Plant Protection

Also of interest

Emerging Contaminants.
Remediation Technologies
Aravind, Kamaraj (Eds.), 2022
ISBN 978-3-11-075158-1, e-ISBN (PDF) 978-3-11-075172-7

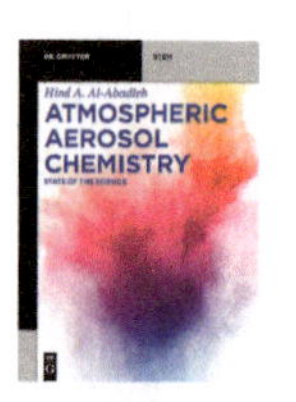

Atmospheric Aerosol Chemistry.
State of the Science
Al-Abadleh, 2022
ISBN 978-1-5015-1936-9, e-ISBN (PDF) 978-1-5015-1937-6

Microbiology of Food Quality.
Challenges in Food Production and Distribution During and After the Pandemics
Hakalehto (Ed.), 2021
ISBN 978-3-11-072492-9, e-ISBN (PDF) 978-3-11-072496-7

Water Resource Technology
Dubey, Mishra, Michalska-Domańska, Deshpande (Eds.), 2021
ISBN 978-3-11-072134-8, e-ISBN (PDF) 978-3-11-072135-5

Photosynthesis.
Biotechnological Applications with Microalgae
Roegner (Ed.), 2021
ISBN 978-3-11-071691-7, e-ISBN (PDF) 978-3-11-071697-9

Plant Protection

From Chemicals to Biologicals

Edited by
Ravindra Soni, Deep Chandra Suyal and Reeta Goel

DE GRUYTER

Editors
Dr. Ravindra Soni
Department of Agricultural Microbiology
College of Agriculture
Indira Gandhi Krishi Vishwa Vidyalaya
Raipur 492012, Chhattisgarh
India

Dr. Reeta Goel
Institute of Applied Sciences and Humanities
GLA University
Chaumuhan
Mathura 281406, Uttar Pradesh
India

Dr. Deep Chandra Suyal
Department of Microbiology
Eternal University
Baru Sahib 173101
Himachal Pradesh
India

ISBN 978-3-11-077147-3
e-ISBN (PDF) 978-3-11-077155-8
e-ISBN (EPUB) 978-3-11-077163-3

Library of Congress Control Number: 2022940995

Bibliographic information published by the Deutsche Nationalbibliothek
The Deutsche Nationalbibliothek lists this publication in the Deutsche Nationalbibliografie; detailed bibliographic data are available on the Internet at http://dnb.dnb.de.

Cover image: Gettyimages/batuhan toker
Typesetting: Integra Software Services Pvt. Ltd.
Printing and binding: CPI books GmbH, Leck

www.degruyter.com

Contents

Sanjay Kumar Joshi, Rashmi Upadhyay

Chapter 1 Agrochemical industry: a multibillion industry

Abstract: The market size of agrochemicals is expected to reach $336.4 billion by the end of 2026, registering a compound annual growth rate of 4.2%. India, being the fourth largest producer of agrochemicals, is succeeded globally by the USA, Japan, and China. United Phosphorus Ltd, Bayer CropScience Ltd, Rallis India Ltd, Gharda Chemicals Ltd, Syngenta India Ltd, BASF India Ltd, and so on are some of the top players who have dominated the market. These companies control almost 80% of the market share. If one is to look at the global population trend, it is of extreme importance that in order to feed the global population, it becomes vital to increase the production of crops within the present arable land. This is where the agrochemicals can play an important role in this sector by assisting farmers in enhancing their crop quality as well as quantity. New developments such as integrated pest management, precision farming, and off-patent products are shaping the outlook of agricultural chemical industry. Not only these are threat to agrochemical industries' long-term revenue targets, long-term disruptions like genome editing, food waste management, improvements in animal feed digestibility, decreasing biofuel demand, and the growing popularity of indoor farming also put a question mark on the brighter future of the sector. Having said that, the future of agrochemicals still looks bright and worth the risk, considering global population growth, growing imperative to protect against crop losses and increase yields, and rising consumer demand for sustainably produced foods. Will leading agrochemical companies eventually increase their exposure to long-term disruptive trends and unfamiliar markets? There are no easy answers, but with growth and opportunity on the horizon, agrochemical companies will have many strategic options to consider as a promising future for the agricultural chemistry industry unfolds.

1.1 Introduction

Agrochemicals are chemically or biologically engineered formulations made available to improve the outcoming yield as well as the quality of crops produced. The two major categories classified under agrochemicals are fertilizers (which are used to add required nutrients to enhance plant as well as soil health) and pesticides (which protect and support the plants to fight against biological stresses, namely, weeds, pests, and insects compromising their optimum growth) [1].

https://doi.org/10.1515/9783110771558-001

With the spread of agriculture throughout various world geographies, agrochemical application has often altered with the upward and downward shift observed in different physical and nonphysical factors affecting agrochemical application. The worldwide agrochemical sector was valued at $231 billion in 2020, and it is predicted to increase at a compound annual growth rate (CAGR) of 4.0% over the next 5 years (2021–2026). The market for herbicides is a largely contributing market in the pesticide segment. The use of agrochemicals is found to be exceptionally high for cereals and grains since they cover a vast area for cultivation across the globe [2] (Figure 1.1).

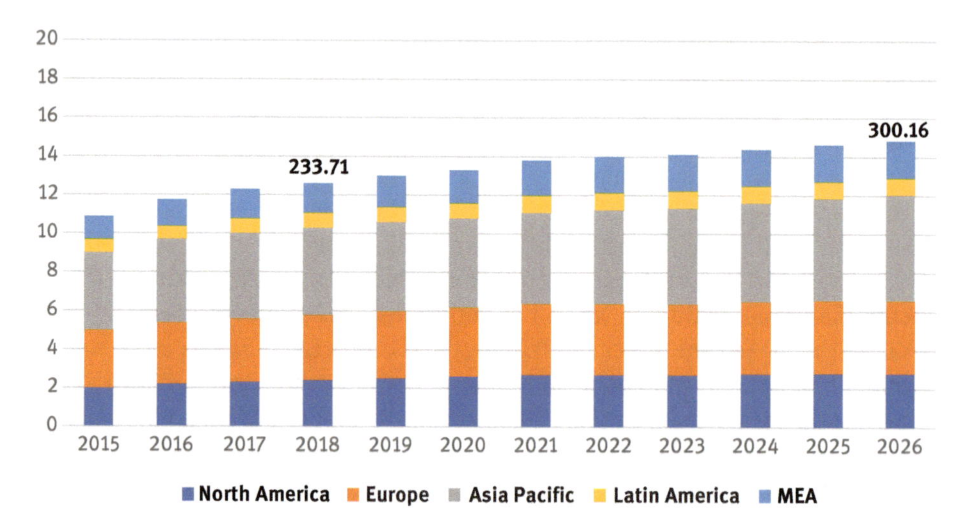

Figure 1.1: Regional agrochemical market growth from 2015 to 2026 (source: Polaris Market Research data [3]).

This can be accounted to a number of factors, like the ever-increasing population giving rise to an increasing number of mouths to feed, which safely directs us toward the limitation of land availability and how agrochemical application can assure both better quality and quantity of crops with the application of high-value fertilizers and pesticides available in the market, keeping in mind the land limitations. Other factors driving agrochemical application are soil degradation, weather anomalies, and growing awareness among consumers and farmers toward the benefits of applying high-quality chemicals in their fields.

1.2 Agrochemical market – life cycle

Figure 1.2 depicts different growth phases occurring in the life cycle of the global agrochemical market and what can be expected from the future.

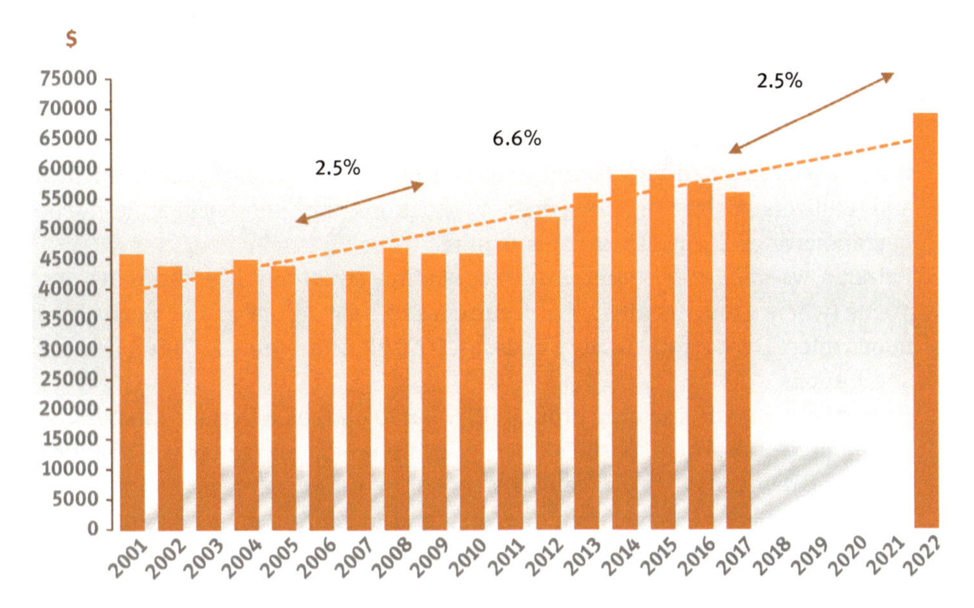

Figure 1.2: Agrochemical market's historical development and forecast (source: author's own compilation of market research data).

Due to poor agricultural economics and the introduction of genetically modified (GM) crops onto the market, the late 1990s saw a period of decline. This was followed by a period of consistent expansion from 2006 to 2010, during which the use of agrochemicals expanded due to higher agricultural revenue and crop prices. From 2010 to 2014, rapid growth was observed with strengthening commodity markets and improving developing markets. This growth was tempered by the *Helicoverpa* outbreak, which incidentally improved the agrochemical sales in Brazil, but that too came to an end by the end of 2014.

The years 2014 to 2017 were traced by the El Nino effect giving rise to adverse weather, resulting in an increased number of crops, and hence reduced prices. The decreased farm incomes led to suppressed investments toward agrochemical applications in fields. Also, as the *Helicoverpa* outbreak too came to an end, the market in Brazil reached its peak of maturity.

In 2018, the first signs of recovery born out of factors like holistic approach toward crop protection targeting all biology, chemistry, as well as precision agriculture-related approach for protection were observed. Other than that, volume growth in developing markets, cheaper distribution facilities for farmers, targeted research and development (R&D) activities, GM adoption leading to better crop prices, and relying on older chemistry instead of new and costly developments.

1.3 Factors affecting agrochemical industry

Some of those factors affecting agrochemical application include:

1) Economic condition of farmers: As farmers face adversity, they rely on absolutely necessary agri-inputs and avoid substantial purchases like pesticides and fertilizers. Instead of taking preventive actions, they indulge more in rescue operations as and when their crops require.
2) Extreme weather conditions: Adverse weather situations like droughts and floods (where droughts affect the occurrence of pests, disease, and weeds; and floods affect the overall viability of crops, leaving no grounds for agrochemical applications).
3) Consumer demands and crop profitability: Based on the harvests, demands and profitability of previous season's crop cycle, crop rotations take place, resulting in varying demands for agrochemicals.
4) Developing markets: Developing markets tend to express increasing demands as their economic conditions move toward the better.
5) Technology development: As newer technologies and methods like GM crops, precision agriculture, and integrated pest Management, the need for chemicals to enhance plant health tends to decrease.
6) Bad press: The potential danger of using chemicals like 2,4-D and atrazine herbicide has gained bad attention from consumers, leading to many averting toward organic, chemical-free products, which are rather costlier than nonorganic products.
7) Policy regulations: Changing laws for using the older and cheaper versions of agrochemicals from time to time create a need for newer developments. This also pushes the farmers to step out of their comfort zones again and again, in turn decreasing their dependability on chemicals for crop protection purposes.
8) Chemical resistance: Agrochemicals, especially single-site active types, are more prone to resistance development in plants, creating a need for newer alternatives which are both costly and difficult to avail locally.

1.4 Leading agrochemical markets

The agrochemical market can be divided into four major geographical locations, namely North America, Europe, Asia-Pacific, and LAMEA [1].

North America:
- USA
- Canada
- Mexico

Europe:
- France
- Germany
- Italy
- Spain
- Rest of Europe

Asia-Pacific:
- Japan
- India
- China
- Australia
- Rest of Asia-Pacific

LAMEA:
- Brazil
- Argentina
- Rest of LAMEA

1.5 Market maturity

North America, EU-15, and Japan are some of the leading markets when we follow agrochemical development as well as consumption. But in recent years, the leading agrochemical markets have been facing a slump. This can be accounted to the recent shift of these markets toward a maturity phase, where the consumer base has almost ceased to increase. Any revenue being generated in such markets is by the way of resistance development, which automatically generates a need for new and improved active ingredients that are high in both quality and price.

Recent statistics highlight the increasing share of markets emerging from the Asia-Pacific region, but this too is a contribution of developing countries, whereas larger markets like Japan, South Korea, and Australia are losing share due to maturity [4].

1.6 Developing markets – volume versus value

As the economic conditions improve for developing countries, pressures like increasing population as well as urbanization are forcing farmers to achieve higher as well as better yields from the same amount of land. This is usually not possible through conventional methods and hence agrochemical application has risen to be necessary (Figure 1.3).

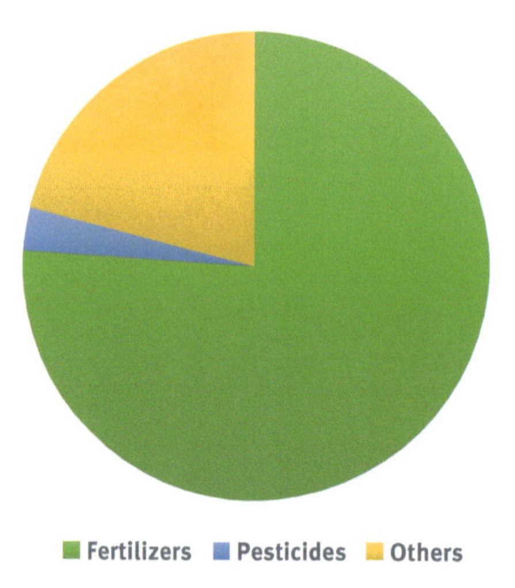

Figure 1.3: Product-wise agrochemical consumption (KT) – 2018 (source: compiled data from various reports).

One important factor leading to an increasing growth for developing markets is the fact that due to previously adhered to poor economic conditions as well as lack of proper advisory services, farmers in such countries have been applying agrochemicals in lesser than optimum quantities. This can also be accounted for with the fact that farmers in such countries prefer generic molecules available in the market for a long time as they are fairly cheap.

And as it goes for generic molecules, they tend to be bulky in nature and need to be applied in quantities much higher than the new-age agrochemicals. This huge gap in application amounts gives a good margin for them to improve, hence, increasing the agrochemical requirements in such countries. This qualifies as volume growth for crop protection-based chemicals [5].

Also, with improving economic conditions, the farmers in developing nations have access to high-quality, registered, and costly products instead of their cheaper, local counterparts. This too stands to increase the demand of agrochemicals in such markets, qualifying it as value growth for these markets.

This is the major difference in the mention of prime agrochemical markets like North America, Japan, and EU-15. For these markets, since volume growth has already reached its maximum threshold due to the usage of optimum quantities of active ingredients, only value growth is possible. This happens either with old chemicals facing resistance from plant varieties creating a need for newer, costlier chemistries, or due to regulatory actions [4] (Figure 1.4).

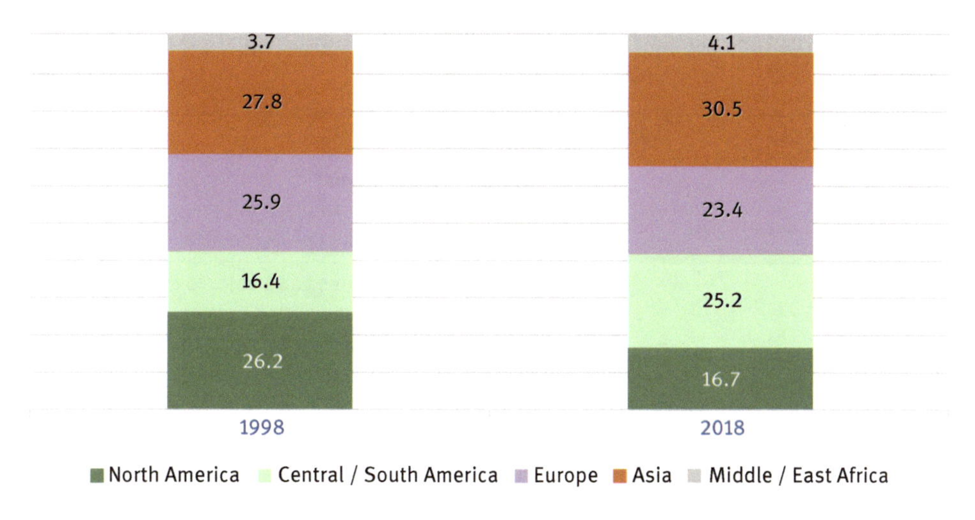

Figure 1.4: Development trends comparisons for 1998 and 2018 for leading agrochemical markets (source: author's own compilation from various reports).

As it is evident in Figure 1.4, major markets like North America and Europe have a decreasing rate of market development for agrochemicals, whereas developing markets like Middle East/Africa, Asia, and Central/South America have an increasing rate of market development for agrochemicals.

1.7 Asia-Pacific market

Asia-Pacific market is one of the fastest growing market segments as far as agrochemicals are concerned, and this is being one of the largest as well as most populated region globally. The region consists of almost 30% of land available and 60% of human population in the entire world. Pesticide consumption has increased in this region in order to increase food production, and India, Japan, Australia, and Thailand stand to be the most pesticide-consuming countries all over the world [2].

1.7.1 Indian agrochemical market

The Indian agrochemical industry is one of the fastest growing markets for agrochemicals, especially in the Asia-Pacific region due to a number of reasons, and some of which include the tropical climate giving rise to many biological threats like pests and diseases, large human population to feed, and the like [7] (Table 1.1).

Table 1.1: Trends and new introduction in Indian crop protection market, FY 20 (source: internal estimates).

Category (3 year CAGR)	Trends and new introductions (FY 20)
Biostimulants 11%	– **New introduction – 3 (2.5 crores)** – Increasing thrust from organized players – Increasing awareness for lesser use of chemical fertilizers
Herbicides 16%	– **New introduction – 5 (4 mixtures and 1 solo; 15 crores)** – Timely and desired monsoon-supported preemergent segment – Growth is driven by the preemergence in rice and wheat, and postemergence in corn and sugarcane – Increase in nonselective herbicide consumption supported by favorable monsoon across the country
Fungicides 13%	– **New introduction – 3 (7.5 crores)** – Scale up of strobe mixtures in rice and vegetables – Gradual shift of low-value to high-value fungicides – Rapid shift from curative to preventive spray habit
Insecticides 10%	– **New introductions – 8 (3 mixtures, 2 formulation change, and 3 new molecules; ₹79 crores)** – Continuous growth of a high-value molecule in BPH-Pexlan – Stem borer segment continuous replacement of cartap by CTPR (Chlorantraniliprole) – Cotton growth from white fly segment, DTF and afidopyropen

Figure 1.5 presents the past and present industry growth for agrochemicals in India, as well as the current trends observed in the Indian agrochemical market.

The industry is estimated to grow at ~8% over the next 3 years with multiple new products in fungicides and herbicide segments, as well as mixtures of high-value insecticides. Figure 1.6 and Table 1.2 represent the industry growth for the next three CAGR for agrochemicals in India.

Figure 1.7 explains different trends and expected introductions in the Indian agrochemical market with respect to pesticides, as well as fertilizers for the next 3-year CAGR.

As is evident from Figure 1.7 for crop protection chemicals in India, herbicides are expected to show maximum growth followed by fungicides and insecticides. The fertilizer sector is expected to grow at higher rates in the Indian market as compared to crop protection chemicals or pesticides.

Figure 1.7 shows the respective shares of key market players prevalent in the Indian agrochemical industry.

Table 1.3 represents the market value, growth figures, as well as trends held by the key market players present in the Indian agrochemical industry.

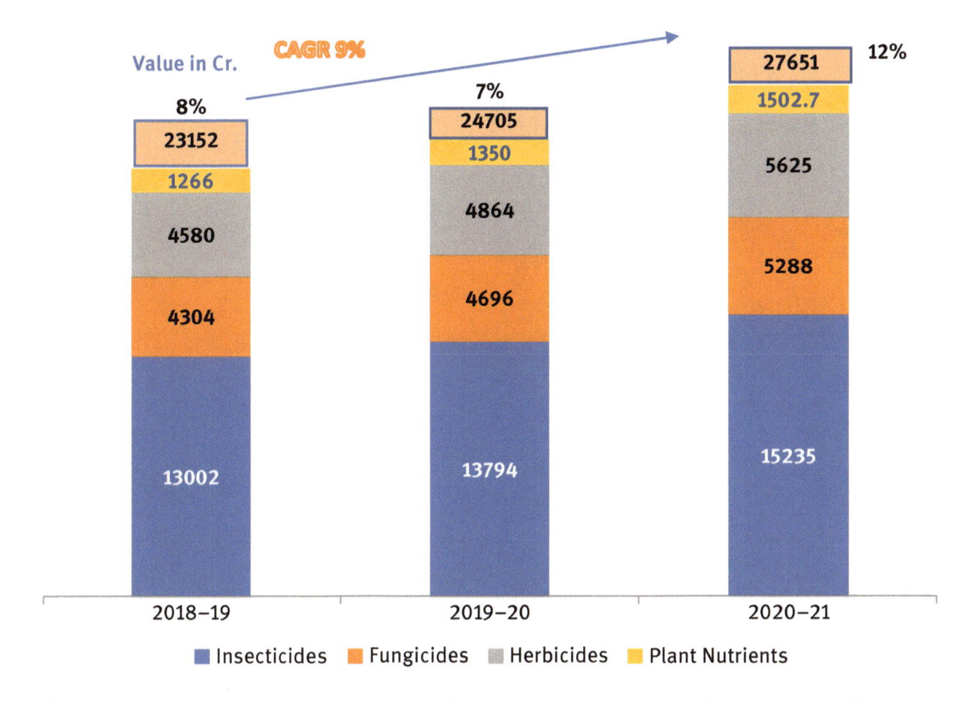

Figure 1.5: Former year-on-year growth in Indian agrochemical industry (source: internal estimates).

1.8 Agrochemical market – by product type

Herbicides (to control undesired plants), insecticides (to control insects), fungicides (to control fungi), biopesticides (are biologically derived pesticides), and plant growth regulators (PGRs) are the most common agrochemical products (to improve plant growth and yields).

Figures 1.8, 1.9, and 1.10 show how these key agrochemical products are broken down by volume of use in agricultural productivity. Rodenticides, miticides, and acaricides are some of the lesser products that fall under this category.

1.9 Research and development in agrochemical industry

Agrochemicals, whether generic or newly developed, are of mainly two types: (a) technical and (b) formulations. Technical are the primary chemicals or active ingredients that form the basis of an insecticide, herbicide, fungicide, or PGR. They are

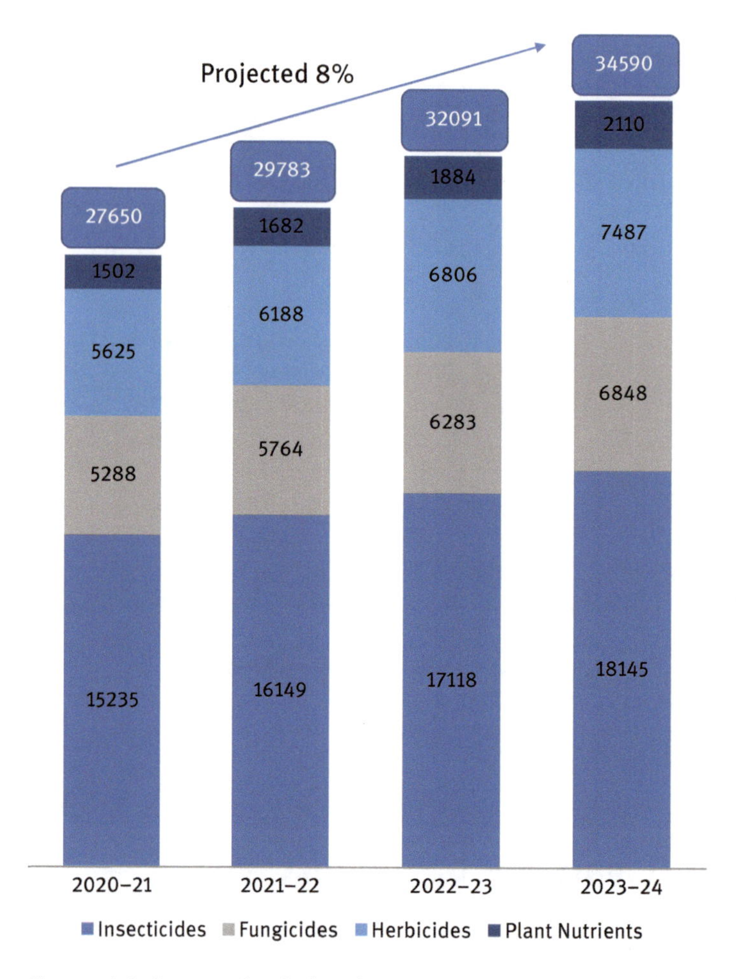

Figure 1.6: Indian agrochemical market's present and future trends (source: author's own compilation of internal estimates).

Table 1.2: Indian agrochemical market trends and expected introductions (3-year CAGR) (source: internal estimates).

Category (3-year CAGR)	Trends and expected introductions
Plant nutrients 12%	– Entry of MNCs in the segment to continue – Network expansion by established players – Increasing acceptance of superior offerings

Table 1.2 (continued)

Category (3-year CAGR)	Trends and expected introductions
Herbicides **10%**	– **Eight new launches in rice herbicide segment in the next 2 years** – **Corn/sugarcane herbicide will keep the momentum, mixture launches** – High-value postemergent herbicide will pick in cotton – Regulation by states will hamper nonselective herbicide growth – **Growth of preemergent herbicide in wheat**
Fungicides **9%**	– High-value fungicides in key crop segments will boost the growth – More premix combinations are expected to launch with high value – Increased awareness on preventive spray
Insecticides **6%**	– **Value shift in BPH segment from 900 to 1,070 crores (Corteva, Syngenta, and PI)** – **Afidopyropen, OP, and DFN mixtures scale up in the cotton sucking complex segment** – Rice stem borer segment: replacement of cartap by CTPR and mixtures – Scale-up of new AI mixtures for sucking complex in vegetables

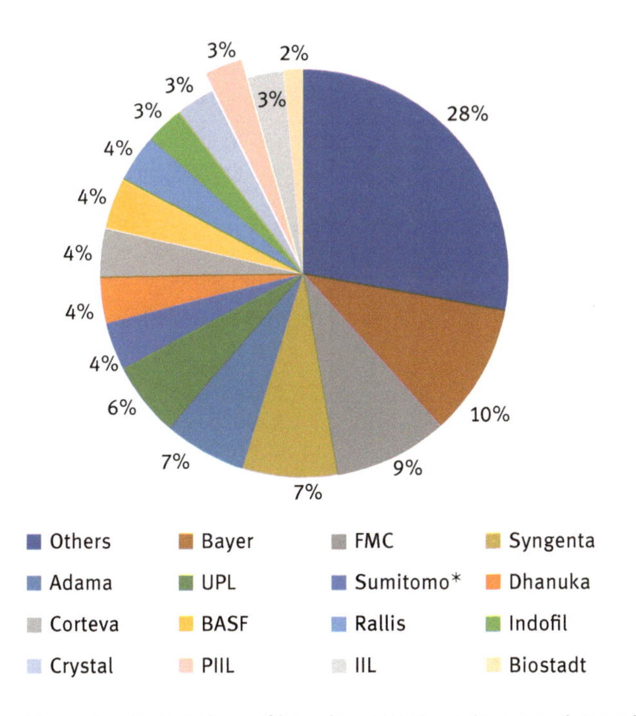

Figure 1.7: Market share of key players in the Indian agrochemical industry (source: internal estimates).

Table 1.3: Indian agrochemical industry trend (2020–21) (source: internal estimates).

S. no.	Companies	FY 2019–20		FY 2020–21		Trend (growth/ decline/ flat)	Remarks
		Revenue	% Share	Revenue	% Share	YOY	
1	Bayer	2700	11%	2940	11%	9%	Continuous growth of timbotrione and glyphosate.
2	FMC	2300	9%	2500	9%	9%	New launches in the herbicide segment and continuous growth of CTPR
3	Syngenta	1850	7%	2100	8%	**14%**	New launches in seed care and corn herbicide, scale-up in Chess Co marketing and Ampligo sales
4	Adama	1598	6%	1850	7%	**16%**	Increased herbicides (Shaked, Barazide, Agil, Tamar) and new launches (Slamberg – PGR and scale-up of Nimrod, Shamir, and Nimitz)
5	UPL	1362	6%	1655	6%	**22%**	Growth in rice fungicides and flexible commercial policy
6	Sumitomo	894	4%	1075	4%	**20%**	Growth through BS and glyphosate.
7	Dhanuka	788	3%	1025	4%	**30%**	Two new launches in fungicide
8	Corteva	915	4%	1080	4%	**18%**	Scale-up of Pexalon, rice fungicides, and herbicides
9	BASF	830	3%	1150	4%	**39%**	Thrust on Sefina and recently launched fungicides
10	Rallis	983	4%	1060	4%	8%	Growth from six new launches in FY'20
11	Indofil	712	3%	850	3%	**19%**	Sustained growth in rice and horticulture fungicides

Table 1.3 (continued)

S. no.	Companies	FY 2019–20		FY 2020–21		Trend (growth/ decline/ flat)	Remarks
		Revenue	**% Share**	**Revenue**	**% Share**	**YOY**	
12	Crystal	788	3%	875	3%	**11%**	Newly acquired Syngenta brands and new launches
13	PIIL	645	3%	840	3%	**30%**	Growth in Awkira, Nominee Gold, Osheen, and Header
14	IIL	735	3%	800	3%	9%	
15	Others	7605	31%	7850	28%	3%	
Total		**24705**		**27651**		**12%**	

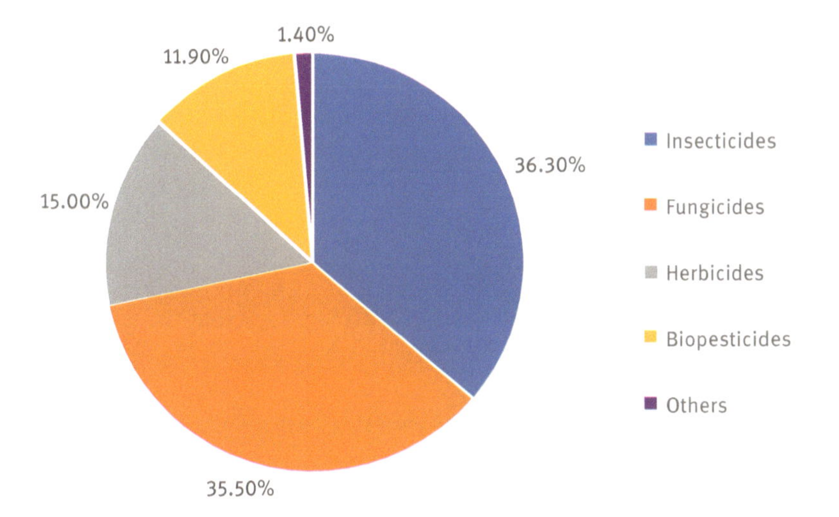

Figure 1.8: In 2018–19, the share of different agrochemical products by volume of use (technical grade) (source: Directorate of Plant Protection, Quarantine, and Storage statistical database [6]).

available in raw form; hence, they are not suitable for direct use. This is where formulations come to the play. Formulations are usually a different type of combination of various proportions and varieties of technical products, mixed with a suitable and compatible carrier. The carrier helps the concentrated mix of technical to dilute and make the technical available as per the recommended dosage of the same, to be used by the farmer [8].

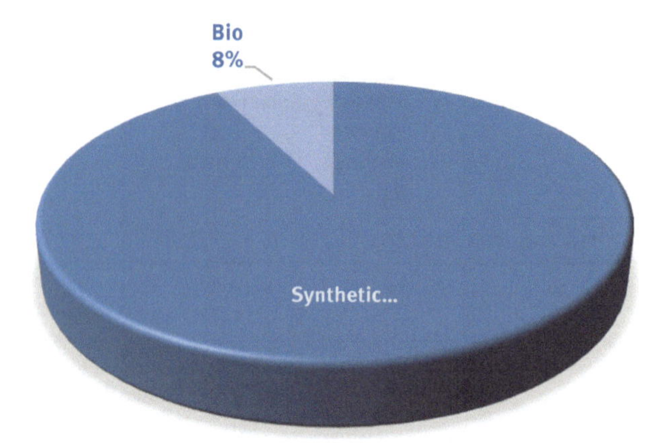

Figure 1.9: Global agrochemical market revenue share (2020) – product origin (source: author's own compilation of International Fertilizer Association).

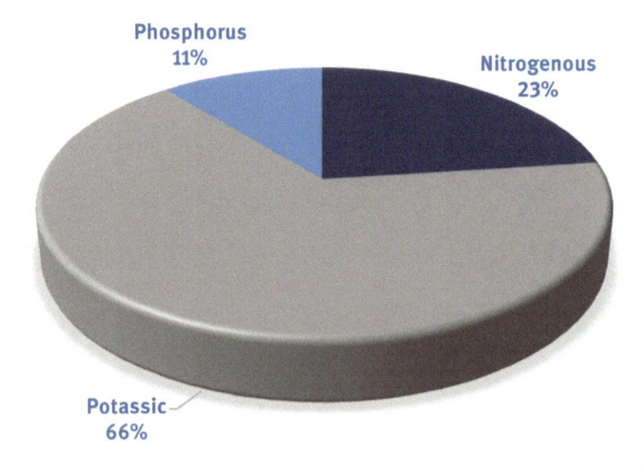

Figure 1.10: Global agrochemical market share (2018) – product type (source: International Fertilizer Association).

As old crop protection formulations reach the stage of resistance, or new regulations tend to reject the legal manufacturing of already available agrochemicals, the need for focused R&D to develop new chemical formulations arises. Earlier this trend has been recurring only in the case of mature markets being propelled by value growth, but with recent developments observed in developing markets, the similar need has been arising there as well.

Better formulations with easier application are being made available even in developing markets with not much R&D backing, highlighting the importance of continuous research into the matter of providing cheaper and more efficient alternatives to farmers. Continuous R&D also has the potential to keep in touch with the emerging needs of farmers and provides viable solutions to them without lagging. This addresses the problem of dependability and reliability usually faced by farmers and protects them from having to rely upon old and nonefficient options available locally [9] (Figure 1.11).

Some important factors affecting agrochemical R&D are:

1. Changing demands of regulatory bodies
2. Development of resistance in plants against previously developed agrochemicals
3. Competition, risks, and consolidation among different companies involved in the agrochemical industry
4. Development of new formulations in order to safeguard the patents that have been obtained for previously developed items
5. Insect resistance and herbicide tolerance may be achieved by genetic manipulation, gene silencing, and gene deletion in GM crops
6. Alternate technologies like precision agriculture and integrated pest management [4]

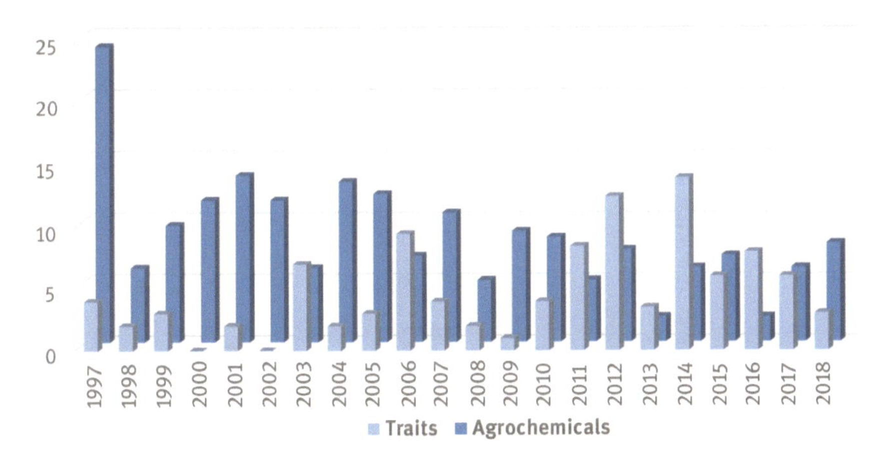

Figure 1.11: Count representation of new active ingredient and GM trait introductions (source: authors' own compilation).

As is evident from Figure 1.11, the development of agrochemicals has been decreasing as years progressed, whereas the trait development through genetic modification has been advancing. As specified earlier, genetic modification in plant varieties is one of the factors affecting the development of new agrochemicals. Recently in the year 2018, due to the issue of glyphosate resistance in weed varieties, there has been a change in trend from the previously suggested pattern.

There has been a noticeable drop in the introduction of newly developed herbicides, particularly in the Americas, where herbicide tolerance characteristics dominate the maize, soybean, cotton, and canola sectors. Other than that, the decrease in agrochemical development can also be attributed to high requirements for monetary as well as resource-based investment, and only a few large companies have enough backing to put forward their efforts into the required field of development.

Also, as more and more awareness is created among consumers regarding environment safety, technologies like GM crops seem more suitable in terms of both health and sustainability. Not only this, such technologies are more long-lasting and reliable than agrochemicals, which tend to retire either due to resistance-related or restriction-related constraints. In contrast to this, developing new chemicals or formulations and getting them registered is a costly as well as a lengthy process [7].

1.10 Consolidation in agrochemical industry

As is evident in Figure 1.12, the agrochemical industry has undergone a number of ups and downs due to a number of controllable as well as uncontrollable factors. The adverse effects of El Nino, which resulted in dryness in Brazil and variable monsoons in Asia, as well as the expanding use of GM crops to combat the underlying stresses, slowed the industry's growth from 2014 to 2018, and the factors contributing to this slump were the adverse effects of El Nino, which resulted in dryness in Brazil and variable monsoons in Asia, and the expanding use of GM crops to combat the underlying stresses.

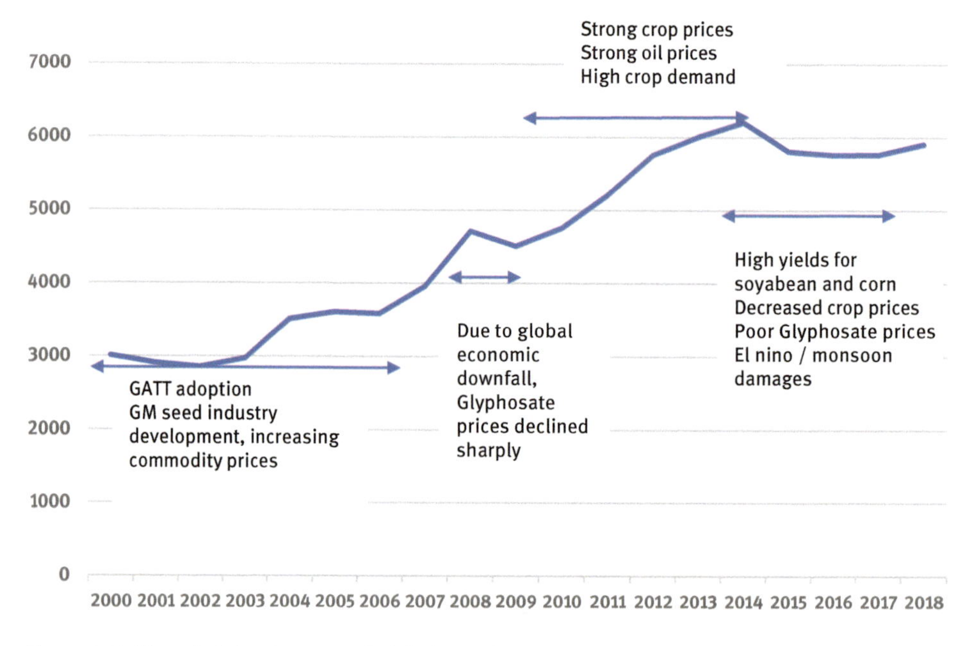

Figure 1.12: Historical development of global agrochemical market (source: internal estimates).

As a result of the decline, the agrochemical industry has seen a wave of corporate consolidation, with Bayer purchasing Monsanto, ChemChina acquiring Syngenta, and Dow combining with DuPont to become Corteva. The consolidation served as a buffer against currency, crop, and crude oil price volatility, which had a little impact on individual enterprises' sales and profit margins [10].

Due to antitrust requirements, much of Bayer's seed and trait operations, as well as some agrochemicals, were divested to BASF, transforming the company into a major seed competitor; DuPont had to divest its agrochemical R&D assets, including research products, to FMC; and Syngenta and Adama had to divest a range of agrochemicals, mostly to Nufarm and Amvac (Figures 1.13 and 1.14).

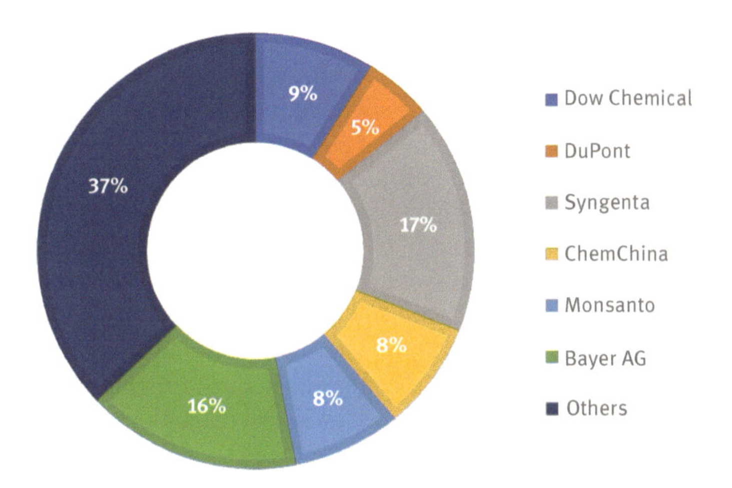

Figure 1.13: Market share before consolidation (CY 2015) (source: Bloomberg; ICICI Securities).

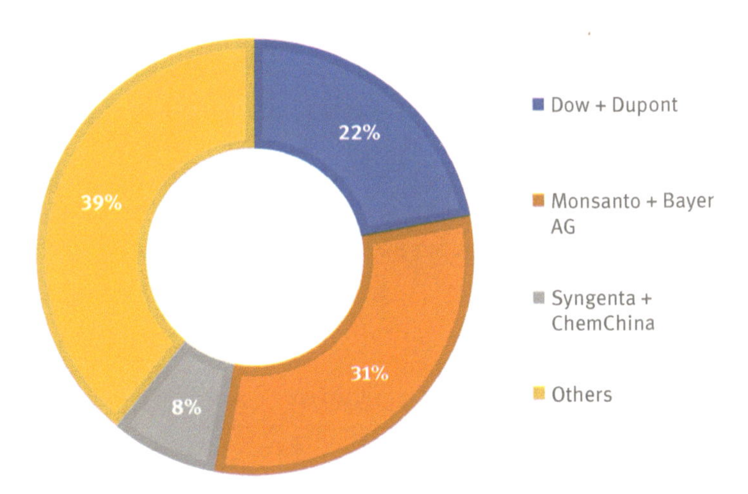

Figure 1.14: Market share after consolidation (source: Bloomberg; ICICI Securities).

Seeds, R&D, and agricultural chemicals are currently handled by the united businesses. This was a strategy aimed at transitioning these firms from pure agro-inputs to agri-science companies, allowing them to serve the farmer's end-to-end demands beyond agro-inputs.

The goal of DuPont's divestiture was to maintain the number of firms engaged in new agrochemical R&D, in order to keep new agrochemical R&D continuing and keep the problem of plant resistance under control. The consolidation has resulted in the agrochemical industry focused on fewer but larger companies controlling almost 80% of market share for agrochemicals, strengthening the overall operations of product supply and demand [7].

Figures 1.13 and 1.14 depict the pre- and post-consolidation market share for the major players in the agrochemical industry.

The sales of the 15 largest agrochemical businesses in 1998 and 2017 are depicted in Figure 1.14. As for the year 1998, there is a slight drop-off in sales achieved by the leading company and the company standing at the fourth position. Beyond that, all companies up to the one in the tenth position are achieving similar levels of sales output.

But for the year 2017, taking the corporate consolidation into account, it can be easily understood that as the cost and difficulty of new agrochemical development have gone up mainly due to strict regulations, only the above five companies have the capacity to indulge in R&D activities for the development of new and improved chemical compositions to sustain crop cycles. This presents a much different picture for the industry from what it was 20 years ago, where most companies were at the same level of output generation.

1.11 Agrochemical sectoral development

Figure 1.15 represents the development trends for the crop protection insecticide sector.

Figure 1.16 represents the development trends for the crop protection herbicide sector.

Figure 1.17 represents the development trends for the crop protection fungicide sector.

1.12 Problem of resistant development

There is an increasing issue of resistance development in weed varieties against particular herbicides. This has been a major issue with single-site active herbicides, whereas broad-spectrum herbicides are less prone to such problems. As most generic

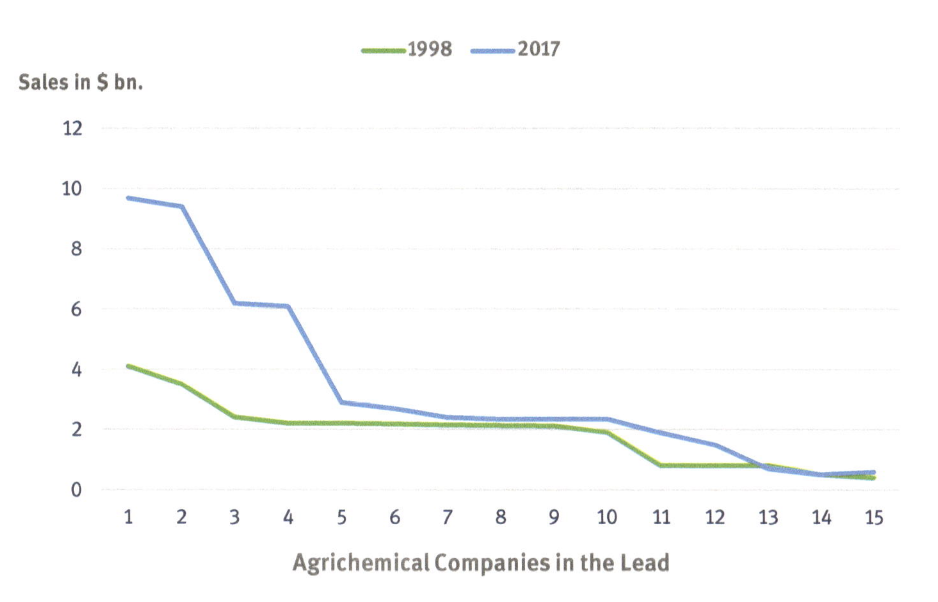

Figure 1.15: Consolidation impacts on the sales and competition of agrochemical companies (source: AgbioInvestor).

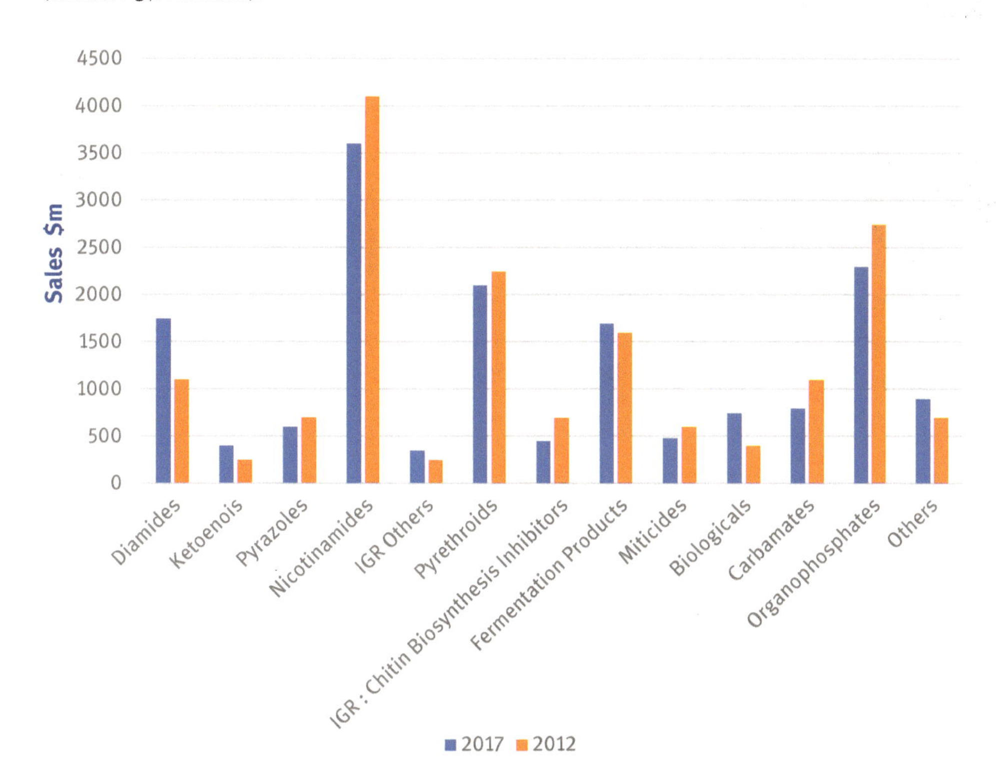

Figure 1.16: Global insecticide market development from 2012 to 2017 (source: AgbioInvestor [11]).

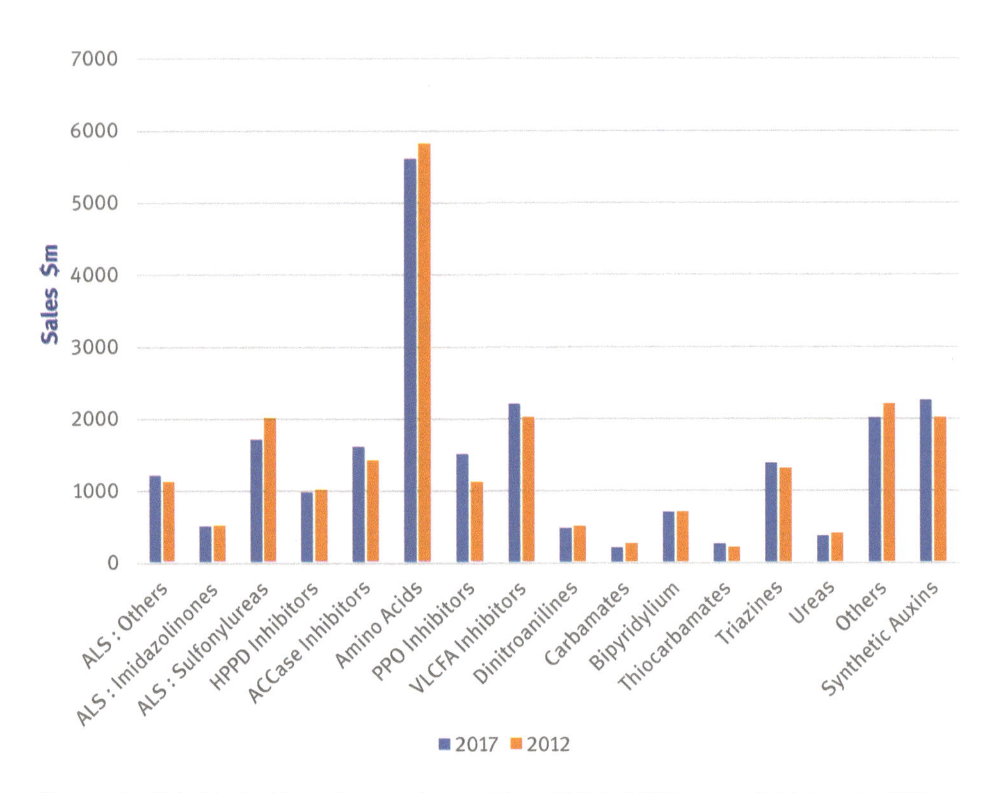

Figure 1.17: Global herbicide market development from 2012 to 2017 (source: AgbioInvestor [11]).

herbicides are single-site active type, this problem tends to emerge with farmers opting for cheaper and locally available solutions in the market.

Figure 1.18 supports this argument by depicting some single-site active herbicides for which resistance has been developed in weed biologies. The graph shows the number of weed species that have gained resistance to each herbicide class versus the year when the first member of that class was created.

1.13 Impact of regulations – registration and reregistration processes

The EU has had the largest impact on agrochemical availability as a result of legislation. The necessity to reregister existing agrochemicals under current-day criteria was introduced by EU Council Directive 91/414, and this was improved by Regulation 1107/2009, which changed the basis of registration criteria to be regulated by hazard rather than risk. It also established the concept of "comparative assessment and safer alternative substitution."

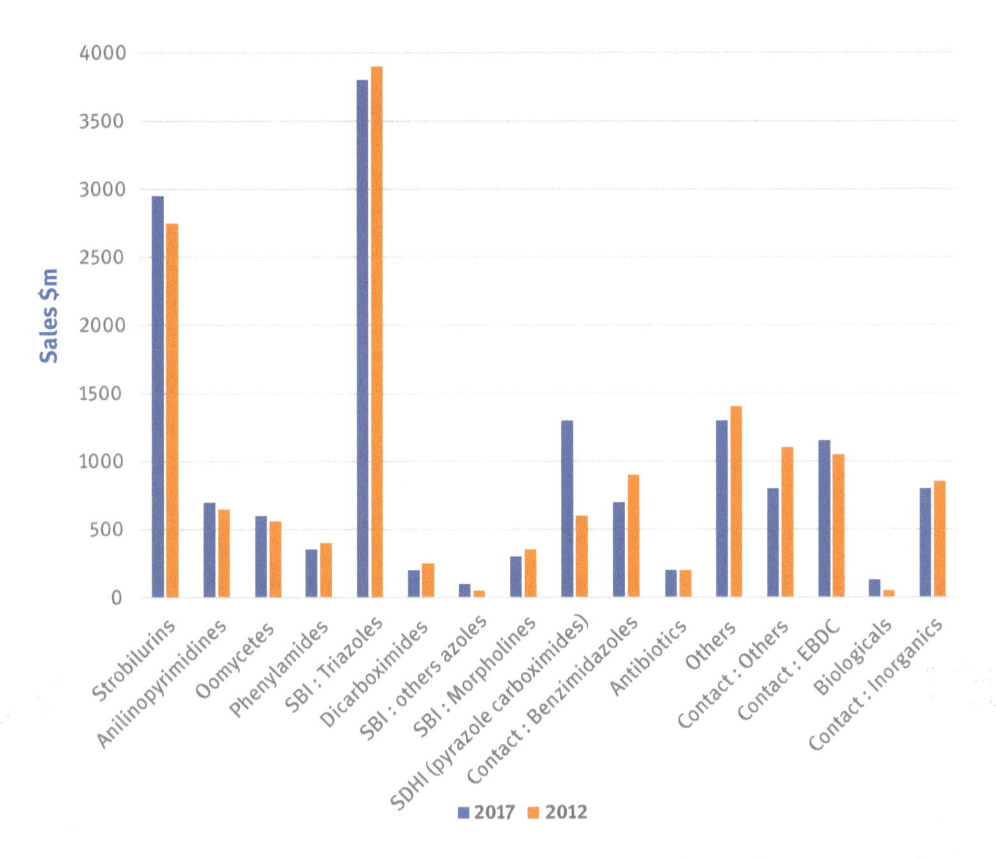

Figure 1.18: Global fungicide market development from 2012 to 2017 (source: AgbioInvestor [11]).

As a result, a list of agrochemicals previously registered in the EU is considered "candidates for substitution." This list includes agrochemicals that fail to meet two of three criteria: persistence, bioaccumulation, or toxicity. Each agrochemical's registration must be evaluated every 10 years, and the product has to abide by the new registration criteria in order for it to remain on the market.

Alternatives to the candidates for substitution will be sought at that time. Existing registrants have the option of continuing to support a molecule during the reregistration process. If agrochemicals are not allowed reregistration or are not supported by existing registrants, they may be removed from the market [4].

For agrochemical classes now accessible in the EU and the USA, Figure 1.19 depicts the year when the agrochemicals were first launched, as well as the fraction of those on the worldwide market in 2018. The dotted line represents the remaining share of active ingredients in EU if all products that are in line for substitution lose their registration (Figure 1.20).

The registration system carried down by the US Environmental Protection Agency (EPA) was defined by risk. It functions a registration review program similar to the

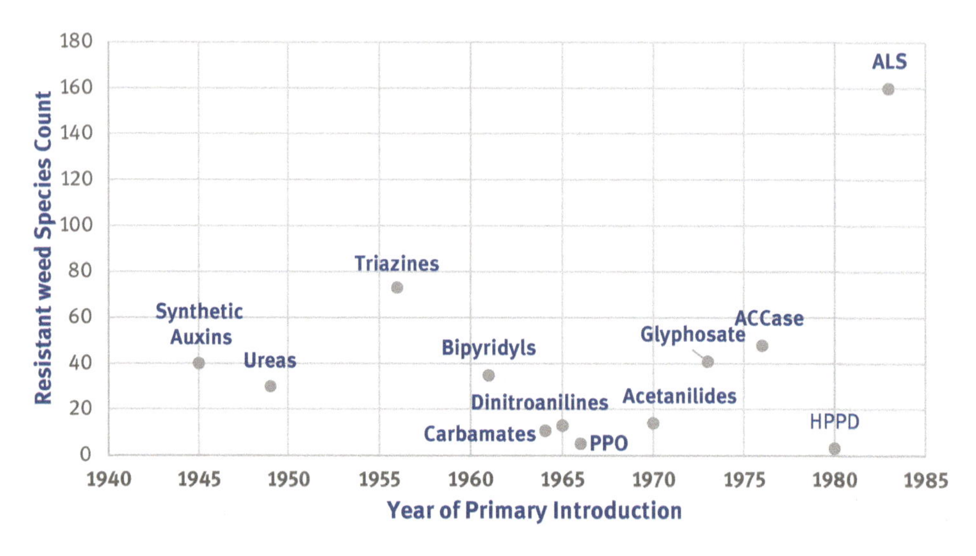

Figure 1.19: Classification of herbicide resistance in weed species based on the year of first introduction (source: Heap, I. The International Survey of Herbicide Resistant Weeds; available online at www.weedscience.org and AgbioInvestor [11]).

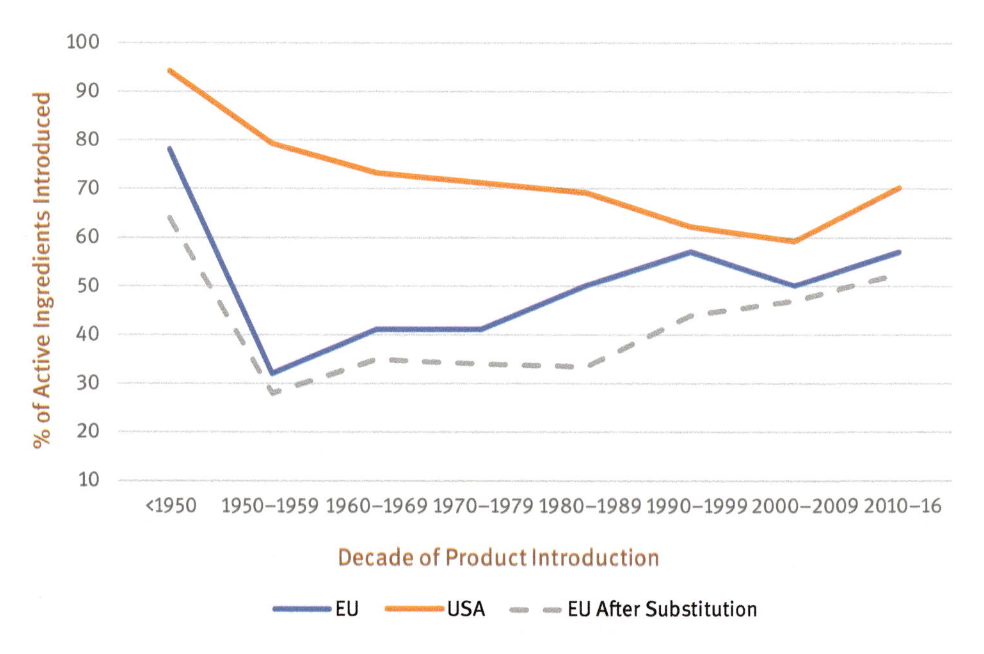

Figure 1.20: Percentage of currently registered agrochemicals in the EU and the USA (source: AgbioInvestor [11]).

one running the EU. Most existing agrochemicals are now being evaluated, but few decisions have been taken, and the requirements for ongoing registration are obviously different. With the changed system, many significant agrochemicals have come under review. Due to this, a number of chemistries might be removed but that would not be as high on the basis of scientific based risk assesment system in the EU.

To support the highly restrictive policies in the EU market, the example of grass weed in small-grain cereal crops is worth mentioning. The herbicides need to be highly specific due to the similarity between weed and crop biologies, but due to rejection in reregistration process because of regulatory limitations, the only few options left now to control these weeds are of postemergence type. The newly appointed herbicides are too candidates for substitution as of now, and hence prone to be removed undergoing the reregistration process, leaving the farmers with very limited options.

Although reregistration might be considered as a chance for new developments through targeted R&D, the stringent EU hazard-based approach is sometimes seen as a barrier to new developments. Only a limited fraction of new active ingredients are eligible to be introduced in the EU market, compared to prior eras. Due to registration failure, if a prospective active component demonstrates any persistence, bioaccumulation, or toxicity effects, the molecule would be unlikely to be proceeded for development in the EU market. In contrast, if such an active ingredient is produced in the US market and kept within acceptable levels of persistence, bioaccumulation, or toxicity, it is very likely to be granted authorization for successful development, registration, and eventual commercialization [4].

1.14 Toward a growing agrochemical industry

To ensure a thriving agrochemical market, some important factors are mentioned for careful consideration:

1) Trade channels: Creating simple and efficient trade channels by way of simplifying export registration and compliance procedures, entering into strategic trade agreements with prime importing nations, infrastructure availability, single-window handling of license, and other registration-related matters are some of the ways in which intranational as well as international trade channels can be created.
2) Technology interventions: Like digital advisory services, artificial chemical application equipment (to address the lack of labor during peak times) and models like pay-per-use technology with postharvest payments and product-as-a-service type of alternatives address the underlying financial concerns of farmers.

3) Extension services: These are required to create consumer awareness regarding the latest products and technologies available in the market and methods of proper application to achieve optimum results.
4) Research and development: This is required to develop cheaper and less resistant active ingredients so as to create an atmosphere of dependability, trust and reliability for farmers, where they can count on particular high-quality products and do not have to make adjustments from time to time. Government subsidies and support to smaller companies can disbar the financial restrictions faced by small developers and encourage them to develop best possible alternatives for farmers.
5) Relaxed regulatory frameworks: Markets like EU find problems not only in reregistering the products lined for substitution but also for the development of new products, leading to a very restrictive environment. Such heavy restrictions may ultimately lead to disarming the agrochemical manufacturing sector in the EU and push the government itself to import such products from outside. Instead, by changing the registration criteria to a hazard-based system only, and not both risk and hazard-based system, more products can be allowed in the market. This will also create an open platform for creativity and scientific innovations.

References

[1] Prihar Y, Prasad E. Agrochemicals Market. 2021, September, Retrieved from Allied Market Research: https://www.alliedmarketresearch.com/agrochemicals-market.

[2] Agrochemicals Market. 2021, March. Retrieved from Markets and Markets: https://www.marketsandmarkets.com/Market-Reports/global-agro-chemicals-market-report-132.html.

[3] Agrochemicals Market size & Forecast, 2015-2026. Polaris Market Research. 2019, Retrieved from https://www.polarismarketresearch.com/industry-analysis/agrochemicals-market

[4] Phillips MW. Agrochemical industry development, trends in R&D and the impact of regulation, Society of Chemical Industry, 2020.

[5] Sharma AD, Mullick H. The role of agrochemicals: Achieving the vision of a USD 5 trillion economy by 2025. FICCI, PwC India, 2020. Retrieved from https://ficci.in/spdocument/23379/Knowledge-Report-Agrochemicals-Conference-20.pdf.

[6] Statistical database. 2019, Retrieved from Directorate of Plant Protection, Quarantine and Storage: http://ppqs.gov.in/statistical-database

[7] Jagasheth UH. Indian agrochemicals industry: Insights and outlook. Care Ratings, 2019.

[8] Aggarwal R. R&D is Vital for the Growth of Crop Protection Sector in India. 2021, December 7, Retrieved from Krishi Jagran: https://krishijagran.com/featured/rd-is-vital-for-the-growth-of-crop-protection-sector-in-india/.

[9] Nishimoto R. Global trends in the crop protection industry. Journal of Pesticide Science. 2019, August 20, Retrieved from https://www.ncbi.nlm.nih.gov/pmc/articles/PMC6718354/.

[10] Bansal S. The Impact of Mergers in the Agrochemical Industry. 2017, May 9. Retrieved from Linked in: https://www.linkedin.com/pulse/impact-mergers-agrochemical-industry-sanjay-bansal.

[11] Special Reports. 2020, Retrieved from AgbioInvestor: https://agbioinvestor.com/.

Himani, Pringal Upadhyay, Sonu Kumar Mahawer, Ravendra Kumar, Om Prakash

Chapter 2 Plant protection through agrochemicals and its consequences

Abstract: Agrochemicals, commonly referred to as crop protection chemicals or pesticides, play a key role in controlling harmful pests that are responsible for damaging crops, thereby significantly reducing the quality and yield of crops. Agrochemicals comprise insecticides that protect crops either by killing insects or by preventing their attacks; antimicrobials that protect plants against plant pathogens such as nematodes, fungi, bacteria, and viruses; herbicides that are used to kill undesirable plants; biopesticides, which are new age crop protection products of natural origin; and other groups such as fumigants, fertilizers, soil conditioners, and plant growth regulators. In India, about 30–35% of the annual crop yield gets wasted due to pest attack; therefore, nowadays agrochemicals are being widely used for crop protection and increasing productivity. Besides the major benefits of pesticides and their role in crop production, there are many harmful effects associated with pesticides such as residues on food commodities, biomagnification and bioaccumulation, acute or chronic toxicity to humans and other nontarget organisms, and the emergence of pest resistance. The purpose of this publication is to explain the protection mechanism of different groups of pesticides, their impact, advantages and risks to human health and environment.

2.1 Introduction

The term “agrochemical” refers to a wide range of compounds, including pesticides, fertilizers, plant growth regulators, and soil conditioners, which are now an integral part of modern agriculture production system all over the world. Pesticides are chemically manufactured compounds that are used in agriculture for the management and control of pests, pathogens, and unwanted plants (weeds). Modern industrial agriculture basically promotes the reliance on synthetic agrochemicals, while neglecting their negative health and environmental consequences. However, their balanced use, prominent doses, right method, and application at the right time provide higher crop output. The damage caused by agricultural pests is a major limiting factor for global food production, and over the last 50 years, the consumption of pesticides used has increased several folds, and their adverse impact on environment and human health came into light. The Food and Agriculture Organization estimated that about 40% of the global crop production destroyed by agricultural insect pests, weeds, and diseases

https://doi.org/10.1515/9783110771558-002

collectively results in economic losses. Every year, plant diseases cost the world economy over $220 billion and invasive insects at least $70 billion; thus, due to great loss in crop production, pesticides are frequently used to tackle the problem. Over the years, agrochemicals with special reference to pesticides had several benefits and also have caused many problems. Benefits of pesticides include increased and improved production and productivity of agriculture commodities, increased farm profits by helping the farmers save money on labor costs, reduce the time required to manually remove weeds and pests from fields, and also prevent transmission of disease by pests or insect vectors [1]. However, there are numerous problems associated with the pesticide usage that surpass their benefits as they possess health and environmental risks such as bioaccumulation and biomagnification of their residues in the food chain; agroecosystem affects major environmental components such as air, water, and soil, potentially resulting in biodiversity extinction; pesticide resistance, as some pests develop genetic resistance to pesticides; adverse effects of pesticides on nontarget species; water pollution through runoff, drift, and leaching; and soil contamination. Alternatively, there are proven strategies that are emerging to replace the conventional synthetic pesticides such as the use of botanicals, which are effective against a limited number of specific target species, biodegradable and acceptable for use in integrated pest management programs [2]. This chapter focuses on the uses of different classes of pesticides, their protection mechanisms, alternative strategies, and, most importantly, the consequences of using pesticides.

2.1.1 Agrochemicals and their classification

Agrochemicals are broadly categorized as pesticides, plant growth regulators, fertilizers, soil conditioners, and chemicals used in animal husbandry. Further, they are again classified into different categories as shown in Figure 2.1.

2.2 History of pesticide usage

The first account of use of insecticide was recorded about 4,500 years ago, where Sumerians used sulfur compounds for the control of insects and mites. The Rigveda, which is 4,000 years old, also mentions the use of poisonous plants against pests [3]. While about 3,200 years ago, the Chinese used arsenical and mercury compounds to control the body lice. In addition, nicotine sulfate extracted from tobacco leaves also came into use as an insecticide during the seventeenth century. Two more natural pesticides were introduced in the nineteenth century , one of which was pyrethrum (derived from chrysanthemums) and the other was rotenone (derived from the roots of tropical crops) [4]. In 1867, an impure form of copper or arsenic was employed to

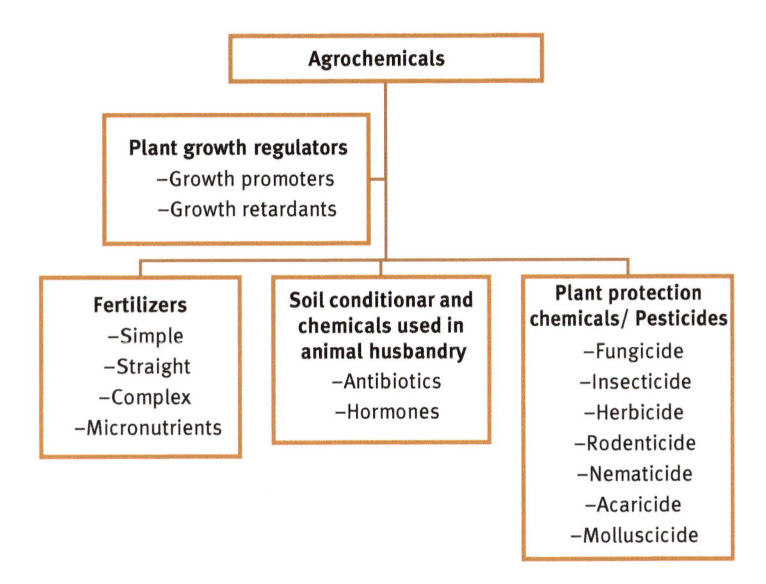

Figure 2.1: Classification of agrochemicals.

combat the infestation caused by the Colorado potato beetle in the USA. Several effective and economical pesticides were synthesized and produced during and after World War II, which marked a huge step forward in pesticide development. The impacts of some organochlorine (OC) insecticides (parathion, captan, and 2,4-D) encouraged and promoted the growth of synthetic pesticides in the 1940s. Out of which, DDT was most popularly employed to its broad-spectrum usage. Between 1950 and1955, fungicides, such as captan and glyodin, as well as the organophosphate insecticide "malathion" were introduced followed by the discovery of triazine herbicides in the years 1955–1960. During the Vietnam War, an experimental wartime herbicide named "Agent Orange" was used, which was introduced by Monsanto in 1961–1971. However, in 1962, an American scientist Rachel Carson stated in her book *Silent Spring* that spraying DDT in the field causes sudden death of nontargeted organisms through direct or indirect toxicity [5, 6]. This highlighted the potential consequences of indiscriminate usage of pesticides and paved the path for safer, more ecologically, and friendly alternatives. The era observed the frequent and large-scale use of well-known agrochemicals, namely, sulfonylurea, imidazolinone herbicide, glyphosate, cyclohexanediones, dinitroanilines, and aryloxyphenoxypropionate. Moreover, third-generation insecticides, namely, benzoylureas, pyrethroids, insecticides, avermectins, and various others were also got popularized. This era also observed triazole, dicarboxamide, imidazole, pyrimidine morpholine, and other similar fungicides. Most of those agrochemicals had single target site, and thus they are selective. Nowadays, researches are being conducted to combat their negative impacts and discover green alternatives.

2.3 Mechanism of action of plant protection chemicals/pesticides

Based on the target sites, pesticides are classified into different categories. The way in which a pesticide causes physiological disruption at the target site is known as its mode of action.

2.3.1 Insecticides

Insecticides are pesticides that are formulated to kill, harm, repel, and control one or more insect species. They penetrate into the insect's body through different routes: they take effect upon ingestion (stomach poisons) and inhalation (fumigants), or act as contact poison by penetrating to the insect's body. Most of the insecticides cause the death of insects by acting mainly on the three target sites in their nervous system: acetylcholinesterase (enzyme responsible in the transmission of nerve impulse), voltage-gated Na^+ channels across the nerve membranes, and the acetylcholine receptors. DDT is a potent broad-spectrum OC insecticide which acts as nerve poison, and its main site of action is the nervous system. The major mechanism by which DDT works is by changing the permeability of nerve axons to Na^+ and K^+ ions and by disturbing the ionic balance, causing nerve axons to fire repetitively which eventually leads to tremors and death [7]. The organophosphorus (OP) and carbamate insecticides act by inhibiting the enzyme cholinesterase (pivotal role in hydrolyzing acetylcholine). The enzyme is said to be phosphorylate (OP) or carbamoylate (carbamates) in the OP and carbamate moieties. The binding is irreversible, when the enzyme gets attached to the phosphorus moiety, while it is reversible for the carbamate moiety. This inhibition causes acetylcholine (Ach) to accumulate at neuron/neuron and neuron/muscle synapses, producing fast twitching of voluntary muscles and eventually paralysis [7, 8]. The neonicotinoids are a newer class of insecticides with a new mode of action. The first commercial neonicotinoids were imidacloprid and acetamiprid. The major target of neonicotinoid insecticides is nicotinic acetylcholine receptors (nAChRs). They mimic the action of acetylcholine (excitatory neurotransmitter in the central nervous system) and act as an agonist by readily interacting with the nAChRs. The interaction results with excitation and paralysis [9, 10]. The mode of action of some important insecticides has been tabulated in Table 2.1.

Table 2.1: Mode of action of some common insecticides.

Groups	Mode of action
Nicotine	Interacts with Ach receptors at the central nervous system, resulting in twitching, convulsions, and death
Rotenone	Respiratory enzyme inhibitor, interferes with NAD^+ (a coenzyme involved in redox reaction in metabolic pathways) and coenzyme Q (a respiratory enzyme responsible for carrying electrons in some electron transport chains), resulting in failure of the respiratory functions
Pyrethrum	Rapid knockdown action, acts as axonic poison
Neem-based insecticides	Inhibits biosynthesis or metabolism of ecdysone, the juvenile molting hormone, and disrupts molting
DDT	Acts as nerve poison, causing tremors, convulsions, and eventually death
Lindane	Interferes with GABA-gated chloride complex, and the treated insect shows tremors and convulsions within hours
OPs	Inhibit cholinesterase, and inhibition results in the accumulation of Ach at the neuron/neuron and neuron/muscle junctions or synapses, causing rapid twitching of voluntary muscles and finally paralysis
Carbamates	Same as OPs, but the binding is reversible when the enzyme gets attached to the carbamate moiety
Pyrethroids	Similar modes of action resembling that of DDT, rapid paralytic action on insect
Neonicotinoids	Mimics the action of acetylcholine and acts as an agonist by readily interacting with the nAChRs, excitation, and paralysis
Nereistoxin	Interacts with nAChRs and acts as an antagonist by blocking cholinergic transmission
Limonene	Affects the sensory nerves of the peripheral nervous system. Used to control fleas, lice, mites, and ticks, while remaining virtually nontoxic to warm-blooded animals and only slightly toxic to fish
Spinosyns	Disrupts binding of acetylcholine with nAChRs at the postsynaptic cells that prevent the repolarization of the receptor and cause hyperactivity
Avermectins	In insects and mites, avermectins work as an agonist of the GABA-gated chloride channel, which eliminates the GABA inhibitory postsynaptic potential as well as excitatory postsynaptic potentials at the neuromuscular junction resulting in behavioral changes such as delay in egg laying and feeding
Benzoylureas	Inhibits or blocks the synthesis of the main component of insect's integument, chitin, during their larval stage. Thus, because of deformed exoskeleton, hatchability of eggs gets impaired

Ach, acetylcholine; OPs, organophosphates; nAChR, nicotinic acetylcholine receptors.

2.3.2 Fungicides

Fungal plant pathogens are among the most important biotic agents that cause disastrous crop disease. They have evolved by a variety of techniques to colonize plants, and these interactions result in a wide range of consequences, from favorable interactions to host death [11]. The fungal disease appeared to be one of the leading causes of crop losses ever since humans started to cultivate plants [12]. Pathogenic fungi can infect plants at any stage of growth and development, from seedling to seed maturation, in both natural and artificial environments, either alone or in combination with other phytopathogens [13]. Phytopathogenic fungal species can cause tremendous loss in the quantity and quality of crop yields, making this a serious economic issue in the global agricultural sector. To promote the effective disease control, precise and rapid detection and identification of plant infecting fungus is required [14]. Fungicides are either synthetic or natural chemical compounds or biological organisms that inhibit or retard the growth of fungi or fungal spores. Modern fungicides do not kill fungi; they simply inhibit the growth for a period of days or weeks. Fungi are the causal organisms which can cause significant damage in agriculture, resulting in critical losses of yield, quality, and profit. Fungicides are used both in agriculture and to fight fungal infections in animals [15].

There are some common modes of actions involved in the fungicidal action [16]:
- Inhibition of ergosterol biosynthesis
- Inhibition of protein biosynthesis
- Inhibition of mitochondrial electron transport

Inhibition of ergosterol biosynthesis: Ergosterol is the main and necessary component of fungal cell membrane structure and integrity. Its inhibition is an important target for fungicides. Ergosterol biosynthesis inhibitors block the cytochrome P450 (CYP-51) or lanosterol-14α demethylase present in the fungi. Ergosterol is synthesized in a multistep pathway and its precursor is squalene. Demethylation inhibitors are a group of systemic fungicides that targets the cell membrane integrity by inhibiting C14-demethylation during ergosterol biosynthesis. This group includes imidazoles, pyrimidines, and triazoles having primary curative activity. These groups target a single metabolic site which promotes the risk of acquiring resistance unless they are used in combination with other fungicides with different and multiple sites of action. These group members are mostly effective against rust pathogens, followed by septoria and, to a lesser extent, against powdery mildews.

Inhibition of protein biosynthesis: Inhibition of protein biosynthesis can also be a factor responsible for fungicidal actions. By modifying the sulfhydryl groups of many proteins, the fungicide "dithianon" plays a significant role of multisite inhibitor of protein formation which prevents or inhibits the germination of spores and germ tube growth.

Inhibition of mitochondrial electron transport: Another major group of fungicides is inhibitors of mitochondrial respiration. As mitochondrial respiration is a common metabolic process across all organisms, inhibition of respiration may lead to problems with selectivity. Common members of this group are strobilurins, which inhibit mitochondrial respiration at the Qo site of cytochrome b which in turn leads to hinder the spore germination and mycelial growth in fungal pathogens. The pyridine group of fungicides is another example of these group members that are also termed as proton gradient uncouplers, which inhibit the mitochondrial respiration in fungal pathogens by uncoupling the proton gradient and reducing the proton-driven synthesis of ATP via the enzyme ATP synthase. Folpet and Captan are some common examples that block the mitochondrial respiration and as a result inhibit the spore germination.

2.3.3 Herbicides

Herbicides, commonly known as weed killers are those compounds that are used in the management and control of unwanted plants (weeds) in the area of agriculture, forestry, gardening, and landscaping. These chemicals are utilized instead of adopting manual methods to remove the weeds. Herbicide action is surpassing the benefits of the plant's metabolic pathways such as photosynthesis, phytohormones function etc. Herbicide must meet a number of criteria to be effective, the target weed must come in contact with the compound, be adequately absorbed by the weed, translocate to the site of action in the weed, and accumulate in enough amounts at the site of action to kill or retard the growth the target plant.

There are some common modes of action involved in the herbicidal action:
- Growth regulators
- Seedling growth regulators
- Inhibition of photosystem I (PSI) and photosystem II (PSII)
- Lipid biosynthesis inhibitors
- Amino acid biosynthesis inhibitors
- Inhibitors of pigment biosynthesis

Growth regulators: Plant growth regulators are primarily used to control/kill or control broadleaf weeds in grass crops. They mainly impact on the growth hormones and mimic the natural growth hormone, indole acetic acid (IAA), that regulates the root growth, cell elongation, protein synthesis, cell division, and stem callus formation. These herbicides belong to the family of phenoxy carboxylic acid, benzoic acid, quinoline carboxylic acid, pyridine carboxylic acid, and the mechanisms involve behind their activities depend on their auxin-like capacity; thus, they are also known as synthetic auxins [17].

Seedling growth regulators: Some herbicides including thiocarbamates and acid amides are effective root and shoot growth inhibitors. They obstruct the plant's growth, primarily at its growing points.

PSII inhibitors: These group herbicides promote the inhibition of photosynthetic pathway, specifically PSII. They bind with the Q_B-binding site of the D1 protein complex present in the chloroplast thylakoid membrane which disrupts the electron transport system from Q_A to Q_B and leads to stop the CO_2 fixation and production of NADPH and ATP in the cells. They form toxic hydroxyl radicals that destroy unsaturated lipid membrane and chlorophyll. Eventually, the plant dies because of the highly reactive molecules which damage the cell membranes. Triazines, phenylureas, and uracils are the common examples of PSII inhibitors.

PSI inhibitors: These group members are also known as electron diverters and are represented by bipyridilium family of herbicides such as diquat and paraquat. They generates radicals by accepting electrons from PSI, which on reacting with O_2 forms superoxide radicals. The superoxide radicals are catalyzed in the presence of the enzyme, dismutase; it forms H_2O_2 and OH radicals. These radicals damage the cell membrane's unsaturated fatty components, lipids, plant pigments, and proteins. As a result, the cell membrane is damaged beyond repair, allowing the leakage of cytoplasm, causing wilting and which eventually leads to the death of plant.

Lipid biosynthesis inhibitors: These are primarily used as postemergence herbicides for controlling broad leaf grasses. They block the production of lipid and fatty acids as a result plants gradually turn purple, brown, and ultimately die. Common examples of this group of herbicides are fluazifop and sethoxydim [18].

Amino acid biosynthesis inhibitors: Some herbicides can also act as inhibitors of amino acid biosynthesis.

- **Aromatic amino acid inhibitors:** Glyphosate (roundup) is a common example of this group of herbicide. It is nonselective and strongly bound to the soil, preventing the root uptake. These group herbicides inhibit the enzyme EPSP (5-enolpyruvylshikimate-3-phosphate) synthase, which is the main enzyme in the shikimate pathway. These group members inhibit the conversion of shikimate-3 phosphate to 5-enol pyruvyl shikimate-3-phosphate and prevent the biosynthesis of aromatic amino acids including tryptophan, tyrosine, and phenylalanine [19].
- **Glutamine synthesis inhibitors:** There are different compounds like glufosinate and bialophos that act as strong inhibitors of glutamine synthase enzyme which is responsible for the conversion of ammonia into glutamate. Accumulation of ammonia leads to the disturbance of numerous cell functions, especially close down the PSI and PSII.
- **Branched chain amino acid inhibitors:** Herbicides that inhibit branched chain amino acids are also accepted as acetolactase synthase inhibitors and are also known as acetohydroxy acid synthase inhibitors which catalyze the

formation of branched chain amino acids such as leucine, isoleucine, and valine in the first step that are important for protein synthesis and plant growth. These groups comprise the chemical family of imidazolinones, sulfonylureas, pyrimidinylthiobenzoates, sulfonylaminocarbonyltriazolinones, and triazolopyrimidines.

Inhibitors of pigment biosynthesis: These group herbicides are also acknowledged as carotenoid biosynthesis inhibitors that destroy the plant pigment, chlorophyll, which plays a vital role in photosynthesis in plants. Carotenoids play a crucial role in protecting chlorophyll from light, and if carotenoids are not present in the plant, chlorophyll will be damaged, thus preventing plants from performing photosynthesis. These group members also act as bleachers as they impart white color in plant tissues, when they come in contact with the foliage. Their action leads to cell and tissue injury which ultimately kills the weeds. These groups are represented by the chemical family of amides, anilidex, furanones, phenoxybutanamides, pyridiazinones, pyridines, and isoxazoles.

2.4 Advantages of using pesticides

Pesticides play a vital role in reducing threats caused by pathogens, weeds, and insects which drastically reduce the crop outcome and thereby improving agriculture commodity production and productivity to feed the population which is growing day by day. Malaria, which is a vector-borne disease, can be avoided by eliminating the vector and only by using insecticides it is possible to restrain the insects that cause mortal diseases. In addition, insecticides protect buildings and other wooden structures from damage caused by termites and other wood-boring insects. Numerous advantages of pesticides were classified by Cooper and Dobson in 2007 into primary and secondary benefits. According to them, primary benefits include better crop and livestock quality, as well as better crop and livestock yields, reduced labor and fuel use for weeding, protection of turf and controlling invasive species. Secondary advantages are those that come as a result of the primary benefits but are less immediately or visibly apparent. Food safety and security, increased export revenues, and reduced global disease spread are examples of secondary benefits [1].

2.5 Harmful effects of using pesticides

Besides having several advantages, pesticides can cause several adverse health problems in people and livestock and can also damage the environment. Pesticide residue absorption can cause acute and chronic toxicities by means of inhalation, ingestion, or penetration via dermal contact [20]. It primarily affects the human body by disrupting the metabolic and systemic activities, and can harm the immunological and endocrine systems. [21]. Further notable impacts of pesticides in the human health and ecosystem are presented in Figure 2.2.

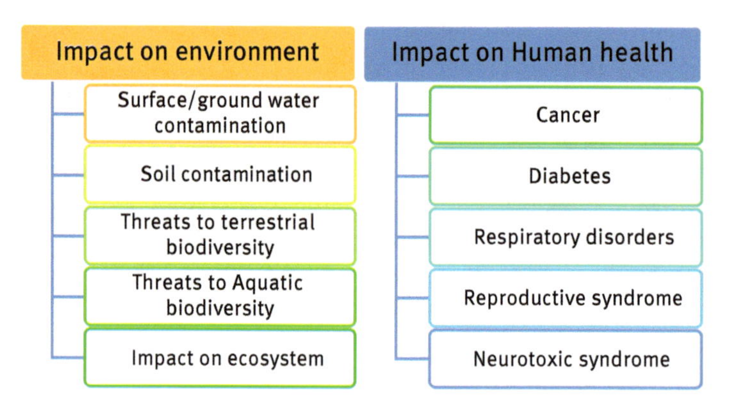

Figure 2.2: Impact of pesticides on human health and environment.

2.5.1 Adverse health effects in humans

Pesticides can enhance human health by reducing vector-borne diseases; still, their continuous and indiscriminate usage has developed major health consequences. As per the WHO, 3,000,000 cases of pesticide poisoning and about 220,000 deaths are reported in developing nations every year [22]. Some people are more vulnerable to the hazardous effects of pesticides than others, like the workers in the pesticide manufacturing organizations, infants, young children, and pesticide applicators [23]. Pesticides have a wide range of consequences on human health which are produced by direct exposure, pesticide handling, or their residues in the food stuffs. Three types of toxicity can be caused by pesticide: acute, delayed, or chronic and allergic. The effect of acute toxicity may occur within 24 h of pesticide contact, and causes headache, allergic reactions in nose and throat, stinging of the eyes and skin, skin irritations, breathing difficulties, sleeplessness, diarrhea, cramps, sickness and vomiting, impaired vision, and leads to death in rare cases. Chronic effects are long-term effects that may not appear even for years but cause prolonged illness, such as neurological disorders, memory loss problems, hypersensitivity, leukemia, brain cancer, and reproductive disorders, and long-term pesticide exposure may also damage the organs and may cause blood diseases. Allergic reactions are hazardous reactions that some people have in response to chemicals that do not have the same impact on others. Asthma, skin irritation, open sores, watery eyes, sneezing, and so on are all allergic reactions in response to pesticides.

2.5.2 Diabetes

Insecticidal exposure triggers gut microbes to produce glucose in the liver, leading to high blood sugar or hyperglycemia in the body, leading to diabetes. A study conducted on degradation of OP pesticides suggests that upon degradation, OPs produce short chain fatty acids, like acetates that via by gluconeogenesis converted to glucose in the liver and contributes to high glucose level in body and results in diabetes [24]. Several investigations have confirmed the correlation between exposure of pesticide and type-2 diabetes. A positive correlation between the OC pesticides and type-2 diabetes mellitus (T2DM) was observed in a study conducted in North India [25]. The study suggested that lipophilic OCs accumulate in the body as a result of taking contaminated water with high OCs, potentially increasing the risk of T2DM. A study also confirmed the risk of gestational diabetes in pregnant women due to contact with herbicides and insecticides [26]. Thus, with prolonged exposure with the pesticides increased the risk of diabetes.

2.5.3 Respiratory disorders

Respiratory disorders usually occur on the exposure of organic dusts, toxic gases, and highly volatile pesticide products, which mainly happen without using respiratory protective equipment. Respiratory disorders include asthma, bronchitis, oversensitivity pneumonitis, and lung cancer. Many pesticides can affect the bronchial lining by causing irritation, inflammation, immunological suppression, and other stimuli. Among these, OP pesticides are linked to cause asthma to a greater extent. They act by inhibiting acetylcholinesterase enzyme which hydrolyzes the neurotransmitter acetylcholine. The inhibition process leads to accumulation of acetylcholine and its persistence in nicotinic and muscarinic receptors in target tissues, resulting in cholinergic overexpression. Cholinergic overexpression of muscarinic receptors M3 on the smooth muscle of the airways causes bronchoconstriction. [27, 28]. Previous studies have revealed that occupational exposure to organophosphates, carbamates, and herbicides with phenoxy acids also lead to development of lung cancer with a high mortality rate. Chlorpyrifos, diazinon, metolachlor, pendiemethalin, dieldrin organochlorine, and carbofuran carbamates have been extensively studied in relation to the lung cancer risks [29, 30].

2.5.4 Neurotoxicity

Some pesticides, such as organophosphates, OC, and carbamates, produce toxic effects on the central and peripheral nervous systems, leading to long-term neurological illnesses like Alzheimer's and Parkinson's disease.

- **Alzheimer's disease**

Alzheimer's disease is a type of dementia that causes gradual memory loss and power loss owing to cerebral cortex deterioration. Pesticide disrupts the neuron function by disturbing the normal functioning of microtubules and causes hyperphorylation which leads to Alzheimer's disease. OP and OC pesticides have been reported to affect the function of acetylcholinesterase in the nervous system synaptic junctions, which can lead to disease, especially in people who are exposed in their later years [31].

- **Parkinson's disease**

It is a central nervous system disorder that arises when dopamine is not secreted by the substantia nigra neuron (dopaminergic neuron) in the brain resulting in trembling, lack of coordination, and loss of muscle control. Some investigations showed that some pesticides like rotenone and paraquat disrupt the dopaminergic neuron and inhibit the secretion of dopamine leading to Parkinson's disease [32].

2.5.5 Cancer

Pesticide exposure may cause cancer, and investigations have showed that many pesticides such as sulfallate, OCs, and sulfates are carcinogenic in nature while other pesticides lindane and chlordane are tumor causing. The carcinogenic effects of pesticides are the most dangerous for both adults and children. Increasing evidence derived from epidemiological and agricultural health studies suggested that continuous pesticide exposure can cause different types of cancers as enlisted in Table 2.2 [33–36].

Table 2.2: Immune alteration-related cancer associated with pesticides [34–36].

Cancer	Pesticide
Leukemia	Chlordane, chlorpyrifos, diazinon, EPTC, fonofos
Non-Hodgkin lymphoma (NHL)	Lindane, chlordane
Brain cancer	Chlorpyrifos
Prostate cancer	Methyl bromide, DDT, lindane, and simazine
Colon cancer	Aldicarb, dicamba, imazethapyr
Pancreatic cancer	EPTC and pendimethalin
Lung cancer	Chlorpyrifos, diazinon, dicamba
Rectum cancer	Chlordane, chlorpyrifos
Bladder cancer	Imazethapyr

2.5.6 Reproductive disorders

Certain pesticides can cause reproductive problems in men and women, including sperm deformity, low fertility, lack of male child, abortion, birth malformations, and retardation in fetal growth. Several OP pesticides such as chlorpyrifos, diazinon, malathion, parathion are linked with adverse reproductive effects such as sperm aneuploidy, sperm DNA fragmentation in men, reduces birth weight and gestational age in women.

2.6 Impact on environment

Pesticides have a number of environmental consequences, which are discussed in the following sections.

2.6.1 Surface and ground water contamination

Through drainage, drift, runoff, and leaching, pesticide contaminates the surface water, causing harm to the living organism. The two principal mechanisms by which pesticide causes ecological impact are bioconcentration and biomagnification.

- **Bioconcentration:** Bioconcentration is the accumulation of a chemical in or on an organism when the source of chemical is solely water. This is the movement of chemicals from the source (surrounding medium) into the organism. The primary "sink" for some pesticides is fatty tissue ("lipids"). DDT, for example, is a lipophilic insecticide that accumulates in the fatty tissue, such as in edible fish tissue and human fatty tissue. Other pesticides such as glyphosate are metabolized and excreted.
- **Biomagnification:** It is also known as bioamplification and it describes how a chemical or pesticide's concentration rises as food energy is transformed within the food chain in each tropic level. In the higher tropic level, when an organism in the food chain consumes the lower organism containing such chemicals, the chemicals can get accumulated in the organism present in higher tropic level. The process of biomagnification can be described as follows (Figure 2.3):
  1. Small concentration of pesticide gets into the bodies of animal that are present in lower tropic level in the food chain, such as the small fishes (primary consumer).
  2. Large fishes (secondary consumer) will eat many small fish and therefore the concentration of pesticide will increase in their body.
  3. When the high-level predator such as the vulture eats large fishes and other prey, the pesticide content in its body increases several times.

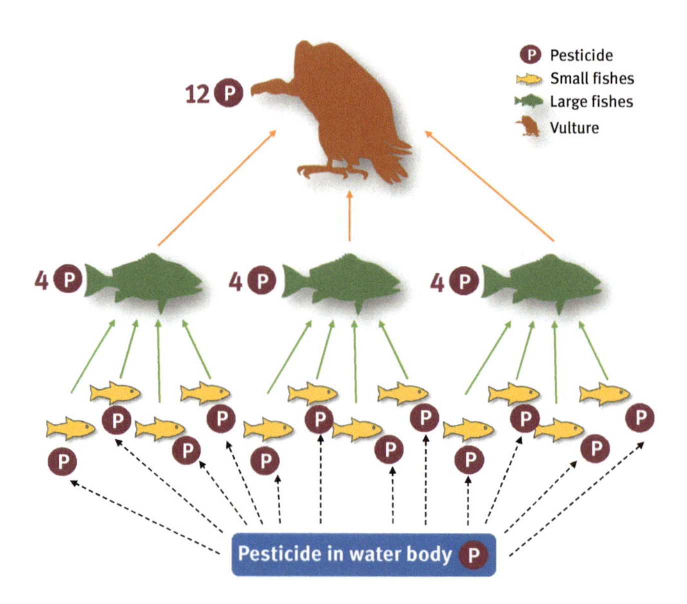

Figure 2.3: Biomagnification of pesticide in the environment.

The whole ecosystem disrupts by this process as the concentration of pesticide will go on increasing in each tropic level, and as a result, the population of secondary consumer will increase and primary consumer will decrease [37].

Pesticide from treated plants and soil reaches the surface water through runoff and leaching and contaminate it. Similarly, the ground water contamination through pesticide and other heavy metals is also a global problem. According to the US Geological Survey, at least 143 different pesticides and 21 transformation products have been found to occur in ground water, out of which pesticide is the major chemical class and once the ground water is polluted with toxic chemicals, it may take many years to dissipate or remove the contamination.

2.6.2 Soil contamination

Pesticides and fertilizers used in agriculture over long periods of time may impair the soil microbial activity and processes, which negatively affect the soil nutrient cycle and crop productivity. The pesticides used in soil treatment get converted to different transformation products which might result in declination in the population of beneficial microorganisms. Microbial biomass is an important component of soil organic matter and plays a key function in the soil nutrient cycle [38]. The declining population of beneficial microorganisms in soil reduces soil fertility which can also lead to poor crop output and productivity, lowering farm income and increasing hunger and poverty. Pesticides adversely affect the activity of soil enzymes that plays an important role of biocatalysts to enhance the soil quality, to manage nutrient cycle, and also involved in fertilization. Pesticides adversely affect the metabolic activity of several enzymes such as acid phosphatase, phosphatase, dehydrogenase, fluorescein diacetate hydrolase, β-glucosidase, urease, aryl-sulfatase, and alkaline phosphatase that are involved in the nutrient cycles [39, 40].

2.6.3 Risk to aquatic species

Water contaminated with pesticides poses a significant threat to aquatic form of life by means of lowering the dissolved oxygen level in water bodies. Pesticides and other toxic chemicals enter water bodies through a variety of routes, including spillage, industrial effluent, surface runoff, and pesticide-treated soils that affect the aquatic species at different tropic levels, ranging from zooplanktons to large fishes, causing physiological and behavioral changes in fish populations [41]. Aquatic lives get exposed to pesticides by three different means:

- Dermally: Penetrate to the skin via direct absorption
- Breathing: Reach to the gills during respiration
- Orally: Enter the body by taking pesticide-contaminated water

Aquatic plants provide about 80% of the total dissolved oxygen to the aquatic species. Killing aquatic plants by using herbicides drastically lowers the oxygen level in water bodies, which leads to suffocation of fish and ultimately leads to death [42]. Atrazine is the most common herbicide that causes negative effects on few fish species, and has a big impact on the immunity of few amphibians. Pesticide-contaminated water contributes to overexploitation and habitat loss of amphibians and all of which disrupts the entire aquatic ecosystem.

2.6.4 Threats to terrestrial biodiversity

Pesticide exposure can cause sublethal effects on terrestrial plants, in addition to affecting nontarget plants. Some herbicides like glyphosate make plants more prone to disease and deteriorate the quality of seeds [43]. Many other herbicides such as imidazolinones, sulfonylureas, and sulfonamides have disastrous effects on nontargeted crops, natural plant communities, and wildlife. Many poisonous pesticide groups such as organophosphates, carbamate insecticides, and anticoagulant rodenticides are directly involved in declining and affecting bird populations and can lead to detrimental changes in their behavior. Rare and endangered bird species are threatened by broad-spectrum herbicides as these herbicides reduce the abundance of weeds consumed by birds. Similarly, the use of insecticides declines the number of insects which are the primary source of feeding for birds that can produce a significant drop in rare species populations [44]. Use of broad-spectrum chemicals significantly declines the populations of beneficial insects like bees and beetles. Many of these species are important pollinators, food web components, and natural insect pest antagonists. Some carbamate pesticides including aldicarb, carbofuran, benomyl, methiocarb, organophosphates (chlorpyrifos, diazinon, dimethoate, and fenitrothion), pyrethroids (cyfluthrin and cyhalothrin), and neonicotinoids (imidacloprid, carbofuran, and thiamethoxam) are all highly poisonous to bees, bumblebees, and other useful insects [44, 45]. Herbicides such as trifluralin and oryzalin also cause considerable damage to inhibit the growth of symbiotic mycorrhizal fungi which help in nutrient uptake. The atmospheric nitrogen fixation to nitrates involves several soil microorganisms. Fungicides like chlorothalonil and dinitrophenyl inhibit bacteria-dependent nitrification and denitrification processes and similarly the herbicide, triclopyr, prevents ammonia from being converted to nitrite by inhibiting soil bacteria involved in the transformation [46, 47]. Glyphosate, a nonselective herbicide, reduces the growth and activity of nitrogen-fixing bacteria in soil, whereas 2,4-D inhibits the transformation of ammonia into nitrates carried out by the soil bacteria [48, 49]. Glyphosate and chlorpyrifos have a negative effect on earthworms at the cellular level, damaging the DNA. Earthworms are important members of the soil ecosystem because they serve as bioindicators of soil pollution, models for soil toxicity testing, and contribute to soil fertility. Many other pesticides produce neurotoxic effects in earthworms and after a long time of exposure, they get physiologically damaged [50].

2.6.5 Risk of resistance development in insect pests and weeds

Insect pest species may acquire resistance to pesticides if they are used continuously and indiscriminately. Insects that have more susceptibility to the pesticide die quickly, whereas some resistant individuals stay. These resistances may occur by the alteration in the genetic makeup with respect to the insecticides. Continuous breeding of such insect species produces resistant insect population against them and the insecticides become ineffective. For instance, in the 1950s, housefly population became resistant against DDT and pyrethroids as they both have similar types of mode of action. As the resistance was developed in housefly population against both DDT and pyrethroids because of continuous exposure to DDT only, this phenomenon is called as cross-resistance. A closely related phenomenon, multiple resistance, is also reported, in which insects became resistant against two or more classes of insecticides with unlike mode of actions [51]. A similar phenomenon is also seen in weed species and is also difficult to control by the preexisting herbicides.

2.7 Alternative preventive strategies for pesticide management

Because of the hazardous effects of pesticides toward the environment and human health, there must be execution of remedial strategies that facilitate the less use of pesticides like banning most toxic pesticides, adopting integrated pest management, and using botanicals and ecofriendly agents against plant protection. Integrated Pest Management (IPM) emerged as a result of efforts to decline the total reliance on synthetic pesticides for pest management. IPM is "a pest management system that, in the context of the associated environment and the pest species population dynamics, employs all the appropriate methods and techniques in as compatible a manner as possible, and maintains the pest populations below the level at which they cause economical damage or loss" [52]. IPM seeks to combine the use of biological, cultural, and mechanical practices to manage or control insect pests in agricultural production. It allows the use of chemical pesticides only when it is necessary or when the pest population cannot be controlled with natural means. Botanical pesticides are safer for the environment, with a novel mode of action, and are rich source of biologically active components. As they are biodegradable, ecofriendly in nature, have the ability to change the behavior of target pests, and possess favorable safety profile, they can be recommended as an ecochemical and sustainable strategy for the management of pests and diseases.

2.8 Conclusion

Agrochemicals are considered as key elements in agriculture production in the form of fertilizers, pesticides, soil health amendments, soil conditioners, plant growth regulators, and many more. Among these pesticides are the agrochemicals which are used for plant protection against different types of pests and diseases. As plant protection agrochemicals have a nonforgettable role in agricultural crop production by increasing the quality and production by minimizing the damages caused by pest and diseases, these also have several consequences related to human health, livestock, nontarget organisms along with tremendous environmental impacts. These consequences need to be taken into consideration for minimizing the health and environmental risks of pesticide uses. There is a need to reduce the harmful impacts of pesticides by either minimizing the uses or by strengthening the alternative options such as botanicals, biopesticides, and integrated pest management practices. There is a need of in-depth study about the consequences related to plant protection chemicals and to find out the strategies for their mitigation.

References

[1] Cooper J, Dobson H. The benefits of pesticides to mankind and the environment. Crop Protection 2007, 26(9), 1337–1348.
[2] Gupta S, Dikshit AK. Biopesticides: An eco-friendly approach for pest control. Journal of Biopesticides 2010, 3(1), 186–188.
[3] Rao GV, Rupela OP, Rao VR, Reddy YV. Role of biopesticides in crop protection: Present status and future prospects. Indian Journal of Plant Protection 2007, 35(1), 1–9.
[4] Miller GT. Living in the environment. 12th edn, Belmont, Wadsworth/Thomson Learning, 2002.
[5] Jabbar A, Mallick S. Pesticides and environment situation in Pakistan (Working Paper Series No. 19). Available from Sustainable Development Policy Institute, 1994.
[6] Delaplane KS. Pesticide usage in the United States: History, benefits, risks, and trends. In: Cooperative extension service. 2000, 1–6.
[7] Ware GW, Whitacre DM. An introduction to insecticides. The pesticide book, 2004, 6.
[8] Fukuto TR. Mechanism of action of organophosphorus and carbamate insecticides. Environmental Health Perspectives 1990, 87, 245–254.
[9] Ishaaya I, Degheele D. Insecticide with novel mode of action-mechanism and application. Springer Verlag, 1998.
[10] Taillebois E, Cartereau A, Jones AK, Thany SH. Neonicotinoid insecticides mode of action on insect nicotinic acetylcholine receptors using binding studies. Pesticide Biochemistry and Physiology 2018, 151, 59–66.
[11] Doehlemann G, Okmen B, Zhu W, Sharon A. Plant pathogenic fungi. Microbiology Spectrum 2017, 5(1), 5–1.
[12] Cornelissen BJ, Melchers LS. Strategies for control of fungal diseases with transgenic plants. Plant Physiology 1993, 101(3), 709.
[13] Ndayihanzamaso P, Karangwa P, Mostert D, Mahuku G, Blomme G, Beed F, Viljoen A. The development of a multiplex PCR assay for the detection of *Fusarium oxysporum* f. sp. cubense lineage VI strains in East and Central Africa. European Journal of Plant Pathology 2020, 158(2), 495–509.
[14] Hariharan G, Prasannath K. Recent advances in molecular diagnostics of fungal plant pathogens: A mini review. Front Cell Infect Microbiol 2021, 10, 829.
[15] Rachid R. Introduction and Toxicology of Fungicides, Fungicides, Odile Carisse (Ed.), ISBN: 978-953-307-266-1. Intechopen, 2010.
[16] Rani L, Thapa K, Kanojia N et al. An extensive review on the consequences of chemical pesticides on human health and environment. Journal of Cleaner Production 2021, 283, 124657.
[17] Grossmann K. Auxin herbicides: Current status of mechanism and mode of action. Pest Management Science 2010, 66, 113–120.
[18] Konishi T, Sasaki Y. Compartmentalization of two forms of acetyl-CoA carboxylase in plants and the origin of their tolerance toward herbicides. Proceedings of the National Academy of Sciences USA 1994, 91, 3598–3601.
[19] Herrmann KM, Weaver LM. The shikimate pathway. Annual Review of Plant Physiology and Plant Molecular Biology 1999, 50, 473–503.
[20] Spear R. Recognized and possible exposure to pesticides. In: Handbook of pesticide toxicology. 2011, 245–275.
[21] Wesseling C, McConnell R, Partanen T, Hogstedt C. Agricultural pesticide use in developing countries: Health effects and research needs. International Journal of Health Services : Planning Administration Evaluation 1997, 27, 273–308.

[22] Lah K. Effects of pesticides on human health. Toxipedia. Available from http://www.toxipedia.org/display/toxipedia/Effectsof+Pesticides+on+Human+Health.2014.
[23] Maroni M, Fanetti AC, Metruccio F. Risk assessment and management of occupational exposure to pesticides in agriculture. La Medicina Del Lavoro 2006, 97(2), 430–437.
[24] Velmurugan G, Ramprasath T, Swaminathan K et al. Gut microbial degradation of organophosphate insecticides-induces glucose intolerance via gluconeogenesis. Genome Biology 2017, 18(1). 1–18.
[25] Tyagi S, Siddarth M, Mishra BK, Banerjee BD, Urfi AJ, Madhu SV. High levels of organochlorine pesticides in drinking water as a risk factor for type 2 diabetes: A study in north India. Environmental Pollution 2021, 271, 116287.
[26] Saldana TM, Basso O, Hoppin JA et al. Pesticide exposure and self-reported gestational diabetes mellitus in the agricultural health study. Diabetes Care 2007, 30(3). 529–534.
[27] Hernández AF, Parrón T, Alarcón R. Pesticides and asthma. Current Opinion in Allergy and Clinical Immunology 2011, 11(2), 90–96.
[28] Shaffo FC, Grodzki AC, Fryer AD, Lein PJ. Mechanisms of organophosphorus pesticide toxicity in the context of airway hyperreactivity and asthma. American Journal of Physiology-Lung Cellular and Molecular Physiology 2018, 315(4), L485–L501.
[29] Pesatori AC, Sontag JM, Lubin JH, Consonni D, Blair A. Cohort mortality and nested case-control study of lung cancer among structural pest control workers in Florida (United States). Cancer Causes & Control 1994, 5(4), 310–318.
[30] Bonner MR, Lee WJ, Sandler DP, Hoppin JA, Dosemeci M, Alavanja MC. Occupational exposure to carbofuran and the incidence of cancer in the agricultural health study. Environmental Health Perspectives 2005, 113, 285–289.
[31] Hayden KM, Norton MC, Darcey D et al. Occupational exposure to pesticides increases the risk of incident AD: The cache county study. Neurology 2010, 74(19). 1524–1530.
[32] Le Couteur DG, McLean AJ, Taylor MC, Woodham BL, Board PG. Pesticides and Parkinson's disease. Biomedicine and Pharmacotherapy 1999, 53(3), 122–130.
[33] Alavanja MC, Bonner MR. Occupational pesticide exposures and cancer risk: A review. Journal of Toxicology and Environmental Health Part B Critical Reviews 2012, 15(4), 238–263.
[34] Purdue MP, Hoppin JA, Blair A, et al. Occupational exposure to organochlorine insecticides and cancer incidence in the agricultural health study. International Journal of Cancer 2007, 120, 642–649.
[35] Lee WJ, Hoppin JA, Blair A, et al. Cancer incidence among pesticide applicators exposed to alachlor in the agricultural health study. American Journal of Epidemiology 2004, 159, 373–380.
[36] Mokarizadeh A, Faryabi MR, Rezvanfar MA, Abdollahi M. A comprehensive review of pesticides and the immune dysregulation: Mechanisms, evidence and consequences. Toxicology Mechanisms and Methods 2015, 25(4), 258–278.
[37] Katagi T, Tanaka H. Metabolism, bioaccumulation, and toxicity of pesticides in aquatic insect larvae. Journal of Pesticide Science 2016, 41(2), 25–37.
[38] Azam F, Farooq S, Lodhi A. Microbial biomass in agricultural soils-determination, synthesis, dynamics and role in plant nutrition. Pakistan Journal of Biological Sciences 2003, 6(7), 629–639.
[39] MdMeftaul I, Venkateswarlu K, Dharmarajan R, Annamalai P, Megharaj M. Pesticides in the urban environment: A potential threat that knocks at the door. The Science of the Total Environment 2019, 711, 134612.
[40] Riah W, Laval K, Laroche-Ajzenberg E, Mougin C, Latour X, Trinsoutrot-Gattin I. Effects of pesticides on soil enzymes: A review. Environmental Chemistry Letters 2014, 12, 257–273.

[41] Scholz NL, Fleishman E, Brown L, Werner I, Johnson ML, Brooks ML, Mitchelmore CL. A perspective on modern pesticides, pelagic fish declines, and unknown ecological resilience in highly managed ecosystems. Bioscience 62(4), 428–434.
[42] Helfrich LA, Weigmann DL, Hipkins P, Stinson ER. Pesticides and aquatic animals: A guide to reducing impacts on aquatic systems. In: Virginia Polytechnic Institute and State University, 2009.
[43] Brammall RA, Higgins VJ. The effect of glyphosate on resistance of tomato to Fusarium crown and root rot disease and on the formation of host structural defensive barriers. Canadian Journal of Botany 1988, 66, 1547–1555.
[44] Isenring R. Pesticides and the loss of biodiversity. London, Pesticide Action Network Europe, 2010, 26.
[45] Yang EC, Chuang YC, Chen YL, Chang LH. Abnormal foraging behavior induced by sublethal dosage of imidacloprid in the honey bee (Hymenoptera: Apidae). Journal of Economic Entomology 2008, 101, 1743–1748.
[46] Lang M, Cai Z. Effects of chlorothalonil and carbendazim on nitrification and denitrification in soils. Journal of Environmental Sciences 2009, 21, 458–467.
[47] Pell M, Stenberg B, Torstensson L. Potential denitrification and nitrification tests for evaluation of pesticide effects in soil. Ambio 1998, 27, 24–28.
[48] Frankenberger WT, Tabatabai MA, Tabatabai MA. Factors affecting L-asparaginase activity in soils. Biology and Fertility of Soils 1991, 11, 1–5.
[49] Santos A, Flores M. Effects of glyphosate on nitrogen fixation of free-living heterotrophic bacteria. Letters in Applied Microbiology 1995, 20, 349–352.
[50] Schreck E, Geret F, Gontier L, Treilhou M. Neurotoxic effect and metabolic responses induced by a mixture of six pesticides on the earthworm Aporrectodea caliginosa nocturna. Chemosphere 2008, 71(10), 1832–1839.
[51] Buhler W. (n.d.). Introduction to Insecticide Resistance. retrieved from https://pesticidestewardship.org/resistance/insecticide-resistance/on30, December, 2021.
[52] FAO. Report of the first session of the FAO panel of experts on integrated pest control. Rome, Food and Agriculture Organization of the United Nations, 1967.

Parul Chaudhary, Shivani Singh, Anuj Chaudhary, Upasana Agri, Geeta Bhandari

Chapter 3 Agrochemicals and their effects on soil microbial population

Abstract: Recently, the agriculture system depends on the usage of agrochemicals, which enhance the crop productivity by providing nutrition to plants to meet the global food requirements. A prolonged extensive use of agrochemicals is a principal source of contamination, which persists and biomagnifies in nature and affects the soil characteristics and sustainability. Soil microbes are key components of the agricultural ecosystem to improve crop productivity, and their actions are crucial for maintaining the fertility and health of the soil. Agrochemicals directly or indirectly lead to a shift in the diversity, richness, and evenness of nontargeted beneficial microorganism which orderly decreases the availability of plant nutrition and increases the fate of disease in crops. This chapter focusses on agrochemicals, their classification, and their impact on the soil microflora, especially bacterial and fungal population of cultivated soils.

3.1 Introduction

Biological agents, chemical agents, and equipment used during agricultural production are referred to as agricultural inputs [1]. Agrochemical or agrichemical refers to chemicals used in agriculture. These chemicals are fertilizers, soil conditioners, liming agents, plant growth hormones, and pesticides [2]. The present status of world's population is 7.2 billion, which may increase up to 9.3 billion in 2050, thereby increasing the global demand of food to a greater extent [3]. Boxall et al. [4] also reported that the crop production is lost around 25% worldwide because of pests, weeds, and pathogenic microbes. Therefore, weeds and insects are major reducing biotic sources of agriculture [5]. Before the industrial revolution, organic farming system was used, which protects the natural environment without using agrochemicals, whereas conventional or modern farming system completely depends on the use of agrochemicals [6]. These agrochemicals are generally used worldwide to enhance the crop production by using pesticide, which is controlling pest, disease, pathogens, and weeds, and on the other side, by using fertilizers, nutrient is added to the soil [7]. These chemicals are moreover applied directly to the upper layer of soil or sprayed over the crop later. Due to constant usage of agrochemicals and their

https://doi.org/10.1515/9783110771558-003

degraded products, these enter the water bodies as agricultural runoff and accumulate in soil which severely affects soil microbes and so detrimental for soil health [8].

Agrochemicals absorbed by plants get transferred to the ecosystem through runoff in water, by leaching in soil and by volatilization in air. Alongside contaminating the environment, agrochemicals, especially pesticides, are also detrimental to microflora and microfauna. They also hindered the absorption of important minerals and nutrients from soil to plants [9]. Leached agrochemicals pollute the drinking water and ground water. The leached chemicals can affect both aquatic and terrestrial organisms and also get biomagnified through food chain to affect the human health. The use of agrochemicals is also affected by climatic factors. The Intergovernmental Panel on Climatic Change (IPCC) assessed that the agrochemicals account for 24% of global greenhouse gas emission. Volatilization of pesticides and fertilizers will occur when temperature increases, resulting in the contamination of air [10]. Heavy rainfall causes the leaching of chemicals in the soil. Accumulation of these chemicals in the soil has direct consequence on soil microbiota. Soil microflora are the crucial basics that respond fast to any alteration in the soil system due to the entrance of agrochemicals. Intensification in agricultural efficiency is usually related to the usage of agrochemicals [11]. Long-term application of NPK fertilization decreased the bacterial and fungal abundance and diversity, while potassium along with straw improved the bacterial population and the rice crop yield [12].

Soil microorganisms are widely accepted bioindicators of soil health because they play an imperative part in maintaining soil nutrient, recycling of chemicals, and biodegradation of harmful chemicals, and also act as a biocontrol agent of phytopathogens [13–15]. Evaluation and analysis of microbial community of a system provide the active health status of that system, especially soil. Numerous rhizospheric microbes are identified to secrete enzymes such as alkaline phosphatase, proteases, lipases, and dehydrogenases to get essential nutrients via the mineralization of composite polymers [16, 17]. Diverse microorganisms respond according to the environmental changes. Growing human activities, industrial pollution, and injudicious use of chemicals have led to soil degradation and thus fertility [18, 19]. More effective and accessible strategies are required by the farmers for precision farming in agricultural sector for enhanced crop production and maintenance of soil fertility [20]. There are various culturable and unculturable methods that are suitable for bacterial community analysis. Metagenomics has facilitated bacterial diversity analysis in the environment which reflected the long-term effects of any chemical.

China is at the top in using agrochemicals (about 1,763,000 tons), whereas India ranks 13 (52,750 tons/year). India ranks second for the largest use of pesticides in Asia. According to Mathur and Tannan [21], majority of agrochemicals used in India are pesticides and 76% of these pesticides are insecticides. Among the 28 states of India, consumption of agrochemicals, especially pesticides, is highest in Maharashtra and then in Uttar Pradesh. Deposition of agrochemicals in the soil directly affects microorganisms in the soil and also increases the risk to ecosystem, water bodies, and plants, and is also associated with human health disorders such as cancers, kidney disease, and arthritis [22]. Keeping in mind the extensive use of agrochemicals here, we discussed on the classification of agrochemicals and its impact on soil's microflora.

3.2 Different types of agrochemicals used in agriculture

Agrochemicals are broadly classified on the bases of their purposes into pesticides (plant-protecting chemicals), plant growth regulators (PGRs, plant hormones, stimulants, retardants, and additives), fertilizers (plant growth-promoting chemicals), soil conditioners (antibiotics and hormones), and acidifying agents (chemicals used to maintain soil pH) [23]. These chemicals are required for different purposes at diverse phases of plant growth to protect from various infections, as growth stimulant, and for maintaining the soil nutrient and pH [24].

3.2.1 Pesticides

Pesticides are classified as a plant protector which protects plants from infections and diseases caused by pests and enhances the crop yields. Also, these are defined as complex chemical substances which are used to inhibit or kill the pest. An ideal pesticide should act on the target harmful organisms, which is biodegradable and does not leach down the ground water [25]. Most of the pesticides used in agriculture produce vapor of pesticides which become air pollutant and also get absorbed in the soil through surface runoff from treated plants, and their deposition in soil directly influences the soil microorganism and can severely affect the surrounding ecosystem, plants, human health, and water bodies [26]. On the bases of the mode of target organisms, pesticides are further classified into algicides, fungicides, herbicides, insecticides, molluscicides, nematicides, and rodenticides [27].

Biocide that is used for killing or preventing the growth of algae is known as algicides. Benzalkonium chloride, copper sulfate, dichlone, diuron, simazine, and endothal are some common algicides. Any chemical which has the tendency to inhibit or kill the harmful fungus is known as fungicides. Common fungicides are Bordeaux mixture, copper sulfate, lime sulfur, captan, and formaldehyde. Chemicals that are used for killing unwanted plants (agriculture weeds and unnecessary weeds) are known as herbicides. In 1896, the first herbicide "Sinox" was developed. Glyphosate, glufosinate, dicamba, atrazine, 2,4,5-tri dichlorophenoxyacetic acid, and Vietnam orange are some examples [28].

Chemical agents such as (insecticides) are used to destroy insects at different stages such as eggs (ovicides), larvae (larvicides), and adults (adulticides) which disrupt molting, and maturation stages (insect growth regulators) disrupt the mating behaviors (pheromones). The common examples are organic nitrogen, organic phosphorus, organic chlorine, pyrethroids, and carbamates. Chemicals which are used to control gastropod pests (snails and slugs) and damage plants by feeding on them are known as molluscicides, and metaldehyde, niclosamide, Slug-Tox, and limatox are some examples. Chemicals which are used to kill plant parasitic nematodes are known as nematicides. Aldicarb, Nimitz, velum prime, and Mocap are some common examples [29]. Chemical agents which are used to either inhibit the growth or kill rodents, including rats, mice, and household pests (like chipmunks, woodchucks, and squirrels), are known as rodenticides. Some names of rodenticides are difenacoum, bromadiolone, warfarin, and zinc phosphide.

3.2.2 Plant growth regulators and retardants

Small, simple molecules other than nutrients produced by plants in small quantities to regulate growth, differentiation, morphogenesis, and metabolisms are known as PGRs [30]. Substances produced within plants are known as plant hormone or phytohormones. All phytohormones are derived from common metabolic pathways and they regulate the processes like defense, development, and plant growth [31]. All phytohormones are growth regulators but all growth regulators are not phytohormones. These hormones are auxins, cytokinins, gibberellins, ethylene, and abscisic acid [32–34]. Plant growth retardants are natural growth inhibitors that have the activities of antagonists to the plant growth-promoting hormones. They generally retard the growth of root and shoot elongation, bud opening, and seed germination [35].

3.2.3 Liming or acidifying agent

Due to the continuous use of chemicals in fields and acid rains, soil becomes either too acidic (pH 4–5) or too alkaline (pH 8–9). Crops cannot utilize proper nutrients when they grow on acidic or alkaline soil. Lime stone ($CaCO_3$) powder and crushed mussel shells are used to neutralize the acidic soils, whereas calcium sulfate (gypsum) and S_2 are used to neutralize the basic soils [36]. Lime is expensive for lowland farmers, so they prefer to use charcoal, biochar, and plant ash to neutralize (pH 6–7) the soil in a slow manner (Figure 3.1).

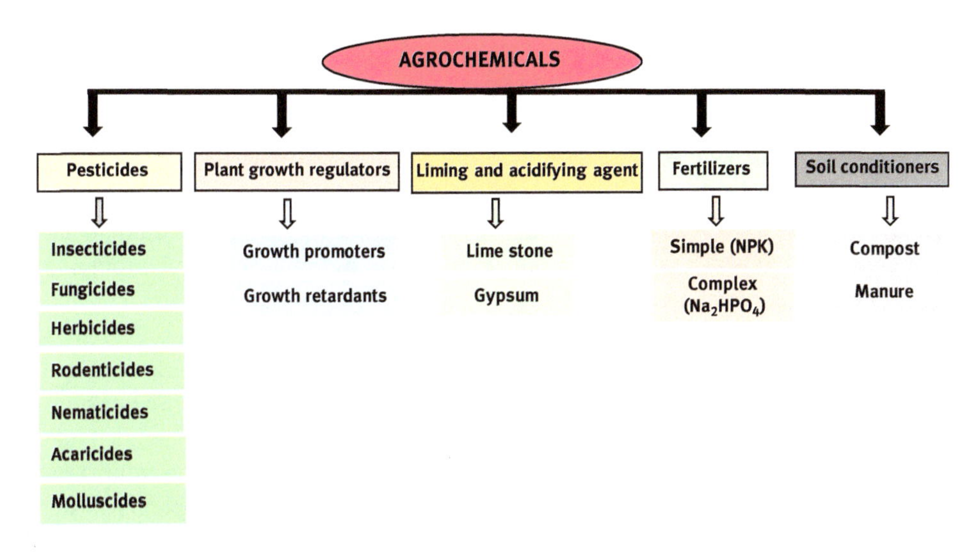

Figure 3.1: Types of agrochemicals used in agriculture practices.

3.2.4 Fertilizers

Plant growth-promoting substances are generally denoted as fertilizers. Any organic or inorganic mixture that supplies essential nutrients to crops in soluble forms is known as fertilizers. Organic or natural fertilizers include animal manure, compost, human manure, recycled waste, and by-products of industries, while synthetic fertilizers are nitrogen fertilizers (urea and ammonium sulfate), phosphate fertilizers, potassic fertilizers, and sulfate of potash (K_2SO_4) [37]. Fertilizers are classified as straight fertilizers, which supply one plant's primary nutrient at a time (either N or P or K fertilizers), whereas complex fertilizers provide two or three nutrients at a time like diammonium phosphate.

3.2.5 Soil conditioners

Soil conditioners are materials that are added to the soil to improve physiochemical properties of soil. Thus, these enhance the crop production by increasing the soil's ability to hold water and oxygen and to rebuild poor soil fertility, in alkali soil reclamation, for better root development, for high yields, and for the quality and release of locked nutrients. Materials used as soil conditioners are peat, compost, manure, sewage, bone meal, and sphagnum moss [38].

3.3 Effect of agrochemicals on soil fungal and bacterial population

Soil enzymes are considered as key “engines” in operating biogeochemical cycles in soil, which are used as an activity index of microorganisms and for the evaluation of soil fertility index [39, 40]. Soil bacteria are one of the largest functional microbial populations in soil which take part in transformation of nutrient, soil respiration, and decomposition of organic matter. Earlier studies have shown that soil microbes and enzyme activities are extremely sensitive to fluctuations in properties of soil such as pH, organic matter, moisture content, and mineral [41, 42]. Still, soil microbes are similarly observed as dynamic aspects driving the creation and turnover of enzymes and their actions in soil. Chemicals such as pesticides badly disturb the microbial metabolism and modify the enzymatic activities of soil [43, 44]. Various literatures reported that the evolution of different bacterial and fungal population depends on the concentration of applied agrochemicals, type of chemicals, incubation time, physicochemical properties, and the type of soil [45, 46]. Elaine Ingham, American microbiologist, stated that “if we lose both bacteria and fungi, then the soil degrades.” Bacterial and fungal population of the soil is important for maintaining soil fertility by utilizing large complex agrochemicals as sole energy or carbon sources [47]. Accumulation of agrochemicals in soil may undergo various transport, degradation, adsorption, and desorption which depend on the soil property and its microflora [48]. These chemicals affect the soil microorganism by disturbing the cell growth, cell division, photosynthesis, respiration, biochemical process, and so on. The utmost plentiful microbes existing in the soil are bacteria, and then actinomycetes, fungi, algae and protozoa are associated with the fertility of soil, cycling of nutrients, and breakdown of complex substances. Agrochemicals have both positive and negative consequences toward the microbes in soil (Figure 3.2).

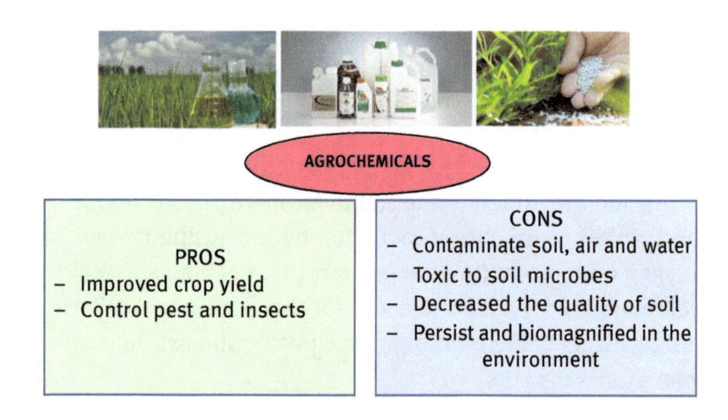

Figure 3.2: Positive and negative effects of agrochemicals.

Arbuscular mycorrhizal fungi (AFM) form mutualistic association with the terrestrial plant and enhance nutrient attainment from soil, which also has better quality tolerance toward stress conditions [49]. Several studies reported that persistence of agrochemical residues affects the growth and colonization of AFM, root colonizing bacteria, and other microbes [50]. Herbicides along with fertilizers decrease the functioning of soil microbes. Application of fomesafen inhibits the growth of mycorrhizal colonization and soil microbial biomass after 12 days of the herbicide treatment [51, 52]. Higher level of phosphorus decreases the mycorrhizal colonization and also decreases the diversity of AMF [53]. Fungal endophytes colonize the plant without producing the disease symptoms, increase the plant defense against pests (phytopathogens), and also provide nutrients to plants [54]. Due to the effect of agrochemicals, the nature of soil gets changed, which can in turn affect the fungal endophytes, and other associated microflora as a result directly or indirectly affect the plant health.

Application of pesticides reduces the bacterial and fungal biomass by 76% and 47%, respectively, on average after 9 days [55]. Long-term exposure of different chemical fertilizers usually shifts the soil microbial diversity. Agrochemicals and their degradation products inhibit the microbial diversity which in turn responsible for soil infertility [56]. Fungal population of *Gibberella*, *Candida*, and *Saitozyma* was susceptible toward pesticides (imidacloprid and metribuzin) along with starch mixture [57]. Application of tebuconazole and clotrimazole strongly reduced the fungal diversity by targeting on reproduction, whereas terbinafine stimulate the fungal growth at very low concentration [58]. Herbicides like clopyralid, ioxynil, mecoprop, and bifenox inhibit the spore germination at low concentration but act as growth stimulator on higher concentrations [59]. Fenpropimorph (0.02 mg/L) shows negative effect on the spore production, root colonization, and hyphal length of fungus [60]. The herbicides oryzalin, oxadiazon, and trifluralin inhibit the spore formation and propagation of mycorrhizae [61]. Glycophosphates inhibited the root mycorrhization growth (40%) and also influenced the metabolite production in plants [62]. Azadirachtin is an insecticide that shows a negative effect on the growth of arbuscular fungus [63]. The pesticide “isoproturon” affects the fungal growth and favors the growth of bacterial populations. Captan is a harmful pesticide, which affects the microbial population such as fungi, and nitrifying and nitrogen-fixing species [64]. Chlorpyrifos is the organophosphorus insecticide, which is widely used to control insects, termites, and beetles throughout the world. The soil treated with chlorpyrifos showed a decrease in the colony-forming count of bacteria and fungi after 14 days of treatment [65]. Application of pesticides influences the species richness of endophytic fungi such as *Colletotrichum*, *Phyllosticta*, and *Pleosporales* [66].

Nitrogenase enzyme involved in nitrogen fixation is inhibited by the application of endosulfan and carbendazim under pot and field condition [67]. Isoproturon and profenofos inhibited the urease activity which is involved in the nitrogen cycle.

Herbicides can also affect the nitrogen fixation process and influence the nodulation in legumes by distressing rhizobacterial infection process. They also reduce the nitrogenase enzyme, plant's dry matter, and ATP synthesis in *Rhizobium* and *Bradyrhizobium*. The application of starch mineral mixture along with pesticide (carbendazim) decreases the microbial population of *Burkholderia* and *Mycobacterium* [68]. Application of inorganic fertilizers enhanced the abundance of Bacteroidetes, Acidobacteria, and phylum of fungi belonging to Zygomycota population, while organic fertilization enhanced the Proteobacteria phylum and fungal classes such as Orbiliomycetes classes [69]. Nitrogen fertilization decreased the P-solubilizers, alkaline phosphatase activity, and Gammaproteobacteria [70]. Kumar et al. [71] described that potassium and farmyard manure enhanced the soil enzyme activity and microbial population under wheat cultivation. NPK fertilization (450 kg C/ha) along with soybean litter under paddy cultivation enhanced the microbial biomass and Actinobacteria population [72]. Chemical fertilization of N causes a decrease in the bacterial alpha diversity and also reduced the microbial biomass [73]. Mineral fertilization (N) increases the nitrification rate but lessens the abundance of *Nitrospira* and increases the richness of *Nitrosomonas* [74]. On the other hand, N fertilization improved the population of *Azotobacter, Actinomycetes*, and phosphate-solubilizing microbes [75]. Butachlor inhibited the enzyme activities of protease, urease, and dehydrogenase under flooded conditions [76]. The usage of paclobutazol affects the fertility and quality of soil, which causes ecosystem unevenness [77]. Therefore, to protect the beneficial natural microflora which is very important for maintaining the soil fertility, nowadays agrochemicals are substituted by biopesticides, bioherbicides, and transgenic plants.

3.4 Conclusion

Scientists tried to find the impact of agrochemicals on soil microbes which are involved in the maintenance of soil fertility. The practice of agrochemicals in the agriculture field for a long time may generate a lethal effect on activities of soil microbes that affect nutrient cycling and crop productivity. Thus, it is important to use agrochemicals at optimal concentration to maintain the environmental balance and sustenance of soil health. Current methods for better usage of agrochemicals would be suitable in reducing lethal possessions and avoiding human well-being threats. Furthermore, to protect the beneficial natural microflora which is very important for maintaining soil fertility, nowadays, the application of biopesticides, nanofertilizers, nanobiofertilizers, and nanopesticides should be emphasized to circumvent the inconsiderate usage of agrochemicals.

References

[1] Zhang L, Yan C, Guo Q, Zhang J, Ruiz-Menjivar J. The impact of agricultural chemical inputs on environment: Global evidence from informetrics analysis and visualization. International Journal of low-Carbon Technologies 2018, 13, 338–352.
[2] Mandal A, Sarkar B, Mandal S, Vithanage M, Patra AK, Manna MC. Impact of agrochemicals on soil health. In: Agrochemicals detection, treatment and remediation. Butterworth-Heinemann, 2021, 161–187.
[3] FICCI (Federation of Indian Chambers of Commerce and Industry). Ushering in the 2nd green revolution: Role of crop protection chemicals. Fine Chemicals Corporation. What are Agrochemicals? Jayhaws News. 2016. https://www.jayhawkchem.com/2018/05/11/what-areagrochemicals/
[4] Boxall AB, Hardy A, Beulke S, Boucard T, Burgin L, Falloon PD, Williams RJ. Impacts of climate change on indirect human exposure to pathogens and chemicals from agriculture. Environment Health Perspectives 2009, 117, 508–514.
[5] Oliveira CM, Auad AM, Mendes SM, Frizzas MR. Crop losses and the economic impact of insect pests on Brazilian agriculture. Crop Protection 2014, 56, 50–54.
[6] Singh D, Singh SK, Modi A, Singh PK, Zhimo VY, Kumar A. Impacts of agrochemicals on soil microbiology and food quality. In: Agrochemicals detection, treatment and remediation. Butterworth-Heinemann 2020, 101–116.
[7] Varma D, Meena RS, Kumar S. Response of mungbean to fertility and lime levels under soil acidity in an alley cropping system of Vindhyan Region, India. International Journal of Chemical Studies 2017, 5, 1558–1560.
[8] Prashar P, Shah S. Impact of fertilizers and pesticides on soil microflora in agriculture. In: Sustainable agriculture reviews. Cham, Springer, 2016, vol. 19, 331–361.
[9] Van der Werf HM. Assessing the impact of pesticides on the environment. Agriculture Ecosystems Environment 1996, 60, 81–96.
[10] Yeo HG, Choi M, Chun MY, Sunwoo Y. Concentration distribution of polychlorinated biphenyls and organochlorine pesticides and their relationship with temperature in rural air of Korea. Atmospheric Environment 2003, 37, 3831–3839.
[11] Ismael LL, Rocha EMR. Estimate of the contamination of groundwater and surface water due to agrochemicals in the sugar-alcohol area, Santa Rita, State of Paraíba, Brazil. Ciência & Saúde Coletiva 2019, 24, 4665–4676.
[12] Liu Z, Xie W, Yang Z, Huang X, Zhou H. Effects of manure and chemical fertilizer on bacterial community structure and soil enzyme activities in North China. Agronomy 2021, 11, 1017.
[13] Canet R, Birnstingl JG, Malcolm DG, Lopez-Real JM, Beck AJ. Biodegradation of polycyclic aromatic hydrocarbons (PAHs) by native microflora and combinations of white-rot fungi in a coal-tar contaminated soil. Bioresource Technology 2001, 76, 113–117.
[14] Agri U, Chaudhary P, Sharma A. In vitro compatibility evaluation of agriusable nanochitosan on beneficial plant growth-promoting rhizobacteria and maize plant. National Academy Science Letters 2021, 44, 555–559.
[15] Agri U, Chaudhary P, Sharma A, Kukreti B. Physiological response of maize plants and its rhizospheric microbiome under the influence of potential bioinoculants and nanochitosan. Plant Soil 2022. https://doi.org/10.1007/s11104-022-05351-2.
[16] Kukreti B, Sharma A, Chaudhary P, Agri U, Maithani D. Influence of nanosilicon dioxide along with bioinoculants on *Zea mays* and its rhizospheric soil. 3 Biotech 2020, 10, 345.
[17] Chaudhary P, Sharma A, Chaudhary A, Khati P, Gangola S, Maithani D. Illumina based high throughput analysis of microbial diversity of maize rhizosphere treated with nanocompounds and *Bacillus* sp. Applied Soil Ecology 2021, 159, 103836.

[18] Chaudhary A, Parveen H, Chaudhary P, Khatoon H, Bhatt P. Rhizospheric microbes and their mechanism. In: Bhatt P, Gangola S, Udayanga D, Kumar G, eds. Microbial technology for sustainable environment. Singapore, Springer, 2021, 79–93.
[19] Kumar G, Suman A, Lal S, Ram RA, Bhatt P, Pandey G, Chaudhary P, Rajan S. Bacterial structure and dynamics in mango (Mangifera indica) orchards after long term organic and conventional treatments under subtropical ecosystem. Scientific Reports 2021, 11, 20554.
[20] Chaudhary P, Khati P, Gangola S, Kumar A, Kumar R, Sharma A. Impact of nanochitosan and *Bacillus* spp. on health, productivity and defence response in *Zea mays* under field condition. 3 Biotech 2021, 11, 237.
[21] Mathur SC, Tannan SK. (1999) Future of Indian pesticides industry in next millennium. Pesticide Information 1999, 24, 9–23.
[22] Jayasumana C, Fonseka S, Fernando A, Jayalath K, Amarasinghe M, Siribaddana S, Paranagama P. Phosphate fertilizer is a main source of arsenic in areas affected with chronic kidney disease of unknown etiology in Sri Lanka. Springer Plus 2015, 4, 1–8.
[23] FICCI (Federation of Indian Chambers of Commerce and Industry). Report on use of agrochemicals for sustainable farming. 8th Agrochemicals Conference 2019 – July 16, 2019, New Delhi, India. http://ficci.in/past-event-page.asp?evid¼24230
[24] Gupta PK. Toxic effects of pesticides and agrochemicals. In: Concepts and applications in veterinary toxicology. Cham, Springer, 2019, 59–82.
[25] Johnsen K, Jacobsen CS, Torsvik V, Sørensen J. Pesticide effects on bacterial diversity in agricultural soils–a review. Biology and Fertility of Soils 2001, 33, 443–453.
[26] Agnihotri NP, Gajbhiye VT, Kumar M, Mohapatra SP. Organochlorine insecticide residues in Ganga river water near Farrukhabad, India. Environmental Monitoring and Assessment 1994, 30, 105–112.
[27] Dhananjayan V, Jayakumar S, Ravichandran B (2020). Conventional methods of pesticide application in agricultural field and fate of the pesticides on the environment and human health. Eds (Thomas S et al.), In: Controlled Release of Pesticides for Sustainable Agriculture. ISBN 978-3-030-23395-2, Cham, Springer.
[28] Gunstone T, Cornelisse T, Klein K, Dubey A, Donley N. Pesticides and soil invertebrates: A hazard assessment. Frontiers in Environmental Science 2021, 9, 643847.
[29] Sasanelli N, Konrat A, Migunova V, Toderas I, Iurcu-Straistaru E, Rusu S, Bivol A, Andoni C, Veronico P. Review on control methods against plant parasitic nematodes applied in southern member states (C Zone) of the European Union. Agriculture 2021, 11, 602.
[30] Le VN, Nguyen QT, Nguyen TD, Nguyen NT, Janda T, Szalai G, Le TG. The potential health risks and environmental pollution associated with the application of plant growth regulators in vegetable production in several suburban areas of Hanoi, Vietnam. Biologia Futura 2020, 71(3), 323–331.
[31] Zhao C, Kleiman DE, Shukla D. Optimization of hydration sites in plant hormone receptors for agrochemical design. Biophysical Journal 2022, 121, 190.
[32] Nisler J, Kopečný D, Pěkná Z, Končitíková R, Koprna R, Murvanidze N, Werbrouck SPO, Havlíček L, De Diego N, Kopečná M, Wimmer Z, Briozzo P, Moréra S, Zalabák D, Spíchal L, Strnad M. Diphenylurea-derived cytokinin oxidase/dehydrogenase inhibitors for biotechnology and agriculture. Journal of Experimental Botany 2020, 72, 355–370.
[33] Konstantinova N, Korbei B, Luschnig C. Auxin and root gravitropism: Addressing basic cellular processes by exploiting a defined growth response. International Journal of Molecular Science 2021, 22, 2749.
[34] Mishra BS, Sharma M, Laxmi A. Role of sugar and auxin crosstalk in plant growth and development. Physiologia Plantarum 2022, 174, e13546.

[35] Kefeli VI, Kadyrov CS. Natural growth inhibitors, their chemical and physiological properties. Annual Review of Plant Physiology 1971, 22, 185–196.
[36] Chen L, Lee YB, Ramsier C, Bigham J, Slater B, Dick WA. Increased crop yield and economic return and improved soil quality due to land application of FGD-gypsum. In: Proceedings of the World of Coal Ash. Lexington, Ky, 2005.
[37] Sharma P, Meena RS, Kumar S, Gurjar DS, Yadav GS, Kumar S. Growth, yield and quality of cluster bean (*Cyamopsis tetragonoloba*) as influenced by integrated nutrient management under alley cropping system. Indian Journal of Agricultural Sciences 2019, 89, 1876–1880.
[38] Kacprzak M, Kupich I, Jasinska A, Fijalkowski K. Bio-based waste substrates for degraded soil improvement – advantages and challenges in European context. Energies 2022, 15, 385.
[39] Chaudhary P, Khati P, Chaudhary A, Maithani D, Kumar G, Sharma A. Cultivable and metagenomic approach to study the combined impact of nanogypsum and *Pseudomonas taiwanensis* on maize plant health and its rhizospheric microbiome. PLoS One 2021, 16, e0250574.
[40] Chaudhary P, Parveen H, Gangola S, Kumar G, Bhatt P, Chaudhary A. Plant growth-promoting rhizobacteria and their application in sustainable crop production. In: Bhatt P, Gangola S, Udayanga D, Kumar G, eds. Microbial technology for sustainable environment. Singapore, Springer, 2021, 217–234.
[41] Kumari S, Sharma A, Chaudhary P, Khati P. Management of plant vigor and soil health using two agriusable nanocompounds and plant growth promotory rhizobacteria in Fenugreek. 3 Biotech 2020, 10, 1–11.
[42] Chaudhary P, Chaudhary A, Bhatt P, Kumar G, Khatoon H, Rani A, Kumar S, Sharma A. Assessment of soil health indicators under the influence of nanocompounds and *bacillus* spp. Field Condition. Frontiers in Environmental Science 2022, 9, 769871.
[43] Li J, Gan G, Chen X, Zou J. Effects of long-term straw management and potassium fertilization on crop yield, soil properties, and microbial community in a rice–oilseed rape rotation. Agriculture 2021, 11, 1233.
[44] Chaudhary P, Chaudhary A, Parveen H, Rani A, Kumar G, Kumar A, Sharma A. Impact of nanophos in agriculture to improve functional bacterial community and crop productivity. BMC Plant Biology 2021, 21, 519.
[45] Kumar A, Sharma A, Chaudhary P, Gangola S. Chlorpyrifos degradation using binary fungal strains isolated from industrial waste soil. Biologia 2021, 76, 3071–3080.
[46] El Hussein AA, Mohamed AT, El Siddig MA, Sherif AM, Osman AG. Effects of oxyfluorfen herbicide on microorganisms in loam and silt loam soils. Research Journal of Environmental Sciences 2012, 6, 134–145.
[47] Wesley B, Ajugwo G, Adeleye S, Ibegbulem C, Azuike P. Effects of agrochemicals (insecticides) on microbial population in soil. EC Microbiology 2017, 8, 211–221.
[48] Hussain S, Siddique T, Saleem M, Arshad M, Khalid A. Impact of pesticides on soil microbial diversity, enzymes, and biochemical reactions. Advances in Agronomy 2009, 102, 159–200.
[49] Xu X, Chen C, Zhang Z, Sun Z, Chen Y, Jiang J, Shen Z. The influence of environmental factors on communities of arbuscular mycorrhizal fungi associated with *Chenopodium ambrosioides* revealed by MiSeq sequencing investigation. Scientific Reports 2017, 7, 45134.
[50] Tien CJ, Chen CS. Assessing the toxicity of organophosphorous pesticides to indigenous algae with implication for their ecotoxicological impact to aquatic ecosystems. Journal of Environmental Science and Health 2012, 47, 901–912.
[51] Santos JB, Jakelaitis A, Silva AA, Costa MD, Manabe A, Silva MCS. Action of two herbicides on the microbial activity of soil cultivated with common bean (*Phaseolus vulgaris*) in conventional-till and no-till systems. Weed Research 2006, 46, 284–289.

[52] Zhang Q, Zhu L, Wang J, Xie H, Wang J, Wang F, Sun F. Effects of fomesafen on soil enzyme activity, microbial population, and bacterial community composition. Environmental Monitoring and Assessment 2014, 186, 2801–2812.
[53] Cheng Y, Ishimoto K, Kuriyama Y, Osaki M, Ezawa T. Ninety-year-, but not single, application of phosphorus fertilizer has a major impact on arbuscular mycorrhizal fungal communities. Plant Soil 2013, 365, 397–407.
[54] Stuart AK, Stuart RM, Pimentel IC. Effect of agrochemicals on endophytic fungi community associated with crops of organic and conventional soybean (*Glycine max L. Merril*). Agriculture and Natural Resources 2018, 52, 388–392.
[55] Xie H, Gao F, Tan W, Wang SG. A short-term study on the interaction of bacteria, fungi and endosulfan in soil microcosm. Science of the Total Environment 2011, 412, 375–379.
[56] Munoz-Leoz B, Ruiz-Romera E, Antiguedad I, Tebuconazole GC. Application decreases soil microbial biomass and activity. Soil Biology & Biochemistry 2011, 43, 2176–2183.
[57] Streletskii R, Astaykina A, Krasnov G, Gorbatov V. Changes in bacterial and fungal community of soil under treatment of pesticides. Agronomy 2022, 12, 124.
[58] Pimentão AR, Pascoal C, Castro BB, Cássio F. Fungistatic effect of agrochemical and pharmaceutical fungicides on non-target aquatic decomposers does not translate into decreased fungi-or invertebrate-mediated decomposition. Science of the Total Environment 2020, 712, 135676.
[59] Dodd JC, Jeffries P. Effects of herbicides on three vesicular-arbuscular fungi associated with winter wheat (*Riticum aestivum* L.). Biology and Fertility of Soils 1989, 7, 113–119.
[60] Campagnac E, Fontaine J, Sahraoui ALH, Laruelle F, Durand R, Grandmougin-Ferjani A. Differential effects of fenpropimorph and fenhexamid, two sterol biosynthesis inhibitor fungicides, on arbuscular mycorrhizal development and sterol metabolism in carrot roots. Phytochemistry 2008, 69, 2912–2919.
[61] Pasaribu A, Mohamad RB, Hashim A, Rahman ZA, Omar D, Morshed MM, Selangor DE. Effect of herbicide on sporulation and infectivity of vesicular arbuscular mycorrhizal (*Glomus mosseae*) symbiosis with peanut plant. Journal of Animal and Plant Sciences 2013, 23, 1671–1678.
[62] Zaller JG, Heigl F, Ruess L, Grabmaier A. Glyphosate herbicide affects belowground interactions between earthworms and symbiotic mycorrhizal fungi in a model ecosystem. Scientific Reports 2014, 4, 1–8.
[63] Ipsilantis I, Samourelis C, Karpouzas DG. The impact of biological pesticides on arbuscular mycorrhizal fungi. Soil Biology and Biochemistry 2012, 45, 147–155.
[64] Martinez-Toledo MV, Salmero´n V, Rodelas B, Pozo C, Gonzalez-lópez J. Effects of the fungicide Captan on some functional groups of soil microflora. Applied Soil Ecology 1998, 7, 245255.
[65] Supreeth M, Chandrashekar MA, Sachin N, Raju NS. Effect of chlorpyrifos on soil microbial diversity and its biotransformation by *Streptomyces* sp. HP-11. 3 Biotech 2016, 6, 1–6.
[66] Win PM, Matsumura E, Fukuda K. Effects of pesticides on the diversity of endophytic fungi in tea plants. Microbial Ecology 2021, 82, 62–72.
[67] Khan MS, Zaidi A, Rizvi PQ. Biotoxic effects of herbicides on growth, nodulation, nitrogenase activity, and seed production in chickpeas. Communications in Soil Science and Plant Analysis 2006, 37, 1783–1793.
[68] Fang H, Han L, Cui Y, Xue Y, Cai L, Yu Y. Changes in soil microbial community structure and function associated with degradation and resistance of carbendazim and chlortetracycline during repeated treatments. Science of the Total Environment 2016, 572, 1203–1212.

[69] Wang J, Song Y, Ma T, Raza W, Li J, Howland JG, Huang Q, Shen Q. Impacts of inorganic and organic fertilization treatments on bacterial and fungal communities in a paddy soil. Applied Soil Ecology 2017, 112, 42–50.
[70] Dai Z, Liu G, Chen H, Chen C, Wang J, Ai S, Wei D, Li D, Ma B, Tang C. Long-term nutrient inputs shift soil microbial functional profiles of phosphorus cycling in diverse agroecosystems. The ISME Journal 2020, 14, 757–770.
[71] Kumar S, Dhar S, Barthakur S, Rajawat MVS, Kochewad SA, Kumar S, Kumar D, Meena LR. Farmyard manure as K-fertilizer modulates soil biological activities and yield of wheat using the integrated fertilization approach. Frontiers in Environmental Science 2021, 9, 764489.
[72] Wu L, Wang Y, Zhang S, Wei W, Kuzyakov Y, Ding X. Fertilization effects on microbial community composition and aggregate formation in saline-alkaline soil. Plant Soil 2021, 463, 523–535.
[73] Cui J, Yuan X, Zhang Q, Zhou J, Lin K, Xu J, Zeng Y, Wu Y, Cheng L, Zeng Q. Nutrient availability is a dominant predictor of soil bacterial and fungal community composition after nitrogen addition in subtropical acidic forests. PLoS One 2021, 16, e0246263.
[74] Zou W, Lang M, Zhang L, Liu B, Chen X. Ammonia-oxidizing bacteria rather than ammonia-oxidizing archaea dominate nitrification in a nitrogen-fertilized calcareous soil. Science of the Total Environment 2021, 811, 151402.
[75] Cinnadurai C, Gopalaswamy G, Balachandar D. Diversity of cultivable Azotobacter in the semi-arid alfisol receiving long-term organic and inorganic nutrient amendments. Annals of Microbiology 2013, 63, 1397–1404.
[76] Rasool N, Reshi ZA, Shah MA. Effect of butachlor (G) on soil enzyme activity. European Journal of Soil Biology 2014, 61, 94–100.
[77] Kumar G, Lal S, Maurya SK, Bhattacherjee AK, Chaudhary P, Gangola S. Exploration of *Klebsiella pneumoniae* M6 for paclobutrazol degradation, plant growth attributes, and biocontrol action under subtropical ecosystem. PLoS One 2021, 16, e0261338.

Nitika Negi, Siya Sharma, Nitika Bansal, Aditi Saini,
Ratnaboli Bose, M.S. Bhandari, Amit Pandey, Shailesh Pandey

Chapter 4
Effect of abiotic stresses on plant systems and their mitigation

Abstract: Terrestrial plants are exposed to adverse environments from their very birth. A multitude of physical and chemical factors pose threats to them, namely water deficit, flooding, extremes of temperature, saline loads, heavy metals, and ultraviolet radiation, among others. Together, these stresses come under abiotic stresses severely affecting agroecosystems resulting in massive crop yield loss. Abiotic stresses generally arise collectively, thus necessitating crops to be fortified with multistress tolerance. The world's population is rapidly increasing, yet, agricultural productivity is not at par with the demand for food. A considerable portion of the population in poor nations, where agriculture is still practiced at a subsistence level, is continually challenged by abiotic stress factors and their interaction with biotic stress factors. Furthermore, when plants are challenged with multiple stresses simultaneously, coping becomes extremely difficult. Crop plants being sessile cannot escape, hence, must either adapt or die. This chapter aims at looking into the major plants' abiotic stresses and their possible methods of mitigation.

4.1 Introduction

The world's population is rapidly increasing with an estimated 9 billion people by the year 2050. Agricultural productivity, on the other hand, is not progressing at the required rate to meet rising food demand [1, 88]. Plants being primary producers support the world population but a declining trend of productivity has been observed probably due to climate change, low fertility, water scarcity, and other abiotic stressors. Climate and the geographical distribution are the major determinants of the type of plantation they support, and as plants can sense changes in their environment, they respond accordingly in order to maximize their chances of survival. Furthermore, as urbanization progressed, humans have acquired all of the conveniences for their everyday living, but in return nature has deteriorated severely, exposing plant systems to numerous changing environments. These changing environments alter plant responses as decreased productivity, flowering, seed formation, and changes in metabolism and growth rates that eventually cause senescence (Table 4.1). In 2014, Rhodes defined plant stress as "Any external factor that negatively influences plant growth, productivity, reproductive capacity or survival." Findings from several studies state

https://doi.org/10.1515/9783110771558-004

that under extreme stress, the structure of biological macromolecules such as lipids, proteins, and nucleic acids is vulnerable to damage and/or destruction [62]. Although the plant can recover from these stresses if the source of stress is mild but if they are severe or left untreated, whole plantations or fields may be wiped out all at once (Figure 4.1). Furthermore, when plants are challenged with multiple factors at once, it becomes extremely difficult for them to cope with the stress, and the only alternative left for them in the midst of these stresses is to adapt or to die since they are sessile and cannot escape [47]. The combined effect of temperature and drought is more detrimental than when they occur independently [73]. However, the combined effect is not necessarily cumulative but is dependent on plant–stress factor relationship. In general, when a plant is exposed to unfavorable environmental or biological conditions, it is known as "plant stress."

Table 4.1: Major plant system processes affected by stressors and their defense mechanisms.

Stress	Key process affected	Plant defense mechanism	References
Salinity	– Altered flowering – Limited water uptake – Reduced growth and metabolism – Senescence – Limited CO_2 uptake – Reduced photosynthetic capacity	– Ion exclusion – Tissue tolerance	[65, 87]
Heavy metal	– Inhibit photosynthesis – Growth retardation – Progressive chlorosis – Fewer branching – Less fruiting	– Constitutive mechanisms – Adaptive mechanisms	[90]
Drought	– Early senescence – Reduced photosynthetic capacity – Less CO_2 uptake – Prolonged closure of stomata – Disturbed TCA – Less ATP formation	– Production of antioxidants – Production of phytohormones – Production of osmoprotectants	[2, 79]

Table 4.1 (continued)

Stress	Key process affected	Plant defense mechanism	References
Temperature	– Seed germination – Photosynthetic efficiency – Lipid modification – Reduced photosynthesis	– Hormonal homeostasis – Production of HSPs – Cold acclimatization – Production of antioxidants – Presence of ice-binding proteins	[11, 12, 34, 37, 51, 102, 103]
pH	– Low nutrient availability – Less mobility – Limited plant water intake – Oxidative stress – Electrolyte leaks – Less CO_2 absorption	– Production of polyamines – Increased hormonal activity – Induction via plant microflora	[28, 61, 112]
Waterlogging	– Less root cell ATP – Oxygen deficit – Altered stomatal conductance and CO_2 assimilation – Reduced nutrient as well as water uptake – Production of secondary metabolites	– Plant adaptation mechanisms – Hormonal induced resistance	[37, 71, 98]

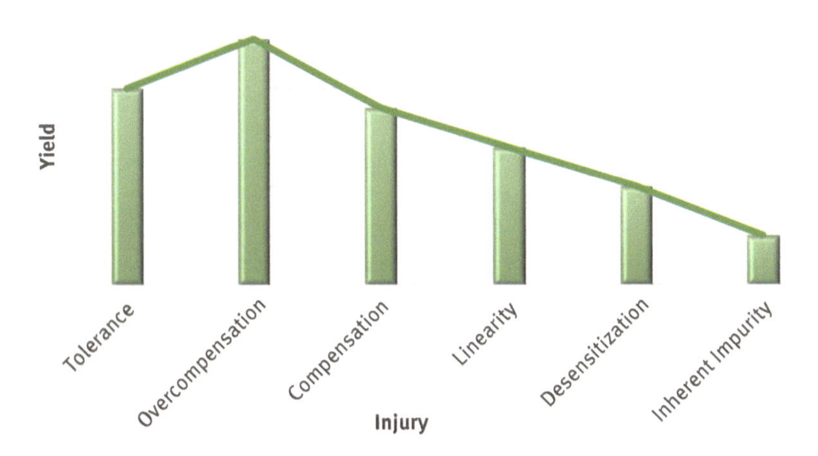

Figure 4.1: The damage curve is divided into six segments, each of which represents a different response of yields to injuries. (1) Tolerance: no loss of yields, (2) overcompensation: high yields, (3) compensation: low yields, (4) linearity: lower yields, (5) desensitization: lowest yields, and (6) inherent impunity: yields with injury.

The stresses in which plants are exposed to are caused by biotic and abiotic factors. Biotic stresses are caused by the number of organisms, especially fungi, bacteria, viruses, insects, and pests. The damage caused may be partial or complete, which can be overcome or cannot be tolerated by the plant. Causal organisms penetrate the plasma membrane and deprive their host of the nutrient, causing the plant to lose vitality and, in certain cases, death while the abiotic stresses occur as drought, waterlogging, temperature, pH, salinity, toxins, radiation, sunlight, heavy metals, and nutrient deficiency (Figure 4.2). Abiotic stressors can have either beneficial (eustress) or detrimental consequences (distress), but according to literature, they have been identified as the principal cause of crop damage in over half of all cases worldwide [80]. The ability of plants to survive such unfavorable environments is determined by a variety of processes and factors, including a wide range of molecular processes and regulatory mechanisms, and because of the economic implications, knowing how plants withstand poor environmental or biological conditions has become a prominent priority in agricultural research [63]. Furthermore, microbial communities have also been identified as a key factor in disease prevention and can be used as a biological marker in regulating stress (Table 4.2). However, since India gained independence, human population has increased manifold and with this

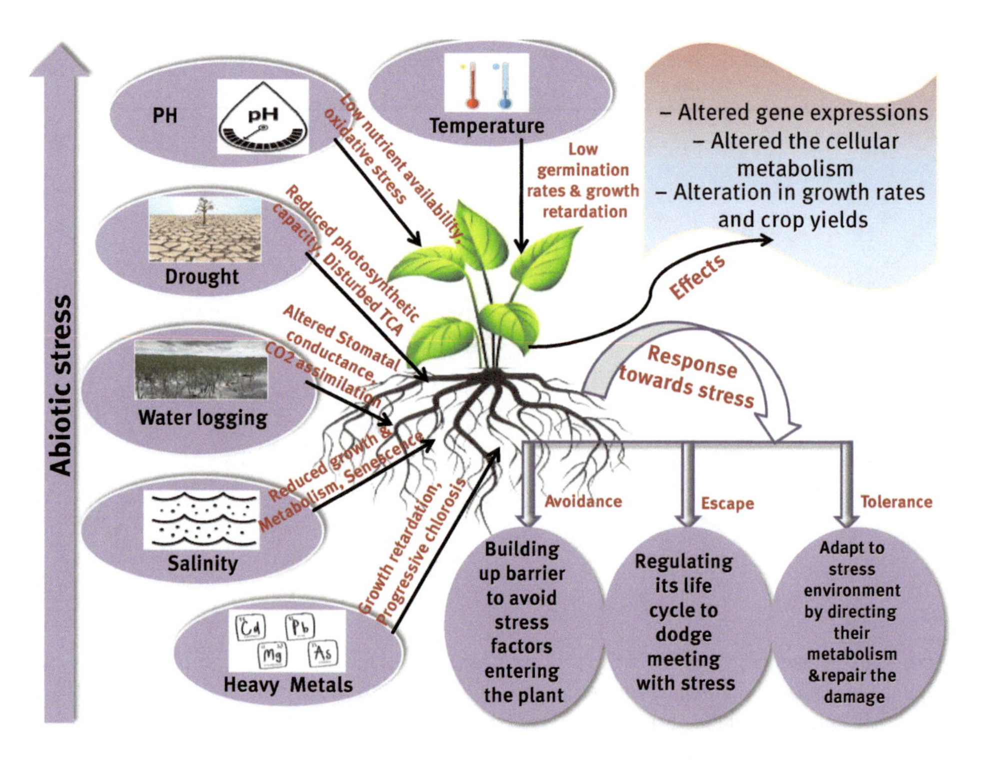

Figure 4.2: Effects of different abiotic stresses on plants.

Table 4.2: Microbial management of abiotic stresses.

Stress	Test plant	Microorganism	Mechanism	Reference
Salinity	Barley (*Hordeum vulgare*)	*Sesuvium portulacastrum*	Roots facilitate Na^+ substitution by Ca^{2+} at the adsorption sites and ameliorate soil's physical properties by their expansion leading to a leaching of sodium	[75]
	Pearl millet (*Cenchrus americanus*) and wheat (*Triticum aestivum*)	*Nostoc ellipsosporum* HH-205 and *N. punctiforme* HH-206	Cyanobacteria are reported to increase the phosphate availability to rice plants by mobilizing different insoluble forms of inorganic phosphate, possibly by synthesizing chelator(s) and/or organic acids as well as total Kjeldahl nitrogen and total organic carbon	[70]
	Peas (*Pisum sativum*)	*Streptomyces lydicus* WYEC108	Colonization of the organism in the roots increases the size of the nodules, improving iron and nutrient assimilation, which benefits the nitrogen-fixing bacteria (*Rhizobium* sp.) that live inside them	[81]
Heavy metal	-	*Propionibacterium freudenreichii shermanii* JS, *Bifidobacterium breve* Bbi99/E8, *Actinomyces*,	Physical adsorption to bacterial surface is the major mechanism	[4]
	-	*Corynebacterium glutamicum*	Overexpression of ars operons (ars1 and ars2)	[59]

Table 4.2 (continued)

Stress	Test plant	Microorganism	Mechanism	Reference
Drought	Thale cress (*Arabidopsis thaliana*)	*Phyllobacterium brassicacearum*	Synchronized changes in transpiration, abscisic acid content, photosynthesis, and development resulted in higher water-use efficiency and a better tolerance to drought of inoculated plants	[13]
	Red rice (*Oryza sativa*)	*Gluconacetobacter diazotrophicus* Pal5	Elicitation of induced systemic tolerance	[26]
	Rice (*O. sativa*)	*Bacillus* sp.	Presence of mineral solubilization and other plant growth-promoting traits	[6]
Temperature	Tomato (*Solanum lycopersicum*)	*Trichoderma harzianum* (AK20G)	Increased expression of transcription factor NAC1 and dehydrin TAS14, while P5CS expression was decreased	[31]
	Tomato seedlings (*S. lycopersicum*)	*Funneliformis mosseae* and *Paraburkholderia graminis* C4D1M	Increased efficiency of photosystem II	[14]
	Tomato (*S. lycopersicum*)	*T. asperellum* (HK703)	Inhibition of cell division in the meristems or by suppressing cell elongation and expansion due to low turgor pressure	[19]

Table 4.2 (continued)

Stress	Test plant	Microorganism	Mechanism	Reference
pH	*Pinus massoniana* Lamb. and *Cinnamomum camphora*	*Bradyrhizobium canariense* and *Terracidiphilus* sp.	Production of large amounts of external exopolysaccharides Improved tricarboxylic acid cycle, gluconeogenesis, and pentose phosphate pathway not only affect soil carbon but are also involved in reactive oxygen species detoxification	[17]
	Maize (*Zea mays*)	*T. asperellum*	Reducing soil pH and salt ion concentration and controlling the fungal microbial population, resulting in higher crop yields	[28]
Waterlogging	–	1-Aminocyclopropane-1-carboxylic acid-producing bacteria (α-proteobacteria, β-proteobacteria, and γ-proteobacteria; Actinobacteria, Firmicutes, Bacteroidetes, and Flavobacterium)	Lowering ethylene concentration in plants	[86]

increasing trend, food demands have also been increased drastically and in order to sustain the population, suitable mitigation measures should be undertaken.

4.2 Plant stress factors

4.2.1 Salinity

Salinity is one of the most serious factors that cause severe osmotic, ionic, and oxidative stress in the plants. Saline soil simply makes up 3.1% (397 million hectare) of the world's total land area, of which more than 45 million hectares of irrigated land have been affected by salt globally; however, in India nearly 23 million hectares of land is affected due to high saline levels [66, 89]. The damages caused by salt stress are growth inhibition, accelerated development, senescence, decreased agricultural crop productivity, affecting germination, plant vigor, and ultimately death due to prolonged exposure. Growth inhibition is the primary harm that causes secondary symptoms; however, programmed cell death can also occur under severe salt shock. The reason behind salinity is the presence of excessive amounts of water-soluble salts such as sodium sulfate (Na_2SO_4), sodium nitrate ($NaNO_3$), sodium chloride (NaCl), sodium carbonates ($NaHCO_3$ and Na_2CO_3), potassium sulfate (K_2SO_4), calcium sulfate ($CaSO_4$), magnesium sulfate ($MgSO_4$), and magnesium chloride ($MgCl_2$). The majority of these salts are beneficial to the plant and aid in its development and metabolism. However, if present in high enough amounts, they can become toxic [87]. Salinity could occur naturally due to weathering of parent material, fossil salts, and transport of salt in river but human intervention has disturbed natural ecosystem to drastic levels via clearing land, incorrect irrigation, overextraction of groundwater, canal water seepage, and overuse of chemicals [43]. World soils may lose 6.8 Pg soil organic carbon by 2100 due to salinity [89].

4.2.1.1 Remediation of salt

4.2.1.1.1 Salt tolerance

Plants are classified into two groups according to their ability to grow in saline soil as halophytes and glycophytes. Halophytes comprise 2% of those plants that can tolerate high level of salt concentration while glycophytes comprise 98% of incapable tolerating high salinity level, ultimately making it extremely difficult for such plants to thrive in these conditions [76]. Every plant, however, responds differently to the concentration of salt to which it is exposed. A two-phase model has been described for the probable effects of salt stress (Figure 4.3). The presence of osmotic stress outside the roots slows the development of all plants and requires metabolic management during phase 1, but some plants adapt to the toxicity while sensitive plants perish during phase 2.

In general, salt-tolerant species have tolerance to large salt concentrations in their systems, whereas sensitive plants have none or only a few salts. The salt tolerance might be due to a cellular mechanism of the plant system. Reactive oxygen species (ROS) acts as a signal during salt stress, although there are various genes that can help plants detoxify the ROS and cope with the stress. For instance, a transcriptomic study of *Sorghum* under salt stress [73], revealed differentially expressed genes (DEGs) to be the major group of genes involved in many metabolic processes such as carbohydrate and lipid metabolism, cell wall synthesis, photosynthesis, and hormone transduction, thereby alleviating the adverse effects of salinity by arresting oxidative stress (H_2O_2) and invigorating enzymatic and nonenzymatic antioxidant

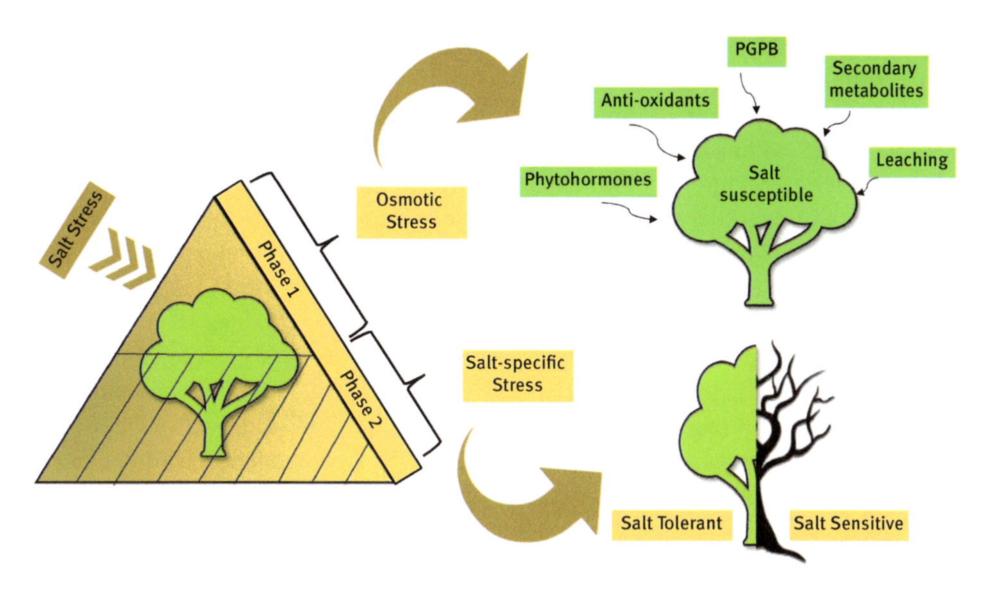

Figure 4.3: Hypothetical two-phase growth response of salinity.

activities (superoxide dismutase (SOD), ascorbate peroxidase (APX), catalase (CAT), peroxidase (POD), glutathione reductase (GR), glutathione, ascorbate (ASC), proline, and glycine betaine), as well as protecting the cell membrane structure by reducing the electrolyte leakage [87]. Furthermore, rhizospheric fungi and plant growth-promoting rhizobacteria (PGPR) could also be employed to the stressed soil to increase plant yields. Under salinity stress, salt-tolerant PGPR can lower Na^+ levels in shoots, enhance the expression of stress-responsive transcription factors, increase proline synthesis, promote ROS scavenging, and improve plant biomass. Salt tolerance is further aided by the root system, which improves access to water and nutrients while minimizing the salt accumulation [65]. However, factoring in the climate change scenario, these aspects of the major genes could be utilized to cope with the stress in future studies.

4.2.1.1.2 Reclamation of saline soils

Subsurface drainage and salt leaching with fresh ponded water are some of the well-known technological interventions to address the problem of soil salinity. These approaches necessitate good-quality water with low electrolyte content that percolates through the soil profile, allowing salts to dissolve and travel lower, preventing undue salt accumulation in the root zone. The amount of water necessary to reclaim a soil is governed by the texture of the soil, the presence of salts in it, the volume of soil to be reclaimed, the desired salt level in the rhizosphere, and the plants that will be produced following reclamation [91]. Haryana, Rajasthan, Gujarat, Punjab, Andhra Pradesh, Maharashtra, Madhya Pradesh, and Karnataka have successfully implemented subsurface drainage technology, rehabilitating about 110,000 hectares of waterlogged saline soils [92]. However, there are some other methods as well to reclaim the saline soil: flushing, scrapping, biomimicry, and phyto-salinization. These methods could be implemented alone or in conjunction with other methods. Biomimicry is a new technology that aims to replicate the capillary movement of vascular plants in order to gather salt on the soil surface while phyto-salinization uses plants to reclaim the soil [91]. Plants with economic value have also been reported to be useful in the reclamation of salty and sodic soils. For instance, *Portulaca oleracea* was successfully used as a salt removal crop [52].

However, in regions where water is scarce, the availability of good seeding water is a significant barrier to desalination systems. Furthermore, these conventional processes are extremely costly and labor intensive. Phyto-salinization with halophytes might be a promising alternative in this context. They have been exploited extensively as they are found in saline soils and have the potential to restore salt in their systems [54]. They have also been shown to form symbiotic relationships with arbuscular mycorrhizal fungi and PGPR, and enhance tolerance to saline soil [54, 91].

4.2.1.1.3 Microbial remediation

Soil is an important habitat for microorganisms where plants and microbes interact continuously in a direct or indirect way. Bioreclamation via microbiota offers itself as a new and ecologically acceptable approach in the case of salt disturbed soils (naturally occurring or manmade), which are normally cleansed or modified with calcium salts. Halophilic bacteria, sometimes known as "salt-loving" microorganisms, thrive in high-salt conditions that would lower the salt content of the soil [70]. Halophilic individuals may be found in all branches of the Proteobacteria, and they often have nonhalophilic relations. They may be found in the Cyanobacteria, the Flavobacterium-Cytophaga branch, Spirochetes, and Actinomycetes [81]. The potential of halophilic bacteria and fungi has been reported by many researchers. The ability of the obligate halophyte, *Sesuvium portulacastrum* [75] to desalinize an experimentally salinized soil was tested. Also, the consortium of indigenous cyanobacterial species was reported to be used successfully for phytoremediation and for better productivity of saline soils in semiarid conditions [70]. They have been shown to stimulate plant development by improving the availability of N and P and by releasing a range of plant growth-promoting phytohormones. Not only Cyanobacteria but members of the bacterial genus *Streptomyces* due to their low pathogenicity, resistance to harsh environmental conditions, and synthesis of a wide variety of secondary metabolic compounds are known to be promising candidates for bioreclamation of salt-affected soils [81].

4.2.2 Heavy metal toxicity

Heavy metals are defined as the elements having density greater than 4 g/cm^3 [47]. Plants, like other forms, are typically sensitive to both deficiency and an excess of certain heavy metal ions. They require metals, such as cobalt (Co), copper (Cu), iron (Fe), manganese (Mn), molybdenum (Mo), nickel (Ni), and zinc (Zn), but excessive amounts of these, and nonessential metals such as cadmium (Cd), mercury (Hg), silver (Ag), and lead (Pb) could be hazardous, and as the world becomes more industrialized, the problem of disposing of garbage containing heavy metals has become a major concern [27,47]. Industrial effluents and sewage sludge laden with heavy metal waste are dumped into soil leading to the buildup of metal toxicity in the soil. These metals directly influence the developmental processes of plant due to their high reactivity. It causes photosynthesis inhibition, stunted or retarded growth, chlorosis, reduced branching, and less fruiting [90]. Although the plant is able to cope with a significant level of toxicity by bioaccumulating heavy metals in its system, as a consequence, heavy metals circulate within the food chain that ultimately harms humans. These are implicated in various health disorders in humans like cardiovascular diseases, cancer, cognitive impairment, chronic anemia, problems of kidneys, nervous system, brain, skin, and bones [53].

Plants have developed varied mechanisms to maintain physiological metal ion concentrations while minimizing exposure to nonessential heavy metals [53]. Defense against metal toxicity is governed by an intertwined network of physiological and molecular processes of the plant system. Plants possess both constitutive and adaptive mechanisms to cope with the stress. This has so far been accomplished through a variety of physical, chemical, and biological methods. The physical and chemical methods are expensive and generate by-products while the biological processes are cheaper but time-consuming [1]. In contrast to any other stress, heavy metal toxicity is highly resistant to remediation measures as it persists for long in the environment. Therefore, it is critical to employ innovative, site-specific remediation measures that can safely and effectively treat heavy metal polluted soils. Some conventional bioremediation methods involve in situ vitrification, soil incineration, excavation and landfill, nanoremediation, electroremediation, soil washing, soil flushing, solidification, chemical leaching, and stability of electrokinetic systems [1]. The majority of these conventional clean-up procedures are expensive to install and create more disturbance to an already polluted area. In these circumstances, plant-based bioremediation is employable where some plants termed "metallophytes" are introduced to the contaminated region that are able to absorb heavy metals in their system several times faster than a typical plant [32]. In addition, certain microorganisms are also known to remediate organic contaminants along with heavy metals by converting the organic contaminants into CO_2, water, and other metabolites [93]. For instance, *Pseudomonas putida* strains, namely, E41 and E46, have been documented to express heavy metal tolerance (Zn, Cd, and Pb) and possess few PGPR traits such as indole acetic acid synthesis, siderophores, HCN production, and phosphate solubilization [68].

4.2.2.1 Remediation of heavy metals

The primary goal of heavy metal remediation is to reduce the dangers that these hazardous substances present to the human and environmental health. Discussed below are some of the methods of heavy metal bioremediation.

4.2.2.1.1 Physical methods

Physical removal procedures may be used to eliminate almost all contaminants. However, there are certain drawbacks. Physical methods, in general, need more processing and have a somewhat high application cost when compared to other techniques [93]. Majority of them are based on pollutant particle size distribution. The following are some of the most often employed physical remediation techniques.

Soil replacement

This procedure, sometimes known as the "dig-and-haul" method, entails the partial or total removal of metal-contaminated soil and subsequent replacement with clean soil. The major drawbacks of this method are that it is quite costly and is practicable for only a small region that may be heavily polluted. It is also not feasible since it moves contaminated soil from a high-risk area to a low-risk one, increasing the likelihood of contamination at the new location [93].

Soil washing

Soil washing is the most widely used method, though it is time-consuming. In this method, the soil contaminated with heavy metals is flushed out with the water containing chelators, inorganic washing agents, and surfactants. The technique can be used in situ, where the solution is forced into the soil, or ex situ, when the soil is physically excavated and washed in reactors. Heavy metals dissolve in these solutions, which can then be readily removed [25]. Heavy metals were efficiently removed from polluted soil through the application of ferric nitrate and ethylenediamine disuccinate [108]. Triton X-100 and Tween 80 with ethylenediaminetetraacetic acid (EDTA) proved to be effective in removing nickel, lead, and zinc from the soil [84]. Natural chelators, on the other hand, degrade quickly but are less effective, while synthetic chelators have resulted in secondary pollution

[83]. Thus, accordingly, a suitable method should be applied that is efficient and also cost-effective.

Vitrification technology

Vitrification is the process of heating the contaminated soils at high temperatures (1,400–2,000 °C) until they liquidize and freeze rapidly, in order to break down or volatilize organic matter and generate vitreous materials like solid oxide. This technique may be employed in ex situ as well as in situ procedures. In "in situ" vitrification, electric rods are inserted directly into the contaminated soil to heat it, whereas in ex situ vitrification, the contaminated soil is first excavated from the site and then subjected to pretreatments such as screening, dewatering, and blending before being transported to the reactors. This is the most efficient method but it requires large energy investments [82].

4.2.2.1.2 Chemical methods

Chemical method involves the use of immobilization and extraction of heavy metals from the soil utilizing various chemical agents. These methods allow remediation of contaminated soil by chemically transforming the chemical characteristics of the contaminant but the major problem lies in the formation of certain by-products. Some methods are discussed further.

Precipitation

Precipitation being simple and inexpensive is the most effective and widely used method for the effective elimination of heavy metals. Chemical agents that are used react with heavy metals and form insoluble precipitates. The two most common chemical methods of precipitation include hydroxide precipitation and sulfide precipitation. The mechanisms involved are as follows [27, 93]:

$$M^{2+} + 2(OH)^{-} \rightarrow M(OH)_2 \downarrow$$

$$M^{2+} + H_2S \rightarrow MS \downarrow + 2H^{+}$$

Chemical leaching

Chemical leaching is the process of dissolving heavy metal ions from the soil in the presence of water, chemical reagents, and other fluids or gases. Acidic leaching solutions comprising H_2SO_4, HCl, or HNO_3 at a pH of 1.5–2.0 is used to extract heavy metals. Leaching is efficient, applicable in situ, yet cost prohibitive if large amounts of chemicals are later required for neutralization [93].

Chemical extraction

Chelating agents are chemicals that bind tightly to metal ions that are employed to remove heavy metals from the site. Several chelating agents are used for the purpose of remediation such as EDTA, nitrilotriacetic acid, *n*-hydroxyethylethylenediaminetriacetic acid, hemoglobin, chlorophyll, as well as several simple organic acids like oxalic acid and malic acid. Among all, EDTA is widely reported as the most effective chelating agent [57, 83]. Although common soil elements (such as Ca^{2+}, Fe^{2+}, Mg^{2+}, and Al^{3+}) compete with toxic metals for chelating agent binding sites, an additional quantity of chelate is required to assure sufficient contaminant removal.

4.2.2.1.3 Biological methods

Biological methods involve the use of plants, animals, and microbes. The process is considered to be economical and eco-friendly. A few techniques utilizing biological agent remediation are listed below.

Phytoremediation

Phytoremediation employs plants to remediate contaminated soil. The idea of employing these plants dates back to 1983, as they were thought to be an ecologically acceptable, appealing, aesthetically pleasant, noninvasive, energy-efficient, and cost-effective technology. The process is based upon the ability of plants to take up and degrade the pollutant. There are certain plants that accumulate a large quantity of heavy metals around 100–500 times higher in their system compared to a typical plant with no effect on yield, known as metallophytes [51]. They are categorized as metal excluders, metal indicators, and metal hyperaccumulators which work via the process of rhizofiltration, phytostabilization, phytoextraction, and phytotransformation [93].

- Rhizofiltration: Plant root remediates contaminated water through adsorption, concentration, and precipitation.
- Phytostabilization: Plant reduces the mobility of the contaminant.
- Phytoextraction: Plant species extract metals from the soil by accumulating it in various plant parts.
- Phytotransformation: Enzymatic breakdown of pollutant into simpler products is employed.
- Benefits of phytoremediation include in situ application, passive, solar-driven "green" technology, easy applicability, and pertinence to a wide range of metals, radionuclides, and organic substances. However, the integrated use of plants with microbes could also be employed and, in some researches, has also shown to be synergistic in nature. Despite the fact that phytoremediation has several advantages over traditional soil remediation procedures, a comprehensive assessment of phytoremediation reveals some severe drawbacks; for example, hyperaccumulators accumulate specific metals only, if grown in a site

where multiple metal toxicity is observed, then the hyperaccumulator may not be able to remediate the soil. Also, the phytoremediation is relatively slower than any other traditional method [60].

Microbial remediation

In this process, microorganisms are employed to remediate contaminated soil by transforming the toxic form of metal contaminant to a nontoxic or less toxic form. Recently, this method employing different microbial species has emerged as an effective approach to clean up the contaminated site. Several studies have shown the use of Actinobacteria in remediating the soil from numerous organic chemicals and heavy metals [4]. Among the most researched genera are *Streptomyces*, *Rhodococcus*, and *Amycolatopsis* [4, 59]. They have important ecological functions in the environment, including recycling chemicals, degrading complex polymers, and producing bioactive compounds. The ability of Actinobacteria to remove organic and inorganic contaminants proved their biotechnological potential in the environment. It is now also feasible to develop microorganisms with desirable traits such as the capacity to endure metal stress, overexpression of metal chelating proteins and peptides, and metal accumulation ability using modern genetic engineering methods. For example, *Corynebacterium glutamicum*, a biotechnological workhorse, has been employed as an arsenic biocontainer and is genetically modified to clean As-contaminated sites via overexpression of ars operons (ars1 and ars2) [59]. Several fungi are also known to remediate soil by accumulating heavy metal toxicity in their system such as *Aspergillus*, *Rhizopus*, and *Penicillium*. Algae and yeast also perform well in the field of bioremediation [93].

Vermiremediation

Vermiremediation utilises earthworms for the clean-up of contaminated soils. They perform a major role in the functioning of soil ecosystems, and are thus known as engineers of soil ecosystem. Earthworms survive in heavy-metal-contaminated soils sequester metals via dermal and intestinal routes to efficiently acquire high-tissue metal concentrations such as Pb, Cd, and Zn. Some of the reported earthworms utilized for bioremediation are *Eisenia andrei*, *Eisenia fetida*, *Metaphire guillelmi*, *Eudrilus eugeniae*, and *Lumbricus rubellus* [20].

4.2.3 Drought

Plants are subjected to a variety of environmental factors, from rising temperatures to abrupt changes in atmospheric gas levels, resulting in sporadic rainfall and its dispersal. Although the temperature of the air rises, ice melts, resulting in waterlogging in the area and a decrease in water in storage systems, limiting the availability of water to crops in the near future. Due to continuing climate change, annual precipitation in rain-fed agricultural areas has decreased, resulting in severe drought conditions [76]. Drought stress is considered to be one of the major threats to plants as it affects nearly 25% of world's agriculture [36]. However, the severity is uncertain as it depends upon infrequent rainfalls, poor irrigation, soil moisture retention, and high evaporation rate, which leads to decrease in precipitation with increased transpiration, lowering moisture content. Plants under drought stress have compromised morphological, biochemical, and physiological traits and have declined crop yield and number of changes, including a reduction in leaf size, a decrease in cell division, elongation of stems, root proliferation, and plant nutrition absorption [97]. Annual rainfall patterns have an important role in the management of plant water resources, since high rainfall areas have abundant water resources, while low rainfall areas are more prone to experience drought [2].

Drought stressed plants can be managed by using stress-resistant cultivars, foliar application of plant growth hormones (ABA (abscisic acid), gibberellin (GA), cytokinin, and salicylic acid (SA)), antioxidants (ascorbic acid and H_2O_2), and seed treatment with osmoprotectants (proline, phenol, and soluble sugars) and foliar extracts [3]. Another strategy is to use plant growth-promoting bacteria (PGPB), such as *Gluconacetobacter diazotrophicus*, to elicit "induced systemic tolerance" to drought stress, which has been proven to help reduce the detrimental effects of drought stress while also improving biomass and seed yield [26]. Although plants are able to defend themselves to the stressors to a limited extent by activating their resistance mechanisms to recoup vigor and ensure optimal development, they induce their system for transduction and transcription pathways in order to manage the stress [48]. Plant response to cope with stress is usually by the expansion of their roots and formation of the ramified root system, and this enhancement results in accumulation of biomass under water scarcity. The expanded root system generally helps plants to uptake more water, and instead of reducing the growth of the shoot, plants reduce the growth of root and manage the stress conditions [85]. In higher plants, reduction in photosynthesis rate is observed. The decline in photosynthesis influences the expansion of leaves, shortening of leaf surface area, and early senescence of leaves [85]. The photosynthetic rate is supposed to be affected by stomatal and nonstomatal factors due to the prolonged closure of stomata and restriction of CO_2 concentration in the plants. The disturbed plant system ultimately declines the molecular oxygen and accumulates ROS such as super oxides and hydroxyl radicals, leading to oxidative damage in chloroplast and disruption of light and dark reactions of plants [50]. The reduction in photosynthesis and respiration leads to changes in the metabolism of carbon under water scarcity. Under water-deficit environment, there is a decline in respiration rate because of disturbed tricarboxylic acid cycle and biosynthesis of ATP that affects the plant health [79].

4.2.3.1 Remediation of drought

4.2.3.1.1 Stress tolerance

When plants are exposed to drought, their resistance mechanisms evolve strategies that help them avoid or survive stress. Plant ROS buildup occurs due to oxidative stress. When the concentration of ROS in plant system rises, a chain of cellular signaling starts in the plant system that initiates the stress responses [42]. Plants have a potent antioxidant repertoire that detoxifies prooxidants like ROS and lipid peroxyl radicals to keep proper cellular redox equilibrium. Antioxidant activities of enzymatic (SOD, PODs, and CAT) and nonenzymatic antioxidants (carotenoids, tocopherol, and ASC) are produced in response to drought stress to scavenge ROS by decomposing ROS into oxygen, hydrogen peroxide, and water while osmoprotectants like proline have been known to play a key role in reducing the negative effects of ROS by increasing the biosynthesis of proline or decreasing the consumption of proline by plant systems [7]. Nonetheless, overexpression of tomato carotenoid ε-hydroxylase gene has revealed improvement in drought tolerance of transgenic tobacco [104]. However, ROS also induce base deletions, pyrimidine dimers, strand breakage, and base alterations including alkylation and oxidation in plants [42].

4.2.3.1.2 Seed priming

Seed priming is considered as one of the best approaches to reduce the effect of stress on plants that allows the pregermination metabolisms without actual germinations. The germination process of primed seeds is crucial since it leads to a higher rate of germination and uniformity, which are not present in non primed seeds. The technique includes hydro-priming (water), halo-priming (salts), osmo-priming (KNO_3, KH_2PO_4, KCl, NaCl, $CaCl_2$, or $MgSO_4$), hormonal priming (plant growth regulators and polyamines), and bio-priming (PGPB) [23]. The method is adopted in various crops of rice, wheat, maize, soybean, and chickpea to mitigate the drought stress. Osmo-priming seeds with saturated $CaHPO_4$ and 4% KCl solution reveal increment in production even under stress conditions [23]. Primed rice seeds germinate effectively in drought-prone areas, and seedlings emerge quicker and more uniform, resulting in higher yields [10]. Hydro-priming and hormonal priming with gibberellic acid improved the performance of soybean under stress conditions in terms of germination, biochemical traits, and seed yield [56]. Likewise, priming with ascorbic acid in wheat has shown improved leaf emergence, leaf elongation, and shoot and root growth under drought [23]. The benefits of seed priming include stand establishment, growth, and increased yields. Furthermore, natural plant extracts (sorghum, brassica, sunflower, and moringa) are promising and cost-effective ways to develop drought tolerance, especially in terminal drought [24, 36]. Besides, inoculation with arbuscular mycorrhizal fungus (AMF) and biochar amendment has also been reported to promote the growth of a range of agricultural plants. Increased growth and physiological features owing to biochar additions and AMF inoculation under drought stress might be attributed to their roles in nitrogen and phosphorus intake, chlorophyll synthesis, and photosynthesis [39]. Hormonal priming with phytohormones such as ethylene and brassinolide is also a great method to cope with stress condition and enhance the plant tolerance through various pathways that regulate the defense mechanism by activating BZR1/BES1 transcription factors [97].

4.2.3.1.3 Plant growth-promoting bacteria (PGPR)

Microorganisms can be used to mitigate the effects of drought stress and improve plant productivity by regulating oxidative damage. They could be employed for both direct and indirect processes, such as phytohormone synthesis, organic acids, siderophores, fixation of atmospheric nitrogen, phosphate solubilization, production of antibiotics, chitinase production, and plant defense through induced systemic resistance and tolerance [13]. Microorganisms, especially the phyllosphere bacteria, seem to have the capacity of changing the plant ecology and biogeography by altering plant performance under different environmental conditions, and targeting such important microbes could be helpful in improving the plant development. Microbes such as *Bacillus* species accumulate solutes in rice plants, allowing them to cope with drought and avoid damage [6]. The PGPR *Phyllobacterium brassicacearum* is reported to induce a reproductive delay and physiological changes leading to improved drought tolerance in *Arabidopsis*. The PGPRs colonize the root and promote stress-induced development, as well as assist in the solubilization of nutrients and making them accessible to plants [13]. Strains of rhizobacteria containing ACC deaminase and rhizobia (*Rhizobium leguminosarum*) have been reported [110] for their potential to improve root elongation, nodulation, and consequently improved growth of lentil under axenic conditions.

4.2.4 pH

pH or hydrogen potential is regarded as “a master variable,” which is a vital factor for the plant growth and development. The ideal pH for good crop yields is between 6 and 8; pH less than 4 is caused by acidic sulfate generated from sulfur in sediments, which is oxidized to sulfuric or pyrite oxidation, and pH greater than 9 is caused by alkaline soil, which can create stress [8]. The leaching of basic cations such as Ca, Mg, K, Mn, P, Mo, and Na from weathered minerals regulates soil pH, leaving H^+ and Al^{3+} ions as the major exchangeable cations in the soil [58]. The effect of soil pH affects nutrient availability, mobility, plant water intake, and causes oxidative stress and electrolyte leaks by increasing the generation of active oxygen species (ROS), and prevents CO_2 absorption and alters soil biological processes. Also, the microbial community residing in soil is altered by the shift in pH that ultimately affects the growth of many plant systems. For instance, *Trichoderma* prefers acid soils, whereas *Actinomycetes* thrive in alkaline soils. It has been observed that fungal respiration is frequently higher than bacterial respiration under low pH settings, and vice versa, since fungus is better acclimated to acidic soil conditions than bacteria [45].

Low soil pH stress is the major limitation of crop production as acidic soils make up around 30% of the world's ice-free territory, while only about 12% of crops is grown on acidic soils. Low pH conditions increase the solubility of heavy metals in soil such as Al, Mn, and Fe, and with accumulation of these heavy metals in the soil, toxicity in the plant system increases manifolds. However, lacking of certain essential minerals like Ca, Mg, Mo, and P can also limit water uptake [58]. Alkaline pH in the soil also disrupts metabolic conditions by increasing oxidative stress, causing protein and nucleic acid imbalances, chlorophyll breakdown, and reduced photosystem II performance, all of which ultimately leads to photosynthesis failures in plants [18]. Furthermore, excessive use of chemical fertilizers, pesticides, and soils with high heavy metal solubility all contribute to acid rain in the region and raise the pH of the soil, limiting agricultural output.

Mitigation of stress is accomplished by soil conditioning and adoption of agriculture protection techniques for crop conservation and protection from excessive soil alkalinity and acidity. The addition of organic materials or modified soils, as well as the planting of acid-tolerant tree species, is a measure used to control pH stress [99]. Some interventions are discussed below.

4.2.4.1 Remediation of pH

4.2.4.1.1 Mitigation of soil acidity

Acidic soil is characterized by low pH, low organic matter, and aluminum and iron toxicity. A key cause of soil acidity is the excessive use of nitrogen fertilizers that deteriorate agricultural productivity. The most common and convenient way of reducing soil acidity is to utilize lime ($CaCO_3$) in agriculture and forestry [99]. Liming, by lowering the pH of the soil, alleviates the solubility of different elements in the soil. Not only does pH rise, but so do HCO^{3-} or Ca^{2+} ion concentrations, enhancing mobility of the elements [100]. Plant uptake of 56 elements (Ag, Al, As, B, Ba, Be, Bi, Br, Ca, Cd, Ce, Co, Cr, Cs, Cu, Dy, Er, Eu, Fe, Gd, Ge, Hf, Hg, Ho, K, La, Li, Lu, Mg, Mn, Mo, Na, Nb, Nd, Ni, P, Pb, Pr, Rb, S, Sb, Sc, Se, Si, Sm, Sr, Tb, Th, Tl, U, V, W, Y, Yb, Zn, and Zr) has been demonstrated to improve when the pH is raised by adding fine-grained precipitated calcium carbonate [99]. However, the application of a blend of charcoal, sago bark ash, and urea is reported to raise soil pH and base cations while lowering exchangeable acidity, exchangeable Al^{3+}, and exchangeable H^+ [35]. Furthermore, planting *Cinnamomum camphora* (Linn) Presl, or a blend of *C. camphora* and *P. massoniana*, gave the highest exchangeable calcium

(Ca) and magnesium (Mg) concentration, as well as a greater soil pH. Increased relative presence and abundance of beneficial taxa (e.g., *Bradyrhizobium canariense* and *Terracidiphilus* sp.) may have resulted in soil aggregate formation and increased environmental stress tolerance (e.g., Gaiellales), whereas soil linked to *C. camphora* seems to have fewer acid-producing and acidophilic species, which aids in acidification rehabilitation [17]. In addition, biochar has also been reported to lower metal toxicity and increase soil pH. It was discovered that applying biochar to soil can improve soil fertility and agricultural production by preventing nutrient leaching and even supplying nutrients to plants [94].

4.2.4.1.2 Mitigation of soil alkalinity

Alkaline stress is defined as the presence of alkaline salts (Na_2CO_3 or $NaHCO_3$) in the soil, which is said to be more dangerous than saline stress because of the high pH [112]. Evident alkaline stress symptoms were found to be in tomato [112], maize [61], zucchini [15], cucumber [69], and others. Polyamines are reported to be the best alternative for the mitigation of soil alkalinity that promotes growth, cell division, DNA replication, and protein synthesis. Spermidine (Spd) and spermine (Spm) and their precursor, putrescine (put), are the most common polyamines found in plant systems. They act as secondary messengers, mediating responses to environmental stressors such as osmotic stress, salinity changes, drought, ozone, heavy metals, and ultraviolet. These exogenous polyamines could enhance plant tolerance to alkalinity stresses [112]. Furthermore, hormonal activity has also been correlated to the reduction of alkalinity stress. Pretreating plant systems with jasmonic acid (JA) can considerably reduce the hazardous effects of alkalinity in soil. The decrease in ROS and malondialdehyde might be to blame for the better tolerance [61]. However, SA revealed reduction in Na^+ accumulation, maintaining ionic homeostasis and normal photosystem operation, improved the ROS scavenging system, reduced oxidative damage, and reduced lipid peroxidation, thereby improving alkaline tolerance of plants [69]. Plant microbiomes, particularly rhizospheric microorganisms found in a wide variety of natural settings, have the capacity to increase plant growth and soil health. Inoculation of AMF to the plant system was reported to enhance plant nutrient acquisition, overcome the detrimental effect of salinity, improve drought tolerance, and enhance plant tolerance to alkalinity [15]. Nonetheless, *T. asperellum* exposure reduced saline–alkaline stress for maize development by lowering the soil pH and salt ion concentration and controlling the fungal microbial population, resulting in higher crop yields [28].

4.2.5 Temperature

Climate change, which is driven by global warming, has become one of the major environmental issues. The most notable change is an increase in atmospheric temperature as a result of higher greenhouse gas concentrations in the atmosphere that directly affects the agriculture sector [16]. The rate of plant growth and development is influenced by the temperature around it, and each species has its own temperature range, which is represented by a minimum, maximum, and optimal temperature. Occurrence of extreme fluctuations in temperature has the most detrimental impact on plant productivity as most physiological processes normally function between 0 and 40 °C [41]. The effect of temperature extremes (frost and heat) on wheat (*Triticum aestivum*) indicated that frost, which induced sterility and abortion of produced grains, while extreme heat caused a drop in grain number and shortened the grain filling time [9]. The incidence of extreme events might reduce the productivity of three major crops (rice, wheat, and maize) in the near future [17].

High-temperature stress is a warming trend beyond a critical threshold that substantially reduces crop productivity. It affects numerous species including legumes that affect seed germination and photosynthetic efficiency. Physiological processes of the plant system are affected the stress to some extent, and if exposed for longer duration may cause injury or ultimately death [38]. Excessive levels of ROS, which lead to oxidative stress, are the most serious outcomes of high-temperature stress. When plants are subjected to extremely high temperatures, various changes occur at the cellular level, including lipid modification (become more liquidy and develop higher osmotic pressure) and ROS formation, which disrupt plasma membrane metabolism. Proteins execute their functions well at room temperature, but when the temperature is raised over their optimal range, the protein deactivates, causing alterations in enzymatic activity, thereby increasing the formation of ROS and AOS (active oxygen species) [51]. Plant damage and mortality are caused by a reduction in photosynthesis, changes in assimilate movement, increased oxidative stress, and changes in the AOS effect [34].

Another important environmental factor that impacts plant development is low temperature or cold stress. The cold stress is categorized as chilling and freezing. Chilling happens when plants are exposed to temperatures below 10–20 °C without ice crystal formation, whereas freezing occurs when plants are subjected to temperatures that cause ice crystal formation [38]. When plants are exposed to low temperatures, the integrity of cell membrane shifts by changing the state of lipid from liquid to solid, which is mostly determined by the amount of unsaturated fatty acids present [51]. Several symptoms appear when fruits and vegetables are stored at a low temperature and then brought to room temperature. The peeling and superficial burn-like symptoms of the peel, as well as browning of the skin, are all symptoms of cold stress [33]. Furthermore, plants that have evolved in temperate environments can withstand cold temperatures, although the degree of sensitivity varies from species to species.

However, resistance of plants to extreme freezing is not natural, since, at low temperatures plants activate their various biochemical and physiological processes in order to cope with freezing. This process is known as "cold acclimatization." Plants of *Secale cereale*, for example, did not survive when exposed to −5 °C without prior cold acclimatization; nevertheless, when cold acclimatized at 2 °C for 1–2 weeks, the plants were not able to endure temperatures as low as −30 °C [51].

Cold stress or heat stress cause cellular alterations that result in an excess accumulation of harmful chemicals, particularly ROS which causes oxidative stress. Phosphorylation is a mechanism giving rise to free radicals in the plant cellular environment. O_2 is reduced to H_2O by the addition of four electrons during the oxidative phosphorylation process, resulting in superoxide anion radicals, which are subsequently converted to hydrogen peroxide (H_2O_2) by the SOD enzyme. Plants have evolved a number of enzymatic and nonenzymatic reactions to cope with extreme temperatures, protecting cells from oxidative damage and sustaining cellular homeostasis [5, 38].

4.2.5.1 Remediation of temperature

4.2.5.1.1 Heat stress

Heat avoidance

Plants have a variety of systems that allow them to endure heat stress: short-term avoidance or acclimatization, and long-term phenological and morphological alterations. Leaf rolling or changes in leaf position, lipid membrane composition, and cooling as a result of transpiration are some of the short-term processes while increased density of leaf stomata and hair, as well as bigger vasculature, are long-term morphological alterations in response to heat stress [40]. Other processes exist in plants, including aggressive transpiration; for example, great millet (*Sorghum* spp.), maize (*Zea mays* subsp. mays), poplar (*Populus* spp.), pine (*Pinus* spp.), wild jujube (*Ziziphus lotus*), and *Eucalyptus* [40]. The probable mechanisms contribute to changes in membrane permeability, an increase in cuticle permeability and a decrease in water viscosity that eventually led to increase in transpiration rate [83]. Furthermore, *Arabidopsis thaliana* can respond to spring heat and drought stress as reported by Wolfe and Tonsor [106]. Variable freezing tolerance, vernalization demands, sensitivity to light quality, heat shock protein (HSP) expression, growth rate, seed dormancy, season of germination, and age at the beginning of flowering are all factors that contribute to the adaptation [106].

Heat tolerance

Plants can detect and adapt to changes in their environment, albeit the degree of adaptation or tolerance to certain stressors, and hormones play an essential role in this regard. Hormonal homeostasis, stability, and biosynthesis are all impacted when a plant is stressed. Phytohormones viz. play essential roles in the response of plants to heat stress, including ABA, auxin, GAs, cytokinins, SA, JA, ethylene, and brassinosteroids (BRs) [102]. For instance, treatment of BRs to *Solanum melongena* is reported to increase stomatal conductance and quantum efficiency of PSII, indicating that BRs can induce intrinsic heat tolerance [107]. However, exogenous application of protectants such as osmoprotectants, signaling molecules, trace elements, phytoprotectants, and polyamines improve heat tolerance in plants cultivated under high temperature since these protectants have growth-promoting and antioxidant abilities [40].

Furthermore, molecular techniques are assisting in the clear understanding of the idea of heat stress tolerance in plants, in addition to other physiological and biochemical pathways. Plants tolerate stress by modulating multiple. For instance, *Populus simoni* was investigated for heat and drought stress. RNA-sequencing analysis revealed that a large chunk of genes showed differential expression in its roots and leaves in response to high temperature and desiccation. However, a small number of these genes showed overlapping functioning for both heat and drought responses. These genes were mainly concerned with RNA regulation, transport, phytohormone physiology, and stress. For instance, high temperature and/or drought-induced ABA accumulation and decreases in auxin and other phytohormones corresponded well with the differential expression of a few genes involved in hormone metabolism. Their conclusions were that heat-drought-responsive genes play important roles in the transcriptional and physiological metamorphosis of poplars to high temperature and/or drought when exposed to changing climatic contexts [46].

Further, HSPs are regarded as the "master players" to maintain cellular homeostasis [38]. When cells are triggered by many types of stress, such as oxidative stress, heat, cold, inflammation, and oxygenation abnormalities, plants start producing unique proteins called HSPs. Normally, they are present as companion cells in the body but when the plant is stressed, cells survive by acting as chaperones [5]. In response to rising temperatures, plants produce HSPs of various sizes. It has been demonstrated that heat resistance is directly linked to the synthesis of HSP accumulation [38].

4.2.5.1.2 Chilling stress

Plant growth regulators

The predominant impact of chilling stress is the accumulation of ROS, that is, hydrogen peroxide (H_2O_2), superoxide anion radicals (O_2^-), and hydroxyl radicals (OH) that interrupt electron transport chain of chloroplast, ultimately causing cell dysfunction [103]. Different strategies have been evolved by the plant system to cope with the stress: for instance, enzymatic antioxidants (SOD, CAT, POD, and APX), nonenzymatic antioxidants (glutathione, ascorbic acid, proline, silicon, and caffeic acid), sugars (fructan, trehalose, galactinol, mannitol, D-inositol, and raffinose), and amino acids (melatonin, proline, and tryptophan) [37, 103]. Various compounds like SA, JA; moringa leaf extract, sorghum water extract, and thiourea have been documented to be involved in the protecting plants from chilling injury by the overproduction and enhanced activity of CAT enzyme. Also, the allelopathic effects of the plant extract such as moringa leaf extract and sorghum water extract increased the photosynthetic activity of plant systems under chilling stress by acting as an ROS scavenger [105]. Nonetheless, seeds hydroprimed with SA or JA at 10^{-2}, 10^{-3}, and 10^{-4} M concentration, and 24-epibrassinolide at 10^{-6}, 10^{-8}, and 10^{-10} M concentration followed by heat shock treatment markedly reported reduced root inhibition as well as lateral root establishment by permitting the germinating seeds to cope with the growth-retarding chilling effects [96]. Furthermore, increase in CAT activity and sugar metabolism stimulated the development of seedlings and lessened the decrease of F_v/F_m (maximum photochemical efficiency of photosystem II, a measure of plant stress) caused by chilling conditions [96].

Fungicides

Fungicides, particularly the triazoles, have been employed from many decades for seed treatments. They help by regulating the production, especially under low temperatures [111]. Azole fungicides reduce the rate of chilling injury as well as control postharvest deterioration caused by green mold and others. They are used as a hot dip at 40 °C to alleviate the effects of chilling symptoms. Studies designate that in some citrus cultivars, the usage of thiabendazole is efficient in decreasing the rate of chilling injury. A comparative study [44] revealed the susceptibility of "Washington Navel" oranges (*Citrus sinensis*) to chilling conditions. Fungicides, that is, benzimadazole, Benazid®, thiabendazole (Tecto®), and Thiabendazole (ICA-TBZ®) have been employed, and thiabendazole has been concluded to be the most convenient and efficient alternative in decreasing the rate and severity of chilling injury. Furthermore, fungicides of the strobilurin class (pyraclostrobin, azoxystrobin, trifloxystrobin) have been shown to delay senescence by lowering the amount of ACC which is an ethylene precursor. Pyraclostrobin works by increasing the activity of SOD and reducing ozone toxicity, whereas azoxystrobin is known to promote antioxidant enzyme activities such as SOD, CAT, and APX, and trifloxystrobin has been shown to improve abiotic stress tolerance [95].

Mycoremediation

Fungal usage to mitigate temperature induced stress is mycoremediation. *Trichoderma* species are widely employed to mitigate the negative impacts of biotic and abiotic stressors on agricultural plants. The colonization controls the endogenous plant hormones, plant enzymes, antioxidants, compatible solutes, and compounds like phytoalexins and phenolics. For instance, *T. harzianum* (AK20G) confers tolerance to cold, leading to enhanced photosynthetic rate, growth rates, water content, and proline accumulation while reducing lipid peroxidation rate and electrolyte leakage [31]. Nonetheless, inoculation of plant system with *T. asperellum* (HK703) improved some of the traits of chilling stressed plants [19]. However, the potential for the use of AMF and PGPR (*Funneliformis mosseae* and *Paraburkholderia graminis*), both alone and in combination, has been exploited to protect the processed tomato seedlings from damage caused by severe chilling stress (1 °C). Results revealed best performances when inoculated alone with *F. mosseae* [14].

4.2.5.1.3 Freezing stress

Avoidance

Growers can employ passive measures such as optimal site selection, land clearance, soil management, or choosing crop types that are less temperature sensitive to avoid frost damage. Alternatively, they might utilize active methods like crop cover, smoke clouds, wind devices, water spraying, or heating [51]. However, some growers may not be able to afford to implement these safeguards; therefore, they may choose to sow crops later in the season instead [11].

Cold acclimatization

Plants can adapt to cold stress by exposing them to it for shorter periods of time. The ability of a plant to acclimatize is primarily determined by its genetic background as well as environmental factors such as minimum temperature and cold exposure periods [11, 51]. For instance, frost-sensitive mustard *Arabidopsis thaliana*, has a maximum freezing tolerance of roughly −12 °C after cold acclimation, compared to −5 °C in nonacclimated plants. On the other hand, cold-acclimated frost-hardy winter wheat and forage grasses can endure temperatures below −25 °C, while certain plants from the Arctic and alpine areas can withstand temperatures below −50 °C [11]. There is no particular molecule, transcript, or mechanism responsible for cold acclimatization but rather it appears that a multidimensional reorganization of metabolic homeostasis is involved. However, regulation of mRNA synthesis plays a pivotal role in the process. Furthermore, photosynthesis and carbohydrate metabolism that prevents ROS generation is tightly regulated under low temperatures [29].

Ice-binding proteins (IBPs)

Ice-binding proteins (IBPs) are the family of low-temperature-associated proteins found in certain plants that facilitate survival in extreme conditions. Seeds, for example, can prevent freezing by overwintering with limited water content or by accumulating sugars and polyols, which lowers the freezing point. Plants may also withstand freezing by acclimating to the cold and employing IBPs to protect cells [11]. The IBP method relies on their capacity to adsorb to ice crystals in an irreversible manner, resulting in the "shaping" of ice as they get integrated into the ice crystal lattice by the process of ice adhesion and ice structuring [11]. For instance, ryegrass (*Lolium perenne*) can withstand temperature below −13 °C. IBP presumably is responsible for preventing the damage in frost conditions [12].

Many industrial- and biotechnological-based enterprises exploit cold-active biomolecules derived from cold-dwelling organisms. They have been categorized as ice-nucleation proteins (INPs), antifreeze proteins (AFPs), and antinucleation proteins (ANPs) [11]. INPs have been discovered in bacteria such as *Pseudomonas*, *Pantoea*, and *Xanthomonas*, in fungi such as *Fusarium*, while ANPs have been found in bacteria such as *Acinetobacter* and *Bacillus*. AFPs have been first discovered in plants; in filamentous fungi such as *Penicillium*, *Typhula*, *Coprinus*, and *Flammulina*; in yeasts such as *Leucosporidium*, *Glaciozyma*, and *Rhodotorula*; and in bacteria such as *Pseudomonas*, *Moraxella*, *Flavobacterium*, and *Marinomonas* [49]. This can be extrapolated to the agricultural industry as well for developing freeze-tolerant crop cultivars favoring the wide adaptation of such crops, thus contributing to greater crop yield.

4.2.6 Waterlogging

Waterlogging is a situation where the surface of farmland is completely saturated with water. It is estimated that around 12% of world's cultivated area might be flooded on a regular basis, leading to 20% reduction in agricultural output [98]. Soil waterlogging is becoming more prevalent due to climate change in the near future, particularly in irrigated areas and areas with unexpected rainfalls, such as the Yangtze Watershed, the plains of Huang-Huai-Hai, Sanjiang, and Songnen in China, as well as irrigated areas in India, Pakistan, the United States, Argentina, and Europe [98]. Simultaneously, owing to the continued use of agricultural equipment, a considerable amount of clay is heavily compacted, resulting in poor drainage and increased waterlogging conditions [67, 72]. The waterlogging begins with hypoxic (oxygen shortage) conditions and progresses to anoxic (absence of oxygen) conditions. Thus, the major consequence is the reduction in oxygen level that leads to accumulation of ethylene and carbon dioxide, thereby shifting root respiration from aerobic to anaerobic. This will result in a significant drop in root cell ATP generation and have an impact on the plant's metabolic pathway [37, 98].

Waterlogging causes oxygen deficit, light deprivation, altered stomatal conductance, and CO_2 assimilation, resulting in reduced nutrient as well as water uptake, accumulation of ROS such as superoxide ($O_2^{\bullet}$), hydrogen peroxide (H_2O_2), and hydroxyl radical ($OH^{\bullet}$), and the production of a variety of secondary metabolites, all of which contribute to reduction in the ability of the plant to grow and show productivity under stress [37]. Studies reveal that waterlogging causes a substantial reduction in crop yield at the seedling, joining, and tillering stages. Crop varieties that are mostly affected include cotton (23.66–34.79%), wheat (7.75–16.30%), and rice (7.48–57.42%) [98]. Photooxidative damage has also been induced, causing wilting of leaf, lowering biomass, thus decreasing the weight of seed and production of grain [98].

Plant adaptation to water stressors such as waterlogging has recently been widely researched in a variety of plant species. Plants use a variety of responses when they are stressed by water, albeit the degree of acclimation varies by species and stress threshold. Plants that are subjected to excessive water stress have a depleted oxygen supply and undergo a variety of physiological, morphological, and metabolic changes in order to survive in submerged or waterlogged conditions: for instance, production of hormone and antioxidants.

4.2.6.1 Remediation of waterlogging

Excessive water in the soil causes an increase in ROS, resulting in oxidative stress and cellular damage such as lipid peroxidation, protein oxidation, nucleic acid damage, enzyme inhibition, and programmed cell death, and in consequence, it employs two strategies, that is, plant adaptation mechanisms and hormonal-induced resistance to cope with the stress. Plant adaptation mechanisms include morphological and anatomical adaptations, photosynthetic adaptations, respiratory adaptations, and antioxidant defense mechanisms, while hormonal-induced resistance comprises several hormones such as ethylene, gibberellic acids, ABA, SA, JA, brassinosteroids, and melatonin [71].

Antioxidants play a significant role in stress tolerance since they scavenge ROS from stressed plant systems. They are generally classified into two types: enzymatic and nonenzymatic antioxidants. Nonenzymatic antioxidants include APX, SOD, POD, CAT, and GR, whereas enzymatic antioxidants include ascorbic acid, glutathione, tocopherols, and carotenoids [103]. Several investigations have found a significant change in the endogenous levels of several enzymatic and nonenzymatic antioxidants. For instance, when the pigeon pea (*Cajanus cajan*) is water stressed, antioxidant enzymes such as SOD, APX, GR, and CAT increased under waterlogging condition [55]. Similarly, when the legume plant, *Vigna sinensis*, is subjected to water stress, it showed a marked increase in the production of antioxidants such as proline, ascorbic acid, CAT, SOD, and POD [22]. Likewise, in soybean, the activity of

CAT, SOD, POD, and malondialdehyde has been reported to increase [21]. As a conclusion of these findings, it is obvious that when plants are exposed to waterlogged conditions, their antioxidant defense mechanism kicks in to combat the harmful effects of oxidative stress generated by ROS.

Application of exogenous hormones has a vital role in stress tolerance as well. Previous studies showed that the application of ethylene, melatonin, auxin, GAs, ACC, ABA, and cytokinin might help plants recover from the adverse effects of waterlogging [71]. For instance, exogenous spraying of 6-benzyladenine (6-BA) to waterlogged summer maize could promote grain filling and photosynthesis, resulting in a considerable increase in grain productivity by delaying leaf senescence, increasing chlorophyll content, and improving photosynthetic efficiency [78]. Furthermore, following priming with ethylene gas, the potential of okra (*Abelmoschus esculentus*) and maize (*Zea mays*) to endure waterlogging was evaluated, and the results revealed greater stress tolerance [101]. Melatonin likewise promotes waterlogging tolerance by modulating polyamines and ethylene production through ethylene repression and polyamine stimulation, resulting in more stable cell membranes, improved photosynthesis, and reduced ethylene-responsive senescence [64]. Nonetheless, transcriptomic study of the transgenic maize under normal and waterlogged conditions revealed that ZmERFVII gene mediated AR production and regulation of ROS, thereby regulating waterlogged tolerance [109]. However, the capacity of ACC deaminase-producing bacteria (α-proteobacteria, β-proteobacteria, and γ-proteobacteria, and Actinobacteria, Firmicutes, Bacteroidetes and Flavobacterium) to stimulate the development of terrestrial plants under waterlogged settings has been the focus of research. Ethylene is produced in large levels inside the plant tissue during flooding due to the obvious enhanced activity of ACC synthase in the waterlogged roots. PGPRs that have ACC deaminase activity can reroute ACC away from the ethylene biosynthesis pathway in host plants' roots, resulting in decreased ethylene production [86].

4.3 Conclusion

Abiotic stressors are one of the major problems limiting global food productivity. Plants have successfully inhabited the Earth, so it is no wonder that they have figured out how to contend with abiotic stresses. They can withstand stress up to a threshold, but when that threshold is surpassed significantly, they become susceptible by causing a cascade of events, including reduced leaf number and size, increased leaf rolling and abscission, changes in stomatal size and resistance, and increased root suberization, thereby causing low yields. Furthermore, reduction in agricultural production due to abiotic stress has been positively correlated with economic losses. Many studies have revealed that greater than 30% reduction is caused by abiotic factors in major food crops [30]. However, if left untreated, the yield loss may go up to 100%. Plants, on the contrary, utilize sophisticated processes (stress recognition, signal transduction, and transmission) as well as many nonenzymatic and enzymatic systems such as SOD, POD, CAT, and APX to activate the cell and adjust its metabolism to stress. Fortunately, as highlighted by numerous plant abiotic stress researchers, silicon from the Earth has compensated the ill effects of both abiotic and biotic stresses [37, 103]. However, these concepts are under scrutiny. Furthermore, taking rising population and food demands into consideration, finding techniques to improve crop tolerance to abiotic stress factors will be critical if agricultural productivity and food security are to be improved further.

References

[1] Ali H, Khan E, Sajad MA. Phytoremediation of heavy metals – Concepts and applications. Chemosphere 2013 91(7), 869–881.
[2] Ali N, Anjum MM. Drought stress: Major cause of low yield and productivity. Austin Environmental Sciences 2016, 1, 10–12.
[3] Ali Z, Basra SMA, Munir H, Mahmood A, Yousaf S. Mitigation of drought stress in maize by natural and synthetic growth promoters. Journal of Agriculture and Social Science 2011 7(2), 56–62.
[4] Alvarez A, Saez JM, Costa JSD, Colin VL, Fuentes MS, Cuozzo SA . . . Amoroso MJ. *Actinobacteria*: Current research and perspectives for bioremediation of pesticides and heavy metals. Chemosphere 2017, 166, 41–62.
[5] Argosubekti N. A review of heat stress signalling in plants. In: IOP Conference Series: Earth and Environmental Science, IOP Publishing 2020, 484(1), 012041.
[6] Arun KD, Sabarinathan KG, Gomathy M, Kannan R, Balachandar D. Mitigation of drought stress in rice crop with plant growth-promoting abiotic stress-tolerant rice phyllosphere bacteria. Journal of Basic Microbiology 2020 60(9), 768–786.
[7] Azmat R, Moin S. The remediation of drought stress under VAM inoculation through proline chemical transformation action. Journal of Photochemistry and Photobiology. B, Biology 2019, 193, 155–161.
[8] Balks MR, Zabowski D. Celebrating soil: Discovering soils and landscapes. Switzerland, Springer, 2016.
[9] Barlow KM, Christy BP, O'leary GJ, Riffkin PA, Nuttall JG. Simulating the impact of extreme heat and frost events on wheat crop production: A review. Field Crops Research 2015, 171, 109–119.
[10] Bitew JM. The impacts of drought stress on crop production and productivity. GPH-IJAR International Journal of Agriculture and Research 2021 4(02), 18–38.
[11] Bredow M, Walker VK. Ice-binding proteins in plants. Frontiers in Plant Science 2017, 8, 2153.
[12] Bredow M, Vanderbeld B, Walker VK. Ice-binding proteins confer freezing tolerance in transgenic Arabidopsis thaliana. Plant Biotechnology Journal 2017 15(1), 68–81.
[13] Bresson J, Varoquaux F, Bontpart T, Touraine B, Vile D. The PGPR strain Phyllobacterium brassicacearum STM 196 induces a reproductive delay and physiological changes that result in improved drought tolerance in Arabidopsis. New Phytologist 2013 200(2), 558–569.

[14] Caradonia F, Francia E, Morcia C, Ghizzoni R, Moulin L, Terzi V, Ronga D. Arbuscular mycorrhizal fungi and plant growth promoting rhizobacteria avoid processing tomato leaf damage during chilling stress. Agronomy 2019 9(6), 299.

[15] Cardarelli M, Rouphael Y, Rea E, Colla G. Mitigation of alkaline stress by arbuscular mycorrhiza in zucchini plants grown under mineral and organic fertilization. Journal of Plant Nutrition and Soil Science 2010 173(5), 778–787.

[16] Chauhan BS, Mahajan G, Randhawa RK, Singh H, Kang MS. Global warming and its possible impact on agriculture in India. Advances in Agronomy 2014, 123, 65–121.

[17] Chen Z, Maltz MR, Zhang Y, O'Brien BJ, Neff M, Wang Y, Cao J. Plantations of Cinnamomum camphora (Linn) Presl with distinct soil bacterial communities mitigate soil acidity within polluted locations in southwest China. Forests 2021 12(6), 657.

[18] Choudhury FK, Rivero RM, Blumwald E, Mittler R. Reactive oxygen species, abiotic stress and stress combination. The Plant Journal 2017 90(5), 856–867.

[19] Cornejo-Ríos K, Osorno-Suárez MDP, Hernández-León S, Reyes-Santamaría MI, Juárez-Díaz JA, Pérez-España VH, Saucedo-García M. Impact of Trichoderma asperellum on chilling and drought stress in tomato (Solanum lycopersicum). Horticulturae 2021 7(10), 385.

[20] Das A, Osborne JW. Bioremediation of Heavy Metals. In: Gothandam K, Ranjan S, Dasgupta N, Ramalingam C, Lichtfouse E. (eds) Nanotechnology, Food Security and Water Treatment. Environmental Chemistry for a Sustainable World, Cham, Springer, 2018.

[21] Da-Silva CJ, do Amarante L. Time-course biochemical analyses of soybean plants during waterlogging and reoxygenation. Environmental and Experimental Botany 2020, 180, 104242.

[22] El-Enany AE, Al-Anazi AD, Dief N, Al-Taisan WAA. Role of antioxidant enzymes in amelioration of water deficit and waterlogging stresses on Vigna sinensis plants. Journal of Biology and Earth Science 2013 3(1), 144–153.

[23] Farooq M, Irfan M, Aziz T, Ahmad I, Cheema SA. Seed priming with ascorbic acid improves drought resistance of wheat. Journal of Agronomy and Crop Science 2013 199(1), 12–22.

[24] Farooq M, Rizwan M, Nawaz A, Rehman A, Ahmad R. Application of natural plant extracts improves the tolerance against combined terminal heat and drought stresses in bread wheat. Journal of Agronomy and Crop Science 2017 203(6), 528–538.

[25] Feng W, Zhang S, Zhong Q, Wang G, Pan X, Xu X, Zhang Y. Soil washing remediation of heavy metal from contaminated soil with EDTMP and PAA: Properties, optimization, and risk assessment. Journal of Hazardous Materials 2020, 381, 120997.

[26] Filgueiras L, Silva R, Almeida I, Vidal M, Baldani JI, Meneses CHSG. Gluconacetobacter diazotrophicus mitigates drought stress in Oryza sativa L. Plant and Soil 2020 451(1), 57–73.

[27] Fu F, Wang Q. Removal of heavy metal ions from wastewaters: A review. Journal of Environmental Management 2011 92(3), 407–418.

[28] Fu J, Xiao Y, Wang YF, Liu ZH, Zhang YF, Yang KJ. Trichoderma asperellum alters fungal community composition in saline–alkaline soil maize rhizospheres. Soil Science Society of America Journal 2021 85(4), 1091–1104.

[29] Fürtauer L, Weiszmann J, Weckwerth W, Nägele T. Dynamics of plant metabolism during cold acclimation. International Journal of Molecular Sciences 2019 20(21), 5411.

[30] Gharde Y, Singh PK, Dubey RP, Gupta PK. Assessment of yield and economic losses in agriculture due to weeds in India. Crop Protection 2018, 107, 12–18.

[31] Ghorbanpour A, Salimi A, Ghanbary MAT, Pirdashti H, Dehestani A. The effect of Trichoderma harzianum in mitigating low temperature stress in tomato (Solanum lycopersicum L.) plants. Scientia Horticulturae 2018, 230, 134–141.

[32] Ghosh M, Singh SP. A review on phytoremediation of heavy metals and utilization of its by-products. Asian Journal of Energy and Environment 2005 6(4), 18.

[33] Golding J. A review of chilling injury causes and control, 2019. (Accessed at https://citrusaustralia.com.au/news/latest-news/a-review-of-chilling-injury-causes-and-control).
[34] Hall AE. Crop responses to the environment. Boca Raton, Florida, CRC Press, 2001.
[35] Hamidi NH, Ahmed OH, Omar L, Ch'ng HY. Combined use of charcoal, sago bark ash, and urea mitigate soil acidity and aluminium toxicity. Agronomy 2021 11(9), 1799.
[36] Hanafy R. Using Moringa olifera leaf extract as a bio-fertilizer for drought stress mitigation of Glycine max L. plants. Egyptian Journal of Botany 2017 57(2), 281–292.
[37] Hasanuzzaman M, Al Mahmud J, Nahar K, Anee TI, Inafuku M, Oku H, Fujita M. Responses, Adaptation, and ROS Metabolism in Plants Exposed to Waterlogging Stress. In: Khan M, Khan N, eds. Reactive Oxygen Species and Antioxidant Systems in Plants: Role and Regulation under Abiotic Stress. Singapore, Springer, 2017, 257–281.
[38] Hasanuzzaman M, Nahar K, Fujita M. Extreme temperature responses, oxidative stress and antioxidant defense in plants. Abiotic Stress-plant Responses and Applications in Agriculture 2013, 13, 169–205.
[39] Hashem A, Kumar A, Al-Dbass AM, Alqarawi AA, Al-Arjani ABF, Singh G, AbdAllah EF. Arbuscular mycorrhizal fungi and biochar improves drought tolerance in chickpea. Saudi Journal of Biological Sciences 2019 26(3), 614–624.
[40] Hassan MU, Chattha MU, Khan I, Chattha MB, Barbanti L, Aamer M, Aslam MT. Heat stress in cultivated plants: Nature, impact, mechanisms, and mitigation strategies – A review. Plant Biosystems-An International Journal Dealing with All Aspects of Plant Biology 2021 155(2), 211–234.
[41] Hatfield JL, Prueger JH. Temperature extremes: Effect on plant growth and development. Weather and Climate Extremes 2015, 10, 4–10.
[42] Hernández I, Cela J, Alegre L, Munné-Bosch S. Antioxidant Defenses Against Drought Stress. In: Aroca, R, eds. Plant Responses to Drought Stress. Berlin, Heidelberg, Springer, 2012, 231–258.
[43] Hoang TML, Tran TN, Nguyen TKT, Williams B, Wurm P, Bellairs S, Mundree S. Improvement of salinity stress tolerance in rice: Challenges and opportunities. Agronomy 2016 6(4), 54.
[44] Hordijk J, Cronjé PJR, Opara UL. Postharvest application of Thiabendazole reduces chilling injury of citrus fruit. In: II All Africa Horticulture Congress 1007. Acta Hortic, 2012, 119–125.
[45] Husson O. Redox potential (Eh) and pH as drivers of soil/plant/microorganism systems: A transdisciplinary overview pointing to integrative opportunities for agronomy. Plant and Soil 2013 362(1), 389–417.
[46] Jia J, Zhou J, Shi W, Cao X, Luo J, Polle A, Luo ZB. Comparative transcriptomic analysis reveals the roles of overlapping heat-/drought-responsive genes in poplars exposed to high temperature and drought. Scientific Reports 2017 7(1), 1–17.
[47] Kalaivanan D, Ganeshamurthy AN. Mechanisms of Heavy Metal Toxicity in Plants. In: Rao N, Shivashankara K, Laxman R, eds. Abiotic Stress Physiology of Horticultural Crops. New Delhi, Springer, 2016, 85–102.
[48] Kaur G, Asthir B. Molecular responses to drought stress in plants. Biologia Plantarum 2017 61(2), 201–209.
[49] Kawahara H. Cryoprotectants and ice-binding proteins. In: Margesin R, eds. Psychrophiles from biodiversity to biotechnology. Cham, Springer, 2017, 237–257.
[50] Keyvan S. The effects of drought stress on yield, relative water content, proline, soluble carbohydrates and chlorophyll of bread wheat cultivars. Journal of Animal and Plant Science 2010 8(3), 1051–1060.

[51] Khalid MF, Hussain S, Ahmad S, Ejaz S, Zakir I, Ali MA, Anjum MA. Impacts of abiotic stresses on growth and development of plants. In: Hasanuzzaman M, Fujita M, Oku H, Tofazzal MI, eds. Plant tolerance to environmental stress. CRC Press, 2019, 1–8.
[52] Kiliç CC, Kukul YS, Anaç D. Performance of purslane (Portulaca oleracea L.) as a salt-removing crop. Agricultural Water Management 2008 95(7), 854–858.
[53] Kumar Das S, Singh Grewal A, Banerjee M. A brief review: Heavy metal and their analysis. Organization 2011 11(1), 003.
[54] Kumar P, Sharma PK. Soil salinity and food Security in India. Frontiers in Sustainable Food Systems 2020, 4, 174.
[55] Kumutha D, Ezhilmathi K, Sairam RK, Srivastava GC, Deshmukh PS, Meena RC. Waterlogging induced oxidative stress and antioxidant activity in pigeonpea genotypes. Biologia Plantarum 2009 53(1), 75–84.
[56] Langeroodi ARS, Noora R. Seed priming improves the germination and field performance of soybean under drought stress. The Journal of Animal and Plant Sciences 2017 27(5), 1611–1621.
[57] Leštan D, Luo CL, Li XD. The use of chelating agents in the remediation of metal-contaminated soils: A review. Environmental Pollution 2008 153(1), 3–13.
[58] Long A, Zhang J, Yang LT, Ye X, Lai NW, Tan LL, Chen LS. Effects of low pH on photosynthesis, related physiological parameters, and nutrient profiles of citrus. Frontiers in Plant Science 2017, 8, 185.
[59] Mateos LM, Villadangos AF, Alfonso G, Mourenza A, Marcos-Pascual L, Letek M, Gil JA. The arsenic detoxification system in Corynebacteria: Basis and application for bioremediation and redox control. Advances in Applied Microbiology 2017, 99, 103–137.
[60] McIntyre T. Phytoremediation of heavy metals from soils. Phytoremediation, 2003, 78, 97–123.
[61] Mir MA, John R, Alyemeni MN, Alam P, Ahmad P. Jasmonic acid ameliorates alkaline stress by improving growth performance, ascorbate glutathione cycle and glyoxylase system in maize seedlings. Scientific Reports 2018 8(1), 1–13.
[62] Miranda H. Stress response in the cyanobacterium Synechocystis sp. PCC 6803. Doctoral dissertation, Department of Chemistry, Umeå University, 2011.
[63] Mosa KA, Ismail A, Helmy M. Introduction to plant stresses. In: Stuber M, eds. Plant stress tolerance. Cham, Springer, 2017, 1–19.
[64] Moustafa-Farag M, Mahmoud A, Arnao MB, Sheteiwy MS, Dafea M, Soltan M, Ai S. Melatonin-induced water stress tolerance in plants: Recent advances. Antioxidants 2020 9(9), 809.
[65] Munns R, Gilliham M. Salinity tolerance of crops – What is the cost? New Phytologist 2015 208(3), 668–673.
[66] Munns R, Tester M. Mechanisms of salinity tolerance. Annual Review of Plant Biology 2008, 59, 651–681.
[67] Najeeb U, Bange MP, Tan DK, Atwell BJ. Consequences of waterlogging in cotton and opportunities for mitigation of yield losses. AoB Plants, 2015, 7, plv080.
[68] Nesme J, Cania B, Zadel U, Schöler A, Płaza GA, Schloter M. Complete genome sequences of two plant-associated Pseudomonas putida isolates with increased heavy-metal tolerance. Genome Announcements 2017, 5(47), 01330, 17.
[69] Nie W, Gong B, Chen Y, Wang J, Wei M, Shi Q. Photosynthetic capacity, ion homeostasis and reactive oxygen metabolism were involved in exogenous salicylic acid increasing cucumber seedlings tolerance to alkaline stress. Scientia Horticulturae 2018, 235, 413–423.
[70] Nisha R, Kiran B, Kaushik A, Kaushik CP. Bioremediation of salt affected soils using cyanobacteria in terms of physical structure, nutrient status and microbial activity. International Journal of Environmental Science and Technology 2018 15(3), 571–580.

[71] Pan J, Sharif R, Xu X, Chen X. Mechanisms of waterlogging tolerance in plants: Research progress and prospects. Frontiers in Plant Science, 2020, 11, 627331.
[72] Ploschuk RA, Miralles DJ, Colmer TD, Ploschuk EL, Striker GG. Waterlogging of winter crops at early and late stages: Impacts on leaf physiology, growth and yield. Frontiers in Plant Science 2018, 9, 1863.
[73] Prasad PVV, Pisipati SR, Momčilović I, Ristic Z. Independent and combined effects of high temperature and drought stress during grain filling on plant yield and chloroplast EF-Tu expression in spring wheat. Journal of Agronomy and Crop Science 2011 197(6), 430–441.
[74] Punia H, Tokas J, Malik A, Sangwan S, Rani A, Yashveer S, El-Sheikh MA. Genome-wide transcriptome profiling, characterization, and functional identification of NAC transcription factors in sorghum under salt stress. Antioxidants 2021 10(10), 1605.
[75] Rabhi M, Ferchichi S, Jouini J, Hamrouni MH, Koyro HW, Ranieri A, Smaoui A. Phytodesalination of a salt-affected soil with the halophyte Sesuvium portulacastrum L. to arrange in advance the requirements for the successful growth of a glycophytic crop. Bioresource Technology 2010 101(17), 6822–6828.
[76] Radyukina NL, Kartashov AV, Ivanov YV, Shevyakova NI, Kuznetsov VV. Functioning of defense systems in halophytes and glycophytes under progressing salinity. Russian Journal of Plant Physiology 2007 54(6), 806–815.
[77] Ray DK, West PC, Clark M, Gerber JS, Prishchepov AV, Chatterjee S. Climate change has likely already affected global food production. PloS One 2019, 14, 5.
[78] Ren B, Zhu Y, Zhang J, Dong S, Liu P, Zhao B. Effects of spraying exogenous hormone 6-benzyladenine (6-BA) after waterlogging on grain yield and growth of summer maize. Field Crops Research 2016, 188, 96–104.
[79] Ristvey AG, Belayneh BE, Lea-Cox JD. A Comparison of irrigation-water containment methods and management strategies between two ornamental production systems to minimize water security threats. Water 2019 11(12), 2558.
[80] Rodríguez M, Canales E, Borrás-Hidalgo O. Molecular aspects of abiotic stress in plants. Biotecnología Aplicada 2005 22(1), 1–10.
[81] Romano-Armada N, Yañez-Yazlle MF, Irazusta VP, Rajal VB, Moraga NB. Potential of bioremediation and PGP traits in Streptomyces as strategies for bio-reclamation of salt-affected soils for agriculture. Pathogens 2020 9(2), 117.
[82] RoyChowdhury A, Datta R, Sarkar D. Heavy metal pollution and remediation. In: Török B, Dransfield T, eds. Green chemistry. Elsevier United States, 2018, 359–373.
[83] Sadok W, Lopez JR, Smith KP. Transpiration increases under high-temperature stress: Potential mechanisms, trade-offs and prospects for crop resilience in a warming world. Plant, Cell & Environment 2021 44(7), 2102–2116.
[84] Saeedi M, Li LY, Grace JR. Simultaneous removal of polycyclic aromatic hydrocarbons and heavy metals from natural soil by combined non-ionic surfactants and EDTA as extracting reagents: Laboratory column tests. Journal of Environmental Management 2019, 248, 109258.
[85] Salehi-Lisar SY, Bakhshayeshan-Agdam H. Drought stress in plants: Causes, consequences, and tolerance. In: Hossain M, Wani S, Bhattacharjee S, Burritt D, Tran LS, eds. Drought stress tolerance in plants. Cham, Springer, 2016, 1, 1–16.
[86] Sapre S, Gontia-Mishra I, Tiwari S. ACC deaminase-producing bacteria: A key player in alleviating abiotic stresses in plants. In: Kumar A, Meena V, eds. Plant growth promoting rhizobacteria for agricultural sustainability. Singapore, Springer, 2019, 267–291.
[87] Sazzad K. Exploring plant tolerance to biotic and abiotic stresses. Uppsala, Swedish University of Agricultural Sciences, 2007.

[88] Searchinger T, Hanson C, Ranganathan J, Lipinski B, Waite R, Winterbottom R, Ari TB. Creating a sustainable food future. A menu of solutions to sustainably feed more than 9 billion people by 2050. World resources report 2013–14: interim findings (Doctoral dissertation, World Resources Institute (WRI); World Bank Groupe-Banque Mondiale; United Nations Environment Programme (UNEP); United Nations Development Programme (UNDP); Centre de Coopération Internationale en Recherche Agronomique pour le Développement (CIRAD); Institut National de la Recherche Agronomique (INRA), 2014).

[89] Setia R, Gottschalk P, Smith P, Marschner P, Baldock J, Setia D, Smith J. Soil salinity decreases global soil organic carbon stocks. Science of the Total Environment 2013, 465, 267–272.

[90] Shah FUR, Ahmad N, Masood KR, Peralta-Videa JR. Heavy metal toxicity in plants. In: Ashraf M, Ozturk M, Ahmad M, eds. Plant adaptation and phytoremediation. Dordrecht, Springer, 2010, 71–97.

[91] Shankar V, Evelin H. Strategies for reclamation of saline soils. In: Giri B and Varma A. (eds) Microorganisms in saline environments: Strategies and functions. Cham, Springer 2019, 439–449.

[92] Sharma PC, Kaledhonkar MJ, Thimmappa K, Chaudhari SK. In: ICAR-CSSRI. Reclamation of waterlogged saline soils through subsurface drainage technology, 2016.

[93] Sharma S, Tiwari S, Hasan A, Saxena V, Pandey LM. Recent advances in conventional and contemporary methods for remediation of heavy metal-contaminated soils. 3 Biotech 2018 8(4), 1–18.

[94] Shetty R, Vidya CSN, Prakash NB, Lux A, Vaculík M. Aluminum toxicity in plants and its possible mitigation in acid soils by biochar: A review. Science of the Total Environment 2021, 765, 142744.

[95] Takahashi N, Sunohara Y, Fujiwara M, Matsumoto H. Improved tolerance to transplanting injury and chilling stress in rice seedlings treated with orysastrobin. Plant Physiology and Biochemistry 2017, 113, 161–167.

[96] Tan L, Xu W, He X, Wang J. The feasibility of F_v/F_m on judging nutrient limitation of marine algae through indoor simulation and in situ experiment. Estuarine, Coastal and Shelf Science 2019, 229, 106411.

[97] Tanveer M, Shahzad B, Sharma A, Khan EA. 24-Epibrassinolide application in plants: An implication for improving drought stress tolerance in plants. Plant Physiology and Biochemistry 2019, 135, 295–303.

[98] Tian LX, Zhang YC, Chen PL, Zhang FF, Li J, Yan F, Dong Y, Feng BL. How does the waterlogging regime affect crop yield? A global meta-analysis. Frontiers in Plant Science 2021, 12, 634898.

[99] Tyler G, Olsson T. Plant uptake of major and minor mineral elements as influenced by soil acidity and liming. Plant and Soil 2001 230(2), 307–321.

[100] Uchida R, Hue NV. Soil acidity and liming. In: Silva JA, Uchida R, eds. Plant nutrient management in Hawaii's soils: Approaches for tropical and subtropical agriculture. Honolulu, College of Tropical Agriculture & Human Resources University of Hawai'i at Mānoa, 2000, 101–111.

[101] Vwioko E, Adinkwu O, El-Esawi MA. Comparative physiological, biochemical, and genetic responses to prolonged waterlogging stress in okra and maize given exogenous ethylene priming. Frontiers in Physiology 2017, 8, 632.

[102] Wahid A, Gelani S, Ashraf M, Foolad MR. Heat tolerance in plants: An overview. Environmental and Experimental Botany 2007 61(3), 199–223.

[103] Wang M, Zhang S, Ding F. Melatonin mitigates chilling-induced oxidative stress and photosynthesis inhibition in tomato plants. Antioxidants 2020 9(3), 218.

[104] Wang S, Zhuang K, Zhang S, Yang M, Kong F, Meng Q. Overexpression of a tomato carotenoid ε-hydroxylase gene (SlLUT1) improved the drought tolerance of transgenic tobacco. Journal of Plant Physiology 2018, 222, 103–112.
[105] Waqas MA, Khan I, Akhter MJ, Noor MA, Ashraf U. Exogenous application of plant growth regulators (PGRs) induces chilling tolerance in short-duration hybrid maize. Environmental Science and Pollution Research International 2017 24(12), 11459.
[106] Wolfe MD, Tonsor SJ. Adaptation to spring heat and drought in northeastern Spanish Arabidopsis thaliana. New Phytologist 2014 201(1), 323–334.
[107] Wu X, Yao X, Chen J, Zhu Z, Zhang H, Zha D. Brassinosteroids protect photosynthesis and antioxidant system of eggplant seedlings from high-temperature stress. Acta Physiologiae Plantarum 2014 36(2), 251–261.
[108] Yoo JC, Beiyuan J, Wang L, Tsang DC, Baek K, Bolan NS, Li XD. A combination of ferric nitrate/EDDS-enhanced washing and sludge-derived biochar stabilization of metal-contaminated soils. Science of the Total Environment 2018, 616, 572–582.
[109] Yu F, Liang K, Fang T, Zhao H, Han X, Cai M, Qiu F. A group VII ethylene response factor gene, ZmEREB180, coordinates waterlogging tolerance in maize seedlings. Plant Biotechnology Journal 2019 17(12), 2286–2298.
[110] Zafar-ul-Hye M, Ahmad M, Shahzad SM. Short communication synergistic effect of rhizobia and plant growth promoting rhizobacteria on the growth and nodulation of lentil seedlings under axenic conditions. Soil Environment 2013, 32, 79–86.
[111] Zhang C, Wang Q, Zhang B, Zhang F, Liu P, Zhou S, Liu X. Hormonal and enzymatic responses of maize seedlings to chilling stress as affected by triazoles seed treatments. Plant Physiology and Biochemistry 2020, 148, 220–227.
[112] Zhang Y, Hu XH, Shi Y, Zou ZR, Yan F, Zhao YY, Zhao JZ. Beneficial role of exogenous spermidine on nitrogen metabolism in tomato seedlings exposed to saline-alkaline stress. Journal of the American Society for Horticultural Science 2013 138(1), 38–49.

Hari Narayan, Pragati Srivasatava, Santosh Chandra Bhatt,
Divya Joshi, Ravindra Soni

Chapter 5
Plant pathogenesis and disease control

Abstract: The development of any disease or disorder is termed pathogenesis. With respect to plants, any pathogen whether it is fungi, bacteria, or viruses which interfere with normal functioning of the plants is termed plant pathogenesis. After the onset of infection, the plant's normal structure and metabolic functions are disturbed, hence, leading to unbalanced physiological process. There are various factors which contribute in the disease progression like soil pH, relative humidity, temperature, soil fertility, and moisture. The susceptibility and disease proneness of the plant host, multiplication rate of the infectious entity, and the adequate inoculum potential along with a biotic component mediate the successful infection process in the host plants. Plant's pathogenic microorganisms affect its growth and thereby reduce the agricultural productivity. The proper understanding of pathogenicity and basis for plant disease control is essential. The first step involved in controlling the disease is to identify the mechanism of disease outbreak, next is to derive new control measures by natural phenomenon which is inbuilt within the plant parasite interaction.

5.1 Introduction

Plant pathogenesis is referred to as the process of developing diseases in plants. From the initial connection between the plant pathogen and its host through the completion of the syndrome, the host sequences or development in disease development or the capacity of the pathogen to interfere with one or more of the plant's vital processes are covered in it [1].

A plant becomes infected when its normal structure, growth, function, or other activities are disrupted by a causative agent on a regular basis, resulting in an aberrant physiological process. Plant disease is actually a phenomenon in which the physiological and/or biochemical attributes of the plants are disrupted, and undesirable symptoms arise.

On the basis of a primary causative agent, they can be categorized as follows [2]:

Infectious: Plant pathogenic organism, like a virus, bacteria, fungus, nematodes, mycoplasma, or viroid, causes infectious plant diseases. An infectious agent has the ability to reproduce within or on its host, as well as transfer from one vulnerable host to another [3].

https://doi.org/10.1515/9783110771558-005

Noninfectious: Unfavorable growth circumstances, such as temperature extremes, lack of moisture–oxygen relationships, poisonous compounds in the soil or environment, and an excess or shortage of an important mineral, cause noninfectious plant diseases. As noninfectious causal agents are incapable of reproducing within a host, they are not dispersed [3].

5.2 Plant disease development and transmission

5.2.1 Saprogenesis and pathogenesis

The stage of plant disease pathogenesis occurs when the pathogen is in close contact with the living host cell/tissue. Further, they involve virulence.

Virulence is one of the most essential features of pathogenic organisms. A pathogen's capacity to spread through and damage tissue is influenced by a variety of factors, such as toxins, enzymes, extracellular polysaccharides, and certain chemicals. These virulence factors may contribute to the killing of the cells, disrupt the cell walls, or affect cell development processes. In terms of virulence, not all pathogenic species contribute equally. In other words, same amounts of compounds are not produced by them for plant tissue invasion, succession, and damage [4]. Furthermore, not every disease has all of the virulence factors activated. Toxins that kill cells, for example, are crucial in necrotic disease, whereas enzymes that break down cell walls are vital in soft rots in the plants [5].

Several plant pathogens including fungi, mycoplasma and bacteria live as pathogens as well as saprotrophs during their different life spans [6]. When a pathogen is no longer in crucial interaction with the living host tissue, it enters saprogenesis, where it either gets dormant or continues to grow on dead host cells. Some fungus generates their sexual fruiting bodies during this stage. In fallen apple leaves, for example, *Venturia inaequalis* (apple scab) forms flask-shaped structures called perithecia; *Rhizoctonia solani* and *Claviceps purpurea* (ergot fungi) develop sclerotia. In the absence of a live host, the pathogen may survive in soil and plant debris for months or years due to these resistant dormant bodies.

5.2.2 Epiphytotics

A disease is considered epidemic when the number of persons affected increased drastically. Epiphytotic ("on plants") is a more exact phrase for plants, whereas epizootic ("on animals") is the analogous term for animals.

Downy mildews (*Sclerospora* species) and rusts (*Puccinia* species) of maize in Africa during 1950s, chestnut blight (*Endothia parasitica*) in the United States during 1900s, and coffee rust in Brazil during 1960s are the well-known examples for epidemics. Conditions that favor a fresh epiphytotic might tip the balance substantially. Weather conditions (mainly temperature and moisture) that may favor pathogen proliferation, dissemination, and infection; introduction of new and more vulnerable hosts; changes in cultural and socioeconomic practices that produce better environment for the disease; as well as the emergence of a more aggressive pathogen race are examples of such situations [7].

5.2.3 Factors affecting plant disease development

Environmental elements that can influence the development and progression of plant diseases are mentioned further [8, 9].

5.2.3.1 Temperature

Each pathogen has a growing temperature that it favors. Furthermore, several stages of fungal growth and development, such as the development of reproductive units (spores), followed by their germination, and ultimately the growth of filamentous mycelium, required ideal temperatures.

Few examples are *Pseudoperonospora cubensis* (downy mildews of vine crops), *Peronospora tabacina* (blue mold of tobacco), *Phytophthora infestans* (late blight of potato), *Phytophthora phaseoli* (wheat leaf rust), and *Cercospora beticola* (sugar beet leaf spot). Temperature effects, on the other hand, can obscure the symptoms of several viral and mycoplasmal infections, making them more difficult to identify [10, 11].

5.2.3.2 Relative humidity

In the germination of fungal spores and the development of storage rots, relative humidity is crucial. Even if the optimal temperature is used for pathogenic growth, and the relative humidity is kept between 85% and 90%, a storage disease does not develop. It has been observed that suberized (corky) tissues of sweet potato root can block by *Rhizopus stolonifer*. under such conditions.

Guzman-Plazola et al. [12] have observed that 90–95% relative humidity is best for germinating the spores of powdery mildew. Further, they have suggested that controlling the air humidity is a good alternative to reduce the infection of molds in the leaves, roots, flowers, and stems under greenhouse conditions.

5.2.3.3 Soil moisture

High soil moisture encourages the growth of damaging water mold fungus such as *Phytophthora*, *Pythium*, and *Aphanomyces*. Watering houseplants too much is a regular issue. Overwatering makes roots more vulnerable to root-rotting organisms by reducing oxygen and increasing carbon dioxide levels in the soil.

Low soil moisture levels make diseases like cereal take-all (*Ophiobolus graminis*), charcoal rot of sorghum, soybean, and maize (*Macrophomina phaseoli*), common onion white rot (*Sclerotium cepivorum*), and scab of potato (*Streptomyces scabies*) more severe [11].

5.2.3.4 Soil pH

It has been observed that several diseases, including common scab of potatoes and clubroot of crucifers, are influenced significantly by soil pH, which is a measure of acidity or alkalinity (*Plasmodiophora brassicae*). At a pH of 5.2 or slightly below, the growth of the potato scab organism is inhibited (pH 7 is neutral; numbers <7 indicate acidity, and those >7 indicate alkalinity). When the natural soil pH is about 5.2, scab is usually not a concern. To keep the pH of their potato soil at 5.0, some growers inject sulfur. Clubroot of crucifers (mustard family members such as cabbage, cauliflower, and turnips) can typically be managed by properly putting lime into the soil until the pH reaches 7.2 or above.

5.2.3.5 Soil fertility

The excess or limitation of the soil nutrients affects the plant growth and development significantly. Moreover, they also affect the development and progression of the diseases, namely, stalk rots of corn and sorghum, fire blight of the fruits, botrytis blights, S powdery mildew of the cereals, septoria diseases, and so on. These diseases, as well as a slew of others, become more harmful if an overabundance of nitrogen fertilizer is applied. Potash, a potassium-based fertilizer, can typically be used to alleviate this problem [13].

5.2.4 Phenomenon of infection/infection process

After inoculum survival and dissemination, it is the third step in the infection chain. Infection refers to the pathogen's establishment in the host plant. Establishment refers to the pathogen's entry and colonization of host tissues, whereas inoculum refers to the infective propagules that come into contact with the host. Potential inoculum is the inoculum required for infection to be successful. It is determined by the density and capacity of the inoculum.

This process is the result of pathogen vigor, host susceptibility, environmental actions, and inoculum concentration. In other words, it can also be explained as the pathogen vigor for developing an infection over the host surface.

The development of the infection depends on the following factors:
- **Host factors:** Susceptibility of host and disease proneness of the host
- **Pathogen factors:** Virulence/aggressiveness of the pathogen, high multiplication rate of the pathogen, and proper inoculum potential
- **Environmental factors:** Temperature, relative humidity, moisture, and so on

Further, the process of infection can be arranged into three stages [14–18]:

(a) Prepenetration
- **Active invaders**: Phytopathogenic fungi and phanerogamic parasites
- **Passive invaders**: Plant viruses and phytopathogenic bacteria

(b) Penetration
- *Penicillium*, *Colletotrichum*, *Diplodia*, etc.) or natural openings (stomata: *Puccinia graminis tritici*; lenticels: *Sclerotinia fructicola*; hydathodes: *Xanthomonas campestris* pv. *campestris*).
- **Direct penetration**: Breakdown of physical and chemical (*Colletotrichum circinans*)

(c) Postpenetration: Invasion and colonization (*Diplocarpon rosae*, black spot of rose), exit of the pathogen (viruses, bacteria ooze, and fungi) [19, 20].

5.3 Enzymes and their role in pathogenesis

Enzymes are big protein molecules that catalyze all of the events that take place in a live cell. The majority of infections get their energy from the enzymatic breakdown of food items in the host tissue [21, 22] (Figure 5.1).

Composition of the cell wall: In general, cell wall has three regions, namely, primary cell wall (made up of cellulose and pectic substances), secondary cell wall (made up of entirely cellulose), and middle lamella (made up of pectins).

Cutin: It is primarily found at the tip of the germ tube and at the infection peg of aspersorium forming fungi; for example: *Helminthosporium victoriae*, *Sphaerotheca pannosa*, *Venturia inaequalis*, and *Colletotrichum gloeosporioides*.

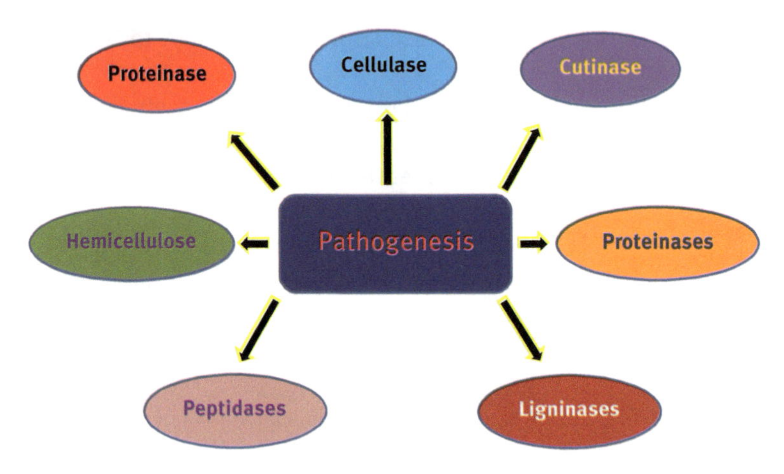

Figure 5.1: Enzymes involved in plant pathogenesis.

Cellulose: Degrading enzymes of cellulose play a major role in softening and decomposition of cell wall material and help for easy penetration and disperse of pathogen in the host; for example: basidiomycetes fungi.

Hemicellulose: Hemicelluloses are degraded by hemicellulase enzymes, which are classified as glucanase, galactanase, mannose, arabinase, or xylanase based on the monomer synthesized from the polymer on which they work; for example: *Sclerotinia fructigena* and *Sclerotinia sclerotiorum*.

Lignin: One or more ligninases were secreted by white rot fungus (basidiomycetes), allowing them to use lignin; for example: *Chaetomium*, *Alternaria*, *Xylaria*, and *Cephalosporium*.

Proteins (cell wall): Proteases, proteinases, and peptidases are enzymes that break down proteins.

5.4 Role of toxins in plant pathogenesis

Toxin is a microbial compound expelled (exotoxin) or produced by lysed cells (endotoxin) that are immediately poisonous to the suspect's cells at extremely low concentrations (host). Toxins differ from enzymes in that they do not damage the host tissues' structural integrity, but they do alter the host's metabolism since they operate on the cell's protoplast [23].

Toxin hypothesis

- All of the symptoms linked with the plant disease should be caused by a toxin.
- Toxic sensitivity will be linked to plant pathogen susceptibility.

- The pathogen's potential to cause disease will be exactly proportional to its toxin production.
- Except for victorin, a toxic metabolite of *Cochliobolus victoriae*, the great majority of toxins linked to plant diseases lack any of the characteristics listed above.

5.4.1 Types of toxins

Toxins are divided into three categories based on their source of origin [23] (Figure 5.2).

(a) Pathotoxins: These are the toxins that play a crucial role in disease transmission and are responsible for all or most disease symptoms in susceptible plants. Pathogens create the majority of these poisons during pathogenesis. The Victoria blight of oats, *Cochliobolus victoriae*, is caused by Victorin (*Helminthosporium victoriae*). This is a poison that is particular to the host.

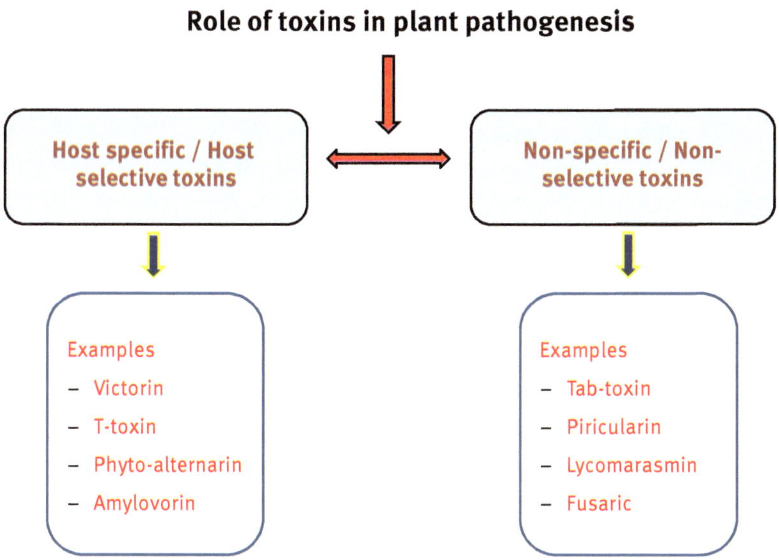

Figure 5.2: Role of toxins in plant pathogenesis.

Other examples:

- **Selective**
  Phytoalternarin: *Alternaria kikuchiana*
  HS toxin: *Helminthosporium sacchari*
  T toxin: *Helminthosporium maydis* race T
  PC toxin: *Periconia circinata*
  HC toxin: *Helminthosporium carbonum*

- **Nonselective**
  Phaseolotoxin: *Pseudomonas syringae* pv. *phaseolicola*
  Tabtoxin or wildfire toxin: *Pseudomonas tabaci*
  Tentoxin: *Alternaria tenuis*

(b) Phytotoxins: These are compounds produced in the host plant as a result of host–pathogen interactions, but their role in disease is only hypothesized, not confirmed. These are parasite products that produce little, if any, of the symptoms induced by the live infection. They are nonspecific, and there is no association between toxin production and the disease-causing agent's pathogenicity; for example: alternaric acid – *Alternaria solani.*

(c) Vivotoxins: These are the compounds produced by the diseased host by the pathogen and/or its host, which aid in the disease's development but are not the disease's initial inciting cause; fusaric acid – wilt causing *Fusarium* sp.

5.4.2 Impact of the toxins on host tissues

- Alteration in the cell permeability
- Disruption of normal metabolic processes
- Affect the growth and development of the host plant; for example: restricted root growth by *Fusarium moniliforme*.

5.5 Growth regulators and their role in plant pathogenesis

Growth regulators can be classified as follows:
- growth-promoting regulators and
- growth-inhibiting regulators.

Excessive cell growth (hypertrophy) and atrophy are caused by an imbalance of growth-promoting and growth-inhibiting chemicals (decrease in cell size). The important symptoms are knots, tumors, stunting, witches, broom, galls, defoliation, suppression of bud development, and excessive root branching [24–26].

5.5.1 Growth-promoting substances

5.5.1.1 Auxins

The enzyme IAA (indole acetic acid) oxidase reduces the concentration of auxin if it is too high. IAA influences membrane permeability, increases respiration, and stimulates mRNA production in addition to regulating differentiation and cell elongation. Hypertrophy occurs when IAA levels are high, while atrophy occurs when IAA levels are low. Inhibition of IAA oxidase might cause an increase in IAA; for example: *Ralstonia solanacearum* (*Pseudomonas solanacearum*). IAA levels have been shown to be higher in diseased plants with the following pathogens: *Sclerospora graminicola* (downy mildew), *Plasmodiophora brassicae* (club root of crucifers), *Phytophthora infestans* (late blight of potato), *Ustilago maydis* (maize smut), *Meloidogyne* (root knot nematode), and *Agrobacterium tumefaciens* (crown gall of apple).

5.5.1.2 Gibberellins

Flowering and fruit growth are promoted by cell elongation, as well as stem and root extension. It also stimulates the production of IAA. IAA and GA have a synergistic effect. *Sclerospora sacchari*, the cause of sugarcane downy mildew, stimulates GA production.

5.5.1.3 Cytokinins

Cytokinins are required for cell division and proliferation. They have the ability to regulate the flow of amino acids and other nutrients toward high cytokinin concentrations, inhibiting the breakdown of proteins and amino acids and so inhibiting senescence. In rust-infected bean leaves, crown galls, and club root, the cytokinin activity rises.

5.5.2 Growth-inhibiting substances

5.5.2.1 Ethylene

Ethylene is important in the rapid ripening of banana fingers infected with *Pseudomonas solanacearum*, the pathogen that causes Moko disease. Leaf epinasty, a sign of vascular wilt syndrome, was also shown to contain ethylene; for example: *Fusarium oxysporum* f. sp. *lycopersici* (wilt in tomato).

5.5.2.2 Abscisic acid

It promotes seed dormancy, seed germination suppression, stomata closure and development, and fungal spore germination stimulation. It is one of the variables that cause plant stunting.

5.5.2.3 Dormin/abscission II

It is known to induce dormancy by converting bud scales from growing leaf primordia. It works as a gibberellin antagonist and hides the effects of IAA.

5.6 Importance of polysaccharides in developing pathogenesis

>Bacteria, fungi, and nematodes are well known to produce mucilaginous compounds that cover their bodies and serve as a contact between the microbe's exterior surface and its surroundings [27, 28]. In wilt diseases, the involvement of slimy polysaccharides is critical. In vascular wilts, pathogens are known to produce large polysaccharides within the xylem tissues and thus induce their mechanical blockage. It results in wilting of the plant, for example, bacterial wilt by *Ralstonia solanacearum*.

5.7 Defense mechanisms in the plants

Plants fight themselves against diseases in two ways, in general [29].

5.7.1 Structural defense mechanisms

These structures can be preexisting, which means that they develop in the plant tissues even earlier than the pathogen entry. Sometimes they can be induced, which means they form after the disease has pierced the preformed defensive structures to defend the plant against further pathogen invasion.

5.7.1.1 Preexisting structural defense structure

The quality as well as quantity of cuticle and wax covering epidermal cells are among these parameters. Moreover, their size, position, and form of natural holes (stomata and lenticels) and the existence of thick-walled cells in plant tissues are known to slow down the pathogen's progress.

5.7.1.2 Induced-type structural barriers (postinfection)

Defensive histological barriers (abscission layers, cork layer, and tyloses) and cellular defense structures can be examined (hyphal sheathing); for example, *Prunus domestica* infect potato tubers; *Coccomyces pruniphorae* attack leaves; *Closterosporium carpophylum* and *Xanthomonas pruni* attack peach leaves.

5.7.1.3 Tyloses

Tyloses are protoplast overgrowths from neighboring live parenchymatous cells that protrude into xylem vessels via pits. Tyloses have cellulose walls and develop swiftly ahead of the infection, clogging the xylem arteries and preventing the disease from progressing further in resistant types. Prior to pathogen invasion, little or no tyloses are produced in sensitive cultivars [30]; for example, most vascular wilt infections cause tyloses to develop in the xylem arteries of most plants.

5.7.2 Biochemical defense mechanisms

5.7.2.1 Preexisting chemical defenses

(a) Inhibitors synthesis: Exudates from the leaves and roots of plants contain flavones, sugars, glycosides, alkaloids, organic acids, amino acids, enzymes, growth factors, inorganic ions, and poisonous substances. Inhibitory compounds either directly influence microorganisms or promote specific groups to take control of the environment, perhaps acting as pathogen antagonists; for example, *Botrytis cinerea* is inhibited by leave exudates of the tomato. Further, *Ceratocystis fimbriata* can be inhibited by chlorogenic acid present in the carrot and potato. Similarly, sweet potato and apple contain caffeic acid and phloretin, respectively.

(b) Inhibitors develop in plant tissues before infection: Sulfur-containing chemicals, phenols, lactones, saponins, cyanogenic glycosides, and phenolic glycosides are antimicrobial molecules found in plant cells. Saponins have antifungal membranolytic action; hence, fungal infections lacking saponinases are excluded [31]. The example is avenacin present in oats and tomatine present in tomatoes.

5.7.2.2 Induced or postinfection defenses

(a) Phytoalexins: These are antimicrobial and toxic compounds generated in significant numbers in plants only after they have been stimulated by phytopathogenic microorganisms or have been injured chemically or mechanically.

(b) Hypersensitive response: This concept was termed "hypersensitivity," which was developed to explain wheat that is infected with the rust fungus *Puccinia graminis*. The hypersensitive response limits pathogen growth by causing localized induced cell death of the plant tissues at the pathogen infected site. HR appears as large water-soaked regions that get necrotic and disintegrate in the damaged plant part. HR develops only when the host and pathogen are incompatible. It can happen when virulent pathogen strains are introduced into nonhost plants or resistant variations, as well as when avirulent pathogen strains are introduced into susceptible cultivars. Further, it is triggered by the identification of elicitors, which are pathogen-produced signal molecules. The host's recognition of the elicitors causes changes in cell activity, which leads to the synthesis of defense-related chemicals [32].

(c) Plantibodies: Plants that have been genetically modified to integrate foreign genes into their genome and can express such genes. For example, mouse genes are known to make antibodies against particular plant diseases that have been created. Plantibodies are antibodies that are encoded by animal genes but generated in and by plants [33]; for example, plantibodies against viral coat proteins, such as artichoke mottle crinkle virus, have been produced in transgenic plants [33].

5.8 Principles of disease control

The majority of pathogen control strategies are based on the concepts of exclusion and avoidance, eradication, protection, host resistance and selection, and treatment [33–35].

5.8.1 Exclusion and avoidance

The goal of these processes is to keep the pathogen from infecting the host plant while it is developing. Pathogens are often eliminated by disinfecting plants, seeds, or other components using chemicals or heat. Seed and other planting stock are inspected and certified to guarantee that they are disease free. Gardeners can do this by sifting bulbs or corms before planting and discarding sick plants. Plant quarantines, sometimes known as embargoes, have been imposed by the federal and state governments to prevent the spread of potentially harmful infections into regions that are now free of the illness. Quarantine rules have now been enacted in more than 150 nations.

5.8.2 Eradication

Eradication refers to the process of eliminating a disease agent once it has established itself in the growing host's environment or has pierced the host. Crop rotation, eradication of unhealthy plants, removal of alternate hosts, disinfection of the field, trimming of the affected portions, and heat treatments of the infected tissues are examples of such interventions.

Many infections, on the other hand, are unaffected by rotation because they establish themselves as saprotrophs in the soil (e.g., *Pythium* and *Fusarium* species, *Rhizoctonia solani*, and *Streptomyces scabies*). To combat diseases like tomato leaf blights, farmers utilize burning, thorough ploughing of plant waste, and fall spraying. Cucumber mosaic and curly top are two viral diseases that can be controlled by destroying weed hosts. Pruning and removing a diseased area of the plant has helped to reduce inoculum sources for shade tree canker and wood-rot diseases, as well as pome fruit fire blight.

5.8.3 Protection

It is the process to maintain a safe distance between host plants and the pathogen by means of physical and other barriers. This can be achieved by environmental regulation, handling practices, management of insect carrier, cultural practices, and the use of agrochemicals.

5.8.4 Environmental regulation

Disease may be controlled by manipulating the environment by choosing outdoor growth sites where the weather is unfavorable for disease. Controlling viral infections in potatoes, for example, can be achieved by cultivating the seed crop in northern locations where the aphid carriers are harmed by cold temperatures. The storage and in-transit environment is another environmental component that may be controlled. In storage and shipping, a number of postharvest diseases of potato, sweet potato, onion, cabbage, apple, pear, and other crops are managed by maintaining low humidity and temperature, as well as limiting the amount of ethylene and other natural gases in storage buildings.

5.8.5 Cultural practices

A cultural approach that lowers disease burden is choosing the ideal time and depth for sowing and planting. *Rhizoctonia* canker can be avoided by growing potatoes shallowly. Seeding winter wheat in the early fall may make seedling infection by wheat bunt teliospores more difficult. In soils afflicted with root knot nematode, cool-temperature crops can be produced and collected before nematode activity becomes favorable. Adjusting soil moisture is another cultural activity that has a wide range of applications.

Excessively moist soils, for example, favor the killing of seedlings (damping-off) seed degradation, and other diseases. The pH of the soil can be adjusted to 5.2 or lower to reduce common potato scab; however, alternative acid-tolerant plants must be utilized in crop rotation.

5.8.6 Fertility-level regulation

Several bacterial, viral, nematodal, and fungal diseases of the crops, namely, cotton, tobacco, sugar beet, and corn, may be influenced by potassium and nitrogen levels, as well as the balance between the two. Many crop and ornamental plants are susceptible to noninfectious diseases caused by microelements such as manganese, molybdenum, boron, copper, iron, sulfur, and zinc. They can be balanced through various methods; for example, addition of chelating compounds (that can bound organic substances), pH regulation, addition of similar salts, and spraying of inorganic salts.

5.8.7 Handling practices

To prevent late blight on potato tubers, wait until the foliage has been damaged by frost, pesticides, or mechanical beaters before harvesting. Disease incidence is reduced by avoiding bruising and wounds when during crop digging, grading, and packing certain crops.

5.8.8 Control of insect vectors

Controlling leafhoppers, aphids, thrips, beetles, and other carriers of bacteria, viruses, and mycoplasma-like disease agents can minimize losses caused by bacteria, viruses, and mycoplasma-like disease agents in many cases. Organic or synthetic pesticides, as well as biological control, can be used to control insect vectors [36].

5.8.9 Chemical control

There are a number of pesticides on the market that are meant to prevent plant diseases by blocking or eliminating the microorganisms that cause them. Bactericides, fungicides, and nematodes (nematicides) are chemicals that can be applied to seeds, leaves, flowers, fruit, or soil to control bacteria, fungi, and nematodes. They use several disease control principles to prevent or minimize infections. Eradicants are chemicals that are used to destroy pathogens that are found in the soil, on seeds, or on vegetative propagative organs like bulbs, corms, and tubers. Between the plant and the pathogen, protectants create a chemical barrier. In order to battle an illness that is already present, therapeutic drugs are used.

Soil treatments are used to kill nematodes, fungi, and bacteria that live in the soil. Steam or chemical fumigants can be used to achieve this elimination. Nematicides, either granular or liquid, can be used to kill soil-borne nematodes. The majority of soil is treated thoroughly before planting; however, some fungicides can be placed into the soil at the time of planting.

Chemicals are often used to kill harmful bacteria, fungi, and nematodes in seeds, bulbs, corms, and tubers, as well as to protect seeds against soil organisms that cause decay and damping-off, primarily fungi. Systemic fungicides and other agrochemicals are widely used to treat the seeds for their protection.

Several agrochemicals are available in the market that can be sprayed or may be used as dusts for the agricultural crops. They just protect the plant from the surface generated and localized infections because they cannot penetrate the cells. Further, their second application is always required for effective results because one-time application can be diminished via rain, sunshine, wind, or irrigation. As new infections are being discovered regularly, therefore, there will always be a demand for novel agrochemicals [37].

5.8.10 Biological control

The use of microorganisms other than humans to minimize or prevent infection by a pathogen is known as biological control of plant diseases. These organisms are known as antagonists, and they can be found in the host's natural environment or administered to certain sections of the potential host plant where they can operate directly or indirectly on the pathogen.

The processes by which antagonists accomplish control are not totally understood, despite the fact that the consequences of biological control have long been documented. Some antagonists manufacture antibiotics that kill or diminish the quantity of closely related infections; others are pathogen parasites; and still others just compete with other for food and shelter.

One important method to reduce disease incidences is to adopt safe cultural practices in which naturally present enemies of the pathogens are employed to control them. Further, in another method, green manure can be applied in the field so that saprotrophs can grow and compete with pathogens for the nutrient and other resources. Suppressive soils can also be recommended for this purpose. In such soils, competitors naturally present and suppress the pathogenic population by limiting the resources.

Antibiosis is another major process that can be employed for biological control of the pathogens. Several organisms produce cell-killing substances which can be utilized for this purpose. For example, terthienyls, a chemical, is secreted by *Tagetes* and is very effective against fungal and nematode pathogens.

Cultural practices that favor and utilize a naturally existing antagonist's positive function are typically successful in lowering sickness. Incorporating green manure, such as alfalfa, into the soil is one method. Potential pathogens are deprived of accessible nitrogen by saprotrophic bacteria that feed on green manure. Another strategy is to employ suppressive soils, which are the ones in which a disease is known to survive but it does not harm the crop. The presence of antagonists in suppressive soils, which compete with the pathogen for food and hence limit pathogen population growth is a plausible explanation for this phenomena.

Other antagonists create chemicals that hinder or kill germs that are in close contact. Marigold (*Tagetes* species) roots, which emit terthienyls, compounds that are harmful to various species of nematodes and fungus, are an example of this phenomenon known as antibiosis.

Only a few antagonists have been created evidently for the treatment of plant diseases. Citrus trees are injected with a weaker strain of the tristeza virus, which successfully suppresses the disease-causing strain. To prevent crown gall induced by *Agrobacterium tumefaciens* infection, an avirulent strain of *Agrobacterium radiobacter* (K84) can be administered to plant wounds. Many more particular antagonists are being researched, and they offer a lot of potential for disease management in the future.

5.8.11 Therapy

Plant pathology has utilized therapeutic techniques far less frequently than human or animal medicine. Growers may now treat many plants once an infection has started because to the recent invention of systemic fungicides such as oxathiins, benzimidazoles, and pyrimidines. Systemic chemicals are absorbed by the plant and translocated inside it, limiting pathogenic growth and development by direct or indirect harmful effects or by enhancing the host's ability to fight infection.

Antibiotics were created to combat a variety of plant diseases. The majority of these medications are absorbed by the plant and translocated throughout it, resulting in systemic treatment. Streptomycin is effective against a wide range of bacterial infections, tetracycline inhibits the growth of some mycoplasmas, and cycloheximides are useful against fungi-caused diseases.

5.8.12 Host resistance and selection

Many agricultural diseases may be controlled with disease-resistant plant cultivars, which are efficient, safe, and very affordable. Most commercial agricultural plant types are resistant to at least one disease, and frequently many infections. For low-value crops, resistant or immune cultivars are necessary since alternative controls are either unavailable or too expensive. Disease-resistant cultivars of field crops, vegetables, fruits, turf grasses, and ornamentals have made significant progress. Pathogens, like most economically significant plants, have a lot of plasticity and the ability to modify their genetic makeup. Occasionally, a novel plant variety is generated that is very sensitive to a disease that was previously inconsequential.

5.8.13 Variable resistance

Plant disease resistance can be absolute (a plant is resistant to a specific pathogen) or partial (a plant is susceptible to a specific infection; a plant is tolerant to a pathogen, suffering minimal injury). Vertical (specific) and horizontal (general) resistance to plant diseases are the two major kinds (nonspecific). Vertical resistance refers to a plant variety that has a high level of resistance to a single pathogenic race or strain; this capacity is generally regulated by the specific genes of the plants. Further, another mechanism, that is, horizontal resistance, protects plants against various pathogenic strains, but not as well. Horizontal resistance is more prevalent, and it is caused by a combination of genes.

5.8.14 Obtaining disease-resistant plants

A variety of methods for developing disease-resistant plants are routinely used, either separately or in combination. These include outside-in introduction, selection, and variation. Further, all of these can be employed at different phases of a continuous process. For example, insect- and disease-free varieties can be introduced for comparison with local types. The most effective and potent strains are then identified and selected for continued replication and enhanced by encouraging as much variety as possible through specific and modern techniques. Finally, the plants that have the most potential are chosen. Plants that are disease-resistant are still being developed.

5.9 Conclusion

Plant pathogenesis and disease control have a significant role in the prosperity of agriculture-based countries. In recent years, different techniques have been tested to control the disease development and progression starting from conventional breeding techniques to the latest genetic engineering methods. Further, it is always critical to understand as much as about the nature of the pathogen so that those techniques can be employed alone or in combination with each other.

References

[1] Oku H. Plant pathogenesis and disease control. Boca Raton, Florida, CRC Press, vol. 1, 2020, 208.

[2] Duncan JM, Torrance L. Techniques for the rapid detection of plant pathogens. New Jersey, United States, Published for the British Society of Plant Pathology by Blackwell Scientific Publications,1991.

[3] Keen NT. A century of plant pathology: A retrospective view on understanding host-parasite interactions. Annual Review of Phytopathology 2000, 8, 31–48.

[4] Dollet M. Plant diseases caused by flagellate protozoa (Phytomonas). Annual Review of Phytopathology 1984, 22, 115–132.

[5] Nenadić M, Grandi L, Mescher MC, De Moraes CM, Mauck KE. Transmission-enhancing effects of a plant virus depend on host association with beneficial bacteria. Arthropod-Plant Interactions 2022, 15, 1–7.

[6] Dhagat S, Jujjavarapu SE. Microbial pathogenesis: Mechanism and recent updates on microbial diversity of pathogens. In: Kumar V, Shriram V, Paul A, Thakur M (eds.), Antimicrobial resistance. Singapore, Springer, vols 71-111, 2022, 71–111.

[7] Nalçacı N, Kafadar FN, Özkan A, Turan A, Başbuğa S, Anay A, Mart D, Öğut E, Sarpkaya K, Atik O, Can C. Epiphytotics of chickpea Ascochyta blight in Turkey as influenced by climatic factors. Journal of Plant Diseases and Protection 2021, 128, 1121–1128.

[8] Barbedo JG. Factors influencing the use of deep learning for plant disease recognition. Biosystems Engineering 2018, 172, 84–91.

[9] Suyal DC, Soni R, Singh DK, Goel R. Microbiome change of agricultural soil under organic farming practices. Biologia 2021, 76, 1315–1325.

[10] Wosula EN, Tatineni S, Wegulo SN, Hein GL. Effect of temperature on wheat streak mosaic disease development in winter wheat. Plant Disease 2017, 101, 324–330.

[11] Leharwan M, Gupta M, Shukla A. Effect of temperature and moisture levels on disease development of stem gall of coriander. Agricultural Science Digest-A Research Journal 2018, 38, 307–309.

[12] Guzman-Plazola RA, Davis RM, Marois JJ. Effects of relative humidity and high temperature on spore germination and development of tomato powdery mildew (Leveillula taurica). Crop Protection 2003, 22, 1157–1168.
[13] Silva MG, Pozza EA, Vasco GB, Freitas AS, Chaves E, Paula PV, Dornelas GA, Alves MC, Silva ML, Pozza AA. Geostatistical analysis of coffee leaf rust in irrigated crops and its relation to plant nutrition and soil fertility. Phytoparasitica 2019, 47, 117–134.
[14] Cobb NA. Rhabditin: Contribution to a science of nematology. The Journal of Parasitology 1914, 1, 1–40.
[15] Romantschuk M. Attachment of plant pathogenic bacteria to plant surfaces. Annual Review of Phytopathology 1992, 30, 225–243.
[16] Stanley MS, Callow ME, Perry R, Alberte RS, Smith R, Callow JA. Inhibition of fungal spore adhesion by zosteric acid as the basis for a novel, nontoxic crop protection technology. Phytopathology 2002, 92, 378–383.
[17] Emmett RW, Parbery DG. Appressoria. Annual Review of Phytopathology 1975, 13, 147–165.
[18] Hillocks RJ, Waller JM. Soilborne diseases of tropical crops. Surrey, UK, Cabi, 1997.
[19] Perfect SE, Green JR. Infection structures of biotrophic and hemibiotrophic fungal plant pathogens. Molecular Plant Pathology 2001, 2, 101–108.
[20] Meredith DS. Significance of spore release and dispersal mechanisms in plant disease epidemiology. Annual Review of Phytopathology 1973, 11, 313–342.
[21] Pandey BP. A textbook of plant pathology: Pathogen and plant disease. New Delhi, India, IK International Pvt Ltd 1992.
[22] Dubey RC, Maheshwari DK. Text book of microbiology. New Delhi, India, S. Chand & Company Limited, 1999.
[23] Durbin R. editor. Toxins in plant disease. Amsterdam, Netherlands, Elsevier, 2012.
[24] Gaspar T, Kevers C, Penel C, Greppin H, Reid DM, Thorpe TA. Plant hormones and plant growth regulators in plant tissue culture. In Vitro Cellular & Developmental Biology-Plant 1996, 32, 272–289.
[25] Denancé N, Sánchez-Vallet A, Goffner D, Molina A. Disease resistance or growth: The role of plant hormones in balancing immune responses and fitness costs. Frontiers in Plant Science 2013, 24, 4–155.
[26] Flors V, Ton J, Van Doorn R, Jakab G, García-Agustín P, Mauch-Mani B. Interplay between JA, SA and ABA signalling during basal and induced resistance against Pseudomonas syringae and Alternaria brassicicola. The Plant Journal 2008, 54, 81–92.
[27] Moradali MF, Rehm BH. Bacterial biopolymers: From pathogenesis to advanced materials. Nature Reviews Microbiology 2020, 18, 195–210.
[28] Andersen EJ, Ali S, Byamukama E, Yen Y, Nepal MP. Disease resistance mechanisms in plants. Genes 2018, 9, 339.
[29] Clérivet A, Déon V, Alami I, Lopez F, Geiger JP, Nicole M. Tyloses and gels associated with cellulose accumulation in vessels are responses of plane tree seedlings (Platanus × acerifolia) to the vascular fungus Ceratocystis fimbriata f. sp platani. Trees 2000, 15, 25–31.
[30] Tomlinson JA, Walker VM, Flewett TH, Barclay GR. The inhibition of infection by cucumber mosaic virus and influenza virus by extracts from Phytolacca americana. Journal of General Virology 1974, 22, 225–232.
[31] Mur LA, Kenton P, Lloyd AJ, Ougham H, Prats E. The hypersensitive response; the centenary is upon us but how much do we know? Journal of Experimental Botany 2008, 59, 501–520.
[32] Gibbs WW. Plantibodies. Scientific American 1997, 277, 44.
[33] Collett SR, Smith JA, Boulianne M, Owen RL, Gingerich E, Singer RS, Johnson TJ, Hofacre CL, Berghaus RD, Stewart-Brown B. Principles of disease prevention, diagnosis, and control. Diseases of Poultry 2020, 13, 1–78.

[34] Singh RS. Introduction to principles of plant pathology. New Delhi, India, Oxford and IBH Publishing, 2017, 30.
[35] Chaube HS, Singh US. Plant disease management: Principles and practices. Boca Raton, Florida, CRC Press, 2018.
[36] Raymaekers K, Ponet L, Holtappels D, Berckmans B, Cammue BP. Screening for novel biocontrol agents applicable in plant disease management–a review. Biological Control 2020, 144, 104240.
[37] Elderfield JA, Lopez-Ruiz FJ, van den Bosch F, Cunniffe NJ. Using epidemiological principles to explain fungicide resistance management tactics: Why do mixtures outperform alternations? Phytopathology 2018, 108, 803–817.

Sonu Kumar Mahawer, Sushila Arya, Tanuja Kandpal, Ravendra Kumar, Om Prakash, Manoj Kumar Chitara, Puspendra Koli

Chapter 6 Plant defense systems: mechanism of self-protection by plants against pathogens

Abstract: Plants respond against biotic and abiotic stresses through various morphological, biochemical, and molecular mechanisms. Biochemical mechanisms of self-defense against plant pathogens are widespread, extremely dynamic, and mediated by both direct and indirect defenses. Physical or morphological and biochemical mechanisms such as formation of different protective layers, modified natural openings, synthesis of secondary metabolites, and sometimes primary metabolites are known to protect to some extent from pathogens and other biotic and abiotic stresses. Both morphological and biochemical mechanisms are further of two types, namely, preexisting such as special morphological and biochemical characteristics of plants and postexisting which are developed in plants in response to different pathogenic attacks. Apart from the production of secondary metabolites and others, two types of resistance such as systemic acquired resistance and induced systemic resistance also have a significant role in the self-protection of plants against different pathogenic attacks.

6.1 Introduction

Plants respond by evolving intricate defense systems against biotic and abiotic stresses as natural systems pose adequately contrasting forces on plants. One such type of stress is the recurrent attack on plants by insect pests and microbial pathogens. Plant pathogenic microorganisms are considered and reported to cause drastic losses to the plants and trees from long ago. To cope with such conditions, the defense mechanism can be performed either by production of toxic secondary metabolites or can be induced. Plants sense biotic stress conditions, trigger the regulatory or transcriptional mechanism, and finally generate a proper response. Among these mechanisms, one of the well-known is the production of secondary metabolites which are synthesized in the plant cell via metabolic pathways derived from the primary metabolic pathways and the synthesis of these metabolites in the plants is often under stress (abiotic and/or biotic) conditions, primarily intervened by different signaling molecules [1]. Around 100,000 secondary metabolites exist in

https://doi.org/10.1515/9783110771558-006

the plant kingdom, which are linked to different chemical classes [2]. Based on the biosynthetic pathway, these chemical compounds are broadly divided into three categories: nitrogen-containing compounds (amines, cyanogenic glycosides, alkaloids, and glucosinolates), phenolic compounds (flavonoids and phenyl propanoids), and terpenes (essential oils/volatile constituents and isoprenoids) [3].

Apart from chemical mechanisms, there are some resistance mechanisms also reported in plants in response to pathogenic attacks. These include acquired and induced resistances. In both, the resistances of different types of signaling are involved. In this chapter, we provide a detailed comprehensive view of different types of defense systems in plant systems against pathogenic attacks along with a brief of the mechanism involved in response to pathogen attacks.

6.2 Mechanism of plant defense

The largest and the most important group of autotrophic groups of life forms on the Earth is constituted by plants. The nutritional requirement of all the heterotrophic organisms, including animals, insects, and microbes, is served by the abundant organic material of these plants. Plants are exposed to a varied range of pathogens. Through their virulent functions, the plant pathogens colonize and damage the host plant [4].

With time, following the evolution of the pathogens, plants have also evolved defense mechanisms in them to protect themselves and prevent potential pathogens from entering and colonizing, derived from different phyla.

Plants' perceptibility and defense response are primarily controlled by two branches of active immune scheme. The gene plant disease resistance encodes the R proteins (R). This protein regulates plant defense by detecting the presence of avirulence proteins or effectors produced by the pathogen. The same can take place through straight binding of the effector or avr protein, by binding of effector-modified target, or by recognizing an effector/target complex [5, 6]. In addition to these possible binding interactions, according to a recent model, the R proteins can guard key cellular hubs that can be common targets for the effectors from different pathogen origins [7].

On the other hand, the transmembrane pattern recognition receptors establishing the subsequent arm of active immune system recognize the molecular patterns of the microbe or the pathogen (MAMPs) and respond to intruding pathogen. MAMPs generally occur in all members of a class of pathogens which are required for the vitality of pathogens. Examples of MAMPs include flagellin in bacterial pathogens and chitin in fungal pathogens [8].

A variety of defense mechanisms are triggered by the plants, also including a hypersensitive response (HR) for a quick breakdown of confronted host cells, biosynthesis of enzymes having the capability to decompose pathogen cell walls, production of phytoalexins having antimicrobial potential, and modifications in plant cell walls especially the papillae deposition loaded with the (1,3)-β-glucan, callose, which is the cell wall polymer. At the site of the microbial attack, cell wall thickenings are formed, which are found to act as a physical barrier slowing down the invasion of pathogens [8].

Two systems of induced resistances are seen among plants: systemic acquired resistance (SAR) and induced systemic resistance (ISR). Combination of the two increases the defense of plants against pathogens. In the case of ISR, if any antagonist is found at the site exposure, the biological control agent could synthesize an antimicrobial substance that is transported through the plant and ultimately inhibits the pathogen directly. In the case of SAR, mobile signals are generated at the site of induction and travel within the plant, creating an induction state in tissues far away from the site exposed to the trigger. It provides long-term protection against a wide range of microorganisms [9].

Oxidative burst is another defense mechanism found in plants. Oxidative burst is related to the emission of local and systemic signals that induce the expression of genes and oxidative cross-linking of the components of the host cell wall. A sufficient amount of reactive species of oxygen accumulates in the source of infection in vitro to kill microorganisms. The suppression of oxidative bursts in the laboratory suggests that it is involved in the onset of later defense responses [10]. Details of different defense mechanisms are as follows:

6.3 Physical barriers against plant pathogens/plant defense

Passive defense mechanisms are those that are present prior to contact with the pathogen also known as constitutive defense (Figure 6.1), while active defense mechanisms are activated only after pathogen detection also known as autonomous/dynamic defense. These defense mechanisms can also be grouped into preexisting and postexisting.

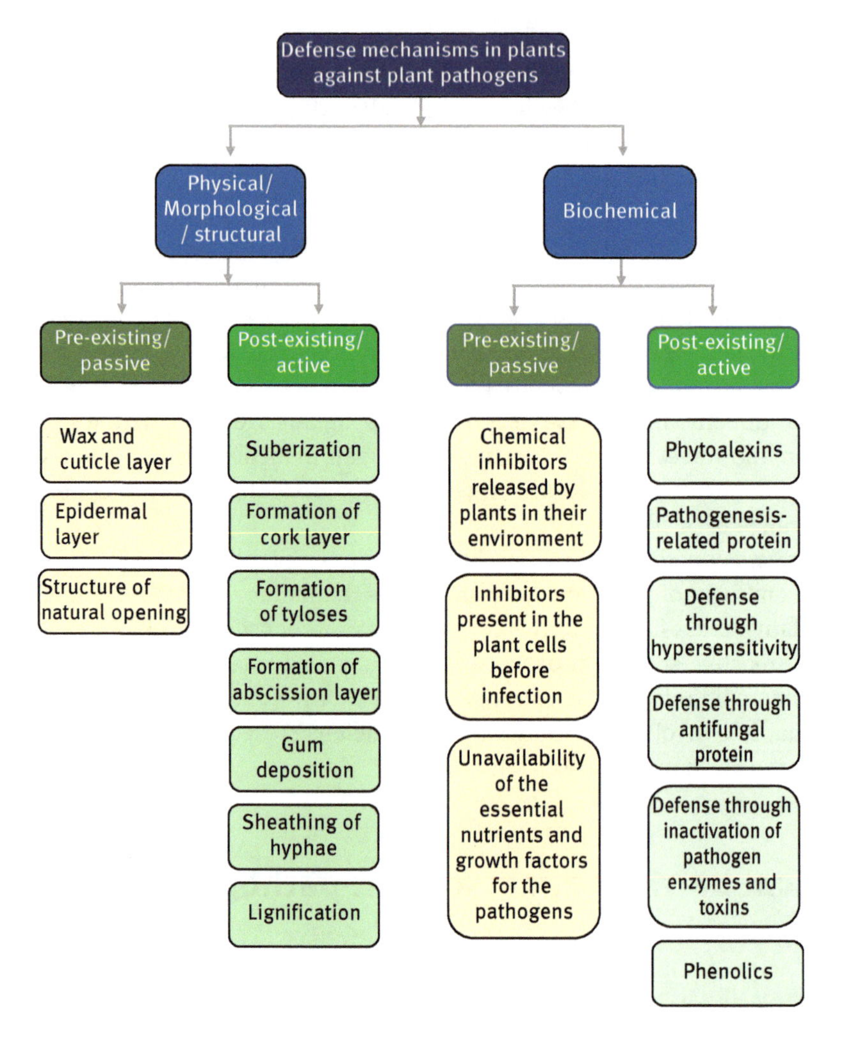

Figure 6.1: Different types of mechanisms are involved in the plant defense system against pathogens.

6.3.1 Preexisting or passive defense mechanism as physical barriers

6.3.1.1 Preexisting structural defenses

The pathogens penetrate the crop through the epidermal, cuticles, and chitinous wax, and the amount of natural pores present before pathogenesis can prevent invasion. It impacts preexisting interior barriers after penetrating. Preexisting defensive structures, also known as passive or anti-infection structures, are exterior and interior structural barriers that exist before a pathogen attack (Figure 6.1).

a. Wax and cuticle: Waxes are a combination of aliphatic side chemicals that inhibit water from remaining on the leaf surfaces, which is necessary for sporulation. Cutin is recognized to be dissolved enzymatically by a select range of pathogens, preventing infection. For example, *Monilinia fructicola* enters the cherry leaf cuticle but not the *Gingko biloba* leaf cuticle; the latter contains more cutin than the former.

b. Epidermis layer: The epidermis is the first membrane of living host tissue that comes in contact with the invading microorganisms. It is the best example of a structural barrier that keeps infections and pests out of the living tissue beneath it. *Pythium debaryanum*-resistant potato tubers have more fibers. Resistance to fungal infections is provided by silicon accumulation in epidermal membranes.

c. Structure of natural opening: Different types of natural opening provide a basal line of defense against pathogens. For instance, hydathodes are natural apertures on the corners of leaf that allow extra water from the core to be expelled. Pathogens are repelled by the filaments on the leaf. Chickpeas with thick hairlines on their leaves and stems have a natural resistance *to Ascochyta rabei.* On the leaf surface, groundnut types resistant to *Cercospora* leaf have a dense epidermis, cuticles, and dense palisade coating, as well as fewer stomata and trichomes. Likewise, lenticels form in the external walls and participate in respiration. Except if the cork cells underneath them are suberized, they are weak spots in defenses. Lenticels seem to be more resistant to microbial penetration after suberization and periderm development [11]. Horns are adapted twigs that defend plants from farm animals. Barrel cactus has thorn-like projections that are modified leaves termed spines that provide similar features. Another kind is guard cells, which are found among the many undifferentiated cells of epidermis and monitor respiration through minute apertures known as stomata. Plants adjust the size of stomata pores, and guard cells can help defend themselves by closing in reaction to MAMPs [12]. Idioblasts, often referred to as "crazy cells," are immunological plant cells with a high level of specialization. Because they possess toxic compounds or spiky crystals, primarily calcium oxalate, which hurt the powerful jaws of insects and mammals as they graze, they aid in the protecting crops against herbivory. Pigmented cells, sclereids, crystalliferous cells, and silica cells are among the many types of idioblasts. Plant components with parenchymal cells typically comprise bitter-tasting tannins, making them unsuitable as a feed ingredient [13]. Sclereids are randomly oriented cells with strong secondary walls that are difficult to bite down: the harsh surface of pear fruit (*Pyrus* spp.) is induced by numerous sclereid stone cells which can roust grazing animals' teeth. Trichomes are also another example of a plant that provides both physical and pharmacological resistance to insect predators. Lots of small trichomes occur on the surface of dusty miller (*Senecio cineraria*), giving it a silky texture. Insect eggs are prevented from entering the epidermis by trichomes on the surface of soybeans (*Glycine max*).

6.3.2 Postexisting morphological and structural defense

Postexisting morphological and structural defenses which are well known in plants against different types of pathogens and other biotic stresses are as follows:

(1) Suberization: Infectious cells in some crops are surrounded by suberized tissues. As a result, they are separated from normal tissues. Plants' natural defense system includes the production of tough outer layers. The tough outer layer formation is used to defend against common potato scab and sweet potato rot.
(2) Cork surface formation: The generation of corky layers is a part of plants' rejuvenation system. Cork barriers appear in some circumstances, such as when potato tuber tissue is infected with the powdery scab pathogen, to hinder the infection from colonizing the area.
(3) Tyloses: Tyloses are outgrowths of surrounding live parenchymatous cells' protoplasts, which protrude into the vascular system via partial pits.
(4) Formation of abscission layer: It is a space between the host layers of cells and the mechanisms that allow infected leaves and ripened fruits to fall off. This is also used by the plants as a defensive measure to fall sick off or invade plant portions, as well as pathogens.
(5) Gum deposition: The gums and vascular gels quickly aggregate and fill the cell membrane or inside the cell surrounds of the infected string and haustoria, leaving them with nothing to eat or death.
(6) Hyphae sheathing: Fungal hyphae that penetrate the cell membrane are frequently unsheathed when the cellular membranes expand. This stops hyphae from coming into contact with protoplasm. Following that, the hyphae penetrate the membrane and kill the cell nucleus lumen.
(7) Lignifications: Lignified cell membranes operate as an impenetrable shield to hyphal invasion and unfettered transport of nutrients, starving pathogens; for example, radish (*Peronospora parasitica* and *Alternaria japonica*) and potato (*Phytophtora infestans*).

6.4 Chemical barriers against plant pathogens

Chemical barriers are among the passive defense responses elicited by plants against pathogens. Exudates on the surface of the plants or certain other compounds present in the cells may excite or inhibit the growth and development of pathogens. Plants occasionally do not provide the required nutrients to the pathogens resulting in resisting infection. Many a times plants produce some compounds during while their natural growth which inhibit the development of pathogens. Phytoanticipins are preformed antimicrobial chemicals in plants that have the remarkable virtue of being created in plants during normal growth even before pathogens or illnesses attack. They can be released into the environment, passively sequestered in vacuoles, or accumulate in dead cells. For example, quinones, catechol, and protocatechuic acid are found in the dead cells of brown onion skins which inhibit the germination of spores of smudge pathogen, *Colletotrichum circinans*, and the neck rot pathogen, *Botrytis cinerea*.

The saponins constituting the group of phytoanticipins are the plant glycosides possessing surfactant properties, bind sterols in cell membranes of the pathogen, and thus destroying membrane integrity and function. Thus, saponins are toxic to the organisms, with sterols in their membranes. Molecules of inactive saponin precursors seem to be stored in the vacuoles of intact plant cells which get converted to their active, antimicrobial form by the action of hydrolase enzyme released following wounding or infection. There are some plant peptides named as plant defensins which inhibit the development of bacteria, viruses, fungi, and insects. These molecules act as proteinase and polygalacturonase inhibitors, as ribosome inhibitors or lectins. The defensins interfere with the growth and development of the pathogens and also retard their nutritional requirement and finally resist the disease. Up to 10% of the total protein in cereals, solanaceous seeds, and legumes is constituted by defensins. The antifeedant activity of defensins provides a defense against insect-transmitted viruses [10].

6.4.1 Preexisting biochemical defense

In the preexisting biochemical defense, plants release chemical substances against pathogen infection in their surrounding environments or present in the plant cells before infection which prevents growth and induces resistance against the pathogen (Figure 3.1). These chemicals and the biochemical conditions that develop may act either directly through a toxic effect on the intruder or indirectly through stimulating antagonistic plant surface microflora. The compounds are preexisting in plants as constitutive antibiotics and those that are produced in response to wounds as wound antibiotics. These chemical modes of action are discussed in detail further.

6.4.1.1 Chemical inhibitors released by plants in their environment

Some plants species have antifungal compounds which are generally phenolic compounds. These phenolic compounds released by plants in their surrounding environments tend to be associated with preventing pathogen spore germination before infection, for example, red scale variety of onion resistance in contrast to white onion variety against onion smudge caused by *Colletotrichum circinans*. The resistance of the red scales variety of the onion is due to the presence of the phenolic compounds, namely, catechol and protocatechuic acid, which prevents the fungal spore germination [14]. Similarly, blight-resistant varieties of the gram have been associated with larger production of the malic acid compared to susceptible variety [15, 16].

6.4.1.2 Inhibitors present in the plant cells before infection

Some plants can form inhibitor compounds themselves, which prevents the pathogen growth, for example, the peel of the potato tubers has phenolic compounds, namely, chlorogenic acids, which gives resistance against scab caused by *Streptomyces scabies* [17]. Oat leaves and roots contain glucoside, namely avenacin, which prevents the infection of the root disease like take-all disease caused by *Ophiobolus graminis* var. *avenae* [18]. The presence of the phenolic glucoside arbutin in pear gives resistance against fire blight caused by *Erwinia amylovora* [19].

6.4.1.3 Unavailability of the essential nutrients and growth factors for the pathogens

Some pathogens are host-specific or can grow and reproduce on only specific varieties due to the presence of the essential nutrients and growth factors such as vitamins, amino acids, polypeptides, and enzymes. Certain times, in the case of mutant varieties, lack essential nutrients and growth factors, so these situations can be unfavorable for the pathogen growth or may prevent the pathogenic infection on these particular host varieties. In case of low nitrogen, the impact of diseases decreases and high nitrogen increases the impact of the disease on plants [20].

6.4.2 Postinfection biochemical defense

Postinfection biochemical defense is considered the last barrier against pathogen infection. In this regard, plants are showing resistance against pathogens employing chemical production in the form of toxins, namely, phenol and phytoalexins (Figure 3.1). These toxins restrict or slow down the pathogen growth as well as trigger the defense cascades in plants.

6.4.2.1 Phytoalexins

Phytoalexins are toxic antimicrobial constituents synthesized in plants only after stimulus by several kinds of plant pathogenic microbes or by chemical and physical damages. These are low-molecular-weight secondary metabolites with antimicrobial properties induced after stress in plants, which are also known as post-inhibitors [21]. Accumulation of phytoalexins at the infected site occurs; therefore, they reduce the growth of fungi and bacteria in vitro. Thus, it is considered a possible plant defense compound against fungal and bacterial diseases [22]. These inhibit the development of the fungal infection in the host cell through hypersensitive response to stop the movement of the fungal mycelium to the adjacent cells [23]. Phytoalexin fungitoxicity is evidenced by the radial mycelial growth, hampering the germ tube elongation and increasing the mycelial dry weight. Phytoalexins may also have some distinct influences on the cytological, morphological, and physiological characteristics of fungal cells. Phytoalexins, significantly important in plant disease management, have been enlisted in Table 6.1.

6.4.2.2 Pathogenesis-related protein

Pathogenesis-related (PR) is defined as proteins that are developed in plants as a result of pathogenic attack. These proteins are antimicrobial, attacking molecules in fungal or bacterium cell wall [24]. The main role of most PRs is to exert antifungal action. Some PRs also have antibacterial, insecticidal, or antiviral properties [24].

6.4.2.3 Defense through hypersensitivity

Hypersensitive response (HR) is a robust resistance action of plants against pathogens. Plant HR is the result of an "unsuited reaction," wherein the "R" gene of the nonhost plant corresponds to the "AVR" gene of the pathogen, whereas in a "suited reaction" the "R" gene of the host plant does not go with the "AVR" gene of the pathogen resulting in the extend of pathogen at the time of plant diseases arise [24]. HR is a process of programmed cell death (PCD) associated with plant reaction to pathogens. This response happens only in incompatible host–parasite combinations. HR in plants is observed by increased reactive oxygen species like hydrogen peroxide (H_2O_2) and nitric oxide (NO), and afterward a systemic response is termed as systemic acquired resistance [24].

Table 6.1: Phytoalexin and their effect on plant disease management.

Phytoalexin	Chemical group	Host	Pathogen	Effect on plants	References
Phaseollin	Isoflavanoid	French bean	*C. lindemuthianum*	Reduced the radial growth of the pathogen (74–92%) under in vitro condition	[25]
Ipomeamarone	Sesquiterpene	Sweet potato	*C. fimbriata*	Induced fungal infection on plants	[26]
Rishithin	Terpenoids	Tomato	*B. cinera*	Increased resistance against pathogen	[27]
Pisatin	–	Pea	*F. oxysporum* f. sp. *pisi*	Inhibition of spore germination of the fungus	[28]
Glyceollin	–	Soybean	*P. sojae*	Increased resistance to fungus	[29]
Isocoumarin	–	Kiwi	*P. syringae* pv. *actinidiae*	Inhibit bacterial activities/ antibacterial activity	[30]
Medicarpin	–	*Dalbergia congestiflora* Pittier tree	*Trametes versicolor*	Inhibit fungal activities/antifungal activities	[31]
Colletotrichumine A	Alkaloid	Chilli	*C. capsici*	Increased virulence of the pathogen	[32]

6.4.2.4 Defense through antifungal protein

Plants need huge quantity of proteins to produce constituents like phenolics, terpenoids, and alkaloids. Hence, several defensive proteins are only produced in substantial amounts only when pathogenic invasion occurs in plants. The effect of these variations may be limited to the infection site or nearby cells. Increased production and activity of phenyl ammonia-lyase (PAL) has been described in various bacterial and viral pathogens in resistant mechanisms. PAL plays an important role in the synthesis of phenols, phytoalexins, and lignin. The efficiency of resistance depends on the rapidity and number of produced products and their movements to nearby healthy tissues to develop defensive hurdles.

6.4.2.5 Defense through inactivation of pathogen enzymes and toxins

The proactive approach of resistant plants, via the action of phenols, tannins, and protein as enzyme inhibitors make pathogen ineffective by deactivating their enzymes. In immature grape fruits, catechol-tannin is recognized to minimize *Botrytis cinerea*-produced enzymes. Toxins are known to be involved in pathogenesis to diverse levels (pathotoxins/vivotoxins). The resistance to toxins, in host, will be resistance to pathogens. This can be attained by detoxification or absence of receptor sites for such toxins. *Pyricularia oryzae* produce three toxins: a-picolinic acid, pyricularin, and piricularin-binding protein. Piricularin is poisonous to both the rice plant and fungi. Piricularin-binding protein detoxicates piricularin in contradiction of the fungus, and chlorogenic and ferulic acids detoxicate it in contrast to the rice plant.

6.4.2.6 Phenolics

Phenols are secondary metabolic substances that have played an important role in plant defense against pathogen infection [33]. In olive (*Olea europaea* L.) plants, phenolic compounds quercetin and luteolin aglycons, followed by rutin, oleuropein, luteolin-7-glucoside, tyrosol, *p*-coumaric acid, and catechin, give natural resistance against the fungus *Verticillium dahliae* Kleb., which has been associated with xylem dysfunction of the plants [34].

6.5 Primary metabolites against the plant pathogen

Primary metabolism is indispensable for the growth, development, and reproduction of cells. They contribute to the primary action by regulating lipids, carbohydrates, and protein and to pathogenic infections. Plant growth is influenced by primary and secondary metabolism all through pathogenic infection.

6.5.1 Primary metabolites against plant virus

Various primary metabolites are released by plants such as lipids, carbohydrates, and proteins; nonetheless, antiviral properties are only in proteins and polysaccharides. Defense responses are formed with the aid of using proteins against specific viral, fungal, and bacterial pathogens. A total of 17 families belong to defense-related proteins (DRPs). During viral infection in sugar beet leaves, a DRP beetin 27 was produced which can take greater action toward many pathogens. Signals released by salicylic acid (SA), hydrogen peroxide, and RNA polynucleotides produced by viral infection react to beetin 27 proteins [35]. The release of DRPs causes anti-tobacco mosaic virus in *Bougainvillea buttiana* [36]. Antiviral properties are noticed in other primary metabolic constituents like polysaccharides belonging to carbohydrate components present in plants. These constituents take part in various biological activities, thus, creating their investigation sympathetic because of their broad target specificity, little residual effects, and lower toxic levels with broader action including anti-aging, hypoglycemia, antioxidation, immune strengthening, and anticancer [37].

6.5.2 Primary metabolites against bacteria

Bacterial pathogens go through the host by mechanical wounding, lenticels, and stomata. The various immune responses can be observed defensive as crops against bacterial pathogens. Different pathogen-associated molecular pattern-prompted immune response is helpful and untimely to encounter the infection before the establishment of disease in the host [38]. Effector-triggered immunity (ETI) leads the establishment of different signaling cascades like SA pathway activation, SAR initiation, and PR protein production [12]. PR proteins are eminent weapons against bacterial pathogens, utilized to increase resistant crops against bacteria.

6.5.3 Primary metabolites against insect pests

Primary and secondary metabolism rearrangement occurs due to signal integrations because of elicitors and wounding, specific to insect. Carbohydrates perform the direct function of defense [39]. Numerous plants have vegetative storage proteins (VSPs) in their vegetative tissues, acting as amino acid reservoirs and providing source–sink connections. Arabidopsis VSPs (AtVSPs) are known to be induced by insect attack, JA applications, and other nearby stresses. There are positive associations between the expression of AtVSPs and insect resistance. Studies reported that AtVSPs have been found active against wide range of insect species. The anti-insect properties of AtVSPs were bound for mutagenesis and phosphatase activity, indicating the enzymatic nature of VSPs [40]. Despite the fact that AtVSP2 targets in the insect digestive system are unknown, it is clear that AtVSP2 interferes with herbivores' phosphate metabolism.

6.5.4 Primary metabolites against fungi

By increasing the resistance in constitutive expression, tobacco and tomato plants get protected against the pathogenic fungi such as *Alternaria longipes* and *A. solani* [41]. Resistance to blackleg disease caused by *Leptosphaeria maculans* was improved by constitutively expressing the defensin of *Brassica napus* [42]. The fungus *Rhizictonia solani* is suppressed in transgenic tobacco and canola by overexpressing PR (PR-3) [43]. The most important feature of a plant's basic defense mechanisms is the formation of physical barriers at potential fungal penetration sites. These structures prevent the growth of the pathogen in plant tissues. Some reports of physical and biochemical mechanisms have been depicted in Table 6.2.

6.6 Secondary metabolites in plant defense

Secondary plant metabolites are the chemical compounds derived from the primary metabolites and produced during the several metabolic pathways in the plant system. In general, the categories of secondary plant metabolites comprise lipids, alkaloids, saponins, terpenes, phenolics, and carbohydrates [45]. These compounds have prominent biological functions as antifungal, antibiotic, antiviral, ant-insecticide, anti-nematicide, antifeedant, toxic or precursors to the physical defense systems. The numerous secondary metabolites have been described with their potential role as a repellent, oviposition deterrent, feeding deterrent, acutely toxic, developmental disruptor, and growth inhibitor in bioassay at the laboratory level [46]. The synthesis of secondary metabolites in the plant is an inductive effect by biotic constraints which causes problems to plants and as a result synthesize some chemicals to defend themselves. The most defensive metabolites are resultant of shikimic acid or aromatic amino acids [47]. The production of these compounds is influenced by genetics of plants and environmental factors. In recent

Table 6.2: Plant defense mechanism against various plant pathogens (adopted and modified from Verma et al. [44]).

S. no.	Defense mechanism	Pathogen	Host	Types of physical barrier
Preexisting defense mechanism				
1	**Preexisting morphological and structural defense**			
	Waxes and cuticle	*Melampsoralini*	Linseed	Waxes and cuticle
		Colletotrichum coffeanum	Coffee	–
		Albugo candida	Brussels sprout	–
		Puccinia graminis	Barberries	–
	Epidermal layer	*Pyriculariaoryzae*	Rice	Thick and tough outer cell wall (presence of silicic acid and lignified epidermal cells)
		Pythium debaryanum	Potato	High-fiber silicon accumulation
	Structure of natural opening	*Xanthomonas campestris* pv. *citri*	Citrus	Broad cuticular ridge projection over the stomata
		Puccinia graminis tritici	Wheat	Late opening of stomata (Functional resistance)
		Entyloma oryzae	Rice	Mechanical tissues (sclerenchymatous tissue)
		Xanthomonas campestris pv. *campestris*	Cabbage	Hydathodes
		Streptomyces scabies	Potato	Lenticels (suberization)
		Ascochyta rabei	Chickpea	leaf hairs

2.	**Preexisting biochemical defense**			
	Inhibitors released by the plant in the environment	*Gloeosporium limetticola*	Citrus	Cutin acid
		Ascochyta rabiei	Gram	Malic acid
		Colletotrichum circinans	Red scale of onion	Catechol and protocatechuic acid
	Inhibitory substances present in the plant cells	*Streptomyces scabies*	Potato	Chlorogenic acid
		Ophiobolus graminis	Wheat	Avenacin
		Erwinia amylovora	Pear	Arbutin
	Phenolic substances	*Venturia inaequalis*	Apple	Polyphenols
		Pyricularia oryzae	Rice	–
		Venturia pirina	Pear	–
	Absence of nutrients required by the pathogen	*Venturia inaequalis*	Apple	Riboflavin or choline
		Erysiphe cichoracearum	Lettuce	High osmotic pressure and permeability effect
		Alternaria solani	Potato	Sugars
Postexisting defense mechanism				
1	**Postexisting morphological and structural**			
	Suberization	*Rhizoctonia solani*	Potato	Formation of cork layer
		Coccomyces prunophorae	European plum	–
	Formation of tyloses	*Fusarium oxysporum* f. sp. *batatas*	Sweet potato	Blocking the spread of pathogen

(continued)

Table 6.2 (continued)

S. no.	Defense mechanism	Pathogen	Host	Types of physical barrier
	Formation of abscission layer	*Verticillium alboatrum*	Tomato	–
		Cladosporium carpophilum	Sour cherry	–
	Gum deposition	*Stereum purpureum*	Plum	Gums and vascular gels
		Drechslera oryzae	Rice	
	Sheathing of hyphae	*Phytophthora infestans*	Potato	Delays contact between hyphae and protoplasm
2	**Postexisting biochemical defense**			
	Phytoalexins	*Botrytis cinerea*	Grapevine	Inhibition of germ tube elongation, radial mycelial growth
	Defense through hypersensitivity	*Puccinia graminis*	Oat, wheat and barley	–
		Pseudomonas syringae pv. *Syringae*	Tobacco	Programmed cell death (PCD) associated with plant reaction to pathogens
	Defense through inactivation of pathogen enzymes and toxins	*Botrytis cinerea*	Grapes	Catechol-tannin
		Pyricularia oryzae	Rice	Picolinic acid, pyricularin, and piricularin

studies, it is found that the abiotic factors have determinable effects on secondary metabolism during the in vitro and in vivo growth of plants. Nowadays, the role and production of the secondary metabolites during the biosynthetic pathways can be easily understood at the cellular, subcellular, organ and whole plant system by applying metabolic engineering [48]. The biological importance of secondary metabolites brought them into the current research interest, but the extraction of secondary metabolites from the plant system is challenging and it depends on the type of metabolites and plants [49]. The highly successful methods depend on the proper selection and preparation of samples and it is very important to minimize the interference from coextracts, avoid contamination, and protect from decomposition.

Secondary metabolites have several advantages over synthetic pesticides and the foremost one is less resistant due to their novel mode of action. Therefore, globally there is an increasing concern in the utilization of these compounds as safer molecules in agriculture food production to sustain the continually growing human population under a regularly fluctuating environment.

6.6.1 Secondary metabolites against fungi

Fungal diseases are key threats to the most agriculturally important crops. The fungal infections result in a significant reduction in both quality and quantity of agricultural produce and often lead to the loss of an entire plant. To minimize these losses, it is essential to apply fungicide to control these fungal pathogens at an early stage. Several synthetic fungicides are available in the market but they have some negative effects on nontarget organisms and the human environment. There are many reports where several fungal pathogens acquired resistance over these synthetic fungicides; therefore, nowadays, the use of safer and eco-friendly chemicals such as secondary metabolites are increasing to overcome such problems. The secondary metabolites have a different kind of mode of action to kill and protect from fungal infections [50]. Interference with the molecular target site of cells and tissue is a well-known mechanism. Other major targets include biomembrane, nucleic acids, and protein disruption. A list of a few secondary metabolites produced to control fungus are depicted in Table 6.3.

6.6.2 Secondary metabolites against bacteria

Bacterial diseases caused by plant pathogenic bacteria (PPB) are too diverse and spread worldwide. About 150 species out of 7,100 classified bacteria are accountable for various kinds of plant diseases [51]. The main classification of these PPB contains tree families known as Xanthomonadaceae, Pseudomonaceae, and Enterobacteriaceae. These bacteria infect plants by performing various actions, which include the release of cell wall deteriorating enzymes such as cellulases, xylanases, pectinases, or

Table 6.3: List of secondary metabolites, their source, and use.

Compound	Source	Target	References
Macrolactin A	Bacillus sp.	Potato scab-*Streptomyces scabies*; *Fusarium oxysporum* causing dry rot disease	[54]
Syringomycin E	*Pseudomonas syringae*	Citrus green mold *Penicillium digitatum*	[55]
Blasticidin-S	*Streptomyces griseochromogenes* I	*Pyricularia oryzae*	[56]
Kusagamycin	*Streptomyces kasugaensis*	*Pyricularia oryzae, Cercospora* spp.	[57]
Cytochalsins	*Phomopsis* sp.	*Sclerotinia sclerotium*, *Fusarium oxysporum*, *Botrytis cineria*, *Bipolaris sorokiniana*, and *Rhizoctonia cerealis*	[58]
Colletotric acid	*Colletotrichum gloeosporioides*	*Helminthosporium sativum*	[59]
Rufuslactone	*Lactarius rufus*	*Alternaria brassicae*, *Fusarium graminearum*, *Botrytis cineria*and *Alternaria alternata*, *Alternaria brassicae*	[60]
Oxytetracycline	*Streptomyces rimosus*	Fire blight caused by *Erwinia amylovora*	[61]
Esters of chrysanthemic acid and pyrethric acid (pyrethrins I and II, cinerins I and II, jasmolins I and II)	*Tanacetum cinerariifolium* (Trevisan) Schultz-Bip	Insecticide, acaricide	[62–64]
Azadirachtin, dihydroazadirachtin, triterpenoids (nimbin, salannin, and others)	*Azadirachta indica* A. Juss	Insecticide, acaricide, fungicide	[62–64]
Rotenone, deguelin (isoflavonoids)	*Derris*, *Lonchocarpus*, and *Tephrosia* species	Insecticide, acaricide	[62–64]

Nicotine sulfate	*Nicotiana* spp.	Insecticide	[62–64]
Ryanodine, ryania, 9,21-didehydroryanodine (alkaloids)	*Ryania* spp. (*Ryania speciosa* Vahl)	Insecticide	[62–64]
Mixture of alkaloids (cevadine, veratridine)	*Schoenocaulon* spp. (*Schoenocaulon officinale* Gray)	Insecticide	[62–64]
Quassin (triterpene lactone)	*Quassia*, *Aeschrion*, *Picrasma*	Insecticide	[62]
Cinnamaldehyde	*Cassia tora* L., *Cassia obtusifolia*	Fungicide, insect attractant	[63, 65]
Physcion, emodin	*Reynoutria sachalinensis* (Fr. Schm.) Nakai	Fungicide, bactericide	[65, 66]
Alkaloids, sanguinarine chloride, and chelerythrine chloride	*Macleaya cordata* R. Br.	Fungicide	[65, 66]
Karanjin	*Derris indica* (Lam.) Bennet	Insecticide, acaricide	[63]
Phenethyl propionate	Component of peppermint oil (*Mentha piperita* L.) and peanut oil	Insecticide, insect repellent, herbicide	[62, 63, 65]
Citric acid	Plant-derived acid herbicide	Insecticide, acaricide, fungicide	[63, 65]
Straight-chain wax esters	*Simmondsia californica* Nutt., *S. chinensis* Link.	Fungicide, insecticide	[63, 65]
Capsaicin	*Capsicum* spp. (*Capsicum frutescens* Mill.)	Repellent, fungicide, nematicide, bactericide	[63, 65]
Eugenol (mixture of several predominantly terpenoid compounds)	*Syzygium aromaticum*, *Eugenia caryophyllus* Spreng	Insecticide, herbicide	[46, 62, 63, 65, 67]
Thymol, carvacrol	*Thymus vulgaris* L., *Thymus* spp.	Insecticide, fungicide, herbicide	[43, 63, 65]

(continued)

Table 6.3 (continued)

Compound	Source	Target	References
1,8-Cineole (borneol, camphor, monoterpenoids)	*Rosmarinus officinalis*	Insecticide, acaricide, fungicide	[43, 65, 67]
Cinnamaldehyde	*Cinnamomum zeylanicum*	Insecticide, herbicide	[43, 65]
Citronellal, citral	*Cymbopogon nardus*, *Cymbopogon citratus* Stapf., *Cymbopogon flexuosus* D.C	Insecticide, herbicide	[43, 65]
Menthol	*Mentha* species (mint)	Insecticide	[43, 65]
Citronellal, geraniol, other terpenes	*Cymbopogon* spp.	Repellent, herbicide	[65]
Omphalotin A	*Omphalotus olearicus*	Nematicides	[68]
Caryospomycin	*Caryospora callicarpa*	Pinewood nematode *Bursaphelenchus xylophilus*	[69]
Dicarboxylic acid	*Paecilomyces* sp.	*Meloidogyne incognita* and *Bursaphelenchus xylophilus*	[69]

by injecting chemicals like hrp harpins, and Avr proteins related to plant diseases [52]. They cause leaf spots and blights, soft rots of fruits, roots, and storage organs, wilts, overgrowths, scabs, and cankers. The several antimicrobial secondary metabolites are listed in Table 6.3.

6.6.3 Secondary metabolites against nematodes

The global annual loss due to the plant-parasitic nematodes is estimated at more than US $100 billion [53]. Several numbers of chemical nematicides have been registered to control nematodes but due to their harmful effects and other environmental detriment effects these nematicides are banned and eventually, their uses are declined in last 30 years. Therefore, to keep the environment safe, other alternatives have been explored which include plant- and animal-derived secondary metabolites. Nowadays, the use of secondary metabolites in the management of nematodes is increasing and the other benefits associated with these metabolites are cost-effective. Concerning the production of nematocidal compounds, ascomycetes fungal group is prominent.

6.7 Systemic acquired resistance (SAR) against plant pathogens

SAR is an inbuilt nature or capability of the plants, in which plants increased resistance against wide ranges of the pathogenic attack. SAR is also known as induced resistance, acquired resistance, acquired immunity, and immunization. The outcome of the SAR is the elevated level of the SA and expression of the pathogen-related (PR) proteins. SA is considered as a signaling molecule, which is accumulated exogenously in plants and expressed during HR reaction during pathogen attack and developing SAR [70]

6.8 Induced systemic resistance (ISR) against plant pathogens

ISR is an indirect type of defense, which is activated through nonpathogenic, root-colonizing, plant growth-promoting rhizobacteria (PGPR). The PGPR reduced the plant disease incidence and severity. The seed treatment through the PGPR has improved the structural changes in the cell wall and physiological/biochemical changes leading to the production of protein, peptides, and chemicals during plant defense mechanisms against pathogenic attack [71] (Figure 6.2).

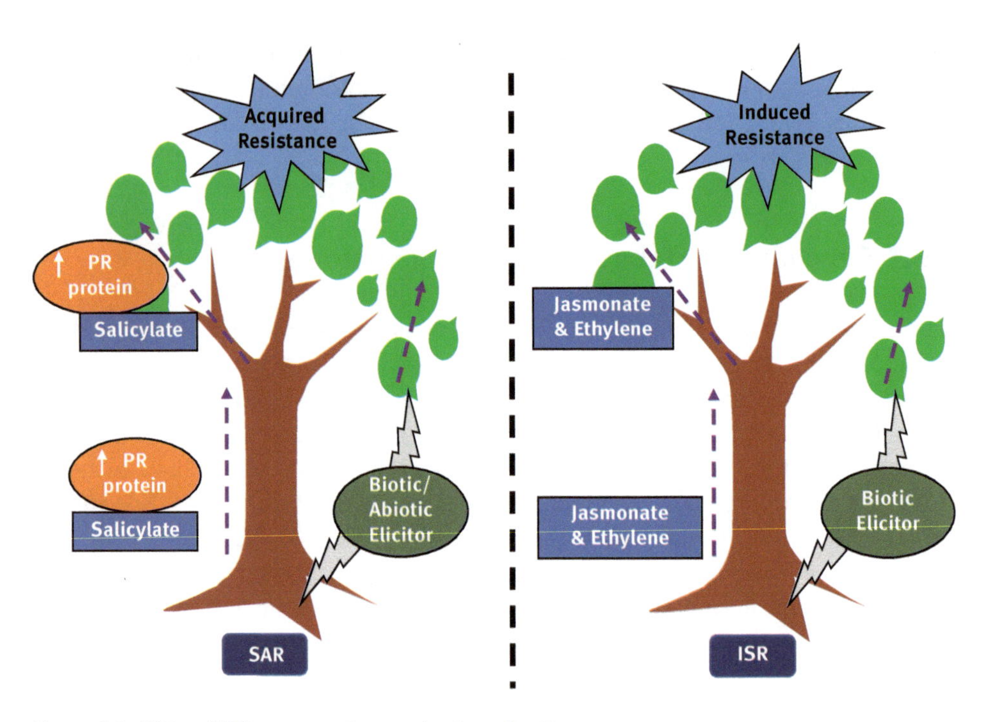

Figure 6.2: SAR and ISR: comparative mechanism of action.

6.9 Conclusion and future prospectus

Increasing human population and increased use of chemical pesticides to control plant diseases and by knowing their dangerous consequences ask for a substitute of chemical pesticides to improve to the quality and production of crop plants. By the available research findings, it can be achieved by exploration of plant pathogenic interactions and different mechanisms of plant defense systems to a greater extent. Alteration and utilization of plant defense mechanisms such as the artificial induction of resistance and utilization of potential plant secondary metabolites in pest control strategies in crop protection can be a game-changing step in the field of plant protection. There is a need for in-depth research about the actual mechanisms involved in plant defense and to make systematic strategies to utilize the natural defense mechanisms in plant protection.

References

[1] Hussein RA, El-Anssary AA. Plants secondary metabolites: The key drivers of the pharmacological actions of medicinal plants. Herbal Medicine 2019, 1, 13.
[2] Li Y, Kong D, Fu Y, Sussman MR, Wu H. The effect of developmental and environmental factors on secondary metabolites in medicinal plants. Plant Physiology and Biochemistry 2020, 148, 80–89.
[3] Fang X, Yang C, Wei Y, Ma Q, Yang L, Chen X. Genomics grand for diversified plant secondary metabolites. Plant Diversity and Resources 2011, 33(1), 53–64.
[4] Kombrink E, Somssich IE. Defense responses of plants to pathogens. Advances in Botanical Research 1995, 21, 1–34.
[5] Flor HH. Current status of the gene-for-gene concept. Annual Review of Phytopathology 1971, 9(1), 275–296.
[6] Dodds PN, Rathjen JP. Plant immunity: Towards an integrated view of plant–pathogen interactions. Nature Reviews Genetics 2010, 11(8), 539–548.
[7] Mukhtar MS, Carvunis AR, Dreze M et al. Independently evolved virulence effectors converge onto hubs in a plant immune system network. science 2011, 333(6042), 596–601.
[8] Voigt CA. Callose-mediated resistance to pathogenic intruders in plant defense-related papillae. Frontiers in Plant Science 2014, 5, 168.
[9] Prasannath K. Plant defense-related enzymes against pathogens: A review. AGRIEST journal of Agricultural Sciences 2017, 11(1), 38–48.
[10] Guest D, Brown J. Plant defences against pathogens. Plant Pathogens and Plant Diseases 1997, 263, 286.
[11] Melotto M, Underwood W, He SY. Role of stomata in plant innate immunity and foliar bacterial diseases. Annual Review of Phytopathology 2008, 46, 101–122.
[12] Jones JD, Dangl JL. The plant immune system. Nature 2006, 444(7117), 323–329.
[13] Doughari J. An overview of plant immunity. Journal of Plant Pathology & Microbiology 2015, 6(11), 10–4172.
[14] Jones HA, Walker JC, Little TM, Larson RH. Relation of color-inhibiting factor to smudge resistance in onion. Journal of Agricultural Research 1946, 72, 259–264.
[15] Singh PJ, PAL M, Devakumar C. Role of malic acid in pycnidiospore germination of *Ascochyta rabiei* and chickpea blight resistance. Indian Phytopathology 1998, 51(3), 254–257.
[16] Cagirgan MI, Toker C, Karhan M, Mehmet AK, Ulger S, Canci H. Assessment of endogenous organic acid levels in Ascochyta blight [Ascochyta rabiei (pass.) Labr.] susceptible and resistant chickpeas (Cicer arietinum. Turkish Journal of Field Crops 2011, 16(2), 121–124.
[17] Singhai PK, Sarma BK, Srivastava JS. Phenolic acid content in potato peel determines natural infection of common scab caused by *Streptomyces spp*. World Journal of Microbiology & Biotechnology 2011, 27(7), 1559–1567.
[18] Türk FM, Egesel CÖ, Gül MK. Avenacin A-1 content of some local oat genotypes and the in vitro effect of avenacins on several soil-borne fungal pathogens of cereals. Turkish Journal of Agriculture and Forestry 2005, 29(3), 157–164.
[19] Smale BC, Keil HL. A biochemical study of the intervarietal resistance of *Pyrus communis* to fire blight. Phytochemistry 1966, 5(6), 1113–1120.
[20] Amtmann A, Troufflard S, Armengaud P. The effect of potassium nutrition on pest and disease resistance in plants. Physiologia Plantarum 2008, 133(4), 682–691.
[21] Ahuja I, Kissen R, Bones AM. Phytoalexins in defense against pathogens. Trends in Plant Science 2012, 17(2), 73–90.
[22] Cruickshank IA, Perrin DR. Isolation of a phytoalexin from Pisum sativum L. Nature 1960, 187 (4739), 799–800.

[23] Noman A, Aqeel M, Qari SH et al. Plant hypersensitive response vs pathogen ingression: Death of few gives life to others. Microbial Pathogenesis 2020, 145, 104224.
[24] Dodds PN, Lawrence GJ, Catanzariti AM et al. Direct protein interaction underlies gene-for-gene specificity and coevolution of the flax resistance genes and flax rust avirulence genes. Proceedings of the National Academy of Sciences 2006, 103(23), 8888–8893.
[25] Pozza Junior MC, Pandini JA, Hein DP et al. Phaseolin induction on common-bean cultivars and biological control of *Colletotrichum lindemuthianum* 89 race by *Baccharis trimera* (Less.) Dc. Brazilian Archives of Biology and Technology 2021, 64.
[26] Lewthwaite SL, Wright PJ, Triggs CM. Sweetpotato cultivar susceptibility to infection by *Ceratocystis fimbriata*. New Zealand Plant Protection 2011, 64, 1–6.
[27] Charles MT, Mercier J, Makhlouf J, Arul J. Physiological basis of UV-C-induced resistance to *Botrytis cinerea* in tomato fruit: I. Role of pre-and post-challenge accumulation of the phytoalexin-rishitin. Postharvest Biology and Technology 2008, 47(1), 10–20.
[28] Bani M, Cimmino A, Evidente A, Rubiales D, Rispail N. Pisatin involvement in the variation of inhibition of *Fusarium oxysporum* f. sp. pisi spore germination by root exudates of *Pisum spp*. germplasm. Plant Pathology 2018, 67(5), 1046–1054.
[29] Jahan MA, Harris B, Lowery M, Infante AM, Percifield RJ, Kovinich N. Glyceollin transcription factor GmMYB29A2 regulates soybean resistance to Phytophthora sojae. Plant Physiology 2020, 183(2), 530–546.
[30] Chen Q, Yu JJ, He J, Feng T, Liu JK. Isobenzofuranones and isocoumarins from kiwi endophytic fungus*Paraphaeosphaeriasporulosa* and their antibacterial activity against *Pseudomonas syringae* pv. actinidiae. Phytochemistry 2022, 195, 113050.
[31] Martínez-Sotres C, López-Albarrán P, Cruz-de-león J et al. Medicarpin, an antifungal compound identified in hexane extract of *Dalbergia congestiflora* Pittier heartwood. International Biodeterioration and Biodegradation 2012, 69, 38–40.
[32] Chitara MK, Keswani C, Varnava KG et al. Impact of the alkaloid colletotrichumine A on the pathogenicity of Colletotrichum capsici in *Capsicum annum* L. Rhizosphere 2020, 16, 100247.
[33] Lattanzio V. Phenolic Compounds: Introduction 50. Natural Product 2013, 1543–1580.
[34] Báidez AG, Gómez P, Del Río JA, Ortuño A. Dysfunctionality of the xylem in *Olea europaea* L. plants associated with the infection process by *Verticillium dahliae* Kleb. Role of phenolic compounds in plant defense mechanism. Journal of Agricultural and Food Chemistry 2007, 55(9), 3373–3377.
[35] Rosario I, Lucìa C, Antimo DM, Jose MF. Biological activities of the antiviral protein BE27 from sugar beet (Beta vulgaris L.). Planta 2015, 241(2), 421–433.
[36] Nandlal C, Harish CK, Madan LL. Cloning and expression of antiviral/ribosomein activating protein from Bougainvillea x buttiana. Journal of Biosciences 2008, 33(1), 91–101.
[37] Chen X, Fang Y, Nishinari K, We H, Sun C, Li J, Jiang Y. Physicochemical characteristics of polysaccharide conjugates prepared from fresh tea leaves and their improving impaired glucose tolerance. Carbohydrate Polymers 2014, 112, 77–84.
[38] Ausubel FM. Are innate immune signaling pathways in plants and animals conserved?. Nature Immunology 2005, 6(10), 973–979.
[39] Zou J, Cates RG. Role of Douglas fir (*Pseudotsuga menziesii*) carbohydrates in resistance to budworm (*Choristoneura occidentalis*). Journal of Chemical Ecology 1994, 20(2), 395–405.
[40] Liu Y, Ahn JE, Datta S et al. Arabidopsis vegetative storage protein is an anti-insect acid phosphatase. Plant Physiology 2005, 139(3), 1545–1556.
[41] Terras FR, Eggermont K, Kovaleva V et al. Small cysteine-rich antifungal proteins from radish: Their role in host defense. The Plant Cell 1995, 7(5), 573–588.
[42] Wang Y, Nowak G, Culley D, Hadwiger LA, Fristensky B. Constitutive expression of pea defense gene DRR206 confers resistance to blackleg (*Leptosphaeria maculans*) disease in

transgenic canola (Brassica napus). Molecular Plant-microbe Interactions 1999, 12(5), 410–418.
[43] Broekaert WF, Terras FR, Cammue BP, Osborn RW. Plant defensins: Novel antimicrobial peptides as components of the host defense system. Plant Physiology 1995, 108(4), 1353–1358.
[44] Verma S, Meena AK. Chapter-2 Integrated Defense Response of Plant against Pathogen. 2020
[45] Hussein RA, Amira A. Plants Secondary Metabolites: The Key Drivers of the Pharmacological Actions of Medicinal Plants. 2018
[46] Fischer D, Imholt C, Pelz HJ, Wink M, Prokop A, Jacob J. The repelling effect of plant secondary metabolites on water voles, *Arvicola amphibius*. Pest Management Science 2013, 69(3), 437–443.
[47] Bennett RN, Wallsgrove RM. Secondary metabolites in plant defence mechanisms. New Phytologist 1994, 127(4), 617–633.
[48] Isah T. Stress and defense responses in plant secondary metabolites production. Biological Research 2019, 52.
[49] Jones WP, Kinghorn AD. Extraction of plant secondary metabolites. Natural Products Isolation 2012, 341–366.
[50] Keswani C, Bisen K, Chitara MK, Sarma BK, Singh HB. Exploring the role of secondary metabolites of Trichoderma in tripartite interaction with plant and pathogens. Agro-environmental Sustainability 2017, 63–79.
[51] Aguilar-Marcelino L, Mendoza-de-gives P, Al-Ani LKT, López-Arellano ME, Gómez-Rodríguez O, Villar-Luna E, Reyes-Guerrero DE. Using molecular techniques applied to beneficial microorganisms as biotechnological tools for controlling agricultural plant pathogens and pest. In: Molecular aspects of plant beneficial microbes in agriculture, Academic Press, 2020, 333–349.
[52] Alfano JR, Collmer A. The type III (Hrp) secretion pathway of plant pathogenic bacteria: Trafficking harpins, Avr proteins, and death. Journal of Bacteriology 1997, 179(18), 5655–5662.
[53] Degenkolb T, Vilcinskas A. Metabolites from nematophagous fungi and nematicidal natural products from fungi as an alternative for biological control. Part I: Metabolites from nematophagous ascomycetes. Applied Microbiology and Biotechnology 2016, 100(9), 3799–3812.
[54] Han JS, Cheng JH, Yoon TM et al. Biological control agent of common scab disease by antagonistic strain *Bacillus sp.* sunhua. Journal of Applied Microbiology 2005, 99(1), 213–221.
[55] Bull CT, Wadsworth ML, Sorensen KN, Takemoto JY, Austin RK, Smilanick JL. Syringomycin E produced by biological control agents controls green mold on lemons. Biological Control 1998, 12(2), 89–95.
[56] Fukunaga K, Misato T, Ishii I, Asakawa M. Blasticidin, a new anti-phytopathogenic fungal substance. Part I. Journal of the Agricultural Chemical Society of Japan 1955, 19(3), 181–188.
[57] Umezawa H, Okami Y, Hashimoto T, Suhara Y, Hamada M, Takeuchi T. A new antibiotic, kasugamycin. The Journal of Antibiotics Series A 1965, 18(2), 101–103.
[58] Fu J, Zhou Y, Li HF, Ye YH, Guo JH. Antifungal metabolites from *Phomopsis sp.* By254, an endophytic fungus in *Gossypium hirsutum*. African Journal of Microbiology Research 2011, 5(10), 1231–1236.
[59] Zou WX, Meng JC, Lu H, Chen GX, Shi GX, Zhang TY, Tan RX. Metabolites of *Colletotrichum gloeosporioides*, an endophytic fungus in *Artemisia mongolica*. Journal of Natural Products 2000, 63(11), 1529–1530.

[60] Luo DQ, Wang F, Bian XY, Liu JK. Rufuslactone, a new antifungal sesquiterpene from the fruiting bodies of the basidiomycete *Lactarius rufus*. The Journal of Antibiotics 2005, 58(7), 456–459.
[61] Finlay AC, Hobby GL, P'an SY et al. Terramycin, a new antibiotic. Science 1950, 111(2874), 85.
[62] Isman MB. Botanical insecticides, deterrents, and repellents in modern agriculture and an increasingly regulated world. Annual Review of Entomology 2006, 51, 45–66.
[63] Copping LG, Duke SO. Natural products that have been used commercially as crop protection agents. Pest Management Science: Formerly Pesticide Science 2007, 63(6), 524–554.
[64] Isman MB, Paluch G. Needles in the haystack: Exploring chemical diversity of botanical insecticides. Green trends in insect control, Cambridge, London, Royal Society of Chemistry, 2011, 248–265.
[65] Dayan FE, Cantrell CL, Duke SO. Natural products in crop protection. Bioorganic & Medicinal Chemistry 2009, 17(12), 4022–4034.
[66] Regnault-Roger C. Trends for commercialization of biocontrol agent (biopesticide) products. In Plant Defence: Biological Control 2012, 139–160.
[67] Isman MB, Machial CM. Pesticides based on plant essential oils: From traditional practice to commercialization. Advances in Phytomedicine 2006, 3, 29–44.
[68] Mayer A, Anke H, Sterner O. Omphalotin, a new cyclic peptide with potent nematicidal activity from *Omphalotus olearius* I. Fermentation and biological activity. Natural Product Letters 1997, 10(1), 25–32.
[69] Dong J, Zhu Y, Song H, Li R, He H, Liu H, Huang R, Zhou Y, Wang L, Ceo Y, Zhang K. Nematicidalresorcylides from the aquatic fungus *Caryospora callicarpa* YMF1.01026. Journal of Chemical Ecology 2007, 33(5), 1115–1126.
[70] Ryals JA, Neuenschwander UH, Willits MG, Molina A, Steiner HY, Hunt MD. Systemic acquired resistance. The Plant Cell 1996, 8(10), 1809.
[71] Meena M, Swapnil P, Divyanshu K et al. PGPR-mediated induction of systemic resistance and physiochemical alterations in plants against the pathogens: Current perspectives. Journal of Basic Microbiology 2020, 60(10), 828–861.

Amir Khan, Mohd Shahid Anwar Ansari, Irsad, Touseef Hussain, Abrar Ahmad Khan

Chapter 7 Role of Beneficial Microbes for Plant growth Improvement

Abstract: Plant growth-promoting microbes comprise of microorganisms in the plant system including rhizospheric bacteria, fungi, mycorrhiza, actinomycetes, endophytic fungi, or those having either symbiotic or nonsymbiotic relationship with plants. Promotion of plant growth by microorganisms is associated with mechanisms such as phytohormone production, siderophore production, nitrogen fixation, solubilization of mineral phosphates, and release of potent secondary metabolites (SMs) that affect the plant health easily. Some fungi such as arbuscular mycorrhizal (AM) fungi have been found in almost all land plant species as obligate root symbiont and increase the uptake of mineral elements in their host plant in exchange of carbon and enhance the plant growth by increasing biomass and chlorophyll content. Such microorganisms are also involved in the management of both biotic and abiotic stresses on plants. Management of biotic stress involves the interaction between pathogenic and nonpathogenic microbes that includes various changes like twisting of hyphae of beneficial microbes around the hyphae of pathogenic microbes as well as certain changes at the cellular level like secretion of lytic enzymes or secretion of antimicrobial compounds and dissolution of host cytoplasm, and such changes inhibit the reproduction and growth of harmful microbes in the nearby places. The exact understanding of the mechanism by which these microbes promote growth of plants will help evolve approaches against damages by several biotic and abiotic stress conditions and help in sustainable agriculture at global scale.

7.1 Introduction

To enhance the total agricultural outputs and production, the modern agricultural system depending on chemicals, namely, fertilizers, pesticides, and weedicides. Excessive and long-term uses of such chemicals cause the loss of soil fertility, environment pollutions, and severe health hazards. Nontarget soil-borne microbes are also affected by excessive utilization of chemicals as well as development of resistance among

Acknowledgments: All the authors are very thankful to Department of Botany, AMU, Aligarh; Faculty of Agricultural Sciences, AMU, Aligarh; SERB, New Delhi; and UGC, New Delhi, for all the support and encouragements. None of the authors have any conflict of interest.

https://doi.org/10.1515/9783110771558-007

target pests [1]. Therefore, it is very important to develop an eco-friendly approach to reduce the uses of chemicals in agriculture production. From the last decades, the demand for biopesticides has been increased, which maintains soil fertility, environment, human health, and nontarget organisms unaffected. The indigenous soil microbiota with their huge communities linked with plants and soil forming an elemental part of vegetated agroecosystem [2]. Several ecosystem services such as soil nutrient cycling, decomposition of organic materials, nitrogen fixation, removal of degradable and biodegradable pollutants, yield and growth promotion of plants (direct and indirect mechanisms), and disease suppression are efficiently done by such microbes present in the vicinity of rhizosphere [3]. Researchers engaged in the findings of microbial mechanism for plant growth provide resistant in plants against harmful microbes in their vicinity [4, 5].

In recent years, several scientists studied on beneficial soil microorganisms, and their two-way association with each other made this study more interesting and enthusiastic for further [6]. These soil microbes affect both quality and quantity of agriculture crops in harmful and beneficial manner. The soil microbes which influence growth and yield of plant may be harmful or deleterious saprophytic one. While the advantageous microorganisms enhance plant growth by many mechanisms, plant growth-promoting microorganisms (PGPMs) are such type of soil microorganisms which can cooperate plants in disease suppression, induction of resistance, nutrient transport, and mobilization [7]. Therefore, the plant–microbiome interaction is an important association for improving growth, yield, and health of the plant. However, helpful microbes found near the rhizospheric area are known as plant growth-promoting rhizobacteria (PGPRs) and the other ones are plant growth-promoting bacteria (PGPB) [8, 9]. Few other PGPMs consist of actinomycetes, protists, cyanobacteria, mycorrhizal fungi, and nonpathogenic saprophytic fungi.

7.2 Need of beneficial microbes for plant growth

The soil biome consists of several components like substrates (dead and decaying leaves, roots, woods, and organic materials), minerals (sulfur, iron, calcium, oxides, nitrates, phosphates, ash, and stone particles), microbes, plants, animals, water, and air. The presence of microorganisms in the soil ecology affects the texture of soil. The speed of soil development is controlled by different factors, in which the microorganisms are the major ones [10]. The quality of microbes and their identification in the soil ecology determined the nutritional status of soil [11]. Microbes play a major role in soil development, health, and improving fertility. Plant growth-promoting microbes are helpful, enhancing the growth and yield of crops under natural and stressful conditions. They play a major role in improving the physiological constraints in retort to external provocative through different procedures. Enhancement of numerous metabolites, production of phytohormones, and atmospheric nitrogen fixation (transformation of atmospheric nitrogen into ammonia) are few modes of actions that help in plant growth promotion and offer resistance against pathogens through systemic acquired resistance (SAR) and induced systemic resistance (ISR) [12, 13].

Microbes enhance nutrients' absorption, yield, and nodulation in chickpea [14]. Furthermore, numerous bacterial species have been employed to mineralize organic toxins in the soil, a process known as bioremediation or bioprocesses of soil contaminants [15]. Phytohormone signal helps [16] the microbial strains for outcompeting and repelling as well as increase the soil nutrients. The three types of mechanisms or steps involved are typically studied and put forward for explaining how the microbial community improves the plant growth [17].

7.3 Plant growth-promoting rhizobacteria (PGPRs)

PGPRs are soil inhabitants in nature, and they actively penetrate the plant roots and boost the strength of plants. Its application in crop cultivation can assist the moderate use of agrochemicals, enhance the long-term food production, and sustain the environment. PGPR is responsible for improved growth of seedling, early nodulation and function, as well as increased surface area of the leaf, vigor, biomass, phytohormone, nutrient, water, and air uptake, and stimulated carbohydrate build up and yield in different plant species [18]. Therefore, several soil microbes that have been described as PGPRs belong to the genera which impact as a beneficial role in plant growth, including *Azospirillum*, *Bacillus*, *Pseudomonas*, *Agrobacterium*, *Azotobacter*, *Alcaligenes*, *Clostridium*, *Beijerinckia*, *Rhizobium*, *Arthrobacter*, *Serratia*, *Enterobacter*, *Phyllobacterium*, *Burkholderia*, *Klebsiella*, *Vario-vovax*, and *Xanthomonas* [19].

Lundberg et al. [20] reported that the growth of plants in field condition is not a simple one but it is very complex community with relatively constant partner relationships. A well-mannered large community of microbes is always symbionts with the crops [21, 22]. Since their earliest history, this microbial community has been connected with land plants to aid early land plants in overcoming problems such as nutrition availability, unfamiliar and frequently stressful circumstances, and diseases [23]. All plant parts are connected with phytomicrobiome components, including bacteria and fungi [24]. The rhizomicrobiome (microbes associated with the roots) is the most abundant and complex of those attached with higher plants. Microbes that fix nitrogen in the legumes are the best examples for understanding symbiosis between microbes and roots of plants [25]. The rhizosphere also harbors more than 8,000 species, living symbiotically or causing diseases in plants.

Agriculture production can be improved and sustained by using PGPRs as biofertilizer inoculants [26]. Generally, bacteria increase the growth of plants in three modes [27]: by enhancing phytohormones production for plant growth [28], by enhancing nutrients' uptake from the soil [29], and by protecting the plants from harmful diseases causing pathogens [30].

7.3.1 Characteristics of an ideal PGPR

If a rhizobacterial strain contains certain plant growth-promoting properties and may boost plant development after inoculation, it is termed as putative PGPR [31]. The following are the characteristics of an optimal PGPR strain:
- It should be rhizosphere-friendly and rhizosphere-competent.
- Upon inoculation, it should colonize the plant roots in substantial quantities.
- It should be able to aid in the development of plants.
- It should be capable of a wide range of actions.
- It has to get along with the other microorganisms in the rhizosphere.
- It must be resistant to physicochemical variables such as heat, dehydration, radiation, and oxidants.
- It should outperform current rhizobacterial communities in terms of competitive abilities.

7.3.2 Types of PGPR

Plant growth-promoting microbes are strongly associated with root cells of the plants and are classified as follows:
- Intracellular PGPR (iPGPR/symbiotics)
- Extracellular PGPR (ePGPR/free living)

7.3.2.1 Extracellular plant growth-promoting rhizobacteria

It can be found on the rhizosphere or in the gaps between root cortex cells. The genera included as ePGPR are *Serratia*, *Micrococcus*, *Azotobacter*, *Bacillus*, *Azospirillum*, *Caulobacter*, *Chromobacterium*, *Burkholderia*, *Erwinia*, *Flavobacterium*, *Arthrobacter*, *Pseudomonas*, and *Agrobacterium* [32].

7.3.2.2 Intracellular plant growth-promoting rhizobacteria

It is mainly found inside the root cells, where a unique nodular structure contains a bacterial population capable of fixing atmospheric nitrogen, which is particularly beneficial to terrestrial plants [33]. The genera included as iPGPR are , *Bradyrhizobium*, *Rhizobium*, *Mesorhizobium*, *Allorhizobium*, and *Frankia*.

The iPGPR may be found inside root cells, usually in specialized formations called nodules. ePGPR is located around the root surface (rhizoplane), or in the root cortex's intercellular spaces, populating the plant tissue intercellularly [25].

7.3.3 Mechanism of action of PGPR

PGPR has complex mechanism by which it helps the increased growth of plants either directly or indirectly. The direct mechanism of PGPR stimulates the resource utilization such as nitrogen, phosphorus, key nutrients through phosphate solubilization, iron sequestration by siderophores, and nitrogen fixation. iPGPRs regulate the plant hormonal balance like auxin, gibberellin (GA3), and cytokinin (CK), or stimulate the rhizospheric competition and ISR.

7.3.3.1 Direct mechanism

In the absence of pathogens, these PGPRs promote plant development. According to Vessey, the soil microbial community which is grown in, on, or around the rhizoplane of plant tissue encourages the development of plants through a variety of ways [34]. Microbial community which is present around the rhizospheric zone of plants influences the rooting pattern as well as transportation of water and nutrients to the plants and also provides the mechanical support.

7.3.3.1.1 Nitrogen fixation

The most important element for plant health and growth is nitrogen (N). In spite of 78% availability of N_2 in the atmosphere, it is inaccessible to developing plants. Biological nitrogen fixation converts N_2 present in the atmosphere into plant-usable forms by nitrogen-fixing bacteria which converts nitrogen into ammonia with the help of nitrogenase enzyme [35]. Bacteria that fix the nitrogen are extensively dispersed in nature, and fix nitrogen biologically at mild temperatures [36]. N_2-fixing bacteria are mainly categorized as (a) bacteria which fix N_2 symbiotically in leguminous plants (e.g., *Rhizobia*) and in nonleguminous plants (e.g., *Frankia*), and (b) bacteria which fix N_2 nonsymbiotically (as free living, endophytes, and associative manner) in nitrogen-fixing forms consist of cyanobacteria (e.g., *Nostoc* and *Anabaena*) [33]. On the other hand, nonsymbiotic nitrogen-fixing bacteria fix a little bit amount of nitrogen to linked host plants [37]. Nitrogenase (nif) genes are compulsory for nitrogen fixation as well as iron (Fe) protein activation, iron–molybdenum cofactor activation, electron donation, and the regulatory gene essential for enzyme synthesis and activity. It is found in groups of roughly 20–24 kb in diazotrophic (nitrogen-fixing) bacteria with 20 coding of distinct proteins with 7 operons [37].

7.3.3.1.2 Phosphate solubilization

Phosphorus (P) is also the most vital macronutrients after nitrogen (N) for the plants and also responsible for its growth. Commonly, it is available in the form of both organic and inorganic in the soil [38]. Plants can take soluble P only in two forms: monobasic (H_2PO_4) and dibasic (H_2PO_4) [33]. Plants absorb just a small portion of broadcasting phosphate fertilizers, while the remainder is swiftly transformed into insoluble complexes in the soil [39]. Phosphate fertilizer application is not only expensive but also ecologically unfavorable. This has sparked a search for a method of enhancing crop yield in low-phosphorus soils that are both eco-friendly and commercially viable. Phosphate-solubilizing microbes (PSMs) are organisms that have phosphate-solubilizing activity and can provide accessible fertilizers in this situation [40]. Phosphate-solubilizing bacteria (PSB) are regarded as potential biofertilizers among different PSMs populating the rhizosphere because they may feed plants with phosphorus from sources that are otherwise inaccessible to plants through diverse ways [41] (Figure 7.1). Some bacterial genera that have been identified as phosphate solubilizers are *Azotobacter*, *Bacillus*, *Burkholderia*, *Beijerinckia*, *Erwinia*, *Microbacterium*, *Flavobacterium*, *Pseudomonas*, *Enterobacter*, *Serratia*, and *Rhizobium* [42]. Rhizobacteria have the ability to solubilize inorganic phosphate sources, hence, increasing agricultural plant growth and yield. The capacity of PGPRs has been used to solubilize mineral phosphate, so piqued the agromicrobiologists' interest because it has the potential to improve phosphorus availability for optimal plant development. Plants with PGPRs have already been demonstrated to solubilize accumulated phosphates, indicating a potential means of field plant growth promotion [43].

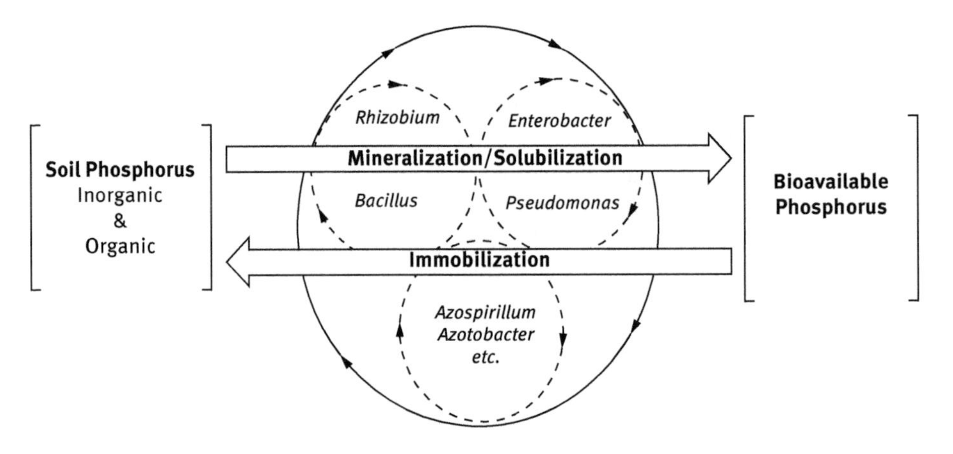

Figure 7.1: Schematic diagram of soil phosphorus mobilization and immobilization by bacteria.

7.3.3.1.3 Siderophore production

Siderophores are low-molecular-weight compounds which bind the iron protein molecules that precipitate in the chelation of Fe^{3+} ion from the environment. Iron (Fe) is the most vital mineral present in the soil, but it is not utilized and accessible by the plants. It is usually present in the Fe^{3+} form. When Fe is scarce, microbial siderophores supply it to plants, boosting their development, and also solubilize Fe from organic compounds in Fe-deficient situation. Siderophores often form 1:1 complexes with Fe^{3+}, which are subsequently taken up by the cell membrane of bacteria, where Fe^{3+} is reduced to Fe^{2+} and released into the cell from the siderophore.

PGPR has been shown to improve plant development by creating highly effective extracellular siderophores that allow management of numerous plant diseases by denying pathogens of Fe feeding, resulting in enhanced crop output. Plants cultivated in metal-contaminated soils are frequently iron deficient, and bacteria may assist plants in obtaining sufficient iron. Metal chelating agents such as microbial siderophores are employed to manage the availability of iron in the plant rhizosphere. This, in turn, aids plants in reducing metal toxicity, as evidenced by arsenic absorption by numerous plants.

7.3.3.1.4 Production of phytohormone

The microbial manufacture of the plant's hormone, auxin (indole-3-acetic acid/indole acetic acid (IAA)), has long been recognized. According to reports, 80% of microbes isolated from the vicinity of different crops can synthesize and secrete auxins as SMs [44]. In general, IAA released by such bacteria participates in improvement in plant's health and growth because the acquisition of IAA secreted by soil bacteria alters the endogenous pool of plant IAA. As a result, IAA is crucial in rhizobacteria–plant interactions [45]. Furthermore, bacterial IAA improves the root surface area and length, giving the plant better access to soil nutrients. Furthermore, rhizobacterial IAA loosens plant cell walls, allowing for increased root exudation, which offers extra nutrients to support the proliferation of rhizosphere bacteria [37]. As a result, rhizobacterial IAA has been discovered as an effector molecule in plant–microbe interactions, including disease and phytostimulation. Tryptophan, an amino acid, is considered as the major IAA precursor and influenced the biosynthesis of IAA as a key molecule that modulates the level of IAA production [46]. IAA production has been demonstrated in the majority of *Rhizobium* species because IAA is involved in a variety of activities, including cell differentiation, cell division, and formation of vascular bundle, all three are also required for nodule formation.

1-Aminocyclopropane-1-carboxylate (ACC) deaminase

Ethylene is the plant hormone, synthesized in plants, and plays a significant role in its growth and development [47]. This plant growth hormone is generated endogenously by nearly all plants and is also created in soils by several biotic and abiotic processes. It is vital for triggering a variety of changes in the physiological function of plants. In addition, it is worked as a growth inhibitor for plants and also identified as a stress hormone [48]. It inhibits the plant growth and development in water logging, salt, heavy metals, drought, and pathogenicity under stress condition. High ethylene content, for example, induces defoliation and other biological functions, which might lead to poor crop growth [49]. For lowering the ethylene level in plants, rhizobacteria producing enzyme 1-aminocyclopropane-1-carboxylate (ACC) deaminase helps in plant growth development and protects them from stress condition [50]. Some bacterial genera which are identified as an ACC deaminase producer are *Achromobacter, Agrobacterium, Acinetobacter, Azospirillum, Bacillus, Burkholderia, Alcaligenes, Enterobacter, Rhizobium, Ralstonia, Serratia*, and *Pseudomonas* [51]. These rhizobacteria consume the ethylene precursor ACC and convert it to 2-oxobutanoate and NH_3. As a consequence, the most visible impacts of seed/root inoculation with ACC deaminase-producing rhizobacteria include plant root elongation, shoot growth stimulation, and augmentation in rhizobial nodulation, N, P, and K absorption, as well as mycorrhizal colonization in diverse crops [52].

7.3.3.2 Indirect mechanism

The use of microorganisms in disease suppression or control is a type of management through biological control, which are safe and eco-friendly in favor of environment [53]. The rhizobacteria promote the plant growth and development indirectly, that is why it is considered as biocontrol agents (BCAs) [37]. In general, PGPR triggers the ISR of plants, nutritional competition, and also stimulates the synthesis of antifungal metabolites. Antifungal compounds produced by several rhizobacteria have been identified, including pyrrolnitrin, HCN, 2,4-diacetylphloroglucinol, phenazines, viscosinamide, pyoluteorin, and tensin [33]. Resistance in plants from pathogenic microbes (fungus, bacteria, and virus) is a result of rhizobacterial association with plant roots, which is termed ISR. Furthermore, the connection of ISR from jasmonate and ethylene signaling hormone boosts the innate immunity inside the plants against different plant diseases [37].

7.3.4 Microbial secondary metabolites for plant growth and its mechanism of action

Beneficial microorganisms produce several natural compounds that are heterogeneous in nature and play a major role in some basic functions like symbiosis with microbes in their vicinity, competition against harmful pathogens, and metal transport to plants [54]. These compounds produced by beneficial soil microbes during their cell cycle event's late idiophase are not helpful for development and growth of plants but also boost resistance, enhance signaling with nearby beneficial microbes, and increase adaptation in plants for adverse condition [55]. These microbial SMs are also important for human beings as these metabolites are being utilized in various fields such as medicine, agriculture, chemical industry, food industry, and related fields [56, 57]. Among all metabolites, the primary microbial metabolites, amino acids, ethanol, and lactic acid are essential for reproduction, growth, and development of organisms/producers. However, SMs (antibiotics and pigments) formed by microbes, namely, bacteria and fungi are needed for other living organisms for their metabolic processes [58]. Seven distinct metabolic pathways are engage in the biosynthesis of active SMs: the polyketide synthase pathway, the peptide pathway, the hybrid pathway, the nonribosomal polypeptide synthase pathway, the b-lactam synthetic pathway, the shikimate pathway, and the carbohydrate pathway. Previous study suggests that PGPRs associated with plant roots boost their SM production during stress conditions [59–62]. According to Jain et al. [63], pea plant with beneficial microbes, namely, *Bacillus subtilis* and *Pseudomonas aeruginosa*, produces more amounts of phenolic compounds and gallic acid compared to uninoculated plants that protect pea plants from harmful pathogens. The findings indicated that root ingredients are responsible for enhancing the advantageous microbial communities in the area of roots that protect and increase defense in plants against harmful soil microbes, or other abiotic stress [64–66] (Table 7.1).

7.3.4.1 Categorization of microbial secondary metabolites

7.3.4.1.1 Water-soluble secondary metabolites

Water-soluble SMs with higher polarity acquire high degree of functionality [86]. Some examples are given further.

Polyketides

Polyketides are considered as most ample SMs formed by several bacteria, which are used as antibiotics; for example, erythromycin is a good example of antibiotics used in the cure of many infectious diseases [72]. Phloroglucinol is another SM used as antibiotics having 2,4-diacetylphloroglucinol (DAPG) by *Pseudomonas* species.

Phloroglucinol

Phloroglucinol (1,3,5-trihydroxybenzene or 1,3,5-benzenetriol) and its derivatives are phenolic compounds. These compounds possess antibacterial, antihelminthic, antiviral, antifungal, and other phytotoxic properties [87]. DAPG is a potent antimicrobial SM produced by fluorescent pseudomonads having the ability to manage phytopathogens [88, 89].

Table 7.1: Production of secondary metabolites by various microorganisms and their effect on plant growth.

Name of microorganisms	Name of secondary metabolites	Activity/target organisms	Observed effect on plants	References
Trichoderma spp.	Viridin	Inhibit egg hatching of *Meloidogyne* spp., *Rhizoctonia solani*, and *Fusarium* spp.	Increase growth of plants	[67, 68]
Bacillus subtilis	Zwittermicin A, kanosamine, and lipopeptides	Antifungal	Stimulate plant growth promotion	[69, 70]
B. subtilis	Bac 14B	Antimicrobial agent against *Agrobacterium tumefaciens*	Protect plants against *A. tumefaciens*	[71]
Pseudomonas spp.	2,4-Diacetylphloroglucinol (DAPG)	Antibacterial	Stimulate plant growth promotion	[72]
Azospirillum brasilense	Salicylic acid	Reduce biotic stress	Enhance growth of strawberry plant	[73]
P. fluorescens	Salicylic acid	*F. oxysporum*	Induction of systemic resistance in plants	[74]
PGPR (*Rhizobium*, *Azospirillum*, and *Frankia*)	*N*-Acylhomoserine lactone (AHL)	*Golovinomyces orontii* and *P. syringae*	Induce systemic resistance (ISR) in *Arabidopsis thaliana*	[75]
Streptomyces spp. RP1A-12	Siderophores	Reduce biotic stress, namely, stem rot disease	Enhance growth of groundnut plant	[76]
T. herzianum, *T. aureoviride*, and *T. koningii*	Koninginins	Antibiotic activity against *Gaeumannomyces graminis* var. *tritici*	Enhance growth of wheat plant	[77]

Table 7.1 (continued)

Name of microorganisms	Name of secondary metabolites	Activity/target organisms	Observed effect on plants	References
P. fluorescens	DAPG	*Pythium* spp.	Protect sugar beet seedling from damping off and enhance growth	[78]
B. subtilis and *B. amyloquefaciens* strain GB03, IN937a	Volatile organic compounds, namely, ammonia, phenazine-1-carboxylic acid, butyrolactonas, and HCN	*Erwinia carotovora* subsp. *carotovora*	ISR pathway in *Arabidopsis* seedlings	[79]
T. virens	Auxin	Antifungal and antibacterial activity	Increase the yield of *Zea mays* and *Arabidopsis thaliana*	[80]
P. aeruginosa	Salicylic acid	*Rhizoctonia solani*	Induce pathogen-related genes and boost the immune system of plants	[81]
P. fluorescens strain (Pf4-92)	DAPG	*F. oxysporum* f. sp. *ciceri*	Enhance growth and yield of chickpea plant	[82]
P. aurantiaca strain SR1	DAPG	Antifungal activity against *Macrophomina phaseolina*	Enhance growth and yield of plant	[83]
Pseudomonas aeruginosa and *Burkholderia gladioli*	Antioxidant	Inhibit root-knot nematode	Enhance growth and yield of tomato plant	[84]
Trichoderma spp.	Jasmonic acid	Inhibit root-knot nematode	Enhance growth and yield of plants by inducing resistance	[85]

Nonribosomal peptides

Siderophore and lipopeptide are nonribosomal metabolites produced by microbes against harmful pathogens. Siderophores (i.e., desferrioxamine, pyoverdines, bacillibactin, etc.) are iron-chelating compounds having the ability to solubilize iron under iron-limited conditions [90, 91]. Lipopeptides exhibit as both antimicrobial and surfactant agents [92].

Ribosomal peptides

Bacteriocin is an antimicrobial ribosomal peptide that protects plants from harmful pathogens. Bac. 14B is a bacteriocin that acts as antimicrobial produced by *B. subtilis* against *Agrobacterium tumefaciens*, and causes crown gall disease [71]. Putidacin is the other best antimicrobial metabolite obtained from *P. putida* to suppress plant pathogens [93].

7.3.4.1.2 Volatile organic compounds (VOCs)

Volatile organic compounds (VOCs) are the best compounds for antagonistic and synergistic interactions among microorganisms. Followings are VOCs.

Nitrogen-containing VOCs

Pyridines, thiazoles, pyrazines, pyrroles, indole, and aniline derivatives are the nitrogen-containing compounds that possess the ability to kill harmful pathogens and protect inoculated plants [94, 95].

Indole

Both gram-positive and gram-negative bacteria are actively involved in the production of indole metabolites against harmful pathogens [96–98].

Pyrazines

Pyrazine-type metabolites obtained from *Bacillus* and *Pseudomonas* species are popular for their antimicrobial activity [99, 100].

Sulfur-containing VOCs

Dimethyl sulfide (a peculiar soil fumigant against phytonematodes and harmful soil microbes), dimethyl disulfide, and trisulfide are well-known sulfur-containing VOCs that act as inhibitory compounds against several harmful fungi and also in plant–microbe interaction [101, 102].

Terpenes

Various types of terpenes such as albaflavenone, geosmin, germacrene D-4-ol, a-pinene, and g-terpinene are formed by microorganisms having antimicrobial activities [103, 104].

7.3.5 Plant growth-regulating secondary metabolites

Such type of metabolic compounds formed in excessively low amount and manages development and growth of plant also protect from abiotic and biotic stresses [105–107]. These are CKs, auxins (glucosinolate), abscisic acid (ABA), gibberellins (GAs), jasmonic acid (alkaloids), brassinosteroids (polyhydroxy steroids), salicylic acids (phenolics), and terpenes. CKs and auxins are the best growth enhancers that help in cell division, cell expansion, apical dominance, and stimulate the synthesis of protein and RNA [98, 108, 109]. Flowering, seed dormancy breaking, and cell expansion functions are attributed to GAs. Salicylic acids provide defense against phytopathogens and also play a role in photosynthesis, nutrient transport, and transpiration. Brassinosteroids are polyhydroxylated steroidal growth regulators that regulate vascular germination immunity, stomatal aperture, and photomorphogenesis. ABA in plants regulates leaf abscission signaling and defense against pathogens while jasmonic acid protects plants from phytopathogens.

7.3.5.1 Fungal secondary metabolites

Among microbial SMs, fungal metabolites are important to improve metabolism and growth of plants, while other metabolites attack on different fungal stages such as hyphal elongation and sporulation [110]. Fungal inoculants have more potential compared to bacterial inoculants in response to plant growth and development within the soil. Beneficial fungi promote plant growth and yield by antibiotic production, competition with harmful pathogens, and enhancement of defense response in inoculated plants [111]. In addition, several other beneficial fungi have the ability to parasitize sclerotia, spores, or hyphae of harmful fungi, resulting in biocontrol. Mycoparasitism is the mechanism through which beneficial fungus, namely, *Trichoderma* spp., kill or inhibit the growth of other harmful fungi. Mycoparasitism in fungi is started by host sensing of recognition than penetration and degradation. Various degradative enzymes like proteases, chitinases, and glucanases enhance the mycoparasitism in beneficial fungus and engage in biocontrol process [111, 112].

7.3.5.2 Secondary metabolites from *Trichoderma* spp

Trichoderma spp. is considered as good BCAs and used as biopesticides worldwide. Several species of genus *Trichoderma* are producers of many SMs with antibiotic activity [113]. The different strains of *Trichoderma* spp. produced SMs that differ each other and have nonvolatile and volatile antifungal substances such as viridin, gliotoxin, harzianopyridone, 6-*n*-pentyl-6H-pyran-2-one (6PP), peptaibols, and harziandione [113].

Pyrone (6PP)

Culture filtrate of different species of *Trichoderma* (e.g., *T. koningii*, *T. viride*, *T. harzianum*, and *T. atroviride*) produces a metabolite known as pyrone 6-pentyl-2H-pyran-2-one (6-pentyl-α-pyrone or 6PP). Both in vitro and in vivo antifungal activities against numerous phytopathogenic fungi possessed 6PP. It is applied exogenously at very low concentration (by 1 µm shoot spray or 1 ppm root application) to tomato or wheat seedlings and etiolated pea plants after application of auxin-like effect was observed in plant growth [114].

Koninginins

Koninginins are complex pyranes isolated from *T. koningii*, *T. harzianum*, and *T. aureoviride*. Koninginins A, B, D, E, and G demonstrate in vitro antibiotic activity against take-all fungus *Gaeumannomyces graminis* var. *tritici* and enhance growth and yield of wheat plant [77, 115].

Viridins

Viridin is another antifungal compound isolated from various *Trichoderma* spp. (*T. virens*, *T. koningii*, and *T. virideae*). This compound prevents spore germination of some fungal pathogens such as *Colletotrichum lini*, *Botrytis allii*, *Penicillium expansum*, *Fusarium caeruleum*, *Stachybotrys atra*, and *Aspergillus niger* [113].

Recently, a new metabolite named harzianic acid comprising pyrrolidinedione ring system has been isolated from *T. harzianum* strain. This tetramic acid derivative expressed in vitro antibiotic activity against *Sclerotinia sclerotiorum*, *Pythium irregulare*, and *Rhizoctonia solani* [116].

Metabolites formed by beneficial fungi participated in the induction of plant resistance such as (i) avirulence-like gene products that increase defense reactions in plants, (ii) increase proteins with enzymatic activity, that is, xylanase [111], and (iii) low-molecular-weight compounds secreted from either fungal or plant cell walls with enzymatic activities [111, 117]. Few low-molecular-weight degradation compounds secreted from cell wall of fungi having short oligosaccharides comprised two types of monomers, with and without an amino acid residue [117]. These compounds facilitate certain defense reactions when applied to roots and leaves, or injected into leaf tissues. Instead of defense reactions in plants, these compounds also increased the biocontrol ability of *Trichoderma* spp. by switching on the mycoparasitic gene expression "cascade" [117]. Vinale et al. [116] reported the reduction of disease symptoms caused by *Leptosphaeria maculans* or *B. cinerea* on canola and tomato seedlings treated with 6PP metabolites. Moreover, "soil drench applications of 6PP four days before inoculation with *Fusarium moniliforme* showed considerable suppression of seedling blight and substantial plant growth promotion, compared with the untreated control" [118]. Application of 6PP on maize seedlings increased the activities of polyphenol oxidase, peroxidase, and β-1,3-glucanase in both root and shoot tissues that illustrate an induction of defense activities in maize plants [118].

7.3.5.3 Bacterial secondary metabolites

Production of SMs in bacteria occurred at static growth phase. Large numbers of metabolites are released in the growth media from which they are easily extracted due to which their production can be maximized or minimized by changing the media composition and growth conditions. These metabolites protect plants from other harmful microbes and improve their yield and growth. Among bacteria, *Pseudomonas* spp. and *Bacillus* spp. are important genera that release SMs and act as BCAs. Various *Bacillus* species have been considered as PGBR since they inhibit pathogens [119, 120]. *Bacillus* spp. are also involved in the production of peptide antibiotics and some other compounds that are toxic to plant pathogens [121]. Several antifungal compounds, namely, iturins, mycobacillins, surfactins, bacillomycins, mycosubtilins, subsporins, and fungistatins are produced by *Bacillus* spp. having the ability to kill pathogens in soil as well as enhance plant growth. Bacterial metabolites are categorized into volatile organic metabolites and soluble SMs.

7.3.5.3.1 Volatile organic secondary metabolites

Various VOCs such as terpenes, sesquiterpenes, nitrogen-containing pyrazines and indole, and sulfur-containing dimethyl disulfide (DMDS), dimethyl sulfide (DMS), and dimethyl trisulfide (DMTS) are potent against harmful pathogens.

Terpenes

The terpene-building units are dimethylallyl pyrophosphate and isopentenyl pyrophosphate through deoxyxylulose phosphate pathway or mevalonate pathway [122]. These are considered as important bacterial metabolites that protect plants against harmful soil pathogen and improve growth and yield of plants.

Sesquiterpene

With another name albaflavenone, it is produced by *Streptomyces albidoflavus* which shows antimicrobial and antibacterial activities [123]. A monoterpene known as β-pinene is isolated from bacteria *Collimonas pratensis* which exhibited inhibition against *Rhizoctonia solani* and *Staphylococcus aureus* [104].

Pyrazines

Pyrazines are produced by bacteria *Chondromyces*, *Pseudomonas*, *Bacillus*, and *Streptomyces* which act as antimicrobial compounds and enhance plant growth and yield [124, 125].

Indole

This is released from gram-positive and gram-negative bacteria. Indole inhibits bacteria, and its pathogenesis by suppressing virulence factor from which virulent bacteria cause disease and control plant defense systems, growth, and shoot and root development [126, 127].

Other volatile sulfur-containing compounds like DMTS, DMS, and DMDS also act as SMs from beneficial bacteria that inhibit the growth of harmful soil microorganism as well as improve plant yield and growth.

7.3.5.3.2 Soluble secondary metabolites

Due to their high polarity, these metabolites are soluble in water. They possess strong biocontrol activities for short duration as antibiotics or toxins against harmful soil microbes and have high degree of functionalization as well as plant growth stimulators. Several soluble SMs such as bacteriocins, nonribosomal peptides, polyketides, and lipopeptides act as soluble SMs.

Bacteriocins

Bacteriocins act as antimicrobial compounds obtained from *Pseudomonas* spp. Their antimicrobial activities against other microorganisms are either from across genera (broad spectrum) or same species (narrow spectrum) [128]. Putidacine is an example of bacteriocin obtained from *Pseudomonas putida* strain BW11M1, a potent inhibitor for *P. putida* GR12-2R3 [93]. *B. subtilis* 14B, involved in the production of an important bacteriocin Bac 14B, shows inhibitory action against crown gall disease caused by *Agrobacterium tumefaciens* [71].

Nonribosomal peptides

Nonribosomal peptide synthetase enzymes are involved in the synthesis of compounds, nonribosomal peptides. These include two groups of SMs: lipopeptides and siderophores.

Lipopeptides

These compounds consist of surfactants' antipredation, antimicrobial, and cytotoxic properties against other harmful pathogens and enhance the growth of plants [92, 129, 130].

On the other hand, there are many genera of fungi and bacteria involved in the production of SMs for plant growth. Few of them are rhizobacterium *B. tequilensis* SSB07 formed by numerous plant growth-promoting compounds including GA1, GA3, GA5, GA8, GA19, GA24, and GA53, ABA, and IAA. Inoculation of *B. tequilensis* SSB07 in seedlings of Chinese cabbage increased growth, while in soybean, there is enhancement in biomass, shoot length, photosynthetic pigment, and leaf development.

7.3.6 Biological control

Biological control is defined as the depletion of deleterious action of one or more microorganisms regularly through the living agents [131, 132]. The biocontrol of soil-borne phytopathogens with the help of living organism continued from more than 80 years ago. It has been found that microorganisms in the vicinity of roots are considered as BCAs which enhance resistant in plants against plant pathogens [133]. From last decades, farmers utilized chemicals as a fast remedy to manage harmful soil-borne pathogens, and such chemical-based control leads to environmental issues. An alternative of chemical and eco-friendly approaches evolved, such as biocontrol, which is considered as potential control approaches in recent years against harmful phytopathogens [134–138]. Scientists are engaged in to find out the mode of action of BCAs against harmful pathogens and control of plant diseases through which they are able to change the soil environment conducive for successful biocontrol or to improve biocontrol strategies [139]. Biocontrol activities are also involved to enhance plant growth without disturbing flora and fauna as well as increase soil fertility [140].

7.3.6.1 Beneficial microbes as biocontrol agents against phytopathogens

The beneficial roles of bacteria are fixation of N_2, reduction in toxic compounds, development of plant growth and yield, and biocontrol of soil-borne phytopathogens. Several genera of beneficial bacteria are considered as good BCAs that control diseases of plants caused by plant parasitic nematodes (PPNs), bacterial pathogens, and fungal pathogens. An efficient utilization of BCAs to control nematodes and other harmful microbiota also influenced the host plant and surrounding organisms. In vitro mortality of second-stage juveniles (J2s) of *Meloidogyne javanica* is increased by the application of *Pseudomonas* spp. [141]. Soil infestation of root-knot nematodes is reduced effectively by using the bioagent *Bacillus* isolates [142–144]. *P. fluorescence* isolates activate the enzymes for defense in tomato against root-knot nematodes [145]. At present, more than 50% losses to fruits and other agricultural crops by phytopathogenic fungi are found at the postharvest stage [146]. Biocontrol of phytopathogenic organisms at present gives an efficient and alternative of chemicals against phytopathogens at the postharvest stage [147]. Jisha et al. [148] determined the biocontrol activity of *P. aeruginosa* against *Colletotrichum capsici*, which causes anthracnose on chilli plant. *P. aeruginosa* also induces the systemic resistance on chilli plant against anthracnose. Among fungi, *Trichoderma* spp. are considered as potent BCAs, and other fungi such as *Penicillium* [149], *Gliocladium* [150], *Aspergillus* [151], and *Saccharomyces* [152] also possess antagonistic activities against few fungal pathogens, including *Fusarium, Alternaria, Pythium, Gaeumannomyces, Rhizoctonia, Aspergillus, Phytophthora, Pyricularia*, and *Botrytis* [132]. *Trichoderma* functions as a mycoparasite against several soil-borne and aeronautical phytopathogens and is utilized as biopesticides in both greenhouse and field experiments. AM fungi and nematophagous fungi were also considered as potent BCAs. In AM fungi, inoculated plants' activation of SAR and pathogenesis-related genes were found, whereas nematophagous fungi trap PPNs including root-knot nematodes by forming trapping structures [153, 154]. Nematodes' egg-parasitizing fungus *Paecilomyces lilacinus* is the most efficient and important BCAs for PPNs and a promising alternative to pesticides at both planting and preplanting applications [155]. Reduction in egg hatching and juvenile's mortality of root-knot nematode,*M. incognita*, in tomato crop were increased by the application of *P. lilacinus* [156]. Effective and most extreme control of root-knot nematodes *was* achieved by the combined application of *B. firmus* and *P. lilacinus* 2 weeks before transplantation of tomato crops [157]. Extreme research in the previous two decades has concentrated on different aspects of beneficial microbes as their antifungal activity, rhizosphere colonization, and beneficial effect on plant health. The mechanisms of biocontrol basically depend on antibiosis, parasitism, competition, systemic resistance, hydrolytic enzyme production, and rhizosphere competence (Figure 7.2 and Table 7.2).

7.3.7 Toxicity of heavy metals to the environment and human health

Environmental pollution with heavy metals has been accepted as an environmental health threat in sustainable residences as a result of long-term industrial advancement and development [172, 173]. Heavy metals have been potentially causing serious

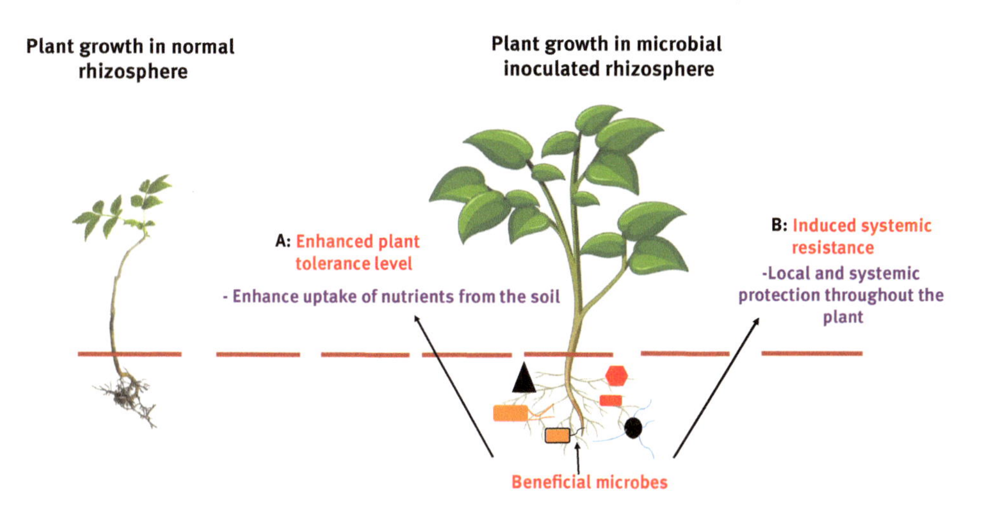

Figure 7.2: Growth of plant with beneficial microbes and without beneficial microbes.

Table 7.2: Effect of various biocontrol agents against plant pathogenic microbes on several crops.

Name of biocontrol agents	Test plants	Test diseases	Target pathogens	References
Pseudomonas fluorescens	Apple	Mucor rot	*Mucor piriformis*	[158]
Pseudomonas putida BP25	Rice	Blast disease	*Magnaporthe oryzae*	[159]
Rhizobium japonicum	Soybean	Root rot	*Fusarium solani*, *Macrophomina phaseolina*	[160]
Bacillus thuringiensis	*Brassica campestris* L.	Scleretiniose	*Sclerotinia sclerotiorum*	[161]
Aspergillus fumigates	Cocoa plant	Black pod	*Phytophthora palmivora*	[151]
Paecilomyces lilacinus	Tomato	Root-knot disease	*Meloidogyne javanica*	[162]
Penicillium oxalicum	Tomato	Wilt	*Fusarium oxysporum* f. sp. *lycopersici*	[163]
Pochonia chlamydosporia	Carrot	Root knot disease	*Meloidogyne incognita*	[164]
Trichoderema asperellum T8a	Mango	Anthracnose	*Colletotrichum gloeosporioides*	[165]

Table 7.2 (continued)

Name of biocontrol agents	Test plants	Test diseases	Target pathogens	References
Trichoderma spp.	Tobacco	Root rot	*Rhizoctonia solani*	[166]
Trichoderma harzianum	Rice	Brown spot	*Bipolaris oryzae*	[167]
Penicillium sp. EU0013	Tomato and cabbage	Wilt	*Fusarium oxysporum*	[149]
Azospirillum brasilense	Strawberry	Anthracnose	*Colletotrichum acutatum*	[73]
Bacillus methylotrophicus	Maize	Stalk rot	*Fusarium graminearum*	[168]
Brevibacillus brevis	Strawberry	Gray mold	*Botrytis cinerea*	[169]
Purpureocillium lilacinum	*Vigna radiata*	Root-knot disease	*Meloidogyne incognita*	[170]
Trichoderma virens	Okra	Root-knot disease	*Meloidogyne incognita*	[171]

health issues due to their persistency and are known from a long time ago. Heavy metal contamination of agricultural soil is a significant environmental issue that can mitigate both the productivity and the safety of plants and their products used as foods and feeds [174]. Plants are primary producers that convert energy into food from sunlight. They uptake advantageous nutrients/elements and water from soil ecosystem. Iron, manganese, copper, cobalt, molybdenum, and vanadium are among the metals that show physiological functions in organisms. Furthermore, high level of these metals can cause different degree of toxic effects in organisms [172]. Numerous other elements have been accumulated and regarded as highly toxic such as mercury (Hg), cadmium (Cd), lead (Pb), and arsenic (As) [175] (Table 7.3).

Plant–microbe interaction enhances heavy metal accumulation, resistance, and tolerance, which has major implications for heavy metal detoxification [187, 188]. During stressful condition, microbes and plants use various defense mechanisms, including compartmentalization, complex formation, exclusion, and release of metal-binding proteins such as phytochelatins and metallothioneins [189, 190]. Toxic metals are deposited in the environment by means of industrial wastes, waste disposal, agricultural intervention, and atmospheric deposition [191] (Figure 7.3).

Table 7.3: Effect of heavy metals on human health and plants.

Heavy metals	Impact on plant and human health	References
Copper (Cu)	Retards the plant development, influences biomass community, reveals chlorosis, chances of kidney and brain damage to humans, and intestinal and gastric pain	[176, 177]
Mercury (Hg)	Interrupts the mitochondrial functions, causes oxidative stress, affects the immune system, loss of hair, visual ailment, and failure of lung and kidney	[178, 179]
Lead (Pb)	Reduces the seed germination, disturbs the nutrition of minerals, affects cell division, and has chances of memory loss, coordination, and learning difficulties	[180]
Nickel (Ni)	Inhibits growth and development, leaf chlorosis, skin allergy, neurotoxic and immune toxic effect, and loss of hairs	[181]
Chromium (Cr)	Retards growth, reduces the growth of roots, leaves, and shoots, and shows breathing problem, nasal congestion, and hair damage and hair fall	[182, 183]
Cobalt (Co)	Reduces biomass, growth of shoots, and interferes the protein concentration	[184]
Arsenic (As)	Disturbs the cell mechanism	[185]
Barium (Ba)	Cardiac disease, gastrointestinal disorder, and respiratory issue	[186]

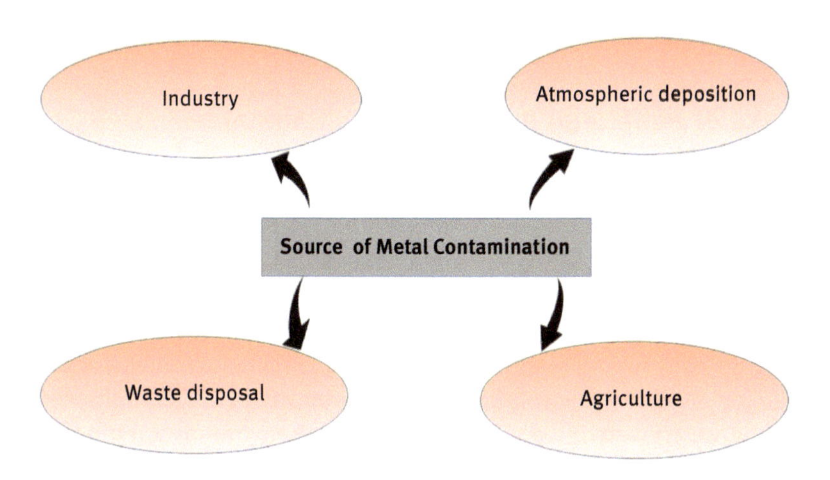

Figure 7.3: Source of metal deposition.

7.3.7.1 Role of beneficial microbes in detoxifying the heavy metals and plant growth enhancement

Several soil-inhabiting microbes have been recognized as a potential tool in remediation of contaminated environment. Microbes are present in plenty in rhizosphere that counters the devastating pathogen and provide strength to plants to develop defense mechanism against pathogens. Improving the metal solubility by boosting plant rhizosphere microbiota allows better polysaccharides as well as organic acid synthesis for receiving competitive advantage during phytoremediation [192, 193]. Detoxification of metals is considered a unique category of contaminants, while phytoremediation technology for wastewater treatment can be employed in a cost-effective manner [194]. Because of unique characteristics such as enzymes and antioxidant activity against reactive oxygen species and biomass, plants can tolerate diverse contaminants available within the environment. Plant metal-binding protein-like genes have been isolated from a variety of microorganisms and plants, including *Brassica*, rice, maize, tobacco, soybean, and wheat for their role in the degradation of metal [195]. The phytoremediation effectiveness of *Helianthus annus* L. for the restoration of heavy metal-polluted industrial sites was recently reported. Other plants suitable for phytoremediation include *Zea mays*, *Erato polymnioides*, *Solanum lycopersicum*, *Hibiscus cannabinus*, *Paxillus involutus*, *Festuca arundinacea*, *Helichrysum italicum*, and *Poulus canescens* [196]. Translocation, chelation, volatilization, solubilization, precipitation, complexation, and immobilization are some of the biochemical processes involved in plant–microbe interaction and heavy metal uptake.

7.3.8 Importance of beneficial microbes for Sustainable agriculture and environment

The world's population is steadily increasing at a rate of 1.05% per year, and this has prompted the development of malnutrition in approximately 2.4 billion people [197]. To feed the entire world's population, synthetic pesticides are often used to significantly boost crop productivity with the objective of intensifying agriculture [198]. However, continued and disproportionate use of pesticides exacerbated threats such as increased soil salinity and toxicity, soil hardening, significantly reduced nutrient-carrying capacity, and waterlogging [199]. This has influenced environmental disruption, necessitating the use of farming practices to meet increased food demand without detrimentally impacting the environment. As a result, a pursuit to develop a model for sustainable agriculture began. Microbes are recognized for their enormous potential in sustainable agriculture. Plants are home to a multitude of microbial communities, including archaea, fungi, and bacteria, which are referred to as the plant–microbiome.

Plant–microbiome contributes to the plant health and survival while posing no damage to the environment [200].

The architecture of the plant–microbiome and its interactions with the host may now be studied due to the technological progress [201]. As an endophyte or epiphyte, microbes inhabited numerous locations within and on the plant body [202]. Plant health is affected by the plant–microbiome in pleiotropic ways [203]. The life span of plants is increased by microbes through promoting health and yield [204], protecting against diseases [205], and strengthening immunity [206]. Site-specific microbes can be employed as BCAs to mitigate the environmental damage caused by pesticides since they enhance crop productivity and longevity. Despite the various advantages, does the microbes contain all of the ideal properties for long-term viability development? Agrochemicals that are being used in the past are still needed. They promote crop production to ensure food security, but they cause havoc on agricultural resources and on our ecosystem. To maintain the balance between environment protection initiatives and increasing food demand, the plant microbiota must be utilized. Several plant microorganisms are already being used in agriculture as sustainable approaches for higher yield [206]. This may be an effective alternative for sustainable crop production and environment protection (Figure 7.4).

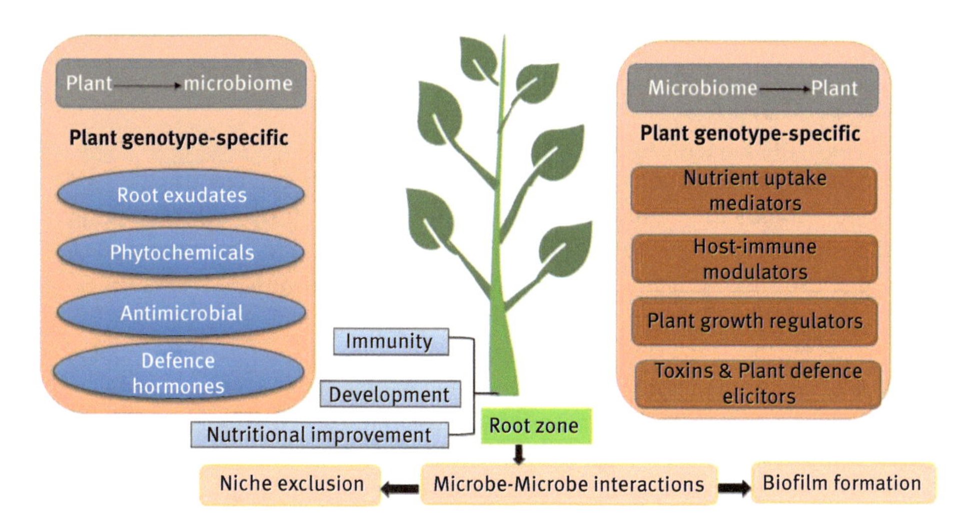

Figure 7.4: Beneficial features of microbes for plant growth and development.

7.3.9 Beneficial attributes of plant–microbiome

Plant growth is encouraged by rhizospheric microbes, which also protect the plant from pathogenic microorganisms. Rhizobacteria create compounds that restrict harmful microorganism's growth and activity, and rhizospheric fungi can produce antibiotics. Harzianic acid, terpines, glovirin, gliotoxin, polyketides, massoilactone, tricholin, peptaibols, and viridian are compounds that are produced by *Trichoderma* sp.

These metabolites have antimicrobial properties and can be used to combat pathogens. Phyllosphere microorganisms can control these pathogens. Microorganisms in the phyllosphere can protect plants from pathogens by eliciting an immune response in the plant by supporting nonpathogenic organisms, antibiotic production, enhancing competition among microorganisms, and thus outcompeting the pathogens (Figure 7.5).

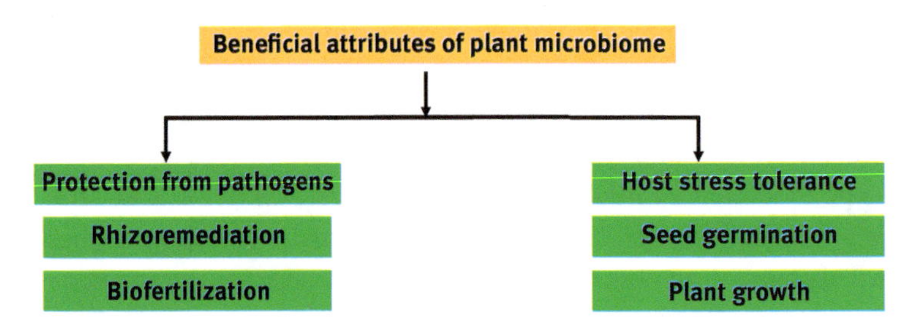

Figure 7.5: Advantage of microbes in response to plant development.

7.4 Conclusion

In the present scenario, food security is the major challenge to meet the global demand of food. Plants face numerous challenges and cope up by their own natural defense mechanism. Plant–microbe interaction occurred at different magnitudes to develop defense mechanisms against pathogens, improve nutrient uptakes, and boost the soil fertility. Plants uptake several nutrients with complex toxic substances from soil ecosystem regarded as heavy metals. These heavy metals are the potential constraints of crop productivity and cause severe health issues to humans and cattle. To combat abiotic stresses without negatively impacting soil and groundwater resources, PGPB and mycorrhizal fungi are recognized as beneficial microbes in many ways. Thus, reduction or elimination of toxicants from the environment by biological resources are the explanation of bioremediation mechanisms (microorganism and plants). Recent advances in heavy metal bioremediation have extensively studied, and microbial-based heavy metals processing provide environment-friendly solution than the traditional approach. Plant-associated beneficial microbes increase the effectiveness of phytoremediation by modifying abundance and distribution in plant tissues and indirectly by boosting root and shoot biomass production. Moreover, bioremediation necessitates the use of genetically engineered microbes with improved efficiency to ramify or remove the toxic effect of heavy metals in the contaminated area. Bioremediation is a cost-effective and novel approach in mitigating abiotic stress for sustainable agriculture. We need to explore such more possible us of soil microbiome to combat the abiotic stress in relation to better plant's growth and development which can ultimately contribute to the sustainable development.

7.5 Future prospects

In today's world, the scientific community aims to demonstrate the role of the microbiome in protecting environment, improving plant health, and enhancing crop yield for achieving sustainable agriculture goals and mitigating the environmental stresses. Plant–microbe interactions have been found to be significant in enhancing crop productivity through improving nutrient translocation, improving soil health condition, and detoxifying the toxic metals. So far, observations have been made to isolate and determine the efficacy of microbes in suppressing pathogens and improving plant health and crop productivity. Such knowledge will enable us to improve the microbe performance under different geographic locations and conditions. There is a need to develop strategies that can integrate available knowledge in formulating ecological principles that can translate this knowledge into increased crop production in order to attain the ultimate goal of sustainable agriculture. Government and Policy makers should try to boost up the startup based manufactures and their product developed, can be more widely utilized under collaborative research programs.

References

[1] Sujatha M, Vimala Devi PS, Reddy TP. Insect pests of castor (Ricinus communis L) and their management strategies, Hyderabad, India, BS Publications; Leiden, The Netherlands, CRC Press, 2011.

[2] Aislabie J, Deslippe JR, Dymond J. Soil microbes and their contribution to soil services. Ecosystem Services in New Zealand – Conditions and Trends *Manaaki Whenua Press, Lincoln, New Zealand* 2013, 1(12), 143–161.

[3] Prasad R, Kumar M, Varma A. Role of PGPR in soil fertility and plant health. In: Egamberdieva et al., eds. Plant-Growth-Promoting Rhizobacteria (PGPR) and Medicinal Plants, Soil Biology 42, Cham, Springer, 2015, 247–260.

[4] Hussain T, Akhtar N, Aminedi R, Danish M, Nishat Y, Patel S. Role of the potent microbial based bioagent and their emerging strategies for the eco-friendly management of Agricultural Phytopathogens. In: Singh J, Yadav AN. Natural Bioactive Products in Sustainable Agriculture. Singapore, Springer, 2020a, 45-66. ISBN-978-981-15-3023-4.

[5] Hussain T, Singh S, Danish, M, Pervez, R, Hussain K, Husain R. Natural Metabolites an eco-friendly approach to manage plant diseases and for better agricultural farming. In: Singh J, Yadav AN, eds. Natural Bioactive Products in Sustainable Agriculture, Singapore, Springer, 2020b, 1-13. doi.org/10.1007/978-981-15-3024-1_1, e-ISBN-978-981-15-3023-4.

[6] Bulgarelli D, Schlaeppi K, Spaepen S, Van Themaat EVL, Schulze-Lefert P. Structure and functions of the bacterial microbiota of plants. Annual Review of Plant Biology 2013, 64, 807–838.

[7] Lugtenberg B, Kamilova F. Plant-growth-promoting rhizobacteria. Annual Review of Microbiology 2009, 63, 541–556.

[8] Compant S, Clément C, Sessitsch A. Plant growth-promoting bacteria in the rhizo-and endosphere of plants: Their role, colonization, mechanisms involved and prospects for utilization. Soil Biology & Biochemistry 2010, 42(5), 669–678.

[9] Mandal SD, Sonali, Singh S, Hussain K, Hussain T. Plant Microbe Association for the mutual benefits for plant growth and soil health. In: Yadav AN, et al., eds. Current Trends in Microbial Biotechnology for Sustainable Agriculture, Environmental and Microbial Biotechnology, Singapore, Springer Nature, 95-121, 2021. https://doi.org/10.1007/978-981-15-6949-4_5

[10] Egli M, Wernli M, Burga C, Kneisel C, Mavris C, Valboa G, Mirabella A, Plötze M, Haeberli W. Fast but spatially scattered smectite-formation in the proglacial area Morteratsch: An evaluation using GIS. Geoderma 2011, 164(1–2), 11–21.

[11] Lombard N, Prestat E, van Elsas JD, Simonet P. Soil-specific limitations for access and analysis of soil microbial communities by metagenomics. FEMS Microbiology Ecology 2011, 78(1), 31–49.

[12] Ahmad F, Ahmad I, Khan MS. Screening of free-living rhizospheric bacteria for their multiple plant growth promoting activities. Microbiological Research 2008, 163(2), 173–181.

[13] Ahamd G, Nishat Y, Haris M, Danish M, Hussain T. Efficiency of soil, plant and microbes for the healthy plant immunity and sustainable agricultural system. In: Varma A, Tripathi S, Prasad R, eds. Plant-Microbe Interface. Cham, Springer. 2019. ISBN 978-3-03019830-5. doi.org/10.1007/978-3-030-19831-2_15.

[14] Verma JP, Yadav J, Tiwari KN, Kumar A. Effect of indigenous Mesorhizobium spp. and plant growth promoting rhizobacteria on yields and nutrients uptake of chickpea (Cicer arietinum L.) under sustainable agriculture. Ecological Engineering 2013, 51, 282–286.

[15] Zaidi S, Usmani S, Singh BR, Musarrat J. Significance of Bacillus subtilis strain SJ-101 as a bioinoculant for concurrent plant growth promotion and nickel accumulation in Brassica juncea. Chemosphere 2006, 64(6), 991–997.

[16] Verbon EH, Liberman LM. Beneficial microbes affect endogenous mechanisms controlling root development. Trends in Plant Science 2016, 21(3), 218–229.

[17] Van Der Heijden MGA, Bardgett RD, Van Straalen NM. The unseen majority: Soil microbes as drivers of plant diversity and productivity in terrestrial ecosystems. Ecology Letters 2008, 11(3), 296–310.

[18] Podile AR, Krishna Kishore G. Plant growth-promoting rhizobacteria. In: Gnanamanickam SS, ed. Plant-associated bacteria, Dordrecht, Springer, 2007, 195–230. © 2006 Springer. Printed in the Netherland.

[19] Kalita M, Bharadwaz M, Dey T, Gogoi K, Dowarah P, Unni BG, Ozah D, Saikia I. Developing novel bacterial based bioformulation having PGPR properties for enhanced production of agricultural crops. Indian Journal of Experimental Biology 2015 Jan, 53(1), 56–60. PMID: 25675713.

[20] Lundberg DS, Lebeis SL, Paredes SH. Yourstone Scott. *Gehring J., Malfatti S., Tremblay J., Engelbrektson A., Kunin V., del Rio TG, Edgar RC, Eickhorst T., Ley RE, Hugenholtz P., Tringe SG, Dang JL Defining the core Arabidopsis thaliana root microbiome* Nature 2012, 488(7409), 86–90.

[21] Turner TR, James EK, Poole PS. The plant microbiome. Genome Biology 2013, 14(6), 1–10.

[22] Chaparro JM, Badri DV, Vivanco JM. Rhizosphere microbiome assemblage is affected by plant development. The ISME Journal 2014, 8(4), 790–803.

[23] Smith DL, Praslickova D, Ilangumaran G. Inter-organismal signaling and management of the phytomicrobiome. Frontiers in Plant Science 2015, 6, 722.
[24] Berg G, Rybakova D, Grube M, Köberl M. The plant microbiome explored: Implications for experimental botany. Journal of Experimental Botany 2016, 67(4), 995–1002.
[25] Gray EJ, Smith DL. Intracellular and extracellular PGPR: Commonalities and distinctions in the plant–bacterium signaling processes. Soil Biology & Biochemistry 2005, 37(3), 395–412.
[26] Schippers B, Scheffer RJ, Lugtenberg BJJ, Weisbeek PJ. Biocoating of seeds with plant growth-promoting rhizobacteria to improve plant establishment. Outlook on Agriculture 1995, 24(3), 179–185.
[27] Glick BR. Phytoremediation: Synergistic use of plants and bacteria to clean up the environment. Biotechnology Advances 2003, 21(5), 383–393.
[28] Dobbelaere S, Vanderleyden J, Okon Y. Plant growth-promoting effects of diazotrophs in the rhizosphere. Critical Reviews in Plant Sciences 2003, 22(2), 107–149.
[29] Çakmakçi R, Dönmez F, Aydın A, Şahin F. Growth promotion of plants by plant growth-promoting rhizobacteria under greenhouse and two different field soil conditions. Soil Biology & Biochemistry 2006, 38(6), 1482–1487.
[30] Saravanakumar D, Lavanya N, Muthumeena B, Raguchander T, Suresh S, Samiyappan R. Pseudomonas fluorescens enhances resistance and natural enemy population in rice plants against leaf folder pest. Journal of Applied Entomology 2008, 132(6), 469–479.
[31] Vejan P, Abdullah R, Khadiran T, Ismail S, Boyce AN. Role of plant growth promoting rhizobacteria in agricultural sustainability – A review. Molecules 2016, 21(5), 573.
[32] Martínez-Viveros O, Jorquera MA, Crowley DE, Gajardo GMLM, Mora ML. Mechanisms and practical considerations involved in plant growth promotion by rhizobacteria. Journal of Soil Science and Plant Nutrition 2010, 10(3), 293–319.
[33] Bhattacharyya PN, Jha DK. Plant growth-promoting rhizobacteria (PGPR): Emergence in agriculture. World Journal of Microbiology & Biotechnology 2012, 28(4), 1327–1350.
[34] Vessey JK. Plant growth promoting rhizobacteria as biofertilizers. Plant and Soil 2003, 255(2), 571–586.
[35] Kim J, Rees DC. Nitrogenase and biological nitrogen fixation. Biochemistry 1994, 33(2), 389–397.
[36] Raymond J, Siefert JL, Staples CR, Blankenship RE. The natural history of nitrogen fixation. Molecular Biology and Evolution 2004, 21(3), 541–554.
[37] Glick BR. Plant growth-promoting bacteria: Mechanisms and applications. Scientifica 2012, 2012. https://doi.org/10.6064/2012/963401.
[38] Khan MS, Zaidi A, Wani PA, Oves M. Role of plant growth promoting rhizobacteria in the remediation of metal contaminated soils. Environmental Chemistry Letters 2009, 7(1), 1–19.
[39] Bruto M, Prigent-Combaret C, Muller D, Moënne-Loccoz Y. Analysis of genes contributing to plant-beneficial functions in plant growth-promoting rhizobacteria and related Proteobacteria. Scientific Reports 2014, 4(1), 1–10.
[40] McKenzie RH, Roberts TL. Soil and fertilizers phosphorus update. In: Proceedings of Alberta Soil Science Workshop Feb. 20–22, Edmonton Alberta, 1990. pp. 84–104
[41] Khan MS, Zaidi A, Wani PA. Role of phosphate-solubilizing microorganisms in sustainable agriculture – A review. Agronomy for Sustainable Development 2007, 27(1), 29–43.
[42] Zaidi A, Khan M, Ahemad M, Oves M. Plant growth promotion by phosphate solubilizing bacteria. Acta Microbiologica et Immunologica Hungarica 2009, 56(3), 263–284.
[43] Kumar V, Behl RK, Narula N. Establishment of phosphate-solubilizing strains of Azotobacter chroococcum in the rhizosphere and their effect on wheat cultivars under greenhouse conditions. Microbiological Research 2001, 156(1), 87–93.
[44] Neubauer U, Nowack B, Furrer G, Schulin R. Heavy metal sorption on clay minerals affected by the siderophore desferrioxamine B. Environmental Science & Technology 2000, 34(13), 2749–2755.

[45] Patten CL, Glick BR. Bacterial biosynthesis of indole-3-acetic acid. Canadian Journal of Microbiology 1996, 42(3), 207–220.
[46] Spaepen S, Vanderleyden J. Auxin and plant-microbe interactions. Cold Spring Harbor Perspectives in Biology 2011, 3(4), a001438.
[47] Brandl MT, Lindow SE. Environmental signals modulate the expression of an indole-3-acetic acid biosynthetic gene in Erwinia herbicola. Molecular Plant-microbe Interactions 1997, 10(4), 499–505.
[48] Khalid A, Akhtar MJ, Mahmood MH, Arshad M. Effect of substrate-dependent microbial ethylene production on plant growth. Microbiology 2006, 75(2), 231–236.
[49] Saleem M, Arshad M, Hussain S, Bhatti AS. Perspective of plant growth promoting rhizobacteria (PGPR) containing ACC deaminase in stress agriculture. Journal of Industrial Microbiology & Biotechnology 2007, 34(10), 635–648.
[50] Zahir ZA, Munir A, Asghar HN, Shaharoona B, Arshad M. Effectiveness of rhizobacteria containing ACC deaminase for growth promotion of peas (Pisum sativum) under drought conditions. Journal of Microbiology and Biotechnology 2008, 18(5), 958–963.
[51] Kang BG, Kim WT, Yun HS, Chang SC. Use of plant growth-promoting rhizobacteria to control stress responses of plant roots. Plant Biotechnology Reports 2010, 4(3), 179–183.
[52] Nadeem SM, Zahir ZA, Naveed M, Arshad M. Rhizobacteria containing ACC-deaminase confer salt tolerance in maize grown on salt-affected fields. Canadian Journal of Microbiology 2009, 55(11), 1302–1309.
[53] Arshad M, Saleem M, Hussain S. Perspectives of bacterial ACC deaminase in phytoremediation. TRENDS in Biotechnology 2007, 25(8), 356–362.
[54] Demain AL, Fang A. The natural functions of secondary metabolites. History of Modern Biotechnology I. Advances in Biochemical Engineering/Biotechnology. 2000, 69, 1–39. doi:10.1007/3-540-44964-7_1. PMID: 11036689.
[55] Zitouni A, Boudjella H, Lamari L, Badji B, Mathieu F, Lebrihi A, Sabaou N. Nocardiopsis and Saccharothrix genera in Saharan soils in Algeria: Isolation, biological activities and partial characterization of antibiotics. Research in Microbiology 2005, 156(10), 984–993.
[56] Berdy J. Bioactive metabolites, a personal view. Journal of Antibiotics 2005, 58(1), 1–26.
[57] Ruiz B, Chávez A, Forero A, García-Huante Y, Romero A, Sánchez M, Rocha D, et al. Production of microbial secondary metabolites: Regulation by the carbon source. Critical Reviews in Microbiology 2010, 36(2), 146–167.
[58] Desire MH, Bernard F, Forsah MR, Assang CT, Denis ON. Enzymes and qualitative phytochemical screening of endophytic fungi isolated from Lantana camara Linn. leaves. Journal of Applied Biology and Biotechnology 1970, 2(6), 001–006.
[59] Singh JS. Plant growth promoting rhizobacteria. Resonance 2013, 18, 275–281.
[60] Ryffel F, Helfrich EJN, Kiefer P, Peyriga L, Portais J-C, Piel J, Vorholt JA. Metabolic footprint of epiphytic bacteria on Arabidopsis thaliana leaves. The ISME Journal 2016, 10(3), 632–643.
[61] Etalo DW, Jeon J-S, Raaijmakers JM. Modulation of plant chemistry by beneficial root microbiota. Natural Product Reports 2018, 35(5), 398–409.
[62] Vimal SR, Patel VK, Singh JS. Plant growth promoting Curtobacterium albidum strain SRV4: An agriculturally important microbe to alleviate salinity stress in paddy plants. Ecological Indicators 2019, 105, 553–562.
[63] Jain A, Singh A, Singh S, Singh HB. Phenols enhancement effect of microbial consortium in pea plants restrains Sclerotinia sclerotiorum. Biological Control 2015, 89, 23–32.
[64] Maruvada P, Leone V, Kaplan LM, Chang EB. The human microbiome and obesity: Moving beyond associations. Cell Host & Microbe 2017, 22(5), 589–599.

[65] Vogel C, Bodenhausen N, Gruissem W, Vorholt JA. The Arabidopsis leaf transcriptome reveals distinct but also overlapping responses to colonization by phyllosphere commensals and pathogen infection with impact on plant health. New Phytologist 2016, 212(1), 192–207.
[66] Vimal SR, Singh JS, Arora NK, Singh S. Soil-plant-microbe interactions in stressed agriculture management: A review. Pedosphere 2017, 27(2), 177–192.
[67] Sharon E, Chet I, Viterbo A, Bar-Eyal M, Nagan H, Samuels GJ, Spiegel Y. Parasitism of Trichoderma on Meloidogyne javanica and role of the gelatinous matrix. European Journal of Plant Pathology 2007, 118(3), 247–258.
[68] Gajera HP, Vakharia DN. Production of lytic enzymes by Trichoderma isolates during in vitro antagonism with Aspergillus niger, the causal agent of collar rot of peanut. Brazilian Journal of Microbiology 2012, 43, 43–52.
[69] Emmert EAB, Klimowicz AK, Thomas MG, Handelsman J. Genetics of zwittermicin a production by Bacillus cereus. Applied and Environmental Microbiology 2004, 70(1), 104–113.
[70] Ongena M, Jacques P, Touré Y, Destain J, Jabrane A, Thonart P. Involvement of fengycin-type lipopeptides in the multifaceted biocontrol potential of Bacillus subtilis. Applied Microbiology and Biotechnology 2005, 69(1), 29–38.
[71] Hammami I, Rhouma A, Jaouadi B, Rebai A, Nesme X. Optimization and biochemical characterization of a bacteriocin from a newly isolated Bacillus subtilis strain 14B for biocontrol of Agrobacterium spp. strains. Letters in Applied Microbiology 2009, 48(2), 253–260.
[72] Shen B. Polyketide biosynthesis beyond the type I, II and III polyketide synthase paradigms. Current Opinion in Chemical Biology 2003, 7, 285–295.
[73] Tortora ML, Díaz-Ricci JC, Pedraza RO. Azospirillum brasilense siderophores with antifungal activity against Colletotrichum acutatum. Archives of Microbiology 2011, 193(4), 275–286.
[74] Wang C, Ramette A, Punjasamarnwong P, Zala M, Natsch A, Moënne-Loccoz Y, Défago G. Cosmopolitan distribution of phlD-containing dicotyledonous crop-associated biocontrol pseudomonads of worldwide origin. FEMS Microbiology Ecology 2001, 37(2), 105–116.
[75] Schikora A, Schenk ST, Stein E, Molitor A, Zuccaro A, Kogel K-H. N-acyl-homoserine lactone confers resistance toward biotrophic and hemibiotrophic pathogens via altered activation of AtMPK6. Plant Physiology 2011, 157(3), 1407–1418.
[76] Jacob S, Sajjalaguddam RR, Vijay Krishna Kumar K, Varshney R, Sudini HK. Assessing the prospects of Streptomyces sp. RP1A-12 in managing groundnut stem rot disease caused by Sclerotium rolfsii Sacc. Journal of General Plant Pathology 2016, 82(2), 96–104.
[77] Ghisalberti EL, Rowland CY. Antifungal metabolites from Trichoderma harzianum. Journal of Natural Products 1993, 56(10), 1799–1804.
[78] Ortíz-Castro R, Contreras-Cornejo HA, Macías-Rodríguez L, López-Bucio J. The role of microbial signals in plant growth and development. Plant Signaling & Behavior 2009, 4(8), 701–712.
[79] Ryu C-M, Farag MA, Hu C-H, Reddy MS, Kloepper JW, Paré PW. Bacterial volatiles induce systemic resistance in Arabidopsis. Plant Physiology 2004, 134(3), 1017–1026.
[80] Contreras-Cornejo HA, Macías-Rodríguez L, Cortés-Penagos C, López-Bucio J. Trichoderma virens, a plant beneficial fungus, enhances biomass production and promotes lateral root growth through an auxin-dependent mechanism in Arabidopsis. Plant Physiology 2009, 149(3), 1579–1592.
[81] Saikia R, Kumar R, Arora DK, Gogoi DK, Azad P. Pseudomonas aeruginosa inducing rice resistance against Rhizoctonia solani: Production of salicylic acid and peroxidases. Folia Microbiologica 2006, 51(5), 375–380.
[82] Saikia R, Varghese S, Singh BP, Arora DK. Influence of mineral amendment on disease suppressive activity of Pseudomonas fluorescens to Fusarium wilt of chickpea. Microbiological Research 2009, 164(4), 365–373.

[83] Andrés JA, Rovera M, Guinazú LB, Pastor NA, Rosas SB. Role of Pseudomonas aurantiaca in crop improvement. In: Maheshwari DK, ed. Bacteria in agrobiology: Plant growth responses, Germany, Springer, Berlin-Heidelberg, 2011, 107–122.
[84] Forghani F, Hajihassani A. Recent advances in the development of environmentally benign treatments to control root-knot nematodes. Frontiers in Plant Science 2020, 11, 1125. doi:10.3389/fpls.2020.01125.
[85] Martínez-Medina A, Fernandez I, Lok GB, Pozo MJ, Pieterse CMJ, Saskia CM Van Wees, Wees V. Shifting from priming of salicylic acid-to jasmonic acid-regulated defences by Trichoderma protects tomato against the root knot nematode Meloidogyne incognita. New Phytologist 2017, 213(3), 1363–1377.
[86] Schmidt R, Cordovez V, De Boer W, Raaijmakers J, Garbeva P. Volatile affairs in microbial interactions. The ISME Journal 2015, 9(11), 2329–2335.
[87] Loper JE, Hassan KA, Mavrodi DV, Davis EW, Lim CK, Shaffer BT, Elbourne LDH, et al. Comparative genomics of plant-associated Pseudomonas spp.: Insights into diversity and inheritance of traits involved in multitrophic interactions. PLoS Genetics 2012, 8(7), e1002784.
[88] Troppens DM, Dmitriev RI, Papkovsky DB, O'Gara F, Morrissey JP. Genome-wide investigation of cellular targets and mode of action of the antifungal bacterial metabolite 2, 4-diacetylphloroglucinol in Saccharomyces cerevisiae. FEMS Yeast Research 2013, 13(3), 322–334.
[89] Sonnleitner E, Haas D. Small RNAs as regulators of primary and secondary metabolism in Pseudomonas species. Applied Microbiology and Biotechnology 2011, 91(1), 63–79.
[90] Chu BC, Garcia-Herrero A, Johanson TH, Krewulak KD, Lau CK, Sean Peacock R, Slavinskaya Z, Vogel HJ. Siderophore uptake in bacteria and the battle for iron with the host; a bird's eye view. Biometals 2010, 23(4), 601–611.
[91] Hider RC, Kong X. Chemistry and biology of siderophores. Natural Product Reports 2010, 27(5), 637–657.
[92] Raaijmakers JM, De Bruijn I, de Kock MJD. Cyclic lipopeptide production by plant-associated Pseudomonas spp.: Diversity, activity, biosynthesis, and regulation. Molecular Plant-Microbe Interactions 2006, 19(7), 699–710.
[93] Parret AHA, Schoofs G, Proost P, De Mot R. Plant lectin-like bacteriocin from a rhizosphere-colonizing Pseudomonas isolate. Journal of Bacteriology 2003, 185(3), 897–908.
[94] Dickschat JS, Wickel S, Bolten CJ, Nawrath T, Schulz S, Wittmann C. Pyrazine biosynthesis in Corynebacterium glutamicum. European Journal of Organic Chemistry 2010, 2687–2695. https://doi.org/10.1002/ejoc.201000155.
[95] Groenhagen U, Maczka M, Dickschat JS, Schulz S. Streptopyridines, volatile pyridine alkaloids produced by Streptomyces sp. FORM5. Beilstein Journal of Organic Chemistry 2014, 10(1), 1421–1432.
[96] Lee J-H, Lee J. Indole as an intercellular signal in microbial communities. FEMS Microbiology Reviews 2010, 34(4), 426–444.
[97] Pandey R, Swamy KV, Khetmalas MB. Indole: a novel signaling molecule and its applications. Indian Journal of Biotechnology 2013, 12(3), 297–310.
[98] Vimal SR, Gupta J, Singh JS. Effect of salt tolerant Bacillus sp. and Pseudomonas sp. on wheat (Triticum aestivum L.) growth under soil salinity: A comparative study. Microbiology Research 2018, 9(1), 26–32.
[99] Dickschat JS, Reichenbach H, Wagner-Döbler I, Schulz S. Novel pyrazines from the myxobacterium Chondromyces crocatus and marine bacteria. European Journal of Organic Chemistry 2005, 4141–4153. https://doi.org/10.1002/ejoc.200500280.
[100] Rajini KS, Aparna P, Sasikala C, Ramana CV. Microbial metabolism of pyrazines. Critical Reviews in Microbiology 2011, 37(2), 99–112.

[101] Garbeva P, Hordijk C, Gerards S, Boer WD. Volatiles produced by the mycophagous soil bacterium Collimonas. FEMS Microbiology Ecology 2014, 87(3), 639–649.
[102] Tyc O, Zweers H, de Boer W, Garbeva P. Volatiles in inter-specific bacterial interactions. Frontiers in Microbiology 2015, 6, 1412.
[103] Moody SC, Zhao B, Lei L, Nelson DR, Mullins JGL, Waterman MR, Kelly SL, Lamb DC. Investigating conservation of the albaflavenone biosynthetic pathway and CYP170 bifunctionality in streptomycetes. The FEBS Journal 2012, 279(9), 1640–1649.
[104] Song C, Schmidt R, de Jager V, Krzyzanowska D, Jongedijk E, Cankar K, Beekwilder J, et al. Exploring the genomic traits of fungus-feeding bacterial genus Collimonas. BMC Genomics 2015, 16(1), 1–17.
[105] Davies PJ ed. Plant hormones: Physiology, biochemistry and molecular biology. Dordrecht, The Netherlands, Springer Science & Business Media, 2013.
[106] Park J, Lee Y, Martinoia E, Geisler M. Plant hormone transporters: What we know and what we would like to know. Bmc Biology 2017, 15(1), 1–15.
[107] Vimal SR, Singh JS. Salt tolerant PGPR and FYM application in saline soil paddy agriculture sustainability. Climate Change and Environmental Sustainability 2019, 7(1), 61–71.
[108] Kramer EM, Bennett MJ. Auxin transport: A field in flux. Trends in Plant Science 2006, 11(8), 382–386.
[109] Piotrowska A, Czerpak R. Cellular response of light/dark-grown green alga Chlorella vulgaris Beijerinck (Chlorophyceae) to exogenous adenine-and phenylurea-type cytokinins. Acta Physiologiae Plantarum 2009, 31(3), 573–585.
[110] Keller NP, Turner G, Bennett JW. Fungal secondary metabolism – From biochemistry to genomics. Nature Reviews. Microbiology 2005, 3(12), 937–947.
[111] Harman GE, Howell CR, Viterbo A, Chet I, Lorito M. Trichoderma species – Opportunistic, avirulent plant symbionts. Nature Reviews: Microbiology 2004, 2(1), 43–56.
[112] Haris M, Shakeel A, Ansari MA, Hussain T, Khan AA, Dhankar R. Sustainable crop production and improvement through bio-prospecting of fungi. In: Sharma VK, Shah MP, Parmar S, Kumar A. Fungi Bio-Prospects in Sustainable Agriculture, Environment and Nano-Technology, Vol-1, AP U.K., Elsevier, 2020, 4-7-428. https://doi.org/10.1016/B978-0-12-821394-0.00016-0.
[113] Reino JL, Guerrero RF, Hernandez-Galan R, Collado IG. Secondary metabolites from species of the biocontrol agent Trichoderma. Phytochemistry Reviews 2008, 7(1), 89–123.
[114] Vinale F, Sivasithamparam K, Ghisalberti EL, Marra R, Woo SL, Lorito M. Trichoderma–plant–pathogen interactions. Soil Biology & Biochemistry 2008, 40(1), 1–10.
[115] Almassi F, Ghisalberti EL, Narbey MJ, Sivasithamparam K. New antibiotics from strains of Trichoderma harzianum. Journal of Natural Products 1991, 54(2), 396–402.
[116] Vinale F, Flematti G, Sivasithamparam K, Lorito M, Marra R, Skelton BW, Ghisalberti EL. Harzianic acid, an antifungal and plant growth promoting metabolite from Trichoderma harzianum. Journal of Natural Products 2009, 72(11), 2032–2035.
[117] Woo SL, Scala F, Ruocco M, Lorito M. The molecular biology of the interactions between Trichoderma spp., phytopathogenic fungi, and plants. Phytopathology 2006, 96(2), 181–185.
[118] El-Hasan A, Buchenauer H. Actions of 6-pentyl-alpha-pyrone in controlling seedling blight incited by Fusarium moniliforme and inducing defence responses in maize. Journal of Phytopathology 2009, 157(11–12), 697–707.
[119] Lee Y-J, Lee S-J, Kim SH, Lee SJ, Kim B-C, Lee H-S, Jeong H, Lee D-W. Draft genome sequence of Bacillus endophyticus Journal of Bacteriology 2012 Oct, 194(20), 5705–5706. doi:10.1128/JB.01316-12. PMID: 23012284; PMCID: PMC3458666.
[120] Mukherjee S, Pandey V, Parvez A, Qi X, Hussain T. *Bacillus* as a Versatile Tool for Crop Improvement and Agro-Industry. In: Islam MT, Rahman M, Pandey P, eds. Bacilli in

Agrobiotechnology. Bacilli in Climate Resilient Agriculture and Bioprospecting. Cham, Springer. 2022, 429-452. https://doi.org/10.1007/978-3-030-85465-2_19.
[121] Yu GY, Sinclair JB, Hartman GL, Bertagnolli BL. Production of iturin A by Bacillus amyloliquefaciens suppressing Rhizoctonia solani. Soil Biology & Biochemistry 2002, 34(7), 955–963.
[122] Dickschat JS. Isoprenoids in three-dimensional space: The stereochemistry of terpene biosynthesis. Natural Product Reports 2011, 28(12), 1917–1936.
[123] Gurtler H, Pedersen R, Anthoni U, Chritophersen C, Nielsen PH, Wellington EMH, Pedersen C, Bock K. Albaflavenone, a sesquiterpene ketone with a zizaene skeleton produced by a streptomycete with a new rope morphology. The Journal of Antibiotics 1994, 47(4), 434–439.
[124] Rajini KS, Aparna P, Sasikala C, Ramana Ch V. Microbial metabolism of pyrazines. Critical Reviews in Microbiology 2011, 37(2), 99–112.
[125] Braña AF, Rodríguez M, Pahari P, Rohr J, García LA, Blanco G. Activation and silencing of secondary metabolites in Streptomyces albus and Streptomyces lividans after transformation with cosmids containing the thienamycin gene cluster from Streptomyces cattleya. Archives of Microbiology 2014, 196(5), 345–355.
[126] Lee J-H, Wood TK, Lee J. Roles of indole as an interspecies and interkingdom signaling molecule. Trends in Microbiology 2015, 23(11), 707–718.
[127] Erb M, Veyrat N, Robert CAM, Hao X, Frey M, Ton J, Turlings TCJ. Indole is an essential herbivore-induced volatile priming signal in maize. Nature Communications 2015, 6(1), 1–10.
[128] Cotter PD, Hill C, Paul Ross R. Bacteriocins: Developing innate immunity for food. Nature Reviews. Microbiology 2005, 3(10), 777–788.
[129] Raaijmakers JM, Mazzola M. Diversity and natural functions of antibiotics produced by beneficial and plant pathogenic bacteria. Annual Review of Phytopathology 2012, 50, 403–424.
[130] Hussain T, Khan AA. *Bacillus subtilisHussainT-AMU*and its antifungal activity against Potato black scurf caused by *Rhizoctonia solani*. Biocatalysis and Agricultural Biotechnology 2020, 23, 101433.
[131] Nega A. Review on concepts in biological control of plant pathogens. Journal of Biology, Agriculture and Healthcare 2014, 4(27), 33–54.
[132] Pal KK, Gardener BM. Biological control of plant pathogens. The Plant Health Instructor. 2006. doi:10.1094/PHI-A-2006-1117-02.
[133] Suprapta DN. Potential of microbial antagonists as biocontrol agents against plant fungal pathogens. Journal ISSAAS 2012, 18(2), 1–8.
[134] Ahmad G, Nishat Y, Ansari MS, Khan A, Haris M, Khan AA. Eco-Friendly Approaches for the Alleviation of Root-Knot Nematodes. In: Mohamed HI, El-Beltagi HEDS, Abd-Elsalam KA, eds. Plant Growth-Promoting Microbes for Sustainable Biotic and Abiotic Stress Management. Cham, Springer. 2021. https://doi.org/10.1007/978-3-030-66587-6_20
[135] Hussain T, Khan A, Khan M. Biocontrol of Soil Borne Pathogen of Potato Tuber Caused by *Rhizoctonia solani* through Biosurfactant based *Bacillus* strain. *Journal of Nepal Agricultural Research Council*, 2021, 7(1),54-66. https://doi.org/10.3126/jnarc.v7i1.36921
[136] Hussain T, Khan AA. Determining the Antifungal activity and Characterization of *Bacillus siamensis AMU03* against *Macrophomina phaseolina* (Tassi) Goid. *Indian Phytopathology* 2020. doi.org/10.1007/s42360-020-00239-6.
[137] Hussain T, Khan AA. Biocontrol prospective of *Bacillus siamensis-AMU03* against Soil-borne fungal pathogens of potato tubers. *Indian Phytopathology* 2022a. doi.org/10.1007/s42360-021-00447-8
[138] Hussain T, Khan AA, Ibrahim H. Potential efficacy of biofilm-forming biosurfactant *Bacillus firmus HussainT-Lab.66* against *Rhizoctonia solani* and mass spectrometry analysis of their metabolites. *International Journal of Peptide Research and Therapeutics*, Springer 2022b, 28, 3.

[139] Ahmad A-GM, Abo-Zaid GA, Mohamed MS, Elsayed HE. Fermentation, formulation and evaluation of PGPR Bacillus subtilis isolate as a bioagent for reducing occurrence of peanut soil-borne diseases. Journal of Integrative Agriculture 2019, 18(9), 2080–2092.
[140] Ahmad G, Khan A, Khan AA, Ali A, Mohhamad HI. Biological control: A novel strategy for the control of the plant parasitic nematodes. Antonie van Leeuwenhoek 2021 Jul, 114(7), 885–912. doi:10.1007/s10482-021-01577-9. Epub 2021 Apr 24. PMID: 33893903.
[141] Nasima AI, Siddiqui IA, Shaukat SS, Zaki MJ. Nematicidal activity of some strains of Pseudomonas spp. Soil Biology & Biochemistry 2002, 34(8), 1051–1058.
[142] Lee YS, Kim KY. Antagonistic potential of Bacillus pumilus L1 against root-Knot nematode, Meloidogyne arenaria. Journal of Phytopathology 2016, 164(1), 29–39.
[143] Hussain T, Haris M, Shakeel A, Ahmad G, Khan AA, Khan MA. Bio-nematicidal activities by culture filtrate of *Bacillus subtilisHussainT-AMU*: new promising biosurfactantbioagent for the management of Root Galling caused by *Meloidogyne incognita*. *Vegetos* 2020, 33, 229-223.
[144] Hussain T, Khan AA. *Bacillus firmus HussainT:Lab.66*, a new biosurfactantbioagent having potential bio-nematicidal activity against root knot nematode *Meloidogyneincognita*. National Conference on Recent Advances in Biological Science (NCRABS-2020) held on 05th March, 2020, organized by Dept. of Biosciences, Faculty of Natural Sciences, JamiaMillialslamia University, New Delhi, OP-25, 2020, 42.
[145] Kavitha PG, Jonathan EL, Nakkeeran S. Effects of crude antibiotic of Bacillus subtilis on hatching of eggs and mortality of juveniles of Meloidogyne incognita. Nematologia Mediterranea 2012, 40, 203–206.
[146] Zhang H, Mahunu GK, Castoria R, Apaliya MT, Yang Q. Augmentation of biocontrol agents with physical methods against postharvest diseases of fruits and vegetables. Trends in Food Science & Technology 2017, 69, 36–45.
[147] Ghazanfar MU, Hussain M, Hamid MI, Ansari SU. Utilization of biological control agents for the management of postharvest pathogens of tomato. Pakistan Journal of Botany 2016, 48(5), 2093–2100.
[148] Jisha MS, Linu MS, Sreekumar J. Induction of systemic resistance in chilli (Capsicum annuum L.) by Pseudomonas aeruginosa against anthracnose pathogen Colletotrichum capsici. Journal of Tropical Agriculture 2019, 56, 2.
[149] Alam SS, Sakamoto K, Amemiya Y, Inubushi K. Biocontrol of soil-borne Fusarium wilts of tomato and cabbage with a root colonizing fungus, Penicillium sp. EU0013. In *19th World Conference. Soil Science Proceedings*, pp. 20–22, 2010.
[150] Agarwal T, Malhotra A, Trivedi PC. Isolation of seed borne mycoflora of chickpea and its in vitro evaluation by some known bioagents. International Journal of Pharmacy and Life Sciences 2011, 2(7), 899–902.
[151] Adebola MO, Amadi JE. Screening three Aspergillus species for antagonistic activities against the cocoa black pod organism (Phytophthora palmivora). Agriculture and Biology Journal of North America 2010, 1(3), 362–365.
[152] Nally MC, Pesce VM, Maturano YP, Muñoz CJ, Combina M, Toro ME, De Figueroa LIC, Vazquez F. Biocontrol of Botrytis cinerea in table grapes by non-pathogenic indigenous Saccharomyces cerevisiae yeasts isolated from viticultural environments in Argentina. Postharvest Biology and Technology 2012, 64(1), 40–48.
[153] Vos C, Claerhout S, Mkandawire R, Panis B, De Waele D, Elsen A. Arbuscular mycorrhizal fungi reduce root-knot nematode penetration through altered root exudation of their host. Plant and Soil 2012, 354(1), 335–345.
[154] Pendse MA, Karwande PP, Limaye MN. Past, present and future of nematophagous fungi as bioagent to control plant parasitic nematodes. The Journal of Plant Protection Sciences 2013, 5(1), 1–9.

[155] Atkins SD, Clark IM, Pande S, Hirsch PR, Kerry BR. The use of real-time PCR and species-specific primers for the identification and monitoring of Paecilomyces lilacinus. FEMS Microbiology Ecology 2005, 51(2), 257–264.
[156] Kalele DN, Affokpon A, Coosemans J, Kimenju JW. Suppression of root-knot nematodes in tomato and cucumber using biological control agents. African Journal of Horticultural Science 2010, 3, 72-80.
[157] Anastasiadis IA, Giannakou IO, Prophetou-Athanasiadou DA, Gowen SR. The combined effect of the application of a biocontrol agent Paecilomyces lilacinus, with various practices for the control of root-knot nematodes. Crop Protection 2008, 27(3-5), 352–361.
[158] Wallace RL, Hirkala DL, Nelson LM. Efficacy of Pseudomonas fluorescens for control of Mucor rot of apple during commercial storage and potential modes of action. Canadian Journal of Microbiology 2018, 64(6), 420–431.
[159] Ashajyothi M, Kumar A, Sheoran N, Ganesan P, Gogoi R, Subbaiyan GK, Bhattacharya R. Black pepper (Piper nigrum L.) associated endophytic Pseudomonas putida BP25 alters root phenotype and induces defense in rice (Oryza sativa L.) against blast disease incited by Magnaporthe oryzae. Biological Control 2020, 143, 104181.
[160] Al-Ani RA, Adhab MA, Mahdi MH, Abood HM. Rhizobium japonicum as a biocontrol agent of soybean root rot disease caused by Fusarium solani and Macrophomina phaseolina. Plant Protection Science 2012, 48(4), 149–155.
[161] Wang M, Geng L, Sun X, Shu C, Song F, Zhang J. Screening of Bacillus thuringiensis strains to identify new potential biocontrol agents against Sclerotinia sclerotiorum and Plutella xylostella in Brassica campestris L. Biological Control 2020, 145, 104262.
[162] Hanawi MJ. Tagetes erecta with native isolates of Paecilomyces lilacinus and Trichoderma hamatum in controlling root-knot nematode Meloidogyne javanica on tomato. International Journal of Application or Innovation in Engineering and Management 2016, 5(1), 81–88.
[163] Sabuquillo P, De Cal A, Melgarejo P. Biocontrol of tomato wilt by Penicillium oxalicum formulations in different crop conditions. Biological Control 2006, 37(3), 256–265.
[164] Bontempo AF, Lopes EA, Fernandes RH, De Freitas Leandrograssi, Dallemole-Giaretta R. Dose-response effect of Pochonia chlamydosporia AGAINST Meloidogyne incognita on carrot under field conditions. Revista Caatinga 2017, 30, 258–262.
[165] Santos-Villalobos DL, Sergio, Guzmán-Ortiz DA, Gómez-Lim MA, Délano-Frier JP, De-folter S, Sánchez-García P, Peña-Cabriales JJ. Potential use of Trichoderma asperellum (Samuels, Liechfeldt et Nirenberg) T8a as a biological control agent against anthracnose in mango (Mangifera indica L.). Biological Control 2013, 64(1), 37–44.
[166] Gveroska B, Ziberoski J. The influence of Trichoderma harzianum on reducing root rot disease in tobacco seedlings caused by Rhizoctonia solani. International Journal of Pure And Applied Sciences Technology 2011, 2(2), 1–11.
[167] Khalili E, Sadravi M, Naeimi S, Khosravi V. Biological control of rice brown spot with native isolates of three Trichoderma species. Brazilian Journal of Microbiology 2012, 43, 297–305.
[168] Cheng X, Ji X, Ge Y, Li J, Qi W, Qiao K. Characterization of antagonistic Bacillus methylotrophicus isolated from rhizosphere and its biocontrol effects on maize stalk rot. Phytopathology 2019, 109(4), 571–581.
[169] Haggag WM. Isolation of bioactive antibiotic peptides from Bacillus brevis and Bacillus polymyxa against Botrytis grey mould in strawberry. Archives of Phytopathology and Plant Protection 2008, 41(7), 477–491.
[170] Khan A, Tariq M, Asif M, Khan F, Ansari T, Siddiqui MA. Research article integrated management of Meloidogyne incognita infecting Vigna radiata L. using biocontrol agent Purpureocillium lilacinum. Trends Appl. Scientific Research 2019, 14, 119–124.

[171] Tariq M, Khan A, Asif M, Siddiqui MA. Interactive effect of Trichoderma virens and Meloidogyne incognita and their influence on plant growth character and nematode multiplication on Abelmoschus esculentus (L.) Moench. Current Nematology 2018, 29, 1–9.
[172] Lajayer BA, Najafi N, Moghiseh E, Mosaferi M, Hadian J. Removal of heavy metals (Cu 2+ and Cd 2+) from effluent using gamma irradiation, titanium dioxide nanoparticles and methanol. Journal of Nanostructure in Chemistry 2018, 8(4), 483–496.
[173] Kumar A, Hussain T, Susmita C, Maurya DK, Danish M, Farooqui SA. Microbial remediation and detoxification of heavy metals by plants and microbes. In: Shah M, et al., eds. The future of effluent treatment plants-Biological Treatment Systems. UK, Elsevier, 2021, 589–614. https://doi.org/10.1016/B978-0-12-822956-9.00030-1.
[174] Zheljazkov VD, Craker LE, Xing B. Effects of Cd, Pb, and Cu on growth and essential oil contents in dill, peppermint, and basil. Environmental and Experimental Botany 2006, 58(1–3), 9–16.
[175] Järup L. Hazards of heavy metal contamination. British Medical Bulletin 2003, 68(1), 167–182.
[176] Yruela I. Copper in plants. Brazilian Journal of Plant Physiology 2005, 17, 145–156.
[177] Wuana RA, Okieimen FE. Heavy metals in contaminated soils: A review of sources, chemistry, risks and best available strategies for remediation. International Scholarly Research Notices 2011, 2011, Article ID 402647. https://doi.org/10.5402/2011/402647.
[178] Gulati K, Banerjee B, Lall SB, Ray A. Effects of diesel exhaust, heavy metals and pesticides on various organ systems: Possible mechanisms and strategies for prevention and treatment. Indian Journal of Experimental Biology 2010 Jul, 48(7), 710–721. PMID: 20929054.
[179] Messer RLW, Lockwood PE, Tseng WY, Edwards K, Shaw M, Caughman GB, Lewis JB, Wataha JC. Mercury (II) alters mitochondrial activity of monocytes at sublethal doses via oxidative stress mechanisms. Journal of Biomedical Materials Research Part B: Applied Biomaterials: An Official Journal of the Society for Biomaterials, the Japanese Society for Biomaterials, and the Australian Society for Biomaterials and the Korean Society for Biomaterials 2005, 75(2), 257–263.
[180] Abdul G. Effect of lead toxicity on growth, chlorophyll and lead (Pb+) contents of two varieties of maize (Zea mays L.). Pakistan Journal of Nutrition 2010, 9(9), 887–891.
[181] Chen C, Huang D, Liu J. Functions and toxicity of nickel in plants: Recent advances and future prospects. Clean–soil, Air, Water 2009, 37(4-5), 304–313.
[182] Salem HM, El-Fouly AA. Minerals reconnaissance at Saint Catherine area, Southern Central Sinai, Egypt and their environmental impacts on human health. In: *Proceedings of the International Conference for Environmental Hazard Mitigation (ICEHM 2000)*, Cairo University, Egypt, pp. 586–598, 2000.
[183] Shanker AK, Cervantes C, Loza-Tavera H, Avudainayagam S. Chromium toxicity in plants. Environment International 2005, 31(5), 739–753.
[184] Nagajyoti PC, Lee KD, Sreekanth TVM. Heavy metals, occurrence and toxicity for plants: A review. Environmental Chemistry Letters 2010, 8(3), 199–216.
[185] Tripathy RD, Srivastava S, Mishra S, Singh N, Tuli R, Gupta DK, Maathuis FJ. Arsenic hazard, strategies for tolerance and remediation by plant. Trends in Biotechnology 2007, 25, 158–165.
[186] Jacobs IA, Taddeo J, Kelly K, Valenziano C. Poisoning as a result of barium styphnate explosion. American Journal of Industrial Medicine 2002, 41(4), 285–288.
[187] Haris M, Shakeel A, Hussain T, Ahmad G, Khan AA. New Trends in Removing Heavy Metals from Industrial Wastewater Through Microbes. In: Shah MP, eds. Removal of Emerging Contaminants Through Microbial Processes. Singapore, Springer. 2021. https://doi.org/10.1007/978-981-15-5901-3_9.
[188] Kumar A, Hussain T, Susmita C, Maurya DK, Danish M, Farooqui, SA. Microbial remediation and detoxification of heavy metals by plants and microbes. In: Shah M, et al., eds. The Future

of Effluent Treatment Plants-Biological Treatment Systems. U.K, Elsevier, 2021, 589-614. doi.org/10.1016/B978-0-12-822956-9.00030-1
[189] Pal R, Rai JPN. Phytochelatins: Peptides involved in heavy metal detoxification. Applied Biochemistry and Biotechnology 2010, 160(3), 945–963.
[190] Gautam N, Verma PK, Verma S, Tripathi RD, Trivedi PK, Adhikari B, Chakrabarty D. Genome-wide identification of rice class I metallothionein gene: Tissue expression patterns and induction in response to heavy metal stress. Functional & Integrative Genomics 2012, 12(4), 635–647.
[191] Hussain T, Dhanker R. Science of Microorganisms for the Restoration of Polluted sites for Safe and Healthy Environment. In: Shah M, Susana Rodriguez-Couto, eds. Microbial Ecology of Wastewater Treatment Plants. U.K, Elsevier, 2021, 127-144. DOI: 10.1016/C2019-0-04695-X
[192] Wang L, Ji B, Hu Y, Liu R, Sun W. A review on in situ phytoremediation of mine tailings. Chemosphere 2017, 184, 594–600.
[193] Verma S, Kuila A. Bioremediation of heavy metals by microbial process. Environmental Technology & Innovation 2019, 14, 100369.
[194] Cameselle C, Gouveia S. Phytoremediation of mixed contaminated soil enhanced with electric current. Journal of Hazardous Materials 2019, 361, 95–102.
[195] Duan G, Kamiya T, Ishikawa S, Arao T, Fujiwara T. Expressing ScACR3 in rice enhanced arsenite efflux and reduced arsenic accumulation in rice grains. Plant & Cell Physiology 2012, 53(1), 154–163.
[196] Liu S, Yang B, Liang Y, Xiao Y, Fang J. Prospect of phytoremediation combined with other approaches for remediation of heavy metal-polluted soils. Environmental Science and Pollution Research 2020, 27(14), 16069–16085.
[197] Singh P, Hussain T, Patel S, Akhtar N. Impact of Climate Change on Root–Pathogen Interactions. In: Giri B, Prasad R, Varma A, eds. Root Biology. Soil Biology, Vol 52. Cham, Springer, 2018. doi.org/10.1007/978-3-319-75910-4_16.
[198] Pingali PL. Green revolution: Impacts, limits, andthe path ahead. Proceedings of the National Academy of Sciences of the United States of America 2012, 109(31), 12302–12308.
[199] Babin D, Deubel A, Jacquiod S, Sørensen SJ, Geistlinger J, Grosch R, Smalla K. Impact of long-term agricultural management practices on soil prokaryotic communities. Soil Biology & Biochemistry 2019, 129, 17–28.
[200] Brown SP, Grillo MA, Podowski JC, Heath KD. Correction to: Soil origin and plant genotype structure distinct microbiome compartments in the model legume Medicago truncatula. Microbiome 2021, 9(1), 1–1.
[201] Song C, Zhu F, Carrión VJ, Cordovez V. Beyond plant microbiome composition: Exploiting microbial functions and plant traits via integrated approaches. Frontiers in Bioengineering and Biotechnology 2020, 8, 896.
[202] Ravanbakhsh M, Kowalchuk GA, Jousset A. Root-associated microorganisms reprogram plant life history along the growth–stress resistance tradeoff. The ISME Journal 2019, 13(12), 3093–3101.
[203] Tkacz A, Poole P. Role of root microbiota in plant productivity. Journal of Experimental Botany 2015, 66(8), 2167–2175.
[204] Turner TR, Ramakrishnan K, Walshaw J, Heavens D, Alston M, Swarbreck D, Osbourn A, Grant A, Poole PS. Comparative metatranscriptomics reveals kingdom level changes in the rhizosphere microbiome of plants. The ISME Journal 2013, 7(12), 2248–2258.
[205] Rout ME. The plant microbiome. Advances in Botanical Research 2014, 69, 279–309.
[206] Pandey A, Tripathi A, Srivastava P, Choudhary KK, Dikshit A. Plant growth-promoting microorganisms in sustainable agriculture. In: Kumar A, Singh AK, Choudhary KK. Role of plant growth promoting microorganisms in sustainable agriculture and nanotechnology, Woodhead Publishing, Sawston, United Kingdom, 2019, 1–19.

Hemant Sharma, Binu Gogoi, Arun Kumar Rai

Chapter 8 Microbial bioproducts for plant growth and protection: trends and prospective

Abstract: Agriculture sector has been under tremendous pressure for merely producing enough food products to fulfill the considerable demand of progressively expanding population in this planet. Although chemical inputs in productive agriculture is imperative for proper plant growth and good yield, negative impacts of such compounds have undoubtedly increased the apparent importance of environmentally friendly pesticides and fertilizers of plants or of microbial origin. Microorganisms efficiently perform a significant role in improving plant health, increasing plant yield, effective promotion of plant growth under stressful and adverse conditions, mitigating plant diseases, and inducing plant defenses. Impressive array of specific functionalities from the beneficial microorganism could significantly increase the necessary sustenance of plants under hostile and adverse conditions as well as increase crop yields with least adverse effect to human health and the environment. Extensive exploration of beneficial microorganisms from poorly explored habitats is still in its nascent stage and the active search for better microbial isolates from exotic habitats with prospective source of bioproducts could divulge some novel and promising isolates with unique functionalities suitable for use in sustainable agriculture practices.

8.1 Importance of microorganisms in agriculture

Microorganisms naturally carry out an intrinsic role in the conducive environment. The beneficial microorganisms efficiently deliver vital applications in almost all domains right from key industries to pharmaceutical companies, food industries to bioremediation. In productive agriculture, beneficial microbes are known for promoting growth factors like nitrogen fixation, increased nutrient uptake by plants, control of phytopathogens and plant diseases, and increase soil fertility. Disease resistance capacity of plants is undoubtedly increased by specific microorganisms. Microorganisms can be invariably found in a remarkable variety of complex environments, including soil, water, air, volcanoes, and hot springs. Every capable microorganism is typically identified to properly discharge unique functions. Microorganisms typically use a comprehensive range of organic substances and have the ability to efficiently convert a broad range of valuable products via various metabolic processes. Primary metabolites, secondary metabolites, active enzymes, and microbial biomass are all known products of various capable microorganisms. The most substantial

https://doi.org/10.1515/9783110771558-008

need is to adequately meet the growing population's food demands. Plant pathogens must be properly managed in successful cultivation and crop improvement techniques along with an increase in successful production without relying on chemical fertilizers so that crops produce an excellent yield, have high protein content, and are resistant to wide factors such as prevalent diseases, pests, or droughts. This possible type of successful production can be realistically accomplished with the considerable help of several beneficial microorganisms known scientifically for their various agricultural functions. The extensive use of beneficial microbes in sustainable agriculture accurately represents an environmentally friendly approach that will powerfully aid in the effective conservation of ecological balance. Nitrogen fixation, crop disease control, bioproducts, and nutrient uptake by cultivated plants in common are some of the specific domains where microorganisms are popularly known to typically perform significant roles. *Rhizobia*, *Mycorrhizae*, *Azospirillum*, *Bacillus*, *Pseudomonas*, *Trichoderma*, and abundant *Streptomyces* species are some of the recognized genera of capable microorganisms that are considered extremely valuable in modern agriculture [1].

8.1.1 Plant growth promoters

Plant growth-promoting rhizobacteria (PGPR) thrive on the plant roots and naturally induce plant growth by supplying essential nutrients like nitrogen, phosphorus, and potassium. Successful resistance of the cultivated plants to abiotic stresses such as drought has been reasonably found to be appreciably reduced along with successful promotion of plant growth with the considerable help of specific microorganisms typically belonging to the recognized genera of *Azotobacter*, *Flavobacterium*, *Bacillus*, *Burkholderia*, *Methylobacterium*, *Pseudomonas*, *Serratia*, and so on [2]. PGPR also decreases the considerable effects of plant pathogens that hamper the overall growth and development of the cultivated plant. The PGPR benefit in sustainable agriculture is well depicted by the symbiotic association of *Rhizobium* sp. and legume plants in nitrogen fixation. Many PGPR are now commercially available, where the successful reports sufficiently indicate toward crop improvement after the extensive application of PGPR.

8.1.2 Biological nitrogen fixation

The complex process of typically fixing atmospheric nitrogen that invariably involves microorganisms has been considered under biological nitrogen fixation. Nitrogen is one of the most essential nutrients universally required for the continuous growth of plants. As the plants are unable to utilize the atmospheric nitrogen directly, some of the microorganisms present in the soil are able to fix atmospheric nitrogen and make it available to the plants in their assessable form. Microorganisms such as bacteria and cyanobacteria are the ones that have been known to fix atmospheric nitrogen.

Some of the bacterial genera such as *Azotobacter*, *Bacillus*, *Clostridium*, and *Klebsiella* have been reliably reported to fix atmospheric nitrogen. Apart from bacterial species, nitrogen fixation is also carried out by cyanobacterial species such as *Anabaena*, *Cylindrospermum*, *Gloeocapsa*, *Nostoc*, *Oscillatoria*, *Plectonema*, and *Trichodesmium* [3].

8.1.3 Phosphate solubilization

Phosphorus is an element that constitutes an essential part of nutrients required for plant growth. To fulfill the requirement of phosphorus by plants, chemical fertilizers containing phosphates are commonly applied to the fields, which increase the productivity of plants but have adverse effect on the surrounding environment. Phosphate-solubilizing microorganisms (PSM) have been reported to solubilize insoluble phosphorus present in the rocks through the secretion of organic acids. Microbial species such as *Bacillus*, *Streptomyces*, *Pseudomonas*, *Rhizobium*, *Penicillium*, *Aspergillus*, *Trichoderma*, actinomycetes, arbuscular mycorrhiza (AM), and cyanobacteria solubilize phosphorus, and naturally make them easier for the plants to efficiently absorb. These specific groups of beneficial microorganisms are being efficiently utilized in the sustainable agriculture practices to amply supply the required amount of phosphorus to the cultivated plants without adverse effect on the ecological habitat [4].

8.1.4 Plant waste decomposition

Microorganisms decompose the raw organic matter to produce compost. Bacteria including actinomycetes and some of the fungi feed on decaying materials and turn it into compost. Actinomycetes efficiently convert the dead plant matter in the soil into peat-like materials. Carbon, nitrogen, and ammonia are typically released during the decomposition processes of microorganisms, naturally making the soil fertile and nutrient-rich. Some of the microbial species such as *Pleurotus sajor-caju* and *Trichoderma harzianum* typically promote waste decomposition. *Rhodanobacter spathiphylli*, *Moraxella osloensis*, *Lysobacter* sp., *Corynebacterium* sp., *Pigmentiphaga kullae*, and Firmicutes were frequently found in the compost [5].

8.1.5 Phytohormone production

Plants typically secrete specific growth regulators to efficiently perform a remarkable variety of specific functions such as fruit ripening, dormancy of mature seeds, cell division, possible prevention of senescence, and opening and closing of stomata. There are published reports of phytohormones being secreted by some of the microbial species. Some of the phytohormones such as gibberellin, abscisic acids, salicylic acids, auxin, and cytokinins were invariably found to be secreted by microorganisms. Apart from bacterial and fungal species, endophytes belonging to rare actinomycetes have also been reported to secrete the plant hormone, namely, indole acetic acid (IAA). In one such study, six isolates of actinomycetes which were endophytes obtained from mandarin plant secreted IAA of which one of the isolates belonging to the recognized genus *Nocardiopsis* secreted the most notable amount of IAA [6].

8.1.6 Biocontrol agents

Microorganisms typically belonging to the recognized genera of actinomycetes such as *Streptomyces* and *Micromonospora* carefully secrete secondary metabolites that act against pathogenic microorganisms. Some of the species of *Trichoderma* have been known to act as an antagonistic agent against some of the phytopathogens such as microorganisms responsible for naturally causing damping off of seedlings and rotting of roots. Environment-friendly biopesticides are being produced using such microorganisms that effectively control the phytopathogens under sustainable agriculture practices. Some of the specific strains of *Bacillus thuringiensis* and *Lysinibacillus sphaericus*, and viruses belonging to Baculovirus have been reported to possess insecticidal properties and a possible approach for controlling insect pests in a sustainable way [7] (Figure 8.1).

Figure 8.1: Potential microbial functionalities useful in sustainable farming system.

8.2 Characteristics of microorganisms suitable for plant growth promotion

8.2.1 Phosphate solubilization

As discussed earlier, specific microorganisms are responsible for solubilizing phosphorus in the soil, and such microorganisms are categorized under PSMs. Solubilizing of phosphates has been observed in the bacterial, fungal, and algal species. Some of the bacterial species with the unique ability to solubilize phosphates typically belong to the following species, that is, *Rhizobium* sp., *Bacillus* sp., *Pseudomonas* sp., and actinomycetes such as *Streptomyces* sp., while some of the notable phosphate-solubilizing fungi include *Penicillium* sp., *Trichoderma* sp., and *Aspergillus* sp. These beneficial microorganisms have been positively identified to bioconvert organic as well as inorganic phosphorus to much more usable soluble phosphorus through different processes that are readily available for the plants to efficiently utilize them [4].

It has been observed that the phosphorus is naturally available in bound form with other metallic ions such as calcium, aluminum, and iron. The PSMs ideally have the unique capability to release various types of acids from the outer cytoplasmic membranes that are organic in nature, namely, formic acids, lactic acids, acetic acids, propionic acids, and succinic acids. These acids naturally possess ion-chelating property that releases bound phosphorus by lowering the pH of soil. The released phosphorus is thus made available to the plants [8].

A recognized species of nematofungus, *Arthrobotrys oligospora*, has also been reported to possess phosphate solubilization potential. It has been correctly observed that the growth of plants and yields of cultivated crops increased significantly after the application of PSMs for the solubilization of phosphates.

In a published study conducted using phosphate-solubilizing species of *Halobacillus* and *Halomonas* obtained from soils with alkaline nature, a significant growth was observed in the plants which were equally capable of withstanding high alkaline conditions [9].

Insertion of phosphate-solubilizing genes in engineered microorganisms typically belonging to rhizosphere using biotechnological tools has been reported to significantly increase its phosphate-solubilizing ability, thus making it a suitable organism for use as a suitable biofertilizer [10]. Plant growth-promoting bacteria have been identified to efficiently convert insoluble phosphates to orthophosphates which can be readily used by cultivated plants. A significant improvement in the growth of plants was observed when AM fungi (AMF) and phosphate-solubilizing bacteria (PSBs) were coinoculated. Phosphorus uptake increases in the plants when inoculated with specific microorganisms possessing phosphate-solubilizing potential. PSMs with exceptional phosphate-solubilizing ability are reasonably considered as a more suitable choice for utilizing as a commercial biofertilizer [11]. PSM strains that have amply demonstrated the remarkable ability to solubilize phosphate under a variety of stress conditions are a better choice for commercial biofertilizers.

8.2.2 Nitrogen fixation

Nitrogen accounts to one of the most abundant elements in the atmosphere which is mostly available in the gaseous form. However, plants can only utilize the reduced form of nitrogen. Microorganisms are known to convert gaseous nitrogen into organic compounds through various biological reactions. Nitrogen fixation is naturally carried out by heterotrophic bacteria that are present in the soil. Different microbial species, respectively, belonging to the recognized genera *Azotobacter, Bacillus, Clostridium*, and *Klebsiella* typically fix atmospheric nitrogen without any established association with another host/organism.

Another form of nitrogen fixation, that is, associative nitrogen fixation is carried out by certain species of *Azospirillum*. The microorganisms present in the rhizosphere form an associative link with plants belonging to the Poaceae family, wherein these microbes establish a symbiotic relationship with various parts of plants, resulting in the formation of notable structures that serve as nitrogen fixation sites. Symbiotic nitrogen fixation has been carried out by distinct species typically belonging to the genera of *Azospirillium, Frankia*, and *Rhizobium. Rhizobium* sp. is known to develop a symbiotic relationship with leguminous plants, forming a root nodule structure that facilitates gas diffusion to the nitrogen-fixing microbe under low oxygen conditions. Nitrogen fixation occurs in the root nodules formed by the bacteria *Rhizobium* or *Bradyrhizobium* which are associated with leguminous plants and actinorhiza formed by *Frankia* in nonleguminous plants. Cyanobacterial species such as *Anabaena, Cylindrospermum, Gloeocapsa, Nostoc, Oscillatoria, Plectonema*, and *Trichodesmium* are cyanobacteria that are popularly known to typically fix atmospheric nitrogen. Heterocysts are specialized cells that contain the enzyme nitrogenase in cyanobacteria formed during nitrogen deficiency. Heterocysts typically carry out the nitrogen fixation process in cyanobacteria and serve as the nitrogen fixation site.

Nitrogenase, an enzyme found in all nitrogen-fixing microbes, aids in the conversion of nitrogen to ammonia. Two metalloproteins, the larger molybdenum–iron (Mo–Fe) protein and the smaller iron–protein (Fe–protein), mediate the fixation of nitrogen. The Fe–protein interacts with ATP and Mg^{2+}, as well as receives electrons from ferredoxin or flavodoxin. The larger Mo–Fe protein binds to N_2 and generates two NH_3 molecules. The reduction of N_2 occurs without the breakage of nitrogen bonds, resulting in the release of two molecules of ammonia from the enzyme [12].

Leghemoglobin (LHb) typically comprises another crucial factor that contributes positively in the fixation of atmospheric nitrogen. This active molecule is structurally and functionally similar to hemoglobin that is naturally found in the human blood. It is a pink-red pigment that is typically found in healthy nodules of leguminous plants. The specific function of LHb is to adequately regulate the necessary concentration of adequate oxygen as well as the concentration of nitric oxide (NO). LHb aids in the transfer of oxygen molecules to the deficient areas of nodules for the aerobic respiration to continue and properly throttle the adequate supply of oxygen. Therefore, the nitrogenase activity that requires an anaerobic environment and is involved in the fixation of atmospheric nitrogen remains active. The active presence of LHb naturally allows oxygen molecules to diffuse through the nodules, satisfactorily completing the formation of required number of ATPs. The effective concentration of LHb can be properly used to carefully calculate the necessary amount of nitrogen fixation by nodules [13].

Symbiotic genes in the rhizobia genome include *nod* genes, *nif* genes, and *fix* genes. These three specific genes are collectively referred to as *sym* genes. The *nod* gene is responsible for nodule formation and represents nodulation which is further classified into four distinct types: nod A, B, C, and D. These genes are responsible for coding proteins for different amino acid residues. The *nif* gene, which stands for nitrogen-fixing genes, is in charge of encoding the Mo–Fe-dependent nitrogenase complex enzyme. Many rhizobia *nif* genes have been known to be present in the plasmids, but *Bradyrhizobium nif* genes are located on the chromosomes. *nif* genes also encode a considerable variety of regulatory proteins ordinarily required for nitrogen fixation [14].

There are several other methods employed during the artificial fixation of atmospheric nitrogen and ammonia production. Haber-Bosch is one of the processes in which nitrogen is carefully extracted from the atmosphere and combined with hydrogen at extremely high pressures and moderately high temperatures to produce ammonia. Plasma-induced nitrogen fixation includes both thermal plasma nitrogen fixation and nonthermal nitrogen fixation.

8.2.3 Siderophore production

Iron-chelating compounds having low molecular weight are known as siderophores which are produced by beneficial microorganisms such as bacteria and fungi. Certain species of actinomycetes are equally recognized to produce siderophores. Siderophore mediates the uptake of iron by microbial cells from its local surroundings. Siderophores, in addition to chelating iron, have also been found to chelate other elements that include Cu, Zn, Ni, and Mn [15]. A recent study has convincingly demonstrated the successful production of siderophore by plant growth-promoting bacteria. Six different bacteria such as *Bacillus subtilis*, *B. circulans*, *B. coagulans*, *B. licheniformis*, *Pseudomonas fluorescens*, and *P. koreensis* were carefully tested for the siderophore activity. Among these six isolates, *B. subtilis* and *P. koreensis* exhibited siderophore production and antagonistic activity against the plant pathogen *Cephalosporium maydis* which is responsible for causing late wilt disease in maize [16].

There are precisely two specific types of siderophores based on their structure which has been classified as catechol and hydroxamate.

8.2.3.1 Catechol siderophore

Catechol siderophore is present only in bacteria that has been characterized by the presence of catecholate and has hydroxyl groups where these ends are known to bind to Fe^{3+} [17].

8.2.3.2 Hydroxamate siderophore

These types of siderophores consist of acylated and hydroxylated alkylamines usually present in bacteria and in fungi and based on ornithine which is hydroxylated and alkylated [18].

Based on the production of siderophores by the microorganisms, siderophores are grouped into fungal siderophores, bacterial siderophores, and cyanobacterial siderophores.

8.2.3.3 Fungal siderophore

Fungi are known to produce hydroxamate and carboxylate siderophores. The fungus *Aspergillus nidulans* is known to utilize hydroxamate fusarinine C and triacetyl fusarinine C [19].

8.2.3.4 Bacterial siderophore

Different types of siderophore are produced by several species of gram-positive and gram-negative bacteria. *Streptomyces* sp. are known to produce G, B, and E desferrioxamine siderophores. *Mycobacterium* produces carboxymycobactins and exochelin siderophores [20]. Pyoverdine and pyochelin siderophores are produced by *Pseudomonas aeruginosa* [21].

8.2.3.5 Cyanobacterial siderophore

A large number of cyanobacterial species produce siderophores such as schizokinen, synechobactin, anachelin, and cyanochelin A [22].

In siderophore production, the outer membrane receptors such as FepA, FecA, and FhuA are typically required for iron uptake which then binds to the cognate ferric siderophore complex. For the active transport of the said complex, binding proteins like FhuD are required, which promptly takes the complex and directs it to ABC permease which is present in the outer membrane [23]. Every siderophore consists precisely of a functional unit bound to ligate with transferrin and lactoferrin molecules. The peptide bond present in them is known to interact with the outer membrane receptors present at the cell surface [21]. Apart from iron-chelating activity, siderophores are known to perform other roles too. The two siderophores, pyoverdine and pyochelin, secreted by *Pseudomonas aeruginosa* stop the production of exotoxin A and the protease prpL [24].

8.2.4 Production of plant hormones

Plant hormones or phytohormones are essential for the continuous growth and development of a plant. In every completed phase of growth and formation of a plant, phytohormones have multiple roles to play, which maybe antagonistic or synergistic. Many PGPR have been known to produce phytohormones which perform various specific functions like plant growth, senescence, seed germination, elongation of root tips and root hairs, maturation, and ripening of fruits [25]. The exogenous use of phytohormones has been carefully studied to rigorously test the unique capability of the cultivated plant to survive stress conditions.

Some of the important phytohormones secreted by microorganisms include auxins, gibberellins, cytokinin, and ethylene.

8.2.4.1 Auxin

Rhizospheric and epiphytic bacteria are widely known to produce IAA. This phytohormone typically promotes the abscission of older leaves and fruits and naturally induces parthenocarpy. Auxins are naturally produced by the growing apices of the plant part. Published studies have indicated the notable production of IAA by a certain group of beneficial microorganisms. The secretion of indole-3-acetic acid (IAA) and indole-3-acetamide have been reported from *Fusarium* sp. [26].

The production of IAA has been observed to be mediated by *trp* gene via the indole-3-pyruvate pathway present in some microorganisms. This method of production is well depicted by *Ustilago maydis*, which produces IAA from tryptophan [27]. The plant pathogen *Colletotrichum gloeosporioides* was reported to use tryptophan for the production of IAA [28]. Auxin is additionally used as an herbicide to kill dicotyledonous weeds. IAA was noted to decrease the infection caused by *Fusarium oxysporum* and the presence of IAA increased plant growth [29]. *Bacillus siamensis* was noted for the production of IAA [30]. Many microorganisms have an ability to alter the phytohormones and promote drought resistance capacity in cultivated plant species [2]. Number of cyanobacteria species like *Nostoc*, *Calothrix*, and *Anabaena* have been found to secrete IAA [25].

8.2.4.2 Gibberellins

Strains of *Gibberella fujikuroi* were analyzed for the production of plant hormone gibberellins, and the study revealed the similarity between the number of gibberellic acids produced and the virulence of strain [31].

8.2.4.3 Cytokinin

An array of microorganisms is known to secrete plant hormone cytokinins. In one study, *Azospirillum brasilense* was found to secrete zeatin in its supernatant along with other phytohormones [32].

8.2.4.4 Ethylene

Ethylene, a gaseous phytohormone, helps in breaking seed dormancy, promotes fruit ripening, increases root growth, and helps in the formation of root hairs. In the plants, the biosynthesis of ethylene originates from the formation of methionine. A large number of bacterial and fungal species are known to produce ethylene [25].

The potential application of phytohormones and secondary metabolites is popularly known to typically induce plant development under drought conditions [2]. Phytohormones are known to appreciably reduce heavy metal stress without affecting the growth of the plants [33].

8.3 Biofertilizers and biopesticides

Microbial inoculants as biofertilizer sources remain an effective way to eliminate the use of chemical biofertilizers. Biofertilizers are nonpolluting, easy to apply, and can be used by small farmers. Biofertilizers are known to increase soil nutrient content, soil texture, and plant resistance. The use of biofertilizers and biopesticides should be maximized to improve soil quality and reduce the negative impacts of chemical fertilizer and pesticide application and production.

Excessive use of chemical fertilizers and pesticides has affected the environment and human health by causing soil, air, and water pollution, decreasing soil fertility, and releasing greenhouse gases. Biofertilizers in sustainable agriculture have favorably received a lot of special attention recently as a better approach to improve the agriculture sector. Biofertilizers are complex substances containing live microbial inoculants that colonize within the roots of the rhizosphere positively enhancing the soil quality and nutrient uptake, resulting in crop improvement. The extensive use of chemical fertilizers and pesticides disrupts the established equilibrium and leads to biodiversity loss. Biofertilizers and biopesticides undoubtedly represent environmentally beneficial solutions to the ever-increasing environmental challenges produced by toxic chemical fertilizers and insecticides. Biofertilizers are organic and biodegradable, and they progressively increase the soil quality while also stimulating plant development by efficiently converting atmospheric nitrogen into ammonia and organic derivatives, as well as hydrolyzing organic and inorganic insoluble phosphorus compounds into soluble form and production of phytohormones [34]. Biofertilizers are typically comprised of two compounds: a microbial strain and a carrier material. A carrier molecule should be able to hold water and also able to gradually release the viable cells in the soil. Agroindustrial wastes, peat, organic compost, vermiculture, charcoal combined with suitable soil, bentonite, and perlite are some of the carrier molecules [35]. In 1896, Nobbe and Hiltner in the USA prepared the first biofertilizer, Nitragin, from *Rhizobium* sp. Many other biofertilizers, such as Azotogen and Phosphobactin, were developed later. *Rhizobium* fertilizers were produced first in India in 1934. Today, a considerable variety of other biofertilizers are commercially produced and available [3].

8.3.1 Classification of biofertilizers

Biofertilizers are basically grouped into three types [36]:
1. Nitrogen-fixing biofertilizers include *Azospirillum*, *Azotobacter*, *Nostoc*, and *Rhizobium*.
2. Phosphorus-solubilizing biofertilizers such as *Aspergillus*, *Bacillus*, *Pseudomonas*, and *Penicillium*.
3. Phosphorus-mobilizing biofertilizers such as AMF.
4. Compost fertilizer, that is, *Azotobacter*

Biofertilizers like *Rhizobium* are present in suitable soil and in the root nodules of legumes. Their inoculation as biofertilizers has been substantially known to increase the yield of crops. Legume plants are undoubtedly benefitted by the symbiotic association of *Rhizobium*, also aiding in biological nitrogen fixation [3]. *Azospirillum*, nitrogen-fixing bacteria, colonizes the roots and inner cortex without the development of any structures. *Azospirillum lipoferum* and *Azospirillum brasilense* are commonly used as biofertilizers [37].

Azotobacter is a free-living nitrogen-fixing bacteria present in the soil and rhizosphere. Its inoculation has exhibited significant growth of crops, hence, proving it as an efficient biofertilizer. Inoculants of *Acetobacter*, *Rhizobium*, and VAM have demonstrated a considerable increase in the effective yield of *Triticum aestivum* [38].

Cyanobacterial inoculants like *Anabaena*, *Aulosira*, *Cylindrospermum*, *Gleotrichia*, *Nostoc*, *Plectonema*, and *Tolypothrix* species fix atmospheric nitrogen and discharge it in the surroundings as plant growth-promoting substances. Being a source of potential biofertilizer, it has been known to appreciably reduce the ill impacts of chemical fertilizers. Some of the strains of cyanobacteria do not lose their viability up to 3 years.

Azolla is an aquatic heterosporous fern that has an endophytic cyanobacterium, *Anabaena azolla*, which has been used as a significant biofertilizer in rice and other crop fields. *Azolla* has been reported to be a fast grower and even survives at low temperatures [39] and also helps in the phytoremediation of heavy metals [40].

AMF comprise fungi that constitute a symbiotic relationship with the roots of vascular plants. AMF is more common in plants that are deprived of essential nutrients, particularly phosphorus and nitrogen. AMF performs a variety of roles in plants such as growth yield, resistance to climatic stress, pathogens, and pests. In recent years, it is extensively used as a natural biofertilizer [41].

PGPR are rhizospheric bacteria known for improving overall plant growth through the solubilization of phosphates, siderophore production, and phytohormone production [42]. Many PGPR species like *Azotobacter*, *Azospirillium*, *Arthrobacter*, *Bacillus*, *Burkholderia*, *Enterobacteria*, *Klebsiella*, *Pseudomonas*, and *Rhizobium* have induced the nutrient uptake capability of the plants [3]. PGPR has been used all over the world to increase plant yield and improve soil quality.

The use of microalgae *Chlorella vulgaris* and *Spirulina platensis* as a liquid biofertilizer increased the uptake of most essential nutrients of the plant, that is, nitrogen, phosphorus, and potassium as well as increased the overall growth of *Vigna radiata* plant parts [43]. The use of an endophyte *Burkholderia phytofirmans* as a biofertilizer along with *Acacia ampliceps* has been known to increase plant biomass along with remediation of soil when studied in arid region [44].

Negative effects of pesticides on the cultivated plants' and animals' health have led to an extensive search for an alternative method to eliminate pests and plant pathogens that have devastated agriculture in the recent years. The excessive use of chemical pesticides has invariably led to gradual degradation of soil quality as well as groundwater pollution. Microbial biopesticides may contain either of or in specific combination of living specific microorganisms such as beneficial bacteria, fungi, protozoa, mycoplasmas, and, under some cases, viruses laden with insecticidal properties. These microbial insecticides are effective in their preferred mode of action. *Bacillus thuringiensis* is an effective bioinsecticide that produces specific toxins like α-exotoxin, β-exotoxin, γ-exotoxin, and δ-endotoxin [7]. It is the most commercially successful biopesticide in the bioinsecticide market. Viral pesticides used to control insect pests belong to Baculovirus and cytoplasmic polyhedrosis viruses. A notable example of a viral pesticide is Baculovirus which has been genetically engineered for the insertion of foreign genetic material. One such published work is the insertion in *Bacillus thuringiensis* which is known to produce toxins, therefore yielding prominent results as an effective bioinsecticide [45]. Insect pests are also controlled using entomopathogenic fungi. The infective parts of these entomopathogenic fungi, such as conidia and spores, enter the insect's body via the mouth or integument and attach to the epicuticle, where they germinate via germ tubes. The fungus multiplies and carefully secretes mycotoxins, intentionally causing the pest to eventually die. Mycopesticides include *Zoophthora radicans*, which is used to control the diamondback moth, *Plutella xylostella* [46].

8.4 Microbial products in organic farming practices

Chemically derived fertilizers have satisfactorily performed significant roles in typically improving the crop yield and fulfill the insistent demands of the overwhelming population. The fertilizers are undoubtedly the vital source of nitrogen, phosphorus, and potassium at different concentrations. It has been credibly reported that there is consistent and increasing demand for ammonia and phosphorus [47]. Although chemical fertilizers are preferred inputs to increase the attractive yields in typical farms, successful assimilation of essential nutrients is low and one of the primary reasons for severe pollution of the local environment eventually cascades into ecological catastrophes [48]. Microorganisms and microbial products are considered very important for the promotion of plant growth. Some of the microorganisms such as PGPR, PGPF, fungal and bacterial endophytes, mycorrhizae, and certain bioactive agents produced by these microorganisms have been reported to affect the growth of plants as well as plant health [49].

Organic farming is a long-term approach that typically promotes the possible use of manures, particularly green manures, to increase crop productivity. Chemical fertilizers, pesticides, and growth regulators are not used in organic farming system. Crop rotation, crop residues, cow dung, vermicompost, green manures, and biological methods are preferred inputs to manage crop-destroying insects and pests which are considered some of the components of organic farming practices. Plant pathogens constitute a grave threat to cultivated crops, so using beneficial microbes to sufficiently reduce the pathogens and pest activity while increasing plant productivity without injuring the plant is always preferable. In organic farming systems, bioproducts such as Azadirachtin (derived from *Azadirachta indica*), essential oils, and microbial inoculants can be properly used [50]. As the demand for organic products grows tremendously, the prevalent methods and products used in organic farming should also improve. The application of microbial inoculants in the organic farming system represents a sustainable approach in terms of environmental and economic ways. The extensive use of biofertilizers and biopesticides which contains microorganisms are known to boost plant growth and protect the plant from harmful pests and disease-causing microorganisms. The use of microbial products to increase the yield of crops does not cause any adverse impact on the environment.

The extensive use of potassium-solubilizing microbes in organic farming must be efficiently implemented. Inoculation of *Rhizobia* and *Azotobacter* that fix atmospheric nitrogen could be utilized in the form of biofertilizers. These beneficial microorganisms foster a symbiotic relationship with the host plant, generously assisting in nitrogen uptake. Synthetic nitrogen fertilizers can be replaced by growing leguminous plants in the fields which are known to harbor nitrogen-fixing bacteria which eventually helps in supplying considerable amount of nitrogen to the soil.

Some of the bacterial species belonging to the genus *Bacillus*, *Streptomyces*, *Pseudomonas*, *Rhizobium*, and *Streptomyces*, and fungal species belonging to the genera of *Penicillium*, *Aspergillus*, and *Trichoderma*, including AMF and cyanobacteria are examples of PSM [4]. Because of their heterotrophic nature, PSM can be universally used for all types of cultivated crops.

Mycorrhizal fungi inoculants have grown in phenomenal popularity over the years because of their unique ability to promote plant growth, induce resistance to stress in plants, and provide resistance to plant pathogens. Plants growing under nutrient-deprived conditions containing lower amount of phosphorus and nitrogen levels in the soil are known to associate with mycorrhizal fungi. Mycorrhizal inoculation is undoubtedly encouraged as a mycorrhiza that progressively develops into a complex structure which is considered as a mantle in plants lacking root hairs. The hyphae form such structures so that they can cover more surface area of the soil and can be used for the acquisition of essential nutrients from the soil, including available phosphorus, as seen in ectomycorrhiza such as *Laccaria bicolor* and *Amanita mascaria*. Endomycorrhiza, such as *Endogone* and *Rhizophagus*, are mycorrhiza found on the root surface [51].

Crop productivity in organic farming has a relatively high nutrient content. The inoculation of seed with *Rhizobium* and PSB resulted in high nitrogen and phosphorous uptake as well as high protein content. Organic farming efficiently produces high-quality foods while preserving soil quality [52].

Organic compost is produced from decomposed plant parts and organic materials. Compost typically contains many beneficial microorganisms such as *Rhodanobacter spathiphylli*, *Moraxella osloensis*, *Lysobacter* sp., *Corynebacterium*, *Pigmentiphaga kullae*, and other Firmicutes, which increases the humus content of the soil [5]. Compost represents a more suitable input in organic farming practices as it is both environmentally friendly and cost-effective when compared to hazardous and expensive chemical fertilizers.

Certain plant diseases can be controlled by the possible use of various antagonistic microbes. Actinomycetes like *Streptomyces* sp., *Micromonospora*, and *Trichoderma* sp. are examples of antagonistic microbes known to secrete fungicidal metabolites against phytopathogenic fungi [53] (Figure 8.2).

8.4.1 Microbial biostimulants

Organic farming practices envisage elimination of harmful effects of inorganic fertilizers on the environment through the application of more sustainable agricultural inputs. However, yields of the plants are significantly lower when compared to those grown through conventional techniques [54]. Biostimulants are products that improve the growth of plants, through effective and efficient uptake of nutrients, even when applied in minute quantities such as Kelpak SL and Asahi SL. Some of effective

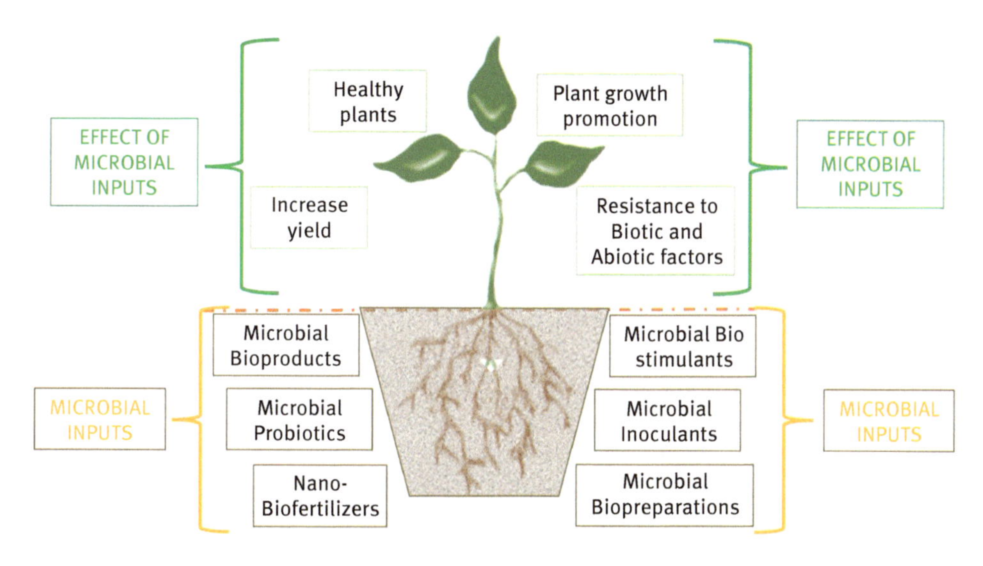

Figure 8.2: Microbial inputs in agriculture promoting plant growth.

biostimulants include extracts of seaweeds, chitosans, derivatives of proteins, humic products, biopolymers, chemical compounds, and microorganisms such as bacteria and fungi [55].

8.4.1.1 Fungal biostimulants

Fungi and fungi-based products that promote the development of plants, resistance to stress and pathogens, and improve yields are considered as biostimulants. AMF are the group of fungi that promote the uptake of macro- and micronutrients such as phosphorus, nitrogen, and zinc through the formation of arbuscules in the roots of plants including improvement in plant health under stressful conditions [50]. Recent findings indicate the possibility of interplant signaling through the hyphal networks [56].

Fungal biostimulant was found to improve yield, growth of plants, and uptake of nitrogen when applied to durum wheat [57]. Fiorentino et al. demonstrated the positive effects of biostimulants prepared using different species of *Trichoderma* on the yield of lettuce and rocket plants. *Trichoderma virens* (GV41) and *Trichoderma harzianum* (T22) significantly increased the total yield of leafy plants when planted in unfertilized soil along with an increase in the uptake of nitrogen from the soil with low nitrogen content as well as affecting the composition of eukaryotic organisms in the soil [58]. Apart from mycorrhiza, fungal endophytes typically residing within a plant without causing any apparent disease in the host plants have been reported to facilitate the transfer of nutrients to the plants along with promoting plant growth and enhancing stress resistance [59]. Endophytes are known to secrete bioactive metabolites that help in promoting plant growth and possess antagonistic activities against phytopathogens [60,61]

8.4.1.2 Bacterial biostimulants

Root-colonizing beneficial bacterial species that belong to a nonpathogenic group are able to bring about an array of effects on the plants inoculated with such microorganisms, including increase in yield, use and uptake of nutrients, and in due resistance to stresses. *Rhizobium* and PGPR are typically used as biofertilizers, whereby bacterial species facilitate the uptake of nutrients by the plants and proper upkeep of the plants. Members of bacterial species belonging to *Azotobacter* sp., *Azospirillum* sp., *Bradyrhizobium* sp., *Rhizobium* sp., *Bacillus* sp., *Bacillus* sp., have been found to act as biostimulants to the plants under stressful conditions [62]. Consortia of *Rhizobium* sp. and rhizobacterium *Enterococcus mundtii* have been reported to increase productivity and tolerate saline conditions by spring mungbeans [59]. Similarly, *Azotobacter* species when inoculated with wheat plants planted in saline soil showed increased yield of wheat and nitrogen content [63].

8.4.2 Plant probiotics

Microbiome of plants forms important factors for plant health, which interacts with the host through array of mechanisms. The exudates from the roots helps in building up the population of microflora around the rhizosphere. The beneficial microorganisms in the soil that help the plants to sustain dynamic environmental conditions may be termed as plant probiotics [64]. Agriculture inputs such as pesticides and fertilizers along with the other environmental factors alter the structure of the microorganisms in the community. Engineering of rhizosphere is being proposed, in which microorganisms are being introduced to the soil that stimulates retrieval of beneficial microorganisms as well as the functionalities that are related to the fertility of the soil, thereby increasing the native microflora. Some of the effects of the introduction of beneficial microorganisms include enhanced fixation of atmospheric nitrogen, solubilization of insoluble phosphates, increase in phytohormone production, as well as increase in the secretion of polysaccharides that are known to provide increased tolerance to abiotic factors such as high or low temperatures, pH, saline conditions of the soil, stress due to scarce water, and soil contaminated with heavy metals and intolerable pesticides. There have been reports of mutual association between normal microflora of the rhizosphere and the inoculated microorganisms. The microbial consortia could be utilized to introduce novel growth-promoting factors apart from the ones being provided by the native microorganisms [65].

8.4.3 Nanobiofertilizers

For the last decade, extensive work was faithfully rendered to eagerly seek viable alternatives for chemical inputs with ecologically reliable products. After green and biotechnological revolutions in the field of modern agriculture, nanotechnology is undoubtedly gaining paramount importance. The technological advances in modern agriculture through the effective implementation of bio- and nanotechnology along with other scientific knowledge have demonstrated the possibility of positively transforming conventional farming to productive ones and to adequately accommodate the overwhelming demands of ever-increasing population through sustainable and less expensive approaches [66].

Nanobiofertilizers can be defined as precise formulations of microbial origin integrated with nanoparticles that positively enhance plant growth through the steady secretion of bioproducts and extending the practical usefulness of the valuable fertilizers [67]. The concoction reduces the rate of release of nutrients and extends the flow of nutrients to the plants over a period of time, thus increasing the efficiency of the nutrients eventually improving the yields from targeted crop plants [68]. The collective effect of biofertilizers and nanoparticles has numerous advantages through slow and sustainable release of necessary nutrients for a longer period of plant growth [69].

As described by Kumari and Singh [66], some of the features of nanobiofertilizers include:

a. Nitrogen fixation, phosphorus solubilization, improved plant growth-promoting hormones, and healthy soil microbiome
b. Increased solubilization and dispersion of available nutrients
c. Fixation and absorption of nutrients by soil is reduced
d. Bioavailability of nutrients to plants is enhanced
e. Continuous and slow release of nutrients promotes proper utilization by the plants
f. Efficient use of available nutrients
g. Requirement of fertilizers is reduced significantly when used over larger area
h. Cost-effective
i. Ecologically sustainable

Zinc is one of the most crucial elements necessary for different types of metabolisms in plants as well as animals. Advantages of nanotechnology in improving the growth and other process in maize plant has been reported through the application of zinc chelate or nanobiofertilizers under water-stressed conditions [70]. Application of nanobiofertilizer-containing iron was found to significantly improve the amount and quality of sepals of roselle plants [71]. Similarly, application of nanobiofertilizer containing acylated homoserine lactone-coated fibers of iron–carbon and spores of *Panebacillus polymyxa* on chickpea and wheat exhibited substantial improvement in the growth of roots, biomass, chlorophyll content, and proteins along with resistance to common fungal phytopathogens [72]. Weight and vigor of plants treated with a nano-biofertilizer developed from the extracts of onions showed significant improvement with possible application in tomato and brinjal plants [73].

8.5 Future prospective

Microbial products like biofertilizers and biopesticides have continuously proven to be suitable for organic farming practices without typically causing any adverse effect on human health as well as environment. However, environmental factors, complex nature and dynamic composition of soil, abiotic and biotic factors, distinct type of microflora, and so on influence the desired effect of beneficial microorganisms and their active metabolites when applied on the chosen field. Foremost importance of such products is to properly maintain uniformity and effectivity under varying conditions which undoubtedly remain a daunting challenge and need a proper redressal. Other challenges which require considerable attention prominently include the fundamental understanding and developing propagation techniques of AMF at industrial level, scientifically studying its host types and distinctive community.

Valuable and advantageous strains of microorganisms could be helpful in complying with the increasing demand of biofertilizers and progressively reducing our potential requirement of inorganic fertilizers. Technological tools could be gainfully employed to understand the specific nutrient requirement for culturing agriculturally valuable microorganisms, carefully following their preferred mode of action, as well as the cross talk between the plants and microorganisms which could be efficiently utilized to our distinct advantage during its practical application in the chosen fields. One of the approaches that should be considered would be to add certain metabolites or compounds that would promote the growth of beneficial bacteria already present in the soil rather than the introduction of bacterial inoculum itself. Along with the studies on microorganism–host studies, reports on microorganism–microorganism interactions are of prime importance to understand the nature and effect of the inoculum during its application in the fields. There are reports of genes being transferred to more than 2,000 genomes of bacteria most likely between bacteria in the same niche than between bacteria belonging to different ecosystems. Managing microorganisms in a community would be better understood by understanding microbiota of a plant which would influence the techniques in agriculture practice [74].

Exploring the rarely explored niches for microorganisms could help in finding novel functionalities with potential plant growth-promoting factors and antimicrobial properties against prevalent phytopathogens which may be suitable for growing crops in harsh and unfavorable conditions in a more sustainable way [75].

8.6 Conclusion

The demand for agricultural produce is increasing day by day due to ever-growing human population. Although chemical fertilizers and pesticides are undoubtedly the most reliable inputs for adequately fulfilling the increasing demand of the agriculture produce, alternative and more sustainable methods of modern agriculture are again gaining interest among the masses due to its eco-friendly and beneficial products. Microorganisms in the soil efficiently perform a substantial role in promoting plant growth, maintain proper health of the soil, carefully help the plants to tolerate stressful environmental conditions, and undoubtedly increase the yields of the cultivated crops. Harnessing the capable microorganisms that are laden with beneficial features would adequately support the cultivated plants to propagate and produce through sustainable approach. Technological innovations such as metagenomic and genetic engineering approaches are helping in steady exploration of diverse microbes in soil and understanding the cross talk between the plants and microorganisms. Plant growth-promoting microorganisms have typically remained a key topic of research for quite some time to properly understand their unique capabilities in solubilization and facilitation of uptake of important plant nutrients from the soil and help in the upkeep of plants. The agriculture sectors are efficiently utilizing technological tools to suitably harness the beneficial microorganisms using different formulations as per the key requirements of agriculture produce. In the last couple of decades, there has been a considerable advancement in the knowledge about the active roles and complex mechanisms played by microorganisms and our remarkable ability to manipulate them, yet the sector requires more understanding for its utilization in situ to significantly improve farm yields. Innovative methods like cleverly engineering the microbiome of the rhizosphere, adequately understanding the interactions between endophytes and plants, and understating the precise mechanisms to make use of different microbial formulations will help in the growth of agriculture industries. Biotechnological tools could be further utilized to divulge diverse array of desired functionalities of microbiota to increase agriculture produce in a sustainable way.

Use and continuous optimization of microbial formulations should be carefully monitored, and research should be efficiently conducted for a prolonged duration to accurately perceive its ecological effects along with its probable interactions. Commercial success of bioformulations will reasonably require the active participation of local farmers, public, and researchers for accurate assessment of considerable risks and necessary modification of the formulations as per the data thus obtained.

References

[1] Gupta VV. Beneficial microorganisms for sustainable agriculture. Microbiology Australia 2012, 33(3), 113–115. doi:10.1071/MA12113.

[2] Chhaya YB, Jogawat A et al. An overview of recent advancement in phytohormones-mediated stress management and drought tolerance in crop plants. Plant Gene 2021, 25, 100264. doi:10.1016/j.plgene.2020.100264.

[3] Swarnalakshmi K, Yadav V, Murugeasn S, Dhar D. Biofertilizers for higher pulse production in India : Scope, accessibility and challenges. Indian Journal of Agronomy 2016, 61, 173–181.

[4] Kalayu G. Phosphate solubilizing microorganisms: Promising approach as biofertilizers. International Journal of Agronomy 2019, 2019, 1–7. doi:10.1155/2019/4917256.

[5] Vaz-Moreira I, Silva ME, Manaia CM, Nunes OC. Diversity of bacterial isolates from commercial and homemade composts. Microbial Ecology 2008, 55(4), 714–722. doi:10.1007/s00248-007-9314-2.

[6] Shutsrirung A, Chromkaew Y, Pathom-Aree W, Choonluchanon S, Boonkerd N. Diversity of endophytic actinomycetes in mandarin grown in northern Thailand, their phytohormone production potential and plant growth promoting activity. Soil Science and Plant Nutrition 2013, 59(3), 322–330. doi:10.1080/00380768.2013.776935.

[7] El-Bendary MA, Moharam ME, Mohamed SS, Hamed SR. Pilot-scale production of mosquitocidal toxins by Bacillus thuringiensis and Lysinibacillus sphaericus under solid-state fermentation. Biocontrol Science and Technology 2016, 26(7), 980–994. doi:10.1080/09583157.2016.1177710.

[8] Kim KY, McDonald GA, Jordan D. Solubilization of hydroxyapatite by Enterobacter agglomerans and cloned Escherichia coli in culture medium. Biology and Fertility of Soils 1997, 24(4), 347–352, doi:10.1007/s003740050256.

[9] Joshi G, Kumar V, Brahmachari SK. Screening and identification of novel halotolerant bacterial strains and assessment for insoluble phosphate solubilization and IAA production. Bulletin of the National Research Centre 2021, 45(1), 1–12. doi:10.1186/s42269-021-00545-7.

[10] Fatima F, Ahmad MM, Verma SR, Pathak N. Relevance of phosphate solubilizing microbes in sustainable crop production: A review. International Journal of Environmental Science and Technology Published online June 12 2021. doi:10.1007/s13762-021-03425-9.

[11] Nacoon S, Jogloy S, Riddech N, et al. Combination of arbuscular mycorrhizal fungi and phosphate solubilizing bacteria on growth and production of Helianthus tuberosus under field condition. Scientific Reports 2021, 11(1), 6501. doi:10.1038/s41598-021-86042-3.

[12] Hoffman BM, Lukoyanov D, Yang ZY, Dean DR, Seefeldt LC. Mechanism of nitrogen fixation by nitrogenase: The next stage. Chemical Reviews 2014, 114(8), 4041–4062, doi:10.1021/cr400641x.

[13] Kosmachevskaya OV, Nasybullina EI, Shumaev KB, Topunov AF. Expressed soybean leghemoglobin: effect on Escherichia coli at oxidative and nitrosative stress. Molecules 2021, 26(23), 7207. doi:10.3390/molecules26237207.

[14] Sheoran S, Kumar S, Kumar P, Meena RS, Rakshit S. Nitrogen fixation in maize: Breeding opportunities. Theoretical and Applied Genetics 2021, 134(5), 1263–1280, doi:10.1007/s00122-021-03791-5.

[15] Dimkpa C. Microbial siderophores: Production, detection and application in agriculture and environment. Endocytobiosis Cell Research 2016, 27(2), 7–16.

[16] Ghazy N, El-Nahrawy S. Siderophore production by Bacillus subtilis MF497446 and Pseudomonas koreensis MG209738 and their efficacy in controlling Cephalosporium maydis in maize plant. Archives of Microbiology 2021, 203(3), 1195–1209, doi:10.1007/s00203-020-02113-5.

[17] Paul A, Dubey R. Characterization of protein involve in nitrogen fixation and estimation of CO factor. International Journal of Advanced Biotechnology Research 2014, 5(4), 582–597.
[18] Baakza A, Vala AK, Dave BP, Dube HC. A comparative study of siderophore production by fungi from marine and terrestrial habitats. Journal of Experimental Marine Biology and Ecology 2004, 311(1), 1–9. doi:10.1016/j.jembe.2003.12.028.
[19] Schrettl M, Beckmann N, Varga J et al. HapX-mediated adaption to iron starvation is crucial for virulence of Aspergillus fumigatus. Cowen LE, ed PLoS Pathogens 2010, 6(9), e1001124. doi:10.1371/journal.ppat.1001124.
[20] Winkelmann G. Microbial siderophore-mediated transport. Biochemical Society Transactions 2002, 30(4), 691–696. doi:10.1042/bst0300691.
[21] Marathe RJ, Phatake YB, Sonawane AM. Bioprospecting of Pseudomonas aeruginosa for their potential to produce siderophore, process optimization and evaluation of its bioactivity. International Journal of Bioassays 2015, 4, 3667–3675.
[22] Galica T, Borbone N, Mareš J et al. Cyanochelins, an overlooked class of widely distributed cyanobacterial siderophores, discovered by silent gene cluster awakening. Nojiri H, ed Applied and Environmental Microbiology 2021, 87(17), 1–13. doi:10.1128/AEM.03128-20.
[23] Köster W. ABC transporter-mediated uptake of iron, siderophores, heme and vitamin B12. Research in Microbiology 2001, 152(3-4), 291–301, doi:10.1016/S0923-2508(01)01200-1.
[24] Lamont IL, Beare PA, Ochsner U, Vasil AI, Vasil ML. Siderophore-mediated signaling regulates virulence factor production in Pseudomonas aeruginosa. Proceedings of the National Academy of Sciences 2002, 99(10), 7072–7077, doi:10.1073/pnas.092016999.
[25] Sharma S, Kaur M. Plant hormones synthesized by microorganisms and their role in biofertilizer – a review article. International Journal of Advanced Research 2017, 5(12), 1753–1762.
[26] Tsavkelova E, Oeser B, Oren-Young L et al. Identification and functional characterization of indole-3-acetamide-mediated IAA biosynthesis in plant-associated Fusarium species. Fungal Genetics and Biology 2012, 49(1), 48–57. doi:10.1016/j.fgb.2011.10.005.
[27] Reineke G, Heinze B, Schirawski J, Buettner H, Kahmann R, Basse CW. Indole-3-acetic acid (IAA) biosynthesis in the smut fungus Ustilago maydis and its relevance for increased IAA levels in infected tissue and host tumour formation. Molecular Plant Pathology 2008, 9(3), 339–355, doi:10.1111/j.1364-3703.2008.00470.x.
[28] Maor R, Haskin S, Levi-Kedmi H, Sharon A. In planta production of indole-3-acetic acid by Colletotrichum gloeosporioides f. sp. aeschynomene. Applied and Environmental Microbiology 2004, 70(3), 1852–1854. doi:10.1128/AEM.70.3.1852-1854.2004.
[29] Sharaf EF, Farrag AA. Induced resistance in tomato plants by IAA against Fusarium oxysporum lycopersici. Polish Journal of Microbiology 2004, 53(2), 111–116.
[30] Suliasih WS. Isolation of Indole Acetic Acid (IAA) producing Bacillus siamensis from peat and optimization of the culture conditions for maximum IAA production. IOP Conference Series: Earth and Environmental Science 2020, 572(1), 012025, doi:10.1088/1755-1315/572/1/012025.
[31] Desjardins AE, Manandhar HK, Plattner RD, Manandhar GG, Poling SM, Maragos CM. Fusarium species from Nepalese rice and production of mycotoxins and gibberellic acid by selected species. Applied and Environmental Microbiology 2000, 66(3), 1020–1025, doi:10.1128/AEM.66.3.1020-1025.2000.
[32] Perrig D, Boiero ML, Masciarelli OA et al. Plant-growth-promoting compounds produced by two agronomically important strains of Azospirillum brasilense, and implications for inoculant formulation. Applied Microbiology and Biotechnology 2007, 75(5), 1143–1150. doi:10.1007/s00253-007-0909-9.

[33] Saini S, Kaur N, Pati PK. Phytohormones: Key players in the modulation of heavy metal stress tolerance in plants. Ecotoxicology and Environmental Safety 2021, 223, 112578. doi:10.1016/j.ecoenv.2021.112578.
[34] Kumar R, Kumawat N, Sahu YK. Role of biofertilizers in agriculture. Popular Kheti 2017, 5(4), 63–66.
[35] Rani U, Kumar V. Microbial bioformulations: Present and future aspects. In: Prasad R, Kumar V, Kumar M, Choudhary D eds. Nanobiotechnology in bioformulations. Springer International Publishing, 2019, 243–258. doi:10.1007/978-3-030-17061-5_10.
[36] Asoegwu CR, Awuchi, Chibueze Gospel Nelson KCT, Orji, Chimaroke Gabriel Nwosu UO, Egbufor, Uchenna Christian Awuchi CG. A review on the role of biofertilizers in reducing soil pollution and increasing soil nutrients. Himalayan Journal of Agriculture 2020, 1(1), 34–38.
[37] Kennedy I, Choudhury ATM, LKecskés M. Non-symbiotic bacterial diazotrophs in crop-farming systems: Can their potential for plant growth promotion be better exploited? Soil Biology & Biochemistry 2004, 36(8), 1229–1244, doi:10.1016/j.soilbio.2004.04.006.
[38] Mazid M, Khan TA. Future of bio-fertilizers in Indian agriculture: An overview. Irish Journal of Agricultural and Food Research 2014, 3(3), 10–23. www.sciencetarget.com.
[39] Bocchi S, Malgioglio A. Azolla-Anabaena as a biofertilizer for rice paddy fields in the Po valley, a temperate rice area in Northern Italy. International Journal of Agronomy 2010, 2010, 1–5. doi:10.1155/2010/152158.
[40] Naghipour D, Ashrafi SD, Gholamzadeh M, Taghavi K, Naimi-Joubani M. Phytoremediation of heavy metals (Ni, Cd, Pb) by Azolla filiculoides from aqueous solution: A dataset. Data Br 2018, 21, 1409–1414. doi:10.1016/j.dib.2018.10.111.
[41] Begum N, Qin C, Ahanger MA et al. Role of arbuscular mycorrhizal fungi in plant growth regulation: Implications in abiotic stress tolerance. Frontiers in Plant Science 2019, 10, 1068. doi:10.3389/fpls.2019.01068.
[42] Maougal RT, Kechid M, Ladjabi C, Djekoun A. PGPR characteristics of rhizospheric bacteria to understand the mechanisms of Faba bean growth. Proceedings 2021, 66(1), 27, doi:10.3390/proceedings2020066027.
[43] Dineshkumar R, Duraimurugan M, Sharmiladevi N et al. Microalgal liquid biofertilizer and biostimulant effect on green gram (Vigna radiata L) an experimental cultivation. Biomass Convers Biorefinery 2020. doi:10.1007/s13399-020-00857-0.
[44] Afzal M, Shabir G, Tahseen R et al. Endophytic Burkholderia sp. strain PsJN improves plant growth and phytoremediation of soil irrigated with textile effluent. Clean Soil Air Water 2014, 42(9), 1304–1310. doi:10.1002/clen.201300006.
[45] Chang JH, Choi JY, Jin BR et al. An improved baculovirus insecticide producing occlusion bodies that contain Bacillus thuringiensis insect toxin. Journal of Invertebrate Pathology 2003, 84(1), 30–37. doi:10.1016/S0022-2011(03)00121-6.
[46] Shah PA, Pell JK. Entomopathogenic fungi as biological control agents. Applied Microbiology and Biotechnology 2003, 61(5-6), 413–423, doi:10.1007/s00253-003-1240-8.
[47] da Silva JG, Nwanze KF, Ertharin C. The state of food insecurity in the world: Meeting the 2015 international hunger targets: Taking stock of uneven progress, Rome FAO, Food Agric Organ United Nations, 2015.
[48] Savci S. Investigation of effect of chemical fertilizers on environment. APCBEE Procedia 2012, 1(January), 287–292, doi:10.1016/j.apcbee.2012.03.047.
[49] Rai AK, Sunar K, Sharma H. Agriculturally important microorganism: Understanding the functionality and mechanisms for sustainable farming. In: Soni R, Suyal DC, Prachi B, Reeta G, eds. Microbiological activity for soil and plant health management. Singapore, Springer, 2021, 35–64. doi:10.1007/978-981-16-2922-8_2.

[50] Pylak M, Oszust K, Frąc M. Review report on the role of bioproducts, biopreparations, biostimulants and microbial inoculants in organic production of fruit. Reviews in Environmental Science and Biotechnology/Technology 2019, 18(3), 597–616, doi:10.1007/s11157-019-09500-5.
[51] Jagnaseni B, Aveek S, Babita S, Siraj D. Mycorrhiza: The oldest association between plant and fungi. Resonance 2016, 21, 1093–1104.
[52] Yadav SK, Babu S, Yadav MK, Singh K, Yadav GS, Pal S. A review of organic farming for sustainable agriculture in Northern India. Barker A V, ed International Journal of Agronomy 2013, 2013, 1–8. doi:10.1155/2013/718145.
[53] Zin NA, Badaluddin NA. Biological functions of Trichoderma spp. for agriculture applications. Annals of Agricultural Sciences 2020, 65(2), 168–178, doi:10.1016/j.aoas.2020.09.003.
[54] Dorais M, Alsanius B. Advances and trends in organic fruit and vegetable farming research. In: Horticultural reviews: Volume 43. vol. 43, John Wiley & Sons, Inc., 2015, 185–268. doi:10.1002/9781119107781.ch04.
[55] du Jardin P. Plant biostimulants: Definition, concept, main categories and regulation. Scientia Horticulturae 2015, 196, 3–14. doi:10.1016/j.scienta.2015.09.021.
[56] Johnson D, Gilbert L. Interplant signalling through hyphal networks. The New Phytologist 2015, 205(4), 1448–1453, doi:10.1111/nph.13115.
[57] Laurent EA, Ahmed N, Durieu C, Grieu P, Lamaze T. Marine and fungal biostimulants improve grain yield, nitrogen absorption and allocation in durum wheat plants. The Journal of Agricultural Science 2020, 158(4), 279–287, doi:10.1017/S0021859620000660.
[58] Fiorentino N, Ventorino V, Woo SL et al. Trichoderma-based biostimulants modulate rhizosphere microbial populations and improve N uptake efficiency, yield, and nutritional quality of leafy vegetables. Frontiers in Plant Science 2018, 9(June), 1–15. doi:10.3389/fpls.2018.00743.
[59] Das T, Dey A, Pandey DK, Panwar JS, Nandy S. Fungal endophytes as biostimulants of secondary metabolism in plants: A sustainable agricultural practice for medicinal crops. In: Singh H, Vaishnav ABTN, FD in MB and B eds. New and future developments in microbial biotechnology and bioengineering. Elsevier, 2022, 283–314. doi:10.1016/B978-0-323-85163-3.00010-7.
[60] Sharma H, Rai AK, Chettri R, Nigam PS. Bioactivites of Penicillium citrinum isolated from a medicinal plant Swertia chirayita. Archives of Microbiology 2021, 203(8), 5173–5182, doi:10.1007/s00203-021-02498-x.
[61] Sharma H, Rai AK, Dahiya D, Chettri R, Nigam PS. Exploring endophytes for in vitro synthesis of bioactive compounds similar to metabolites produced in vivo by host plants. AIMS Microbiology 2021, 7(2), 175–199, doi:10.3934/microbiol.2021012.
[62] Van Oosten MJ, Pepe O, De Pascale S, Silletti S, Maggio A. The role of biostimulants and bioeffectors as alleviators of abiotic stress in crop plants. Chemical and Biological Technologies in Agriculture 2017, 4(1), 1–12, doi:10.1186/s40538-017-0089-5.
[63] Chaudhary D, Narula N, Sindhu SS, Behl RK. Plant growth stimulation of wheat (Triticum aestivum L.) by inoculation of salinity tolerant Azotobacter strains. Physiology and Molecular Biology of Plants 2013, 19(4), 515–519, doi:10.1007/s12298-013-0178-2.
[64] Kumar V, Kumar M, Sharma S, Prasad R. Probiotics and plant health. In: Kumar V, Kumar M, Sharma S, Prasad R, eds. Singapore, Springer, 2017. doi:10.1007/978-981-10-3473-2.
[65] Woo SL, Pepe O. Microbial consortia: Promising probiotics as plant biostimulants for sustainable agriculture. Frontiers in Plant Science 2018, 9(2003), 7–12. doi:10.3389/fpls.2018.01801.

[66] Kumari R, Singh DP. Nano-biofertilizer: An emerging eco-friendly approach for sustainable agriculture. Proceedings of the National Academy of Sciences India Section B: Biological Sciences 2020, 90(4), 733–741, doi:10.1007/s40011-019-01133-6.
[67] Zulfiqar F, Navarro M, Ashraf M, Akram NA, Munné-Bosch S. Nanofertilizer use for sustainable agriculture: Advantages and limitations. Plant Science 2019, 289(August). doi:10.1016/j.plantsci.2019.110270.
[68] Shukla SK, Anand P, Rajesh K, Mishra RK, Anupam D. Prospects of nano-biofertilizer in horticultural crops of Fabaceae. Agriculture Situation India 2013, 70(9), 45–50.
[69] Thirugnanasambandan T. Advances and trends in nano-biofertilizers. SSRN 2018, 59. doi:10.2139/ssrn.3306998.
[70] Farnia A, Omidi MM. Effect of nano-zinc chelate and nano-biofertilizer on yield and yield components of maize (Zea mays L.), under water stress condition. Indian Journal of Natural Sciences 2015, 5(29), 4614–4704.
[71] Hashemi Fadaki SE, Fakheri BA, Mehdi Nezhad N, Mohammad Pour R. Effects of nano and nano bio-fertilizer on physiological, biochemical characteristics and yield of roselle under drought stress. Journal of Crop Improvement 2018, 20(1), 45–66, doi:10.22059/JCI.2018.219078.1568.
[72] Gahoi P, Omar RA, Verma N, Gupta GS. Rhizobacteria and acylated homoserine lactone-based nanobiofertilizer to improve growth and pathogen defense in Cicer arietinum and Triticum aestivum plants. ACS Agricultural Science & Technology 2021, 1(3), 240–252, doi:10.1021/acsagscitech.1c00039.
[73] Gosavi VC, Daspute AA, Patil A et al. Synthesis of green nanobiofertilizer using silver nanoparticles of Allium cepa extract. International Journal of Chemical Studies 2020, 8(4), 1690–1694. doi:10.22271/chemi.2020.v8.i4q.9854.
[74] Berlec A. Novel techniques and findings in the study of plant microbiota: Search for plant probiotics. Plant Science 2012, 193-194, 96–102. doi:10.1016/j.plantsci.2012.05.010.
[75] Rai AK, Sharma H. Cold-adapted microorganisms and their potential role in plant growth. In: Goel R, Soni R, Suyal DC, Khan M, eds. Survival strategies in cold-adapted microorganisms. Singapore, Springer, 2022, 321–342. doi:10.1007/978-981-16-2625-8_14.

Sougata Ghosh, Bishwarup Sarkar, Sirikanjana Thongmee

Chapter 9
Nanopesticides: challenges and opportunities

Abstract: Nanotechnology has touched almost all aspects of life right from energy, pharmaceutics, food, textiles, and cosmetics. Even agriculture is witnessing revolution due to advent of nanofertilizer, nanoherbicides, nanoinsecticides, and many more. Application of nanopesticides has particularly created considerable impact in advancement of agrobiotechnology. However, such extensive application of nanopesticides has recently started to raise concerns about their plausible impact on the health and environment. Hence, this chapter discusses in detail, a number of nanopesticide formulation that include nanoemulsions, nanodispersions, and polymers. Silica-based nanocomposites, metals, and metal oxide nanoparticles consist of elemental silver, selenium, iron, nickel, calcium, and copper that have emerged as promising nanopesticides are also discussed in detail. In spite of the potential applications of the nanopesticides in agriculture, it is still in its infancy that needs to be reconsidered in order to provide amicable solution to address the obstacles employing a proper standardization and regulation. In view of the background, the toxicity and environmental impacts of nanopesticides are also covered along with the regulatory guidelines. The proposed solutions in developing more biocompatible environmentally benign nanopesticides can serve dual purpose of benefitting both agriculture and environment.

9.1 Introduction

Unrestricted application of pesticides has resulted in escalation of pest resistance, reduction in nitrogen fixation, and soil biodiversity. It has also contributed to increased bioaccumulation of pesticides [1]. Such extreme adverse effects due to extensive use of conventional pesticides have therefore reduced the yield and nutritional value of crops and increased environmental pollution. New strategies are required to overcome these problems and hence nanotechnology seems to be a potential solution. Nanotechnology is recently applied in various areas such as medicine, paints, textiles, energy, and even agriculture [2]. Likewise, number of studies has highlighted the

Acknowledgments: SG acknowledges Kasetsart University, Bangkok, Thailand, for postdoctoral fellowship and funding under Reinventing University Program (ref. no. 6501.0207/10870 dated 9 November 2021).

https://doi.org/10.1515/9783110771558-009

advantages of "nanopesticides" over conventional pesticide that include reduction in requirement, soil leaching, and environmental toxicity. Further, nanopesticides are more stable. Befits of using nanopesticides include increase in the solubility, targeted delivery, reduction of toxicity, and sustained release [1]. Silica, chitosan (CS), lipid-based nanostructures with herbicidal, fungicidal, and pesticidal activity are considered as ideal for agricultural applications [3–5]. The core structure of the porous nanoparticles (NPs) can be loaded with the pesticide for ensuring sustained release while the shell structure can protect the active molecule from UV-light-mediated degradation [6]. CS-based formulation in combination with organic, inorganic, and/or copolymers helps to enhance the solubility of the pesticide, better adherence to plant surface, enhancement in contact time, and hence superior uptake of the pesticidal agent [7]. Solid lipid NPs entrapped pesticides; on the other hand, they facilitate the uniform dispersion and sustained release of the lipophilic bioactive molecules [8].

This chapter describes the recent developments of several kinds of nanoproducts that are proposed to be a safer alternative as compared to conventional pesticides. Nanoemulsions are formulations in which pesticide is generally dispersed as nanosized droplets in water along with localization of surfactant molecules at the interface of water and pesticide. Similarly, nanodispersions have pesticide molecules dispersed as solid nanocomposites in water. Mesoporous silica NPs (MSNs) are also extensively reported to be useful as a nanocarrier for pesticides. Apart from these nanoformulations, several metal and metal oxide NPs have potential pesticidal activity. The physicochemical characteristics, features, and applications of all such nanoproducts are summarized in Table 9.1.

However, increasing interest in application of such nanoproducts also raised questions on environmental safety and regulatory guidelines for proper monitoring and management of these nanopesticides. Therefore, the toxic impact of these nanopesticides along with current regulatory guidelines is also discussed in this chapter to provide amicable solutions and standardizations regarding nanopesticide utilization at a large scale in near future.

9.2 Nanoemulsion

Nanoemulsion is considered as an advanced pesticide delivery system that is able to provide a stable release rate along with an appropriate effective concentration of active ingredient over a specified period of time. Hence, it can reduce the wastage and toxicity of the pesticide and improve its bioavailability. Zhang et al. [9] prepared castor oil-based polyurethane (CO-PU) nanoemulsion containing avermectin (AVM) biopesticide [9]. Prepolymer dispersion method was carried out for preparation of CO-PU emulsion which was further loaded with AVM to formulate AVM/CO-PU nanoemulsion. High-speed dispersing method was adopted to fabricate AVM-encapsulated

Table 9.1: Nanoformulations containing pesticides.

Pesticide type	Physicochemical characteristics	Features	Application	References
Nanoemulsions				
Avermectin	Hydrodynamic diameter: 40–50 nm	Controlled release of biopesticide, increased adhesion property, enhanced photostability	Increased bioavailability of pesticides after foliar spray on corn leaves	[9]
Lambda-cyhalothrin	Hydrodynamic size: 42.3 ± 1.4 nm–75.5 ± 0.9 nm, PDI: 0.112 ± 0.01 to 0.0806 ± 0.02, zeta potential: −9.3 ± 0.7 mV	Controlled insecticide release, enhanced adhesion property	–	[10]
Mancozeb	Average diameter: 182 ± 17 nm, zeta potential: −9.3 ± 0.7 mV, PDI: <0.25 ± 0.02	Spherical morphologies, regular and well-defined edges	Enhanced antifungal activity against *G. cingulata* (MIC value: 0.02 µg/mL)	[11]
Nanodispersions				
Avermectin	Mean size: 188 ± 8 nm, PDI: 0.292 ± 0.143, zeta potential: −33 mV	Spherical morphology, better suspensibility and wettability, improved photostability	Efficient toxicity and bioavailability against *P. xylostella* L.	[12]
Chlorantraniliprole	Mean size: 29 ± 1 nm, PDI: 0.26 ± 0.01	Irregular morphologies, amorphous nature, electronegative, enhanced wettability and suspensibility	Efficient toxicity and bioavailability against *P. xylostella* L.	[13]
Lambda-cyhalothrin	Mean size: 21.7 ± 0.1 nm, zeta potential: − 47 mV	Spherical shape, enhanced wettability, storage stability and suspensibility	Efficient toxicity and bioavailability against *P. xylostella* L.	[14]

(continued)

Table 9.1 (continued)

Pesticide type	**Physicochemical characteristics**	**Features**	**Application**	**References**
Polymeric nanoparticles				
Carbendazim	Average diameter: 90 nm, zeta potential: 58 mV	Spherical in shape with agitated edges, sustained release at varying pH, reduced phytotoxicity	Enhanced antifungal activity against *F. oxysporum* and *A. parasiticus*	[15]
Imidacloprid	Average size: 150 nm, encapsulation efficiency: 98.66%	Reduced cytotoxicity	Enhanced insecticidal activity against leafhopper with no harmful effects on crop	[16]
Metachlor	Average size: 97.87 ± 5.5 nm, PDI: 0.128 ± 0.02	Spherical morphology, sustained pesticide release, enhanced stability and root permeation	Reduction in cytotoxicity and improved herbicidal activity	[17]
Silica nanoparticles				
Abamectin	Average diameter: 320 nm, maximum specific surface area: 318.62 m^2/g, maximum pesticide loading capacity: 111.0 mg/g	Sustained release of pesticide, protective effect against photolysis	–	[18]
Prochloraz	Average size: 70.89 nm,	Spherical morphology, sustained pectinase-responsive fungicide release	Enhanced fungicidal activity against *M. oryzae*, increased translocation of fungicide in different parts of rice plant	[19]
Pyraclostrobin	Average diameter: 299 nm, pesticide loading capacity: 41.6%, zeta potential: 28.8 mV	Spherical shape, monodispersed, efficient in vitro release of fungicide in aqueous medium	Efficient fungicidal activity against *P. asparagi*	[20]

nanoemulsion that exhibited a hydrodynamic diameter within a range of 40–50 nm. Aqueous solution of acetone was used as a solvent that increased the size of the NPs and also facilitated penetration of AVM molecules into the swelled CO-PU NPs. Hence, a stable biopesticide-loaded nanoemulsion was prepared because of efficient compatibility between AVM and PU matrix as well as cross-linkage of CO-PU. Transmission electron microscope (TEM) images revealed smooth and flat film of AVM/CO-PU nanoemulsion without the presence of any AVM powder on the surface highlighting uniform encapsulation of AVM molecules inside CO-PU NPs as evident from Figure 9.1. Further, drug loading capacity and encapsulation efficiency of AVM/PO-CU nanoemulsion were 42.3% and >85%, respectively. However, encapsulation efficiency decreased with subsequent increase in AVM content.

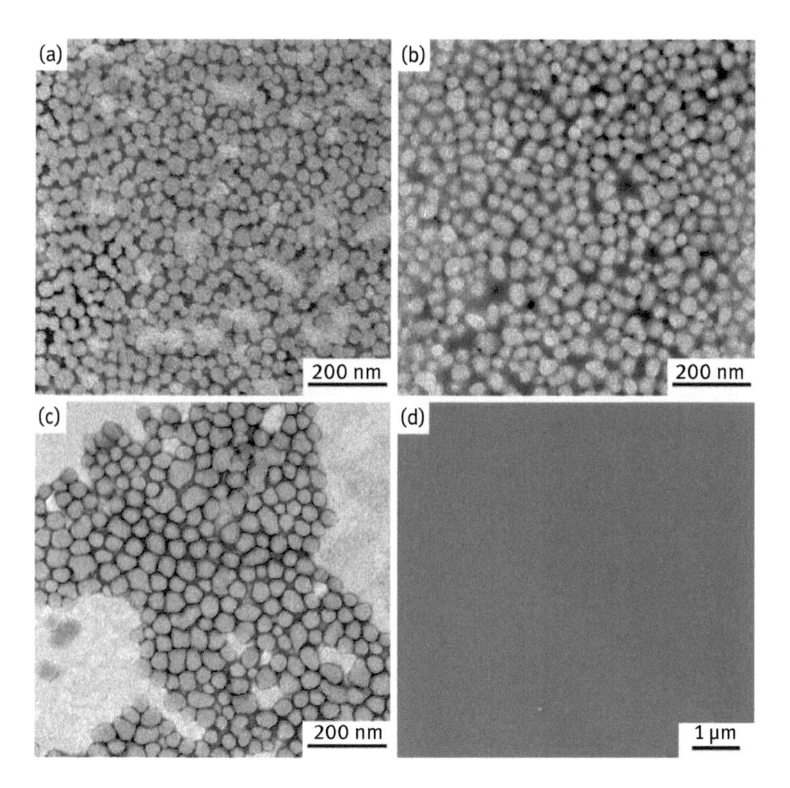

Figure 9.1: TEM images of the nanoparticles: (a) CO-PU, (b) AVM/CO-PU (sample H_1), and (c) AVM/CO-PU (sample H_4). (d) SEM image of AVM/CO-PU (sample H_4) film [9].

Behavior of AVM release from the nanoemulsion was also investigated where increase in temperature over 35 °C showed complete release of AVM molecules within 72 h. On the other hand, 20% of AVM molecules were still present inside the nanoemulsion when the incubation time was increased up to 180 h at 25 °C. Likewise, maximum cumulative release rate of around 99.6% and 94.1% were obtained at pH

4.0 and 10.0, respectively. Moreover, release profiles of AVM/CO-PU nanoemulsions indicated non-Fickian transport controlled by both diffusion and matrix erosion. Foliar retention on *Zea mays* L. (corn plant) and adhesion property of AVM/CO-PU nanoemulsion also was shown to be better as compared to free AVM. Photostability of AVM molecules was also observed to increase wherein 50% of free AVM molecules were degraded within 3.5 min of ultraviolet (UV) light irradiation (1,000 W UV lamp) while AVM molecules encapsulated in CO-PU NPs showed 50% degradation within 11.5 min.

Qin et al. [10] also reported formulation of lambda-cyhalothrin/polyurethane (LC/CO-PU) drug-loaded nanoemulsions [10]. In situ soap-free phase-inversion emulsification method was carried out for preparation of nanoemulsions wherein, CO-PU prepolymer having hydrophilic carboxyl groups was first synthesized by step-growth addition polymerization after which LC molecules were encapsulated into the CO-PU NPs. FTIR analysis of nanoemulsion showed characteristic peaks of both the CO-PU carrier and free LC molecules implying proper loading of LC molecules inside the carrier. Differential scanning calorimetry curves further revealed melting temperature (T_m) of free LC powder at 51.4 °C that was absent in the nanoemulsion indicating conversion of LC from crystal to amorphous state after loading that was suggested to improve its solubility and bioavailability. Hydrodynamic size of four nanoemulsion samples made using varying concentrations of LC were found to range from 42.3 ± 1.4 nm to 75.5 ± 0.9 nm that had uniform spherical shapes. Polydispersity indices (PDI) were in a range from 0.112 ± 0.01 to 0.0806 ± 0.02 while zeta potential had a range of −59.8 ± 1.7 mV to −51.5 ± 2.9 mV. The storage capacity of nanoemulsion samples that had less than 30 wt% loading capacities of LC were stable and maintained homogeneity without any agglomeration. The encapsulation efficiency was more than 85% indicating proper loading of LC molecules into the CO-PU nanocarrier. Additionally, release rate of LC/CO-PU nanoemulsion was rapid only for the first 30 h. Thereafter, stable release behavior was obtained until equilibrium was attained after 80 h. On the other hand, commercial emulsifier concentrates of LC (LC-EC) and wettable powder of LC (LC-WP) showed accumulated release of over 90% after 48 h only. Increase in drug loading capacity from 10 wt% to 30 wt% also showed a subsequent increase in accumulated release of LC from 65.6% to 71.9%, respectively. Furthermore, increase in temperature from 20 to 30 °C resulted in concomitant increase of accumulated release from 66.9% to 76.1% after 72 h. Moreover, foliar adhesion property of LC/CO-PU nanoemulsion was demonstrated on *Gossypium hirsutum* (cotton) and corn foliar surfaces that showed retention of most of the particles after water wash as compared to commercial LC-EC and LC-WP formulations. Volatilization of water was shown to form latex films on the crop leaves that were composed of aggregated and coalesced LC/CO-PU nanoemulsions. The average surface tension of LC/CO-PU latex film was 42.95 mN/m, which was less than the critical surface tension of both cotton and crop leaves.

Mancozeb fungicide nanoemulsion was also prepared by Velho et al. [11] that exhibited antifungal activity against *Glomerella cingulata* [11]. Spontaneous emulsification method >was carried out using an oily phase comprised of eugenol, mancozeb, sorbitan monooleate, and acetone along with an aqueous phase made up of polysorbate 80 and water. The physicochemical characteristics of prepared nanoemulsion was evaluated that showed an average diameter of 182 ± 17 nm, a PDI less than 0.25 ± 0.02, and a negative zeta potential (−9.3 ± 0.7 mV), as well as acidic pH 5.6 ± 0.04. Around 14 mg/mL of eugenol was found in nanoemulsions that represented 40% of the theoretical value of about 33.3 mg/mL. Further, TEM images of prepared nanoformulations demonstrated spherical morphologies, regular and well-defined edges. Stability studies of nanoemulsions showed that refrigeration conditions at 4 °C was optimum as there were negligible changes in the physicochemical parameters. With regards to antifungal activity, 10^3 times higher dose of free fungicide exhibited similar efficacy as that of nanoemulsion. Moreover, the minimum inhibitory concentration (MIC) of nanoemulsion made using mancozeb and eugenol was 0.02 µg/mL as compared to MIC value of free mancozeb that was 9 µg/mL.

9.3 Nanodispersion

Several nanodispersions have also been reported to be formulated using poorly water-soluble pesticides in order to enhance bioavailability. One such study by Cui et al. [12] developed solid nanodispersion containing AVM by microprecipitation and lyophilization [12]. Combination of maleic rosin–polyoxypropylene–polyoxyethylene ether sulfonate (MRES) and polycarboxylate surfactants at a ratio of 1:1 along with 87% w/w of sucrose as water-soluble carrier and 1% w/w of 2-(2-hydroxy-5-*tert*-octylphenyl)benzotriazole (UV 329) as a light stabilizer was used for formulation of nanodispersion containing 10% w/w of AVM and having a particle size of around 46 nm. AVM NPs were spherical morphology while mean size and PDI of NPs were 188 ± 8 nm and 0.292 ± 0.143, respectively. Zeta potential and pH of the redispersed nanosuspension were −33 mV and 7.0, respectively. Suspensibility of the AVM solid nanodispersion was 99.8% which was higher than conventional formulations of AVM. A shorter wetting time of 13 s was also obtained for the solid nanodispersion which was superior than other AVM formulations. Further, the mean size and PDI of the nanoformulations were increased to 180 nm and 0.439, respectively after storing at 25 °C for 14 days. Photolytic stability of the free AVM and prepared nanodispersion were also compared wherein, 13% photolysis of AVM was observed in the solid nanodispersion after 264 h of exposure to simulated light while free AVM revealed 29% degradation under similar conditions. Such slow degradation of the pesticide in case of nanoformulation was suggested to be due to presence of UV 329 and sucrose. Moreover, biological activity of solid nanodispersion of AVM was evaluated against *Plutella xylostella* L. (diamondback moths) wherein, the lethal concentration-50 (LC_{50}) of the nanodispersion was 3.4, 1.5, and 14.2 times higher than that of commercial

AVM products, namely Cuiwei water dispersible granule, Kaiwei water dispersible granule, and Yipaohong wettable powder, respectively.

Cui et al. [13] also reported preparation of solid nanodispersions of chlorantraniliprole. High-pressure homogenization method was carried out for preparation of chlorantraniliprole nanosuspension followed by solidification of the aqueous dispersion through addition of sucrose as a water-soluble carrier and removal of water [13]. Two different surfactants, namely MRES and polycarboxylate, were used in a 1:1 ratio to prepare nanoemulsions having a mean size and PDI of 13 nm and 0.23, respectively. It was suggested that both electrostatic repulsion and steric stabilization effects caused due to adsorption of the two surfactants on the pesticide surface were responsible for maintaining stability of the nanoemulsions which in turn, prevented agglomeration of the particles. Moreover, chlorantraniliprole content of the solid nanodispersion was 91.5% with a mean particle size and PDI of 68 nm and 0.24, respectively. The mean size and PDI of nanodispersion containing 2.5% w/v of chlorantraniliprole were 29 ± 1 nm and 0.26 ± 0.01, respectively. Scanning electron microscopy (SEM) and TEM images showed irregular shape of particles that were suggested to be a result of asymmetric application of shear, cavitation, and collision forces during the homogenization process. The pH and zeta potential of the prepared solid nanodispersions were 7.4 and − 22 mV, respectively that highlighted the electronegative nature of the particles due to adsorption of anionic surfactants on the surface of pesticides. X-ray diffraction (XRD) pattern of the solid nanodispersion further revealed a certain level of amorphous nature imparted by surfactants as compared to pure crystalline nature of chlorantraniliprole. Suspensibility of the solid nanodispersion was 97.32%. Wettability of the prepared solid nanodispersion was much efficient as compared to conventional formulations. Excellent storage stability of the nanodispersion was obtained at 25 °C where the average size and PDI of the nanoformulation remained unchanged. In addition, biological activity of pesticide-loaded nanodispersion was evaluated against diamondback moths. The toxicity as well as bioavailability of solid nanodispersion was highly efficient as compared to aqueous suspension concentrate and technical products of chlorantraniliprole.

In a similar study, Cui et al. [14] also prepared solid nanodispersion of lambda-cyhalothrin by melt-emulsification and high-speed shearing. MRES and 1-dodecanesulfonic acid sodium salt (SDS) surfactants were combined in a 3:1 ratio to prepare nanosuspension with a mean size and PDI of 16.2 ± 0.1 nm and 0.29 ± 0.01, respectively [14]. Shearing speed of 10,000 rpm for 10 min was observed to be optimum for preparation of lambda-cyhalothrin nanosuspension. Moreover, the mean size of pesticide particles decreased to 28.7 nm when incubated at 49.2 °C due to melting of lambda-cyhalothrin. Hence, 80 °C was considered for preparation of pesticide nanosuspensions as smaller droplets of lambda-cyhalothrin was proposed to be easily encapsulated and dispersed. Later on, solid nanodispersion was obtained using lyophilization. Dynamic light scattering (DLS) results demonstrated the mean size of redispersed solution of nanoformulations to be 21.7 ± 0.1 nm while SEM images revealed its spherical morphology.

Further, zeta potential and pH of the nanodispersion were −47 mV and 7.0, respectively indicating strong stabilization due to electrostatic effects. XRD patterns of nanodispersion in turn, demonstrated diffraction peaks at 12.8°, 15.6°, 18.9°, 20.9°, and 26.0° that indicated preservation of crystalline nature of the pesticide during the preparation of nanodispersion. Suspensibility of 99.5% of the solid nanodispersion was obtained in water. Additionally, wettability of the solid nanodispersion containing 0.5% w/w pesticide was investigated on leaves of *Cucumis sativus* L. (cucumber) and *Oryza sativa* L. (rice) plants. The contact angles of the nanodispersion were 58 ± 2° and 120 ± 4° whereas contact angles of pure water were 84 ± 4° and 136 ± 3° on cucumber and rice leaves, respectively which highlighted better wetting properties of prepared nanoformulation. Further, the particle size increased from 21.7 nm to 57.8 nm after 5 days of storage at 25 °C and thereafter, remained unchanged for 14 days. The suspensibility and wettability were 99.4% and 23 s after 14 days, respectively. The biological activity of the synthesized lambda-cyhalothrin solid nanodispersion was also evaluated against diamondback moths. The LC_{50} of solid nanodispersion was lower than technical (TC), emulsion in water (EW), and emulsifiable concentrate (EC) formulations of lambda-cyhalothrin. Similarly, the toxicity of nanodispersion was 1.4, 1.2, and 1.1 times higher than that of conventional TC, EW, and EC formulations, respectively.

9.4 Polymeric nanoparticles

Polymeric NPs are effective carriers of pesticides. Sandhya et al. [15] reported formation of carbendazim nanoformulation using pectin and CS by ionic interaction [15]. The resulting carbendazim-loaded CS-pectin NPs showed 0.5% w/v of CS and 0.15% w/v of pectin with an average size of 129.4 nm and zeta potential of 58 mV. Entrapment efficiency and pesticide loading were 99.2% and 3.2%, respectively. TEM images of optimized nanoformulations indicated nearly spherical shape having agitated edges with an average diameter of about 90 nm. FTIR spectra demonstrated loading of pesticide in the NPs as the peaks at 1,628, 1,589, and 1,018 cm^{-1} corresponding to C=N stretching and C–H vibration of aromatic rings, respectively were observed. These peaks resembled the spectral pattern of pure carbendazim as well. Sustained release of the pesticide from carbendazim-loaded CS-pectin NPs was also demonstrated at varying pH conditions as compared to free carbendazim. For instance, 62.8 ± 0.13% of carbendazim was released from the NPs at pH 10.0 whereas pure carbendazim showed 86.8 ± 0.2% release under similar parameters. Furthermore, germination of *Cucumis sativa*, *Zea mays* L., and *Lycopersicum esculantum* seeds were around 96% after treatment with carbendazim-loaded CS-pectin NPs. In addition, seeds treated with the nanoformulations showed higher growth with respect to average length of roots and shoots as compared to pure carbendazim treatment. Antifungal activity of the nanoformulation was also compared with pure carbendazim and commercial formulation against *Fusarium oxysporum* and *Aspergillus parasiticus*. Both fungi were inhibited up to 100% in presence of 0.5 and 1.0 ppm of prepared nanoformulation, whereas 0.5 and 1.0 ppm of pure carbendazim showed 80 ± 0 and 97.20 ± 1.1% inhibitions against *F. oxysporum* along with 85.97 ± 0.6% and 100% inhibition against *A. parasiticus*, respectively. Likewise, commercial formulation was shown to be less effective than carbendazim-loaded CS-pectin NPs.

In a similar study, Kumar et al. [16] synthesized sodium alginate NPs loaded with imidacloprid insecticide [16]. Emulsion cross-linking method was optimized for preparation of sodium alginate NPs in which 1% w/v of sodium alginate and 15% w/v of dioctyl sodium sulfosuccinate were used which resulted in obtaining particles with average size and encapsulation efficiency of 150 nm and 98.66%, respectively. The size of the particles was in a range from 50 to 100 nm. In vitro cytotoxicity analysis of insecticide-loaded NPs was also investigated against Vero cell lines wherein, 2.46% imidacloprid pesticide-loaded sodium alginate NPs showed reduced toxicity as compared to original insecticide. On field studies, polymeric NPs containing pesticide were also studied. Treatment with nanoencapsulated insecticide was efficient in controlling leafhopper population up to 1–7 per 3 leaves for 15 days as shown in Figure 9.2. Moreover, no harmful effects in crops were observed after application of insecticide-loaded NPs.

Tong et al. [17] also fabricated polymeric NPs that acted as a carrier for metachlor pesticide [17]. Dialysis method was carried out for synthesis of water-based metachlor-loaded methoxy polyethylene glycol-poly(lactic-co-glycolic acid) (mPEG-PLGA) NPs. The organic phase comprised of mPEG-PLGA while pesticide was mixed with aqueous phase under stirring conditions which resulted in immediate synthesis of NPs due to swift diffusion of organic solvent as well as deposition of polymer on the water-organic solvent interphase followed by dialysis in order to remove organic solvent and free pesticide. The optimal ratio of copolymer:pesticide was 4:1 with a concentration of 20 mg/mL of mPEG-PLGA. TEM images then revealed spherical morphologies with an average size and PDI of 97.87 ± 5.5 nm and 0.128 ± 0.02, respectively for the NPs comprised of 5 mg/mL of metachlor. FTIR spectra of pesticide-loaded NPs further showed peaks at 2,883 cm^{-1}, 1,760, and 1,668 cm^{-1} that corresponded with methyl stretching, carbonyl group vibrations from mPEG-PLGA, and the aromatic group of metachlor, respectively. In-vitro release of metachlor was investigated using Murashige and Skoog (MS) culture medium supplemented with 10% acetonitrile and 0.1% SDS and incubated at 30 °C. Release of 48%, 51%, and 60% of metachlor was observed within 24 h in only MS culture medium, MS culture medium containing 10% acetonitrile, and MS culture medium supplemented with 0.1% SDS, respectively after which a slow release was seen in all three mediums. Hence, an overall sustained release of metachlor from mPEG-PLGA NPs was obtained. Stability of pesticide-loaded NPs was evident as no significant alteration in particle size was observed after 3 days of incubation when suspended in MS culture

Figure 9.2: Leaf infestation in (a) control, (b) dummy, (c) pesticide-encapsulated nanoformulation, and (d) plain/normal plot at different (1st, 7th, and 15th day) time of observation [16].

medium and Dulbecco's Modified Eagle Medium at 25 and 37 °C, respectively. Absorption capacity of prepared nanoformulations was evaluated on rice seedlings wherein, NPs were labeled with Cyanine 5 (Cy5) fluorescent dye and treated on the roots of rice seedlings. A distinct fluorescence was observed at 633 nm only when Cy5 was loaded in the NPs indicating its permeation in the root cross-sections. Furthermore, bioassay tests demonstrated less inhibition of metachlor-loaded NPs on germination of *Oryza sativa* and *Digitaria sanguinalis* as compared to other nanoformulations such as metachlor-loaded microparticles along with better herbicidal activity. Cytotoxicity of metachlor-loaded NPs was also demonstrated against preosteoblast cell line (MC3T3) in which cell viability was found to be higher after treatment with pesticide–NP conjugate as compared to free pesticide.

9.5 Silica nanoparticles

Porous silica NPs are extensively studied for its utilization as nanocarriers. In one such study, Wang et al. [18] fabricated porous silica NPs loaded with abamectin pesticide [18]. The porous silica NPs were uniformly spherical having an average diameter of 320 nm. The porous nature of the NPs was confirmed in SEM micrographs after NaOH etching treatment wherein, the specific surface area of the particles increased from 11.31 to 318.62 m^2/g. Furthermore, polyvinylpyrrolidone (PVP) was indicated to be useful as a surface protecting agent during NP synthesis as the carbonyl groups of PVP may form strong hydrogen bonds with the hydroxyl groups on the surface of NPs. The pesticide-loaded silica NPs showed rough and porous surface which indicated adsorption of some pesticide molecules on the surface of nanocarriers. Increase in NaOH etching time from 45 to 120 min further demonstrated subsequent increase in abamectin loading capacity from 81.0 to 111.0 mg/g. Moreover, porous nature of abamectin-loaded silica NPs showed faster release rate of pesticides adsorbed on the surface of nanocarrier at the beginning which gradually reduced eventually maintaining a slower rate of release. Additionally, photolytic rate of pure abamectin and abamectin-loaded silica NPs were 77% and 22.5%, respectively, after 72 h of UV light irradiation thus, demonstrating protective effect of silica nanocarrier.

In another recent study, Abdelrahman et al. [19] fabricated MSNs loaded with prochloraz pesticide and surface coated with pectin [19]. Tetraethoxysilane (TEOS) was used as a silica precursor along with hexadecyltrimethylammonium bromide (CTAB) that acted as a template for the formation of NPs. MSNs obtained were then loaded with prochloraz pesticide and coated with pectin through amide bond formation. Mesoporous prochloraz-loaded pectin-coated MSNs (Pro@MSN-Pec) were spherical in shape with an average size of 70.89 nm. Controlled release of prochloraz was monitored from Pro@MSN-Pec when suspended in phosphate buffer solution (PBS) wherein, 59.62% and 77.36% of fungicide was released at pH 5.0 and 9.0, respectively, after 30 days of treatment. Further, the mechanism of responsive release of prochloraz fungicide from Pro@MSN-Pec was proposed in which production of pectinases by plant pathogens was suggested to break the pectin surface coat on Pro@MSN-Pec which would in turn trigger the release of fungicide. In order to prove this hypothesis, fungicide release behavior was investigated from Pro@MSN-Pec in presence of pectinase enzyme at a pH value of 7.0 and temperature of 25 °C. Addition of pectinase showed increase in prochloraz release from 12.48% to 74.55% after 2 days. Hence, pectinase-mediated hydrolysis indeed resulted in increased release of prochloraz from Pro@MSN-Pec. Fungicidal activity was further evaluated against *Magnaporthe oryzae* P131 strain. Effective concentration-50 (EC_{50}) of Pro@MSN-Pec, commercial prochloraz EC and technical prochloraz were 0.113, 0.151, and 0.196 mg/L, respectively after seventh day of treatment while EC_{50} values were 0.248, 0.453, and 0.725 mg/L after 14 days of treatment, respectively. Additionally, translocation of MSNs in roots and leaves of rice plants were examined by fluorescein isothiocyanate (FITC) labeling on the surface of MSN. Confocal microscopy images of cross-sections of different parts of rice plant revealed efficient translocation of FITC-labeled MSNs through leaves and proper penetration through roots as well.

Cao et al. [20] capped MSNs with CS that was further used as a carrier for pyraclostrobin pesticide [20]. TEOS and CTAB were used as silica source and structure-directing agent, respectively for MSN synthesis using liquid crystal templating mechanism that was then capped with *N*-2-hydroxypropyl trimethyl ammonium chloride CS (HTCC) followed by loading of pyraclostrobin fungicide. Optimization of pesticide loading was shown to be increased in presence of 60 mg of HTCC with a loading capacity of 41.6% and zeta potential of 28.8 mV. The morphology of pyraclostrobin-loaded quaternized CS-capped MSNs (Py@MSNs-HTCC) was observed using SEM and TEM as evident from Figure 9.3. The surface of the nearly monodispersed spherical particles was found to be rough after coating with HTCC and loading of pesticide. The average diameter of Py@MSNs-HTCC was found to be 299 nm using DLS. FTIR spectral analysis further showed characteristic peaks of pyraclostrobin, MSNs and HTCC that were observed in Py@MSNs-HTCC. Brunauer–Emmett–Teller (BET) surface analysis showed a low BET specific surface area of 29.02 m^2/g and a total pore volume of only 0.22 cm^3/g which confirmed coating of HTCC on the surface. In vitro release

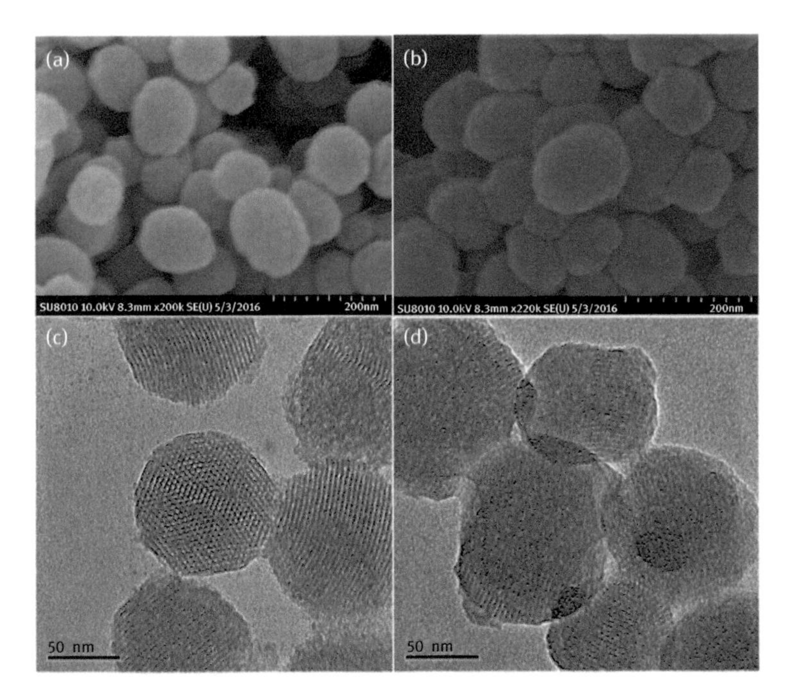

Figure 9.3: (a) Scanning electron microscopy (SEM) images of MSNs and (b) pyraclostrobin-loaded HTCC-capped MSNs; (c) transmission electron microscopy (TEM) images of MSNs and (d) pyraclostrobin-loaded HTCC-capped MSNs [20].

behavior of pyraclostrobin was also evaluated due to its poor water-soluble nature. Py@MSNs-HTCC released 72% of pyraclostrobin within 120 min when suspended in 30% aqueous methanol solution added with 0.5% Tween-80 emulsifier whereas bare MSNs released only 55% under similar conditions. Fungicidal activity of Py@MSNs-HTCC was also investigated against *Phomopsis asparagi* (Sacc.) wherein, 87.72% and 94.74% inhibition were observed at a low concentration of 2.5 and 10.0 mg/L of Py@MSNs-HTCC, respectively after 5 days of treatment. MSNs-HTCC carrier on the other hand, exhibited only 26.32% and 47.37% inhibition under similar conditions at a concentration of 5.0 and 20.0 mg/L, respectively, indicating that fungicidal activity is primarily due to release of pyraclostrobin.

9.6 Metal nanoparticles

Various metal NPs exhibit considerable pesticidal activity and hence, several studies highlighted the potential of these metal NPs to be used as alternative pesticides in near future as listed in Table 9.2. Sankar and Abideen [21] in one such study reported pesticidal activity of biogenic silver NPs (AgNPs) against *Sitophilus oryzae* which is a common grain storage pest [21]. Fresh leaf extract of *Avivennia marina* was prepared using methanol as the solvent. Formation of AgNPs was evident as visible color change from dark grey to dark brown was observed after treatment of $AgNO_3$ with methanolic extract of *A. marina* leaves. UV-vis spectra also confirmed formation of AgNPs with demonstration of surface plasmon resonance band at 425 nm. The crystalline AgNPs were spherical with a size range of 15–25 nm. Further, complete mortality of *S. oryzae* was obtained in presence of 50 mg/mL of biogenic AgNPs after 4 days of treatment whereas only 30% pesticidal activity was observed in presence of 250 mg/mL of aqueous extract of *A. marina* leaves.

AgNPs were also biologically synthesized by Bharani and Namasivayam [22] using *Punica granatum* (pomegranate) peel extract wherein, 1 mM of $AgNO_3$ was added in the supernatant of the prepared extract and reacted at room temperature in darkness under stirring conditions [22]. Synthesis of AgNPs was indicated by visible color change from pale yellow to brown along with absorbance maxima at 458 nm. Extreme stability of AgNPs for up to 6 months was observed without any aggregation or change in pH and color of the solution. Crystalline and face-centered nature of AgNPs was confirmed by XRD diffraction patterns wherein, four different peaks at 38.068°, 44.28°, 64.34°, and 77.41° were obtained that corresponded to four diffraction facets of silver. Hexagonal AgNPs without any aggregation were in a size range of 14–28 nm. *Spodoptera frugiperda*-21 (SF-21) cell lines were used for evaluation of toxicity of biogenic AgNPs that indicated half maximum inhibitory concentration (IC_{50}) of AgNPs was 31.2 µg/mL. Microscopic observations also showed morphological alterations and cell disruption of SF-21 cell lines in presence of AgNPs. Further, 100%

Table 9.2: Metal and metal oxide nanoparticles having pesticidal activity.

Metal nanoparticles	Physicochemical characteristics	Features	Application	References
AgNPs	Average size: 15–25 nm	Spherical NPs	Pesticidal activity against *S. oryzae* with 100% mortality in the presence of 50 mg/mL	[21]
AgNPs	Average size: 14–28 nm	Hexagonal, crystalline, face-centered and monodispersed particles	Pesticidal activity against SF-21 cell lines, larvicidal activity against *S. litura*	[22]
AgNPs	Average size: 3.89–55 nm	Spherical particles showing antioxidant activity	Antibacterial and antifungal activity against known phytopathogens	[23]
CaONPs and CuONPs	Average size of arrangements: 2.0 μm (CaONPs), 1.0 μm (CuONPs), BET surface area- 23.27 m^2/g (CaONPs) 40.54 m^2/g (CuONPs)	CaONPs: polygonal particles with hexagonal shape and a cubic-pure phase, CuONPs: aggregated nanoneedles forming microflower-like structure having monoclinic tenorite phase	Efficient pesticidal activity against *S. littoralis* with LC_{50} value of 129.03 mg/L (CaONPs) and 232.75 mg/L (CuONPs), histopathological changes in cuticle and gut of *S. littoralis*	[24]
FeNPs	Average size: 3.8–11.95 nm	Irregular morphology, crystalline particles	Pesticidal activity against *T. absoluta*	[25]
NiNPs	Average size: 47 ± 2 nm, zeta potential: −53.9 mV	Cubical shaped particles	Larvicidal activity against *Ae. aegypti* and effective pesticidal activity against *C. maculates*	[26]
SeNPs	Average size: 547.2 nm, zeta potential: − 9.16 mV	Oval morphology, smooth surface, and agglomeration	Efficient pesticidal activity against various instars of *Ae. aegypti* exhibiting histopathological changes	[27]

mortality in first and second larval instars of *Spodoptera litura* was obtained in presence of 100 μg of AgNPs.

In an eco-friendly route, AgNPs were also prepared by Amin [23] using aqueous extract of *Ulva lactuca* (seaweed) and 1 mM of $AgNO_3$ wherein, the reaction mixture changed its color from transparent to brown demonstrating successful formation of AgNPs. UV-vis spectral analysis also exhibited peak at 406 nm [23]. FTIR spectroscopy results showed major bands at 3,447.13, 3,227.29, and 1,639 cm^{-1} that depicted O–H/N–H stretching of alcohol group, C–H stretching of amide group, and C=O of carbonyl group, respectively. SEM images of obtained AgNPs showed its spherical morphology with an average size range of 3.89–55 nm. Antioxidant activity of biogenic AgNPs was confirmed by DPPH radical scavenging potential with IC_{50} value of around 0.3 mg/mL. Furthermore, antibacterial activity of AgNPs was observed against gram-positive organisms such as *Bacillus subtilis*, *Staphylococcus aureus*, and *Streptococcus faecalis* with zones of inhibition (ZOI) of 30, 25, and 13 mg/mm of AgNPs. Similarly, antimicrobial activity against gram-negative bacteria such as *Escherichia coli*, *Klebsiella* sp., *Pseudomonas aeruginosa*, and *Neisseria gonorrhoeae* were observed showing ZOI of 26, 17, 27, and 11 mm/mg of AgNPs, respectively. Moreover, antifungal activity was also exhibited against *Alternaria alternate*, *Aspergillus fugigatus*, *Fusarium oxysporum*, and *Penicillium* sp. having ZOI of 0.5 and 1.0 mg/mm of AgNPs, respectively. Hence, such efficient antimicrobial activity of biogenic AgNPs against different microbes that are reported to cause several agricultural diseases was demonstrated and therefore, such biogenic AgNPs were proposed to be useful as a safe alternative to pesticides.

Ayoub et al. [24] in another study synthesized copper oxide (CuO) and calcium oxide (CaO) NPs and investigated its pesticidal activity [24]. XRD patterns of both NPs revealed sharp peaks which highlighted their crystalline nature wherein, monoclinic tenorite phase of CuO and cubic-pure phase of CaO was obtained. FTIR spectra also demonstrated characteristic absorption bands at 477.0, 594.82 cm^{-1} and 551, 659 cm^{-1} that corresponded with vibrational modes of Cu–O and Ca–O bonds, respectively. Further, BET surface area of CuONPs and CaONPs were 40.54 and 23.27 m^2/g, respectively. The CuONPs were made up of aggregated nanoneedles that had a diameter ranging from 200 to 300 nm that were further arranged in a microflower-like structure with an average size of 1.0 μm as evident from Figure 9.4. CaONPs on the other hand, exhibited polygonal particles that were hexagonal in shape and had sharp edges with an average size and thickness of 2.0 μm and 100 nm, respectively. Moreover, *Cinnamomum camphora* (camphor) oil plant leaf-dip bioassay was carried out for analysis of pesticidal activity of CuONPs and CaONPs. The mortality of leaves was evaluated against *Spodoptera littoralis* after treatment with 3 days and 11 days post treatment with CuONPs and CaONPs, respectively. LC_{50} value of CuONPs was 232.75 mg/L after 3 days of treatment highlighting efficient entomotoxic effect whereas CaONPs exhibited slower entomotoxic effect with LC_{50} value of 129.03 mg/L after 11 days of treatment. Histological changes that may have occurred due to NP ingestion were also investigated wherein loss of compact appearance of the muscular layer, exfoliation, and vacuolization were observed after CuONPs and CaONPs treatment. Exoskeleton damage of *S. littoralis* through abrasion of cuticle water barrier was also evident after treatment. Reactive oxygen species generated due to treatment might also attribute in creating pores on the cell membrane leading to intracellular leakage and deterioration of protective waxy layer.

Figure 9.4: SEM images of (a) CuO and (b) CaO nanostructures, and (c, d) TEM micrographs of CuO and (e, f) CaO synthesized by wet-chemical methods [24].

Ramkumar et al. [25] also recently characterized biogenic iron NPs (FeNPs) and investigated its pesticidal activity against *Tuta absoluta* (Lepidoptera) [25]. FeNP was prepared using *Trigonella foenum-graecum* (fenugreek) plant leaf extract and 0.1 M ferric chloride solution. Visible color change of the 24-h-old reaction mixture from transparent yellow to intense black color indicated formation of FeNPs. UV-vis spectroscopy also showed absorption peak at 668 nm while FTIR spectra revealed major peaks at 3,369, 2,927, 2,364, 1,643, 1,394, and 1,037 cm^{-1} corresponding to O–H group, carboxylic acid group, N–H/C–O stretching vibration, N–H bending of amide group, C–C stretching, and C–N stretching of aliphatic amines, respectively along with additional bands at 846, 555, and 428 cm^{-1} highlighting Fe–O stretching. XRD patterns further showed crystalline nature of FeNPs wherein, two prominent peaks at 28.4° and 31.14° corresponded with the crystal planes. SEM images then demonstrated irregular morphologies of FeNPs with size range of 3.8–11.95 nm. Moreover, 100 μg/mL of biogenic FeNPs exhibited a minimum mortality of 50% at 48 h while a maximum mortality of

72% was achieved at 96 h. Likewise, LC_{50} and LC_{90} values of FeNPs against *T. absoluta* pest were 147.32 and 198.0 µg/mL, respectively.

In another study, Elango et al. [26] biosynthesized nickel NPs (NiNPs) using methanolic extract of *Cocos nucifera* (coconut) coir samples and reported its larvicidal and pesticidal activities [26]. UV-vis spectroscopy revealed absorption peak at 252 nm that implied conversion of nickel acetate salt to NiNPs after 4 h of reaction with the secondary metabolites of the extract. XRD pattern analysis then demonstrated average particle size of NiNPs to be 47.39 nm. SEM images exhibited cubical shape of NiNPs while TEM analysis showed an average size of 47 ± 2 nm with overlapping of secondary metabolites on the surface of NiNPs. Zeta potential of NiNPs was −53.9 mV that highlighted effective stability of the NPs. Larvicidal activity of biogenic NiNPs was evaluated against freshly molted larvae of *Aedes aegypti* (dengue) mosquitoes. Larval mortality of 18.4 ± 1.2% was observed in presence of 100 ppm which was increased to a maximum value of 97.5 ± 2.6% with subsequent increase in NiNPs concentration to 500 ppm. Moreover, pesticidal activity against *Callosobruchus maculatus* was also evident with a mortality rate of 96.29 ± 2.3% in presence of 20 mg/L of NiNPs.

Meenambigai et al. [27] also recently fabricated selenium NPs (SeNPs) using leaf extract of *Nilgirianthus ciliatus* and reported its pesticidal activity [27]. Selenious acid with a concentration of 30 mM was used as precursor along with 40 mM of ascorbic acid as a reduction persistence agent. Preliminary color change of the reaction mixture from orange to red was indicative of SeNPs formation that was further confirmed by UV-vis spectroscopy showing SPR band at 265 nm. The nanocrystalline SeNPs showed major peaks in FTIR spectra at 3,417.31, 2,854.19, 2,096.12, 1,714.09, and 1,023.26 cm^{-1} that were specific to O–H group vibrations of dextran, C–H stretching vibration, alkene stretching vibration, carboxyl groups, and C–H bending as well as Se-O stretching vibration, respectively. Oval morphology of smooth-surfaced SeNPs with agglomeration was observed along with a particle size of approximately more than 500 nm. DLS results also depicted an average size of SeNPs to be 547.2 nm whereas zeta potential was −9.16 mV that confirmed stability of the particles. Pesticidal activity of prepared SeNPs was tested against different instars of *Ae. aegypti* larvae. More than 70% of mosquitoes were killed within 24 h when treated with 50 µg/mL of biogenic SeNPs. Consequently, LC_{50} value of SeNPs after 24 h treatment was 3.71, 14.39, 52.14, and 203.92 µg/mL for first to fourth instar stages, respectively. Histology of *Ae. aegypti* tissues revealed complete disorder and destroyed epithelial cells, midgut, and caeca after treatment with 50 ppm of SeNPs.

9.7 Toxicity and environmental impacts

Although nanopesticides are efficient in controlling pests, indiscriminate use of such nanopesticides may result in creating a toxic impact on nontarget organisms as well. In addition, as these nanopesticides have unique physicochemical characteristics, their environmental impacts and interactions with biological systems might be varied from conventional pesticide formulations because of which present ecotoxicity protocols and regulatory guidelines of traditional pesticides may not be applicable to nanopesticides [28].

Soil microbiota plays a vital role in maintaining the integrity of the soil ecosystem. Hence, impact of nanomaterial treatments on soil microbiota needs to be investigated thoroughly. Although reports on soil microbiota alteration due to nanopesticide treatment is not extensively reported, it is demonstrated that repeated exposure of titanium oxide NPs (TiO_2NPs) result in reduction of archaeal and bacterial populations from soil that are responsible for ammonia oxidation and nitrification [29]. Likewise, nanoencapsulated bifenthrin systems were 50% more accumulated in *Eisenia fetida* and *Lumbricus terrestris* worms as compared to traditional bifenthrin [30]. *Caenorhabditis elegans* was affected in terms of growth, reproduction, locomotive behavior, and adenosine triphosphate (ATP) levels after exposure to zinc oxide NPs (ZnONPs) [28]. NPs in association with atrazine, simazine, and paraquat pesticides showed increased toxicity on *C. elegans* model [31].

Plants can also dynamically adsorb and uptake nanopesticides through receptor–ligand interactions followed by translocation to various parts of the plant. Thus, investigation of nanopesticide accumulation by nontarget plants is required to preserve the human food supply. Apart from terrestrial organisms, several aquatic creatures such as algae, *Daphnia* sp. and fishes have also shown toxic effects in presence of several NPs.

9.8 Regulatory guidelines

Environmental risk assessment (ERA) of pesticide-based product is important prior to its commercialization. A four-tiered approach is generally used for risk assessment that focusses on key drivers of impact such as standard toxicity, exposure testing, ecological models, and field monitoring which, in turn, reduce time, costs, and logistics that are required for following regulations [32]. Since nanopesticides exhibit different behavior than conventional pesticides, such ERA may thus be inappropriate. Hence, alternative methods of ERA are required to estimate the ecotoxicological and other effects of nanopesticide. Kookana et al. [32] proposed a rational framework for ERA of nanopesticides which utilizes alteration in toxicokinetics or toxicodynamics for modification in existing tests of ERA of conventional pesticides [32].

Furthermore, the Organization for Economic Co-operation and Development (OECD) has recently established risk assessment guidelines for manufactured nanomaterials. The OECD has also provided standardized testing guidelines for nanopesticide products before its registration and commercialization [33]. However, these regulatory guidelines are applied to only one type of nanopesticide which is nanoscale active ingredient without carrier such as metal NPs. Hence, proper guidelines for other nanocomposites and nanoformulations such as nanoemulsions, nanodispersions, and polymeric nanocarriers are needed to be established. Continuous updates in risk reduction strategies are required in governing nanopesticide and other nano-based formulations applicable in agriculture.

Certain initiatives such as development of user-friendly, web-based decision-support tools which can be suitable to guide stakeholders and other public sectors in order to fulfill the requirements of ERA would also support safe innovation of nanopesticides.

9.9 Conclusion and future perspectives

Nanotechnology-driven solutions for pest management seem to be a powerful strategy that can revolutionize agricultural practices [34]. Nanoformulations of the pesticides can enhance their water solubility and thereby bioavailability. It can provide solutions for agricultural applications and has the potential for targeted delivery and hence reduces nonspecific toxicity. Triggered release is another advantage of nanopesticides as rational alteration of pH or enzymatic actions are required for the release of the functionalized pesticides [35,36]. Further, RNAi molecules based disease management can be achieved using nanocarriers. However, there are many challenges that are needed to be addressed before applying the nanopesticides in the field [37].

The type of formulation should be carefully considered for the distinct type of pesticides. Generalized formulations may fail to deliver certain pesticides. Optimization studies are should be carried out to monitor the parameters required for sustained and effective release of the pesticides from the nanocomposite [38]. Toxicity and retention of the NPs in the environment must be thoroughly investigated. Entry, accumulation, and excretion profile of the nanopesticides within the plant tissue should be established.

Various bacteria, fungi, algae, and medicinal plants can synthesize pesticidal NPs [39–43]. Hence, these NPs which are often biocompatible can be explored as nanocarriers for fabrication of nanopesticides [44–46]. Similarly, natural biopolymers like CS, alginate, cellulose, and gelatine can be used for nanoformulation of pesticides. Bioactive agents from natural resources can be functionalized onto metal-doped oxides, mesoporous, and other nanocarriers for effective pest management [47]. In view of the background, agricultural nanotechnology can have pioneering contribution for developing next generation pesticides with enhanced activity and minimum adverse effects on the human health and environment.

References

[1] Hayles J, Johnson L, Worthley C, Losic D 2017. Nanopesticides: A review of current research and perspectives. In: Grumezescu AH, ed. New pesticides and soil sensors. Cambridge, Massachusetts, USA, Academic Press, Elsevier, 2017, 193–225.

[2] Chhipa H. Nanofertilizers and nanopesticides for agriculture. Environmental Chemistry Letters 2017, 15, 15–22.

[3] Kashyap PL, Xiang X, Heiden P. Chitosan nanoparticle based delivery systems forsustainable agriculture. International Journal of Biological Macromolecules 2015, 77, 36–51.

[4] Mody VV, Cox A, Shah S, Singh A, Bevins W, Parihar H. Magnetic nanoparticle drug delivery systemsfor targeting tumor. Applied Nanoscience 2014, 4, 385–392.

[5] Ekambaram P, Sathali AAH, Priyanka K. Solid lipid nanoparticles: A review. Scientific Reviews & Chemical Communications 2012, 2(1), 80–102.

[6] Barik T, Sahu B, Swain V. Nanosilica – from medicine to pest control. Parasitology Research 2008, 103(2), 253–258.

[7] Malerba M, Cerana R. Chitosan effects on plant systems. International Journal of Molecular Sciences 2016, 17(7), 996.

[8] Borel T, Sabliov C. Nanodelivery of bioactive components for food applications: Types of delivery systems,properties, and their effect on ADME profiles and toxicity of nanoparticles. Annual Review of Food Science and Technology 2014, 5, 197–213.

[9] Zhang H, Qin H, Li L, Zhou X, Wang W, Kan C. Preparation and characterization of controlled-release avermectin/castor oil-based polyurethane nanoemulsions. Journal of Agricultural and Food Chemistry 2017, 66(26), 6552–6560.

[10] Qin H, Zhang H, Li L, Zhou X, Li J, Kan C. Preparation and properties of lambda-cyhalothrin/polyurethane drug-loaded nanoemulsions. RSC Advances 2017, 7, 52684–52693.

[11] Velho MC, de Oliveira DA, da Silva Gündel S, Favarin FR, Santos RCV, Ourique AF. Nanoemulsions containing mancozeb and eugenol: Development, characterization, and antifungal activity against *Glomerellacingulata*. Applied Nanoscience 2019, 9, 233–241.

[12] Cui B, Wang C, Zhao X, Yao J, Zeng Z, Wang Y, Sun C, Liu G, Cui H. Characterization and evaluation of avermectin solid nanodispersion prepared by microprecipitation and lyophilisation techniques. PLoS One 2018, 13(1), e0191742.

[13] Cui B, Feng L, Wang C, Yang D, Yu M, Zeng Z, Wang Y, Sun C, Zhao X, Cui H. Stability and biological activity evaluation of chlorantraniliprole solid nanodispersions prepared by high pressure homogenization. PLoS One 2016, 11(8), e0160877.

[14] Cui B, Feng L, Pan Z, Yu M, Zeng Z, Sun C, Zhao X, Wang Y, Cui H. Evaluation of stability and biological activity of solid nanodispersion of lambda-cyhalothrin. PLoS One 2015, 10(8), e0135953.

[15] Sandhya KS, Kumar D, Dilbaghi N. Preparation, characterization, and bio-efficacy evaluation of controlled release carbendazim-loaded polymeric nanoparticles. Environmental Science and Pollution Research 2017, 24(1), 926–937.

[16] Kumar S, Bhanjana G, Sharma A, Sidhu MC, Dilbaghi N. Synthesis, characterization and on field evaluation of pesticide loaded sodium alginate nanoparticles. Carbohydrate Polymers 2014, 101, 1061–1067.

[17] Tong Y, Wu Y, Zhao C, Xu Y, Lu J, Xiang S, Zong F, Wu X. Polymeric nanoparticles as a metolachlor carrier: Water-based formulation for hydrophobic pesticides and absorption by plants. Journal of Agricultural and Food Chemistry 2017, 65(34), 7371–7378.

[18] Wang Y, Cui H, Sun C, Zhao X, Cui B. Construction and evaluation of controlled-release delivery system of abamectin using porous silica nanoparticles as carriers. Nanoscale Research Letters 2014, 9, 655.

[19] Abdelrahman TM, Qin X, Li D, Senosy IA, Mmby M, Wan H, Li J, He S. Pectinase-responsive carriers based on mesoporous silica nanoparticles for improving the translocation and fungicidal activity of prochloraz in rice plants. Chemical Engineering Journal 2021, 404, 126440.
[20] Cao L, Zhang H, Cao C, Zhang J, Li F, Huang Q. Quaternized chitosan-capped mesoporous silica nanoparticles as nanocarriers for controlled pesticide release. Nanomaterials 2016, 6(7), 126.
[21] Sankar MV, Abideen S. Pesticidal effect of green synthesized silver and lead nanoparticles using *Avicennia marina* against grain storage pest *Sitophilus oryzae*. International Journal of Nanomaterials and Biostructures 2015, 5(3), 32–39.
[22] Bharani RA, Namasivayam SKR. Biogenic silver nanoparticles mediated stress on developmental period and gut physiology of major lepidopteran pest *Spodoptera litura* (Fab.) (Lepidoptera: Noctuidae) – An eco-friendly approach of insect pest control. Journal of Environmental Chemical Engineering 2017, 5(1), 453–467.
[23] Amin HH. Biosynthesized silver nanoparticles using *Ulva lactuca* as a safe synthetic pesticide (in vitro). Open Agriculture 2020, 5, 291–299.
[24] Ayoub HA, Khairy M, Elsaid S, Rashwan FA, Abdel-Hafez HF. Pesticidal activity of nanostructured metal oxides for generation of alternative pesticide formulations. Journal of Agricultural and Food Chemistry 2018, 66(22), 5491–5498.
[25] Ramkumar G, Asokan R, Ramya S, Gayathri G. Characterization of *Trigonella foenum-graecum* derived iron nanoparticles and its potential pesticidal activity against *Tutaabsoluta* (Lepidoptera). Journal of Cluster Science 2021, 32, 1185–1190.
[26] Elango G, Roopan SM, Dhamodaran KI, Elumalai K, Al-Dhabi NA, Arasu MV. Spectroscopic investigation of biosynthesized nickel nanoparticles and its larvicidal, pesticidal activities. Journal of Photochemistry and Photobiology B: Biology 2016, 162, 162–167.
[27] Meenambigai K, Kokila R, Chandhirasekar K, Thendralmanikandan A, Kaliannan D, Ibrahim KS, Kumar S, Liu W, Balasubramanian B, Nareshkumar A. Green synthesis of selenium nanoparticles mediated by *Nilgirianthus ciliates* leaf extracts for antimicrobial activity on foodborne pathogenic microbes and pesticidal activity against *Aedes aegypti* with molecular docking. Biological Trace Element Research 2021, (In Press).
[28] Côa F, Bortolozzo LS, Petry R, Da Silva GH, Martins CH, de Medeiros AM, Sabino CM, Costa RS, Khan LU, Delite FS, Martinez DST. Environmental toxicity of nanopesticides against non-target organisms: The state of the art. In: Fraceto LF, S.s. de Castro VL, Grillo R, Ávila D, Caixeta Oliveira H, Lima R, eds. Nanopesticides. Springer Nature Switzerland AG, 2020, 227–279.
[29] Simonin M, Martins JM, Uzu G, Vince E, Richaume A. Combined study of titanium dioxide nanoparticle transport and toxicity on microbial nitrifying communities under single and repeated exposures in soil columns. Environmental Science & Technology 2016, 50(19), 10693–10699.
[30] Mohd Firdaus MA, Agatz A, Hodson ME, Al-Khazrajy OS, Boxall AB. Fate, uptake, and distribution of nanoencapsulated pesticides in soil–earthworm systems and implications for environmental risk assessment. Environmental Toxicology and Chemistry 2018, 37(5), 1420–1429.
[31] Jacques MT, Oliveira JL, Campos EV, Fraceto LF, Ávila DS. Safety assessment of nanopesticides using the roundworm *Caenorhabditis elegans*. Ecotoxicology and Environmental Safety 2017, 139, 245–253.
[32] Kookana RS, Boxall AB, Reeves PT, Ashauer R, Beulke S, Chaudhry Q, Cornelis G, Fernandes TF, Gan J, Kah M, Lynch I. Nanopesticides: Guiding principles for regulatory evaluation of environmental risks. Journal of Agricultural and Food Chemistry 2014, 62(19), 4227–4240.

[33] Li L, Xu Z, Kah M, Lin D, Filser. Nanopesticides: A comprehensive assessment of environmental risk is needed before widespread agricultural application. Environmental Science & Technology 2019, 53(14), 7923–7924.
[34] Ghosh S, Sarkar B, Thongmee S. Nanoherbicides for field applications. In: Ghosh S, Thongmee S, Kumar A eds. Agricultural nanobiotechnology: Biogenic nanoparticles, nanofertilizers and nanoscale biocontrol agents. Elsevier, 2022a, (In Press).
[35] Ghosh S, Sarkar B, Thongmee S, Mostafavi E. Combinations of nanobiomolecules as next generation antimicrobial agents. In: Joshi S, Lahiri D, Kar R, Nag M, eds. Alternative therapeutics: Lantibiotics and other novel drugs. Elsevier, 2022b, (In Press).
[36] Ghosh S, Sarkar B, Kumar A, Thongmee S. Regulatory affairs, commercialization and economic aspects of nanomaterials used for agriculture. In: Ghosh S, Thongmee S, Kumar A, eds. Agricultural nanobiotechnology: Biogenic nanoparticles, nanofertilizers and nanoscale biocontrol agents. Elsevier, 2022c, (In Press).
[37] Ghosh S, Sarkar B. Microbial enzymes for biodegradation and detoxification of pesticides. In: Singh J, Singh S, Garg VK, Ramamurthy PC, Pandey A, eds. Pesticides: Human health, environmental impacts and management. Elsevie, 2022, (In Press).
[38] Bhattacharya J, Nitnavare R, Shankhapal A, Ghosh S. Microbially synthesized nanoparticles: Aspect in plant disease management. In: Radhakrishnan EK, Kumar A, Raveendran A, eds. Biocontrol mechanisms of endophytic microorganisms, Cambridge, Massachusetts, USA, Academic Press, Elsevier, 2022, 303–325.
[39] Ghosh S, Webster TJ. Nanobiotechnology: Microbes and plant assisted synthesis of nanoparticles, mechanisms and applications. 1st edn, USA, Elsevier Inc, 2021, eBook ISBN: 978-0-12-823115-9; Paperback ISBN: 978-0-12-822878-4.
[40] Rokade S, Joshi K, Mahajan K, Patil S, Tomar G, Dubal D, Parihar VS, Kitture R, Bellare JR, Ghosh S. *Gloriosa superba* mediated synthesis of platinum and palladium nanoparticles for induction of apoptosis in breast cancer. Bioinorganic Chemistry and Applications 2018 Jul 2;2018:4924186. doi: 10.1155/2018/4924186. PMID: 30057593; PMCID: PMC6051271.
[41] Bloch K, Pardesi K, Satriano C, Ghosh S. Bacteriogenic platinum nanoparticles for application in nanomedicine. Frontiers in Chemistry 2021, 9, 624344.
[42] Jamdade DA, Rajpali D, Joshi KA, Kitture R, Kulkarni AS, Shinde VS, Bellare J, Babiya KR, Ghosh S. *Gnidia glauca* and *Plumbago zeylanica* mediated synthesis ofnovel copper nanoparticles as promising antidiabetic agents. Advances in Pharmacological Sciences 2019, 5, 9080279.
[43] Ranpariya B, Salunke G, Karmakar S, Babiya K, Sutar S, Kadoo N, Kumbhakar P, Ghosh S. Antimicrobial synergy of silver-platinum nanohybrids with antibiotics. Frontiers in Microbiology 2021, 11, 610968.
[44] Ghosh S. Copper and palladium nanostructures: A bacteriogenic approach. Applied Microbiology and Biotechnology 2018, 101(18), 7693–7701.
[45] Shende S, Joshi KA, Kulkarni AS, Charolkar C, Shinde VS, Parihar VS, Kitture R, Banerjee K, Kamble N, Bellare J, Ghosh S. *Platanus orientalis* leaf mediated rapid synthesis of catalytic gold and silver nanoparticles. Journal of Nanomedicine and Nanotechnology 2018, 9, 2.
[46] Bhagwat TR, Joshi KA, Parihar VS, Asok A, Bellare J, Ghosh S. Biogenic copper nanoparticles from medicinal plants as novel antidiabetic nanomedicine. World Journal of Pharmaceutical Research 2018, 7(4), 183–196.
[47] Robkhob P, Ghosh S, Bellare J, Jamdade D, Tang IM, Thongmee S. Effect of silver doping on antidiabetic and antioxidant potential of ZnO nanorods. Journal of Trace Elements in Medicine and Biology 2020, 58, 126448.

Bahman Fazeli-Nasab, Ramin Piri, Ahmad Farid Rahmani

Chapter 10
Assessment of the role of rhizosphere in soil and its relationship with microorganisms and element absorption

Abstract: The rhizosphere is the area around the plant root where a large population of microorganisms lives under the influence of chemicals secreted by plant roots. The chemicals released from plant roots are affected by plant species, soil factors, and climatic conditions, all of which together with the microbial population make the rhizosphere. The architecture of root system is very flexible. It is determined by plant species and occurs in response to changes in climatic, biological, and edaphic soil conditions. The distribution of nutrients in the soil is heterogeneous or fragmented. There is evidence that plants can sense the presence of nutrients and allocate more resources to the growth of root system, directing root growth toward these areas. Nevertheless, there is still little information about the effect of low-molecular-weight compounds on rhizosphere processes. A set of scientific efforts are being developed to unveil many of the functions of root secretions as a means of obtaining nutrients (such as iron and phosphorus uptake), creating invasive agents (such as allelopathy), being as chemical signals to attract symbiotic partners (chemotaxis) such as rhizobia and legumes or encouraging beneficial microbial colonies such as *Bacillus subtilis* and *Pseudomonas fluorescens* on the root surface. Focused but limited activities have been carried out to control plants root system (especially the rhizosphere) in order to increase the yield potential of strategic agricultural products at the Global Challenge of Sustainable Production of Food, Fuel and Fiber. However, more extensive and comprehensive research is required to compensate for global demand in the next 50 years. These efforts are taking place despite the changing global climate and increasing global population, which will inevitably require the production of more food and plants on poor and often infertile lands. This is the situation that developing countries are currently facing. Meetings on global challenges of climate change and population growth and focusing on better understanding and controlling rhizosphere processes will be one of the most important frontiers of science in the coming decades, requiring diverse and interdisciplinary trained workforce.

https://doi.org/10.1515/9783110771558-010

10.1 Introduction

Soil is one of the last great scientific frontiers, and the rhizosphere is the most active part of that frontier affecting the biochemical processes of the earth on a global scale. A better understanding of these processes is critical to maintain the health of the planet and to feed its living organisms [1–4].

Soil is an important and effective factor in the emergence and formation of any natural ecosystem. Soil can be defined as a part of limestone or regolith that is able to maintain plant life. By definition, soil is a mixture of minerals, organic matter, gases, liquids, and organisms that have made life possible together. It is the habitat of organisms and an environment to promote plant growth, to collect, supply, and purify water and to modify air temperature. All these factors together alter the properties of the soil [5–7]. Plants are an important source of soil organic matter as the major energy source for microbial activity. This issue, in most cases, is the basis of the tendency of microorganisms toward plants roots and the formation of microorganism-plant interaction both in the form of cooperation and coexistence [8–11].

The environment around plants roots is called as rhizosphere. Rhizosphere is a shallow layer of soil that is directly affected by plant roots and related microorganisms. It is actually the interactive environment of plant and soil. In fact, rhizosphere is a volume of soil that is occupied by plant root system or is affected by its activity. It is the place where the interaction between soil mineral particles, plant roots, and microbes takes place. It is an area of soil where a large number of microorganisms are found and soil particles adhering to plant root surface can be easily removed. The rhizosphere is very narrow and its width range is very variable, depending on various factors such as plant type and species, plant growth stage, plant metabolic status, soil type, and conditions as well as environmental conditions. The thickness of the rhizosphere is not constant in different plant species at different stages of plant growth. For example, most studies have shown that the number and activity of rhizosphere microbes will be high when plant growth and development is at a maximum [12–14].

Rhizosphere is a small area with different chemical, biological, and physical properties of the bulk soil. Rhizosphere is considered as the active roots – containing area, affecting the chemical and biological properties of the soil. Therefore, chemical and biological properties of rhizosphere soil can be different from bulk soil. Plant roots constantly release compounds such as sugars, amino acids, organic acids, and vitamins that can be used by microorganisms to change the properties of rhizosphere soil. Different chemical and biological properties of rhizosphere soil compared to nonrhizosphere soil can affect the shapes of elements and their usability [2, 15, 16].

The rhizosphere (common surface of soil and plants) plays an important role in the green purification of heavy metal contaminants. Actually, soil microbial populations are known to be effective in the mobility and usability of metals for plants through releasing chelates, acidification, phosphate dissolution, and changing reactions. So, they have the potential to promote green refinement. In the green refining strategy, the absorption of heavy metals by plants can be boosted more and more using special heavy metal-compatible rhizobacteria. One of the methods of green refinement which has recently been considered is to use species that establish a relationship with plant-growth promoting rhizobacteria [17, 18].

Through root secretions or changing the rhizosphere pH, plants can transform nutrient to absorb them. Studies have shown that measuring ions activity before planting in soil solution cannot well estimate the usability of elements for the plant [19].

Observations that first showed the effect of legume crop rotation on improving nonlegume plants growth go back for centuries to the Romans and Greeks. Yield improvement takes place by converting atmospheric nitrogen to ammonia (a form of nitrogen which can be used by plants) by rhizobacteria. This discovery was an inspiration to Lawrence Hiltner, as he spent his professional life on researching and promoting rhizobia inoculation to improve agricultural productivity, finally developing the concept of rhizosphere. Since then, intricate studies have been carried out on legumes – rhizobia interactions, and its various aspects have been clarified. What we have learned is that when the plant is under severe nitrogen deficiency, the first dialogue between the plant and the rhizobium begins with releasing a chemical signal (flavonoid) [20–22].

Mycorrhiza (from the Greek word meaning fungus and root) is a general term to describe the symbiotic relationship between soil fungus and plant roots. Unlike rhizobia and their legume symbionts, the mycorrhiza community is ubiquitous, being almost nonselective in 80% of all angiosperms and all gymnosperms. The ability of plants to coexist with mycorrhizae occurred earlier (about 450 million years ago) compared to that of legumes (about 60 million years ago). This is probably why mycorrhizae are now ubiquitous throughout plant evolution. Though there are some parasitic and neutral relationships, the majority of these relationships are beneficial for both the host and the fungal colony. Mycorrhiza helps the plant to obtain water, phosphorus, and other soil nutrients such as copper and zinc and in return receive carbon from the plant [23–25].

Plant growth-promoting rhizobacteria (PGPR) were first defined as organisms that, after inoculation on seeds, can successfully use plant roots to promote plant growth. To date, more than 12 nonpathogenic rhizobacterial genera have been identified. By releasing plant growth stimulants (phytohormones such as auxins and cytokines) and improving nutrient uptake (such as siderophores which increase iron availability), it can be shown that in the absence of root pathogens, plant growth stimulant directly affects plant growth. Moreover, plant growth stimulants can promote induced systemic resistance indirectly by controlling pathogens (biocontrol) through the synthesis of antibiotics or secondary metabolic mediators [20, 21, 26].

It is estimated that roots can release 10–40% of the total soil fixed photosynthetic carbon. Free carbon is present in both organic (such as low molecular weight

(LMW) organic acids) and inorganic compounds (such as bicarbonate), though organic forms have the highest diversity, showing the greatest impact on chemico-physical and biological processes in the rhizosphere [27–29].

Root-secreting compounds are classified according to their chemical composition, release method, or action. These substances are released from the root and dispersed either actively or inactively (due to osmotic differences between soil solution and cell or due to cell loss and spontaneous deterioration of the epidermis and cell membrane). Organic compounds released in these processes can be classified to low- or high-weight molecules. High-molecular-weight (HMW) compounds (cellulose and mucilage) are complex molecules that are not easily used by microorganisms. But LMW compounds such as organic acids, amino acids, and proteins are easily used by microorganisms [30–32].

The structure and amount of compounds released are affected by several factors including plant type, climatic conditions, herbivorous insects, nutrient deficiency, or excess and chemical, physical, and biological properties of the soil. The release of material from plant root into the surrounding soil is generally called rhizodeposit. It involves six main parts including sloughed-off cap and marginal cells, secretions of insoluble materials, soulable root secretions, free volatile organic carbon, carbon released by root symbionts, and carbon released due to the death and destruction of epidermal and cortical root cells. Rhizodiposits make the rhizosphere an appropriate medium to propagate microbial communities. One teaspoon of plowed soil contains more microorganisms than all the people on earth. However, the rhizosphere can have 1,000–2,000 times the number of microorganisms. Access to nutrients for microbial growth in the rhizosphere is greater than in the whole soil. Nevertheless, many microorganisms compete for these nutrients, and some are more successful than the others. The plant-microbs crosstalks in the rhizosphere that have received the most attention include rhizobia and its coexisting plant as well as mycorrhizal fungi and PGPRs [17].

The roots apply a huge amount of pressure (>7 kg/cm^2) on their growing tips. In other words, by pushing, they create a way to penetrate the soil. In order to protect and smooth the movement of the growing roots, the root cap and epidermal cells secrete a viscous, HMW, insoluble material, rich in polysaccharides. Besides lubricating, this mucus protects against dehydration and helps absorb nutrients. Most importantly, it keeps soil particles connected to each other and creates aggregates that improve soil quality by increasing the penetration of air and water. Furthermore, to protect the root-tip meristem, the root cap cells are programmed to reduce friction; otherwise, the root will be damaged. Interestingly, the weakened cells also continue to secrete mucus for several days and absorb beneficial microorganisms by accumulating root pathogens and toxic metals isolated from the roots such as AL ions as biat. Root effluents include both mucous secretions that are actively released from the roots due to osmotic differences between cells and soil solution and destroyed cortex and epidermal cells [33–35].

Organic compounds released through these processes can be further subdivided into HMW and LMW compounds. HMW compounds such as mucus and cellulose are complex molecules making up most of the carbon released by roots. They are not readily available to microorganisms. However, LMW compounds are more diverse, showing known or wider potential functions. The list of specific LMW compounds released from the root is very long, but in general they can be classified into organic acids, amino acids, proteins, sugars, phenolic compounds, and other secondary metabolites, which are generally easily consumed by microorganisms [36–38].

Obviously, rhizosphere soil has a large microbial community with high metabolic activity as compared to bulk soil. Inoculation of soil with some of plant growth-promoting bacteria may be used for some of their beneficial effects on plant growth and nutrition through different mechanisms such as nitrogen fixation, plant hormones production, and hydrophores transfer of nutrients. The thickness of the rhizosphere layer in poor soils is more. The thickness of the rhizosphere layer is reported to be about 5 mm in tomatoes and 16 mm in lupine. The rhizosphere layer will be about 1 mm, and if the soil is fertile, it will be a few millimeters. Nonrhizospheric soil (bulk soil or control) is an environment in which there are no roots or where the soil is not in contact with roots or its various characteristics are not affected by root activity. The rhizosphere soil is such that mineral particles are easily separated by shaking it in sterile water. About 2–3% of the total soil volume is rhizosphere. In fact, rhizosphere is a hot spot [39, 40].

10.2 Rhizosphere differentiation

After Hiltner, different terms are introduced for this part of the soil, such as histosphere, edaphosphere, endorhizosphere, and ectorhizosphere. In 1949, Clark coined the term rhizoplane. Based on general rhizosphere differentiation introduced by Whipps and Lynch, the rhizosphere is differentiated into distinct classes including ectorhizosphere (outer rhizosphere) which involves soil attached to living plant root cells, rhizoplane (root surface) which contains microbes that adhere to the root surface, and endorhizosphere (inside the root) which contains microbes that live inside the epidermal cells or root tissue where bacteria and cations can occupy the intercellular space (apoplastic space). The endorosphere itself is sometimes differentiated into two parts including the histosphere (root tissue) and the cortosphere (epidermis or root skin). Some biologists call the volume of soil affected by germinating seeds, along with its microbes as the spermosphere. Rhizosphere is an area that cannot be changed in size or shape, but instead possess a chemical, biological, and physical differentiation which changes both longitudinally and transversely along the root [41–43].

The rhizosphere and the biological interactions within it are concepts that date back to the beginning of the twentieth century. Nonetheless, the measurement of many functional aspects of these relationships has only recently started. Much of the carbon absorbed from the plant by the rhizosphere may be modulated by VAM fungi. The presence of mycorrhizal fungi can increase photosynthesis rate and compensate for carbon depletion caused by endophytes. Carbon flow from plant roots through fungal and microbial organisms can be one of the major processes in soil ecosystems. There are several positive feedbacks between plants and soil-microbial biological communities. Materials secreted from VAM filaments and thalli provide some energy that will support the diversity of soil microorganisms. Mycorrhizae seem to be important in maintaining soil accumulation. Soil accumulation may be attributed to physicochemical interactions between mud surfaces and partially degraded organic matter or to the physical pressures of roots and thalli that are bound together by microbial-originated polysaccharides. In most cases, loss of soil accumulation and structure reduces the soil's ability to store nutrients and water. Soil carbohydrates, accumulation, and VAM fungal populations decrease when lands are fallowed or plants are cultivated in virgin soils. Therefore, soil organisms can affect plant growth by improving soil structure [44–46].

The effect of microbiologically mediated changes on soil structure is important in sustainable agriculture. Soil structure affects nutrient penetration, soil moisture properties, and erosion potential as well as nutrient cycling. Thomas et al. showed that VAM plants could increase the frequency of water-resistant macroparticles compared to non-VAM plants; therefore, the effect of fungus-plant coexistence in soil should be part of evaluating the effects of VAM on plant growth. The plant used to be highly emphasized previously, but attention to the effects of VAM may now place emphasis on the soil. A useful research field can be the evaluation of the long-term effects of different plant and microbial communities on soil structure. If VAM fungal species are found to be effective in transporting carbon to the soil without directly affecting plant nutrition, the issue of "soil nutrition" will be really interesting [44, 47, 48].

In general, in different ecosystems, the rate of nitrogen mineralization and total soil nitrogen content indicate soil fertility. In soils, most nitrogen is organic and stable, and it is converted into absorbable forms of ammonium and nitrate by mineralization processes (ammonification and nitrification). Organic nitrogen mineralization depends on soil texture, pH, temperature, and soil moisture. Soil nitrogen management and the balance and synchronization between plant nitrogen requirements at different stages of growth and its availability in soil are of great importance. This synchronicity depends on the microbial processes of nitrogen mineralization in the soil, especially in the rhizosphere [49, 50].

The intensity of microbial processes in the rhizosphere is higher, and these reactions may be different in the rhizosphere of different plants. Root secretions significantly affect the biological activity of the rhizosphere [51, 52]. Plant roots continuously release compounds such as sugars, organic acids, amino acids, and vitamins into the rhizosphere. Therefore, the chemical and biological properties of the rhizosphere are very different from those of the nonrhizosphere parts. The release of organic carbon by plant roots and the increase of organic resources such as sludge and fertilizers enhance the population and activity of microbes [53, 54].

Moreover, microorganisms affect the usability of trace elements by affecting root growth and morphology, plant physiology and development, element shapes, and plant uptake of elements. Therefore, it is necessary to determine the zinc forms in rhizosphere soils where the absorption of this element takes place in this area, and the activity of roots and microorganisms has important effects on the soil [55, 56].

The properties of the soil which is in contact with the root (rhizosphere soil) are different from those of bulk soil. The secretions and activity of roots and microorganisms can affect different forms of elements and their usability and absorption by plants. Therefore, paying more attention to the rhizosphere environment is very important for plant growth due to its proximity to plant roots and high efficiency of nutrient uptake in this area as well as easy farmers management to provide sufficient amount of nutrients and fertilizers (providing nutrients locally and in the vicinity of the roots, instead of distributing them in the bulk soil). On the other hand, this will help to preserve and sustain the environment due to right consumption of chemical fertilizers by farmers. The specificities of rhizosphere environment in terms of plant nutrition and its uniqueness among various other plants rhizosphere can be explored for agricultural sustainability and enhancing crop production [57, 58].

10.3 Exudates and chemical signals

For nearly 200 years, scientists have found that roots secrete chemicals that stimulate or inhibit the activity of soil organisms. These organisms include microorganisms, root and seed parasites, and so on. The major role of root-secreting substances is well known, especially in case of fungi, some of which germinate only when exposed to such chemicals [59, 60], showing that strigolactones cause the branching of microrhizal fungi in the tissue. Many rhizosphere chemicals are the constituents of root cells that have emerged from living roots. Pure carbon fixation (most importantly, the roots can absorb the carbon of the rhizosphere) by the roots devotes 5–10% of plant total carbon fixation to itself [61]. Rhizosphere organisms also aid rhizosphere chemistry, releasing mineral nutrients from dead cells that are released by roots, antibiotics and antifungal drugs, phytotoxins, and mucilage [62, 63].

10.4 Rhizosphere engineering: improving the sustainable productivity of plant ecosystems

The rhizosphere is differentiated into three parts:
1. endorhizosphere (inner part of the rhizosphere) which is the part of the cortex and endoderm in which microbes and cations occupy the free inter-cellular space (apoplastic space).
2. rhizoplane (middle region), as the middle region next to root epidermal cells and mucilages (directly attached to the root including root epidermis and mucilage);
3. ectorhizosphere (outer region of the rhizosphere), as the outermost region that extends from the rhizoplane into the bullk soil [64, 65].

Root-secreting substances are classified based on their chemical composition, release method, and the action of performance. These substances are released and dispersed from the roots either actively or inactively (due to osmotic differences between soil and cell or due to cell loss and spontaneous deterioration of the epidermis and cell membrane). Organic compounds released in these processes can be classified as low- or high-weight molecules. HMW compounds (cellulose and mucilage) are complex molecules that are not readily used by microorganisms. However, LMW compounds such as organic acids, amino acids, and proteins are easily used by microorganisms [64–66].

Rhizosphere is severely affected by plant metabolism due to the release of carbon dioxide (CO_2) and the secretion of photosynthates as an array of root secretions (mainly from the rhizoplan and ectorhizosphere). Further, root secretions including phytohormones and other chemicals result in rhizosphere reaction by assisting as an energy source for microbes and works as chemical adsorbents and repellents [59]. Besides the physiological and biological interaction between the soil microbiome and plant roots, they act as communication molecules that inhibit the growth of competing plant species and facilitate beneficial symbioses by affecting the chemical, physical, and microbial properties of the soil [67]. Rhizosphere is of particular importance for carbon and water cycles, nutrient accumulation, crop production, and carbon uptake and storage in ecosystems [68]. Global climate change, including rising temperature and destructive climate patterns due to rising atmospheric CO_2 levels, will disrupt the rhizosphere ecosystem and thus affect ecosystem performance through a variety of direct and indirect methods. For example, global warming between 1981 and 2002 is estimated to reduce major grain yields by approximately 5 billion dollars a year [69].

The physical and chemical texture of the rhizosphere results from many competitive and reciprocal processes that depend mainly on soil type, composition of microbial communities, water content, and the plant's own physiology [70]. All three rhizosphere components including plant, soil, and microbes can be adjusted for better plant productivity. By varying the physical and chemical properties, the soil or its overall quality can be improved, microbiomes can be selected for essential traits such as plant growth initiation and root traits, and plants can be selected as useful and novel features of interest. Natural and artificial plant-microbial interactions can be used to improve the bioavailability of nutrients. Key chemical compounds used in root–microbe communication can be explored as genetic engineering targets to enhance these interactions in future.

10.5 Design and engineering of the rhizosphere

Plant ecosystems are valued for a variety of reasons, including nutrients, climate regulation, carbon and water cycles, food and fuel efficiency, carbon storage, wildlife habitat preparation, recreational activities, and nutrient retention. Given the wide variety of genotypes that can be harvested and/or produced in any plant, genetic diversity is a potentially important asset in maintaining or enhancing plant ecosystem values. For example, it can be used in stability control, stress tolerance in native and agricultural ecosystems [71], cultivated ecosystems, and their performance. It suggests that the selection of both species and its genotypes should be considered in breeding programs and ecosystem designing, especially in the use or reduction of potential adverse effects of climate change and the prevention of adverse effects on regional or global climate. The design or engineering of plant ecosystems for enhanced carbon storage involves increasing the allocation of carbon to groundwater or surface biomass to be allocated to structural components or transferring it to soil to be converted into reversible minerals such as calcite or soil materials. Further studies are needed on long-term carbon storage in the soil, the metabolism of rhizosphere microbial communities, and their interactions with the host plant and soil carbon sequestration mechanisms [71].

One of the main methods of modifying rhizosphere plants is through root exudates. In this regard, little effort has been made to perform rhizosphere engineering by changing root H^+ and organic anions in transgenic plants [70]. Since many leachate regulatory genes have been identified, it appears possible to alter the expression levels of these genes in plants and redesign the rhizosphere to improve traits. For example, transgenic rice and tomato plants containing the Arabidopsis AVP1 H-pyrophosphatase gene had about 50% higher citric acid and malic acid content than wild varieties when treated with the $AlPO_4$ compounds. It was interpreted as a tool to promote stress tolerance of Al^{3+} and improve its ability to utilize insoluble phosphorus [72]. However, plant engineering is a very complex process due to its impact on the rhizosphere through the degradation or inactivation of the engineered soil composition, the small amount of rhizosphere-induced exudate, limited knowledge on root exudate composition and its change over time, and exudate releasing level in relation to plant growth and external stimuli. Bioengineering of artificial microbial communities offers unparalleled benefits for increasing plant growth, disease resistance, and stress tolerance. Hundreds of bacterial strains have been identified, many of which have economic implications.

Engineering a sustainable artificial microbial community is an important challenge, requiring the consideration of six factors of environmental interaction as follows [73]:

(1) Commensalism (the relationship between two organisms in which one benefits and the other has no benefit or harm): this is a strain pressure on the other hand, without affecting it, (2) competition, (3) predation: this is in the hunter's favor, (4) no interaction, (5) cooperation: both sides benefit each other, and (6) amensalism (one side has a negative effect while the other sides remain unaffected).

The rhizosphere is an important part of the plant ecosystem. About 2 mm from the soil is considered the rhizosphere. By comparing the population density (colony forming unit) of rhizosphere soil (R) and bulk soil (S) for measuring "R/S ratio," the effect of rhizosphere on soil microbial population is measured. Most rhizosphere effects affect bacteria, fungi, actinomycetes, and protozoa, respectively. In contrast, algae show higher populations in bulk soils than in rhizosphere soils. The type of companion planting can also affect the R/S ratio, which is associated with the amount and type of root secretions. The effect of the rhizosphere is mainly regulated by plant-based organic carbon compounds and microorganism-mediated siderophores. Surface root expansion determines the effect of the rhizosphere. Plant species of solanaceae, leguminoseae, and gramineae families had a higher rhizosphere effect on available soil phosphorus (P) and soil biological properties than plants of the cruciferae and compositae families [74–76].

Phytosiderophore is an organic substance [nicotinamine, magnesium acid (MA), abionic acid, etc.] produced by plants under iron deficiency conditions and forms an organic complex or chelate complex with Fe^{3+} to form iron in the soil. LMW non-protein acids are released by pollen species under stress conditions of iron and zinc. Siderophore species affect the uptake of phytonnutrients from the rhizosphere. *Rhizopus arhizus* has a slightly higher Fe content than the phytosiderophore. The rhizoferin is not a good source of iron molecules, probably due to the exchange of iron from lysoferrin to phytosiderophores. Soil microorganisms regulate the decomposition dynamics of organic matter and the availability of plant nutrients, playing a key role in ecosystems' response to global climate change. Increased CO_2 level indirectly affects soil microorganisms by promoting root growth and root accumulation rate, as the concentration of CO_2 in soil is much higher than that of atmospheric CO_2, and the effect of high CO_2 level in soil ecosystems is primarily focused on microbial and plant processess [74, 77].

10.6 The effect of rhizosphere on the availability of plant nutrients

Plants respond to nutrient deficiencies by altering root morphology, benefiting from microorganisms, and changing the chemical environment of the rhizosphere. The compounds in root secretions help plants gain access to nutrients by altering the acidity or oxidation conditions in the rhizosphere or by direct chelation with the nutrient. These compounds help to release nutrients by dissolving insoluble mineral phases or releasing clay minerals and soil-soluble organic matter. Then these nutrients are absorbed by the plant. Phosphorus and nitrogen nutrients are the most limiting factor for plant growth, as about 78% of the nitrogen in the Earth's atmosphere is made up of N_2 gas, a form of nitrogen that can only be used for nitrogen fixation by microorganisms. The rhizosphere is highly populated with diverse microorganisms. In it, the roots compete with the neighboring plant roots for nutrients, space, water, and beneficial soil microorganisms, including fungi and bacteria. Therefore, root–root, microbial–root, and insect–root relations are likely to be biologically active in this area due to the the dynamic nature of the roots. These fascinating interactions have been largely ignored. The root–root and microbial–root relations can be either positive (including the relation of epiphytes, nitrogen-fixing bacteria, phosphate solubilizing bacteria, mycorrhizal fungi, and with the root) or negative for the plant (including harmful association with the pathogens and the parasitic plants) [78–81].

10.6.1 Nitrogen

This element is one of the most crucial plant nutrients that reduces the yield and quality of the product. Nitrogen uptake efficiency was found to reduce upto 50% evenafter adopting good management practices. In sandy soils with an annual rainfall of 100–120 cm, the N uptake efficiency may not exceed 20–30%. Additional application of nitrogen fertilizers increases cultivation costs. Soil nitrogen-fixing rhizobacteria are organic and can be easily used by plants. Rhizosphere conditions alter N_2 by nifH gene regulation process in soil as heterotrophic bacteria use organic compounds as an electron source to reduce N_2 [82].

The activity of protease, urease, and dehydrogenase in the rhizosphere soil increases with the growth of wheat, reaches a maximum at the jointing and heading stages, and decreases thereafter, respectively. Catalase activity increases with wheat growth, showing the highest value at maturity stage [83]. In other study, the activity of the enzymes viz. catalase, protease, dehydrogenase, and amylase in the rhizosphere soil was boosted for two cultivars having increasing N application [84]. Furthermore, urease activity was found to be increased with increasing nitrogen utilization as well.

Also, its highest activity was observed in 360 kg N/ha [85]. The use of $(NH_4)_2SO_4$ and urea reduces the pH of the rhizosphere by 0.22–0.29 units and 0.8–0.18 units, respectively. KNO_3 treatment increased the pH of rhizosphere in bean plants [86]. Residual nitrogen fertilizers affect the soil, changing the rhizosphere pH through a chemical reaction [87].

The redox potential (Eh) of the rhizosphere is lower than that of soil, and there is no direct relationship between the redox potential and nitrogen. The use of nitrogen fertilizer lowers in the rhizosphere but increases the N concentration in the soil solution. The reaction of potassium (K), calcium (Ca), magnesium, copper, zinc, manganese, and iron concentrations in the root zone to the type of nitrogen fertilizer differs between differently treated apple seedlings [84, 88]. In an old study, the application of different herbicides showed different rhizosphere-related nitrogenase activity in rice seedlings. Pretilachlor had no effect on nitrogenase activity at two levels of treatment, while butachlor and benthiocarb stimulated nitrogenase activity at that level [89]. Changing the rhizosphere pH using N source fertilizers is recommended.

Nitrogen sources affect the pH of the rhizosphere through three mechanisms [90]: (1) the displacement of adsorbed H/OH in the solid phase, (2) nitrification/denitrification reactions, and (3) diffusion/adsorption of H by roots in response to adsorption of the NH_4/NO_3 ratio. First, two mechanisms are not associated with any plant activity, affecting the total volume of fertile soil, while mechanism 3 is directly related to nutrient uptake, affecting a limited volume of soil near the roots [91]. The availability of soil N and its uptake by some plants decrease with increasing CO_2 in the rangeland [92, 93]. This is related to microbial N immobility, changing the quality and quantity of soil exudates [94]. Some grazing-tolerant species of the grasses are known to increase carbon inflow into the soil [95]. This microbial activity in the rhizosphere is found to stimulate the plant [96], affecting the availability of plant nutrients [97]. In a pulse of plant biomass and microbial activity, the potential for nitrogen mineralization in the rhizosphere soil enhances by plant N uptake. Nevertheless, the extent and direction of the effects of herbivores on the rhizosphere process can alter with the identity of plant species and the intensity of plant leaf loss [98]. Root LMW organic compounds were to 17% of the total carbon fixed by *Bacillus* and *Pseudomonas* species in hydroponic conditions [99, 100] and up to 17% in steppes, but their effect could not be analyzed yet.

In this way, inorganic forms of nitrogen (ammonium nitrate) that can be used by plants are added to the soil. Nitrogen availability is low in many soils due to several reasons including leaching of nitrate from rainfall, N fixation in clay particles and soil organic matter as well as bacterial denitrification. Plants show different reactions depending on the type of nitrogen in the soil. The electric charge of ammonium is positive, and plants must lose one proton (H^+) to absorb each ammonium ion, thereby lowering the pH of the rhizosphere. When nitrate is provided to the plant, the plant increases the rhizosphere pH by releasing bicarbonate ions. Such changes in the rhizosphere pH can affect the plant's access to other vital nutrients (such as zinc, calcium, and magnesium). Phosphate ion, which is the consumable form of phosphorus for plants and is highly insoluble in soils, binds strongly with calcium, aluminum, and iron oxide. Soil organic matter components provide a large amount of inaccessible phosphorus to plants [101–103].

10.6.2 Phosphorus

Phosphorus deficiency is among the most crucial nutritional problems of plants, especially in calcareous as well as acidic lands with high phosphorus retention and deposition [104]. Recently, attention has been focused on promoting phosphorus utilization efficiency (PUE) of about 15–20%. Phosphorus uptake by plants depends on the concentration gradient and its diffusion in the soil near the roots [105]. Under such conditions, it has been observed that root–soil interactions in the rhizosphere significantly impact the phosphorus availability toward the plants. Therefore, soil microbes play an important role in the mobilization and uptake of P, especially those that are able to solve the insoluble P form. The mycorrhizosphere is involved in the absorption of nutrients, especially P, copper, and zinc. Mycorrhizae is a natural solution in which fungal hyphae expand the root system to increase soil volume release organic acids, converting insoluble P in plants into an accessible form [106, 107].

Root leachate is composed of LMW organic matter (LMWOS). They are an important source of easily decomposable organic carbon. The addition of small amounts of LMWOS, such as amino acids and monosaccharides, to the soil has been shown to dramatically increase soil phosphorus levels. This is especially true for glucose, which is an energy source for microorganisms and has a significant impact on rhizosphere microbial activity and phosphorus mineralization. In contrast, amino acids, which are predominantly less important amino acids as sources of N and C in microorganisms, have less effect on P-mineralization than organic sources. Studies show that plant species significantly reduce total mineral P (Pi) in root-associated soils compared to bulk soils. The genus *Bacillus*, *Arthrobacter*, *Streptomyces*, *Pseudomonas*, *Enterobacter*, *Penicillium*, *Aspergillus*, *Mycorrhiza*, and other soil fungi are reported to dissolve insoluble phosphate [108–110].

Under phosphorus deficiency conditions, plants adopt special mechanisms for obtaining phosphate based on plant type (monocotyledonous or dicotyledonous), species, and genotypes. Plant roots can secrete organic acids such as malic acid and citric acid into the rhizosphere. This effectively lowers the pH of the rhizosphere and breaks phosphorus bonds in soil minerals [111, 112].

A species of pigeon chickpea reacts to phosphorus deficiency using another mechanism by secreting pecidic acid which releases phosphate from the iron phosphate compound. Plants also release phosphates from soil organic matter by secreting enzymes such as phosphatase. Iron deficiency causes different responses in plants depending on whether the plant is monocotyledonous or dicotyledonous. Dicots respond to iron depletion by releasing protons into the soil environment and increasing the reduction capacity of rhizodermal cells. In monocots, iron deficiency results in the secretion of phytosiderophores, such as muginic acid as a non-protein amino acid with a strong tendency to bind to iron. Phytosiderophores form strong chelates with iron and then transport them through diffusion to cells where specific plasma membrane transporters for chelated iron are present. Plants that absorb iron from the first method are called strategy 1 plants and those using the second method are called strategy 2 plants [113–115].

The processes involved in the microbial phosphate solubilization are especially the production of organic acids and thereby release of associated protons into soil solutions. Further, alkaline phosphatase activity also found to enhance from 102% to 325%. Moreover, acid phosphatase activity was reported to enhance from 205% to 455% in rhizosphere soil compared to the outer part of the rhizosphere. The N concentration and the NH_4/NO_3 ratio significantly lower down the phosphorus concentration. This may be due to an indirect effect of pH in the rhizosphere. The concentrations of K, Ca, and Mg decrease with increasing N and NH_4/NO_3 ratios in solution and increasing competition in the NH_4 adsorption environment. Numerous studies have been performed to modify phosphorites to increase their phosphorus content as well as their solubility by rhizosphere soil. The P solubility of a particular phosphorite varies depending on the type of organic matter and its rate of decomposition [116, 117].

The residual action of soil phosphate reaction products is a major concern for the conversion of phosphorus from unstable to stable forms. However, it is necessary to release phosphorus from the phosphate rock into the solution before the remaining phosphorus appears. Therefore, it may have less residual effect in the short term than when using phosphate rock. Soil has different types of P fertilizer and soil reaction. Acidic soils with rapid leaching of soluble phosphorus from phosphate rock show greater residual effects in the short term [118, 119].

Organophosphorus increases with soil development processess, but in very moist soils, it tends to reduce soil volume again. Calcium phosphate is the main mineral source for Pi only in moist, pH-neutral soils, while ferric phosphate and aluminum (Al) and Pi are bound or blocked by iron oxide and aluminum in acidic soils, dry soils. Will be gradually accepted [118]. This part of phosphorus can be made available to plants by manipulating the rhizosphere environment [120, 121].

Fungi, bacteria, as well as plants are known to evolve various mechanisms to mobilize phosphorus. Further, most studied is proton release, which lowers soil pH, releases organic and inorganic ligands such as bicarbonates and carboxyl anions and chelating agents and releases phosphatases involved in P mineralization. Increasing the concentration of CO_2 in the atmosphere promotes photosynthesis. A significant amount of root exudate is released from the roots of the plant. This increases the dissolved phosphorus in the soil [122].

With more excess phosphorus available, plants can grow faster with less than optimal CO_2. Dry matter production with field CO_2 enrichment (FACE) increases by an average of 17% on the upper soil surface of the plant and by more than 30% on the underground part. This increased growth through FACE has also been observed in crop yields, showing a 12–14% increase in yield under optimal CO_2 conditions for wheat, rice and soybeans. Under normal CO_2 conditions, the rate of photosynthesis increases, and more mineral nutrients are absorbed from the rhizospheric soil, which promotes plant growth [123].

Trifolium repens and *Plantens lanceolata* potentially boost their PUE under normal CO_2 environments by lowering the phosphorus content of the shoot as one of the components of CO_2-induced photosynthetic adaptation. In less phosphorus soils, N biomass does not increase much in meadows, but when soils with high phosphorus are in normal CO_2 conditions, N biomass is increased by 28% [124, 125].

10.6.3 Iron

Iron availability is mostly low in aerobic soils. Microorganisms and plants release LMW compounds (chelates) that increase the availability of iron [126]. In plants, there are two solutions to compensate iron deficiency. In solution 1, plants (dicotyledons and monocotyledons) release organic acid anions (a type of iron chelate), iron solubility increases with decreasing pH of the rhizosphere, and iron uptake increases with decreasing root capacity (conversion of Fe^{3+} to Fe^{2+}). In solution 2, plants (Poaceae) produce phytosiderophores that release Fe^{3+}. This substance is taken as chelate and in the form of Fe phytosiderophore. Phytosiderophores are released at the root tip for only a few hours a day [127, 128].

Iron contained in young rice phloem can be transferred as deoxy-MAFe. Phloem sap has MA deoxy as an ion equivalent. The Yoneshibu has a high pH (about 8), and under these conditions, iron preferentially binds to MA and not to other organic acids such as citric acid and malic acid. Some rhizosphere bacteria, especially *Bacillus*, *Pseudomonas*, and *Arthrobacter* produce chelates (siderophores) that form soluble Fe_2 complexes available in iron-deficient soils [129–131].

Iron and zinc content in both shoots and roots are inversely related to the pH of the rhizosphere. Manganese levels also increase with elevating pH, but a sharp decrease in manganese levels is observed in pH below 5.5. In the branches of *Phaseolus vulgaris* L. (French bean), iron, zinc, and manganese have a significant relationship with the extractable levels analyzed in the nonrhizosphere as well as rhizosphere sections [132–134].

10.6.4 Manganese

This factor plays an important role in plant growth and reduces the incidence of plant diseases. The dynamics and mobility of manganese in the rhizosphere are very similar to those of iron. Its availability in soil solutions increases with increasing acidic conditions. In alkaline soils where manganese is normally insoluble, rhizosphere action is beneficial, but in manganese-rich acidic soils, its excessive reduction can lead to manganese toxicity in sensitive plants [135, 136]. The availability of manganese in the rhizosphere is influenced by key factors such as redox status, pH, humidity, temperature, other nutrients, and heavy metal concentrations in soil solutions. In oxidized soil, manganese is present in the oxidized form (Mn_4) in the sparingly soluble mineral periolosite. Some rhizosphere bacteria, such as *Pseudomonas*, *Arthrobacterr*, *Geobacter*, *Bacillus*, and *Streptomyces*, reduce oxidized Mn^4 to Mn^2, which are beneficial metabolic chemical forms for plants. This reaction occurs in the rhizosphere, where manganese reduction requires the free flow of protons and electrons, provided by the decomposition of root carbon compounds and organic secretion [137–139].

Several manganese-reducing bacteria (e.g., *Pseudomonas*) and fungi (for example *Gaeumannomyces graminis*) play an important role in increasing its availability in the system and product uptake from soil crust. Further, antagonism between zinc and manganese uptake is ideally not predicted under normal physiological conditions. Further, it cannot be assumed that Zn and Mn have the same carrier system. However, manganese may introduce Zn_2 into root cells through a divalent cation channel system. Elevating the atmospheric mean temperature in corn increases nutrients absorption upto three times by increasing the surface area of the root and the diffusion rate of nutrients and water. It has been observed that detoxification of reactive oxygen species reduces phytohormone activity, the proper function of a number of enzymes as well as P toxicity [135, 140, 141].

In general, organic soils as well as sandy loam soils are more Zn deficient than saline and clay soils. Nevertheless, it can be improved by changing the rhizosphere soil and its physiological parameters. Zinc is usually provided to plants through a chelating system (primarily DTPA or EDTA). However, with zinc fertilizer, the concentration of zinc decreases very rapidly and within a minute. It has been reported that Zn uptake is affected by various factors such as chemical composition, soil nutrient content, pH, and temperature [142–144].

The availability of zinc in rice plants is well correlated with root environment pH as well as ions in the soil solution. Further, Zn deficiency is very common in paddy soils, neutral and calcareous soils, sodium and saline soils, poorly drained soils, peat soils, phosphorus and silicon-containig soils, very sandy and highly acidic soils, and coarse-grained soils are common. In the absence of zinc, its efficiency is related to various processess that operate in the rhizosphere system and within the plant. Utilizing Zn from larger soil volumes, conducting higher synthesis, and releasing zinc-stimulating phytosidophores by the roots in the rhizosphere, the plant's well-elongated roots improve absorption as a siderophore-zinc complex [145, 146].

The use of $Ca(NO_3)_2$ increases the concentration of zinc in *Thlaspi caerulescens* because of the higher rhizosphere pH. Further, $Ca(NO_3)_2$ is found to be more effective in comparison to urea/EDDS for stimulating the phytoxidation of zinc in moist soil. Uptake of zinc and manganese by the crops is a complex reaction. Both are positive cations, and the competition among them for adsorption depends on their concentration in the soil. Further, the cause of the lack of difference between zinc and manganese uptake by rice plants is not explained. It is expected that the rhizosphere pH should affect the availability of both the micronutrients in the rhizosphere [147]. The stability of zinc with citric acid-containing ligands, malic acid-containing ligands, or EDTA [148, 149] is higher than that of manganese (II) [150]. Therefore, at equal concentrations and based on thermodynamic considerations, it is assumed that the solubility of the zinc in the root is almost equal to that of manganese and iron. Therefore, the uptake of such metal ions by the plant mainly depends on their respective activity in the solution [151].

The utilization of chelated metal compounds involves two steps: releasing the metal ions from the set and reducing it. Therefore, increased uptake of iron and manganese by rice crops may be associated with the reduction of excess cellular volume and the secretion of organic acids and phenolics compounds by the roots. Further, NH_4OAc extraction method was found satisfactory for predicting the bioavailability process of cadmium and zinc in the rhizosphere soil. While, reducing zinc concentrations (<10 μmol/L) is beneficial for plants, animals, and human societies, high zinc concentrations are a contaminant in many environmental systems [152–155]. This decreses the pH of the rhizosphere by (1) adding acid solutions to the water that is used for irrigation and (2) indirectly changing the soil pH through the application of nitrogen in the form of ammonia nitrogen fertilizers. The first method is not suitable due to the higher input cost, but the second method is cost-effective. The rate of NH_4-mediated pH changes depends on chemical properties and volume of planting medium, irrigation management (chemical properties, dose, and frequency), plant activity, and environmental factors (including moisture, temperature, salinity, and pH.).

pH changes the availability of soil nutrients in the rhizosphere zone. Lowering the pH of the planting medium due to NH_4 nitrification and removal of root protons is an efficient way to overcome the growth problems caused by micronutrient deficiencies. Further, zinc supplied through the rhizosphere region may be correlated with strong organic chelates, and as a result, its concentration (zinc) in plant portions may be very reduced. Zinc foliar application is known to have been very effective in eliminating P toxicity [156–159].

10.6.5 Copper

Copper dynamics and mobility are similar to that of zinc element. Further, Cu is a micronutrient and cation as well. It has been observed that the root exudates and increases the availability of Cu in the rhizosphere and helps in its uptake. Root-secreting materials in dicots increase the mobility and uptake of copper in nutrient solutions and in the calcareous soils as organic-copper ligands [160]. Some copper bacteria are found in soils with high copper concentrations, reducing the concentration of copper in the soil. It is reported that MS12 and ampicillin together can decrease Cu concentration in comparison to the bacterial inoculation [161]. Further, co-inoculation of bacteria and ampicillin had reduced the copper concentration from root and shoot significantly. Addition of ampicillin was found to reduce 24% to 44% Cu in the branches and 44% in the plant roots [162]. In the wheat root region, the copper-containing portions are more affected by root activities because it changes this concentration relative to the bulk soil. The mean residual copper, carbonates, organic matter, iron-manganese oxides, and exchangeable copper were 18.8, 2.1, 1, 0.37, and 0.24 mg/kg in rhizosphere soil compared to 18.1, 2.43, 0.8, 0.42, and 0.3 mg/kg in the bulk soil [163].

The effect of rhizosphere on nutrient mobilization and its uptake by the plants is an essential activity in agriculture production. Further, manipulation of plants root activities, processess, and secretions by biotechnological tools or different types of nutrients under suitable nutrient conditions increases nutrient efficiency through better plant root architecture with roots that are thinner and longer for greater nutrients uptake or utilization. Screening for bacteria containing above-mentioned organic compounds and using them in seed treatment is beneficial. More detailed research is required on plant science challenges in the rhizosphere, including local identification and measurement of root activity which is affected by root-induced biochemical changes in the rhizosphere region, describing basic regulatory processes and mechanisms at the molecular as well as physiological levels. It helps in transforming the knowledge into rhizosphere modeling methods and root rhizosphere management strategies.

10.7 Conclusion

Roots have many functions, including obtaining essential nutrients, growth factors, and water for plant growth. The soil–root–plant relationship is formed in a versatile region in which multiple biochemical processes take place through physical activity, and variety of organic chemicals are released by plant roots and soil microbes. Such processes in turn control a set of reactions regulating soil carbon and elements cycle that maintain plant growth and have a great impact on the function and structure of the plant and bacterial community. In this way, it also affects a variety of ecosystem surface processes. Understanding and exploiting these interactions to produce sustainable fuel, fiber, and food to support the growing global population and to provide arable land will be the next challenge for the next generation.

References

[1] Berendsen RL, Pieterse CM, Bakker PA. The rhizosphere microbiome and plant health. Trends in Plant Science 2012, 17, 478–486.

[2] Simonin M, Dasilva C, Terzi V, Ngonkeu EL, Diouf D, Kane A, Béna G, Moulin L. Influence of plant genotype and soil on the wheat rhizosphere microbiome: Evidences for a core microbiome across eight African and European soils. FEMS Microbiology Ecology 2020, 96, fiaa067.

[3] Ali-Soufi M, Shahriari A, Shirmohammadi E, Fazeli-Nasab B. Investigation of biological properties and microorganism identification in susceptible areas to wind erosion in Hamoun wetlands. In: University of Zabol (ed.) Congress on restoration policies and approaches of Hamoun international wetland. Zabol, Iran, 2017, 1–10.

[4] Ali-Soufi M, Shahriari A, Shirmohammadi E, Fazeli-Nasab B. Seasonal changes biological characteristics of airborne dust in Sistan plain, Eastern Iran. International Conference on Loess Research, 2017. Gorgan University of Agricultural Sciences and Natural Resources, Gorgan, Iran.

[5] Kalev SD, Toor GS. The composition of soils and sediments. In: Torok B, Dransfield T (eds.) Green chemistry. Elsevier, Amsterdam, Netherlands 2018, 339–357.

[6] Schoonover JE, Crim JF. An introduction to soil concepts and the role of soils in watershed management. Journal of Contemporary Water Research & Education 2015, 154, 21–47.

[7] Ali-Soufi M, Shahriari A, Shirmohammadi E, Fazeli-Nasab B. Investigation of dust microbial community and identification of its dominance species in northern regions of Sistan and Baluchestan Province. Journal of Water and Soil Science (Science and Technology of Agriculture and Natural Resources) 2019, 23, 309–320.

[8] Cotrufo MF, Wallenstein MD, Boot CM, Denef K, Paul E. The M icrobial E fficiency-M atrix S tabilization (MEMS) framework integrates plant litter decomposition with soil organic matter stabilization: Do labile plant inputs form stable soil organic matter? Global Change Biology 2013, 19, 988–995.

[9] Vidal A, Watteau F, Remusat L, Mueller CW, Nguyen Tu -T-T, Buegger F, Derenne S, Quenea K. Earthworm cast formation and development: A shift from plant litter to mineral associated organic matter. Frontiers in Environmental Science 2019, 7, 55.

[10] Adamczyk B, Sietiö O-M, Straková P, Prommer J, Wild B, Hagner M, Pihlatie M, Fritze H, Richter A, Heinonsalo J. Plant roots increase both decomposition and stable organic matter formation in boreal forest soil. Nature Communications 2019, 10, 1–9.

[11] Amozadeh S, Fazeli-Nasab B. Improvements methods and mechanisms to salinity tolerance in agricultural crops. The first national agricultural conference in difficult environments, 2012. Islamic Azad University, Ramhormoz Branch.

[12] Hu L, Robert CA, Cadot S, Zhang X, Ye M, Li B, Manzo D, Chervet N, Steinger T. Van Der Heijden MG. Root exudate metabolites drive plant-soil feedbacks on growth and defense by shaping the rhizosphere microbiota. Nature Communications 2018, 9, 1–13.

[13] Gregory PJ. Roots, rhizosphere and soil: The route to a better understanding of soil science?. European Journal of Soil Science 2006, 57, 2–12.

[14] Yu Z, Wang Z, Zhang Y, Wang Y, Liu Z. Biocontrol and growth-promoting effect of Trichoderma asperellum TaspHu1 isolate from Juglans mandshurica rhizosphere soil. Microbiological Research 2021, 242, 126596.

[15] Fazeli-Nasab B, Rahmani AF. Microbial genes, enzymes, and metabolites: To improve rhizosphere and plant health management. In: Soni R, Suyal DC, Bhargava P, Goel R, Hrsg. Microbiological activity for soil and plant health management. Singapore, Springer Singapore, 2021, 459–506.

[16] Hatami N, Bazgir E, Sedaghati E, Darvishnia M. The symbiosis study of *Arbuscular Mycorrhizal* fungi with some annual herbaceous plants and the morphological identification of dominant species of these fungi in Kerman Province. Biological Journal of Microorganism (BJM) 2020, 9, 41–55.

[17] Fazeli-Nasab B, Sayyed RZ. Plant growth-promoting rhizobacteria and salinity stress: A journey into the soil. In: Sayyed RZ, Arora NK, Reddy MS, Hrsg. Plant growth promoting rhizobacteria for sustainable stress management: Volume 1: Rhizobacteria in abiotic stress management. Singapore, Springer Singapore, 2019, 21–34.

[18] Mehrban A, Fazeli-Nasab B. The effect of different levels of potassium chloride on vegetative parameters of sorghum variety KGS-29 inoculated with *mycorrhizal fungi* under water stress. Journal of Microbiology (Seoul, Korea) 2017, 10, 275–288.

[19] Halifu S, Deng X, Song X, Song R. Effects of two Trichoderma strains on plant growth, rhizosphere soil nutrients, and fungal community of *Pinus sylvestris* var. Mongolica Annual Seedlings. Forests 2019, 10, 758.

[20] Fageria N, Stone L. Physical, chemical, and biological changes in the rhizosphere and nutrient availability. Journal of Plant Nutrition 2006, 29, 1327–1356.

[21] Jin J, Tang C, Sale P. The impact of elevated carbon dioxide on the phosphorus nutrition of plants: A review. Annals of Botany 2015, 116, 987–999.

[22] Nuruzzaman M, Lambers H, Bolland MD, Veneklaas EJ. Distribution of carboxylates and acid phosphatase and depletion of different phosphorus fractions in the rhizosphere of a cereal and three grain legumes. Plant and Soil 2006, 281, 109–120.

[23] Helber N, Wippel K, Sauer N, Schaarschmidt S, Hause B, Requena N. A versatile monosaccharide transporter that operates in the arbuscular mycorrhizal fungus Glomus sp is crucial for the symbiotic relationship with plants. The Plant Cell 2011, 23, 3812–3823.

[24] Yeh C-M, Chung K, Liang C-K, Tsai W-C. New insights into the symbiotic relationship between orchids and fungi. Applied Sciences 2019, 9, 585.

[25] Estrada-Navarrete G, Cruz-Mireles N, Lascano R, Alvarado-Affantranger X, Hernández-Barrera A, Barraza A, Olivares JE, Arthikala M-K, Cárdenas L, Quinto C. An autophagy-related kinase is essential for the symbiotic relationship between Phaseolus vulgaris and both rhizobia and arbuscular mycorrhizal fungi. The Plant Cell 2016, 28, 2326–2341.

[26] Annapurna K, Kumar A, Kumar LV, Govindasamy V, Bose P, Ramadoss D. PGPR-induced systemic resistance (ISR) in plant disease management. In: Maheshwari DK (ed.) Bacteria in agrobiology: Disease management. Springer, Berlin, Heidelberg, 2013, 405–425.

[27] Jones DL, Oburger E. Solubilization of phosphorus by soil microorganisms. In: Bunemann E, Oberson A, Frossard E (eds.) Phosphorus in action. Springer, Berlin, Heidelberg, 2011, 169–198.

[28] Manning DA, Renforth P. Passive sequestration of atmospheric CO2 through coupled plant-mineral reactions in urban soils. Environmental Science & Technology 2013, 47, 135–141.

[29] Yuan H, Zhu Z, Liu S, Ge T, Jing H, Li B, Liu Q, Lynn TM, Wu J, Kuzyakov Y. Microbial utilization of rice root exudates: 13 C labeling and PLFA composition. Biology and Fertility of Soils 2016, 52, 615–627.
[30] Benizri E, Baudoin E, Guckert A. Root colonization by inoculated plant growth-promoting rhizobacteria. Biocontrol Science and Technology 2001, 11, 557–574.
[31] Deacon JW. Fungal biology. John Wiley & Sons, Hoboken, New Jersey, 2013.
[32] Roshchina VV, Roshchina VD. The excretory function of higher plants. Springer Science & Business Media, Berlin, Heidelberg, 2012.
[33] Shabala S, Shabala L, Van Volkenburgh E. Effect of calcium on root development and root ion fluxes in salinised barley seedlings. Functional Plant Biology 2003, 30, 507–514.
[34] Hamada E, Hamoud M, El-Sayed M, Kirkwood R, El-Sayed H. Studies on the adaptation of selected species of the family Gramineae A. Juss. to Salinization. Feddes Repertorium 1992, 103, 87–98.
[35] Eastin JD, Sullivan CY. Environmental stress influences on plant persistence, physiology, and production. Physiological Basis of Crop Growth and Development 1984, 8, 201–236.
[36] Xu Q, Wang C, Li S, Li B, Li Q, Chen G, Chen W, Wang F. Cadmium adsorption, chelation and compartmentalization limit root-to-shoot translocation of cadmium in rice (*Oryza sativa* L.). Environmental Science and Pollution Research 2017, 24, 11319–11330.
[37] De Lorenzo G, Ferrari S, Cervone F, Okun E. Extracellular DAMPs in plants and mammals: Immunity, tissue damage and repair. Trends in Immunology 2018, 39, 937–950.
[38] Kumar MS, Swarnalakshmi K, Annapurna K. Exopolysaccharide from Rhizobia: Production and role in symbiosis. In: Hansen et al. (eds.) Rhizobium biology and biotechnology. Springer, Singapore, 2017, 257–292.
[39] Wei X, Ge T, Zhu Z, Hu Y, Liu S, Li Y, Wu J, Razavi BS. Expansion of rice enzymatic rhizosphere: Temporal dynamics in response to phosphorus and cellulose application. Plant and Soil 2019, 445, 169–181.
[40] Zhang X, Kuzyakov Y, Zang H, Dippold MA, Shi L, Spielvogel S, Razavi BS. Rhizosphere hotspots: Root hairs and warming control microbial efficiency, carbon utilization and energy production. Soil Biology & Biochemistry 2020, 148, 107872.
[41] Surówka E, Rapacz M, Janowiak F. Climate change influences the interactive effects of simultaneous impact of abiotic and biotic stresses on plants. In: Hasanuzzaman M (ed.) Plant ecophysiology and adaptation under climate change: Mechanisms and perspectives I. Springer, Singapore 2020, 1–50.
[42] Shaw LJ, Burns RG. Biodegradation of organic pollutants in the rhizosphere. Advances in Applied Microbiology 2003, 53, 1–60.
[43] Sarhan MS, Hamza MA, Youssef HH, Patz S, Becker M, ElSawey H, Nemr R, Daanaa H-SA, Mourad EF, Morsi AT. Culturomics of the plant prokaryotic microbiome and the dawn of plant-based culture media – A review. Journal of Advanced Research 2019, 19, 15–27.
[44] Dal Cortivo C, Barion G, Ferrari M, Visioli G, Dramis L, Panozzo A, Vamerali T. Effects of field inoculation with VAM and bacteria consortia on root growth and nutrients uptake in common wheat. Sustainability 2018, 10, 3286.
[45] Sylvia DM. Distribution, structure, and function of external hyphae of vesicular-arbuscular mycorrhizal fungi. In: Box JE, Hammond LC (eds.) Rhizosphere dynamics. CRC Press, New York, 2019, 144–167.
[46] Dar M, Reshi Z. Vesicular Arbuscular Mycorrhizal (VAM) fungi-as a major biocontrol agent in modern sustainable agriculture system. Russian Agricultural Sciences 2017, 43, 138–143.
[47] Banerjee S, Walder F, Büchi L, Meyer M, Held AY, Gattinger A, Keller T, Charles R. van der Heijden MG. Agricultural intensification reduces microbial network complexity and the abundance of keystone taxa in roots. The ISME Journal 2019, 13, 1722–1736.

[48] Toju H, Kurokawa H, Kenta T. Factors influencing leaf-and root-associated communities of bacteria and fungi across 33 plant orders in a grassland. Frontiers in Microbiology 2019, 10, 241.
[49] Chen H, Shang Z, Cai H, Zhu Y. Irrigation combined with aeration promoted soil respiration through increasing soil microbes, enzymes, and crop growth in tomato fields. Catalysts 2019, 9, 945.
[50] Nazli F, Khan MY, Jamil M, Nadeem SM, Ahmad M. Soil microbes and plant health. In: Haq IU, Ijaz S (eds.) Plant disease management strategies for sustainable agriculture through traditional and modern approaches. Springer, Switzerland, 2020, 111–135.
[51] Sinegani AAS, Rashidi T. The relationship between available P and selected biological properties in the rhizosphere of ten crop species under glasshouse conditions. Spanish Journal of Soil Science: SJSS 2012, 2, 74–89.
[52] Koul B, Taak P. Soil remediation through microbes. In: Koul B, Taak P (eds.) Biotechnological strategies for effective remediation of polluted soils. Springer, Singapore, 2018, 101–128.
[53] Maryani Y, Rogomulyo R. Impact of plant exudates on soil microbiomes. In: Soni R, Suyal DC, Bhargava P, Goel R (eds.) Microbiological activity for soil and plant health management. Springer, Singapore, 2021, 265–284.
[54] Vahedi R, Rasouli-Sadaghiani M, Barin M, Vetukuri RR. Interactions between biochar and compost treatment and mycorrhizal fungi to improve the qualitative properties of a calcareous soil under rhizobox conditions. Agriculture 2021, 11, 993.
[55] Briat J-F, Gojon A, Plassard C, Rouached H, Lemaire G. Reappraisal of the central role of soil nutrient availability in nutrient management in light of recent advances in plant nutrition at crop and molecular levels. European Journal of Agronomy 2020, 116, 126069.
[56] Ali S, Mehmood A, Khan N. Uptake, translocation, and consequences of nanomaterials on plant growth and stress adaptation. Journal of Nanomaterials 2021, 2021, 1–17, 6677616.
[57] Gurbanov R, Kalkanci B, Karadag H, Samgane G. Phosphorus solubilizing microorganisms. Biofertilizers: Study and Impact 2021, 5, 151–182.
[58] Philip PS, Karthika KS, Rajimol RP. Functional nitrogen in rhizosphere. In: Cruz C, Vishwakarma K, Choudhary DK, Varma A, Hrsg. Soil nitrogen ecology. Cham, Springer International Publishing, 2021, 113–138.
[59] Bais HP, Loyola-Vargas VM, Flores HE, Vivanco JM. Root-specific metabolism: The biology and biochemistry of underground organs. Vitro Cellular & Developmental Biology – Plant 2001, 37, 730–741.
[60] Sindhu SS, Sehrawat A, Sharma R, Dahiya A, Khandelwal A. Belowground microbial crosstalk and rhizosphere biology. In: Singh DP, Singh HB, Prabha R, Hrsg. Plant-microbe interactions in agro-ecological perspectives: Volume 2: Microbial interactions and agro-ecological impacts. Singapore, Springer Singapore, 2017, 695–752.
[61] Farrar J, Hawes M, Jones D, Lindow S. How roots control the flux of carbon to the rhizosphere. Ecology 2003, 84, 827–837.
[62] Keel C, Schnider U, Maurhofer M, Voisard C, Laville J, Burger U, Wirthner PJ, Haas D, Défago G. Suppression of root diseases by Pseudomonas fluorescens CHA0: Importance of the bacterial secondary metabolite 2, 4-diacetylphloroglucinol. Molecular Plant-Microbe Interactions 1992, 5, 4–13.
[63] Yang M, Mavrodi DV, Mavrodi OV, Thomashow LS, Weller DM. Exploring the pathogenicity of Pseudomonas brassicacearum Q8r1-96 and other strains of the *Pseudomonas fluorescens* complex on tomato. Plant Disease 2020, 104, 1026–1031.
[64] McNear DH Jr. The rhizosphere-roots, soil and everything in between. Nature Education Knowledge 2013, 4, 1–16.

[65] Joshi SR, Morris JW, Tfaily MM, Young RP, McNear DH. Low soil phosphorus availability triggers maize growth stage specific rhizosphere processes leading to mineralization of organic P. Plant and Soil 2021, 459, 423–440.
[66] Dasila K, Pandey A, Samant SS, Pande V. Endophytes associated with Himalayan silver birch (Betula utilis D. Don) Roots in Relation to Season and Soil Parameters. Applied Soil Ecology 2020, 149, 103513.
[67] Nardi S, Concheri G, Pizzeghello D, Sturaro A, Rella R, Parvoli G. Soil organic matter mobilization by root exudates. Chemosphere 2000, 41, 653–658.
[68] Adl SM. Rhizosphere, food security, and climate change: A critical role for plant-soil research. Rhizosphere 2016, 1, 1–3.
[69] Lobell DB, Field CB. Global scale climate – Crop yield relationships and the impacts of recent warming. Environmental Research Letters 2007, 2, 014002.
[70] Ryan PR, Dessaux Y, Thomashow LS, Weller DM. Rhizosphere engineering and management for sustainable agriculture. Plant and Soil 2009, 321, 363–383.
[71] Ahkami AH, White III RA, Handakumbura PP, Jansson C. Rhizosphere engineering: Enhancing sustainable plant ecosystem productivity. Rhizosphere 2017, 3, 233–243.
[72] Yang H, Knapp J, Koirala P, Rajagopal D, Peer WA, Silbart LK, Murphy A, Gaxiola RA. Enhanced phosphorus nutrition in monocots and dicots over-expressing a phosphorus-responsive type I H+-pyrophosphatase. Plant Biotechnology Journal 2007, 5, 735–745.
[73] Großkopf T, Soyer OS. Synthetic microbial communities. Current Opinion in Microbiology 2014, 18, 72–77.
[74] Dotaniya M, Meena V. Rhizosphere effect on nutrient availability in soil and its uptake by plants: A review. Proceedings of the National Academy of Sciences, India Section B: Biological Sciences 2015, 85, 1–12.
[75] Tomar PC, Arora K. Response of cadaverine on the protein profiling of cultured tissues of Brassica Juncea (Rh-30) under multiple stress. Journal of Microbiology, Biotechnology and Food Sciences 2021, 10, e4002–e4002.
[76] Scheffknecht S, St-Arnaud M, Khaosaad T, Steinkellner S, Vierheilig H. An altered root exudation pattern through mycorrhization affecting microconidia germination of the highly specialized tomato pathogen *Fusarium oxysporum* f. sp. lycopersici (Fol) is not tomato specific but also occurs in Fol nonhost plants. Botany 2007, 85, 347–352.
[77] Marschner P, Crowley D, Rengel Z. Rhizosphere interactions between microorganisms and plants govern iron and phosphorus acquisition along the root axis – Model and research methods. Soil Biology & Biochemistry 2011, 43, 883–894.
[78] Goodarzi MT, Khodadadi I, Tavilani H, Abbasi Oshaghi E. The role of *Anethum graveolens* L. (Dill) in the management of diabetes. Journal of Tropical Medicine 2016, 2016, 1–11, 1098916.
[79] Jevđović D, Todorović N, Marković L, Kostić B, Sivčev L, Stanković R. Effect of fertilization on yield, seed quality and content of essential oil of anise (*Pimpinela anisum* L.) and dill (*Anethum graveolens* L.). Proceedings of the Seventh Conference on Medicinal and Aromatic Plants of Southeast European Countries,(Proceedings of the 7th CMAPSEEC), Subotica, Serbia, 27–31 May, 2012. Institute for Medicinal Plant Research “Dr Josif Pancic”, Belgrade, Serbia, 2012, 428–434.
[80] Li Y, Wang S, Lu M, Zhang Z, Chen M, Li S, Cao R. Rhizosphere interactions between earthworms and arbuscular mycorrhizal fungi increase nutrient availability and plant growth in the desertification soils. Soil and Tillage Research 2019, 186, 146–151.
[81] Wang X, Tang C. The role of rhizosphere pH in regulating the rhizosphere priming effect and implications for the availability of soil-derived nitrogen to plants. Annals of Botany 2018, 121, 143–151.

[82] Vitousek PM, Menge DN, Reed SC, Cleveland CC. Biological nitrogen fixation: Rates, patterns and ecological controls in terrestrial ecosystems. Philosophical Transactions of the Royal Society B: Biological Sciences 2013, 368, 20130119.
[83] Gahoonia TS, Nielsen NE. The effects of root-induced pH changes on the depletion of inorganic and organic phosphorus in the rhizosphere. Plant and Soil 1992, 143, 185–191.
[84] Tai X, Mao W, Liu G, Chen T, Zhang W, Wu X, Long H, Zhang B, Zhang Y. High diversity of nitrogen-fixing bacteria in the upper reaches of the Heihe River, northwestern China. Biogeosciences 2013, 10, 5589–5600.
[85] Guo R, Li X, Christie P, Chen Q, Jiang R, Zhang F. Influence of root zone nitrogen management and a summer catch crop on cucumber yield and soil mineral nitrogen dynamics in intensive production systems. Plant and Soil 2008, 313, 55–70.
[86] Thomson C, Marschner H, Römheld V. Effect of nitrogen fertilizer form on pH of the bulk soil and rhizosphere, and on the growth, phosphorus, and micronutrient uptake of bean. Journal of Plant Nutrition 1993, 16, 493–506.
[87] Gundersen P, Schmidt IK, Raulund-Rasmussen K. Leaching of nitrate from temperate forests effects of air pollution and forest management. Environmental Reviews 2006, 14, 1–57.
[88] Ring E, Högbom L, Jansson G. Effects of previous nitrogen fertilization on soil-solution chemistry after final felling and soil scarification at two nitrogen-limited forest sites. Canadian Journal of Forest Research 2013, 43, 396–404.
[89] Patnaik G, Kanungo P, Moorthy B, Mahana P, Adhya T, Rao VR. Effect of herbicides on nitrogen fixation (C2H2 reduction) associated with rice rhizosphere. Chemosphere 1995, 30, 339–343.
[90] Kopittke PM, Lombi E, Wang P, Schjoerring JK, Husted S. Nanomaterials as fertilizers for improving plant mineral nutrition and environmental outcomes. Environmental Science: Nano 2019, 6, 3513–3524.
[91] Rasool B, Ramzani PMA, Zubair M, Khan MA, Lewińska K, Turan V, Karczewska A, Khan SA, Farhad M, Tauqeer HM. Impacts of oxalic acid-activated phosphate rock and root-induced changes on Pb bioavailability in the rhizosphere and its distribution in mung bean plant. Environmental Pollution 2021, 280, 116903.
[92] Mousavi SM, Motesharezadeh B, Hosseini HM, Alikhani H, Zolfaghari AA. Root-induced changes of Zn and Pb dynamics in the rhizosphere of sunflower with different plant growth promoting treatments in a heavily contaminated soil. Ecotoxicology and Environmental Safety 2018, 147, 206–216.
[93] Yao Z, Wang R, Zheng X, Mei B, Zhou Z, Xie B, Dong H, Liu C, Han S, Xu Z. Elevated atmospheric CO2 reduces yield-scaled N2O fluxes from subtropical rice systems: Six site-years field experiments. Global Change Biology 2021, 27, 327–339.
[94] Dijkstra FA, Blumenthal D, Morgan JA, Pendall E, Carrillo Y, Follett RF. Contrasting effects of elevated CO2 and warming on nitrogen cycling in a semiarid grassland. New Phytologist 2010, 187, 426–437.
[95] Hamilton III EW, Frank DA. Can plants stimulate soil microbes and their own nutrient supply? Evidence from a grazing tolerant grass. Ecology 2001, 82, 2397–2402.
[96] Paterson E, Thornton B, Sim A, Pratt S. Effects of defoliation and atmospheric CO 2 depletion on nitrate acquisition, and exudation of organic compounds by roots of *Festuca rubra*. Plant and Soil 2003, 250, 293–305.
[97] Henry F, Vestergård M, Christensen S. Evidence for a transient increase of rhizodeposition within one and a half day after a severe defoliation of *Plantago arenaria* grown in soil. Soil Biology & Biochemistry 2008, 40, 1264–1267.

[98] Fu S, Cheng W. Defoliation affects rhizosphere respiration and rhizosphere priming effect on decomposition of soil organic matter under a sunflower species: Helianthus annuus. Plant and Soil 2004, 263, 345–352.
[99] Canarini A, Kaiser C, Merchant A, Richter A, Wanek W. Root exudation of primary metabolites: Mechanisms and their roles in plant responses to environmental stimuli. Frontiers in Plant Science 2019, 10, 157.
[100] Biondini M. Carbon and nitrogen losses through root exudation by Agropyron cristatum, a smithii and Bouteloua gracilis. Soil Biology & Biochemistry 1988, 20, 477–482.
[101] Gul I, Manzoor M, Kallerhoff J, Arshad M. Enhanced phytoremediation of lead by soil applied organic and inorganic amendments: Pb phytoavailability, accumulation and metal recovery. Chemosphere 2020, 258, 127405.
[102] Zhao XQ, Shen RF. Aluminum–nitrogen interactions in the soil–plant system. Frontiers in Plant Science 2018, 9, 807.
[103] Zhang M, Song G, Gelardi DL, Huang L, Khan E, Mašek O, Parikh SJ, Ok YS. Evaluating biochar and its modifications for the removal of ammonium, nitrate, and phosphate in water. Water Research 2020, 186, 116303.
[104] Wu H, Xiang W, Ouyang S, Forrester DI, Zhou B, Chen L, Ge T, Lei P, Chen L, Zeng Y. Linkage between tree species richness and soil microbial diversity improves phosphorus bioavailability. Functional Ecology 2019, 33, 1549–1560.
[105] Ikhajiagbe B, Anoliefo GO, Olise OF, Rackelmann F, Sommer M, Adekunle IJ. Major phosphorus in soils is unavailable, yet critical for plant development. Notulae Scientia Biologicae 2020, 12, 500–535.
[106] Lucas R, Klaminder J, Futter M, Bishop KH, Egnell G, Laudon H, Högberg P. A meta-analysis of the effects of nitrogen additions on base cations: Implications for plants, soils, and streams. Forest Ecology and Management 2011, 262, 95–104.
[107] Jiang C, Yu G, Li Y, Cao G, Yang Z, Sheng W, Yu W. Nutrient resorption of coexistence species in alpine meadow of the Qinghai-Tibetan Plateau explains plant adaptation to nutrient-poor environment. Ecological Engineering 2012, 44, 1–9.
[108] Dotaniya M, Datta S, Biswas D, Meena B. Effect of solution phosphorus concentration on the exudation of oxalate ions by wheat (*Triticum aestivum* L.). Proceedings of the National Academy of Sciences, India Section B: Biological Sciences 2013, 83, 305–309.
[109] Blagodatskaya E, Blagodatsky S, Anderson TH, Kuzyakov Y. Contrasting effects of glucose, living roots and maize straw on microbial growth kinetics and substrate availability in soil. European Journal of Soil Science 2009, 60, 186–197.
[110] Safari Sinegani AA, Rashidi T. Changes in phosphorus fractions in the rhizosphere of some crop species under glasshouse conditions. Journal of Plant Nutrition and Soil Science 2011, 174, 899–907.
[111] Kumar A, Teja ES, Mathur V, Kumari R. Phosphate-solubilizing fungi: Current perspective, mechanisms and potential agricultural applications. In: Yadav AN et al. (eds.) Agriculturally important fungi for sustainable agriculture. Springer, Cham 2020, 121–141.
[112] Kumari A, Sharma B, Singh BN, Hidangmayum A, Jatav HS, Chandra K, Singhal RK, Sathyanarayana E, Patra A, Mohapatra KK. Physiological mechanisms and adaptation strategies of plants under nutrient deficiency and toxicity conditions. In: Aftab T, Roychoudhury A (eds.) Plant perspectives to global climate changes. Elsevier, Amsterdam, Netherlands, 2022, 173–194.
[113] Ariel CE, Eduardo OA, Benito GE, Lidia G. Effects of two plant arrangements in corn (*Zea mays* L.) and soybean (*Glycine max* L. Merrill) intercropping on soil nitrogen and phosphorus status and growth of component crops at an Argentinean Argiudoll. American Journal of Agriculture and Forestry 2013, 1, 22–31.

[114] Fall AF, Nakabonge G, Ssekandi J, Founoune-Mboup H, Apori SO, Ndiaye A, Badji A and Ngom K (2022) Roles of Arbuscular Mycorrhizal Fungi on Soil Fertility: Contribution in the Improvement of Physical, Chemical, and Biological Properties of the Soil. Front. Fungal Biol. 3: 723892
[115] Kaur G, Sudhakara Reddy M. Role of phosphate-solubilizing fungi in sustainable agriculture. In: Satyanarayana T, Deshmukh SK, Johri BN, Hrsg. Developments in fungal biology and applied mycology. Singapore, Springer Singapore, 2017, 391–412.
[116] Sinegani AAS, Sharifi Z. Changes of available phosphorus and phosphatase activity in the rhizosphere of some field and vegetation crops in the fast growth stage. Journal of Applied Sciences and Environmental Management 2007, 11, 113–118.
[117] Silber A, Yones LB, Dori I. Rhizosphere pH as a result of nitrogen levels and NH 4/NO 3 ratio and its effect on zinc availability and on growth of rice flower (*Ozothamnus diosmifolius*). Plant and Soil 2004, 262, 205–213.
[118] Dotaniya M, Sharma M, Kumar K, Singh P. Impact of crop residue management on nutrient balance in rice-wheat cropping system in an Aquic hapludoll. The Journal of Rural and Agricultural Research 2013, 13, 122–123.
[119] Meena AL, Jha P, Dotaniya M, Kumar B, Meena B, Jat R. Carbon, nitrogen and phosphorus mineralization as influenced by type of organic residues and soil contact variation in vertisol of Central India. Agricultural Research 2020, 9, 232–240.
[120] Spohn M, Carminati A, Kuzyakov Y. Soil zymography – A novel in situ method for mapping distribution of enzyme activity in soil. Soil Biology & Biochemistry 2013, 58, 275–280.
[121] Spohn M, Kuzyakov Y. Spatial and temporal dynamics of hotspots of enzyme activity in soil as affected by living and dead roots – A soil zymography analysis. Plant and Soil 2014, 379, 67–77.
[122] Uroz S, Calvaruso C, Turpault M-P, Frey-Klett P. Mineral weathering by bacteria: Ecology, actors and mechanisms. Trends in Microbiology 2009, 17, 378–387.
[123] Ainsworth EA, Long SP. What have we learned from 15 years of free-air CO2 enrichment (FACE)? A meta-analytic review of the responses of photosynthesis, canopy properties and plant production to rising CO2. New Phytologist 2005, 165, 351–372.
[124] Loladze I. Rising atmospheric CO2 and human nutrition: Toward globally imbalanced plant stoichiometry? Trends n Ecology & Evolution 2002, 17, 457–461.
[125] Mikan CJ, Zak DR, Kubiske ME, Pregitzer KS. Combined effects of atmospheric CO 2 and N availability on the belowground carbon and nitrogen dynamics of aspen mesocosms. Oecologia 2000, 124, 432–445.
[126] Marschner P, Crowley D, Rengel Z. Interactions between rhizosphere microorganisms and plants governing iron and phosphorus availability. In: 19th World congress of soil science, soil solutions for a changing world, Brisbane, Australia, 2010.
[127] Sah S, Singh N, Singh R. Iron acquisition in maize (*Zea mays* L.) using Pseudomonas siderophore. 3 Biotechnology 2017, 7, 1–7.
[128] Bartucca ML, Di Michele A, Del Buono D. Interference of three herbicides on iron acquisition in maize plants. Chemosphere 2018, 206, 424–431.
[129] Suzuki M, Urabe A, Sasaki S, Tsugawa R, Nishio S, Mukaiyama H, Murata Y, Masuda H, Aung MS, Mera A. Development of a mugineic acid family phytosiderophore analog as an iron fertilizer. Nature Communications 2021, 12, 1–13.
[130] Yoneyama T. Iron delivery to the growing leaves associated with leaf chlorosis in mugineic acid family phytosiderophores-generating graminaceous crops. Soil Science and Plant Nutrition 2021, 67, 1–12.

[131] Ueno D, Ito Y, Ohnishi M, Miyake C, Sohtome T, Suzuki M. A synthetic phytosiderophore analog, proline-2′-deoxymugineic acid, is efficiently utilized by dicots. Plant and Soil 2021, 469, 123–134.
[132] Bityutskii N, Yakkonen K, Petrova A, Nadporozhskaya M. Xylem sap mineral analyses as a rapid method for estimation plant-availability of Fe, Zn and Mn in carbonate soils: A case study in cucumber. Journal of Soil Science and Plant Nutrition 2017, 17, 279–290.
[133] Vítková M, Puschenreiter M, Komárek M. Effect of nano zero-valent iron application on As, Cd, Pb, and Zn availability in the rhizosphere of metal (loid) contaminated soils. Chemosphere 2018, 200, 217–226.
[134] Khoshgoftarmanesh AH, Afyuni M, Norouzi M, Ghiasi S, Schulin R. Fractionation and bioavailability of zinc (Zn) in the rhizosphere of two wheat cultivars with different Zn deficiency tolerance. Geoderma 2018, 309, 1–6.
[135] Millaleo R, Reyes-Díaz M, Ivanov A, Mora M, Alberdi M. Manganese as essential and toxic element for plants: Transport, accumulation and resistance mechanisms. Journal of Soil Science and Plant Nutrition 2010, 10, 470–481.
[136] Surgun-Acar Y, Zemheri-Navruz F. Exogenous application of 24-epibrassinolide improves manganese tolerance in Arabidopsis thaliana L. via the modulation of antioxidant system. Journal of Plant Growth Regulation 2021, 41, 1–12.
[137] Liu P, Huang R, Hu X, Jia Y, Li J, Luo J, Liu Q, Luo L, Liu G, Chen Z. Physiological responses and proteomic changes reveal insights into Stylosanthes response to manganese toxicity. BMC Plant Biology 2019, 19, 1–21.
[138] González-Villagra J, Escobar AL, Ribera-Fonseca A, Cárcamo MP, Omena-Garcia RP, Nunes-Nesi A, Inostroza-Blancheteau C, Alberdi M, Reyes-Díaz M. Differential mechanisms between traditionally established and new highbush blueberry (*Vaccinium corymbosum* L.) cultivars reveal new insights into manganese toxicity resistance. Plant Physiology and Biochemistry 2021, 158, 454–465.
[139] Li J, Jia Y, Dong R, Huang R, Liu P, Li X, Wang Z, Liu G, Chen Z. Advances in the mechanisms of plant tolerance to manganese toxicity. International Journal of Molecular Sciences 2019, 20, 5096.
[140] Clair SBS, Lynch JP. The opening of Pandora's Box: Climate change impacts on soil fertility and crop nutrition in developing countries. Plant and Soil 2010, 335, 101–115.
[141] Fernando DR, Dyer F, Gehrig S, Capon S, Fernando AE, George A, Campbell C, Tschierschke A, Palmer G, Davies M. Nutritional traits of riverine eucalypts across lowland catchments in southeastern Australia. Australian Journal of Botany 2021, 69, 565–584.
[142] White PJ, Broadley MR. Biofortifying crops with essential mineral elements. Trends in Plant Science 2005, 10, 586–593.
[143] Rouphael Y, Kyriacou MC. Enhancing quality of fresh vegetables through salinity eustress and biofortification applications facilitated by soilless cultivation. Frontiers in Plant Science 2018, 9, 1254.
[144] Graham RD, Welch RM, Bouis HE. Addressing micronutrient malnutrition through enhancing the nutritional quality of staple foods: Principles, perspectives and knowledge gaps. Advances in Agronomy 2001, 70, 77–142.
[145] Singh B, Natesan SKA, Singh B, Usha K. Improving zinc efficiency of cereals under zinc deficiency. Current Science 2005, 88, 36–44.
[146] Verma PK, Verma S, Chakrabarty D, Pandey N. Biotechnological approaches to enhance zinc uptake and utilization efficiency in cereal crops. Journal of Soil Science and Plant Nutrition 2021, 21, 2412–2424.
[147] Kim HS, Lee DS. Proximity to chemical equilibria among air, water, soil, and sediment as varied with partition coefficients: A case study of polychlorinated dibenzodioxins/furans,

polybrominated diphenyl ethers, phthalates, and polycyclic aromatic hydrocarbons. Science of the Total Environment 2019, 670, 760–769.
[148] Vishwakarma K, Mishra M, Jain S, Singh J, Upadhyay N, Verma RK, Verma P, Tripathi DK, Kumar V, Mishra R. Exploring the role of plant-microbe interactions in improving soil structure and function through root exudation: A key to sustainable agriculture. In: Singh DP, Singh HB, Prabha R (eds.) Plantmicrobe interactions in agro-ecological perspectives. Springer, Singapore 2017, 467–487.
[149] Malhotra H, Sharma S, Pandey R. Phosphorus nutrition: Plant growth in response to deficiency and excess. In: Hasanuzzaman M et al. (eds.) Plant nutrients and abiotic stress tolerance. Springer, Singapore, 2018, 171–190.
[150] Zhang R, Yang Z, Wang Y, Wang J, Wang Y, Zhou Z. Root morphology and physiology responses of two subtropical tree species to NH4+-N and NO3−-N deposition in phosphorus-barren soil. New Forests 2021, 1–20. https://doi.org/10.1007/s11056-021-09875-w.
[151] Souri MK, Hatamian M. Aminochelates in plant nutrition: A review. Journal of Plant Nutrition 2019, 42, 67–78.
[152] Pantigoso HA, Yuan J, He Y, Guo Q, Vollmer C, Vivanco JM. Role of root exudates on assimilation of phosphorus in young and old Arabidopsis thaliana plants. PloS One 2020, 15, e0234216.
[153] Liu L, Wu L, Li N, Luo Y, Li S, Li Z, Han C, Jiang Y, Christie P. Rhizosphere concentrations of zinc and cadmium in a metal contaminated soil after repeated phytoextraction by *Sedum plumbizincicola*. International Journal of Phytoremediation 2011, 13, 750–764.
[154] Rajendra P Zinc malnutrition and its alleviation through zinc fortified cereal grains. Proceedings of the Indian National Science Academy 2009, 75, 89–91.
[155] Shivay YS, Kumar D, Prasad R. Effect of zinc-enriched urea on productivity, zinc uptake and efficiency of an aromatic rice–wheat cropping system. Nutrient Cycling in Agroecosystems 2008, 81, 229–243.
[156] Elbana TA, Selim HM, Akrami N, Newman A, Shaheen SM, Rinklebe J. Freundlich sorption parameters for cadmium, copper, nickel, lead, and zinc for different soils: Influence of kinetics. Geoderma 2018, 324, 80–88.
[157] Sacristán D, González–Guzmán A, Barrón V, Torrent J, Del Campillo MC. Phosphorus-induced zinc deficiency in wheat pot-grown on noncalcareous and calcareous soils of different properties. Archives of Agronomy and Soil Science 2019, 65, 208–223.
[158] Piri M, Sepehr E, Rengel Z. Citric acid decreased and humic acid increased Zn sorption in soils. Geoderma 2019, 341, 39–45.
[159] Suri I, Prasad R, Shivay Y. Techniques for making value added coated fertilisers and their economics. Indian Journal of Fertilisers 2009, 5, 49–58.
[160] Degryse F, Verma V, Smolders E. Mobilization of Cu and Zn by root exudates of dicotyledonous plants in resin-buffered solutions and in soil. Plant and Soil 2008, 306, 69–84.
[161] Chen YX, Wang YP, Lin Q, Luo YM. Effect of copper-tolerant rhizosphere bacteria on mobility of copper in soil and copper accumulation by Elsholtzia splendens. Environment International 2005, 31, 861–866.
[162] Tarnawski S, Hamelin J, Jossi M, Aragno M, Fromin N. Phenotypic structure of *Pseudomonas* populations is altered under elevated pCO2 in the rhizosphere of perennial grasses. Soil Biology & Biochemistry 2006, 38, 1193–1201.
[163] Motaghian H, Hosseinpur A. Effect of wheat (*Triticum aestivum* L.) rhizosphere on fractionations of copper in some sewage sludge amended soils. In: E3S Web of Conferences. EDP Sciences, 2013, 04009. doi:https://doi.org/10.1051/e3sconf/20130104009

Viphrezolie Sorhie, Pranjal Bharali, Alemtoshi

Chapter 11
Biosurfactant: an environmentally benign biological agent for sustainable agroecological agriculture

Abstract: With rising food demand, the agricultural sector invests a substantial amount of capital every year in research and development as well as other productivity-boosting measures. Although using fertilizers, pesticides, and other synthetic compounds including surfactants to boost productivity has been successful, it has also resulted in substantial environmental damage by threatening the surrounding ecosystems and eroding soil quality, resulting in a decline in cultivable acreage. Hence, the use of sustainable and greener compounds for the safer production and enhancement of agricultural productivity is the need of the hour. During the metabolic processes of certain microorganisms, biosurfactants are produced as a secondary metabolite. A variety of rhizospheric and plant-associated microbes have been shown to produce such amphiphilic biomolecules, which could be used to perform bioremediation of heavy metals and other anthropogenic pollutants by enhancing their bioavailability in polluted soil. Biosurfactants are currently gaining a lot of attention because of various advantages over synthetic counterparts, such as bioavailability, biocompatibility, biodegradability, environmental friendliness, high foaming, and stability over a broad temperature, pH, and salinity range. The most notable virtue is that they may be made from renewable agricultural waste or residues as feedstock. Many biosurfactant usages featuring agricultural benefits have been identified in recent years. Due to their antibacterial, antifungal, antiviral, larvicidal, nematocidal, and insecticidal qualities, biosurfactants are regarded useful for treating a variety of plant diseases caused by a number of phytopathogens in agronomically important plants. Further, biosurfactants can also be used to promote plant growth and development in a number of crop varieties and hydroponic systems. These broad-scale potential applications of biosurfactant make them a valuable asset and a suitable alternative for the chemically synthesized fertilizers and pesticides, which have been shown to have negative environmental consequences. Due to such unique properties, biosurfactants hold a lot of promise for their broad use in agricultural technology and enhanced sustainable agricultural output. This book chapter analyzed and conferred the potential of biosurfactant as a viable and competent multifaceted candidate in the modern agroecological agriculture practices.

https://doi.org/10.1515/9783110771558-011

11.1 Introduction

Numerous synthetic chemical agents have provided a huge majority of beneficial outcomes in controlling various biotic factors such as fungus, weeds, insects, and other plant-related diseases since the dawn of time, occasionally resulting in severe reductions in production and quantities [1, 2]. Increased crop production is essential to achieving human food demands in today's challenging environment. However, the overuse of chemical pesticides has resulted in a number of environmental issues, including hazardous chemical exposure to humans and aquatic and marine ecosystems, which has resulted in the emergence of resistant pathogens, cancellation of chemical pesticide production due to general environmental issues, substantial environmental damage, pest recurrence, pesticide tolerance, and fatal impacts on nontarget creatures [3]. As a result, consumers and government authorities are calling for a limitation in the employment of synthetic pesticides [4]. One of the most serious ecological challenges facing microbiologists and plant pathologists is the finding of environmentally acceptable options to the routinely used synthetic pesticides for combating a number of crop infections [4]. Furthermore, growing public consciousness of the importance of using environmentally benign and sustainable green products necessitates the development of innovative techniques to reduce production costs by replacing hazardous commercial amphiphilic compounds with biological amphiphiles [5].

Bacterial biosurfactants play an essential part in modern agriculture since they are both environmentally benign and cost-effective [6, 7]. In agriculture, the genera *Bacillus*, *Enterobacter*, *Erwinia*, *Pseudomonas*, and *Streptomyces* [8] have gotten a lot of interest because they generate unique compounds like biosurfactants that have antibacterial properties against plant infections. The most well-studied biosurfactants are glycolipid secreted by *Pseudomonas aeruginosa* [9] and surfactin secreted by *Bacillus subtilis* species [10–13]. Biosurfactants have become popular in commercial uses as substitutes to industrially synthesized ones due to their biodegradability, good biocompatibility, and broad spectrum of functional characteristics. Biosurfactant usage has been expanded to oil recoveries, cleaning of oil spills, and land surface bioremediation, owing to their well-documented tendency to emulsify hydrocarbon-water solutions [14]. Apart from that, a large array of microbes have also been discovered to produce biosurfactant, which has been extensively studied for potential applications including biological control because of its easiness in biodigestion in nature, low ecological toxicity, exceptional stability over a broad range of temperature, pH, and salinity, and target-specific nature [15, 16]. Biosurfactants are multifaceted molecules with a broad spectrum of structures varies from lower molecular mass compounds such as glycolipids, lipopeptides, and fatty acids to higher molecular mass compounds like polysaccharides, lipopolysaccharides, lipoproteins, and proteins, or a complex heterogeneous mixtures of these molecules [17]. Microorganisms and microbially derived metabolites have proven to be excellent biological control agents in crop protection [18]. In many instances, however, attaining considerable pathogen suppression with a single biological control agent may be difficult. The present practice is to integrate different techniques as part of an integrated pest management (IPM) framework, which could improve biopesticide performance [19]. This chapter summarizes recent research on biosurfactant as a bio-control agent and how it can help with sustainable agriculture management with minimal environmental impact.

11.2 Biosurfactant

Biosurfactants, like synthetic surfactants, are amphiphilic molecules with both hydrophilic and hydrophobic motifs which are produced or secreted by a variety of biological systems [20]. Monosaccharide, oligosaccharide or polysaccharides, short peptide or protein, phosphate groups, carboxylic acids, alcohols, and other compounds make up the hydrophilic component of a biosurfactant molecule. A single or double-bounded hydrocarbon chain, or a lengthy chain of fatty acids, hydroxy fatty acids, or -alkyl-hydroxy fatty acids, constitutes the hydrophobic group. The hydrophobic moiety is often represented by a C8 to C22 alkyl chain or an alkyl aryl derivative, either straight or branched. Chemical structures of the major classes of biosurfactants generated by diverse microorganisms while growing on water-soluble/hydrophobic substrates are depicted in Figure 11.1a–f. Apart from the ability to lower liquid surface and interfacial tensions, it can also be used to create micelles and micro-emulsions between two phases. These structures can be used to recover hydrophobic compounds, clean and/or treat contact surfaces, disrupt cell biomass, act as antimicrobial and anti-biofilm agents, introduced straightforwardly in to the preparations as an adjuvant or component, and act as wetting and emulsifying agents in the pharmaceutical, cosmetic, and food industries [21–24]. In contrast to their synthetic counterparts' polar grouping-based classification, biosurfactants are categorized depending on the molecular structure, molecular weight, physicochemical properties, mechanism of action, and microbiological source [25]. Depending on the molecular composition, biosurfactants are classed as glycolipid, lipopeptide, polymeric type, fatty acid, and phospholipid [26]. Attention toward microbially derived surfactants has risen over the years due to their natural origins and promise as a long-term substitute for synthetic surfactants [27].

Synthetic surfactants, a common ingredient in pesticide formulations, are utilized as adjuvants in pesticide formulations and are administered at random, not to state their negative environmental impacts [28]. Biological surfactants or biosurfactants are natural surfactants with numerous unique benefits unlike synthetic surfactants, including biodegradability, the capability to be generated from regenerative and inexpensive materials, less toxic or biocompatible, great precision, and consistency across a wide range of environmental variables including temperature, pH, and salinity [29]. Biosurfactants are also referred to as "green chemicals" because they are fully degradable and pose minimal to negligible environmental risk [30]. Such amphiphilic molecules are widely employed in a number of industries, including environmental remediation, pharmaceuticals, cosmetics, and food, due to their distinct desirable attributes [31–34]. Biosurfactants are reported to promote nutrient bioavailability for beneficial plant-associated bacteria and have antagonistic effects against pathogenic microorganisms, in addition to their physicochemical properties.

11.2.1 Types of biosurfactant

11.2.1.1 Glycolipids

They are composed of a carbohydrate and a fatty acid molecule, and they belong to a category that vary in the composition of the carbohydrate and lipid moiety. Based on the type of carbohydrate moiety, rhamnose lipids, trehalose lipids, sophorose lipids, cellobiose lipids (CLs), manno–sylerythritol lipids, lipomannosyl–mannitols, lipomannans and lipoarabinomannanes, diglycosyl diglycerides, monoacylglycerol, and galactosyl–diglycerids are some of the most prevalent types of glycolipids [17]. Rhamnolipid, tetrahalose, and sophorose lipids have been studied extensively and found to have better surface-active properties than other glycolipids. Microbes including *Arthrobacter* spp., *Cornebacterium* spp., *Mycobacterium tuberculosis*, *Nocardia* spp., and *Pseudomonas* spp. are known to produce glycolipids [35–37].

11.2.1.2 Rhamnolipid

Rhamnolipids are considered to be the most efficient biosurfactant producers having surface active properties and biological activities [20, 38, 39]. These are usually produced by *P. aeruginosa*, in which one or two rhamnose molecules are connected to one or two molecules of β-hydroxy decanoic acid [9, 40, 41] and have been shown to reduce the surface tension to 25–30 mN/m and the interfacial tension to 1 mN/m against *n*-hexadecane. Rhamnolipids can be employed as agriculture additives, according to certain studies because they can operate as direct antibacterial as well as a stimulating agent to boost plant immunity [34].

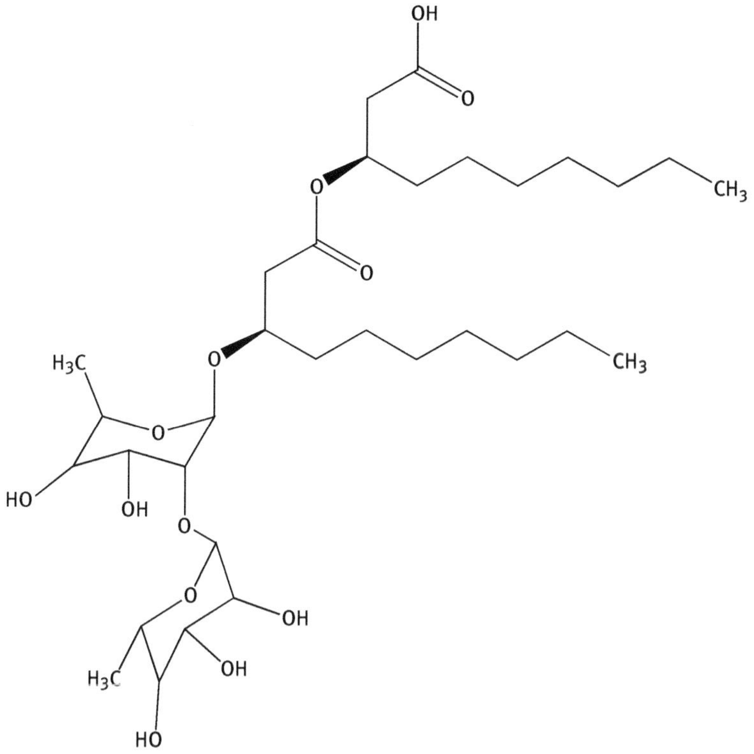

Figure 11.1a: Chemical structure of first identified rhamnolipid; α-L-rhamnopyranosyl-α-L-rhamnopyranosyl-β-hydroxyde-canoyl-β-hydroxydecanoate and symbolized as Rha-Rha-C10-C10; IUPAC names: (*R*)-3-{(*R*)-3-[2-*O*-(α-L-rhamnopyranosyl)-α-L-rhamnopyranosyl]oxydecanoyl} oxydecanoate.

11.2.1.3 Tetrahalolipids

Tetrahalolipids are a type of lipid in which the disaccharide tetrahalose is connected to mycolic acid at C-6 and C-6′, and they have been shown to exhibit surface-active properties. Species of *Mycobacterium*, *Nocardia*, and *Corynebacterium* are some examples producing tetrahalose lipids [42]. The quantity of esterified fatty acids and their overall chain length (C-20 to C-90) vary among these glycolipids. Rhodococcus strain H13-A produces a nonionic trehalose lipid, which is made up of one major and ten minor components [42]. Novel forms of trehalolipids have been identified in several investigations, including mono-, di-, tetra-, hexa-, and octa-acylated trehalose derivatives, trehalosetetraesters, and succinoyltrehalose lipids.

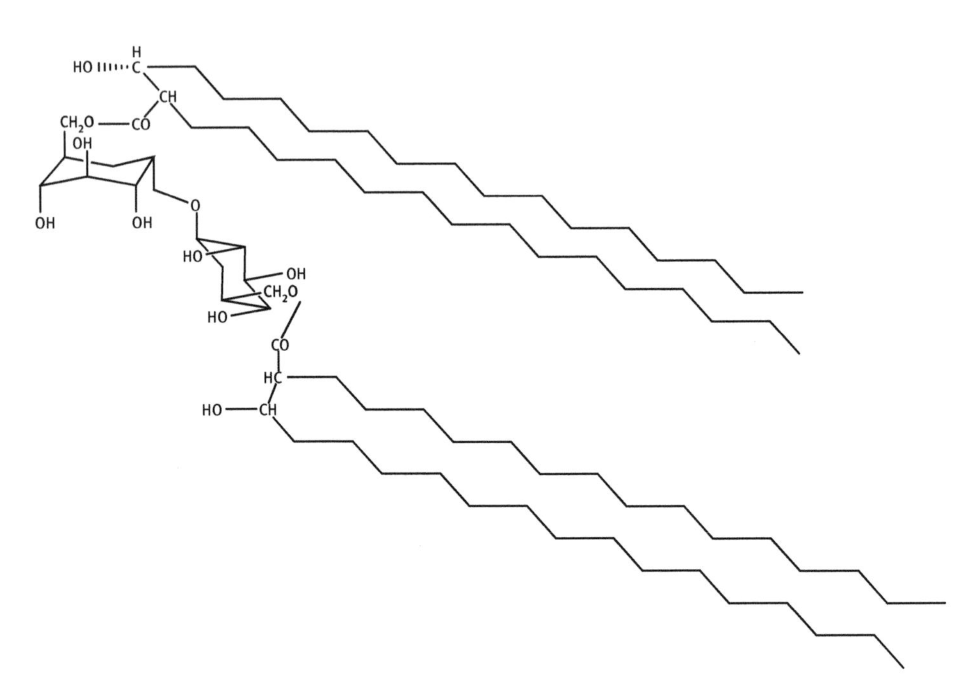

Figure 11.1b: Structure of trehalose lipid with carbon chain upto C-15. IUPAC name; ((1*R*,2*R*,3*R*,4*S*,5*R*)-2,3,4-trihydroxy-5-((((1*R*,2*R*,3*S*,4*R*,5*R*)-2,3,4-trihydroxy-5-(((3-hydroxy-2 pentadecyloctadecanoyl)oxy)methyl)cyclohexyl)oxy)methyl)cyclohexyl)methyl(3*R*)-3-hydroxy-2-pentadecyloctadecanoate.

11.2.1.4 Sophorolipids

Torulopsis bombicola [43], *Candida bogoriensi* [44], and *Torulopsis petrophilum* [45] produce a dimeric carbohydrate sophorose coupled to a long-chain hydroxy fatty acid. Their structures can come in a number of different forms, as a result of a mixture of structurally similar molecules, with up to 40 distinct types and associated isomers. Their structures can come in a number of different forms, as a result of a mixture of structurally similar molecules, with up to 40 distinct types and associated isomers from β-glycosidic bond linked to anomeric carbon of sophorose to terminal and subterminal carbons of fatty acids to the fatty acid chains varying in size with the presence of unsaturation and presence of stereoisomers. Sophorolipids can reduce surface and interfacial tension; however, they are not good emulsifiers [46]. They are environmentally friendly, highly biodegradable, have lesser toxicity, elevated selectivity, and specific action over a wide environmental conditions (pH, temperature, and salinity range) and have a lot of promise for use in fields including agriculture, biomedicine, food, cosmetics, bioremediation, and enhanced oil recovery [47].

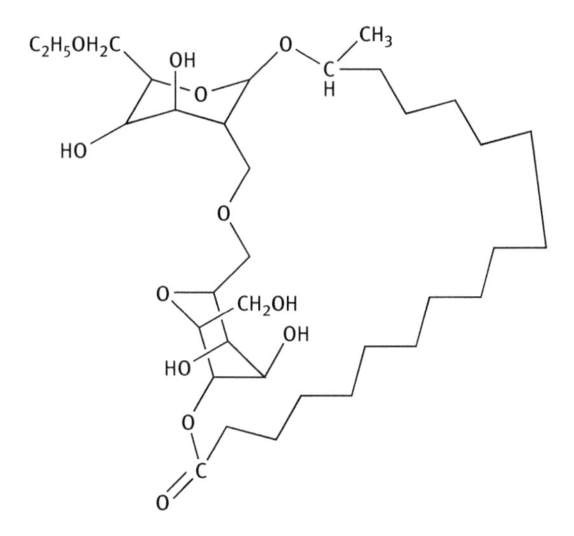

Figure 11.1c: Molecular structure of sophorolipid from *C. combicola*; IUPAC name: (2*R*,4*R*,4a*R*,8*S*,9*R*,11*S*,31*R*)-2-(ethoxymethyl)-3,4,9,10-tetrahydroxy-31-(hydroxymethyl)-29-methyltetracosahydro-2H,5H,7H,13H-8,11-(epoxymethano)pyrano[2,3-b][1, 5, 11] trioxacyclooctacosin-13-one.

11.2.1.5 Mannosylerythritol lipids

Mannosylerythritol lipids (MELs) are amphiphilic molecules categorized as both glycolipids and biosurfactants. MEL is a biosurfactant with 4-*O*-β-D-mannopyranosylmeso-erythritol as the hydrophilic group and a fatty acid and/or an acetyl group as the hydrophobic moiety secreted by *Ustilago* spp. [48], *Kurtzmanomyces* spp. [49], and *Pseudozyma* spp. [50] as a minor component along with CL.

It is important to note that MELs' biological and surface-active properties are closely related to the chemical structure of mannosylerythritol lipids, which is influenced by the microbial fermentation lineage utilized and also the carbon and nitrogen sources in the culture medium. MELs are usually known to be utilized in industrial and medical application; however, some have reported their use as pesticide against certain pathogenic fungi [51]. They are characterized by functional properties, foaming, superior dispersion, emulsification, wetting, solubilization, minimal toxicity, and optimal yields. One of the promising agronomic utilization of MELs, according to Yoshida et al. [51], is as pesticide against a wide array of phytopathogenic bacteria. According to Fukuoka et al. [52], the application of MEL as a potential agrospreader in agriculture has shown to be extremely important due to its high wettability of hydrophobic surfaces and the attachment of microbes to plant surfaces. The effectiveness of MEL produced by yeast strain *Pseudozyma antartica* T-34 [53] was tested against Silwet L77 which is a commercial agrospreader, and the results show a markedly high activity of MEL when used as agrospreader mixed with cells of *B. subtilis* suspension on wheat leaf surfaces.

Figure 11.1d: Molecular structure of mannosylerythritol lipids from *Ustilagomaydis*; IUPAC name: (2*R*,3*S*,4*R*,5*S*,6*R*)-5-(hept-1-en-2-yloxy)-4-(palmitoyloxy)-6-((2*R*,3*S*)-2,3,4-trihydroxybutoxy) tetrahydro-2H-pyran-2,3-diyl diacetate.

11.2.1.6 Phospholipid

Every microorganism has phospholipids in their cellular structure [54]. Examples of considerable extracellular phospholipid synthesis or the measurements of surfactant characteristics of these lipids are less common. All of them have a glycerol unit esterified to two fatty acids and one phosphate group, which is generally substituted further. There is only one fatty acid ester in lysophospholipids. The nature of the phospholipid mixtures formed by microorganisms is influenced by their substrates. When *Candida tropicalis* was cultivated on *n*-alkanes, it produced considerably more phospholipids than when it was fed on glucose [55].

Figure 11.1e: General molecular structure of lysophospholipid; R_1 = alkyl substituents, X = phosphatidic acid, phosphatidyl serine, inositol, phosphatidyl inositol, phosphatidylglycerol, etc.

11.2.1.7 Fatty acids and neutral lipids

Fatty acids and neutral lipids are found throughout all microbes and are commonly generated extracellularly. Surface activity is present in the majority of these lipids, which include alcohols, carboxylic acids, esters, mono-, di-, and triglycerides. The neutral lipids tested, on the other hand, have typically come from nonmicrobial sources [56]. Microbes thriving on hydrocarbons [57] and other species such as *Arthrobacter parafineu*, *Mycobacterium rhodochrou*, and *Corynebacterium lepus* [58] are reported to produce neutral lipids or fatty acids extracellularly in the majority of instances. Holdorn and Turner [59] used ether to extract a surface-active compound from a *Mycobacterium rhodochrous* saturated culture medium enriched with decane (C10). The ether extracts emulsified water and decane and boosted *M. rhodochrous* growth on decane, presumably because it contained neutral lipids. The surface tension was lowered to 44 dyn/cm when the extract was introduced to water (0.1 g/L).

11.2.1.8 Lipopeptides

Lipopeptides have been obtained from a broad array of bacteria and yeasts, but not many have been properly described and investigated for surfactant characteristics. The most potent biosurfactant documented in the literature, however, is a lipopeptide generated by *B. subtilis* [60, 61]. A chain of seven amino acids is covalently connected to the carboxyl function on one end and the hydroxyl function on the other end of a 3-hydroxy fatty acid. Glutamic acid and aspartic acid are the only amino acid residues with free carboxylic acid functions; the others have minor alkyl groups. Many different lipopeptides have been discovered from bacteria, but in spite of their amphipathic architectures, there have been few studies of their surfactant activities. A large number of cyclic lipopeptide antibiotics (polymyxins), produced by *P. polmyxa*, have also been reported to have remarkable surface-active properties showing inhibitory action against single and mixed biofilms [62]. *Bacillus subtilis* produces surfactin, which is a crystalline lipopeptide. It has a molecular weight of 1050 and is composed of L-aspartic acid, L-glutamic acid, L-valine, D-leucine, and fatty acids. Surfactin is an acidic compound that dissolves in alkaline water and a variety of organic solvents. When grown on a glucose carbon source, a strain of *Bacillus lichenifarmis* known as JF2 produces an ionic biosurfactant containing a free amino group and a lipid moiety [63].

The lipopeptide produced by *Corynebacterium lepus* lowers the surface tension of pure water to 52 dyn/cm. This lipopeptide is composed of 35% protein and the rest is carboxylic acids. In such biosurfactant, saturated fatty acids account for 25% of the total, whereas corynomycolic acids account for 75%. Both the types feature a wide spectrum of isomers, which can be attributed to their growth on an alkane-rich substrate [64].

Lipopeptides are used in agriculture to eradicate phytopathogens and arouse defense responses as well as to reduce surface tension, dispersion, stabilization of pesticide and powdered fertilizer suspension, emulsifying pesticide solutions, and enabling the functioning of biocontrol mechanisms [30, 65].

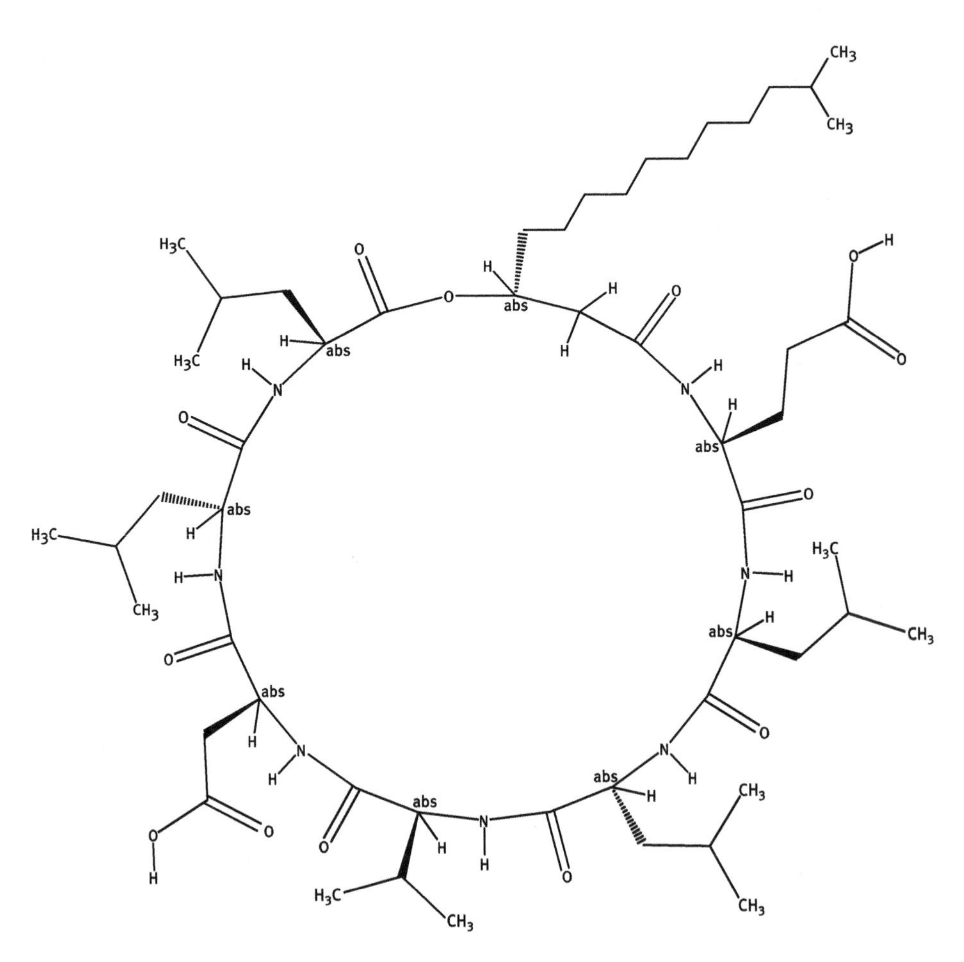

Figure 11.1f: Structure of surfactin; IUPAC anme; 3-((3*S*,6*R*,9*S*,12*S*,15*R*,18*S*,21*S*,25*R*)-9-(carboxymethyl)-3,6,15,18-tetraisobutyl-12-isopropyl-25-(10-methylundecyl)-2,5,8,11,14,17,20,23-octaoxo-1-oxa-4,7,10,13,16,19,22-heptaazacyclopentacosan-21-yl)propanoic acid.

11.3 Role of biosurfactant in agriculture

11.3.1 Biosurfactant as a bioremediation agent for polluted soil

Surfactants have been thoroughly explored and are a well-defined approach for polluted soil cleanup. Surfactants increase the solubility and bioavailability of insoluble pollutants in soil, making them more suitable for bioremediation [66]. Biosurfactants have greater benefits than chemically manufactured surfactants due to their twofold hydrophobic and hydrophilic characteristics. These molecules have a wide range of agricultural applications, including improving pollutant biodegradation, improving soil quality, and indirectly promoting plant growth through antimicrobial activity [32, 66]. They can also be utilized to boost plant-microbe interactions, which is advantageous for plant growth. Since soil-dwelling bacteria use these natural surfactants as a carbon source, they could eventually replace the harsh surfactants employed in pesticide manufacture [67]. Bioremediation is an effective and environmentally acceptable method of removing contaminants from the environment but working with huge quantities of polluted materials requires time and is inefficient. Some approaches, like adopting soil washing to extract pollutants from soil without causing chemical damage, could, however, significantly accelerate biodegradation [68, 69].

Antibiotics and increased number of genes with antibiotic resistance genes (ARGs) and their combinations, which are found in soils conducive to crop production, can endanger human health [65]. As a result, additional research is being focused on developing novel ways for soil restoration and ensuring food security.

According to Ye et al. [70], two consecutive washings with 20 g/L sophorolipid solution along with ultrasonication (35 kHz) efficaciously removed cadmium, roxithromycin, sulfadiazine, and tetracycline by 71.2%, 100%, 96.6%, and 88.2%, respectively. As a result, the levels of ARGs have dropped to 10^{-7} and 10^{-8}. The lettuce cultivation after the second washed soil revealed a decrease in ARG richness in lettuce tissues as well as separate quantities of antibiotic-resistant bacterial endophytes as well as a significant improvement in lettuce growth.

Plant development is boosted by biosurfactant-producing bacteria, which boost adaptive immune responses against organic pollutants in the environment. Furthermore, they are effective in reducing plant stress reactions as well as promoting plant growth and development [71].

11.3.1.1 Sequestration of heavy metals

Heavy metals are found in agricultural soil as pollutants [72]. Heavy metal contamination is caused by the overuse of metal salt-based fungicides as well as sewage and sludge additives in agricultural areas [72, 73]. These heavy metals despite their toxicity are necessary for a variety of crucial physiological processes in plant metabolism and act as vital micronutrients. However, in higher concentrations, it can impair plant growth, causing necrosis of root tissue and purpling of the foliage [74]. Miller [75] has successfully proven the bioremediation of metal-contaminated sewage streams using biosurfactants produced by microbes. To assist in metal removal, entire cells or microbial exopolymers are typically utilized to concentrate and/or precipitate metals in the waste stream. The following are the metal binding methods: (1) anionic biosurfactants produce nonionic complexes with metals via ionic bonding. As the interfacial tension is lowered, these bonds are stronger than the metal's bonds with the soil/sediment, and metal-biosurfactant complexes are desorbed from the soil matrix to the solution and (2) the cationic biosurfactant can compete for some negatively charged surfaces, but not all, and can replace the same charged metal ions (ion exchange). Biosurfactant micelles can remove metal ions from dirt surfaces. The polar head groups of micelles bind metals that have been mobilized in water [76]. In terms of size, biosurfactants have a significant advantage over whole cells or exopolymers; biosurfactant molecular weights are frequently less than 1,500. The literature on the use of biosurfactant as a bioremediation agent for heavy-metal-contaminated soils and wastewater is summarized in Table 11.1. Another advantage is that biosurfactants are available in a variety of chemical configurations, each with its own metal selectivity and, as a result, metal removal efficiency [75]. Mulligan and colleagues [77] conducted batch soil washing studies to assess the possibility of employing surfactin from *B. subtilis*, a lipopeptide biosurfactant, to remove heavy metals from polluted soil and sediments. After five washings with 0.25% surfactin (1% NaOH), 70% of the copper and 22% of the zinc were removed from the soil. In an another study, surfactin was found to be capable of removing metals through sorption and metal complexation at the soil interphase, accompanied with desorption of the metal through lowering of interfacial tension and fluid forces, and finally metal complexation with micelles [77]. San Martn et al. [78] recently revealed that rhamnolipid produced by *P. aeruginosa* PAO1 was found to be effective in removing Vanadium from contaminated sediments.

11.3.1.2 Petroleum-based hydrocarbons

The major causes of environmental damage are accidental spills and the intentional discharge of oil or oily waste [79]. Growth-rate limiting factors include nitrogen deficit in oil-soaked soil and nutritional inadequacies such as phosphorus [80, 81]. Additionally, large levels of biodegradable organics in agricultural soils deplete oxygen stores and impede oxygen diffusion to deeper layers. Since oxygen and nitrogen are limiting factors throughout all types of petroleum degradation, crude-oil contamination tends to persist in soils unless nutrient-based remedial strategies are used [82]. One of the possible solutions is bioremediation, which employs the natural degradative activity of microorganisms, most commonly bacteria and fungi, to transform pollutants toward less hazardous molecules, or preferably carbon dioxide and water [83]. As a result, the US Environmental Protection Agency (EPA) has proposed a number of strategies for cleaning petroleum hydrocarbon-contaminated soil, including chemical, physical, and biological approach. Biosurfactants in PAH-contaminated soil washing solutions lower surface and interfacial tension at the oil-water and air-water interfaces, lessening the capillary force that keeps the oil and soil together and potentially leading to oil mobilization and solubilization [84]. When compared to synthetic surfactants such as Triton X-100 and Tween-80, rhamnolipid biosurfactant derived from *P. aeruginosa* was found to be more effective in the remediation of total petroleum hydrocarbon-contaminated soil [85].

Table 11.1: Recent advances on the use of biosurfactant for bioremediation of heavy-metal-contaminated soils and waters.

Microorganism	Type of biosurfactant	Target pollutants for remediation	Description	References
–	Rhamnolipid	Cu, Cd, Pb, Cr	Rhamnolipid biosurfactant purchased from Huzhou Gemking Biotechnology, Co., Ltd, Zhejiang, China as a washing agent was effective in removing heavy metals from sediment.	[245]
Rahnella **sp.** RM	Not determined	Cu, Pb, Cr	Biosurfactant producing strain from *Rahnella* sp. RM was optimized for heavy metal remediation of Cu, Cr and Pb. The maximum removal rate of Cu was observed 74.3% at the 100 mg/L.	[246]
–	Rhamnolipid	Cu, Zn, Cr, Pb, Ni, Mn	Heavy metal removal efficiency of an electrokinetic decontamination treatment. enhanced by a biodegradable complexing agent Tetrasodium of ***N*,*N***-bis (carboxymethyl) glutamic acid in combination with a biodegradable biosurfactant (rhamnolipid) was investigated for the bioremediation of heavy metals from the sludge.	[247]
Candida **sp.**	Sophorolipids	Fe, Cu, As	Sophorolipids were found to effectively remove toxic heavy metals by increasing the temperature from 22 to 35 °C from mine tailings.	[248]
Burkholderia **sp.** Z-90	Rhamnolipid	Zn, Pb, Mn, Cd, Cu, As	Biosurfactant produced by *Burkholderia* sp. Z-90 was used for the remediation of soils contaminated by heavy metals by combining bioleaching and flocculation.	[249]
Pseudomonas fluorescence RE1 and RE17	Rhamnolipid	Zn, Cr, Cd, Ni	Biosurfactant produced from *Pseudomonas fluroresence* from indigenous rice roots was optimized for heavy metal sequestration at variable pH.	[250]
Bacillus **sp.** H1P3	Surfactin	Cu, Pb, Zn, Cd, Cr	Biosurfactant from *Bacillus* sp. strain isolated from cooking characterized as surfactin was found to be efficient for bioremediation of heavy metals.	[251]

(continued)

Table 11.1 (continued)

Microorganism	Type of biosurfactant	Target pollutants for remediation	Description	References
Pseudomonas aeruginosa MA01	Rhamnolipid	Cd	A sample waste activated by rhamnolipid biosurfactant obtained from *Pseudomonas aeruginosa* MA01 was found to be efficient in the removal of Cd, Zn, Cu, and Pb.	[252]
Bacillus cereus NWUAB01	Lipopeptide	Pb, Cd, Cr	Resistant bacteria were cultured, and the biosurfactant production was carried out which was capable of removing 69% of Pb, 54% of Cd, and 43% of Cr from the batch experiment from the initial concentration of 100 mg/L. The strain removed 83% of Pb, 60% of Cd, and 30% of Cr from polluted soil.	[253]
Bacillus **sp. (MSI 54)**	Lipopeptide	Pb, Hg, Mn, Cd	The synthesized biosurfactant's heavy metal remediation efficacy at a 2.0 critical micelle concentration was 75.5% Hg, 97.73% Pb, 89.5% Mn, and 99.93% Cd in 1000 ppm of the corresponding metal solution, respectively. Surface heavy metal pollutants were effectively removed from farm fresh cabbage, carrot, and lettuce after treatment with 2.0 critical micelle concentration of the lipopeptide.	[254]
Pseudomonas aeruginosa **strain LFM 634**	Rhamnolipid	Ar and transition metals	Di-rhamnolipid congener was used to check the extractive capacity for arsenic and other transition metals through precipitation method.	[255]
Pseudomonas aeruginosa **strain ASW-4**	Rhamnolipid	Pb, Hg	At a critical micelle concentration of 43.73 mg L, 62.50% Pb, and 50.20% Hg were removed from the marine intertidal sediment sample containing 520.32 mg/kg of Pb and 13.15 mg/kg of Hg (dry weight) by the process of leaching.	[256]

Pseudomonas aeroginosa	Rhamnolipid	Cr, Pb	Biosurfactant-rhamnolipid was applied for bioremediation of heavy metals, environmental toxic pollution removal, and stain decolorizer from water sample collected from Muttukadu Lake.	[257]
Pseudomonas aeruginosa **RW9**	Rhamnolipid	Cr VI	Rhamnolipid produced from *Pseudomonas aeruginosa* which can grow well and tolerate up to 40 mg/L Cr(VI), with an OD value ranging from 1.70 to 1.6 has been applied for the bioremediation of hexavalent chromium from wastewater effluent.	[258]
Pseudomonas **sp. CQ2**	Rhamnolipid	Cd, Cu, Pb	*Pseudomonas* sp. CQ2 isolated from the Chongqing oilfield displayed the heavy metal removal efficiencies of 78.7%, 65.7%, and 56.9% for Cd, Cu, and Pb, respectively, at optimized bioleaching conditions of pH: 11, soil/solution ratio: 30:1, and nonsterilized soil.	[259]
*Pseudomonas indica*NBRC 103,045, *Pseudomonas aeruginosa* PAO1	Rhamnolipid	V	Rhamnolipid biosurfactant isolated from an unidentified strain selected with 2.0 critical micelle concentration value used for bioremediation of Vanadium from artificially contaminated sand having a removal efficiency of (240 mg/L) 85.52d ± 4.31.	[78]
Bacillus haynesii E1	Surfactin	Pb	The biosurfactant production process was optimized using raw orange peel extract and the surfactin type biosurfactant effectively remediated Pb^{2+} (high MIC = 2200 mg/L) with a maximum adsorption capacity of 196.08 mg/g	[260]

Along with 2-methylnaphthalene, hexadecane, and pristine, sophorolipid from *Candida bombicola* ATCC 22214 has been evaluated for its bioremediation effectiveness of crude oil polluted soil. The evaluated model compounds deteriorated 95% in 2 days, 97% in 6 days, and 85% in 6 days [86]. *Klebsiella* spp. strain RJ-03, an alkaliphilic bacteria, also produced biosurfactant with lower viscosity and pseudoplastic rheological behavior as well as significant emulsification activity with hydrocarbons and oils [87]. GC-MS analysis confirmed the presence of rhamnose and mannose type of monosaccharides which displayed excellent oil removing efficiency as compared to chemical surfactants like SDS and laundry detergents powders [88].

Recently, Louiza Derguine-Mecheri and coworkers [89] extracted biosurfactant from the yeast *Rhodotorula* sp. YBR which was found to be efficient in remobilization and further favoring the bioavailability of the hydrocarbons from oil contaminated soil. Ahmadi et al. [90] employed a rhamnolipid biosurfactant derived from *P. aeruginosa* strain R4 to remediate pyrene-contaminated soil with an initial pyrene concentration of 200 mg/L. *P. frederiksbergensis* also produced a biosurfactant with a critical micelle concentration (CMC) of 48.3 mg/L, indicating significant benzo(a) pyrene biodegradation efficiency while also increasing soil microbial abundance, water holding capacity (WHC), and dehydrogenase (DH) activity [91]. A lipopeptide secreted by *B. subtilis* SNW3 was found to have a greater prospective for germinating seeds and seedling growth promotion in *Pisum sativum*, *Lactuca sativa*, *Solanum Lycopersicum*, and *Capsicum annuum*, as well as the elimination of crude oil from polluted soil, implying possible applications in the field of environmental and agronomy [92].

11.3.1.3 Other anthropogenic pollutants

There are many pollutants associated with textile industries ranging from inorganic compounds and elements to polymers and organic products which mainly comprise of synthetic dyes, nonbiodegradable pigments, hydrocarbons, and even heavy metals [93]. The release of dyeing wastewater pollutes the environment and constitutes a significant hazard to aquatic ecosystems [94]. Decolorization of dyeing wastewater is now performed primarily through physical, chemical, and biological procedures that reduce the azo dye structure's –N=N– double bond [95]. Azo dyes are difficult and inefficient to remove because of their great structural complexity, toxicity, and recalcitrance [96]. Because of their weak exhaustion qualities, about 10%–15% of the dyestuff employed remains unbound to the fibers and is thus released as color into the environment, posing a risk of ecotoxicity and bio-accumulation [97, 98]. Furthermore, dye in water is highly apparent and has an impact on its transparency and attractiveness, obstructing light penetration and lowering dissolved oxygen concentrations [99]. While various physicochemical methods for removing coloring agents from textile wastewater have been developed, they have several intrinsic downsides such as being unsuitable for dye separation, being expensive, inflexible to a broad array of dye-containing wastewater, being hard to discard, and producing huge amounts of sludge that may contribute to secondary pollution [100].

Many papers address the direct and indirect harmful consequences of dyes and metals. In reality, they can cause tumours, malignancies, and allergies as well as inhibiting the growth of bacteria, protozoa, algae, plants, and various animals, including humans [101, 102]. Consequently, there has recently been an increasing need for their detoxification and/or removal from the environment. Hence use of microorganisms and their consortium appears to be the best alternative which involves the use of bacteria or fungi capable of decolorizing dyes either by direct augmentation or through the employ of their consortia [103, 104]. Biosurfactant is thought to modify cell membranes under aqueous settings by increasing membrane permeability or dye solubility, enabling for far more effective enzyme release, demonstrating its utility in the remediation of anthropogenic toxic dye pollutants [105, 106]. *Bacillus weihenstephanensis* RI12, a lipopeptide biosurfactant isolated from hydrocarbon-contaminated soil, was found to be effective in the biotreatment of Congo red-contaminated wastewater. The treated dyes also displayed germination potencies of tomato seeds under variable conditions showed biotreatment efficiency of the azo dye when added with SPB1 biosurfactant [107]. Nor and colleagues [108] have reported the production of a lipopeptide biosurfactant that can help with dye decolorization by increasing the bioavailability and biodegradability of hydrophobic substances via high emulsification activity and dye solubilization.

Furthermore, the addition of biosurfactant to the culture creates a hydrophobic layer to the bacterial membrane allowing the pollutants with higher potency to penetrate the bacterial cell membranes [109]. In another investigation by Carolin et al. [110], a lipopeptide biosurfactant generated by *Bacillus* sp. was reported to be effective in the degradation of aromatic amine 4-chloroanilline when introduced to the bacterial culture, which further increases the efficiency and delays cell death of bacteria.

11.3.2 Biosurfactants and their role of biosurfactant in solubilization of pesticides

Pesticides are notorious for their low water solubility and tendency for being absorbed in soil, rendering them inaccessible to microbial breakdown [111]. Because of their persistence and recalcitrance, chlorinated pesticides and solvents are common pollutants of surface soil and groundwater [112]. These substances can undergo biomagnification, posing serious health risks to humans and animals. As a result, appropriate remediation processes for these compounds from a variety of environmental compartments, including manufacturing sites, application sites, storage sites, obsolete samples, and accidental spillage sites, are needed. Biosurfactants have previously been proven to assist microorganisms in the degradation of pesticides. Surfactants employed in excess of their CMC can considerably increase the evident bioavailability of hydrophobic organic compounds and can be utilized successfully in conjunction with remediation techniques like soil washing/flushing and pump-and-treat. However, synthetic surfactants sometimes prevent microbial action on pollutants solubilized in their micellar phase [113] leading to the accumulation of surfactant on the subsurface which in turn can cause pollution [114].

Endosulfan biodegradation by co-cultures in the presence of biosurfactant generated by *B. subtilis* MTCC 1427 was described by Awasthi and colleagues [115]. Endosulfan, a chlorinated cyclodiene pesticide widely used on a variety of crops to combat chewing and sucking insect pests, has been designated as a main concern contaminant by several international environmental agencies [116]. The utilization of heavy metal-resistant bacteria that produce rhamnolipid biosurfactant with a CMC value between 45 and 105 mg/L for the solubilization of chlorinated pesticides endosulfan, β-endosulfan, and γ-HCH has been described by Gaur et al. [117]. When comparing to their respective controls, the administration of biosurfactant increased the solubility of 1.18 mg/L (7.2-fold), 0.76 mg/L (2.9-fold), and 13.37 mg/L (1.8-fold) for α-endosulfan, 0.76 mg/L (2.9-fold) for β-endosulfan, and 13.37 mg/L (1.8-fold) for γ-HCH. López-Prieto et al. [118] used lipopeptide biosurfactant generated by *Bacillus mycoide* isolated from maize steep liquor to evaluate the solubilization of Cu-based fungicides. They discovered that when the biosurfactant was introduced at a dose of 20 g/L, the biosurfactant was able to dissolve roughly 90% of Cu-Oxy in 20 min of contact time. Manickam et al. [119] also conducted the solubilization assay on hexachlorocyclohexane (HCH) isomers using three biosurfactants rhamnolipid from *P. aeruginosa*, sophorolipids from *C. bombicola*, and tetrahalose tetraester from *Rhodococcus erythropolis*. They discovered that HCH isomer solubilization improved by 3–9 times, with rhamnolipid and sophorolipid being even more efficient and displaying optimal solubilization of HCH isomers at 40 g/mL compared to trehalose-containing lipid that exhibiting maximum solubilization at 60 g/mL. Payal and Jain [120] discuss the bioremediation, sorption, and solubilization of pentachlorophenol from noxious locations including contaminated soil and water using rhamnolipids from *P. aeruginosa*.

11.3.3 Biosurfactants as antimicrobial agents

Biosurfactants have been proven to have antibacterial properties against pathogenic microorganisms in both plants and animals. The direct toxicity of several classes of biosurfactants to microorganisms has been widely investigated. They have also been reported to be engaged in the stimulation of plant and animal-defensive responses, thanks to breakthroughs in research performed by several microbiologists and in the concept of sustainable agriculture. Rhamnolipids generated by *P. aeruginosa* have been shown to have antibacterial activity against Gram-negative bacteria *viz. B. subtilis, Staphylococcus aureus, Proteus vulgaris, Streptococcus faecalias*, and *Bacillus cereus* as well as against phytopathogenic fungal species of *Alternaria* spp., *Botrytis cinerea, Chaetomium globosum, Colletotrichum capsici, Colletotrichum orbiculare, Fusarium sacchari, Fusarium graminearum, Fusarium moniliforme, Fusarium oxysporum, Fusarium verticillioides, Mucorcircinelloides, Penicillium* spp., *Phytophthora capsica*, and *Phytophthora infestans* [20, 121–128] and antimicrobial properties against destructive zoosporic pathogen *Phytophthora capsica* which causes root rot in cucumber. Table 11.2 summarizes the literature on biosurfactants' antagonistic characteristics and antibacterial capabilities. The possible mode of action of the biosurfactant was by rupturing the plasma membrane, causing the fungus' zoospores to lyse [129]. *Macrophomina phaseolina*, which causes root rot, stem rot, collar rot, charcoal rot, damping off and seedling blight on potato and chickpea plants, was inhibited by a glycolipid-type biosurfactant generated by the yeast *Geotrichum candidum* MK880487 [130]. Bacillus strains have also been utilized to control *Rhizoctonia solani* infections in pepper [131] and potato plants [132, 133]. Rani and her coworkers [134] did a study in which lipopeptide-type biosurfactants were partially recovered from *Bacillus methylotrophicus* OB9 culture supernatants grown in Bushnell–Haas medium broth supplemented with oil using the acid precipitation method. The acid precipitate fraction of the biosurfactant was reported to display antimicrobial activity against the plant pathogenic bacterium *Xanthomonas campestris* as well as *Salmonella serovars* and *Escherichia coli*. Biosurfactants from the plant growth-promoting *Pseudomonas putida* 267 were found to lyse *Phytophthora capsici* zoospores and prevent the growth of the fungal diseases *B. cinerea* and *R. solani*, according to Kruijt et al. [135]. In vitro experiments further demonstrated that the strain requires biosurfactants for swarming and biofilm formation.

Another prospective agronomic application of MELs, according to Yoshida et al. [51], is as pesticides against phytopathogenic microbes. MELs, like other biosurfactants, have been demonstrated in several investigations to have antimicrobial properties [51]. However, the inhibitory mechanisms of MEL have yet to be fully understood [136]. MEL-A was found to prevent the formation of the powdery mildew fungus *Blumeria graminis* f. sp. *tritici* strain T-10 in wheat leaves by Yoshida et al. [51], perhaps due to the suppression of conidial germination.

Certain *Pseudozyma* (Ustilaginales) species have also been attributed to phytopathogenic resistance [137] and biological control [138]. By comparing phenotypic traits on fresh plant surfaces and modified solid surfaces, Yoshida et al. [139] found that Mannosylerythritol lipids generated by the phyllosphere yeast *Pseudozyma antarctica* strain T-34 elicit morphological alterations in bacterial cells on solid surfaces and thereby interfere with the pathogen-host surface interaction. Their analyses suggest that MELs produced on plant surfaces by *P. antarctica* could have a key role in fungal structural development and proliferation on plant surfaces. MELs have been understudied as agricultural supplements to date, with only a limited

Table 11.2: Some recent developments on the antagonistic property of biosurfactants and their antimicrobial properties.

Microorganism	Type	Description	Reference
Candida antarctica	Mannosylerythritol	They have antimicrobial action against Gram-positive bacteria were more prominent	[261]
Pseudomonas aeruginosa	Rhamnolipid	Control of *Phytophthora cryptogea* in chicory witlo in hydroponic forcing system	[225]
Pseudomonas aeruginosa IGB83 *Candida bombicola* ATCC22214	Rhamnolipid Sophorolipid	Biocide for phytopathogenic control of *Pythium* and *Phytophthora* spp. responsible for water-borne damping-off disease	[262]
Pseudomonas putida	Cyclic lipopeptides putisolvin I and II	Control zoospores of the oomycete pathogen *Phytophthora capsici* and impede growth of the fungal pathogens *Rhizoctonia solani* and *Botrytis cinerea*	[135]
Pseudomonas aeruginosa strain B5	Rhamnolipid B	In vivo control and in vitro antifungal activity against *Phytophthora capsici* and *Colletotrichum orbiculare*	[122]
Pseudomonas koreensis	Cyclic lipopeptide-type	Control late blight of potato disease caused by the zoospore-producing pathogen *Phytophthora infestans*	[263]
Bacillus subtilis	Lipopeptides: surfactin, fengycin, and Iturin A	Effective against *Colletotrichum gloeosporioides*, the causative agent for anthracnose on papaya leaves	[264]
Bacillus **spp.**	Lipopeptides	Growth inhibition of *Fusarium* spp., *Aspergillus* spp., and *Biopolaris sorokiniana*	[265]
Pseudomonas aeruginosa ZJU211	Rhamnolipid	They had a broad range of antifungal effects against phytopathogenic fungi, allowing them to be used in agriculture to protect plants.	[124]
Candida bombicoia 1	Sophorolipid	*Phytophthora* spp. and Pythium spp. inhibition for phytopathogenic management	[266]
Pseudomonas aeruginosa	Rhamnolipid	Rhamnolipid biosurfactant against *Fusarium sacchari* – the causal organism of pokkah boeng disease of sugarcane	[125]

Pseudomonas aeruginosa	Rhamnolipid	Rhamnolipid against *Fusarium verticillioides* to control stalk and ear rot disease of maize	[127]
Acinetobacter **sp**. ACMS25	Glycolipid	Biocontrol of *Xanthomonas oryzae*	[267]
Pseudomonas aeruginosa	Rhamnolipid	Application of biosurfactant produced by *Pseudomonas aeruginosa* for biocontrol of *Colletotrichum capsici* responsible for anthracnose disease in chilli	[128]
Bacillus licheniformis	Lipopeptide	Biocontrol of *Rhizoctonia solani* causing root rot in Faba Bean	[268]
Pseudomonas aeruginosa	Rhamnolipid	Glycolipid produced from *Dacryopinax spathularia* was studied for its potential as a natural alternative for the removal of biofilms and as an antimicrobial to control *Listeria monocytogenes* in milk and cheese. The biosurfactant significantly reduced biofilm-associated *Listeria monocytogenes* on both polystyrene and stainless steel at concentrations as low as 45 mg/L.	[269]
Starmerella bombicola	Sophorolipid	The antimicrobial activity of sophorolipids produced by *Starmerella bombicola* against four tomato phytopathogens *Botrytis cinerea*, *Sclerotium rolfsii*, *Rhizoctonia solani*, *and Pythium ultimum* responsible for collar rot and stem rot, causing yellowing and wilting of the leaves, tipping, and early death of the tomato seedling was investigated. The biosurfactant displayed effective antimicrobial activity in *P. ultimum*, with 95% inhibition of mycelial growth, followed by *B. cinerea* with 75.7%, *R. solani* with 64.3%, and *S. rolfsii* with 28.5%	[270]
Lactobacillus plantarum 60 FHE	Glycolipoproteins	Antimicrobial activity against *Pseudomonas aeruginosa* ATCC9027 which can infect the roots of two plants species: Arabidopsis and sweet basil	[271]

(continued)

Table 11.2 (continued)

Microorganism	Type	Description	Reference
Bacillus subtilis	Lipopeptide	*Bacillus subtilis* strains, namely surfactin, fengycin, mycosubtilin, and their mixtures at different purities, were tested against foodborne pathogen and food spoilage microorganisms. Mycosubtilin and mycosubtilin/surfactin mixtures showed for the first time strong antifungal activity, against food-related fungi *Paecilomycesv ariotti* and *Byssochlamys fulva*, with minimum inhibitory concentrations (MICs) of 1–16 mg.	[272]
Bacillus subtilis	Fatty acyl glutamic acid	Antimicrobial mechanism of a fatty acyl glutamic acid biosurfactant against the foodborne pathogens *Listeria monocytogenes* and *Escherichia coli* involving the rupturing and perforation of the cytoplasmic membrane	[273]
Pseudomonas guariconensis LE	Mono- and di-rhamnolipids	Biosurfactant from rhizospheric soil taken from *Lycopersicon esculentum* growing in kitchen waste dumping site was found to be effective in inhibiting the in vitro growth of the phytopathogenic fungus *Macrophomina phaseolina* which is causal agent of charcoal rot in diverse crops	[274]
Bacillus amyloliquefaciens	Lipopeptide Fengycin	Lecithin supplemented biosurfactant displayed as biocontrol agent for diseases affecting pepper and tomato plants – and on the antiviral effect of the PPL strain on Cucumber mosaic virus (CMV)-infected pepper plants	[275]
Bacillus **spp.**	Lipopeptides	The crude extract of synthesized biosurfactant showed antifungal action, inhibiting the mycelial development of *Moniliophthora roreri* and *Moniliophthora perniciosa*. Furthermore, the bacterial cell extract treatment caused swelling, granulation, and fragmentation of both phytopathogens' hyphae.	[276]

Lactobacillus casei TM1B	Rhamnolipid	Biosurfactant producing bacteria isolated from Cameroonian fermented milk "*pendidam*" with antimicrobial activity against gram positive and gram negative food-borne pathogens *P. aeruginosa* PSB2, *P. putida* PSJ1, *Salmonella* sp. SL2, and *Bacillus* spp. BC1	[277]
BS acquired from AGAE Technologies (Corvallis, OR, USA)	Rhamnolipid	Oleoresin extracted from *Apium graveolens* seeds, Limonene, and biosurfactant evaluated as alternative bio-control agent against *Bacillus cereus*, a Gram-positive, toxigenic, and endospore-forming bacterium. From the tests, it was concluded that the presence of biosurfactant with OR and LN inhibited the endospore germination hypothesizing that the structural proteins of the spore coat can be changed due to the binding of their polar and apolar groups to hydrophilic and hydrophobic moieties of the surfactants.	[278]
Bacillus subtilis AKP	Lipopeptide	Lipopeptide biosurfactants extracted from *Bacilllus subtilis* assessed with potential for antagonistic activity against *Colletotrichum capsici*, the causal organism of anthracnose disease of chilli. Crude extract from AKP strain also caused a significant reduction in disease incidence (anthracnose symptoms) by 16.66% ± 0.23% which provides good biocontrol efficiency against *C. capsici* based on in vitro detached fruit assay.	[279]
Bacillus cereus BS14	Cyclic siloxane-type biosurfactant	Biosurfactant produced by Bacillus strain displayed destructive effect on reduction of disease severity index in pulse crop *Vigna mungo* in vitro and in vivo. Also, Scanning electron microscope study revealed deformities at cellular level in the mycelia of *M. phaseolina.*	[280]
Metschnikowia koreensis	Sophorolipid	Biosurfactant produced from an asexual ascomycetes yeast strain CIG-6A^{T} and their inhibition against food spoilage fungi *Fusarium oxysporum* (MTCC9913), *Fusarium solani* (MTCC 350), and *Colletotrichum gloeosporioides* (MTCC 2190) was investigated. MIC value for *Fusarium solani* was 49 µgm/ml confirmed through confocal laser screening microscopy showing effective antifungal properties.	[281]

(continued)

Table 11.2 (continued)

Microorganism	**Type**	**Description**	**Reference**
Bacillus subtilis HussainT-AMU	Lipopeptides	A lipopeptide-type biosurfactant demonstrated antagonistic activity against *Rhizoctonia solani* which is soil-borne pathogen of potato tuber.	[153]
Streptomyces althioticus RG3 and *Streptomyces californicus* RG8	lipids	Streptomyces althioticus RG3 was found to displayed effective antimicrobial activity compared to *Streptomycescalifornicus* RG8. While biosurfactant from both strains showed antimicrobial activity against *Vibrio alginolyticus* MK170250, *Escherichia coli* ATCC 8739, *Pseudomonas aeruginosa* ATCC 4027, and *Staphylococcus aureus* ATTC 25923 which was done using a well diffusion method.	[282]
Bacillus atrophaeus strain B44	Lipopeptides	Lipopeptide biosurfactant produced by *Bacillus atrophaeus* having antimicrobial activity against *Rhizoctonia solani* which causes rhizoctoniosis in cotton plant. Pot tests results suggested that the biocontrol efficacy of B44 strain was related to the lipopeptide production, and furthermore, the strain also significantly increased the size of cotton seedlings due to the GA_3 concentration.	[283]
Bacillus amyloliquefaciens	Fengycin	Lecithin induced fengycin lipopeptide was reported to display antiviral property against CMV-infected pepper plants.	[275]
Pseudomonas aeruginosa	Rhamnolipid	The biosurfactant exhibited in vitro and in vivo antifungal property and found to be suitable as biocontrol agent against Fusarium wilt of *Abelmoschus esculentus*	[284]

data is available in the literature. More research is needed in this area, using diverse plant species and microbial pathogens, to understand the processes behind their antimicrobial effect and interaction with plant surfaces.

11.3.4 Biosurfactant against insecticidal and larvicidal pest in agriculture

Biosurfactants have been used as biopesticides to combat insects and larvae invasion in only a few reports. Glycolipid biosurfactants were effective for the biological control of pests which including eggs, larva, arachnids, grasshoppers, and boxelder bugs, according to a patent [140]. The lethal concentrations, that is, LC_{50} and LC_{90} values for lipopeptide biosurfactant produced by *B. subtilis* SPB1 against *Ectomyelois ceratoniae* Zeller, a kind of destructive moth pest of stored dates in Tunisia, northernmost realm of Africa, were found to be 152 and 641 mg/g, respectively, after 6 days of contact in a report by Mnif and coworkers [18]. Kim et al. [1] discovered that rhamnolipid derived from *Pseudomonas* spp. EP-3 was insecticidal against the green peach Aphid (*Myzus persicae*). The dirhamnolipids killed green peach aphids in a dose-dependent manner, with 50% mortality at 40 g/mL and 100% mortality at 100 g/mL, demonstrating that the rhamnolipid damaged the aphid cuticle membrane. Ghribi et al. [141] discovered the lavicidal effectiveness of *B. subtilis* SPB1 biosurfactant against lepidoptera larvae. Apparently, the bioemulsifier targets the midgut of Olive Moth (*Prays oleae*) larvae, eliciting epithelial cell disintegration and lumen leaking. The death of larvae results from the disintegration of midgut cells. The SPB1 biosurfactant, with LC_{50} and LC_{90} of 142 and 369 g/ml, induces *P. olea* death. In another study of Ghribi et al. [142], they observed that the biosurfactant generated by *B. subtilis* SPB1 strain has a lepidopteran larvicidal potency. The LC_{50} of the biosurfactant against third instar larvae of *Ephestia kuehniella* (Lepidoptera: *Pyralidae*) was found to be 257 mg/g 6 days after treatment. The larvicidal activity of a lipopeptide derived from *Staphylococcus epidermidis* against flour beetle larvae, *Tribolium castaneum*, was recently reported by Fazaeli et al. [143]. The bacterial biosurfactant had highest larval mortality rate at 63.33% at a lipopeptide concentration of 10,000 μg/g.

11.3.5 Nematicidal activity of biosurfactant against root knot nematodes of agriculturally important crops

Rhizobacteria are bacteria that populate in rhizosphere, or the soil under the chemical influence of the root, or are rhizosphere-competent [144]. Plant growth promoting rhizobacteria and their culture filtrates have been applied to seed or planting material in attempts to establish them in the rhizosphere to control cyst nematodes and root-knot nematodes [145]. Nematodes are tiny soil-borne parasites and the symptoms of their infection are usually nonspecific [146]. The four species of *Meleidogyne* spp. include *Meloidogyne hapla*, *Meloidogyne incognita*, *Meloidogyne javanica*, and *Meloidogyne arenaria* that are well known for causing root damage in agricultural crops, resulting in reduced yields [147, 148]. One of the most important species influencing rice production is the root-knot nematode, *M. graminicola*, a stationary endoparasite of rice roots [149].

Gao and colleagues [150] discovered two nematicidal compounds from *Bacillus cereus* strain S2 that have considerable nematicidal activity against root knot nematode *M. incognita*, identified as C16 sphingosine and phytosphingosine. Sphingosine administration resulted in a significant increase in reactive oxygen species (ROS) in the intestinal tract, and the nematode's genital portion disappeared. In addition, *B. cereus* S2 was discovered to generate systemic resistance in tomato and increase the activity of defense-related enzymes for *M. incognita* biocontrol. Hussain and colleagues [151] introduced a novel biosurfactant generated by *B. subtilis* HussainT-AMU that showed nematicidal efficiency against the root galling nematode *M. incognita* found in tomato roots [148]. The bacteria's culture filtrate was effective in causing mortality in second-instar juveniles and lowered egg hatchability in their experiment. They also suggested that biofilm formation over the surface of roots was likely to strengthen plants' systemic resistance and improve host defense.

11.3.6 Biosurfactants as plant growth promoters

Almost all crops around the world are susceptible to some form of pathogenic activity ranging from some kind of fungal diseases to bacterial or viral diseases and the existence of toxic heavy metals in the soil environment which makes the use of pesticides and fungicides a leading demand in the field of agriculture [152]. There are many drawbacks leading to use of synthetic pesticides due to their nonbiodegradability which is a leading cause for environmental pollution. In many cases of crop failures, plants pathogens are the main cause; for example *R. solani*, which is a soil-borne phytopathogen of the potato tuber [153]. Excess nutrients in the soil can contribute to the growth of zoospore-producing plant diseases oomycete *Pythium ultimum* in hydroponic tomato growing, which is another example of pesticide and fungicide overuse [154].

Major portion of the food production comes from the agricultural sector which are under strain due to drought, poor environmental conditions, and phytopathogens in commercial crops [155], and as such, plants growth promoters are also a means for improving the agricultural crisis by enhancing the activity of plants growth, solubilizing toxic metals and elements into plants nutrients that can be utilized by the plants [156]. Plant growth has been observed to be aided by bacteria that produce biosurfactants, such as *Pseudomonas* spp. and *Bacillus* spp. [157–159].

Plants and bacteria have a symbiotic relationship in which the latter produces siderophores and indole-3-acetic acid, along with organic acids such as citric acid, gluconic acid, α-ketobutyric acid, and succinic acid [160]. Biosurfactants' role as plant growth boosters is depicted schematically in Figure 11.2.

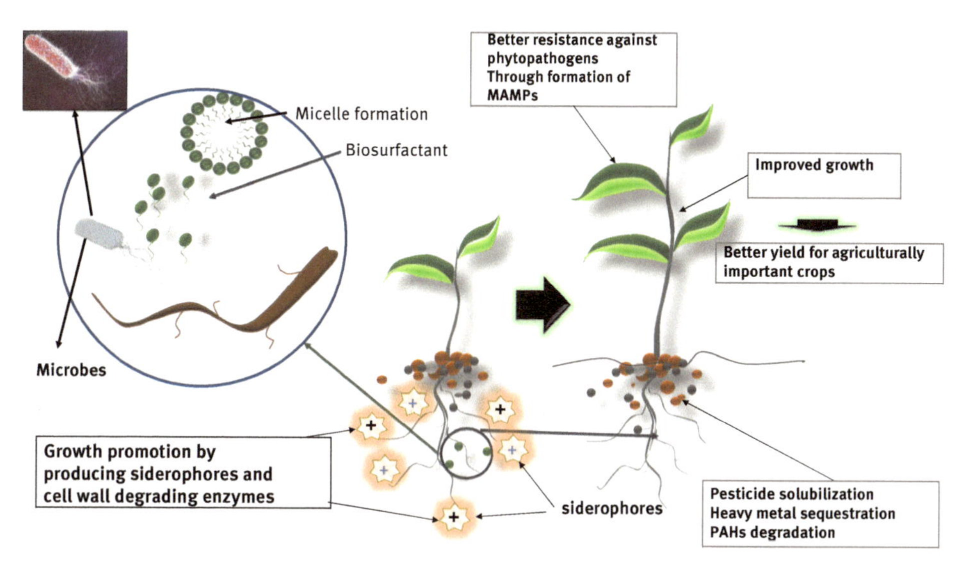

Figure 11.2: A schematic illustration of the role of biosurfactants as plant growth promoters.

Biosurfactant structure consists of hydrophobic and hydrophilic entities [21]. The way biosurfactants interact with plant surfaces and chemicals is determined by their nature and structure. A biosurfactant interacts with plants in three ways: changes in surface properties, changes in compound activities, and collaborations with membranes [1]. The interaction of biosurfactants with plant surfaces is aided by their high surface tension and adhesion [161]. Most biosurfactant-producing bacteria have a high colonization rate, allowing for optimal plant surface coverage [160]. Due to the tendency of biosurfactants to include various plant growth promoting compounds, they can provide protection as well as growth promoting benefits. Hence, externally, most biosurfactants can provide quick access to nutrients and protection from harmful compounds. Internally, they produce siderophores and enzymes that break down cell walls [162]. In this regard, biosurfactant produced by *P. putida* BSP9 obtained from rhizospheric region of *Brassica juncea* was found to show plant growth promoting capabilities which effectively increases the agricultural productivity and the isolate showing effective production of indole acetic acid, siderophores, and had great solubilization capacity [163].

Glycolipid produced by *P. koreensis* JDSCSU15 has been studied for the growth promotion of *Solanum lycopersicum* showing promoted growth in heirloom tomatoes infected with the fungal phytopathogen *F. oxysporum* F. *lycopersici* and the bacterial phytopathogen *X. campestris* pv. *Vesicatoria* [156]. *Enterococcus faecium* LM5.2 was reported to produce biosurfactant effectively with an emulsification index of 45.1 ± 3 and reduction in the surface tension up to 32.98% ± 0.23% [164]. Marchut-Mikoajczyk and coworkers [165] have revealed that glycolipid biosurfactant from endophytic *Bacillus pumilus* 2A promotes growth of *Beta vulgaris* L. (beetroot), *Raphanus* spp. L. (radish), and *Phaseolus vulgaris* L. (bean).

The biocompatibility of lipopeptides with live organisms is one of the favourable effects of their application in agriculture [166]. As a result, seed stimulation tactics can reduce the preliminary dosage of fertilizers, and biosurfactants can ensure that fertilizers are distributed evenly in the soil [167]. Lipopeptides were found to be appropriate for environmental remediation of crude oil (86%) and as an efficacious plant growth-promoting agent which remarkably ($P < 0.05$) increased seed germination rate and support growth of chilli pepper, lettuce, tomato, and pea at a dosage of 0.7 g/100 mL, indicating that they could be used in agriculture and bioremediation operations by reducing financial and environmental burden [92]. The application of a lipopeptide biosurfactant proved successful in germination and seedling growth of maize (*Zea mays* L.) [126]. There are very limited studies available that discuss the employ of lipopeptides that are directly coupled to plant growth. Furthermore, recent research suggests that lipopeptides could be used passively as biostimulants to increase the accessibility of nutrients to plant-associated healthy microbes [30]. However, significant scientific inadequacies regarding biosurfactants' strategies for improved plant growth and development still need to be filled.

The German patent DE102014209346A1 describes the application of sophorolipids to boost crop productivity, even if the plant pathogen is not eradicated or the therapeutic diagnosis is not altered [168]. Sieverding [168] used sophorolipid samples alone or in conjunction with commercial plant protection agents like insecticides, fungicides, and fertilizers to treat plant seeds and leaves in his investigation. He claims that such procedures can protect plants against pest infestation and disease during their early stages of development, based on his field experiments.

11.3.7 Biosurfactant as biostimulant for seed germination and development

Seed coatings, a strategy for covering seeds with neutral substances, homogenizing their size, and protecting them from humidity, radiation, physical traumas, and microbial (bacterial, fungal, and viral) infections, employs surfactants as adjuvants. Seed coating is appealing because it increases the assistance to seeds and could be a secure way to deliver pesticides, nutrients, minerals, fertilizers, or mycorrhizal fungi, among other things, to emerging seedlings [169]. Despite the fact that the coating materials are intended to be inert, the appearance of surfactants near the seeds and in relatively substantial doses may be a reason for issue because they can compromise membrane integrity, granting accessibility to intercellular components [170]. Seeds may be harmed because of such a treatment, with shoots dying or germination failing [167]. Anionic and nonionic surfactants, according to Doige et al. [171], may bind to different membrane proteins, altering their solubility and conformation, as well as impede the biological functions of plant enzymes at micromolar concentrations. According to de Bruin et al. [172], certain surfactants may accelerate electrolyte leakage, diminish chlorophyll concentration, and cause alterations in plant cell ultrastructure. Given the possible toxicity of surfactants and their ubiquitous prevalence in the environment, it is critical to seek for alternate strategies for seed stimulation that include the employment of biopreparation made from natural biological components. Biosurfactant is one such technique [167]. Various investigators have documented the influence of biosurfactants on seed germination. Biopreparations are now commonly utilized to improve seed production and increase germination in polluted soils [173]. The improvement in germination could be attributed to biosurfactant increasing the permeability of the seed coat to water, which speeds up the metabolic activities within the seeds [174]. Increased phytohormone synthesis and increased mineral solubilization in soil could explain the rise in plant biomass in the presence of biosurfactant [175]. Additional factor for the increased root development caused by the use of biosurfactants might be the reduction of soil anoxic conditions [176]. Their use not only protects seeds from microbes but also increases the assimilation of biogenic nutrients by plants, such as phosphorus [173, 177]. Furthermore, these technologies can be used in conjunction with chemical plant protection agents to improve their solubility and uniform distribution [178]. Surface-active agents generated by microbial cells have an undeniable benefit in terms of biodegradability, thermo-stability, and sensitivity to variations in the environmental pH [179]. The majority of biosurfactants are innocuous. They do not build up in live things' tissues; thus, they are not a hazard to the environment [167].

It was observed that seed treatment with sophorolipids biosurfactant quickens and improves the initial growth of plants during sprouting and roots [168]. Krawczyńska et al. [167] reported that biosurfactants, rather than *Azotobacter* spp., were found to stimulate all five investigated seeds, including, *Lupinus luteus*, *Avena sativa*, *Pisum sativum*, *Sinapsis alba*, and *Zea mays* before germination. They proposed that by using a biosurfactant for bioprepration might be used to clean up postflotation waste-contaminated soil. However, combining biosurfactant + *Azotobacter* spp. together did not have a satisfactory effect on germination. At a dosage of 0.7 g/100 mL, the lipopeptide biosurfactant treatments used by Umar and colleagues [92] exhibited a better potential for seed germination and plant growth promotion of *Capsicum annuum*, *Lactuca sativa*, *Solanum lycopersicum*, and *Pisum sativum*, indicating that such biosurfactant could be used in agriculture for biopreparation. Soaked Soybean seeds in lower concentrations (0.5–1 g/L) of rhamnolipid encouraged greater lateral root growth while lowering primary root extensions, without a difference in total root or shoot weight, according to Sancheti and Ju [180]. Marchut

Mikolajczyk et al. [181] discovered that the addition of biosurfactant produced by *B. pumilus* 2A to hydrocarbon-polluted soil contaminated with diesel and waste engine oil improved *Sinapis alba* germination and growth. Adnan et al. [182] revealed that a biosurfactant generated by the endophytic fungus *Xylaria regalis* improved the shoot and root length, dry matter production of the shoot and root, chlorophyll, nitrogen, and phosphorus levels of chilli seedlings, demonstrating its ability to improve agricultural plant growth. Oluwaseun et al. [183] found that the rhamnolipid biosurfactant from *P. aeruginosa* C1501 had no inhibitory effect on the seeds examined (*Sorghum bicolor*, *S. lycopersicum*, *Triticum aestivum*, *Vigna unguiculata*, and *Capsium annuum*) as well as root elongation. On the other hand, the chemical surfactant Tween 20 has been proven to impede germination. Rubio-Ribeaux et al. [184] investigated the biosurfactant from *C. tropicalis* UCP 1613 on seeds of *S. lycopersicum*, *Lactuca sativa* L., and *Brassica oleracea* and found it to be nontoxic and safe.

11.3.8 Biosurfactant for soil micronutrient bioavailability

During the early years, agriculture was focused on adding nutrients like nitrogen and phosphorus to the nutrient deficient soil to solve production problems of otherwise good soils exhausted from years of cultivation and of new, nutritionally marginal lands being introduced for production [185, 186]. Mineral elements like calcium, magnesium phosphorus, nitrogen, and potassium together were known as essential minerals [187], which improved the production to a certain level of expectation but even so there was still a need for extra minerals for boosting the production. Agricultural science then witnessed the discovery of a group of new critical elements required in lesser amounts by all living things, which were collectively known as trace elements or micronutrients in broad terms. The essential micronutrients for growth of higher plants are Fe, Mn, B, Cu, Co, Mo, Ni, Zn, and Cl [188]. Despite the availability of other key micro- and macronutrients, a single micronutrient deficit in soil can impede plant development and productivity [189]. Biosurfactant, a multifaceted microbial by-product, could be a long-term option for raising agricultural productivity by improving nutrient availability [109]. Biosurfactants are less hazardous than conventional surfactants, and their use in agricultural soils is predicted to boost nutritional quality, increase wettability, and ensure that complex nutrients are distributed more uniformly [190]. They have the ability to bind trace metals and form complexes with them, which they then adsorb, desorb, and remove from soil surface contacts, improving micronutrient concentrations and biological availability [191]. Figure 11.3 depicts the mode of action of biosurfactant as a plant growth enhancer via increasing soil micronutrient availability.

Biosurfactants can promote metal mobility by reducing the interfacial tension between metals and soil and creating micelles [192]. When the interfacial tension is low, they can also directly combine to sorbed micronutrients and transport them to

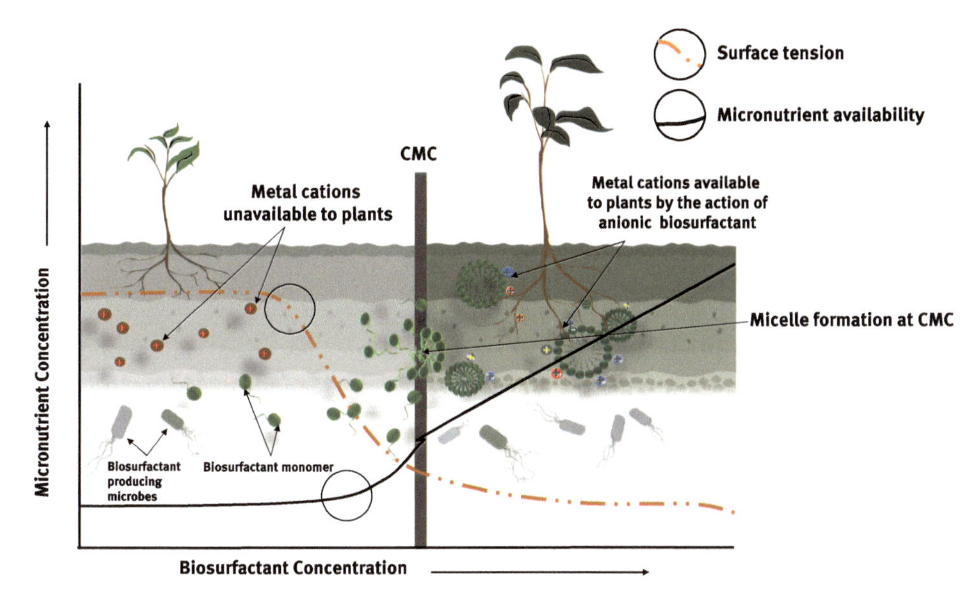

Figure 11.3: Mode of action of biosurfactant as plant growth promoter by enhancing micronutrient availability in soil. Figure is adapted and modified from Singh et al. [190] and Kumar et al. [279].

the region of interface in the roots [193]. *Bacillus* spp. strain J119 was evaluated for its potential to increase plant growth and cadmium uptake in *Brassica napus* (rape), *Zea mays* (maize), *Sorghum drummondii* (sudan grass), and *S. lycopersicum* (tomato) in soil artificially polluted with various concentrations of cadmium [194].

11.3.9 Biosurfactants as potential MAMPs in plant immunity

Plants have an innate immune system that detects and repels potentially harmful microbes [195]. When the innate immune system is activated, it causes local defense responses such as the hypersensitive response (HR) that kills plant cells at the infection site and/or basal defenses. The term "basal defenses" refers to defense responses that occur in the absence of an HR. Both the HR and the basal defenses create a local environment that is less conducive to pathogen growth. Systemic acquired resistance (SAR), a plant-wide condition of increased resistance to a wide spectrum of potential pathogens, can also result from immune activation [196]. Over the last decade, pattern recognition has emerged as a critical component of plant and animal immune responses. Pattern recognition receptors detect chemical signals that actually identify vast classes of microorganisms but are not present in the host, allowing for non-self recognition [197]. Once they have been recognized, these molecular fingerprints, also known as microbe-associated molecular patterns (MAMPs), activate complicated signalling pathways, resulting in the activation of transcription of defense-related genes and the upsurge of antimicrobial compounds in plant cells [198]. Plants' innate immunity is also reported to be stimulated by biological surfactants such as lipopeptides [199]. Biosurfactants from the *B. subtilis* strain were also found to have a considerablly induced systemic resistance (ISR)-mediated protective impact on bean and tomato plants, according to Thonart and colleagues [200]. Surfactin and fengycin biosynthesis genes were overexpressed in *B. subtilis* strain 168, a naturally poor producer, and this was linked to a considerable increase in the derivatives' potential to generate ISR. Figure 11.4 depicting the role of biosurfactant in the formation of MAMPs in plants. Cawoy et al. [201] found that *Bacillus* isolates produce surfactin, which ISR in a concentration-dependent manner. Surfactin, and to a smaller degree fengycin, lipopeptides, are able to trigger defense responses in plants that produce signalling molecules for ISR activation [202]. Waewthongrak et al. [203] used the lipopeptides surfactin and fengycin, both generated by *B. subtilis*, to induce a defense response in *Citrus sinensis* against *Penicillium digitatum* by boosting the transcription of defense genes. Ongena et al. [202] described the application of lipopeptides for triggering ISR in plants, where surfactin and pure fengycin were given straight to bean and tomato plants, resulting in a strong ISR reaction as a protective effect. The findings demonstrate that surfactin reduces infection by 28%, but not fengycin, which only reduces disease by 14% and is identical to reference plants. Yamamoto et al. [204] investigated the usage of lipopeptides in such surfactin and iturin as ISR molecules in plants and revealed the stimulation of defense protein expression that eventually gave resistance to *Colletotrichum gloeosporioides* in strawberry plants. According to López et al. [205], surfactin functions as a signalling molecule that causes cannibalism and matrix formation. Iturin [204], mycosubtilin [206], bacillomycin D [207], and sessilin and orfamide [208] are among the additional lipopeptides that have been described for inducing plant-defensive responses. Gond et al. [126] performed RT-qPCR to look for up-regulation of defense gene expression in an Indian popcorn variety of Maize after treating seedlings with lipopeptide extract produced from *B. subtilis* SG JW.03. The up-regulation of pathogenesis-related (PR) genes, particularly PR-1 and PR-4, which are involved in plant defense toward fungal infections, was observed in the roots of treated host plant seedlings. Further, they discovered that lipopeptide extract alone had no straight influence on the expression of PR genes. They suggest that various endophytic *Bacillus* spp., which are found naturally in many maize varieties, defend hosts by exuding antifungal lipopeptides that suppress fungal pathogens and triggering host plant PR genes up-regulation (systemic acquired resistance).

According to Maget-Dana et al. [209], surfactin is not toxic to fungi on its own, but when mixed with iturin biosurfactant, it has a synergistic fungicidal effect. Fengycins and iturins, on the other hand, are highly fungitoxic [210]. Pure surfactin was also reported to be beneficial in safeguarding up to 70% of wheat against *Zymoseptoria tritici* by Le Mire et al. [211]. According to Waewthongrak et al. [203], Yamamoto et al. [204], and Le Mire et al. [211], surfactin did not demonstrate considerable

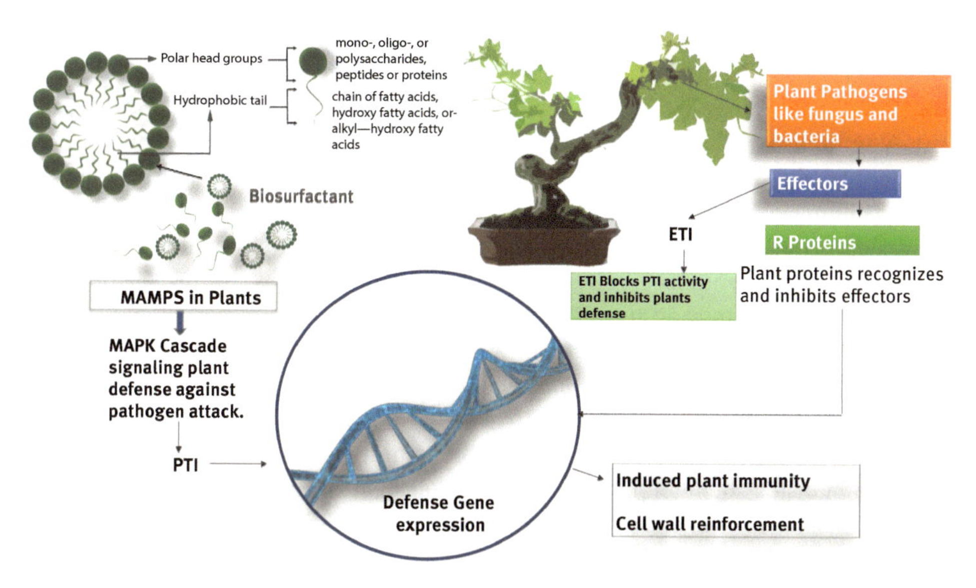

Figure 11.4: A schematic representation of the function of biosurfactant in forming MAMPs in plants. MAMPs: microbe-associated molecular patterns (plant defense responses when plants are introduced to microbial elicitors and are often absent in plants); MAPK: mitogen-activated protein kinase (activation of MAPKs is one of the earliest signaling events after plant sensing of pathogen/ microbe-associated molecular patterns (PAMPs/MAMPs) and pathogen effectors. However, pathogens employ effectors to suppress the activity of MAPKs to induce pathogenesis); effector: Proteins produced by the pathogen to manipulate its host; frequently, a suppressor of PTI: PTI-PAMP triggered immunity (basal resistance induced by MAMPs); ETI: effector triggered immunity (R-gene-mediated resistance induced by a microbial effector, often accompanied by a hypersensitive response); hypersensitive response: a vigorous defense response that culminates in programmed cell death. Figure is adapted and modified from Boller and He [195] and Boller and Felix [197].

antifungal efficacy, but it strongly increased ISR in plants by boosting salicylic acid and jasmonic acid-dependent signalling pathways.

Rhamnolipid biosurfactants may operate as MAMPs in Grapevine, according to the Varnier and collegues [212]. MAMPs, also known as generic elicitors, play a role in plant nonspecific immunity, making them efficient against a broad range of phytopathogens [212, 213]. According to Monnier et al. [214], rhamnolipids from *P. aeruginosa* activate an effective defense of rapeseed foliar tissues against *B. cinerea*. Physical defenses such as callose deposits and stomatal closure are part of the defense activated by rhamnolipids, in addition to chemical processes such as the formation of ROS and the regulation of defense genes. Varnier et al. [212] reported that rhamnolipid biosurfactants act as MAMPs in grapevine, and the specific combination of rhamnolipid effects may contribute to grapevine protection against *B. cinerea* (a phytopathogen for grey mould disease) as well as potentiating defense responses induced by the chitosan elicitor and *Bacillus cinerea* culture filtrate. In the first two minutes following the application, the findings show Ca^{2+} influx, ROS generation, and mitogen-activated protein (MAP) kinase activation. Plant defense mechanisms, such as the activation of a broad range of genes and the HR, can be triggered by these signalling events. Further, Varnier and collegues [212] also found that 0.01 mg/mL of rhamnolipids was sufficient to trigger early signalling cascades and the activation of defense genes including chitinase genes.

According to Sanchez et al. [215], rhamnolipids were found to activate an innate immunological response in Thale cress (*Arabidopsis thaliana*), including the buildup of signalling molecules and the transcription of defense genes. When plants are exposed to *B. cinerea* or *Pseudomonas* syringae pv tomato cells, Sanchez and coworkers [215] found that rhamnolipids amplify plant defense mechanisms involving salicylic acid synthesis. Plant resistance to the oomycete *Hyaloperonospora arabidopsidis*, the bacteria *Pseudomonas syringae* pv tomato, and the fungus *B. cinerea* was found to be improved by using rhamnolipids. Varnier and colleagues [212] and Sanchez and coworkers [215] similarly came to the realization that rhamnolipids can operate as MAMPs in plant cells to elicit nonspecific immunity.

11.3.10 Use of biosurfactants as agrochemical spreaders

The hydrophobicity of plant surfaces is well documented to play a role in pathogen invasion. Surfactants are often used as adjuvants in pesticide formulations to solubilize, suspend, or distribute the pesticide active components homogeneously when administered to the target crop [216]. In a similar fashion, spreaders are used to aid the adhesion and spread of wettable herbicide powders on foliage frequently containing only nonionic surfactants [217]. Spreader active ingredients range from 50% to 100% and have HLB values ranging from 7 to 9. Spreader-surfactants in use include thalestol and blends of various sorbitan surfactants (such as the Tween series). Spreaders can help improve the physical properties of many herbicides, making them more available at the leaf surface [218]. Agrochemical spreaders are typically used in conjunction with agropesticides and insecticides to improve the adhesion of these agrochemicals to plant surfaces, resulting in increased effectiveness of these chemicals on plants and, as a result, a reduction in the amount of chemicals used [219].

Sophorolipids are useful for enhancing or supporting the efficiency of pesticides; however, they have limited spreading or wetting capabilities [65]. Vaughn et al. [220] described the use of sophorolipid as adjuvants in postemergence herbicides to promote plant surface adhesion and cuticle penetration. Similarly, Sachdev and Cameotra [221] proposed employing sorpholipid adjuvants in a number of insecticides and fungicides. Figure 11.5 graphically illustrates the effects of biosurfactants on pesticide application on the leaf surface.

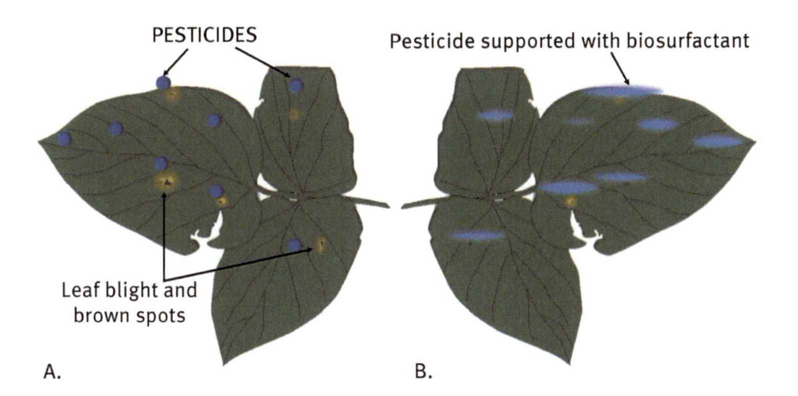

Figure 11.5: Diagrammatic representation of the effects of biosurfactant to pesticidal action: (A) pesticide applied on leaf surface without biosurfactant; (B) pesticide supported by biosurfactant which enables the pesticide to spread over a greater surface area. Figure is adapted and modified from Kumar et al. [280].

Fukuoka et al. [52] investigated the wetting ability of MEL solutions on a variety of plant leaf surfaces, including wheat, rice, strawberry, and mulberry, and found that the contact angle (θ) of MEL solutions was reduced to about 10°. MEL effectively lowered the contact angle (θ) even on plant surfaces of Poaceae family where nonionic surfactant solutions were unable to do so due to their poor wettability. Furthermore, the MELs solution used as a surface pretreatment supported a higher capacity for microbial cells dispersing and anchoring on the plant leaf surface when compared to a variety of standard synthetic surfactants indicating its potential to be employed as agrochemical spreaders, especially for biological pesticides [52]. However, according to Yoshida et al. [51], the capacity of MELs to serve as pesticides by reducing the hydrophobicity of plant surfaces appears to be dependent on the plant and the pathogen being targeted. The effectiveness of MEL produced by yeast strain *Pseudozyma antartica* T-34 was tested against Silwet L77 which is a commercial agrospreader, and the results show a marked high activity of MEL when used as agrospreader mixed with cells of *B. subtilis* suspension on wheat leaf surfaces [53].

11.3.11 Use of biosurfactant in hydroponic system

The release of excess nutrients by the use of agricultural fertilizers and manures into the water bodies through irrigation wastewater has become a global environmental concern owing to the increasing demands in agricultural products [222]. With this challenge in mind, many developed countries have adopted the use of hydroponic systems whereby greenhouse and outdoor nurseries have been built to recapture and recycle the irrigation water [223, 224]. The recycling of nutrient solutions lowers contamination in irrigation wastewater caused by nutrient salt and agricultural fertilizers. In hydroponic growth systems, zoospores are perhaps the most prevalent infectious organisms. In greenhouse and nursery crops, however, recirculation of nutrient solution can result in a large dissemination of root-infecting zoosporic diseases including *Pythium* and *Phytophthora* spp. [222, 225]. Stanghellini and Miller [129] originally reported the application of biosurfactant as a potential combative approach for the biological management of zoosporic phytopathogens in a hydroponic environment. They found that adding rhamnolipid or saponin biosurfactant to the recirculating solution at a dosage of 150–200 g/mL completely stopped pathogen spread and specifically killed the zoospores of *Phytophthora capsica*, the root pathogen. They found that adding rhamnolipid or saponin biosurfactant to the recirculating solution at a dosage of 150–200 g/mL completely stopped pathogen spread and specifically killed the zoospores of *Phytophthora capsica*, the root pathogen [222]. In a hydroponic forcing system of witloof chicory, De Jonghe and colleagues [225] found the effectiveness of rhamnolipids from *P. aeruginosa* in hindering the spreading of brown root rot disease caused by *Phytophthora cryptogea*. Similarly, in hydroponic tomato and lettuce culture, the efficiency of a lipopeptide biosurfactant generated by *P. koreensis* against the oomycete *P. ultimum* has been studied [154, 223]. Folman et al. [226] investigated the impact of biosurfactants in the management of several oomycete infections by *Lysobacter enzymogenes* on *Pythium aphanidermatum* biological control in soil-free environments. Because of its intercalation into the phospholipid bilayer of the zoospore cell, biosurfactants like rhamnolipid rupture the exterior membrane of the zoospore cell, which lacks a cell wall [122, 129]. Aside from antibacterial properties, biosurfactants in hydroponics may improve bioavailability of nutrient and emulsification of water immiscible substances [181].

11.3.12 Valorization of agroindustrial waste

Byproducts and trash from the agriculture and food industries is a huge subject of concern for waste generators and municipal officials, and treating and disposing of them through traditional treatment techniques takes a lot of effort, time, and money [227]. Valorization is a fairly emerging notion in the area of agricultural and industrial waste management that promotes microbial activity as a way of supporting sustainable development [228–230]. The conversion of such waste biomass into a value-added product is a key issue that must be addressed in order to achieve significant gain in energy and cost savings. Biosurfactants, contrary to the synthetic surfactants, can be made from discarded items like leftover food and agriculture-based industrial residue, mill refuse, and waste by-products of food processing industries, which not only save capital but also aid with waste disposal in an environmentally beneficial way [231, 232]. Table 11.3 summarizes the literature on the utilization of various agriculture-based feedstocks for the production of biosurfactant. Due to their rich carbohydrate, fat, oil, and nutritional content, these wastes provide fascinating and less-expensive basic supplies for biosurfactant and other allied metabolite production [229, 233]. Various studies have been published in the past few years on lowering the price of biosurfactant manufacturing by substituting diverse agroindustrial wastes for high-priced medium components. Several renewable resources, for instance, agricultural residues, food-processing industry wastes, and unspent residual edible oils, have been employed as the primary substrate of carbon in the manufacture of biosurfactants, using either solid-state or submerged fermentation [234, 235]. Oil mill effluent, a byproduct of olive oil extraction, was used as a substrate for the extraction of rhamnolipids from *P. aeruginosa* 112 and as a substrate for valorization [236]. Chebbi and coworkers [237] in a recent report utilized agricultural wastes from two different sector of wineries, where *Burkholderia thailandensis* E264 was discovered to exploit the crude dissolved portion of nonfermented waste of white Grape pomace as the primary source of carbon, while concurrently reducing the surface energy to roughly 35 mN/m. *Planomicrobium okeanokoites* IITR52 was also identified for biosurfactant synthesis by means of corncob and pineapple residues, which may lower the surface tension up to 45 Nm/m. The biosurfactant displayed antibacterial activity against oral and gastrointestinal pathogens such as *Neisseria mucosa* (MTCC 1772), *Clostridium perfringens* (MTCC 450), *Streptococcus mutans* (MTCC 497), and *Yersinia enterocolitica* (MTCC 859) and restraining their growth by 77%, 70%, 68%, and 56%, respectively [238]. Umar et al. [92] employed molasses, potato peel powder, white bean powder, and waste frying oil as cheap feedstock for lipopeptide biosynthesis from *B. subtilis* SNW3. Bharali et al. [239] reported the production of rhamnolipid biosurfactant using waste glycerol as substrates. These low-cost substrates can help reduce biosurfactant production costs while also assisting in the recycling of industrial waste.

11.4 Future status and possibilities of biosurfactant

Owing to the efficient and eco-friendly nature of biosurfactants, they can play a very significant part in the agriculture industry. Additionally, there is significant attention in the employment of biosurfactants in a number of fields, including oil recovery, soil bioremediation, pharmaceutical, and cosmetic industries [240]. The manufacturing expense, on the other hand, remains substantial, as it is dependent on raw material supply and downstream processing for industrial scale up [241]. Since primary ingredients account for 10–30% of the entire cost, establishing cost-effective techniques that utilize less-expensive crude feedstock and/or a regenerative resources is critical to the sustainability of metabolite production and application [17, 242]. It is critical to reduce the expense of production substratum, which accounts for around half of the overall price of the product in order to

Table 11.3: Various low-cost agroindustrial feedstocks have been used in recent years for biosurfactant synthesis.

Feedstock	Producer microorganism	Biosurfactant type	Reference
Food and agriculture-based industrial residue	*Bacillus subtilis* ATCC 6051	Rhamolipid	[285]
(brewery waste, sugarcane baggasse, corn steep	*Bacillus subtilis* PC	Surfactin	[286]
liquor, distillers' grains yield, edible paddy straw	*Bacillus subtilis*	Lipopeptide	[287]
mushroom, wheat bran, treated molasses and	*Bacillus amyloliquefaciens*	Glycolipid	[288]
pretreated molasses, cashew nut apple juice,	*Volvariella volvacea*	Anionic polymeric type	[289]
moringa seed residue, cassava waste, whey, orange	*Brachybacterium paraconglomeratum* MSA21		[290]
peel, lime peeling, carrot peel waste, coconut oil	*Pseudomonas aeruginosa* #112		[236]
cake, banana peel, potato peel, pineapple peel, rice	*Pseudomonas aeruginosa* AP 029/GLVIIA		[291]
water, date molasses, sesame peel flour, peanut oil	*Candida tropicalis* UCP 1613		[184]
cake, cassava four wastewater)	*Bacillus licheniformis* (KC710973)		[292]
	Pseudomonas aeruginosa MTCC 2297		[293]
	Bacillus subtilis ANR 88		[294]
	Halobacteriaceae archaeon AS65		[295]
	Bacillus pumilus DSVP18		[260]
	Serratia marcescens UCP 1549		[294]
			[284]
			[296]
			[297]
			[298]

Agriculture-based industrial and mill waste (rice mill polishing residue, two-phase olive mill waste, oil seed cake, oil mill wastewater, palm oil industry waste, olive mill waste, rapeseed cake)	*Bacillus subtilis* MTCC 2423	Lipopeptide	[299]
	Rhodotorula sp.	Surfactin	[300]
	Candida lipolytica UCP0988	Glycolipid	[89]
	Brachybacterium paraconglomeratum MSA21	Rhamnolipid	[301]
	Pseudomonas aeruginosa #112	Surfactin analogues	[290]
	Pseudomonas aeruginosa PAO1		[236]
	P. aeruginosa		[302]
	B. subtilis		[303]
	Pseudomonas aeruginosa AP 029/GLVIIA		[291]
			[304]
Waste residual cooking oil (soybean oil waste, residual frying oil, canola waste frying oil waste frying coconut oil, waste cooking oil, waste soybean oil, soybean post frying oil)	*Bacillus pseudomycoides*	Lipopeptide	[305]
	Staphylococcus spp.	Bioemulsifier	[306]
	Pseodomonas zju.u1M	Cyclic lipopeptide	[307]
	Pseudomonas cepacia CCT6659	Anionic polymeric type	[308]
	Pseudomonas aeruginosa D		[309]
	Pseudomonas SWP-4		[310]
	Candida lipolytica UCP0988		[301]
	Bacillus pseudomycoides BS6		[305]
	Candida tropicalis UCP 1613		[184]

make biosurfactant biosynthesis commercially viable [22, 30]. As a result, media including agroindustrial residual byproducts have been recommended as feedstock to strengthen the feasibility of mass-scale biosurfactant manufacture and increase the competitiveness of such organic products [24, 243]. A better option of feedstock can lead to significant cost saving and create the production more viable [22, 173]. Numerous possible carbon feedstock have been used in the past to reduce production costs, and researchers have been succeeded to a certain extent. Even though, employing alternate sources, the inferior production of biosurfactants remains a constraining issue. Consequently, a thorough investigation into specific parameters, such as biosynthesis process, growth, optimum cultivation parameters, and substratum formulation, is required for the mass-scale manufacture of biosurfactants for industrial applications and economic considerations [244]. However, in the future, it is expected that most of the limiting factors pertaining to it commercial application will be productively resolved through improvements in the fermentation technology of biosurfactants and by developing an effectual strain of biosurfactant producing microbes through genetic manipulation and strain improvement approaches.

11.5 Conclusion

Biosurfactants are amphipathic molecules comprising hydrophilic and lipophilic components in their structure. Plants and microorganisms provide the majority of them. Microbial synthesis, on the other side, has a number of benefits over plant-based surfactants owing to their multifaceted traits, fast and efficient production, and scale-up capability. Currently, biosurfactants are being investigated in a diverse array of disciplines, including oil recovery, soil bioremediation, pharmaceuticals, and cosmetics. Moreover, they have remarkable qualities including proficient surface properties, temperature endurance, pH stability, greater salt tolerance, generally higher disintegration rate, reduced toxicity, and superior selectivity for agricultural applications. In the current chapter, we analyzed and conferred the possibilities of biosurfactants for plant growth, development, and protection in the perspective of sustainable agroecological agriculture. Some of the important attributes of biosurfactants include enhancing the bioavailability, solubility of immiscible chemicals, signalling and differentiating cellular physiological routes, bio-stimulating effect, and antibacterial action for disease prevention. One of the most important aspects of these amphiphilic molecules' agronomic relevance is that they have a similar dual functionality of safeguarding plants via antibacterial properties and boosting local and/or systemic plant immunity, both of which are crucial for biopesticide competency. Experiments on extra pure as well as mixture molecules will be required in the future to gain a thorough grasp of biosurfactant action mechanisms. Although field and laboratory investigations have confirmed the possible application and usefulness of biosurfactants in the agricultural sector in recent years, large-scale manufacture of biosurfactants has proven to be a huge economic challenge, owing to the significant capital investments required. In this context, employing renewable and low-cost alternative substrates can aid to lessen the cost of industrial biosurfactants manufacturing. Aside from economics, their restorative potency in the ground and homogeneity of molecules must also be upgraded in order for them to be used largely in crop protection application. Therefore, owing to their intriguing features, ecologically sound biosurfactants should be seriously considered as biological control options in IPM for greater sustainable agriculture production.

References

[1] Kim SK, Kim YC, Lee S, Kim JC, Yun MY, Kim IS. Insecticidal activity of rhamnolipid isolated from *Pseudomonas sp.* EP-3 against green peach aphid (*Myzus persicae*). Journal of Agricultural and Food Chemistry 2011, 59, 934–938.

[2] Yang SY, Lim DJ, Noh MY, Kim JC, Kim YC, Kim IS. Characterization of biosurfactants as insecticidal metabolites produced by *Bacillus subtilis* Y9. Entomological Research 2016, 471, 55–59.

[3] Nicolopoulou-Stamati P, Maipas S, Kotampasi C, Stamatis P, Hens L. Chemical pesticides and human health: The urgent need for a new concept in agriculture. Frontiers in Public Health 2016, 4, 1–8.

[4] Huang J, Hu R, Pray C, Qiao F, Rozelle S.Biotechnology as an alternative to chemical pesticides: A case study of Bt cotton in China. Agricultural Economics 2003, 29, 55–67.

[5] Shaban SM, Abd-Elaal AA.Studying the silver nanoparticles influence on thermodynamic behavior and antimicrobial activities of novel amide Gemini cationic surfactants. Materials Science and Engineering C 2017, 76, 871–885.

[6] Muthusamy K, Gopalakrishnan S, Ravi TK, Sivachidambaram P.Biosurfactants: Properties, commercial production and application. Current Science 2008, 25, 736–747.

[7] Hafeez FY, Naureen Z, Sarwar A. Surfactin: An emerging biocontrol tool for agriculture sustainability. In: Kumar A, Meena, VS, eds. Plant growth promoting rhizobacteria for agricultural sustainability,1st ed. Singapore, Springer, 2019, 203–213.

[8] Baker J, Levy M, Grewal D.An experimental approach to making retail store environment decisions. Journal of Retailing 1992,684, 445–460.

[9] Robert M, Mercadé ME, Bosch MP, Parra JL, Espuny MJ, Manresa MA, Guinea J.Effect of the carbon source on biosurfactant production by *Pseudomonas aeruginosa* 44T1. Biotechnology Letters 1989, 11, 871–874.

[10] Nitschke M, Pastore GM.Production and properties of a surfactant obtained from *Bacillus subtilis* grown on cassava wastewater. Bioresource Technology 2006, 97, 336–341.

[11] Lima TM, Procópio LC, Brandão FD, Leão BA, Tótola MR, Borges AC.Evaluation of bacterial surfactant toxicity towards petroleum degrading microorganisms. Bioresource Technology 2011,1023, 2957–2964.

[12] Zhou D, Feng H, Schuelke T, De Santiago A, Zhang Q, Zhang J, Luo C, Wei L.Rhizosphere microbiomes from root knot nematode non-infested plants suppress nematode infection. Microbial Ecology 2019,782, 470–481.

[13] Hussain T, Khan AA.*Bacillus subtilis* HussainT-AMU and its antifungal activity against potato black scurf caused by *Rhizoctonia solani* on seed tubers. Biocatalysis and Agricultural Biotechnology 2020, 23, 101443.
[14] Providenti MA, Flemming CA, Lee H, Trevors JT.Effect of addition of rhamnolipid biosurfactants or rhamnolipid-producing *Pseudomonas aeruginosa* on phenanthrene mineralization in soil slurries. FEMS Microbiology Ecology 1995, 17, 15–26.
[15] Chernin L, Ismailov Z, Haran S, Chet I.Chitinolytic *Enterobacter agglomerans* antagonistic to fungal plant pathogens. Applied and Environmental Microbiology 1995, 61, 1720–1726.
[16] Ron EZ, Rosenberg E.Natural roles of biosurfactants. Environmental Microbiology 2001, 3, 229–236.
[17] Mnif I, Ghribi D.Glycolipid biosurfactants: Main properties and potential applications in agriculture and food industry. Journal of the Science of Food and Agriculture 2016, 96, 4310–4320.
[18] Mnif I, Elleuch M, Chaabouni SE, Ghribi D.*Bacillus subtilis* SPB1 biosurfactant: Production optimization and insecticidal activity against the carob moth *Ectomyelois ceratoniae*. Crop Protection (Guildford, Surrey) 2013, 50, 66–72.
[19] Kogan M.Integrated pest management: Historical perspectives and contemporary developments. Annual Review of Entomology 1998, 43, 243–270.
[20] Benincasa M, Abalos A, Oliveira I, Manresa A.Chemical structure, surface properties and biological activities of the biosurfactant produced by *Pseudomonas aeruginosa* LBI from soapstock. Antonie van Leeuwenhoek 2004, 85, 1–8.
[21] Banat IM, Franzetti A, Gandolfi I, Bestetti G.Microbial biosurfactants production, applications and future potential. Applied Microbiology and Biotechnology 2010,872, 427–444.
[22] Jimoh AA, Lin J.Biosurfactant: A new frontier for greener technology and environmental sustainability. Ecotoxicology and Environmental Safety 2019, 184, 109607.
[23] Khanna S, Pattnaik P.Production and functional characterization of food compatible biosurfactants. Applied Food Science Journal 2019, 3, 1–4.
[24] Ribeiro BG, Guerra JMC, Sarubbo LA.Potential food application of a biosurfactant produced by *Saccharomyces cerevisiae* URM 6670. Frontiers in Bioengineering and Biotechnology 2020, 8, 434.
[25] Rosenberg E, Ron EZ.High and low-molecular-mass microbial surfactants. Applied Microbiology and Biotechnology 1999, 52, 154–162.
[26] Mulligan CN, Gibbs B.Types, production and applications of biosurfactants. Proceedings of the Indian National Science Academy Part B Biological Sciences 2004,701, 31–55.
[27] Md F.Biosurfactant : Production and application. Journal of Petroleum & Environmental Biotechnology 2012,34, 124.
[28] Edosa TT, Hun Jo Y, Keshavarz M, Soo Han Y.Biosurfactants: Production and potential application in insect pest management. Trends in Entomology 2018, 14, 79.
[29] Duncan K, Mcinerney M.Comparison of methods to detect biosurfactant production by diverse microorganisms. Journal of Microbiological Methods 2004, 56, 339–347.
[30] Santos DKF, Rufno RD, Luna JM, Santos VA, Sarubbo LA.Biosurfactants: Multifunctional biomolecules of the twenty-first century. International Journal of Molecular Sciences 2016,173, 401.
[31] Chen J, Wu Q, Hua Y, Chen J, Zhang H, Wang H.Potential applications of biosurfactant rhamnolipids in agriculture and biomedicine. Applied Microbiology and Biotechnology 2017, 101, 8309–8319.
[32] Desai JD, Banat IM.Microbial production of surfactants and their commercial potential. Microbiology and Molecular Biology Reviews : MMBR 1997, 61, 47–64.

[33] Lourith N, Kanlayavattanakul M.Natural surfactants used in cosmetics: Glycolipids. International Journal of Cosmetic Science 2009, 31, 255–261.
[34] Vatsa P, Sanchez L, Clement C, Baillieul F, Dorey S.Rhamnolipid biosurfactants as new players in animal and plant defense against microbes. International Journal of Molecular Sciences 2010, 11, 5095–5108.
[35] Abdel-Mawgoud AM, Lépine F, Déziel E.Rhamnolipids: Diversity of structures, microbial origins and roles. Applied Microbiology and Biotechnology 2010, 86, 1323–1336.
[36] Cameotra SS, Makkar RS.Synthesis of biosurfactants in extreme conditions. Applied Microbiology and Biotechnology 1998, 50, 520–529.
[37] Celligoi MAPC, Silveira VAI, Hipólito A, Caretta TO, Baldo C.Sophorolipids: A review on production and perspectives of application in agriculture. Spanish Journal of Agricultural Research 2020, 18, 1–10.
[38] Bharali P, Saikia JP, Ray A, Konwar BK.Rhamnolipid from *Pseudomonas aeruginosa* OBP1 : A novel chemotaxis and antibacterial agent. Colloids and Surfaces B Biointerfaces 2013, 103, 502–509.
[39] Das S, Kalita SJ, Bharali P, Konwar BK, Das B, Thakur AJ.Organic reactions in "green surfactant": An avenue to bisuracil derivative. ACS Sustainable Chemistry & Engineering 2013, 1, 1530–1536.
[40] Bharali P, Konwar BK.Production and physico-chemical characterization of a biosurfactant produced by *Pseudomonas aeruginosa* OBP1 isolated from petroleum sludge. Applied Biochemistry and Biotechnology 2011,1648, 1444–1460.
[41] Haba E, Espuny MJ, Busquets M, Manresa A.Screening and production of rhamnolipids by *Pseudomonas aeruginosa* 47T2 NCIB 40044 from waste frying oils. Journal of Applied Microbiology 2000, 88, 379–387.
[42] Franzetti A, Gandolfi I, Bestetti G, Smyth TJ, Banat IM.Production and applications of trehalose lipid biosurfactants. European Journal of Lipid Science and Technology : EJLST 2010,1126, 617–627.
[43] Göbbert U, Lang S, Wagner F.Sophorose lipid formation by resting cells of *Torulopsis bombicola*. Biotechnology Letters 1984,64, 225–230.
[44] Buchotz ML. The Metabolism of glycolipids in yeast *Candida bogoriensis*. Doctoral dissertation, Florida State University, USA, 1974.
[45] Zhoua QH, Kleknerb V, Kosarica N.Production of sophorose lipids by *Torulopsis bombicola* from Safflower oil and glucose. Journal of the American Oil Chemists' Society 1992, 69, 89–91.
[46] Daverey A, Pakshirajan K.Production, characterization, and properties of sophorolipids from the yeast *Candida bombicola* using a low-cost fermentative medium. Applied Biochemistry and Biotechnology 2009, 158, 663–674.
[47] de Oliveira MR, Magri A, Baldo C, Camilios-Neto D, Minucelli T, Celligoi MAPC.Sophorolipids: A promising biosurfactant and its applications. International Journal of Advanced Biotechnology and Research 2015,62, 161–174.
[48] Hewald S, Linne U, Scherer M, Marahiel MA, Bo M.Identification of a gene cluster for biosynthesis of mannosylerythritol lipids in the basidiomycetous fungus *Ustilago maydis*. Applied and Environmental Microbiology 2006, 72, 5469–5477.
[49] Kakugawa K, Tamai M, Imamura K, Miyamoto K, Miyoshi S, Morinaga Y, Suzuki O, Miyakawa T, Iyamoto KM.Isolation of yeast *Kurtzmanomyces sp*. I-11, novel producer of mannosylerythritol lipid. Bioscience, Biotechnology, and Biochemistry 2014,661, 188–191.
[50] Nimtz M.Formation and analysis of mannosylerythritol lipids secreted by *Pseudozyma aphidis*. Applied Microbiology and Biotechnology 2005,665, 551–559.

[51] Yoshida S, Koitabashi M, Nakamura J, Fukuoka T, Sakai H, Abe M, Kitamoto D.Effects of biosurfactants, mannosylerythritol lipids, on the hydrophobicity of solid surfaces and infection behaviours of plant pathogenic fungi. Journal of Applied Microbiology 2015,1191, 215–224.
[52] Fukuoka T, Yoshida S, Nakamura J, Koitabashi M, Sakai H, Abe M, Kitamoto D, Kitamoto H. Application of yeast glycolipid biosurfactant, mannosylerythritol lipid, as agrospreaders. Journal of Oleo Science 2015, 64, 689–695.
[53] Morita T, Konishi M, Fukuoka T, Imura T, Kitamoto D.Microbial conversion of glycerol into glycolipid biosurfactants, mannosylerythritol lipids, by a basidiomycete yeast, *Pseudozyma antarctica* JCM 10317T. Journal of Bioscience and Bioengineering 2007, 104, 78–81.
[54] Shaw JM, Bottino NR.Endogenous phospholipids of swine liver: Effect of fat deprivation on molecular species of phosphatidylcholine and phosphatidylethanolamine. Journal of Lipid Research 1974, 15, 317–325.
[55] Thorpe RF, Ratledge C.Fatty acids of triglycerides and phospholipids from a thermotolerant strain of *Candida tropicalis* grown on n-alkanes at 30 and 40°C. Microbiol 1973,781, 203–206.
[56] Ledesma-Amaro R, Dulermo R, Niehus X, Nicaud JM.Combining metabolic engineering and process optimization to improve production and secretion of fatty acids. Metabolic Engineering 2016, 38, 38–46.
[57] Spector AA, John K, Fletcher JE.Binding of long-chain fatty acids to bovine serum albumin. Journal of Lipid Research 1969, 10, 56–67.
[58] Cooper DG, Zajic JE.Surface-active compounds from microorganisms. Advances in Applied Microbiology 1980, 26, 229–253.
[59] Holdom RS, Turner AG.A new growth factor and emulsifying agent. The Journal of Applied Bacteriology 1969, 32, 448–456.
[60] Jourdan E, Henry G, Duby F, Dommes J, Barthélemy JP, Thonart P, Ongena M.Insights into the defense-related events occurring in plant cells following perception of surfactin-type lipopeptide from *Bacillus subtilis*. Molecular Plant-microbe Interactions : MPMI 2009,224, 456–468.
[61] Yakimov MM, Timmis KN, Wray V, Fredrickson HL.Characterization of a new lipopeptide surfactant produced by thermotolerant and halotolerant subsurface *Bacillus licheniformis* BAS50. Applied and Environmental Microbiology 1995, 61, 1706–1713.
[62] Quinn GA, Maloy AP, Mcclean S, Carney B, Slater JW.Lipopeptide biosurfactants from *Paenibacillus polymyxa* inhibit single and mixed species biofilms. Biofoul 2012,2810, 1151–1166.
[63] Javaheri M, Jenneman GE, Mclnerney MJ, Knapp RM.Anaerobic production of a biosurfactant by *Bacillus licheniformis* JF-2. Applied and Environmental Microbiology 1985, 50, 698–700.
[64] Cooper DG, Zajic JE, Gerson DF.Production of surface-active lipids by *Corynebacterium lepus*. Applied and Environmental Microbiology 1979, 37, 4–10.
[65] Vasconcelos GMD, Mulinari J, Vko S, Matosinhos RD, Oliveira JV, Oliveira D, Andrade CJ.Biosurfactants as green biostimulants for seed germination and growth. International Journal of Research Studies in Microbiology and Biotechnology 2020,61, 1–13.
[66] Bodour AA, Drees KP, Maier RM.Distribution of biosurfactant-producing bacteria in undisturbed and contaminated arid southwestern soils. Applied and Environmental Microbiology 2003, 69, 3280–3287.
[67] Kosaric N, Sukan FV. Biosurfactants: Production: Properties: Applications, vol. 159. Boca Raton, Florida, USA, CRC Press, 2010.

[68] Chhatre S, Purohit H, Shanker R, Khanna P.Bacterial consortia for crude oil spill remediation. Water Science and Technology : A Journal of the International Association on Water Pollution Research 1996, 34, 187–193.
[69] Pritchard PH, Costa CF.EPA's Alaska oil spill bioremediation project: Final part of a five-part series. Environmental Science & Technology 1991, 25, 372–379.
[70] Ye M, Sun M, Wan J, Feng Y, Zhao Y, Tian D, Hu F, Jiang X.Feasibility of lettuce cultivation in sophoroliplid-enhanced washed soil originally polluted with Cd, antibiotics, and antibiotic resistant genes. Ecotoxicology and Environmental Safety 2016, 124, 344–350.
[71] Almansoory AF, Hasan HA, Abdullah SRS, Idris M, Anuar N, Al-Adiwish WM.Biosurfactant produced by the hydrocarbon-degrading bacteria: Characterization, activity and applications in removing TPH from contaminated soil. Environmental Technology & Innovation 2019, 14, 100347.
[72] Nagajyoti PC, Lee KD, Sreekanth TVM.Heavy metals, occurrence and toxicity for plants: A review. Environmental Chemistry Letters 2010, 8, 199–216.
[73] Klik B, Gusiatin ZM, Kulikowska D.Quality of heavy metal-contaminated soil before and after column flushing with washing agents derived from municipal sewage sludge. Scientific Reports 2021, 11, 1–13.
[74] Bååth E.Effects of heavy metals in soil on microbial processes and populations (a review). Water Air and Soil Pollution 1989, 47, 335–379.
[75] Miller RM.Biosurfactant-facilitated remediation of metal-contaminated soils. Environmental Health Perspectives 1995,1031, 59–62.
[76] Mulligan CN, Kamali M, Gibbs BF.Bioleaching of heavy metals from a low-grade mining ore using *Aspergillus niger*. Journal of Hazardous Materials 2004,1101-3, 77–84.
[77] Mulligan CN, Yong RN, Gibbs BF, James S, Bennett HPJ.Metal removal from contaminated soil and sediments by the biosurfactant surfactin. Environmental Science & Technology 1999,3321, 3812–3820.
[78] San Martín YB, Toledo León HF, Rodríguez AÁ, Marqués AM, López MIS.Rhamnolipids application for the removal of vanadium from contaminated sediment. Current Microbiology 2021, 78, 1949–1960.
[79] Atlas RM.Petroleum biodegradation and oil spill bioremediation. Marine Pollution Bulletin 1995, 31, 178–182.
[80] Ben-Amotz A, Tornabene TG, Thomas WH.Chemical profile of selected species of microalgae with emphasis on lipids. Journal of Phycology 1985,211, 72–81.
[81] Nogueira L, Rodrigues ACF, Trídico CP, Fossa CE, de Almeida EA.Oxidative stress in Nile tilapia (*Oreochromis niloticus*) and armored catfsh (*Pterygoplichthys anisitsi*) exposed to diesel oil. Environmental Monitoring and Assessment 2011,1801, 243–255.
[82] Ayotamuno MJ, Kogbara RB, Ogaji SOT, Probert SD.Bioremediation of a crude-oil polluted agricultural-soil at Port Harcourt, Nigeria. Applied Energy 2006, 83, 1249–1257.
[83] Boopathy R.Factors limiting bioremediation technologies. Bioresource Technology 2000, 74, 63–67.
[84] Urum K, Pekdemir T, Çopur M.Screening of biosurfactants for crude oil contaminated soil washing. Journal of Environmental Engineering and Science 2005, 4, 487–496.
[85] Lai CC, Huang YC, Wei YH, Chang JS.Biosurfactant-enhanced removal of total petroleum hydrocarbons from contaminated soil. Journal of Hazardous Materials 2009, 167, 609–614.
[86] Kang SW, Kim YB, Shin JD, Kim EK.Enhanced biodegradation of hydrocarbons in soil by microbial biosurfactant, sophorolipid. Applied Biochemistry and Biotechnology 2010, 160, 780–790.
[87] Jain RM, Mody K, Mishra A, Jha B.Physicochemical characterization of biosurfactant and its potential to remove oil from soil and cotton cloth. Carbohydrate Polymers 2012,894, 1110–1116.

[88] Mishra A.Physicochemical characterization of biosurfactant and its potential to remove oil from soil and cotton cloth. Carbohydrate Polymers 2012, 89, 1110–1116.

[89] Derguine-Mecheri L, Kebbouche-Gana S, Djenane D.Biosurfactant production from newly isolated *Rhodotorula sp*. YBR and its great potential in enhanced removal of hydrocarbons from contaminated soils. World Journal of Microbiology & Biotechnology 2021, 37, 1–8.

[90] Ahmadi M, Niazi F, Jaafarzadeh N, Ghafari S, Jorfi S.Characterization of the biosurfactant produced by *Pesudomonas areuginosa* strain R4 and its application for remediation pyrene-contaminated soils. Journal of Environmental Health Science & Engineering 2021, 19, 445–456.

[91] Guo J, Wen X.Ecotoxicology and Environmental Safety Performance and kinetics of benzo(α) pyrene biodegradation in contaminated water and soil and improvement of soil properties by biosurfactant amendment. Ecotoxicology and Environmental Safety 2021, 207, 111292.

[92] Umar A, Zafar A, Wali H, Siddique MP, Qazi MA, Naeem AH, Malik ZA, Ahmed S.Low-cost production and application of lipopeptide for bioremediation and plant growth by *Bacillus subtilis* SNW3. AMB Express 2021, 11, 165.

[93] Malik A, Grohmann E, Akhtar R. Environmental deterioration and human health. Switzerland AG, Springer Nature, 2014.

[94] Samchetshabam G, Hussan A, Choudhury TG, Gita S.Impact of textile dyes waste on aquatic environments and its treatment wastewater management view project tribal sub plan view project. Ecology and Environment 2017, 35, 2349–2353.

[95] Mu Y, Rabaey K, Rozendal RA, Yuan Z, Keller J.Decolorization of azo dyes in bioelectrochemical systems. Environmental Science & Technology 2009, 43, 5137–5143.

[96] Feng W, Nansheng D, Helin H.Degradation mechanism of azo dye C.I. reactive red 2 by iron powder reduction and photooxidation in aqueous solutions. Chemosphere 2000, 41, 1233–1238.

[97] Brown MA, De Vito SC.Predicting azo dye toxicity. Critical Reviews in Environmental Science and Technology 1993, 23, 249–324.

[98] Gottlieb A, Shaw C, Smith A, Wheatley A, Forsythe S.The toxicity of textile reactive azo dyes after hydrolysis and decolourisation. Journal of Biotechnology 2003, 101, 49–56.

[99] Ekici P, Leupold G, Parlar H.Degradability of selected azo dye metabolites in activated sludge systems. Chemosphere 2001, 44, 721–728.

[100] Drumond Chequer FM, Junqueira D, de Oliveir DP. Azo dyes and their metabolites: Does the discharge of the azo dye into water bodies represent human and ecological risks? Advances in Treating Textile Effluent 2011, 48, 28–48.

[101] Das A, Mishra S.Removal of textile dye reactive green-19 using bacterial consortium: Process optimization using response surface methodology and kinetics study. Journal of Environmental Chemical Engineering 2017, 5, 612–627.

[102] Lellis B, Fávaro-Polonio CZ, Pamphile JA, Polonio JC.Effects of textile dyes on health and the environment and bioremediation potential of living organisms. Biotechnology Research & Innovation 2019, 3, 275–290.

[103] Pandey A, Singh P, Iyengar L.Bacterial decolorization and degradation of azo dyes. International Biodeterioration & Biodegradation 2007, 59, 73–84.

[104] Selvam K, Swaminathan K, Chae KS.Microbial decolorization of azo dyes and dye industry effluent by *Fomes lividus*. World Journal of Microbiology & Biotechnology 2003, 19, 591–593.

[105] Purkait MK, DasGupta S, De S.Removal of dye from wastewater using micellar-enhanced ultrafiltration and recovery of surfactant. Separation and Purification Technology 2004, 37, 81–92.

[106] Samal K, Das C, Mohanty K.Application of saponin biosurfactant and its recovery in the MEUF process for removal of methyl violet from wastewater. Journal of Environmental Management 2017, 203, 8–16.
[107] Mnif I, Fendri R, Ghribi D.Biosorption of congo red from aqueous solution by *Bacillus weihenstephanensis* RI12; Effect of SPB1 biosurfactant addition on bio-decolorization potency. Water Science and Technology : A Journal of the International Association on Water Pollution Research 2015, 72, 865–874.
[108] Nor FHM, Abdullah S, Yuniarto A, Ibrahim Z, Nor MHM, Hadibarata T.Production of lipopeptide biosurfactant by *Kurthia gibsonii* KH2 and their synergistic action in biodecolourisation of textile wastewater. Environmental Technology & Innovation 2021, 22, 101533.
[109] Pacwa-Płociniczak M, Płaza GA, Piotrowska-Seget Z, Cameotra SS.Environmental applications of biosurfactants: Recent advances. International Journal of Molecular Sciences 2011, 12, 633–654.
[110] Carolin CF, Kumar PS, Chitra B, Jackulin CF, Ramamurthy R.Stimulation of *Bacillus sp.* by lipopeptide biosurfactant for the degradation of aromatic amine 4-Chloroaniline. Journal of Hazardous Materials 2021, 415, 125716.
[111] Joaquin S.Distribution and toxicity of sediment-associated pesticides in agriculture-dominated water bodies of California's central valley. Environmental Science & Technology 2004,3810, 2752–2759.
[112] Kookanaa RS, Aylmoreb LAG.Estimating the pollution potential of pesticides to ground water. Soil Research 1994,325, 1141–1155.
[113] Rouse J, David S.Influence of surfactants on microbial degradation of organic compounds. Critical Reviews in Environmental Science and Technology 1994, 24, 325–370.
[114] Mohan PK, Nakhla G, Yanful EK.Biokinetics of biodegradation of surfactants under aerobic, anoxic and anaerobic conditions. Water Research 2006, 40, 533–540.
[115] Awasthi N, Kumar A, Makkar R, Cameotra SS.Biodegradation of soil-applied endosulfan in the presence of a biosurfactant. Journal of Environmental Science and Health 1999, 34, 793–803.
[116] Keith LH, Lee KW, Provost LP, Present DL. Methods for gas chromatographic monitoring of the environmental protection agency's consent decree priority pollutants. In: Measurement of organic pollutants in water and wastewater, West Conshohocken, PA, USA, ASTM International, 1979.
[117] Gaur VK, Bajaj A, Regar RK, Kamthan M, Jha RR, Srivastava JK, Manickam N.Rhamnolipid from a *Lysinibacillus sphaericus* strain IITR51 and its potential application for dissolution of hydrophobic pesticides. Bioresource Technology 2019, 272, 19–25.
[118] López-Prieto A, Moldes AB, Cruz JM, Pérez Cid B.Towards more ecofriendly pesticides: Use of biosurfactants obtained from the Corn milling industry as solubilizing agent of copper oxychloride. Journal of Surfactants and Detergents 2020, 23, 1055–1066.
[119] Manickam N, Bajaj A, Saini HS, Shanker R.Surfactant mediated enhanced biodegradation of hexachlorocyclohexane (HCH) isomers by *Sphingomonas sp.* Biodegradation 2012, 23, 673–682.
[120] Payal R, Jain A. Influence of biosurfactant in the bioremediation of pentachlorophenol. Green Sustainable Process for Chemical and Environmental Engineering and Science, 2021, 341–352.
[121] Lang S, Katsiwela E, Wagner F. Antimicrobial effects of biosurfactants. Fett wissenschaft technologie-fat science technology 1989, 91, 363–366.

[122] Kim BS, Lee JY, Hwang BK.*In vivo* control and *in vitro* antifungal activity of rhamnolipid B, a glycolipid antibiotic, against *Phytophthora capsici* and *Colletotrichum orbiculare*. Pest Management Science 2000, 56, 1029–1035.
[123] Nitschke M, Costa SGVAO, Contiero J.Structure and applications of a rhamnolipid surfactant produced in soybean oil waste. Applied Biochemistry and Biotechnology 2010, 160, 2066–2074.
[124] Sha R, Jiang L, Meng Q, Zhang G, Song Z.Producing cell-free culture broth of rhamnolipids as a cost-effective fungicide against plant pathogens. Journal of Basic Microbiology 2012, 52, 458–466.
[125] Goswami D, Handique PJ, Deka S.Rhamnolipid biosurfactant against *Fusarium sacchari*-the causal organism of pokkah boeng disease of sugarcane. Journal of Basic Microbiology 2014,546, 548–557.
[126] Gond SK, Bergen MS, Torres MS, White JF Jr.Endophytic *Bacillus spp.* produce antifungal lipopeptides and induce host defense gene expression in maize. Microbiological Research 2015, 172, 79–87.
[127] Borah SN, Goswami D, Sarma HK, Cameotra SS, Deka S.Rhamnolipid biosurfactant against *Fusarium verticillioides* to control stalk and ear rot disease of maize. Frontiers in Microbiology 2016, 7, 1505.
[128] Lahkar J, Goswami D, Deka S, Ahmed G.Novel approaches for application of biosurfactant produced by *Pseudomonas aeruginosa* for biocontrol of *Colletotrichum capsici* responsible for anthracnose disease in chilli. European Journal of Plant Pathology 2018,1501, 57–71.
[129] Stanghellini ME, Miller RM.Biosurfactants: Their identity and potential efficacy in the biological control of zoosporic plant pathogens. Plant Disease 1997,811, 4–12.
[130] Eldin AM, Kamel Z, Hossam N.Purification and identification of surface active amphiphilic candidates produced by *Geotrichum candidum* MK880487 possessing antifungal property. Journal of Dispersion Science and Technology 2020,427, 1082–1098.
[131] Ahmad M, Ahmad I, Hilger TH, Nadeem SM, Akhtar MF, Jamil M, Hussain A, Zahir ZA.Preliminary study on phosphate solubilizing *Bacillus subtilis* strain Q3 and Paenibacillus sp. strain Q6 for improving cotton growth under alkaline conditions. Peer Journal 2018, 6, 5122.
[132] Szczech M, Shoda M.The effect of mode of application of *Bacillus subtilis* RB14-C on its efficacy as a biocontrol agent against *Rhizoctonia solani*. Journal of Phytopathology 2006,546, 370–377.
[133] Ben S, Kilani-feki O, Dammak M, Jabnoun-khiareddine H, Daami-remadi M, Tounsi S.Efficacy of *Bacillus subtilis* V26 as a biological control agent against *Rhizoctonia solani* on potato. Comptes Rendus Biologies 2015,33812, 784–792.
[134] Rani M, Weadge JT, Jabaji S, Saleem M.Isolation and characterization of biosurfactant-producing bacteria from oil well batteries with antimicrobial activities against food-borne and plant pathogens. Frontiers in Microbiology 2020, 11, 64.
[135] Kruijt M, Tran H, Raaijmakers JM.Functional, genetic and chemical characterization of biosurfactants produced by plant growth-promoting *Pseudomonas putida* 267. Journal of Applied Microbiology 2009,1072, 546–556.
[136] Kitamoto D, Yanagishita H, Shinbo T, Nakane T, Kamisawa C, Nakahara T.Surface active properties and antimicrobial activities of mannosylerythritol lipids as biosurfactants produced by *Candida antarctica*. Journal of Biotechnology 1993, 29, 91–96.
[137] Buxdorf K, Rahat I, Gafni A, Levy M.The epiphytic fungus *Pseudozyma aphidis* induces jasmonic acid-and salicylic acid/nonexpressor of PR1-independent local and systemic resistance. Plant Physiology 2013, 161, 2014–2022.

[138] Cheng Y, McNally DJ, Labbé C, Voyer N, Belzile F, Bélanger RR.Insertional mutagenesis of a fungal biocontrol agent led to discovery of a rare cellobiose lipid with antifungal activity. Applied and Environmental Microbiology 2003, 69, 2595–2602.
[139] Yoshida S, Morita T, Shinozaki Y, Watanabe T, Sameshima-Yamashita Y, Koitabashi M, Kitamoto D, Kitamoto H.Mannosylerythritol lipids secreted by phyllosphere yeast *Pseudozyma antarctica* is associated with its filamentous growth and propagation on plant surfaces. Applied Microbiology and Biotechnology 2014, 98, 6419–6429.
[140] Awada SM, Awada MM, Spendlove RS. Compositions and methods for controlling pests with glycolipids. US Patent, US 2012/0058895 A1, 2014.
[141] Ghribi D, Mnif I, Boukedi H, Kammoun R, Ellouze-Chaabouni S.Statistical optimization of low-cost medium for economical production of *Bacillus subtilis* biosurfactant, a biocontrol agent for the olive moth *Prays oleae*. African Journal of Microbiology Research 2011,527, 4927–4936.
[142] Ghribi D, Elleuch M, Abdelke L, Ellouze-chaabouni S.Evaluation of larvicidal potency of *Bacillus subtilis* SPB1 biosurfactant against *Ephestia kuehniella* (Lepidoptera : Pyralidae) larvae and influence of abiotic factors on its insecticidal activity. Journal of Stored Products Research 2012, 48, 68–72.
[143] Fazaeli N, Bahador N, Hesami S.A study on larvicidal activity and phylogenetic analysis of *staphylococcus epidermidis* as a biosurfactant-producing bacterium. Polish Journal of Environmental Studies 2021, 30, 4511–4519.
[144] Schroth MN, Hancock JG.Disease-suppressive soil and root-colonizing bacteria. Science 1982, 216, 1376–1381.
[145] Oostendorp M, Sikora RA.Seed treatment with antagonistic rhizobacteria for the suppression of *Heterodera schachtii* early root infection of sugar beet. Revue de Nématologie 1989, 12, 77–83.
[146] Adomako J, Danso Y, Sakyiamah B, Kankam F, Osei K.Plant-parasitic nematodes associated with sweet potato rhizosphere soil in the semi-deciduous forest and coastal Savannah zones of Ghana. Ghana Journal of Agricultural Science 2020, 55, 1–9.
[147] Ebadi M, Fatemy S, Riahi H.Evaluation of *Pochonia chlamydosporia* var. chlamydosporia as a control agent of *Meloidogyne javanica* on pistachio. Biocontrol Science and Technology 2009, 19, 689–700.
[148] Onkendi EM, Kariuki GM, Marais M, Moleleki LN.The threat of root-knot nematodes (*Meloidogyne spp.*) in Africa: A review. Plant Pathology 2014, 63, 727–737.
[149] Soriano IRS, Prot JC, Matias DM.Expression of tolerance for *Meloidogyne graminicola* in rice cultivars as affected by soil type and flooding. Journal of Nematology 2000, 32, 309–317.
[150] Gao H, Qi G, Yin R, Zhang H, Li C, Zhao X.*Bacillus cereus* strain S2 shows high nematicidal activity against *Meloidogyne incognita* by producing sphingosine. Scientific Reports 2016, 6, 1–11.
[151] Hussain T, Haris M, Shakeel A, Ahmad G, Ahmad Khan A, Khan MA.Bio-nematicidal activities by culture filtrate of *Bacillus subtilis* HussainT-AMU: New promising biosurfactant bioagent for the management of Root Galling caused by *Meloidogyne incognita*. Vegetos 2020, 33, 229–238.
[152] Stukenbrock EH, Mcdonald BA.The origins of plant pathogens in agro-ecosystems. Annual Review of Phytopathology 2008, 46, 75–100.
[153] Hussain T, Khan AA, Khan MA.Biocontrol of soil borne pathogen of potato tuber caused by *Rhizoctonia solani* through biosurfactant based Bacillus strain. Journal of Nepal Agricultural Research Council 2021, 7, 54–66.

[154] Hultberg M, Alsberg T, Khalil S, Alsanius B.Suppression of disease in tomato infected by *Pythium ultimum* with a biosurfactant produced by *Pseudomonas koreensis*. BioControl 2010,553, 435–444.
[155] Sangüesa-Barreda G, Camarero JJ, Oliva J, Montes F, Gazol A.Past logging, drought and pathogens interact and contribute to forest dieback. Agricultural and Forest Meteorology 2015, 208, 85–94.
[156] Perry E. Growth promotion of *Solanum lycopersicum* by addition of a biosurfactant produced by *Pseudomonas koreensis* JDSCSU15. Doctoral dissertation, Southern Connecticut State University, USA, 2020.
[157] Chaurasia LK, Tamang B, Tirwa RK, Lepcha PL.Influence of biosurfactant producing *Bacillus tequilensis* LK5.4 isolate of kinema, a fermented soybean, on seed germination and growth of maize (*Zea mays* L.). 3 Biotechnology 2020,107, 297.
[158] Frommel MI, Nowak J, Lazarovits G.Treatment of potato tubers with a growth promoting *Pseudomonas sp.*: Plant growth responses and bacterium distribution in the rhizosphere. Plant and Soil 1993, 150, 51–60.
[159] Sharma A, Johri BN, Sharma AK, Glick BR.Plant growth-promoting bacterium *Pseudomonas sp.* strain GRP3 influences iron acquisition in mung bean (*Vigna radiata* L. Wilzeck). Soil Biology & Biochemistry 2003, 35, 887–894.
[160] Dey R, Pal KK, Bhatt DM, Chauhan SM.Growth promotion and yield enhancement of peanut (*Arachis hypogaea* L.) by application of plant growth-promoting rhizobacteria. Microbiological Research 2004, 159, 371–394.
[161] D'aes J, De Maeyer K, Pauwelyn E, Höfte M.Biosurfactants in plant–Pseudomonas interactions and their importance to biocontrol. Environmental Microbiology Reports 2010,23, 359–372.
[162] Sheng X, He L, Wang Q, Ye H, Jiang C.Effects of inoculation of biosurfactant-producing *Bacillus sp.* J119 on plant growth and cadmium uptake in a cadmium-amended soil. Journal of Hazardous Materials 2008, 155, 17–22.
[163] Mishra I, Fatima T, Egamberdieva D, Arora NK.Novel bioformulations developed from *Pseudomonas putida* bsp9 and its biosurfactant for growth promotion of *Brassica juncea* (L.). Plants 2020, 9, 1–17.
[164] Chaurasia LK, Tirwa RK, Tamang B. Potential of *Enterococcus faecium* LM5. 2 for lipopeptide biosurfactant production and its effect on the growth of Maize (*Zea mays* L.). Archives of Microbiology 2021, In-press. https://doi.org/10.21203/rs.3.rs-532271/v1.
[165] Marchut-Mikołajczyk O, Drożdżyński P, Polewczyk A, Smułek W, Antczak T.Biosurfactant from endophytic *Bacillus pumilus* 2A: Physicochemical characterization, production and optimization and potential for plant growth promotion. Microbial Cell Factories 2021, 20, 1–11.
[166] Ławniczak Ł, Marecik R, Chrzanowski Ł.Contributions of biosurfactants to natural or induced bioremediation. Applied Microbiology and Biotechnology 2013,976, 2327–2339.
[167] Krawczyńska M, Kolwzan B, Rybak J, Gediga K, Shcheglova NS.The infuence of biopreparation on seed germination and growth. Polish Journal of Environmental Studies 2012,216, 1697–1702.
[168] Sieverding E. Increasing yields by the use of sophorolipids. US Patent, US2017094968A1, 2017.
[169] Gálvez A, López-Galindo A, Peña A.Effect of different surfactants on germination and root elongation of two horticultural crops: Implications for seed coating. New Zealand Journal of Crop and Horticultural Science 2019,472, 83–98.

[170] Bubenheim D, Wignarajah K, Berry W, Wydeven T.Phytotoxic effects of gray water due to surfactants. Journal of the American Society for Horticultural Science. American Society for Horticultural Science 1997, 122, 792–796.
[171] Doige CA, Yu X, Sharom FJ.The effects of lipids and detergents on ATPase-active P-glycoprotein. Biochimica Et Biophysica Acta 1993, 1146, 65–72.
[172] de Bruin W, van der Merwe C, Kritzinger Q, Bornman R, Korsten L.Ultrastructural and developmental evidence of phytotoxicity on cos lettuce (*Lactuca sativa*) associated with nonylphenol exposure. Chemosphere 2017, 169, 428–436.
[173] Mukherjee S, Das P, Sen R.Towards commercial production of microbial surfactants. Trends in Biotechnology 2006,2411, 509–515.
[174] Kaur N, Erickson TE, Ball AS, Ryan MH.A review of germination and early growth as a proxy for plant fitness under petrogenic contamination – knowledge gaps and recommendations. The Science of the Total Environment 2017, 603, 728–744.
[175] Das AJ, Kumar R.Bioremediation of petroleum contaminated soil to combat toxicity on *Withania somnifera* through seed priming with biosurfactant producing plant growth promoting rhizobacteria. Journal of Environmental Management 2016, 174, 79–86.
[176] Shukry W, Al-Hawas G, Al-Moaikal R, El-Bendary M.Efect of petroleum crude oil on mineral nutrient elements, soil properties and bacterial biomass of the rhizosphere of Jojoba. British Journal of Environment and Climate Change 2013,31, 103–118.
[177] Cameotra SS, Makkar RS, Kaur J, Mehta SK.Synthesis of biosurfactants and their advantages to microorganisms and mankind. Advances in Experimental Medicine and Biology 2010, 672, 261–280.
[178] Mata-Sandoval JC, Karns J, Torrents A.Effect of ramnolipids produced by *Pseudomonas aeruginosa* UG2 on the solubilization of pesticides. Environmental Science & Technology 2000,3423, 4923–4930.
[179] Edwards KR, Lepo JE, Lewis MA.Toxicity comparison of biosurfactants and synthetic surfactants used in oil spill remediation to two estuarine species. Marine Pollution Bulletin 2003,4610, 1309.
[180] Sancheti A, Ju LK.Rhamnolipid effects on water imbibition, germination, and initial root and shoot growth of Soybeans. Journal of Surfactants and Detergents 2020,232, 371–381.
[181] Marchut-Mikolajczyk O, Drożdżyński P, Pietrzyk D, Antczak T.Biosurfactant production and hydrocarbon degradation activity of endophytic bacteria isolated from *Chelidonium majus* L. Microbial Cell Factories 2018,171, 1–9.
[182] Adnan M, Alshammari E, Ashraf SA, Patel K, Lad K, Patel M. Physiological and molecular characterization of biosurfactant producing endophytic fungi *Xylaria regalis* from the cones of *Thuja plicata* as a potent plant growth promoter with its potential application. BioMed Research International 2018, 7362148.
[183] Oluwaseun AC, Kola O, Mishra P, Singh JR, Singh AK, Cameotra SS, Oluwasesan MB.Characterization and optimization of a rhamnolipid from *Pseudomonas aeruginosa* C1501 with novel biosurfactant activities. Sustainable Chemistry and Pharmacy 2017, 6, 26–36.
[184] Rubio-Ribeaux D, da Silva Andrade RF, da Silva GS, de Holanda RA, Pele MA, Nunes P, Campos-Takaki GM.Promising biosurfactant produced by a new *Candida tropicalis* UCP 1613 strain using substrates from renewable-resources. African Journal of Microbiology Research 2017, 11, 981–991.
[185] Kilmer VJ.Minerals and agriculture. Journal of Environmental Studies and Sciences 1979, 3, 515–558.
[186] Shive BJW, Robbins WR.Mineral nutrition of plants. Annual Review of Biochemistry 1939, 8, 503–520.

[187] Baligar VC, Fageria NK, He ZL.Nutrient use efficiency in plants. Communications in Soil Science and Plant Analysis 2001, 32, 921–950.

[188] Welch RM, Shuman L.Micronutrient nutrition of plants. Critical Reviews in Plant Sciences 2014,141, 49–82.

[189] Graham RD. Micronutrient deficiencies in crops and their global significance. Micronutr Defic Glob Crop Prod 2008, 41–61.

[190] Singh R, Glick BR, Rathore D.Biosurfactants as a biological tool to increase micronutrient availability in soil : A review. Pedosphere 2018, 28, 170–189.

[191] Abdul AS, Gibson TL, Rai D.Selection of surfactant for the removal of petroleum products from shallow sandy aquifers. Gorund Water 1990, 28, 920–926.

[192] Singh P, Cameotra SS.Enhancement of metal bioremediation by use of microbial surfactants. Biochemical and Biophysical Research Communications 2004, 319, 291–297.

[193] Singh P, Cameotra SS.Potential applications of microbial surfactants in biomedical sciences. Trends in Biotechnology 2004, 22, 142–146.

[194] Sheng X, He L, Wang Q, Ye H, Jiang C.Effects of inoculation of biosurfactant-producing *Bacillus sp.* J119 on plant growth and cadmium uptake in a cadmium-amended soil. Journal of Hazardous Materials 2008, 155, 17–22.

[195] Boller T, He SY.Innate immunity in plants: An arms race between pattern recognition receptors in plants and effectors in microbial pathogens. Science 2009, 324, 742–743.

[196] Métraux JP. Systemic acquired resistance. In: Maloy S, Hughes, K, eds. Brenner's encyclopedia of genetics,2nd edn. Amsterdam, Netherlands, Academic Press, 2013, 627–629.

[197] Boller T, Felix G.A renaissance of elicitors: Perception of microbe-associated molecular patterns and danger signals by pattern-recognition receptors. Annual Review of Plant Biology 2009, 60, 379–407.

[198] Mackey D, McFall AJ.MAMPs and MIMPs: Proposed classifications for inducers of innate immunity. Molecular Microbiology 2006, 61, 1365–1371.

[199] Raaijmakers JM, de Bruijn I, Nybroe O, Ongena M.Natural functions of lipopeptides from Bacillus and Pseudomonas: More than surfactants and antibiotics. FEMS Microbiology Reviews 2010, 34, 1037–1062.

[200] Thonart P, Joris B, Philippe T, Philippe T, Ongena M, Jourdan E, Joris B, Arpigny J.Surfactin and fengycin lipopeptides of *Bacillus subtilis* as elicitors of induced systemic resistance in plants. Environmental Microbiology 2007, 9, 1084–1090.

[201] Cawoy H, Mariutto M, Henry G, Fisher C, Vasilyeva N, Thonart P, Dommes J, Ongena M.Plant defense stimulation by natural isolates of Bacillus depends on efficient surfactin production. Molecular Plant-microbe Interactions 2014,272, 87–100.

[202] Ongena M, Jourdan E, Adam A, Paquot M, Brans A, Joris B, Arpigny JL, Thonart P.Surfactin and fengycin lipopeptides of *Bacillus subtilis* as elicitors of induced systemic resistance in plants. Environmental Microbiology 2007,94, 1084–1090.

[203] Waewthongrak W, Leelasuphakul W, McCollum G.Cyclic lipopeptides from *Bacillus subtilis* ABS-S14 elicit defense-related gene expression in citrus fruit. PLoS One 2014,910, 109386.

[204] Yamamoto S, Shiraishi S, Suzuki S.Are cyclic lipopeptides produced by *Bacillus amyloliquefaciens* S13-3 responsible for the plant defense response in strawberry against *Colletotrichum gloeosporioides*?. Letters in Applied Microbiology 2015,604, 379–386.

[205] López D, Vlamakis H, Losick R, Kolter R.Paracrine signaling in a bacterium. Genes & Development 2009,2314, 1631–1638.

[206] Farace G, Fernandez O, Jacquens L, Coutte F, Krier F, Jacques P, Clément C, Barka EA, Jacquard C, Dorey S.Cyclic lipopeptides from *Bacillus subtilis* activate distinct patterns of defense responses in grapevine. Molecular Plant Pathology 2015,162, 177–187.

[207] Wu L, Huang Z, Li X, Ma L, Gu Q, Wu H, Liu J, Borriss R, Wu Z, Gao X.Stomatal closure and SA-, JA/ET-signaling pathways are essential for *Bacillus amyloliquefaciens* FZB42 to restrict leaf disease caused by *Phytophthora nicotianae* in *Nicotiana benthamiana*. Frontiers in Microbiology 2018, 9, 847.
[208] D'aes J, Kieu NP, Léclère V, Tokarski C, Olorunleke FE, De Maeyer K, Jacques P, Höfte M, Ongena M.To settle or to move? The interplay between two classes of cyclic lipopeptides in the biocontrol strain Pseudomonas CMR 12a. Environmental Microbiology 2014,167, 2282–2300.
[209] Maget-Dana R, Thimon L, Peypoux F, Ptak M.Surfactin/iturin A interactions may explain the synergistic effect of surfactin on the biological properties of iturin A. Biochimie 1992, 74, 1047–1051.
[210] Vanittanakom N, Loeffler W, Koch U, Jung G.Fengycin-A Novel antifungal lipopeptide antibiotic produced by *Bacillus subtilis* F-29-3. The Journal of Antibiotics 1986, 39, 888–901.
[211] Le Mire G, Siah A, Brisset MN, Gaucher M, Deleu M, Jijakli MH.Surfactin protects wheat against *Zymoseptoria tritici* and activates both salicylic acid- and jasmonic acid-dependent defense responses. Agriculture 2018, 8, 11.
[212] Varnier AL, Sanchez L, Vatsa P, Boudesocque L, Garcia-Brugger A, Rabenoelina F, Sorokin A, Renault JH, Kauffmann S, Pugin A, Clement C, Baillieul F, Dorey S.Bacterial rhamnolipids are novel MAMPs conferring resistance to *Botrytis cinerea* in grapevine. Plant Cell Environment 2009, 32, 178–193.
[213] Bent AF, Mackey D.Elicitors, effectors, and R genes: The new paradigm and a lifetime supply of questions. Annual Review of Phytopathology 2007, 45, 399–436.
[214] Monnier N, Furlan A, Botcazon C, Dahi A, Mongelard G, Cordelier S, Clément C, Dorey S, Sarazin C, Rippa S.Rhamnolipids from *Pseudomonas aeruginosa* are elicitors triggering *Brassica napus* protection against *Botrytis cinerea* without physiological disorders. Frontiers in Plant Science 2018, 9, 1–14.
[215] Sanchez L, Courteaux B, Hubert J, Kauffmann S, Renault JH, Clément C, Baillieul F, Dorey S.Rhamnolipids elicit defense responses and induce disease resistance against biotrophic, hemibiotrophic, and necrotrophic pathogens that require different signaling pathways in Arabidopsis and highlight a central role for salicylic acid. Plant Physiology 2012, 160, 1630–1641.
[216] Krogh KA, Halling-Sørensen B, Mogensen BB, Vejrup KV.Environmental properties and effects of nonionic surfactant adjuvants in pesticides: A review. Chemosphere 2003,507, 871–901.
[217] Theodorakis PE, Müller EA, Craster RV, Matar OK.Insights into surfactant-assisted superspreading. Current Opinion in Colloid & Interface Science 2014,194, 283–289.
[218] Wills GD, McWhorter CG. The effects of adjuvants on biological activity of herbicides. In: Doyle P, Fujita, T, eds. Pesticide chemistry: Human welfare and the environment,1st edn. Amsterdam, Netherlands, Pergamon Press, 1983, 289–294.
[219] Abubakar SM.Performance test of a dual-purpose disc agrochemical applicator frame electric motor. Bayero Journal of Pure and Applied Sciences 2013, 6, 105–111.
[220] Vaughn SF, Behle RW, Skory CD, Kurtzman CP, Price NPJ.Utilization of sophorolipids as biosurfactants for postemergence herbicides. Crop Protection (Guildford, Surrey) 2014, 59, 29–34.
[221] Sachdev DP, Cameotra SS.Biosurfactants in agriculture. Applied Microbiology and Biotechnology 2013, 97, 1005–1016.
[222] Nielsen CJ, Ferrin DM, Stanghellini ME.Efficacy of biosurfactants in the management of *Phytophthora capsici* on pepper in recirculating hydroponic systems. Canadian Journal of Plant Pathology 2006,283, 450–460.

[223] Hultberg M, Holmkvist A, Alsanius B.Strategies for administration of biosurfactant-producing pseudomonads for biocontrol in closed hydroponic systems. Crop Protection (Guildford, Surrey) 2011, 30, 995–999.
[224] Paulitz TC, Richard RB.Biological control in greenhouse systems. Annual Review of Phytopathology 2001, 39, 103–133.
[225] De Jonghe K, De Dobbelaere I, Sarrazyn R, Höfte M.Control of *Phytophthora cryptogea* in the hydroponic forcing of witloof chicory with the rhamnolipid-based biosurfactant formulation PRO1. Plant Pathology 2005,542, 219–226.
[226] Folman LB, Postma J, Van Veen JA.Characterisation of *Lysobacter enzymogenes* (Christensen and Cook 1978) strain 3.1T8, a powerful antagonist of fungal diseases of cucumber. Microbiological Research 2003, 158, 107–115.
[227] Rahman PK, Gakpe E.Production, characterisation and applications of biosurfactants-Review. Biotechnol 2008, 7, 360–370.
[228] Pardo G, Moral R, Del Prado A.Simswaste-AD-A modelling framework for the environmental assessment of agricultural waste management strategies: Anaerobic digestion. The Science of the Total Environment 2017, 574, 806–817.
[229] Sharma P, Gaur VK, Kim SH, Pandey A.Microbial strategies for bio-transforming food waste into resources. Bioresource Technology 2020, 299, 122580.
[230] Adesra A, Srivastava VK, Varjani S.Valorization of dairy wastes: integrative approaches for value added products. Indian Journal of Microbiology 2021, 61, 270–278.
[231] Moshtagh B, Hawboldt K, Zhang B.Optimization of biosurfactant production by *Bacillus subtilis* N3–1P using the brewery wastes as the carbon source. Environmental Technology 2018, 40, 3371–3380.
[232] Vea EB, Romeo D, Thomsen MJPC.Biowaste valorisation in a future circular bioeconomy. Procedia Cirp 2018, 69, 591–596.
[233] Lope M, Miranda SM, Belo I.Microbial valorization of waste cooking oils for valuable compounds production-a review. Critical Reviews in Environmental Science and Technology 2020, 50, 2583–2616.
[234] Cruz JM, Hughes C, Quilty B, Montagnolli RN, Bidoia ED. Agricultural feedstock supplemented with manganese for biosurfactant production by *Bacillus subtilis*. Waste Biomass Valorization 2018, 9, 613–618.
[235] Rodríguez A, Gea T, Sánchez A, Font X.Agro-wastes and inert materials as supports for the production of biosurfactants by solid-state fermentation. Waste and Biomass Valorization 2021, 12, 1963–1976.
[236] Gudiña EJ, Rodrigues AI, de Freitas V, Azevedo Z, Teixeira JA, Rodrigues LR.Valorization of agro-industrial wastes towards the production of rhamnolipids. Bioresource Technology 2016, 212, 144–150.
[237] Chebbi A, Tazzari M, Rizzi C, Gomez Tovar FH, Villa S, Sbaffoni S, Vaccari M, Franzetti A.*Burkholderia thailandensis* E264 as a promising safe rhamnolipids' producer towards a sustainable valorization of grape marcs and olive mill pomace. Applied Microbiology and Biotechnology 2021, 105, 3825–3842.
[238] Gaur VK, Gupta P, Tripathi V, Thakur RS, Regar RK, Patel DK, Manickam N.Valorization of agro-industrial waste for rhamnolipid production, its role in crude oil solubilization and resensitizing bacterial pathogens. Environmental Technology & Innovation 2021, 25, 102108.
[239] Bharali P, Singh SP, Dutta N, Gogoi S, Bora LC, Debnath P, Konwar BK.Biodiesel derived waste glycerol as an economic substrate for biosurfactant production using indigenous *Pseudomonas aeruginosa*. RSC Advances 2014,473, 38698–38706.

[240] Patil S, Pendse A, Aruna K.Studies on optimization of biosurfactant production by *Pseudomonas aeruginosa* F23 isolated from oil contaminated soil sample. International Journal of Current Biotechnology 2014,24, 20–30.
[241] Akbari S, Abdurahman NH, Yunus RM, Fayaz F, Alara OR.Biosurfactants-a new frontier for social and environmental safety: A mini review. Biotechnology Research & Innovation 2018, 2, 81–90.
[242] Rufino RD, Sarubbo LA, Campos-Takaki GM.Enhancement of stability of biosurfactant produced by *Candida lipolytica* using industrial residue as substrate. World Journal of Microbiology & Biotechnology 2007,235, 729–734.
[243] De Lima FA, Santos OS, Pomella AWV, Ribeiro EJ, de Resende MM.Culture medium evaluation using low-cost substrate for biosurfactants lipopeptides production by *Bacillus amyloliquefaciens* in pilot bioreactor. Journal of Surfactants and Detergents 2020, 23, 91–98.
[244] Crouzet J, Arguelles-Arias A, Dhondt-Cordelier S, Cordelier S, Pršić J, Hoff G, Mazeyrat-Gourbeyre F, Baillieul F, Clément C, Ongena M, Dorey S.Biosurfactants in plant protection against diseases: Rhamnolipids and lipopeptides case study. Frontiers in Bioengineering and Biotechnology 2020, 8, 1014.
[245] Chen W, Qu Y, Xu Z, He F, Chen Z, Huang S.Heavy metal (Cu, Cd, Pb, Cr) washing from river sediment using biosurfactant rhamnolipid. Environmental Science and Pollution Research International 2017,2419, 16344–16350.
[246] Govarthanan M, Mythili R, Selvankumar T, Choi D, Chang Y.Isolation and characterization of a biosurfactant-producing heavy metal resistant *Rahnella sp.* RM isolated from chromium-contaminated soil. Biotechnology and Bioprocess Engineering 2017, 194, 186–194.
[247] Tang J, He J, Liu T, Xin X, Hu H.Removal of heavy metal from sludge by the combined application of a biodegradable biosurfactant and complexing agent in enhanced electrokinetic treatment. Chemosphere 2017, 189, 599–608.
[248] Arab F, Mulligan CN.An eco-friendly method for heavy metal removal from mine tailings. Environmental Science and Pollution Research International 2018, 25, 16202–16216.
[249] Yang Z, Shi W, Yang W, Liang L, Yao W, Chai L, Gao S, Liao Q.Combination of bioleaching by gross bacterial biosurfactants and flocculation: A potential remediation for the heavy metal contaminated soils. Chemosphere 2018, 206, 83–91.
[250] Karnwal A.Use of bio-chemical surfactant producing endophytic bacteria isolated from rice root for heavy metal bioremediation. Pertanika Journal of Tropical Agricultural Science 2018, 41, 2.
[251] Hisham NHMB, Ibrahim MF, Suraini Ramli N, Abd-Aziz S.Production of biosurfactant produced from used cooking oil by *Bacillus sp.* HIP3 for heavy metals removal. Molecules 2019,2414, 2617.
[252] Shami RB, Shojaei V, Khoshdast H.Efficient cadmium removal from aqueous solutions using a sample coal waste activated by rhamnolipid biosurfactant. Journal of Environmental Management 2019, 231, 1182–1192.
[253] Ayangbenro AS, Babalola OO.Genomic analysis of *Bacillus cereus* NWUAB01 and its heavy metal removal from polluted soil. Scientific Reports 2020, 10, 1–12.
[254] Ravindran A, Sajayan A, Priyadharshini GB, Selvin J, Kiran GS.Revealing the efficacy of thermostable biosurfactant in heavy metal bioremediation and surface treatment in vegetables. Frontiers in Microbiology 2020, 11, 1–11.
[255] Lopes CSC, Teixeira DB, Braz BF, Santelli RE, LVA DC, Gomez JGC, Castro RPV, Seldin L, Freire DMG.Application of rhamnolipid surfactant for remediation of toxic metals of long- and short-term contamination sites. International Journal of Environmental Science and Technology 2021, 18, 575–588.

[256] Chen Q, Li Y, Liu M, Zhu B, Mu J, Chen Z.Removal of Pb and Hg from marine intertidal sediment by using rhamnolipid biosurfactant produced by a *Pseudomonas aeruginosa* strain. Environmental Technology & Innovation 2021, 22, 101456.
[257] Elizabeth Rani C, Balaji Ayyadurai V, Kavitha KK. Bioremediation of heavy metals and toxic chemicals from muttukadu lake, Chennai by biosurfactant and biomass treatment strategies. In: Marimuthu PD, Sundaram, R, Jeyaseelan, A, Kaliannan, T, eds. Sustainable approaches to mitigate environmental impacts,1st edn. Switzerland AG, Springer Nature, 2021, 67–85.
[258] Arisah FM, Amir AF, Ramli N, Ariffin H, Maeda T, Hassan MA, Yusoff MZM.Bacterial resistance against heavy metals in *Pseudomonas aeruginosa* rw9 involving hexavalent chromium removal. Sustain 2021, 13, 1–11.
[259] Sun W, Zhu B, Yang F, Dai M, Sehar S, Peng C, Ali I, Naz I.Optimization of biosurfactant production from *Pseudomonas sp.* CQ2 and its application for remediation of heavy metal contaminated soil. Chemosphere 2021, 265, 129090.
[260] Rastogi S, Kumar R.Statistical optimization of biosurfactant production using waste biomaterial and biosorption of Pb^{2+} under concomitant submerged fermentation. Journal of Environmental Management 2021, 295, 113158.
[261] Kim HS, Jeon JW, Kim SB, Oh HM, Kwon TJ, Yoon BD.Surface and physicochemical properties of a glycolipid biosurfactant, mannosylerythritol lipid, from *Candida antarctica.* Biotechnology Letters 2002, 24, 1637–1641.
[262] Yoo DS, Lee BS, Kim EK.Characteristics of microbial biosurfactant as an antifungal agent against plant pathogenic fungus. Journal of Microbiology and Biotechnology 2005, 15, 1164–1169.
[263] Hultberg M, Bengtsson T, Liljeroth E. Late blight on potato is suppressed by the biosurfactant-producing strain *Pseudomonas koreensis* 2.74 and its biosurfactant. BioControl 2010b, 55, 543–550.
[264] Kim PI, Ryu J, Kim YH, Chi YT.Production of biosurfactant lipopeptides Iturin A, fengycin and surfactin A from *Bacillus subtilis* CMB32 for control of *Colletotrichum gloeosporioides.* Journal of Microbiology and Biotechnology 2010, 20, 138–145.
[265] Velho RV, Medina LFC, Segalin J, Brandelli A.Production of lipopeptides among *Bacillus strains* showing growth inhibition of phytopathogenic fungi. Folia Microbiologica 2011, 56, 297.
[266] Gross RA, Shofield MH. Structure of actoric and acidic forms of sophorolipid mixture produced by *Candida bombicoia.* US Patent, US 2014 / 0024816A1, 2014.
[267] Shalini D, Benson A, Gomathi R, Henry AJ, Jerritta S, Joe MM.Isolation, characterization of glycolipid type biosurfactant from endophytic *Acinetobacter sp.* ACMS25 and evaluation of its biocontrol efficiency against *Xanthomonas oryzae.* Biocatalysis and Agricultural Biotechnology 2017, 11, 252–258.
[268] Akladious SA, Gomaa EZ, El-Mahdy OM.Efficiency of bacterial biosurfactant for biocontrol of *Rhizoctonia solani* (AG-4) causing root rot in faba bean (*Vicia faba*) plants. European Journal of Plant Pathology 2019, 153, 15–35.
[269] Sun L, Forauer EC, Brown SRB, Amico DJD. Application of bioactive glycolipids to control *Listeria monocytogenes* biofilms and as postlethality contaminants in milk and cheese. Food Microbiology 2020, 95, 103683.
[270] de O Caretta T, I Silveira VA, Andrade G, Macedo F, PC Celligoi MA. Antimicrobial activity of sophorolipids produced by *Starmerella bombicola* against phytopathogens from cherry tomato. Journal of the Science of Food and Agriculture 2021, In-press. https://doi.org/10.1002/jsfa.11462.
[271] Sakr EAE, Ahmed HAE, Abo Saif FAA.Characterization of low-cost glycolipoprotein biosurfactant produced by *Lactobacillus plantarum* 60 FHE isolated from cheese samples using food wastes

through response surface methodology and its potential as antimicrobial, antiviral, and anticancer activity. International Journal of Biological Macromolecules 2021, 170, 94–106.
[272] Kourmentza K, Gromada X, Michael N, Degraeve C, Vanier G, Ravallec R, Coutte F, Karatzas KA, Jauregi P.Antimicrobial activity of lipopeptide biosurfactants against foodborne pathogen and food spoilage microorganisms and their cytotoxicity. Frontiers in Microbiology 2021, 11, 3398.
[273] Ren K, Patra P, Lamsal PB, Mendonca A.Antimicrobial mechanism of fatty acyl glutamic acid biosurfactant against *Escherichia coli* O157: H7and *Listeria monocytogenes* involves disruption of the cytoplasmic membrane. ACS Food Science & Technology 2021,110, 1792–1804.
[274] Khare E, Kumar N.Biosurfactant based formulation of *Pseudomonas guariconensis* LE3 with multifarious plant growth promoting traits controls charcoal rot disease in *Helianthus annus*. World Journal of Microbiology & Biotechnology 2021, 37, 1–14.
[275] Kang BR, Park JS, Jung W.Microbial pathogenesis antiviral activity by lecithin-induced fengycin lipopeptides as a potent key substrate against Cucumber mosaic virus. Microbial Pathogenesis 2021, 155, 104910.
[276] Serrano L, Moreno AS, Castillo DSD, Bonilla J, Romero CA, Galarza LL, Coronel-León JR. Biosurfactants synthesized by endophytic Bacillus strains as control of *Moniliophthora perniciosa* and *Moniliophthora roreri*. Scientia Agricola 2021, 78.
[277] Mouafo HT, Mbawala A, Somashekar D, Tchougang HM, Harohally NV, Ndjouenkeu R.Biological properties and structural characterization of a novel rhamnolipid like-biosurfactants produced by *Lactobacillus casei* subsp. casei TM1B. Biotechnology and Applied Biochemistry 2021, 68, 1–34.
[278] Bertuso PDC, Drapp M, Nitschke M.New strategy to control endospore-forming *Bacillus cereus*. Foods 2021, 10, 1–13.
[279] Kumar A, Rabha J, Jha DK.Biocatalysis and agricultural biotechnology antagonistic activity of lipopeptide-biosurfactant producing *Bacillus subtilis* AKP, against *Colletotrichum capsici*, the causal organism of anthracnose disease of chilli. Biocatalysis and Agricultural Biotechnology 2021, 36, 102133.
[280] Kumar S, Dheeman S, Dubey RC, Maheshwari DK, Baliyan N.Cyclic siloxane biosurfactant – producing *Bacillus cereus* BS14 biocontrols charcoal rot pathogen *Macrophomina phaseolina* and induces growth promotion in *Vigna mungo* L. Archives of Microbiology 2021,2038, 5043–5054.
[281] Prasad GS, Pinnaka AK.Production of sophorolipid biosurfactant by insect derived novel yeast *Metschnikowia churdharensis* fa, sp. nov., and its antifungal activity against plant and human pathogens. Frontiers in Microbiology 2021, 12, 1262.
[282] Hamed MM, Alhuzani MR, Youssif AM.Biosurfactant production by marine actinomycetes isolates *Streptomyces althioticus* RG3 and *Streptomyces californicus* RG8 as a promising source of antimicrobial and antifouling effects. Microbiology and Biotechnology Letters 2021,493, 356–366.
[283] Chen L, Zhang H, Zhao S, Xiang B, Yao Z.Lipopeptide production by *Bacillus atrophaeus* strain B44 and its biocontrol efficacy against cotton rhizoctoniosis. Biotechnology Letters 2021, 43, 1183–1193.
[284] Poonguzhali P, Rajan S, Parthasarathi R, Srinivasan R, Kannappan A. Optimization of biosurfactant production by *Pseudomonas aeruginosa* using rice water and its competence in controlling Fusarium wilt of *Abelmoschus esculentus*. South African Journal of Botany 2021, In-press. https://doi.org/10.1016/j.sajb.2021.12.016.

[285] Nazareth TC, Zanutto CP, Maass D, de Souza AAU, Ulson SMDAG. A low-cost brewery waste as a carbon source in bio-surfactant production. Bioprocess and Biosystems Engineering 2021, 44(11), 1–8, 2269–2276.
[286] De Lima AM, De Souzaa RR.Use of sugar cane vinasse as substrate for biosurfactant production using *Bacillus subtilis* PC. Chemical Engineering Journal 2014, 37, 673–678.
[287] Gudiña EJ, Fernandes EC, Rodrigues AI, Teixeira JA, Rodrigues LR.Biosurfactant production by *Bacillus subtilis* using corn steep liquor as culture medium. Frontiers in Microbiology 2015, 6, 59.
[288] Zhi Y, Wu Q, Xu Y.Production of surfactin from waste distillers' grains by co-culture fermentation of two *Bacillus amyloliquefaciens* strains. Bioresource Technology 2017, 235, 96–103.
[289] Ahmadi N, Shafaati M, Mirgeloybayat M, Lavasani PS, Khaledi M, Afkhami H.Molecular identification and primary evaluation of bio-surfactant production in edible paddy straw mushroom. Gene Reports 2021, 24, 101258.
[290] Kiran GS, Sabarathnam B, Thajuddin N, Selvin J.Production of glycolipid biosurfactant from sponge-associated marine actinobacterium *Brachybacterium paraconglomeratum* MSA21. Journal of Surfactants and Detergents 2014, 17, 531–542.
[291] ERB M, Silva FL, Sousa MADSB, Dos Santos ES.Use of different agroindustrial waste and produced water for biosurfactant production. Biosciences, Biotechnology Research Asia 2018, 15, 17–26.
[292] Kumar AP, Janardhan A, Viswanath B, Monika K, Jung JY, Narasimha G.Evaluation of orange peel for biosurfactant production by *Bacillus licheniformis* and their ability to degrade naphthalene and crude oil. 3 Biotechnology 2016,61, 43.
[293] George S, Jayachandran K.Analysis of rhamnolipid biosurfactants produced through submerged fermentation using orange fruit peelings as sole carbon source. Applied Biochemistry and Biotechnology 2009, 158, 694–705.
[294] Rane AN, Baikar VV, Ravi Kumar V, Deopurkar RL.Agro-Industrial wastes for production of biosurfactant by *Bacillus subtilis* ANR 88 and its application in synthesis of silver and gold nanoparticles. Frontiers in Microbiology 2017, 8, 492.
[295] Vieira IMM, Santos BLP, Silva LS, Ramos LC, de Souza RR, Ruzene DS, Silva DP.Potential of pineapple peel in the alternative composition of culture media for biosurfactant production. Environmental Science and Pollution Research International 2021, 28, 68957–68971.
[296] Chooklin CS, Maneerat S, Saimmai A.Utilization of banana peel as a novel substrate for biosurfactant production by *Halobacteriaceae archaeon* AS65. Applied Biochemistry and Biotechnology 2014, 173, 624–645.
[297] Sharma D, Ansari MJ, Gupta S, Al Ghamdi A, Pruthi P, Pruthi V.Structural characterization and antimicrobial activity of a biosurfactant obtained from *Bacillus pumilus* DSVP18 grown on potato peels. Jundishapur Journal of Microbiology 2015,89, 21257.
[298] Araújo HW, Andrade RF, Montero-Rodríguez D, Rubio-Ribeaux D, Alves da Silva CA, Campos-Takaki GM.Sustainable biosurfactant produced by *Serratia marcescens* UCP 1549 and its suitability for agricultural and marine bioremediation applications. Microbial Cell Factories 2019,181, 1–13.
[299] Gurjar J, Sengupta B.Production of surfactin from rice mill polishing residue by submerged fermentation using *Bacillus subtilis* MTCC 2423. Bioresource Technology 2015, 189, 243–249.
[300] Maass D, Moya Ramirez I, Garcia Roman M, Jurado Alameda E, Ulson de Souza AA, Borges Valle JA, Altmajer Vaz D.Two-phase olive mill waste (alpeorujo) as carbon source for biosurfactant production. Journal of Chemical Technology and Biotechnology 2016, 91, 1990–1997.

[301] Souza AF, Rodriguez DM, Ribeaux DR, Luna MA, Lima e Silva TA, Andrade RFS, Campos-Takaki GM.Waste soybean oil and corn steep liquor as economic substrates for bioemulsifier and biodiesel production by *Candida lipolytica* UCP 0998. International Journal of Molecular Sciences 2016, 17, 1608.

[302] Radzuan MN, Banat IM, Winterburn J.Production and characterization of rhamnolipid using palm oil agricultural refinery waste. Bioresource Technology 2017, 225, 99–105.

[303] Moya-Ramírez I, Vaz DA, Banat IM, Marchant R, Alameda EJ, García-Román M.Hydrolysis of olive mill waste to enhance rhamnolipids and surfactin production. Bioresource Technology 2016, 205, 1–6.

[304] Jajor P, Piłakowska-Pietras D, Krasowska A, Łukaszewicz M.Surfactin analogues produced by *Bacillus subtilis* strains grown on rapeseed cake. Journal of Molecular Structure 2016, 1126, 141–146.

[305] Li J, Deng M, Wang Y, Chen W.Production and characteristics of biosurfactant produced by *Bacillus pseudomycoides* BS6 utilizing soybean oil waste. International Biodeterioration & Biodegradation 2016, 112, 72–79.

[306] Hentati D, Cheffi M, Hadrich F, Makhloufi N, Rabanal F, Manresa A, Sayadi S, Chamkha M.Investigation of halotolerant marine *Staphylococcus sp.* CO100, as a promising hydrocarbon-degrading and biosurfactant-producing bacterium, under saline conditions. Journal of Environmental Management 2021, 277, 111480.

[307] Yong Z, Gan JJ, Zhang GL, Yao B, Zhu WJ, Meng Q.Reuse of waste frying oil for production of rhamnolipids using *Pseudomonas aeruginosa* zju. u1M. Journal of Zhejiang University Science B 2007,A8, 1514–1520.

[308] Soares SRCF, de Almeida DG, Brasileiro PPF, Rufino RD, de Luna JM, Sarubbo LA.Production, formulation and cost estimation of a commercial biosurfactant. Biodegra 2018,304, 191–201.

[309] George S, Jayachandran K.Production and characterization of rhamnolipid biosurfactant from waste frying coconut oil using a novel *Pseudomonas aeruginosa* D. Journal of Applied Microbiology 2013, 114, 373–383.

[310] Lan G, Fan Q, Liu Y, Chen C, Li G, Liu Y, Yin X.Rhamnolipid production from waste cooking oil using Pseudomonas SWP-4. Biochemical Engineering Journal 2015,J101, 44–54.

Neela Gayathri Ganesan, Subhranshu Samal, Vinoth Kannan,
Senthil Kumar Rathnasamy, Vivek Rangarajan

Chapter 12
Bacillus lipopeptide-based antifungal agents for plant disease control

Abstract: Plant diseases need to be controlled as they hamper the ecosystem by affecting the quality and availability of food, feed, crops, and fiber produced by plants and farmers worldwide. Different methods and approaches are unraveled to prevent and control plant diseases. In agricultural practice, fungal diseases are the most common crop-devastating disease. The antifungal resistance developed in the crops, their quality degradation, and environmental toxicity caused by present conventional/chemical fungicides prompt the need for new antifungal agents that are considered safe and less toxic. Therefore, researches are focused on finding new alternatives to address the same by finding several derivatives of synthetic pesticides, mixing two or more pesticides, formulating biological originated pesticides, and so on to control/prevent pests and pathogens causing plant diseases. These alternative solutions are referred to as biological controls. A novel biocontrol agent, lipopeptides, produced by microbes possesses peculiar and outstanding properties that show the ability to work against harmful fungi and yeast, which helps in various disease controls in plants. Among various strains, *Bacillus subtilis* and other *Bacillus* species gained attention due to their exceptional properties. *Bacillus* strains have been identified to produce metabolites that serve as antifungal, antimicrobial, and antiviral agents; in addition, these bacterial species improve plants and their products by promoting the plant growth mechanism. Encapsulating all the features in one go, *Bacillus* species have gained importance to be used in the agricultural sector as a biocontrol agent.

12.1 Introduction

Plant disease-causing microorganisms are considered a severe threat to agriculture, and these disease-causing pathogens disturb the flow and consistency inside the ecosystem. Irish potato famine, Dutch elm blight, chestnut blight, and Bengal famine are some notable outbreaks caused due to plant pathogens [1]. Chemical measures used for pest control are generally high in doses to manage their activity depending on the phytopathogens. These chemical antifungal control measures adversely affect soil fertility by altering the microbial environment present in the plant's root, which is responsible for plant growth. Therefore, plant health is adversely affected and deteriorates

https://doi.org/10.1515/9783110771558-012

both the environment and humans. In addition to this, there are a few undesirable effects of using high doses of pest control, such as pathogens developing resistance and impact on nontargeted actions in plants and soil. As the increase in pest control methods generates stress in the disease-causing pathogens, they obtain resistance against the existing methods and chemicals. This situation forges and leads to a global issue for the farmers and consumers. To overcome the undesirable situation, the quest for an alternative method that can suppress the inheritance of occurring resistance against pest control measures and has powerful, sustainable, and less toxic outcomes are in the lookout. Recently, quite a few fungal chemical controls have been banned due to their harmful nature [2]. An alternate source, biological originated antimicrobial agents are considered to control the pathogens for the long term in an environment-friendly way using biodegradable lipopeptides that act as biocontrol agents. The use of microbial treatment can severely act as an alternative to chemical pesticides and prevent crop damage by producing antimicrobial agents, disrupting cell–cell transitions, forming biofilms, and competitive inhibition. Also, they are biodegradable. Mainly *Bacillus* sp. are proven for their properties like the colonization of bacterial community in the root, formation of biofilm, inducing the systematic resistance against pathogens, and producing secondary metabolites that have antimicrobial properties. These secondary metabolites are identified as surfactin, iturin, fengycin, and so on, and their derivatives. Bacillus lipopeptides can act as an antimicrobial agent and perform activities to control the plant disease-causing pathogens that include fungi, bacteria, and yeasts. Iturin is identified as a compound that can work against fungus whereas, surfactin is identified as a compound that can work against bacteria and viruses [3].

The plant disease-causing phytopathogen results in a 15–25% lower yield, increasing the pesticide consumption by nearly 200%. This has caused enormous damage to the soil ecosystem. Eco-friendly microorganisms play an important role in treating the pathogen by inducing the plant's systemic response, mycoparasitism, lytic enzyme production, siderophores, δ-endotoxins, and lipopeptides. The world population is expected to cross nine billion, which can cause tremendous pressure on the agricultural field. And excessive use of chemical fertilizers can add up to environmental pollution [4]. The usage of lipopeptides and their derived products in the global market was increasing day by day and accounted for US $2.2 billion in 2020, and it is expected to reach US $3.1 billion by 2028. Microbial-originated lipopeptides have emerged as a tool to overcome problems humans face without any adverse effects on humankind. Lipopeptides have broad applications in water treatment, detergency, emulsification, lubrification, mineral flotation, petroleum recovery, medical applications, food, bioremediation, cosmetic applications, phytosanitation, and agriculture (as biocontrol). Due to the multinature of lipopeptides, they are considered as versatile weapons in various fields [5].

12.2 Antibiotic producing nature of *Bacillus* species

Different types of antifungal and antimicrobial compounds are produced by both gram-negative and gram-positive bacterial strains that have potential in various fields to target harmful bacteria, fungi, and other microbes. Bacillus has been widely researched in biochemistry, agriculture, and genetics due to their exceptional biofunctionality. More than two dozen antibiotics were reported from these species. About 4–5% of strains were from the *Bacillus subtilis* genome among those antibiotics [6]. Bacillus, a genus of bacteria with great potential in producing Lipopeptides, proteins, volatiles, and peptides [7]. They have high activity against various bacteria, viruses, fungus, mites, oomycetes, insects, nematodes, and phytopathogens [8]. Activity against the plant pathogens is found in the lipopeptides, namely surfactin, iturin, and fengycin. Lipopeptides are proposed as a solution to phytopathogens that has drug resistance to conventional fungal diseases. *Bacillus* species are proven to have excellent cyclic lipopeptides (CLPs), colonization aptitude, fights against plant pathogens. Bacillus has an antibiotic production function which plays a vital role in antifungal and biocontrol activity. *Bacillus* species produce a hydrolytic enzyme that is antimicrobial. These enzymes include chitinases, cellulases, lipases, glucanases, and proteases. These enzymes are proven to hydrolyze the components that are present in the cell walls of phytopathogens in the class of fungi and bacteria [9]. The bacterial lipopeptides are getting more attention due to their less toxic profile, biodegradability, and antibiotic-producing nature [10]. In the *Bacillus* genus, *B. subtilis* and *Bacillus amyloliquefaciens* are the most commonly used strains for lipopeptide production. However, other species like *Bacillus globigii*, *Bacillus mojavensis*, *Bacillus pumilus*, *Bacillus cereus*, *Bacillus thuringiensis*, *Bacillus megaterium*, and *Bacillus methylotrophicus* are also reported for lipopeptide production [11–13].

In a report, a group of different *Bacillus* species were isolated and examined. The result ended in identifying different antimicrobial peptide (AMP) genes, in which about 98.4% produces isoforms of surfactin, 90.6% strain produces isoforms of iturins, and 79.7% strain produces isoforms of fengycins. Among all the *Bacillus* strains, *B. subtilis* and *B. amyloliquefaciens* strains have high antibacterial activity, multiple cyclopeptides (CLP) genes, and the ability to produce various isoforms simultaneously. The presence of the CLP gene determines the antagonistic capacity of a strain; this helps the *Bacillus* strain to fight against plant disease-causing pathogens. These CLP genes are generally present in the *Bacillus* strains like *B. subtilis* and *B. amyloliquefaciens*. Few strains in the group had a very less number to no CLP genes so those strains showed low or no antibacterial activity. The strains including *B. licheniformis*, *B. cereus*, *B. megaterium*, *B. pumilus*, and *B. thuringiensis* show less antagonistic nature [14, 15].

12.3 Lipopeptides: versatile antimicrobial compounds

Lipopeptides are secondary metabolites produced by bacteria, fungi, and viruses. They have a wide range of applications due to their biological activity, including antimicrobial, antifungal, and antibiotic nature and their surface-active nature. A lipopeptide is a molecule that consists of lipids connected to the peptide group. Lipopeptides has a hydrophilic peptide head group attached to the lipid hydrophobic hydrocarbon tail groups; due to this combination, lipopeptides have amphiphilic nature. They can self-assemble into various structures subjected to different conditions [16]. Lipopeptides can affect the cell membrane's integrity and permeability by involving its N-terminal fatty acid tail; it is bactericidal due to inhibition of outer membrane synthesis. Many bacteria produce these lipopeptides as a part of their mechanism. Among the various bacteria, *Bacillus, Pseudomonas, Streptomyces*, and *Serratia* are well known/studied for lipopeptide production. Lipopeptides received antibiotic approval in the United States in 2003. The first antibiotic, lipopeptide, CubicinR (Daptomycin), was approved by the Food and Drug Administration (FDA) in the United States. These antibiotics are used in the medical field to treat infections in blood and skin that are commonly caused by a specific gram-positive microorganism. In the case of gram-negative bacteria, the compound polymyxin binds to lipopolysaccharides due to the electrostatic interaction between the compound and the microbial cell [17–19].

12.3.1 Lipopeptide structure and its role in biocontrol

Bacillus lipopeptides are non-ribosomally synthesized by large and complex multi-enzymes called non-ribosomal peptide synthetases. Various species produce secondary metabolite as an antibiotic when they undergo stress conditions; this includes bacteria and fungi. Lipopeptides attains the properties based on their source. However, these lipopeptides in common have been reported to be biodegradable and less toxic [20]. The molecular structure, that is, the amphiphilic nature of the lipopeptides produced by the bacillus species, adds an advantage in intruding the biological membrane structure. This leads lipopeptides to play a vital role in the activity against pathogens [12]. Lipopeptides consists of cyclic or linear peptide groups with positive or negative charged groups attached covalently to the fatty acid chain to its *N*-terminal. Lipopeptides interacts with the membranes using their lipid tails. Length of the carbon chain (tail) directly relates to the increase in bactericidal activity. Positively charged peptides, owing to their amphiphilic nature, can interact with strong negatively charged bacterial membrane. The affinity of the tail toward the hydrocarbon chain or the interaction of the positively charged peptide with the lipid head group can be related to this nature of lipopeptides. In the case of negatively charged lipopeptides, the presence of the Asp–Gly segment present in their amino acid group tends the lipopeptides to be prone to chemical reactions and degradation under extreme environmental conditions. This nature of lipopeptides prevents the resistance development of the pathogens serving as an antibiotic [18]. These lipopeptides are antibiotics produced by the Bacillus species that are highly active against various microorganisms and multiresistant bacteria. The CLPs produced by the bacillus species have been proven an antifungal agent since the lipopeptides can interact with the fungal cell membrane, resulting in the osmotic difference in the cellular compounds within the cell membrane. However, the mechanism of interaction of lipopeptide and the microbes varies depending on the structure of the lipopeptide and the membrane of the target microbial cell membrane. The combined interaction of different lipopeptides and their derivatives by a different source of strains leads to effective control of plant pathogen as biocontrol [21]. Lipopeptides can act in three ways against pathogens: (1) by directly acting against plant disease-causing pathogens, (2) by inducing the plant's capacity to fight on its own (inducing the systematic response system in the plant), and (3) by forming biofilms in order to control colonization of microorganisms; therefore, they act as the key factor in biocontrol [22]. Lipopeptides possess various properties. Among them, the surface tension reduction property is well established. Due to its surface-active nature, the lipopeptides can bring down the viscosity and surface tension of the exudates produced by the plants, which enhances the cells' swarming process. This helps in uptaking the nutrients for the growth of cells and thereby improving their colonization faster [23].

A new sequence of lipopeptide is extracted from *B. subtilis* (RLID 12.1) and it has the chain Asn-Pro-Tyr-Asn-Gln-Thr-Ser with a slight modification in the acid group (either in the chain length or in acid group branching). Three different and unique homologues are found. Out of three AF (AF 3–5), AF 4 is found to be more effective than the others with 100% inhibition. An interactive study was performed among the three purified lipopeptide homologues/isomers, which shows that they have reduced the cytotoxicity against mammalian cells and proved to be an antibiofilm forming agent [23].

The cecropin A-melittin hybrid peptide BP100 (H-Lys-Lys-Phe-Lys-Lys-Ile-Lys-Lys-Tyr-Leu-NH2) has been reported to act against plant disease-causing pathogens. They possess antimicrobial activity, toxicity, and proteolytic stability. The nature of their activity against microbes and hemolysis were influenced by the length (the number of carbon) and the position where the fatty acid group is attached. Lipopeptides with a butanoyl group or hexanoyl group attached to the chain, in particular, had the best biological activity profile [23, 24].

12.3.2 Lipopeptide production and purification

Living organisms produce lipopeptides by following both ribosomal and non-ribosomal synthesis. In a ribosomal way, small bacteriocins/defensins are produced, whereas in a non-ribosomal way, cyclopeptides, pseudopeptides, and peptaibols are produced. Peptides derived from bacteria are short and contain less than 50 amino acids extracted from living organisms. They are classified structurally as linear peptides, forming helical or open-ended peptides. The cyclopeptides structurally form a peptide ring. Bacteriocins are proteins that are produced by bacteria to kill the other species. Peptides support bacteriocins for the process. Peptaibols are peptides in the linear chain that have amino acids. Cyclopeptides are made up of amino acids and their derivatives. Cyclopeptides are produced by bacteria and fungi as a secondary metabolite. They are antimicrobial in nature. Pseudopeptides are made up of complex molecules of amino acid produced by bacteria [25].

Apart from microbial peptides, solid-phase methods are used to produce AMPs. Other synthesizing methods have been developed to improve the peptides' performance and quality. This development of new molecules is based on the motive to improve the prime compounds, with the goal of increasing or tuning their activity on specific target disease-causing pathogens, minimizing toxicity that it brings to plants and animals, to increase the susceptibility to digest protease. The lipopeptides produced by the *Bacillus* genus are responsible for showing major antipathogenic activity. These low molecular compounds have an amphiphilic nature that protects plants until they are harvested. Though these lipopeptides exhibit a higher tendency for plant protection, their identification, extraction, and quantification require an economic approach. Bacillus has several advantages and usefulness over other microbial strains such as spore formation, various lipopeptides production (total 98 varieties), enzyme production, and so on. Out of which, surfactin, iturin, and fengycin are most effective and widely used. Figure 12.1 presents the general scheme for the upstream and downstream processes involved in the biological production of lipopeptides from bacteria. Although the cost associated with the production of lipopeptide is projected to be comparatively higher than that of their chemical counterparts, it is strongly believed that with the advancement in the processing methods, in particular downstream processing steps along with the use of sustainable resources as substrates, lipopeptides-based antifungal agents would become economically viable green-agents in the near future [26].

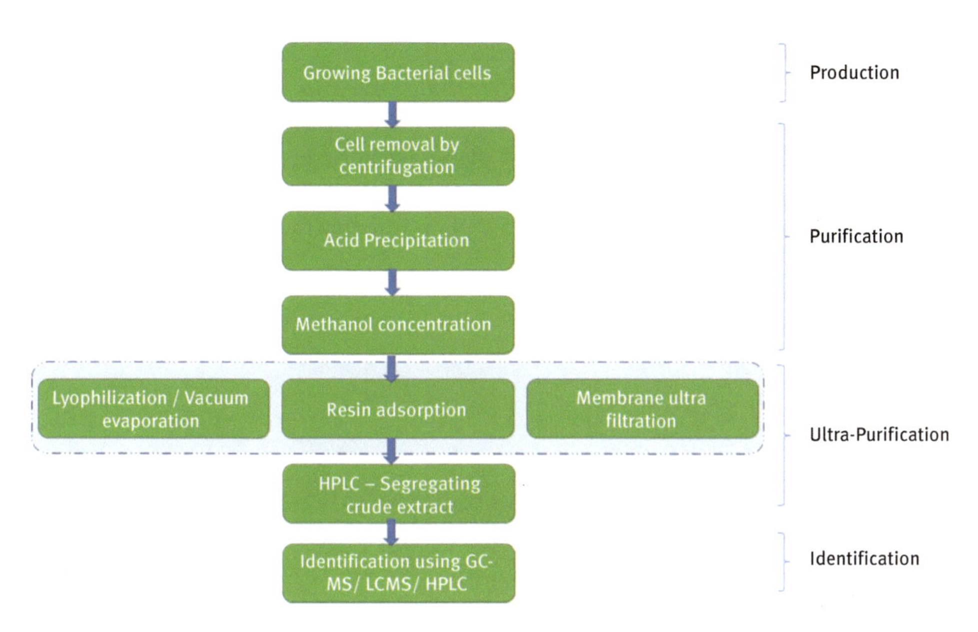

Figure 12.1: General bioprocessing scheme for the production and purification of lipopeptides.

12.4 Microbial lipopeptides and their types

Lipopeptides are classified into different groups based on their amino acid group present in their chemical structure and the lipid moiety length. Among the various lipopeptides produced by bacterial species, iturin, fengycin, surfactin, and kurstakins are notable. The *Bacillus* lipopeptides, namely surfactin, iturin, and fengycin (Figure 12.2a–c), are widely applied for treating pathogenic microbes, including fungi, bacteria, and oomycetes. Surfactin is responsible for exhibiting antimicrobial activity, while iturin and fengycin exhibit antifungal activities.

12.4.1 Iturin

Iturin is a low-molecular-weight CLP with antifungal properties produced by *Bacillus* spp. Iturin is reported to interact with sterol compounds in the fungus cell membrane and has antimicrobial potential against plant pathogens. The molecular structure of iturin has two parts, lipid and a small peptide ring chain of seven amino acids (D- and L- amino acids) connected to a chain of fatty acid (with C14–C17 molecules), due to which they show high polymorphism with distinguishable physicochemical properties. The molecular mass of iturin is ~1.04kDa with about ~38–40 kb size. Iturin has an amphiphilic characteristic that supports the compound in various conditions.

They are neutral or monoanionic lipopeptide in nature, depending on the production source. Iturin A-D, bacillomycin F/D/L, bacillopeptin, and mycosubtilin are some derivatives of iturin. Iturin is known for its powerful antibiotic activity against a broad fungal spectrum [27]. Iturin's structure and the placement of amino acids in the fatty acid chain connected to a peptide group plays a vital role in creating ion-conducting pores inside the cell membranes of fungus. Iturin inhibits the growth of plants pathogens by spore formation. *B. subtilis, B. amyloliquefaciens, B. velezensis, B. siamensis,* and *B. nakamuari* are some sources that produce iturin [28, 29].

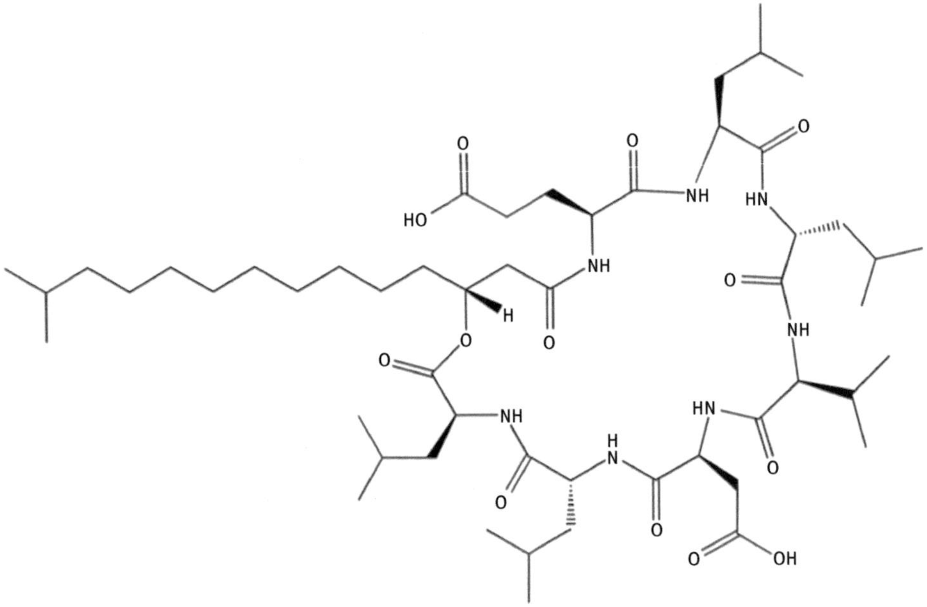

Figure 12.2a: Structure of iturin.

12.4.2 Surfactin

Surfactin is the predominantly known lipopeptide with wide applications due to its antifungal, antibacterial, antibiotic, and outstanding surface-active nature. Surfactin plays a vital role in biofilm formation. Surfactin is proved to fight against bacterial phytopathogen *Pseudomonas syringae,* which causes infection of *Arabidopsis* roots. Surfactin has a molecular mass of ~1.36 kDa. They are amphipathic CLP that has even amino acids (D and L) in the cyclic ring attached to a fatty acid group made of the carbon chain (of C12–C16) that forms a closed ring that contains eight/nine methylene groups. Surfactin is made up of β-hydoxyhepta cyclic peptides, which contains valine, alanine, leucine/isoleucine amino acids in combinations at placed in positions 2/4/7 in the cyclic peptide moiety with a slight variation of carbon chain length of the fatty acid group [30]. Since the chemical structure of surfactin has both hydrophilic and hydrophobic parts, surfactin fits into the heptapeptide type; as a result, surfactin possesses an amphiphilic character. Surfactin, esperin, halobacillin, pumilacidin, bacilysin, and chlorotetaine are some surfactin derivatives. Surfactin is proved to have antiviral, antimycoplasma, and antibacterial activities. *Bacillus velezensis*, *B. amyloliquefaciens*, *B. mojavensis*, and *B. subtilis* are some of the bacillus sources from which surfactin is produced. The primary mode of action of surfactin is attacking the lipid bilayer of cell membranes. Surfactin can act as an antimicrobial agent in different ways like (1) introducing itself into the bilayers, (2) chelating the cations and then solubilizing the membrane walls, and (3) forming pore and thereby lysing the pathogens. Therefore, surfactin can act against both gram-positive/harmful bacterial strains. Surfactin's potential to form a biofilm that disturbs and collapse the biofilm formed by the pathogen can contribute to biocontrol measures even in the absence of pathogen lysis. In plants, surfactin can suppress phytopathogens, namely *P. syringae*, *Sclerotinia sclerotium*, *Colletotrichum gloeosporioides*, *Xanthomonas axonopodis*, and *Botrytis cinerea*, and also against activated induced systemic resistance. Surfactin helps in reducing surface tension, increasing the surface area of hydrophobic compounds, and other generic uses including biofilm and spore formation, quorum sensing, swarming motility, and antimicrobial activity [31].

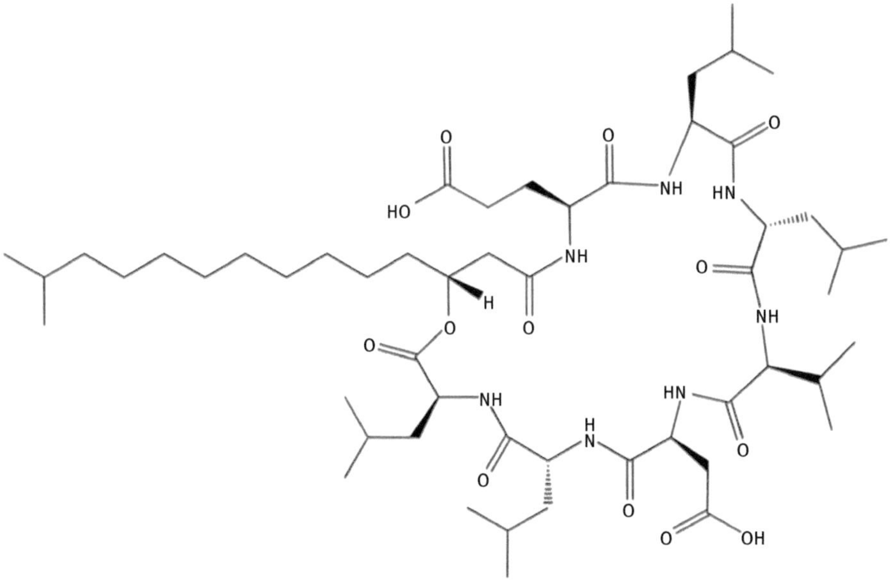

Figure 12.2b: Structure of surfactin.

12.4.3 Fengycin

Fengycin belongs to the lipopeptides that are generally produced by the bacillus and Paenibacillus species. The molecular mass of fengycin is 1,463 Da. Structurally, fengycins are lipodecapeptides with a 14–19 hydroxy saturated/unsaturated fatty acid chain with ten amino acids. Fengycin has a spherical micelle core–shell-like structure. Fengycin plays a vital role in triggering systemic response against plant pathogens in the plants, attracting attention in the bio-agriculture sectors. Fengycin is proved to have activity against filamentous fungi establishing its antifungal activity. They act on membranes of the cells and completely disrupt and collapse them. They also inhibit fungal enzymes like aromatase and phospholipase A2 [32]. Fengycin is reported to possess antimicrobial activity against a range of yeasts. fengycin inhibits spore germination and mycelial growth of plant pathogens. Fengycin A-E and Plipastatin A, B are derivatives of fengycin. Stains that produces fengycin are *B. amyloliquefaciens*, *B. mojavensis*, *B. subtilis*, *Bacillus altitudinis*, and *Bacillus velezensis* [33].

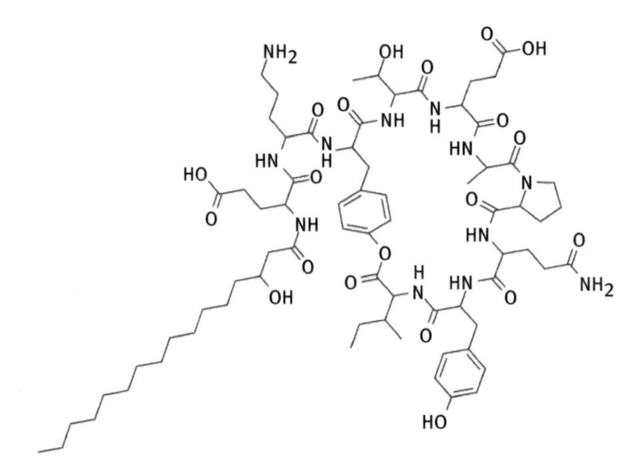

Figure 12.2c: Structure of fengycin.

12.4.4 Kurstakin

Kurstakin is a new lipopeptide with a molecular mass of about 907.4765 to 953.5192 Da. The chemical formula of kurstakin varies from $C_{40}H_{65}O_{13}N_{11}$ to $C_{42}H_{71}O_{14}N_{11}$. It has seven amino groups with a beta-hydroxylated fatty acid chain with 12–14 carbon atoms. Kurstakin is reported to have antifungal activity produced by *B. thuringiensis* [34, 35].

12.4.5 Other types of lipopeptides

The other lipopeptide types are identified as gramicidins, lichenysin, subtilisin, zwittermicins, mycosubtilin, mersacidin, bacilysin, ericin, bacillomycins, and so on are other types of lipopeptides produced by various gram positive and gram negative microbial species. Polymyxins B and E have been used in clinical applications, whereas polymyxins A, C, and D are not commercially used because of their toxicity [36–38]. Daptomycin, viscosins, pseudofactins, and poaeamides are a few lipopeptides produced by *Pseudomonas* species [39].

12.5 Plant diseases

Plant diseases and plant pathogens need to be controlled to maintain the availability of food and maintain the quality of crops. Agriculture is considered the backbone of the country due to the dependence of the entire nation on food, crop, fibers, and other resources. Different approaches, methods, and sources were used to control and prevent the plant diseases and pathogens causing the infections. In the process, excessive use and misuse of chemical sources lead to severe crop damage, and humans consume the crops and their products and food quality. This, in turn, leads to the lookout for an alternate source that has considerably less effect on agriculture. Microbial antagonists, species of actinomycetes, and a few bacterial strains are known for their antifungal properties and thereby inhibits several pathogenic fungi. Hence, these can be used for biocontrol of fungal plant diseases. Bacterial species like *Pseudomonas*, *Bacillus*, *Arthrobacter*, and *Serratia* have been reported in controlling fungal diseases in plants. *B. subtilis* isolates are proven to be effective in plant disease control agents caused by pollutants that are soil borne in nature and the pathogens responsible for postharvest diseases. Iturin produced by *B. subtilis* inhibits *Pythium ultimum*, the plant disease-causing pathogen that causes damping-off disease in beans and *Podosphaera fusca;* this plant pathogen causes powdery mildew disease in cucurbits plants [40, 41]. Surfactin produced by *B. subtilis* and *B. amyloliquefaciens* strains are reported to act against root infection in Arabidopsis caused by *P. syringae*, Foliar or root disease of soybeans caused by the pathogen *X. axonopodis* and white mold disease caused by the plant pathogen *Sclerotinia sclerotiorum* [42–44]. Surfactin produced from *B. subtilis* (SCB-1) was proven to be an antifungal agent that can fight against various fungal infections caused in plants that are caused by the pathogens genera, namely *Saccharicola, Cochliobolus, Alternaria*, and *Fusarium* [1].

In agriculture, the most common crop-devastating diseases that are developed in plants are generally by fungus and its associates. Various strains are refined to create plant disease-causing pathogens. Among those, the important pathogens are different species of *Fusarium* and *Botrytis*. Isolates of *B. megaterium, B. coagulans,*

B. thuringiensis, and *B. amyloliquefaciens* are demonstrated and proved to have excellent biosurfactant activity and can inhibit and suppress the growth of fungus species of *Fusarium* and *Botrytis* [45]. The fungal species *Fusarium* causes a variety of plant diseases that includes rots in plant/fruit/root, wilt and head blights in plants/crops. This plant pathogen can reduce the quality of grains. Mycotoxins and their derivatives produced by this species lead to suppressing immunity and can cause neurological disorders in animals and humans who feed on the plant/plant products. Therefore, *Fusarium* species are considered the most critical economic pathogens. *Fusarium graminearum* causes diseases in various plants like barley, wheat, maize, oats, cotton, rice, kenaf, sweet potato, and others. *F. graminearum* is considered one of the top fungi that destroys crops and grains. *Bacillus* species are identified and reported against *Fusarium* species [44, 46].

Rhizoctonia solani, a major fungal phytopathogen belonging to the Rhizoctonia genus, is responsible for a significant crop loss. *Rhizoctonia solani* is reported to suppress the growth of the plant or the crop and reduce the quality of the crops/products such as cereals, vegetables like potato and lettuce, and fiber crops like cotton. The plant pathogen *P. capsici* causes damages like root, crown, or fruit rot in vegetative plants like different varieties of beans, cucumber, tomato, pepper, and so on. This fungus is a significant cause of soil-borne diseases in crops such as root rot, wilts, and seedling. Furthermore, *Alternaria alternata*, a saprophytic pathogen, causes postharvest diseases and root diseases in plants. Similarly, wide ranges of fungal species are identified to cause plant diseases that lead to a huge loss in agriculture. Lipopeptides like surfactins, echinocandin, fengycin, halobacillin, iturins, mycosubtilin, and plipastatin are produced by microbial sources are under limelight due to their potential activities that include the properties like antimicrobial, anti-inflammatory, antitumor, antiviral, and immune-suppressive activities. Among these lipopeptides, echinocandins, and pneumocandins have the potential to invade into the cell wall and inhibit the glucan produced inside the cell wall of plant disease-causing pathogenic fungal-like *Aspergillus* and *Candida*. New noncytotoxic lipopeptides produced from a marine-originated *B. subtilis* bacterium, "gageopeptides" and its derivatives (A–D:1–4), have been recently discovered by the research community; these new lipopeptides are reported to have a significant role in fighting against pathogenic fungi *R. solani*, *B. cinerea*, and *C. acutatum* [47].

To investigate the mechanisms of biocontrol and the performance of *Bacillus* species, the secondary metabolite (lipopeptides), proteins, and volatile compounds produced from them are extracted and tested. These compounds are identified to contribute to the antagonistic activity against the fungal pathogen *B. cinerea*. Under greenhouse conditions, these compounds that are extracted are proved to prevent the activity of *S. fuliginea* that causes powdery mildew disease in cucumber plants. The isolated lipopeptide is found to be fengycin A. The volatile compound produced is found to be *O*-anisaldehyde (a highly inhibiting volatile compound), among other alcohols, phenols, amines, and alkane amides. MALDI-TOF-MS is used to analyze the product obtained. Cotyledon spraying and flat fumigation methods are used to determine the inhibition of *S. fuliginea* to prevent cucumber cultivation [48]. The disease-control activity of *B. licheniformis* is reported to produce lipopeptides (iturin and surfactin). These are used to treat different plant diseases such as tomato late blight, tomato grey mold, pepper anthracnose, and so on. [49]. Grey mold disease of apple is treated by *B. subtilis* (strain GA1). This strain proved efficient in treating the pathogen for the first five days, followed by 80% efficiency in the next 10 days. This strain resulted in producing a mixture of lipopeptides for treating pathogens, even in unfavorable conditions [45]. Lipopeptides are generally used to control foliar diseases of plants, soil-borne disease, grey mold, and rose grey mold disease. Iturin A produced by *Bacillus marinus* (B-9987) shows activity against *B. cinerea. B. velezensis* (strain 1B-23) produces surfactin, which shows activity against *Clavibacter michiganensis* subsp. strains. Fengycin produced by *B. subtilis* (RLID 12.1), *Bacillus atrophaeus* (CAB-1), and *B. subtilis* (GA1) shows activity against *Sphaerotheca fuliginea, B. cinerea, Candida*, and *Cryptococcus* spp.

Chemical control methods were the main measure to reduce these diseases; due to long usage, the pathogens and their mutants attained resistance that evolved into a new physiological race of pathogens. Due to this reason, many of the chemical pest-control measures became noneffective. This increased the demand for more effective safety measures as well. The control measure should be capable of breaking the resistant developing nature in pathogens plants and secure the crop yield both quantitatively and qualitatively. Natural products produced by microorganisms have a key role in working against plant disease-causing pathogens; researchers are searching for such compounds. The usage of microbial origin pest control can significantly impact plants/crops for a more extended period. Still, in turn, these measures can induce stress on the plant due to the complex interaction between plant and microbial actions. This can be reduced by using or applying a single mode of microbial biocontrol measure. Table 12.1 presents various microbial strains that are used in the treatment of plant diseases. All these eco-friendly biocontrol agents have been said to be more accessible and have less impact on stressing the crop [50].

12.5.1 Post-harvest disease

Postharvest diseases are the diseases that appear or develop after the harvest of crop/plant products. It causes heavy loss during the storage and in the transit of the supply chain. The condition and environment where the products are stored provide suitable factors for pathogenic fungi or bacteria to grow. This causes spoilage of the harvested product. The product of the plants or crops with defects is unmarketable, which leads to the loss of product cost, increase in demand, increase in production cost, and loss of entire product. The defective product/fruit/vegetable that is affected may spread disease from product to product by causing decay and leading to fluid secretion. These fluids can cause damage to other products in the container. The volatiles produced during this

Table 12.1: Plant disease treated by *Bacillus* strains.

Strain	Disease	Plant	Pathogen causing	Activity	Reference
B. subtilis (NH-100) (NH-217)	Bakanae disease (seed-borne fungal infection)	Basmati rice	*Fusarium moniliforme*	Antagonistic activity	[51]
B. amyloliquefaciens	Maize cob disease	maize	*Rhizomucor variabilis*	Inhibitory activity	[52]
B. safensis (B21), *B. polymyxa*, *P. fluorescens*, *Streptomyces* sp. (PM5)	Rice blast	Rice	*Magnaporthe oryzae*	Antifungal, antibacterial, anti-adherent, and cytotoxic activity	[53]
B. vallismortis (ZZ185)	–	–	*P. capsica*, F. *graminearum*, *R. solani*, *A. alternata*, *C. parasitica*	Strong antifungal activity	[54]
Bacillus siamensis S3 and *Bacillus tequilensis* S5	Bayberry twig blight disease	Bayberry	*Pestalotiopsis versicolor* XJ27	Fungal growth inhibition	[55]

infection may lead to undesirable odor, mycotoxin development, and so on. To control these postharvest diseases, synthetic chemical pesticides are applied/sprayed over the harvested products. Other measures to prevent post-harvest diseases are refrigeration, altering the environment, using fungicides and ethanol, irradiation, and so on. Combining one or two methods is also proven to be more efficient. The culture of preharvest measures to prevent postharvest diseases is also followed to control postharvest diseases [56]. One unique type of post-harvest disease is known as latent pathogens, these types infect the plant/crop when they are in cultivation period, and they remain latent/quiescent/subtle during the period, and they start the growth once the product is harvested [57, 58]. Some of the pathogens that cause postharvest diseases are *Penicillium digitatum* that causes green/blue molds, *Geotrichum citri-aurantii* causes sour rot, *Rhizopus stolonifera* causes Rhizopus rot, *Aspergillus niger* causes Aspergillus black rot, *Cladosporium herbarum* causes Cladosporium rot. Some of the pathogens that cause latent diseases are: *Lasiodiplodia theobromae* cause rots in the stem end; *C. gloeosporioides* cause anthracnose that leads to a dark lesion in leaves/stem/twigs in the plant; *Phytophthora* species cause brown rot in the fruits; and *Alternaria* species cause black rot, which is considered as the most important and serious disease in crucifer crops [58].

To date, there are several strategies to overcome post-harvest diseases, as discussed. However, apart from this commercial strategy and approaches, biological control is an alternative and safer option to control postharvest diseases. There are several antagonistic microorganisms that can be used against pathogens causing post-harvest diseases. *Bacillus* species is one of those antagonistic microbes. *Bacillus* lipopeptides exhibit antifungal efficacy which offers an eminent alternate source for controlling postharvest diseases in plants and crops. The inhibitory ability of *Bacillus* spp. is proved to work against *B. cinerea*, a major phytopathogenic fungus that causes grey mold in post-harvest fruits [59]. *B. velezensis* 83 strain is considered foliar biofungicide (Fungifree AB™) in Mexico as the stain can inhibit plant pathogens like *Colletotrichum, Erysiphe, Botrytis, Sphaerotheca,* and *Leveillula.* It is also proved that the strain increases the productivity and quality of tomatoes [60]. The efficacy of lipopeptides produced by *B. amyloliquefacien* against fungi like *A. citri, A. brassicicola, B. cinerea, C. gloeosporioides, F. aromaticum, L. theobromae, M. fructigena P. crustosum, P. expansum, P. persea, R. stolonifer,* and species of *Botryosphaeria* that are responsible for the post-harvest disease has been reported. This proves that lipopeptides can be used as a biocontrol measure to prevent post-harvest disease-causing pathogens [55, 58, 61–63].

12.6 Lipopeptides as biocontrol

The plant-associated microbiome is known to produce a diverse group of compounds with both hydrophilic and hydrophobic moieties that can take up a wide number of structural modifications (depending on the groups attached) that are recognized to have biosurfactant activity. Lipopeptides, phospholipids, neutral lipids, glycolipids, polysaccharide–protein complexes, and fatty acids are examples of such biosurfactants. The ability of these groups of biosurfactants, especially lipopeptides, to show activity against the plant pathogens by biofilm inhibition, pore-forming ability in the pathogen, siderophore activity, colonization prevention, and so on, have resulted in their wide explorations and applications. Recent research has revealed that these lipopeptides can improve the plant's quality by improvising the root colonization (that helps in plant's growth) and also play an important role in stimulating host defense mechanisms by inducing and improving the interaction of *Bacillus* species with plant's hormones. They act as inhibitors and boost host immunity to resist the pathogen [22, 39].

Apart from bacterial, the plant-derived essential oil can also inhibit pathogenic activities. The application of both naturally derived plant essential oil and their volatile constituents and the produced lipopeptides are used for the treatment of various pathogenic activities. The synergistic effect due to the combined effect of bacterial derivatives can result in pathogenic inhibition. They are generally found as preformed compounds, inducible preformed compounds, phytoalexins, and other inhibitory compounds [64].

12.6.1 Lipopeptide plant resistance inducer

Lipopeptides produced in *Bacillus* species (specifically, surfactin and fengycin) can interact with both the plant cells and the bacterial cells. These lipopeptides stimulate the defensive response in the plant through the plant's systematic immune mechanism once the plant detects foreign pathogens invading into the plant tissue. Lipopeptides (surfactin and fengycin) play a prominent role in stimulating the plant's defense mechanism against various diseases caused by multiple pathogenic strains. Table 12.2 shows a glimpse of the defense mechanism played by lipopeptides.

Table 12.2: Lipopeptides inducing resistance in plants.

Culture	Activity	Against	Tested in Plant	Reference
Bacillus spp.	Stimulates the host identification and defense gene expression	*F. moniliforme*	Maize	[65]
B. subtilis	Enhances the tolerance in the plant toward the pathogen	*B. cinerea*	Grapevine and its leaves	[66]
B. amyloliquefaciens (S13–3)	Plant defense response	*C. gloeosporioides*	Strawberry	[67]
B. amyloliquefaciens	Plant defense gene expression	*R. solani*	Lettuce	[68]
B. subtilis	Boosted immune responses triggered by chitin	*Magnaporthe oryzae*	Rice	[69]
B. subtilis (str. SV41) *B. amyloliquefaciens*	Stimulates host defense in plants to suppress wilt caused by *Fusarium* sp.	*F. oxysporum*	Tomato	[70]
Bacillus species	Defense signaling pathways triggered in *Arabidopsis* Inducing acid-dependent (salicylic acid/jasmonic) defense responses	*atmyc2* mutant *npr1–5* mutant	*Arabidopsis*	[71]
B. subtilis (UMAF6639)	Activation of acid-dependent (jasmonate/salicylic) defense responses	Powdery mildew causing fungus: cucurbit *P. fusca*	Melon plants	[72]

12.6.2 Biofilm growth mechanism by lipopetides

Cell colonization is the major challenge faced in biocontrol. The colonization of cells happens when the cells are subjected to extreme conditions like temperature, pressure, pH, nutrient deficiency, exposure to antimicrobial agents, radiations, and so on. This colony formation is a responsive defense mechanism of cells. *Bacillus* species possess a mechanism to overcome this colonization by forming biofilms. *B. subtilis* are generally found in the plant rhizosphere (the area is around the plant's root), from where they can act as a biocontrol agent. *Bacillus* is one of the first successful biocontrol investigated against insects and plant pathogens. *B. subtilis* has evolved several regulatory pathways by triggering biofilm formation. *Bacillus* strain is commercially marketed as biocontrol against fungal disease in plants and crops. Serenade, a biofungicide (contains *B. subtilis* strain) reported as control for *Erwina*, *Pseudomonas*, and *Xanthomonas* strains [73].

Biofilms are assembled on solid surfaces or as pellicles at air or liquid interfaces. Biofilms are complex communities of microorganisms where the cells are bound together by an extracellular matrix containing exopolysaccharides, proteins, and nucleic acids [74]. TasA and BslA are the protein compounds found in the matrix of the colony and pellicle biofilm. epsA-epsO operon is the major exopolysaccharide compound responsible for the synthesis of biofilm. Mutation in the gene of exopolysaccharides leads to defective biofilm formation [75]. The biofilms are formed by the swarming process, which is rapid and helps transport the cells through solid or semisolid phases by forming dendritic patterns. Biofilms form due to the transition from motile planktonic to sessile biofilm cells, which includes the production of secondary metabolites that can withstand a variety of biological, chemical, and physical factors [73]. During this process, the number of motile cells decreases as the biofilm develops; even in mature biofilms, few motile cells remain as a small group. These motile cells change their nature depending on the conditions. The mutants that do not have flagella delay biofilm formation [75]. Due to the structure of Lipopeptides, the formation of dendrites is easier. It is reported that the surfactin with a shorter chained fatty acid is found in the center of the mother colony, and surfactin with longer fatty acid chains are found in the edges of the dendrites [76]. In addition to surfactin, bacillomycin D produced by *B. amyloliquefaciens* SQR9 plays a vital role in inducing biofilm formation by altering the plant gene against *Fusarium oxysporum* that causes vascular wilt in cucumber plants [10]. Biofilms attracted the research communities as they have beneficial applications in microbial development like wastewater treatment, microbial fuel cells and pest control; they have problems in many industrial applications like clogging in pipes and tubings [75].

12.6.3 Bacillus inducing phytohormone production

Bacillus species can increase plant production by inducing plant hormones and stimulating plant growth regulators. Plant hormones have an influence on the development and determination of plant physiology. Auxin, cytokinins (CKs), ethylene, abscisic acid (ABA), and gibberellins (GAs) are some growth regulators that can be stimulated by bacillus species. Auxins are responsible for stimulating plant growth by cell development and tissue generation. GAs are responsible for stem elongation, flowering, seed germination, and fruiting in plants. Cell division in the root and shoot, seed germination, nutrient mobilization, apical dominance, and leaf senescence are all promoted by CKs. ABA affects plant physiological processes such as seed germination and stress tolerance. Ethylene controls the maturation processes along with a response to both abiotic and biotic stresses produced in the plant/crop. Plant hormones that are synthesized by *Bacillus* species directly influence the growth of plants by increasing the total N content, the total segregation and utilizing/uptake of phosphates, increasing the production of amino acids, enzymes, proteins, macro- and microminerals complexes, improving seed germination and seedling vigor, chlorophyll and carotenoids, improving the root's volume, mass, length, along with inducing the vegetative growth, enhancing shoot, and root growth, producing higher levels of hormones, increasing the leaf width and leaf area, promoting stress tolerance. Improving the plant's quality by inducing the plant growth hormone depends on the plant and the bacillus source [76].

12.7 Testing methods

Method 1: The phytopathogen is grown in media; the phytopathogen loaded in the testing plant/fruit/product, in particular, is exposed to various concentrations of the testing source (lipopetides) against the control. The entire setup is kept under a specific temperature and observed throughout the study.

Method 2: The plant to be tested is germinated in a cell tray. After a certain period of growth, the seedling is transferred to the observation pot, and a specific condition is maintained. The plant pathogen at different concentrations is sprayed over the plant. A plant without treatment is considered a control. A specific concentration's conidial suspension (lipopeptide suspension) is sprayed onto the whole plant one day before the bacterial suspension application. Application of lipopeptides and pathogen suspension can be made if needed once in 3 days throughout the observation period. The disease incidence rate of infections can be calculated by this method [77].

Method 3: Agar plates are prepared and inoculated with the standard inoculum of the test pathogen. Sterile discs are placed, and the sample of specific concentration containing test compound (lipopeptide) is loaded and placed on the agar surface. The temperature is maintained according to the test pathogen. The pathogen growth is inhibited when the lipopeptide diffuses and spread into the agar and restricts their growth. The microbial growth and the zone of inhibition can be measured using this method [77].

Other methods used to study the antifungal nature of a compound are the poisoned food technique, till kill test, flow cytofluoric method. These methods provide the data depending on the nature of the inhibitory effect and cell damage to the test microorganism. The antibiogram method offers a qualitative result by grouping the bacteria/pathogen as susceptible, intermediate, and resistant depending on their results. Antimicrobial gradient methods help in determining the MIC values [77].

12.8 Commercial *Bacillus*-based biocontrol

Serenade®, Companion®, Kodiak®, Cease®, Subtilex®, Pro-Mix®, FZB24®, Bio Safe®, Ecoshot®, Biosubtilin®, RhizoVital®42, RhizoVital® 42TB, Ballad®, Plus BioYield®, Rhizocell GC®, Yield Shield®, Sonata®, EcoGuard®, Botrybel®, Symbion-P®, Sublic®, and *Bacillus* SPP® are some of the commercialized *Bacillus* spp. Biocontrol preparation is available in markets of the USA, Canada, France, Denmark, Sweden, Germany, India, and Chile.

12.9 Conclusion

The *Bacillus* species are identified and established as an eminent source of environment-friendly and effective biocontrol methods in plant/crop protection. They work against the pathogen using several mechanisms depending on the plant and the environmental conditions. These bacterial strains not only prevent phytopathogens but also promote plant growth and production. The variety of compounds produced by the *Bacillus* species and their nature that adds advantages in the biocontrol of plant pathogens make the species outstanding and a potential candidate for bioagricultural practices around the world.

References

[1] Hazarika DJ, Goswami G, Gautom T, Parveen A, Das P, Barooah M et al. Lipopeptide mediated biocontrol activity of endophytic Bacillus subtilis against fungal phytopathogens. BMC Microbiology 19(1), 1–13.

[2] Roy S, Chakraborty S, Basu A. In vitro evaluation for antagonistic potential of some bio control isolates against important foliar fungal pathogens of cowpea. International Journal of Current Microbiology and Applied Sciences 2017, 6(9), 2998–3011.

[3] KhemRaj M, Shamsher SK. Lipopeptides as the antifungal and antibacterial agents. BioMed Research International 2015, 1–9. 2015, 2015, 473050. doi: 10.1155/2015/473050.

[4] Valenzuela-Ruiz V, Galvez-Gamboa, Gema Villa-Rodriguez E, Parra-cota F, Santoyo G, de Los Santos-villalobos S. Lipopeptides produced by biological control agents of the genus Bacillus: A review of analytical tools used for their study. Revista Mexicana Ciencias Agrícolas 2020, 11, 419–432.

[5] Bezza FA, Chirwa EMN. Production and applications of lipopeptide biosurfactant for bioremediation and oil recovery by Bacillus subtilis CN2. Biochemical Engineering Journal 2015. 101, 2015, 168–178. ISSN 1369-703X. https://doi.org/10.1016/j.bej.2015.05.007.

[6] Stein T. Bacillus subtilis antibiotics: Structures, syntheses and specific functions. Molecular Microbiology Mei 2005, 56(4), 845–857.

[7] Zhang DD, Guo XJ, Wang YJ, Gao TG, Zhu BC. Novel screening strategy reveals a potent Bacillus antagonist capable of mitigating wheat take-all disease caused by Gaeumannomyces graminis var. tritici. Letters in Applied Microbiology 2017, 65(6), 512–519.

[8] Raaijmakers JM, de Bruijn I, Nybroe O, Ongena M. Natural functions of lipopeptides from Bacillus and pseudomonas: More than surfactants and antibiotics. FEMS Microbiology Reviews 2010, 34(6), 1037–1062.

[9] Miljakovi´cmiljakovi´c D, Marinkovi´c JM, Baleševi´c S, Tubi´c B-T. microorganisms The Significance of Bacillus spp. in Disease Suppression and Growth Promotion of Field and Vegetable Crops. Available at: www.mdpi.com/journal/microorganisms.

[10] Xu Z, Shao J, Li B, Yan X, Shen Q, Zhang R. Contribution of Bacillomycin D in Bacillus amyloliquefaciens SQR9 to Antifungal Activity and Biofilm Formation. 2013, Available at: http://dx.doi.org/10.1128.

[11] Penha RO, Vandenberghe LPS, Faulds C, Soccol VT, Soccol CR. Bacillus lipopeptides as powerful pest control agents for a more sustainable and healthy agriculture: Recent studies and innovations. Planta 251, 70.

[12] Penha RO, Vandenberghe LPS, Faulds C, Soccol VT, Soccol CR. Bacillus lipopeptides as powerful pest control agents for a more sustainable and healthy agriculture: Recent studies and innovations. Planta 251, 70.

[13] Gong A-D, Li H-P, Yuan Q-S, Song X-S, Yao W, He W-J. Antagonistic mechanism of Iturin A and Plipastatin A from Bacillus amyloliquefaciens S76-3 from wheat spikes against Fusarium graminearum. PLoS One 2015, 10(2).

[14] Mora I, Cabrefiga J, Montesinos E. Cyclic lipopeptide biosynthetic genes and products, and inhibitory activity of plant-associated Bacillus against phytopathogenic bacteria. PLoS One 2015, 10(5), e0127738.

[15] Mirkasimovna Mardanova A, Fanisovna Hadieva G, Tafkilevich Lutfullin M, Valer I, Evna K, Farvazovna Minnullina L et al. Bacillus subtilis strains with antifungal activity against the phytopathogenic fungi. Agricultural Science 2017, 8, 1–20.

[16] Geissler M, Heravi KM, Henkel M, Hausmann R. Lipopeptide biosurfactants from bacillus species. In: Hayes DG, Solaiman DKY, Ashby RD, eds. Biobased surfactants, Second Edition,

Cambridge, Massachusetts, USA, AOCS Press, Elsevier, 2019, 205–240. ISBN 9780128127056. https://doi.org/10.1016/B978-0-12-812705-6.00006-X. (https://www.sciencedirect.com/science/article/pii/B978012812705600006X).
[17] Meena KR, Kanwar SS. Lipopeptides as the antifungal and antibacterial agents: Applications in food safety and therapeutics. BioMed Research International 2015, 473050. https://doi.org/10.1155/2015/473050.
[18] Straus SK, Hancock REW. Mode of action of the new antibiotic for Gram-positive pathogens daptomycin: Comparison with cationic antimicrobial peptides and lipopeptides. Biochimica Et Biophysica Acta – Biomembranes 2006, 1758(9), 1215–1223.
[19] Vecino X, Rodríguez-López L, Rincón-Fontán M, Cruz JM, Moldes AB. Nanomaterials synthesized by biosurfactants. Comprehensive Analytical Chemistry 2021, 94, 267–301.
[20] Chen XH, Koumoutsi A, Scholz R, Schneider K, Vater J, Süssmuth R, Piel J, Borriss R. Genome analysis of Bacillus amyloliquefaciens FZB42 reveals its potential for biocontrol of plant pathogens. Journal of Biotechnology 2009, 140(1–2), 27–37.
[21] Liu J, Hagberg I, Novitsky L, Hadj-Moussa H, Avis TJ. Interaction of antimicrobial cyclic lipopeptides from Bacillus subtilis influences their effect on spore germination and membrane permeability in fungal plant pathogens. Fungal Biology 2014, 118(11), 855–861.
[22] Ongena M, Jacques P. Bacillus lipopeptides: Versatile weapons for plant disease biocontrol. Trends Microbiology 2008, 16(3), 115–125.
[23] Li MSM, Piccoli DA, McDowell T, MacDonald J, Renaud J, Yuan ZC. Evaluating the biocontrol potential of Canadian strain Bacillus velezensis 1B-23 via its surfactin production at various pHs and temperatures. BMC Biotechnology 2021, 21(1), 1–12.
[24] Oliveras À, Baró A, Montesinos L, Badosa E, Montesinos E, Feliu L et al. Antimicrobial activity of linear lipopeptides derived from BP100 towards plant pathogens. PLoS One. 2018 Jul 27, 13(7), e0201571. doi: 10.1371/journal.pone.0201571.
[25] Kleinkauf H, Von Döhren H. Peptide Antibiotics. Antimicrobial Agents and Chemotherapy 1999, 43(6), 1317.
[26] Batoni G, de la Fuente-nunez C, Gomes P, Kong Q, Huan Y, Mou H et al. Antimicrobial peptides: classification, design, application and research progress in multiple fields. Frontiers in Microbiology 2020, 11, 582779.
[27] Hsieh F-C, Lin T-C, Menghsiao AE, Ae M, Kao -S-S. Comparing methods for identifying bacillus strains capable of producing the antifungal lipopeptide iturin A. Current Microbiology 2008 Jan, 56(1), 1–5. doi: 10.1007/s00284-007-9003-x.
[28] Dang Y, Zhao F, Liu X, Fan X, Huang R, Gao W et al. Enhanced production of antifungal lipopeptide iturin A by Bacillus amyloliquefaciens LL3 through metabolic engineering and culture conditions optimization. Microbial Cell Factories 2019 Apr 10, 18(1), 68. doi: 10.1186/s12934-019-1121-1.
[29] Chistoserdova L, Espariz M, Dunlap CA, Bowman MJ, Rooney AP. Iturinic lipopeptide diversity in the bacillus subtilis species group – important antifungals for plant disease biocontrol applications. Frontiers in Microbiology 2019 Aug 7, 10, 1794. doi: 10.3389/fmicb.2019.01794.
[30] Khan M. High performance liquid chromatography based characterization of Fengycin produced by Bacillus amyloliquefaciens against Fusarium graminearum and Rhizoctonia solani. Kuwait Journal of Science 2021, 9.
[31] Chen WC, Juang RS, Wei YH. Applications of a lipopeptide biosurfactant, surfactin, produced by microorganisms. Biochemical Engineering Journal 2015, 15(103), 158–169.
[32] Deleu M, Paquot M, Nylander T. Effect of fengycin, a lipopeptide produced by bacillus subtilis, on model biomembranes. Biophysical Journal 2008, 94(7), 2667–2679.

[33] Roy A, Paul D, Korpole S, Franco OL, Mandal SM. Purification, biochemical characterization and self-assembled structure of a fengycin-like antifungal peptide from Bacillus thuringiensis strain SM1, 2013.
[34] Béchet M, Caradec T, Hussein W, Abderrahmani A, Chollet M, Leclère V et al. Structure, biosynthesis, and properties of kurstakins, nonribosomal lipopeptides from Bacillus spp. Applied Microbiology and Biotechnology 2012, 95(3), 593–600.
[35] Hathout Y, Ho Y-P, Ryzhov V, Demirev P, Fenselau C. Kurstakins: A new class of lipopeptides isolated from bacillus thuringiensis. Journal of Natural Products 2000, 63, 1492–1496.
[36] Desai JD, Banat IM. Microbial production of surfactants and their commercial potential. Microbiology and Molecular Biology Reviews. 1997 Mar, 61(1), 47–64. doi: 10.1128/mmbr.61.1.47-64.1997.
[37] Chen WC, Juang RS, Wei YH. Applications of a lipopeptide biosurfactant, surfactin, produced by microorganisms. Biochemical Engineering Journal 2015, 103, 158–169.
[38] Balaji V, Jeremiah SS, Baliga PR. Polymyxins: Antimicrobial susceptibility concerns and therapeutic options. Indian Journal of Medical Microbiology 2011, 29(3), 230–242.
[39] Malviya D, Sahu PK, Singh UB, Paul S, Gupta A, Gupta AR et al. Lesson from ecotoxicity: Revisiting the microbial lipopeptides for the management of emerging diseases for crop protection. International Journal of Environmental Research and Public Health. 2020 Feb 23, 17(4), 1434. doi: 10.3390/ijerph17041434.
[40] Ongena M, Jacques P, Touré Y, Destain J, Jabrane A, Thonart P. Involvement of fengycin-type lipopeptides in the multifaceted biocontrol potential of Bacillus subtilis. Appl Microbiol Biotechnol 2005, 69(1), 29–38.
[41] Romero D, De Vicente A, Rakotoaly RH, Dufour SE, Veening J-W, Arrebola E et al. The iturin and fengycin families of lipopeptides are key factors in antagonism of bacillus subtilis toward podosphaera fusca. Mol Plant-Microbe Interact MPMI 2007, 20(4), 430–440.
[42] Preecha C, Sadowsky MJ, Prathuangwong S. Lipopeptide surfactin produced by bacillus amyloliquefaciens kps46 is required for biocontrol efficacy against xanthomonas axonopodis pv. glycines. Nature Science. 2010, 44, 84–99.
[43] Alvarez F, Castro M, Príncipe A, Borioli G, Fischer S, Mori G et al. The plant-associated Bacillus amyloliquefaciens strains MEP218 and ARP23 capable of producing the cyclic lipopeptides iturin or surfactin and fengycin are effective in biocontrol of sclerotinia stem rot disease. Journal of Applied Microbiology 2012, 112(1), 159–174.
[44] Bais HP, Fall R, Vivanco JM. Biocontrol of Bacillus subtilis against infection of Arabidopsis roots by Pseudomonas syringae is facilitated by biofilm formation and surfactin production. Plant Physiology 2004, 134(1), 307–319.
[45] De Senna A, Lathrop A. Antifungal screening of bioprotective isolates against Botrytis cinerea, Fusarium pallidoroseum and Fusarium moniliforme. Fermentation 2017 3(4), 53.
[46] Ntushelo K, Ledwaba LK, Rauwane ME, Adebo OA, Njobeh PB. The mode of action of bacillus species against fusarium graminearum, tools for investigation, and future prospects. Toxins (Basel) 2019, 11(10), 1–14.
[47] Tareq FS, Lee MA, Lee HS, Lee YJ, Lee JS, Hasan CM et al. Non-cytotoxic antifungal agents: Isolation and structures of gageopeptides A-D from a bacillus strain 109GGC020. Journal of Agricultural and Food Chemistry 2014, 62(24), 5565–5572.
[48] Zhang X, Li B, Wang Y, Guo Q, Lu X, Li S, Ma P. Lipopeptides, a novel protein, and volatile compounds contribute to the antifungal activity of the biocontrol agent Bacillus atrophaeus CAB-1. Applied Microbiology and Biotechnology 2013 97(21), 9525–9534.
[49] Kong HG, Kim JC, Choi GJ, Lee KY, Kim HJ, Hwang EC et al. Production of surfactin and iturin by Bacillus licheniformis N1 responsible for plant disease control activity. Plant Pathology Journal 2010, 26(2), 170–177.

[50] Köhl J, Kolnaar R, Ravensberg WJ. Mode of action of microbial biological control agents against plant diseases: Relevance beyond efficacy. Front Plant Science 2019 Jul 19, 10, 845. doi: 10.3389/fpls.2019.00845.
[51] Sarwar A, Hassan MN, Imran M, Iqbal M, Majeed S, Brader G, Sessitsch A, Hafeez FY. Biocontrol activity of surfactin A purified from Bacillus NH-100 and NH-217 against rice bakanae disease. Microbiological Research 2018, 209, 1–3.
[52] Kulimushi PZ, Arias AA, Franzil L, Steels S, Ongena M. Stimulation of fengycin-type antifungal lipopeptides in Bacillus amyloliquefaciens in the presence of the maize fungal pathogen Rhizomucor variabilis. Frontiers in Microbiology 2017, 8(May), 1–12.
[53] Rong S, Xu H, Li L, Chen R, Gao X, Xu Z. Antifungal activity of endophytic Bacillus safensis B21 and its potential application as a biopesticide to control rice blast. Pesticide Biochemistry and Physiology 2020 162, 69–77.
[54] Revathi N, Kalaiselvi M, Gomathi D, Ravikumar G, Uma C. Antifungal activity of Bacillus species in bio-control of different plant pathogens. Journal Phytopharmacology 2013, 2(6), 14–18.
[55] Qi X, Li B. Agronomy antifungal e ff ects of rhizospheric bacillus species against bayberry twig blight pathogen pestalotiopsis versicolor, 1–16.
[56] Benkeblia N, Tennant DPF, Jawandha SK, Gill PS. Preharvest and harvest factors influencing the postharvest quality of tropical and subtropical fruits. Postharvest Biology and Technology of Tropical and Subtropical Fruits 2011, 1, 112–142e.
[57] Naradisorn M. Effect of ultraviolet irradiation on postharvest quality and composition of foods. In: Galanakis CM, ed. Food Losses, Sustainable Postharvest and Food Technologies, Academic Press, 2021, 255–279. ISBN 9780128219126, https://doi.org/10.1016/B978-0-12-821912-6.00011-0.
[58] Zacarias L, Cronje PJR, Palou L. Postharvest technology of citrus fruits. The Genus Citrus Woodhead Publishing, 2020, 421–446, ISBN 9780128121634, https://doi.org/10.1016/B978-0-12-812163-4.00021-8.
[59] Chen X, Wang Y, Gao Y, Gao T, Zhang D. Inhibitory abilities of *Bacillus* isolates and their culture filtrates against the gray mold caused by *Botrytis cinerea* on postharvest fruit. Plant Pathology Journal 2019 [cited 24 Desember 2021] 35(5), 425–436. Available at. http://www.ppjonline.org/journal/view.php?doi=10.5423/PPJ.OA.03.2019.0064.
[60] Balderas-Ruíz KA, Gómez-Guerrero CI, Trujillo-Roldán MA, Valdez-Cruz NA, Aranda-Ocampo S, Juárez AM et al. Bacillus velezensis 83 increases productivity and quality of tomato (Solanum lycopersicum L.): Pre and postharvest assessment. Current Research in Microbial Sciences 2021, 2, 100076.
[61] Pretorius D, van Rooyen J, Clarke KG. Enhanced production of antifungal lipopeptides by Bacillus amyloliquefaciens for biocontrol of postharvest disease. New Biotechnology 2015, 32(2), 243–252.
[62] Arrebola E, Jacobs R, Korsten L. Iturin A is the principal inhibitor in the biocontrol activity of Bacillus amyloliquefaciens PPCB004 against postharvest fungal pathogens. Journal of Applied Microbiology 2010, 108(2), 386–395.
[63] Rangarajan V, Herbst WJ, Mazibuko S, Clarke KG. Bacillus lipopeptides for a novel postharvest disease control technology. Acta Horticulturae 2021, 1323, 79–86.
[64] Basaid K, Chebli B, Mayad EH, Furze JN, Bouharroud R, Krier F, Barakate M, Paulitz T. Biological activities of essential oils and lipopeptides applied to control plant pests and diseases: A review. International Journal of Pest Management 2021, 67(2), 155–177.
[65] Gond SK, Bergen MS, Torres MS, White JF. Endophytic Bacillus spp. produce antifungal lipopeptides and induce host defence gene expression in maize. Microbiological Research 2015, 172, 79–87.

[66] Farace G, Fernandez O, Jacquens L, Coutte F, Krier F, Jacques P et al. Cyclic lipopeptides from Bacillus subtilis activate distinct patterns of defence responses in grapevine. Molecular Plant Pathology 2015, 16(2), 177–187.
[67] Yamamoto S, Shiraishi S, Suzuki S. Are cyclic lipopeptides produced by Bacillus amyloliquefaciens S13-3 responsible for the plant defence response in strawberry against colletotrichum gloeosporioides? Letters in Applied Microbiology 2015, 60(4), 379–386.
[68] Chowdhury SP, Uhl J, Grosch R, Alquéres S, Pittroff S, Dietel K et al. Cyclic lipopeptides of Bacillus amyloliquefaciens subsp. plantarum colonizing the lettuce rhizosphere enhance plant defense responses toward the bottom rot pathogen Rhizoctonia solani. Molecular Plant-Microbe Interactions 2015, 28(9), 984–995.
[69] Chandler S, Van Hese N, Coutte F, Jacques P, Höfte M, De Vleesschauwer D. Role of cyclic lipopeptides produced by Bacillus subtilis in mounting induced immunity in rice (Oryza sativa L.). Physiological and Molecular Plant Pathology 2015, 91, 20–30.
[70] Aydi Ben Abdallah R, Stedel C, Garagounis C, Nefzi A, Jabnoun-Khiareddine H, Papadopoulou KK et al. Involvement of lipopeptide antibiotics and chitinase genes and induction of host defense in suppression of Fusarium wilt by endophytic Bacillus spp. in tomato. Crop Protection 2017, 99, 45–58.
[71] Kawagoe Y, Shiraishi S, Kondo H, Yamamoto S, Aoki Y, Suzuki S. Cyclic lipopeptide iturin A structure-dependently induces defense response in Arabidopsis plants by activating SA and JA signaling pathways. Biochemical and Biophysical Research Communications 2015, 460(4), 1015–1020.
[72] García-Gutiérrez L, Zeriouh H, Romero D, Cubero J, de Vicente A, Pérez-García A. The antagonistic strain Bacillus subtilisUMAF6639 also confers protection to melon plants against cucurbit powdery mildew by activation of jasmonate- and salicylic acid-dependent defence responses. Microb Biotechnology 2013, 6(3), 264–274.
[73] Morikawa M. Review beneficial biofilm formation by industrial bacteria bacillus subtilis and related species. Journal of Bioscience and Bioengineering 2006 Jan, 101(1), 1–8. doi: 10.1263/jbb.101.1.
[74] Kearns DB, Chu F, Branda SS, Kolter R, Losick R. A master regulator for biofilm formation by Bacillus subtilis. Molecular Microbiology 2005, 55(3), 739–749.
[75] Vlamakis H, Chai Y, Beauregard P, Losick R, Kolter R. Sticking together: Building a biofilm the Bacillus subtilis way. Nature Reviews Microbiology 2013 Mar, 11(3), 157–168. doi: 10.1038/nrmicro2960. Epub 2013 Jan 28.
[76] Debois D, Hamze K, Guérineau V, Le Caër JP, Holland IB, Lopes P et al. In situ localisation and quantification of surfactins in a bacillus subtilis swarming community by imaging mass spectrometry. Proteomics 2008, 8(18), 3682–3691.
[77] Ji SH, Paul NC, Deng JX, Kim YS, Yun BS, Yu SH. Biocontrol activity of bacillus amyloliquefaciens CNU114001 against fungal plant diseases. Mycobiology 2013, 41(4), 234–242.

Pragati Srivasatava

Chapter 13
Use of alkaloids in plant protection

Abstract: Alkaloids derived from plants are among the largest group of plant-derived secondary metabolites having therapeutic effectiveness in both modern and traditional medicines. Major examples including caffeine, nicotine, and emetine are used to fight oral intoxication and the antitumorals: vincristine and vinblastine. Alkaloids serve as a major tool for plant protection against pathogens and predators due to its toxicity. Plants in their natural habitat are circumambient by a different number of adversary including bacteria, fungus, viruses, nematodes, insects, and other herbivores which ultimately leads to reduction in plant growth and development. Alkaloids' basic structure comprise of any nitrogen-containing cyclic compound in negative oxidation state. It has more than 20 different classes including pyrrolidines, pyrrolizidine, quinolizidine, tropanes, piperidines, pyridines, and others. Most alkaloids function as storage reservoirs of nitrogen, defensive elements against predators, especially animals, vertebrates, insects, as well as arthropods due to their general toxic and deterrence effects, and growth regulators, since the structures of some alkaloids are similar to known plant growth regulators. Quick recognition by the attacker (herbivores/insects/microbes) and challenging environmental stress, forwarded by signal transduction via plant upon attack and stimulating alkaloid production is a two-way process; alkaloid mediated plant protection. Toxicity basically depends on specific dosage, exposure time, and aggressor type, such as sensitivity, site of action, and developmental stage.

13.1 Introduction

Many secondary metabolites are plant derived with different physiological functions that vary in structure, amount, site, and activity [1]. Among them alkaloids is naturally occurring chemical compounds containing nitrogenous organic molecules. Alkaloid mainly signifies word alkaline that was used to describe any nitrogen-containing base. Alkaloid classification depends on the presence of a basic nitrogen atom at any position in the molecule, in which nitrogen does not include in peptide bond [2]. Estimates of 12,000 natural products are an example of plant alkaloids. Since ages plant alkaloids have been used in medicines and teas, but the active compounds were not isolated and identified until the nineteenth century. Its chemical nature and structure have been elucidated lately in nineteenth century [3]. Mostly alkaloids are plant derived; however, they have also been present in animals, insects, marine invertebrates, and some microorganisms [4–6]. Inside the cell, they can be present in different

https://doi.org/10.1515/9783110771558-013

organs like mitochondria, vesicles, chloroplasts, and vacuoles. Its precursors (mainly amino acids) are derivatives of metabolic pathways, such as glycolysis. Its general function is to facilitate the survival of plants in the ecosystem, because they are allelopathic compounds having the potential to be a natural herbicide [7].

Plant-defensive alkaloids do not affect the normal vegetative growth of the plant, but makes the tissue less palatable for the feeding herbivore. It is stored constitutively in inactive forms or induced a specific response to predator attack [8]. These defensive alkaloids target organ system unique to herbivores likely nervous, digestive, and endocrine organs and are indulged in plant defense mechanism [9]. Plant families Leguminosae, Liliaceae, Solanaceae, and Amaryllidaceae, that is, 20% of vascular plants synthesize alkaloids [10].

Plants alkaloids are a result of reflex action produced in response to environmental modulations and biotic or abiotic stress. An alkaloid's structural and biological activity makes it suitable for many pharmacological effects on vertebrates. Many updates are reported about alkaloids' therapeutic significance for treating intestinal inflammatory disorders [11, 12]. Nowadays, alkaloids are well investigated as anesthetics, stimulants, antibacterials, antimalarials, analgesics, antihypertensive agents, spasmolysis agents, anticancer drugs, antiasthma therapeutics, vasodilators, antiarrhythmic agents, and so on. Records enumerate different uses of plant alkaloids since ages. In 1400 to 1200 BC, opium poppy (*Papaver somniferum*) was used in the Eastern Mediterranean. Extracts of henbane (*Hyoscyamus*), which contain atropine to dilate pupils, were used by the Egyptian queen Cleopatra [13, 14].

13.2 Alkaloids of plant origin as antimicrobials

Othman et al. [15] studied the antimicrobial activity of polyphenols and alkaloids in Middle Eastern plants. The antibacterial activity of the alkaloids found in the ethanolic extract of *Datura stramonium*, an annual herb commonly found in Baghdad district, was tested using the agar well diffusion method against *E. coli*, *P. aeruginosa*, *S. aureus*, *Proteus mirabilis*, and *K. pneumoniae*. The tested microorganisms were more sensitive to the ethanolic leaf extract as compared to standard antibiotics [16]. Antibiotic resistance is now considered a worldwide problem that puts public health at risk. Casciaro et al. [17] studied naturally occurring alkaloids of plant origin as potential antimicrobials against antibiotic-resistant Infections. Khameneh et al. [18] reviewed mechanistic viewpoint of plant antimicrobials. Thawabteh et al. [19] stated the biological activity of natural alkaloids against herbivores and Pathogens. Yan et al. [20] contributed in the research progress of antibacterial activities and mechanisms of natural alkaloids. In this section, we have discussed about the most effective antimicrobial and anti-insecticidal plant derived alkaloids:

Piperine, a nitrogen-containing pyridine alkaloid. is derived from *Piper nigrum* and *Piper longum* [21]. These alkaloids mainly constitute Palmaceae, Leguminosae, Zingiberaceae, and Solanaceae plant family groups and they are well known for their bacteriostatic or bactericidal activities. Pyridine alkaloids are broad-spectrum antimicrobial agents and have a good antimicrobial activity against a variety of pathogens. When piperine in combination with ciprofloxacin were co-administered, it inhibited growth of a mutant *S. aureus* and MIC values were also reduced for *S. aureus* significantly [22]. Methicillin-resistant *Staphylococcus aureus* growth was inhibited when co-administrated with piperine and gentamicin [23]. Piperine has reported to be a potent efflux proton inhibitor as it prohibited the NorA efflux proton activity of *S. aureus* and MRSA [24].

Reserpine is an indole alkaloid which is a tryptophan derivative having complex structure with significant biological activity. They are mainly grouped in Loganaceae, Apocynaceae, and Rubiaceae plant family. *Rauwolfia serpentine*-derived reserpine has a potent EPI activity [25]. In vitro experiments proved its bactericidal activity against *Staphylococcus* spp. *Streptococcus* spp., and *Micrococcus* spp. and it also increased the antibiotic susceptibility of these abovementioned bacterial species [26].

Berberine is an isoquinoline alkaloid, mainly encompasses of Papaveraceae, Berberidaceae, Ranunculaceae, and Menispermaceae plant family. They constitute the largest family of plant-derived alkaloids and further have four subdivisions including simple isoquinoline alkaloids, benzylisoquinoline alkaloids, bisbenzylisoquinoline alkaloids, aporphine alkaloids, and protoberberine alkaloids [27]. It is also extracted from the roots and stem bark of *Berberis* species, which is also the main active ingredient of *Rhizoma coptidis* and *Cortex phellodendri* and has been widely used in traditional medicine. Berberine has been reported for its various pharmacological effects mainly as anticancer and antibiotic agent. Also well known for its antioxidant, antiobesity, anticancer, anti-inflammatory, antibacterial, and antiviral effects [28–30]. It mainly intercalates between the DNA base pairs, targets RNA polymerase, gyrase, topoisomerase IV, and also inhibits the cell division showing its bactericidal effects. A study reported its antibacterial property against *E. coli* inhibiting cell division protein FtsZ [31]. There are various mechanisms of action reported against bacteria such as: damaging the cell structure, inhibiting protein, and DNA synthesis that ultimately leads to cell death. Feng et al. [32] summarized the biological activity of berberine [33].

Chanoclavin is an ergoline alkaloid consisting of diverse structure with basic ergoline skeleton; due to this, they are also termed as "ergolines" [34]. *Ipomoea muricata* seeds consist of 0.49% clavine alkaloids in which the percentage of chanoclavin is 37% and remaining is lysergol [35]. Previous study demonstrated that alone chanoclavin does not exhibit antibacterial property but when co-administered with tetracycline showed synergistic effect against *E. coli*. Chanoclavin, mediated bactericidal property, involves inhibition of ATPase-dependent bacterial efflux pump. It is also involved in binding with proteins which are drug resistant [36]. Cherewyk et al. [110] validate a sensitive method for the identification of ergot alkaloids from Canadian spring wheat.

Solasodine and tomatodine are steroidal alkaloids having most of its nitrogen contained in the ring structure [37]. They are mostly derived from the Solanaceae genus but are also found in Liliaceae, Apocynaceae, and Buxaceae plant families. Marine invertebrates and amphibians are also major sources of steroidal alkaloids [38–40]. Based on the steroidal skeleton, these alkaloids are divided into three category: pregnane alkaloids, cyclopregnane alkaloids, and cholestane alkaloids [41]. These alkaloids are well known for their starting material for steroidal drug; hence, they have an immense pharmacological value [42]. Tomatodine and solasodine have capability to combat bacterial resistance [43]. Tomatodine inhibited the growth of *S. aureus, E. faecalis, P. aeruginosa*, and *E. coli* when coadministered with ciprofloxacin or ampicillin [44].

Conessine is another example of steroidal alkaloid derived from stem bark of *Holarrhena antidysenterica* (Roth) species belonging to the Apocynaceae family [45]. Conessine plant extract was used as Thai folk medicine to cure diarrhea, bacterial infection, and was also used as astringent [46]. Conessine is a broad spectrum bactericidal agent active against both gram-positive and gram-negative bacteria [47, 48]. When conessine is combined with other conventional antibiotics, such as levofloxacin, it had synergistic effect against *A. baumannii* [49–51]. Also conessine potentiated efflux pump inhibition activity against ade gene cluster Adel JK EP, and efflux multiple antibiotics in *A. baumannii*. A study reported that conessine exerted an overexpression of efflux pump in *Pseudomonas aeruginosa* Mex-oprM conferring antibiotic resistance [52, 53]

Evocarpin is yet an another example of quinoline alkaloids reported as broad spectrum bactericidal agent against *Helicobacter pylori*, gram-positive bacteria *Staphylococcus aureus*, gram-negative *Pseudomonas aeruginosa*, also yeast *Candida albans* [54]. Hochfellner et al. [55] reported the antagonistic effects of indole quinozoline alkaloids on *Mycobacterium tuberculosis*. Evocarpin targets inhibition of peptidoglycan-synthesizing enzyme, ATP-dependent Mur ligase, and hence bacterial cell wall synthesizing machinery is disturbed. Ricine plant derived alkaloid reported to have antibacterial activity against *S. aureus, E. coli, Klebsiella pneumonia, Pseudomonas aeruginosa*, and *Candida albans* [56]. Ricine and its derivatives have been reported as antimicrobial and antiquorum sensing agents [57].

β-carboline is an indole alkaloid, first isolated from Zygophyllaceae plant family (*Peganum harmala* L.) sp. In the Middle East and North America, β-carboline is used as folk medicine and as bactericidal agent and it mainly ruptures the bacterial cell membrane and mediates the release of K^+ ions and other constituents of cytoplasm [58]. β-carboline is effective against different species of *Candida: C. albicans, C. intermedia*, and *C. krusei*, also *E. coli* and *S. aureus*. [58], Tiku [59] reported various β-carboline derivatives and their N_2 alkylated analogs and their role in plants defense in response to growth and stress. Cocaine alkaloid derived from *Erythroxylum coca* is reported to inhibit gram-negative and gram-positive cocci [60].

13.3 Plant-derived alkaloids as insecticides

Plants secrete a range of allelochemicals as the first line of defense against insect herbivores. Some exclusive examples of defensive allelochemicals are alkaloids, saponins, phenols, and terpines. Hence, plant-derived allelochemical is an eco-friendly approach for the control of numerous insects pest-mediated damage, which cause annually more than 15% yield loss [61, 62]. Tlak Gajger and Dar [65] elaborated the role of plant-derived allelochemicals: alkaloids as major source of insecticides. Matsuura and Fett-Neto [64] attributed antiherbivory and pollinator interaction in mainly insects. Mogren and Shikano [63] studied microbiota, pathogens, and parasites as mediators of tritrophic interactions between insect herbivores, plants, and pollinators. Plants mainly give a reflex action in the form of repellents, antinutritive, and toxic compounds. Some alkaloids inhibit the digestion of several sugars, such as sucrose and glycosidase, metabolizing enzymes inside tissues of the mid-gut of the insect which leads to toxic effects and deteriorating growth once the penetrating insects is devoid of using trehalose or sucrose [66].

There are various examples of plant alkaloids which tend to affect insects' nerve transmission. It also disturbs the cytoskeletal and cell membrane structure and leads to leakage of the cell material [67, 68]. Pyrrolizidine alkaloid (PA): Jacobins and erucifoline mediate insect attack in plants belonging to the senecionine type [69, 70]. *Senecio* genus mainly Asteraceae family secretes senecionine pyrolizidine alkaloid which has species-specific structure [70]. The secretion of this compound is induced in roots during insects/pest attack but not in shoots. PA have two major configurations: The tertiary base and *N*-oxides. *N*-oxide forms are nontoxic and is broken down into toxic pyrrole form via enzyme P450 inside the insect's gut *Spodoptera exigua* [71, 72]. But there is a certain example: *Spodoptera littoralis*, a generalist caterpillar, overcomes plant's PA-mediated defense mechanism and excretes PA effectively.

Aphids are responsible for huge economic loss of various vegetables and fruit crops across the globe. They are sap-sucking insects which leads to sudden downfall in the yield. Mainly apple (*Malus domestica*), crab apple (*Malus sylvestris*), and papaya (*Carica papaya*) are affected. Among the 10 *Lycoris radiate* isolated alkaloids, amabiline, deoxytazettine, deoxydihydrotazettine, 3-epimacronine, galanthamine, 11-hydroxygalanthamine, *N*-allylnorgalanthamine, 11b-hydroxygalanthamine, lycorine, and colchicines were reported to have insecticidal property against *Aphis citricola* both in vivo and in vitro experiments [73, 74]. Ricine, a plant alkaloid, also an antimicrobial agent which has insecticidal activity derived from *Ricinus communis* against *Atta sexden* [75, 76]. A three concentration gradient experiment of alkaloid extracted from the roots of *Catalpa ovate* were tested against *Plutella xylostella* and *Oriental armyworm*. The findings suggest potential insecticidal activity [77]. Pellitorine, an alkaloid, present in *Zanthoxylum piperitum* bark has toxicity against third-instar larvae of *Mythimna separata* and *Plutella xylostella* [78]. Kallure et al. [79] characterized constituents of insect herbivore oral secretions and their influence on the regulation of plant defenses. Kortbeek et al. [80] studied the endogenous plant metabolites against insects.

13.3.1 Mechanism

There is a variety of mechanisms involving alkaloids to inhibit the bacterial growth. Its unique chemical structure led gradually to the discovery of their bactericidal property and their site of activity. Bacterial nucleic acid and protein synthesis, modification of the bacterial cell permeability, cell wall, bacterial metabolism, and efflux pumps are the key targets.

13.3.1.1 Inhibition of bacterial nucleic acid and protein synthesis

Bacterial genetic material comprises DNA/RNA. DNA's main function is to store, copy, and transfer the genetic information and RNA's main function is in protein-synthesizing machinery [81]. Therefore, any damage in the DNA/RNA-synthesizing process blocks the expression of virulence genes and halts the bacterial growth and reproduction [82]. Cell division protein (Fts Z), filamentous temperature sensitive protein, is responsible for the formation of diaphragms and ring structure at the site of cell division and directs the bacterial cell division; it can be the target for the certain plant derived alkaloids [83]. Petronio et al. [84] studied the effect of berberine alkaloid on *Galleria mellonella* as an infection model and the results predicted the anti-invasive property of berberine. Chelerythine (CHE), an isoquinoline alkaloid, has a strong bactericidal property against *S. aureus* MRSA (methicillin resistant) and also ESBL *S. aureus* (extended spectrum β-lactamase). CHE effects the protein expression, which was confirmed via SDS page. The ratio of protein leakage was directly proportional to the increasing gradient of CHE [85]. Sanguinarine and metrine are another examples of alkaloids which affect the cell division process in bacteria. It initiates the filamentation in both gram-positive and gram-negative bacteria and disturbs the cytokinetic Z ring and rearrangement of the FtsZ protofilament [86]. Metrine forms a bond with protein inside the cell causing aggregation and disintegration of the cytoplasm leading to cell death [87]. Phenanthroindolizidine plant alkaloids, mainly pergularinine and tylophorinidine, hinder the activity of DHF enzyme which is important for the production of purines and pyrimidine precursors for DNA/RNA, protein, and amino acid synthesis [82, 88].

13.3.1.2 Damage of the cell membrane/cell wall/loss of cell permeability

The bacterial cell membrane serves as a stable environment for bacterial machinery, providing elasticity and shape to the cell. It is basically composed of phospholipid bilayers and proteins. Bacteria by means of quorum sensing forms biofilm under nutrient-deficient condition as a strategy for survival [89]. Any disruption of the cell membrane mediates the leakage of the cell material especially the electrolytes which increases the conductivity of the cell supernatant in the culture broth. Alkaline phosphatase (AKP) is present in between the bacterial cell wall. Increase in cell permeability leads to AKP leakage; hence, AKP gives a direct estimation of integrity of the bacterial cell wall [90, 91]. Sanguinarines have antibacterial property against *Providencia rettgeri* and have an MIC of 7.8 mg/μL determined by agar dilution method. It also inhibited the biofilm formation in *P. rettgeri*, which was confirmed via confocal scanning microscopy and crystal violet staining [92].

13.3.1.3 Efflux pumps

Efflux pumps are membrane spanning proteins, an efficient process for pathogenic bacterial survival [93] making it at least susceptible to the bactericidal alkaloids in the membrane. Efflux pumps enable the expulsion of membrane-permeable bactericidal agents and reduce the exposure of the agent to the pathogenic bacteria [94]. A number of studies have reported *E. coli* to have higher expression of efflux pumps genes involved in biofilm formation [95]. *Callistemon citrinus* alkaloids' bactericidal effect was tested against *S. aureus* and *P. aeruginosa* using MIC sensitivity test, which were grown using broth culture method. The alkaloid extract of *C. citrinus* and *Vernonia adoensis* inhibited the growth of *S. aureus* and *P. aeruginosa* at MIC values of 0.025 and 0.21 mg/μL respectively, whereas minimum bactericidal concentration of *C. citrinus* against *S. aureus* was 0.835 mg/μL. Rhodomine 6 G, a dye, was used to track the sensitivity of the alkaloids against *S. aureus* and *P. aeruginosa*. Two ATP-dependent efflux pumps were involved in the expulsion of 6 G-Rhodomine [96]. *Callistemon citrinus* alkaloids in combination with 6 F rhodomine had highest precipitation value with 121% with that of glucose control. When compared, *S. aureus* was less susceptible to EPI, whereas *P. aeruginosa* was most susceptible to EPI to *C. citrinus* alkaloids [90]. Yu et al. [97] reported jatrorrhizine suppressing the antimicrobial resistance of methicillin-resistant *S. aureus* via interaction of NOR-A with H bond and electrostatic interaction which simultaneously halted the efflux pump and NOR-A expression at transcription level.

13.3.1.4 Inhibition of bacterial metabolism

ATP adenosine triphosphate is among the potential energy units required for the limitless life activities inside the bacterial cell. ATP's role in respiration is significant; apart from this, it is also equally important for carrying out energy-requiring enzyme reaction. ATP synthase is among them. Hence, alkaloids play an active role in interfering with ATP synthase energy-mediated metabolism process which leads to biological death of the cell [99]. A study reported the proteomic investigation into the action mechanism of berberine against *Streptococcus pyogenes* [100]. It basically affects the carbohydrate metabolism of *S. pyogenes* (group A *Streptococcus*, GAS).

13.4 Conclusion

Multiple strategies for defense have evolved in plants against aggressors and environmental stress [111]. Naturally, plant alkaloids provide a massive way to overcome hostile condition generated by microbial and herbivore attack. An extensive study in plant's metabolic efficacy, its modification, and safety of alkaloids will power up plant natural resource utilization. A comprehensive analysis on the genes and their enzymes and its feasibility to clone the gene of interest in transgenic plants or cell cultures will give an idea of complete gene cassette for detailed pathway studies. Sivakumar et al. [108] recently studied the expression analysis of *Ornithine decarboxylase* gene associated with alkaloid biosynthesis in plants. Rai et al. [109] recently updated about the terpenoid indole alkaloid biosynthetic pathway in *Catharanthus roseus*. Plant-derived phytochemical having antimicrobial and insecticidal properties are significant as an eco-friendly approach, as they are less potent as anti-infectives than agents derived microbially such as antibiotics [101]. Alkaloids are unique in their chemical structure having endless pharmaceutical value with very nominal side effects and having a broad range of spectrum against microbes and have a minimum susceptibility to drug resistance. There are many drug-resistant bacteria, such as *S. aureus*, which are resistant to entire range of β-lactam antibiotics: penicillin, amoxicillin, methicillin, oxacillin, and so on. They have gained resistance by obtaining resistance factors via mobile gene elements encoding it through horizontal gene transfer [102]. Health sector is facing an issue of concern, that is, antibiotic resistance of the disease-causing pathogenic bacteria. There is vigorous need for search of agents of plant origin that can eliminate drug resistance. Khadraoui et al. [103] updated about the antibacterial and anti-biofilm activity of *Peganum harmala* seeds extracts against multidrug-resistant *P. aeruginosa* pathogenic isolates and molecular mechanism of action. There are certain points to be reviewed for carrying out experiments for the synthesis of new range of plant-derived alkaloids such as its extraction, optimization, and commercialization. Recently, Han et al. [106] studied the development, optimization, validation, and application of ultra HPLC and TMS tandem mass spectrometry for the analysis of pyrolizidine alkaloids and pyrolizidine alkaloids N oxides in teas and weed. Nowadays, MCA multiple component analysis is the method for alkaloid extraction [104]. Also computational methods and high throughput screening procedure are counted among the new technologies for the enhanced extraction of alkaloids. Alkaloid's bioavailability is also a major concern. The amount of active compound inside the plant is governed by numerous factors, that is, plant's genetics is an internal factor whereas plant's physiology and ecology are external factors [105]. Generally, alkaloids form salt with acids and get dissolved in water. Lipophilic bases mediate absorption from the gastrointestinal tract via lipophilic diffusion [107]. 0.27% to 64.6% is an average range for the bioavailability of different alkaloids. These data are extracted from Traditional Chinese Medicine Systems Pharmacology Database and Analysis Platform (TCMSP; https://tcmspw.com/tcmsp.php, accessed on 2 January 2022) [112]. The discovered drug should evaluate site, factor, link, and mechanism properties of bacterial infection, and optimization experiments should be conducted to obtain permissible dose within acceptable range for toxicity. Since there are a variety of habitats across the globe from hot deserts to cold mountain peaks and densely occupied forests, it is feasible to explore a broad range of plant species in search of new antimicrobials with effective resistance properties and minimal side effects.

References

[1] Matsuura HN, Fett-Neto AG. Plant alkaloids: Main features, toxicity, and mechanisms of action. In: Gopalakrishnakone P, Carlini C, Ligabue-Braun R, eds. Plant toxins. toxinology. Dordrecht, Springer, 2015, 1–5.

[2] Robinson T. Metabolism and function of alkaloids in plants. Science 1974, 184, 430–435.

[3] Gutiérrez-Grijalva EP, López-Martínez LX, Contreras-Angulo LA, Elizalde-Romero CA, Heredia JB. Plant alkaloids: Structures and bioactive properties. In: Plant-derived bioactives. Singapore, Springer, 2020, 85–117.

[4] Lu JJ, Bao JL, Chen XP, Huang M, Wang YT. Alkaloids isolated from natural herbs as the anticancer agents. Evidence-based Complementary and Alternative Medicine 2012, 1–12.

[5] Roberts MF. editor. Alkaloids: Biochemistry, ecology, and medicinal applications. Springer Science & Business Media, New York, Plenum, 2013, 17.

[6] Bribi N. Pharmacological activity of alkaloids: A. Review 2018, 1, 1–6.

[7] Jing H, Liu J, Liu H, Xin H. Histochemical investigation and kinds of alkaloids in leaves of different developmental stages in Thymus quinquecostatus. The Scientific World Journal 2014, 839548.

[8] War AR, Paulraj MG, Ahmad T, Buhroo AA, Hussain B, Ignacimuthu S, Sharma HC. Mechanisms of plant defense against insect herbivores. Plant Signaling & Behavior 2012, 7, 1306–1320.

[9] Fürstenberg-Hägg J, Zagrobelny M, Bak S. Plant defense against insect herbivores. International Journal of Molecular Sciences 2013, 14, 10242–10297.

[10] Yang L, Stöckigt J. Trends for diverse production strategies of plant medicinal alkaloids. Natural Product Reports 2010, 27, 1469–1479.

[11] Zhang YB, Yang L, Luo D, Chen NH, Wu ZN, Ye WC, Li YL, Wang GC. Sophalines E–I, five five quinolizidine-based alkaloids with antiviral activities against the hepatitis B virus from the seeds of sophora alopecuroides. Organic Letters 2018, 20, 5942–5946.

[12] Peng J, Zheng TT, Li X, Liang Y, Wang LJ, Huang YC, Xiao HT. Plant-derived alkaloids: The promising disease-modifying agents for inflammatory bowel disease. Frontiers in Pharmacology 2019, 12, 10–351.

[13] Croteau R, Kutchan TM, Lewis NG. Natural products (secondary metabolites). Biochemistry and Molecular Biology of Plants 2000, 24, 1250–1319.

[14] Evans SR, Hofmann A. Planta de los dioses. Mexico, Fondo de Cultura Econômica, 2006.

[15] Othman L, Sleiman A, Abdel-Massih RM. Antimicrobial activity of polyphenols and alkaloids in middle eastern plants. Frontiers in Microbiology 2019, 15(10), 911.

[16] Al-Matani SK, Al-Wahaibi RN, Hossain MA. Total flavonoids content and antimicrobial activity of crude extract from leaves of ficus sycomorus native to sultanate of Oman. Karbala International Journal of Modern Science 2015, 1, 166–171.

[17] Casciaro B, Mangiardi L, Cappiello F, Romeo I, Loffredo MR, Iazzetti A, Calcaterra A, Goggiamani A, Ghirga F, Mangoni ML, Botta B. Naturally-occurring alkaloids of plant origin as potential antimicrobials against antibiotic-resistant infections. Molecules 2020, 25, 3619.

[18] Khameneh B, Iranshahy M, Soheili V, Bazzaz BS. Review on plant antimicrobials: A mechanistic viewpoint. Antimicrobial Resistance and Infection Control 2019, 8, 1–28.

[19] Thawabteh A, Juma S, Bader M, Karaman D, Scrano L, Bufo SA, Karaman R. The biological activity of natural alkaloids against herbivores, cancerous cells and pathogens. Toxins 2019, 11, 656.

[20] Yan Y, Li X, Zhang C, Lv L, Gao B, Li M. Research progress on antibacterial activities and mechanisms of natural alkaloids: A review. Antibiotics 2021, 10, 318.

[21] Luo YM. Natural medicinal chemistry. Wuhan, China, Huazhong University of Science and Technology, 2011.

[22] Khan IA, Mirza ZM, Kumar A, Verma V, Qazi GN. Piperine, a phytochemical potentiator of ciprofloxacin against Staphylococcus aureus. Antimicrobial Agents and Chemotherapy 2006, 50, 810–812.

[23] Khameneh B, Iranshahy M, Ghandadi M, Ghoochi Atashbeyk D, Fazly Bazzaz BS, Iranshahi M. Investigation of the antibacterial activity and efflux pump inhibitory effect of co-loaded piperine and gentamicin nanoliposomes in methicillin-resistant Staphylococcus aureus. Drug Development and Industrial Pharmacy 2015, 41, 989–994.

[24] Kumar A, Khan IA, Koul S, Koul JL, Taneja SC, Ali I, Ali F, Sharma S, Mirza ZM, Kumar M, Sangwan PL. Novel structural analogues of piperine as inhibitors of the NorA efflux pump of Staphylococcus aureus. Journal of Antimicrobial Chemotherapy 2008, 61, 1270–1276.

[25] Abdelfatah SA, Efferth T. Cytotoxicity of the indole alkaloid reserpine from Rauwolfia serpentina against drug-resistant tumor cells. Phytomedicine 2015, 22, 308–318.

[26] Sridevi D, Shankar C, Prakash P, Park JH, Thamaraiselvi K. Inhibitory effects of reserpine against efflux pump activity of antibiotic resistance bacteria. Chemical Biology Letters 2017, 4, 69–72.

[27] Hagel JM, Facchini PJ. Benzylisoquinoline alkaloid metabolism: A century of discovery and a brave new world. Plant & Cell Physiology 2013, 54, 647–672.

[28] Jamshaid F, Dai J, Yang LX. New development of novel berberine derivatives against bacteria. Mini Reviews in Medicinal Chemistry 2020, 20, 716–724.

[29] Liang Y, Xu X, Yin M, Zhang Y, Huang L, Chen R, Ni J. Effects of berberine on blood glucose in patients with type 2 diabetes mellitus: A systematic literature review and a meta-analysis. Endocrine Journal 2019, 66, 51–63.

[30] Habtemariam S. Berberine and inflammatory bowel disease: A concise review. Pharmacological Research 2016, 113, 592–599.
[31] Boberek JM, Stach J, Good L. Genetic evidence for inhibition of bacterial division protein FtsZ by berberine. PloS One 2010, 5, 13745.
[32] Feng R, Qu J, Zhou W, Wei Q, Yin Z, Du Y, Yan K, Wu Z, Jia R, Li L, Song X. Antibacterial activity and mechanism of berberine on avian Pasteurella multocida. International Journal of Clinical and Experimental Medicine 2016, 9, 22886–22892.
[33] Och A, Podgórski R, Nowak R. Biological activity of berberine – a summary update. Toxins 2020, 12, 713.
[34] Casciaro B, Mangiardi L, Cappiello F, Romeo I, Loffredo MR, Iazzetti A, Calcaterra A, Goggiamani A, Ghirga F, Mangoni ML, Botta B. Naturally-occurring alkaloids of plant origin as potential antimicrobials against antibiotic-resistant infections. Molecules 2020, 25, 3619.
[35] Genest K. A direct densitometric method on thin-layer plates for the determination of lysergic acid amide, isolysergic acid amide and clavine alkaloids in morning glory seeds. Journal of Chromatography A 1965, 19, 531–539.
[36] Maurya A, Dwivedi GR, Darokar MP, Srivastava SK. Antibacterial and synergy of clavine alkaloid lysergol and its derivatives against nalidixic acid-resistant E scherichia coli. Chemical Biology & Drug Design 2013, 81, 484–490.
[37] Luo YM. Natural medicinal chemistry. Wuhan, China, Huazhong University of Science and Technology, 2011.
[38] Li HJ, Jiang Y, Li P. Chemistry, bioactivity and geographical diversity of steroidal alkaloids from the Liliaceae family. Natural Product Reports 2006, 23, 735–752.
[39] Rahman AU, Choudhary MI. Diterpenoid and steroidal alkaloids. Natural Product Reports 1995, 12, 361.
[40] Rahman A, Choudhary MI.Chemistry and biology of steroidal alkaloids. Alkaloids: Chemistry and Biology 1998, 50, 61–108.
[41] Zhang Y, Hu WZ, Chen XZ, Peng YB, Song LY, Shi YS. Bioactive quinolizidine alkaloids from the root of Sophora tonkinensis. Journal of Traditional Chinese Medicine 2016, 41, 2261–2266.
[42] Zhang X, Sun X, Wu J et al. Berberine damages the cell surface of methicillin-resistant *Staphylococcus aureus*. Frontiers in Microbiology 2020 11 621
[43] Chu M, Zhang MB, YC L, Kang JR, Chu ZY, Yin KL, Ding LY, Ding R, Xiao RX, Yin YN, Liu XY. Role of berberine in the treatment of methicillin-resistant Staphylococcus aureus infections. Scientific Reports 2016, 1, 1–9.
[44] Patel K, Singh RB, Patel DK. Medicinal significance, pharmacological activities, and analytical aspects of solasodine: A concise report of current scientific literature. Journal of Acute Disease 2013, 2, 92–98.
[45] Murugan K, Jayakumar K. Solanum alkaloids and their pharmaceutical roles: A review. Journal of Analytical & Pharmaceutical Research 2015, 1–14.
[46] Lamontagne Boulet M, Isabelle C, Guay I, Brouillette E, Langlois JP, Jacques PÉ, Rodrigue S, Brzezinski R, Beauregard PB, Bouarab K, Boyapelly K. Tomatidine is a lead antibiotic molecule that targets Staphylococcus aureus ATP synthase subunit C. Antimicrobial Agents and Chemotherapy 2018, 62, 02197–17.
[47] Kumar N, Singh B, Bhandari P, Gupta AP, Kaul VK. Steroidal alkaloids from holarrhena antidysenterica (L.) W ALL.. Chemical & Pharmaceutical Bulletin 2007, 55, 912–914.
[48] Li-Na ZH, Xiao-Lei GE, Ting-Ting DO, Hui-Yuan GA, Bo-Hang SU. Antibacterial steroidal alkaloids from Holarrhena antidysenteriaca. Chinese Journal of Natural Medicines 2017, 15, 540–545.

[49] Siddiqui BS, Ali ST, Rizwani GH, Begum S, Tauseef S, Ahmad A. Antimicrobial activity of the methanolic bark extract of Holarrhena pubescens (Buch. Ham), its fractions and the pure compound conessine. Natural Product Research 2012, 26, 987–992.
[50] Siriyong T, Voravuthikunchai SP, Coote PJ. Steroidal alkaloids and conessine from the medicinal plant Holarrhena antidysenterica restore antibiotic efficacy in a Galleria mellonella model of multidrug-resistant Pseudomonas aeruginosa infection. BMC Complementary and Alternative Medicine 2018, 18, 1–10.
[51] Damier-Piolle L, Magnet S, Brémont S, Lambert T, Courvalin P. AdeIJK, a resistance-nodulation-cell division pump effluxing multiple antibiotics in Acinetobacter baumannii. Antimicrobial Agents and Chemotherapy 2008, 52, 557–562.
[52] Siriyong T, Chusri S, Srimanote P, Tipmanee V, Voravuthikunchai SP. Holarrhena antidysenterica extract and its steroidal alkaloid, conessine, as resistance-modifying agents against extensively drug-resistant Acinetobacter baumannii. Microbial Drug Resistance 2016, 22, 273–282.
[53] Soltani R, Fazeli H, Najafi RB, Jelokhanian A. Evaluation of the synergistic effect of Tomatidine with several antibiotics against standard and clinical isolates of Staphylococcus aureus, enterococcus faecalis, Pseudomonas aeruginosa and Escherichia coli. Iranian Journal of Pharmaceutical Research 2017, 16, 290–296.
[54] Bligh X, Smith E SA. Quinolone alkaloids from Fructus Euodiae show activity against methicillin-resistant Staphylococcus aureus. Phytotherapy Research 2014, 28, 305–307.
[55] Hochfellner C, Evangelopoulos D, Zloh M, Wube A, Guzman JD, McHugh TD, Kunert O, Bhakta S, Bucar F. Antagonistic effects of indoloquinazoline alkaloids on antimycobacterial activity of evocarpine. Journal of Applied Microbiology 2015, 118, 864–872.
[56] El-Naggar MH, Elgaml A, Abdel Bar FM, Badria FA. Antimicrobial and antiquorum-sensing activity of Ricinus communis extracts and ricinine derivatives. Natural Product Research 2019, 33, 1556–1562.
[57] Ali AH, Abdelrahman M, El-Sayed MA. Alkaloid role in plant defense response to growth and stress. In: Jogaiah S, Abdelrahman M, eds. Bioactive molecules in plant defense. Cham, Springer, 2019, 145–158.
[58] Suzuki K, Nomura I, Ninomiya M, Tanaka K, Koketsu M. Synthesis and antimicrobial activity of b-carboline derivatives with N2-alkyl modifications. Bioorganic & Medicinal Chemistry Letters 2018, 28, 2976–2978.
[59] Tiku AR. Antimicrobial compounds and their role in plant defense. In: Molecular aspects of plant-pathogen interaction. Singapore, Springer, 2018, 283–30.
[60] JK D, Dan WJ, Wan JB. Natural and synthetic β-carboline as a privileged antifungal scaffolds. European Journal of Medicinal Chemistry 2021, 17, 114057.
[61] Oerke EC. Crop losses to pests. The Journal of Agricultural Science 2006, 144, 31–43.
[62] Mendesil E. Plant-resistance to insect herbivores and semiochemicals: Implications for field pea pest management. In: Faculty of landscape architecture, horticulture and crop production science. Swedish University of Agricultural Sciences, Vol. 1, 2014, 1–37.
[63] Mogren CL, Shikano I. Microbiota, pathogens, and parasites as mediators of tritrophic interactions between insect herbivores, plants, and pollinators. Journal of Invertebrate Pathology 2021, 15, 107589.
[64] Matsuura HN, Fett-Neto AG. Plant alkaloids: Main features, toxicity, and mechanisms of action. Plant Toxins 2012, 2, 1–5.
[65] Tlak Gajger I, Dar SA. Plant allelochemicals as sources of insecticides. Insects 2021, 12, 189.
[66] Glendinning JI. (2002) How do herbivorous insects cope with noxious secondary plant compounds in their diet? In: Nielsen JK, Kjær C, Schoonhoven LM, eds. Proceedings of the

11th International Symposium on Insect-Plant Relationships. Series Entomologica, Dordrecht, Springer, 2002, Vol. 57,15–25
[67] Nakagawa A, Matsumura E, Sato F, Minami H. Bioengineering of isoquinoline alkaloid production in microbial systems. Advances in Botanical Research 2013, 68, 183–203.
[68] Mbata GN, Payton ME. Effect of monoterpenoids on oviposition and mortality of Callosobruchus maculatus (F.)(Coleoptera: Bruchidae) under hermetic conditions. Journal of Stored Products Research 2013, 53, 43–47.
[69] Uzor PF. Alkaloids from plants with antimalarial activity: A review of recent studies. Evidence-Based Complementary and Alternative Medicine 2020, 12, 2020.
[70] Nuringtyas TR, Verpoorte R, Klinkhamer PG, van Oers MM, Leiss KA. Toxicity of pyrrolizidine alkaloids to Spodoptera exigua using insect cell lines and injection bioassays. Journal of Chemical Ecology 2014, 40, 609–616.
[71] Hol WG, Macel M, van Veen JA, van der Meijden E. Root damage and aboveground herbivory change concentration and composition of pyrrolizidine alkaloids of Senecio jacobaea. Basic and Applied Ecology 2004, 5, 253–260.
[72] Kopp T, Abdel-Tawab M, Mizaikoff B. Extracting and analyzing pyrrolizidine alkaloids in medicinal plants: A review. Toxins 2020, 12, 320.
[73] Yan H, Xie N, Zhong C, Su A, Hui X, Zhang X, Jin Z, Li Z, Feng J, He J. Aphicidal activities of Amaryllidaceae alkaloids from bulbs of Lycoris radiata against Aphis citricola. Industrial Crops and Products 2018, 123, 372–378.
[74] Ali AH, Abdelrahman M, El-Sayed MA. Alkaloid role in plant defense response to growth and stress. Cham, InBioactive Molecules in Plant Defense Springer, 2019, 145–158.
[75] Bigi MF, Torkomian VL, De Groote ST, Hebling MJ, Bueno OC, Pagnocca FC, Fernandes JB, Vieira PC, Da Silva MF. Activity of Ricinus communis (Euphorbiaceae) and ricinine against the leaf-cutting ant Atta sexdens rubropilosa (Hymenoptera: Formicidae) and the symbiotic fungus Leucoagaricus gongylophorus. Pest Management Science 2004, 60, 933–938.
[76] Santos PM, Batista DL, Ribeiro LA, Boffo EF, de Cerqueira MD, Martins D, de Castro RD, de Souz-neta LC, Pinto EL, Zambotti-Villela E, Colepicolo P, Fernandez LG, Canuto GA, Ribeiro PR. Identification of antioxidant and antimicrobial compounds from the oilseed crop Ricinus communis using a multiplatform metabolite profiling approach. Ind Crop Prod 20182018, 124, 834–844.
[77] Shao J, Zhang Y, Zhu Z, Chen X, He F. Process optimization and insecticidal activity of alkaloids from the root bark of Catalpa ovata G. Don by response surface methodology. Tropical Journal of Pharmaceutical Research 2018, 17, 843–848.
[78] Kim SI, Ahn YJ. Larvicidal activity of lignans and alkaloid identified in Zanthoxylum piperitum bark toward insecticide-susceptible and wild Culex pipiens pallens and Aedes aegypti. Parasites & Vectors 2017, 10, 1–0.
[79] Kallure GS, Kumari A, Shinde BA, Giri AP. Characterized constituents of insect herbivore oral secretions and their influence on the regulation of plant defenses. Phytochemistry 2022, 193, 113008.
[80] Kortbeek RW, van der Gragt M, Bleeker PM. Endogenous plant metabolites against insects. European Journal of Plant Pathology 2019, 154, 67–90.
[81] Salton MR. Structure and function of bacterial cell membranes. Annual Reviews in Microbiology 1967, 21, 417–442.
[82] Othman L, Sleiman A, Abdel-Massih RM. Antimicrobial activity of polyphenols and alkaloids in middle eastern plants. Frontiers in Microbiology 2019, 10, 911.
[83] Barrows JM, Goley ED. FtsZ dynamics in bacterial division: What, how, and why? Current Opinion in Cell Biology 2021, 68, 163–172.

[84] Petronio Petronio G, Cutuli MA, Magnifico I, Venditti N, Pietrangelo L, Vergalito F, Pane A, Scapagnini G, Di Marco R. In vitro and in vivo biological activity of berberine chloride against uropathogenic E. coli strains using Galleria mellonella as a host model. Molecules 2020, 25, 5010.
[85] He N, Wang P, Wang P, Ma C, Kang W. Antibacterial mechanism of chelerythrine isolated from root of Toddalia asiatica (Linn) Lam. BMC Complementary and Alternative Medicine 2018, 18, 1–9.
[86] Beuria TK, Santra MK, Panda D. Sanguinarine blocks cytokinesis in bacteria by inhibiting FtsZ assembly and bundling. Biochemistry 2005, 44, 16584–16593.
[87] Chen DH. In vitro antibacterial activity of alkaloids from Sophora flavescens. China Animal Health 2010, 12, 28–30.
[88] Rao KN, Venkatachalam SR. Inhibition of dihydrofolate reductase and cell growth activity by the phenanthroindolizidine alkaloids pergularinine and tylophorinidine: The in vitro cytotoxicity of these plant alkaloids and their potential as antimicrobial and anticancer agents. Toxicology in Vitro 2000, 14, 53–59.
[89] Mabhiza D, Chitemerere T, Mukanganyama S. Antibacterial properties of alkaloid extracts from callistemon citrinus and vernonia adoensis against staphylococcus aureus and pseudomonas aeruginosa. International Journal of Medicinal Chemistry 2016, 2016.
[90] Lan WQ, Xie J, Hou WF, Li DW. Study on antibacterial activity and action mechanism of compound biological preservative against Staphylococcus squirrel. Natural Product Research and Development 2012, 6, 741–746.
[91] Hara S, Yamakawa M. Moricin, a novel type of antibacterial peptide isolated from the Silkworm, Bombyx mori (*). Journal of Biological Chemistry 1995, 270, 29923–29927.
[92] Zhang Q, Lyu Y, Huang J, Zhang X, Yu N, Wen Z, Chen S. Antibacterial activity and mechanism of sanguinarine against Providencia rettgeri in vitro. Peer Journal 2020, 11, 9543.
[93] Teelucksingh T, Thompson LK, Cox G. The evolutionary conservation of Escherichia coli drug efflux pumps supports physiological functions. Journal of Bacteriology 2020, 202, 00367–20.
[94] Wei JT, Qian J, Su QLY, Liu ZX, Wang XL, Wang YP. Research Progress on the mechanism of bacterial biofilm induced drug resistance and the effect of antimicrobial peptide LL-37 on biofilm. Journal of Hexi University 2020, 36, 38–43.
[95] Ito A, Taniuchi A, May T, Kawata K, Okabe S. Increased antibiotic resistance of Escherichia coli in mature biofilms. Applied and Environmental Microbiology 2009, 75, 4093–4100.
[96] Maesaki S, Marichal P, Bossche HV, Sanglard D, Kohno S. Rhodamine 6G efflux for the detection of CDR1-overexpressing azole-resistant Candida albicans strains. Journal of Antimicrobial Chemotherapy 1999, 44, 27–31.
[97] Yu H, Wang Y, Wang X, Guo J, Wang H, Zhang H, Du F. Jatrorrhizine suppresses the antimicrobial resistance of methicillin-resistant Staphylococcus aureus. Experimental and Therapeutic Medicine 2019, 18, 3715–3722.
[98] Sobti M, Ishmukhametov R, Stewart AG. ATP Synthase: Expression, Purification, and Function. Methods Mol Biol 2020, 2073, 73–84.
[99] Du GF, Le YJ, Sun X, Yang XY, He QY. Proteomic investigation into the action mechanism of berberine against Streptococcus pyogenes. Journal of Proteomics 2020, 215, 103666.
[100] Yamada H. Natural products of commercial potential as medicines. Current Opinion in Biotechnology 1991, 2, 203–210.
[101] Vuong C, Yeh AJ, Cheung GY, Otto M. Investigational drugs to treat methicillin-resistant Staphylococcus aureus. Expert Opinion on Investigational Drugs 2016, 25, 73–93.

[102] Li J, Liu D, Tian X, Koseki S, Chen S, Ye X, Ding T. Novel antibacterial modalities against methicillin resistant Staphylococcus aureus derived from plants. Critical Reviews in Food Science and Nutrition 2019, 59, 153–161.
[103] Khadraoui N, Essid R, Jallouli S, Damergi B, Ben Takfa I, Abid G, Jedidi I, Bachali A, Ayed A, Limam F, Tabbene O. Antibacterial and antibiofilm activity of Peganum harmala seed extract against multidrug-resistant Pseudomonas aeruginosa pathogenic isolates and molecular mechanism of action. Archives of Microbiology 2022, 204, 1–2.
[104] Belwal T, Pandey A, Bhatt ID, Rawal RS. Optimized microwave assisted extraction (MAE) of alkaloids and polyphenols from Berberis roots using multiple-component analysis. Scientific Reports 2020, 10, 1–0.
[105] Djarot P, Utami NF, Veonicha N, Rahmadini A, Iman AN. Antibacterial activity tests of staphylococcus aureus and phytochemical screening in family asteraceae, clusiaceae, phyllanthaceae. Journal of Southwest Jiaotong University 2020, 55, 6.
[106] Han H, Jiang C, Wang C, Wang Z, Chai Y, Zhang X, Liu X, Lu C, Chen H. Development, optimization, validation and application of ultra high performance liquid chromatography tandem mass spectrometry for the analysis of pyrrolizidine alkaloids and pyrrolizidine alkaloid N-oxides in teas and weeds. Food Control 2022, 132, 108518.
[107] Chen YW, Liu HG. Recent advances in pharmacokinetics of alkaloids. Medicine Review 2009, 15, 3489–3491.
[108] Sivakumar HP, Sundararajan S, Rajendran V, Ramalingam S. Genome wide survey, and expression analysis of Ornithine decarboxylase gene associated with alkaloid biosynthesis in plants. Genomics 2022, 114, 84–94.
[109] Rai SK, Rai KK, Kumar S, Rai SP. Functional genomics approaches for gene discovery related to terpenoid indole alkaloid biosynthetic pathway in *Catharanthus roseus*. In: Kole C, eds. The catharanthus genome. Compendium of plant genomes. Cham, Springer, 2022, 155–173.
[110] Cherewyk J, Grusie-Ogilvie T, Blakley B, Al-Dissi A. Validation of a new sensitive method for the detection and quantification of *R* and *S*-Epimers of ergot alkaloids in canadian spring wheat utilizing deuterated lysergic acid diethylamide as an internal standard. Toxins 2022, 14, 22.
[111] Gershenzon J, Ullah C. Plants protect themselves from herbivores by optimizing the distribution of chemical defenses. Proceedings of the National Academy of Sciences 2022, 119.
[112] Traditional Chinese Medicine Systems Pharmacology Database and Analysis Platform, 2022 (TCMSP; https://tcmspw.com/tcmsp.php, accessed on 2 January 2022.

U. M. Aruna Kumara, N. Thiruchchelvan

Chapter 14 Biotechnological approaches for plant stress management

Abstract: Plant protection technologies are aiming to minimize damage caused by main plant pests such as insect pests, plant pathogens, and weeds. Apart from these biotic agents, various other abiotic factors namely, extreme soil pH, drought, salinity, and global climate change have negative impacts on plant growth and development. As a result of that, crop plants would be unable to reach their genetic potential. Therefore, expected yield and postharvest quality of agricultural commodities will be reduced. Damage is continued from preharvest to postharvest. Estimated crop loss in global agricultural production due to pest and disease infestation from field to fork is around 30–40%. Sustainable development goals (SDGs) work toward better future. That try to meet the needs of present generation without harming to fulfill the needs of future generation. A-line with SDGs, plant biotechnology plays a major role as a sustainable tool to fulfill the future food demand without disturbing the natural ecosystem. Biotechnological tools have greatly contributed to many areas in crop production systems. These tools are employed to speed up the multiplication process of vegetatively propagated crops through plant tissue culture. Development of novel technological applications will benefit the farming community through utilization of disease free planting materials, environmental friendly plant protection technologies, and good postharvest qualities in final harvest with higher economic potentials. The use of biotechnological approaches, such as biological and microbial biocontrol, protoplast fusion, marker-assisted selection, RNA interference, cisgenesis/intragenesis, and genome editing tools, like CRISPR/Cas9, has given tremendous support in plant stress management.

14.1 Introduction

Plant protection chemicals have detrimental effects on consumer and environment. Therefore, there is a requirement to develop new approaches or new technologies to exploit genetic diversity in host plant resistance for biotic and abiotic stress. Elucidation of durable host plant disease resistance is perceived as the goal in disease management [1]. Usually, host plant resistance is accompanied with crop wild relatives. These resistance characteristics can be transferred to commercial cultivar having susceptibility to such resistance factor with high yielding properties through conventional breeding approaches such as marker-assisted selection, hybridization,

https://doi.org/10.1515/9783110771558-014

and by biotechnology-based approaches of gene editing and gene silencing. Development of host plant resistance through genetic transformation would be able to minimize the need for use of conventional plant protection chemistries [1]. Development of these technologies takes considerably long period of time. Especially, host phenotyping is a time-consuming and labor-intensive process. In order to overcome these existing problems, new technologies namely optical sensors, artificial intelligence, and machine learning have been proposed by various research groups [2].

As described by Mahlein et al. [2], "digital phenotyping" would be beneficial to develop more accurate biotechnological approaches in plant stress management. This durable field application tool has ability to recognize changes of host plant molecular and physiological status under pathogenic invasion. In addition to that, gradual development of resistance in host population can be detected from this novel application. Progressive development of the biotechnological approaches for induced host resistance has many constrains such as cost of virulence [3]. New gene editing technologies have opened up the opportunities for development of new plant genotypes with disease-resistance abilities under epidemic conditions amalgamate with biotechnological approaches [4]. Most of the plant breeding research are trying to elucidate the resistance gene and their capacity in disease management in resistant or tolerant cultivars. Newton et al. [5] questioning the potential of quick development of plant resistance against biotic and abiotic stress under the catastrophic emergence of pest and diseases. Development of host resistance against biotic and abiotic factors is really a challenge in molecular plant pathology for tree crops or perennials with slow growth and a long generation time [6]. Therefore, conventional approaches will take several decades to develop durability or resistance in perennials. In order to overcome these challenges, development of biotechnological approaches is timely important [1].

Exotic and emergence or reemergence of new plant pathogens and spread to new locations are the great challenges faced by plant protection agencies under global climate change [7]. Exchange of plant materials as planting materials or breeding lines has been accelerated in the last few decades and leading to development of exotic plant diseases worldwide. That is negatively impacted on the sustainability of agricultural and natural production systems [8]. Therefore, development of non-native pathogens and their prolong existence in some of the areas have been observed as a threat to plant health management. Early warning or development of broad host resistance against phytopathogens and pest is prerequisite in plant stress management [1]. The sudden emergence or invasion of Fall Army-worm (*Spodoptera frugiperda*) in Sri Lanka during 2018/2019 is a remarkable example for above situation.

Biochemical profiling of resistant phenotypes against plant pathogens and environmental regulators is another important area in plant stress management. Basically crops diseases management strategies mainly depend on the deployment of host plant resistance and fungicides as mentioned earlier [9]. In cooperation of biotechnological approaches in order to develop resistance, crop varieties are the most economical way of managing brown spot disease [10]. Less abundance of those resistance line is the major bottleneck in practical application of conventional and molecular plant breeding [11]. Also resistance may be unstable under the development of exotic or immerging a new virulence plant pathogens [12]. Therefore, it is necessary to use gene experiment analysis with the help of plant biotechnology such as construction of cDNA libraries and differential display [13].

Complex network of molecular and biochemical reactions is involved in the process of determination of host resistance. Host resistance is mediated by a complex network of molecular and biochemical events that determine a range between the susceptibility and resistance. Therefore, understanding the molecular mechanism of host pathogen interaction is a vital role in the identification of effective and durabale resistance genotypes [14]. The most commonly identified first line of defense compounds through cDNA library screening is peroxidase (POD), mitogen-activated protein kinases (MAPKs), and phenylalanine ammonia lyase [15]. Pathogenesis-related (PR) proteins, like β-1,3-glucanase, catalase, superoxide dismutases, and polyphenol oxidase, contribute immensely for the systemic acquired resistance by accumulating profusely at the site of infection in plants under biotic and abiotic stresses [10]. Hence, this chapter is devoted to elaborate interaction among plant, pest, and their environment to study the stress generated by these biotic and abiotic factors. It ultimately discusses the potential applications of plant biotechnology in order to develop stress tolerant in crop plants.

14.2 Plants as photoautotrophs

Plants or flora are living biological organisms. They do growth and development, reproduction, and response to their environmental stimuli. Plants lose their locomotion otherwise, they totally look similar to the animals in ecosystem. Furthermore, plants are multicellular eukaryotic life form characterized by photosynthetic nutrition [16]. In which chemical energy need for their survival is generated from water, mineral, and carbon dioxide with the help of chlorophyll pigments and radiant energy of the sun. Plants are primary produces or photoautotrophs, and they produce their own energy for their cellular activities. As a result, heterotrophs are highly dependent upon this photoautotrophs. The synthesized carbohydrates through photosynthesis will be stored in various parts of the plants such as root system, shoot system, leaf system, and reproductive organs. These are commonly called as vegetative parts or vegetables. Almost all of the herbivorous animals consume plants to fulfill their energy requirement. Being an omnivores or vegetarians, human and animals consume shoot, leaf, root, tuber, bulbs, seed, and various other plant-derived substances as their primary food. Carbohydrate is the main energy source supplemented with plant materials which provide the energy for cellular functions to maintain biosystem alive.

Therefore, plants are highly important to ensure the life and its consistency of living organisms in our planet [16].

14.3 Early civilization and dawn of agriculture

According to the archeological evidences proposed by Paleoanthropologists, the earliest fossil evidence of *Homo sapiens*, the ancestor of modern humans was found roughly 196,000 years ago [17]. Until the dawn of early civilization, our ancestors were used to acquire their foods by hunting and gathering [18]. They hunt wild animals for carcasses and gather wild plant parts and most of the edible fungi as their staple food. Some of these wild crop relatives are widely cultivated today as agricultural crops [19]. With evolutionary adaptation of early ancestors, several behavioral changes were taken place. Among these, the most significant change was gradual translation of hunter- gatherer lifestyle toward cultivating crops and raising animals for their food requirements. This translation was taken place as early as 11000 BCE. This historical change was independent process which was taken place in several parts of the world, which include North China, Central America, and the Fertile Crescent [18].

Changes in climatic conditions, greater population density, and farming provided more food per acre are some of the main reasons for this evolutionary changes of hunter-gatherer lifestyle toward cultivating crops [20]. In addition to that, technological changes, such as domesticated seeds, would have made agriculture a more viable lifestyle [21]. Agriculture is the process of cultivation of crops to generate food and rearing animals to produce meat, milk, far, wool, and egg through farming. This practice is the major way of acquiring world food needs. It is believed to be practiced sporadically for the past 13,000 years [18]. Our ancestors spent more time nearly 200,000 years in the wild with hunting and gathering lifestyle. Compared to that, agricultural transformation has a short history. But during that time, several radical transformations were taken place in human societies including raising up of the global population from 4 million to 7 billion since 10000 BCE and is still growing [22]. Due to this continuous population growth, production of agricultural commodities had to multiply several folds. As a result, subsistence agriculture turned into commercial agriculture. During this subsistence, agriculture, marginal, or small-scale lands were used for cultivation of crops and raising farm animals. Subsistence agriculture used most conventional and indigenous technologies to manage the crop and livestock. Crop livestock integration, mixed cropping, and forest farming are more sustainable farming systems. Most of the natural enemies and ecofriendly systems limit the pest and disease development. However, with this commercial agricultural revolution, started to cultivate hectares of lands as mono cropping. Where one type of crop covers the entire landscape of several hundreds of acres. As a result, sudden emergence of insect pests and diseases and weeds was taken place. Where not reported earlier became a serious issue in commercial agriculture. To control these insect pests, diseases, and weeds, various plant protection technologies emerged. Among these, chemicals or pesticides played a major role.

The evolution of agricultural development up to date has not been smooth. Limitation of the resource and resources degradation, rapid population growth, emergence of pest and disease, global climatic changes, and several other forces have periodically crippled food supplies [23]. A-line with this statement, International Plant Protection Convention (IPPC) has stipulated emergence of new pests and diseases, global climate change, phytosanitary capacity of inspection and pest reporting systems, phytosanitary capacity of pest surveillance, and overuse of chemicals for pest risk management as five most common emerging issues across globe in their report on global emerging issues [24]. Still today, we faced the same or greater challenges as faced by our ancestors. The global emerging issues identified by IPPC regional workshop in 2016 highlighted that cankers, blights, fungal infections, rots of roots, stems, fruit and pods, phytoplasmas, viruses, and various soil diseases are more destructive and widespread in the agricultural settings. In addition to that, emerging risks include insects, arachnids, nematodes, and mollusks. Insects are specified as the most common plant pests of concern including fruit flies, beetles and weevils, several moth species (larvae and adults), and others [24]. As a result of that, protecting our crops from those enemies or plant protection technologies emerged as a most important segment in agricultural operations.

14.4 Plant protection

Qu Dongyu, FAO Director-General, has pointed out that in his message year 2020 was declared as the International Year of Plant Health (IYPH) in December 2018 [25]. "Protecting plants, protecting life" was the IYPH slogan. That symbolizing the concept of plants are life and contributes immensely to the health and well-being of all living beings. Eighty percentage of the food we eat come from the plant-derived substances. As well as 98% of oxygen we breathe are also produced by plants. Keeping them healthy is key to securing several of the SDGs according to IPPC.

Furthermore, plant health is increasingly at risk. The annual loss incurred by plant pests in food crops is up to 40% globally [25]. This annual crop loss has direct and indirect effect on the millions of smallholder farmers scattered all over the world. These rural communities rely on agriculture as a primary source of income. The global climate change and unsustainable human activities are altering the ecosystems and reducing the biodiversity. As a result of that, new niches are invasive pests to thrive well. Simultaneously, pests and diseases rapidly spread around the world due to international travel and trade as an indirect route. This causes great damage to native plants and the environment [25, 26]. The official year of the plant health has ended but our efforts for the protection of plant health are essential. Therefore, IPPC has been working on setting up of the international standards and phytosanitary regulations for plant health management worldwide [25, 26].

Generally, diseases, insects, and weeds are the major pests of agricultural crops. Therefore, plant protection is the process of reducing the detrimental effect of the pests. The Plant Protection Act No. 35 of 1999 of Sri Lanka use as a legal measure to insure introduction of foreign/exotic or invasive pest, disease, or weed to Sri Lanka from other countries [27]. These legal document further strengthens the plant protection through quarantine approaches. Cultural practices like ploughing, hoeing and basin preparation, hand picking of pests and their destruction, practice of applying cow dung as a thin layer, cow urine as a disinfectant, pruning, and training of fruit trees are the most important indigenous technologies used by the Asians for plant protection [28]. With the emergence of commercial agriculture, physical, chemical, and biological methods are involved in the plant protection approaches. The large-scale application of pesticides, primarily insecticides, plays a major role in the plant protection at the moment. As mentioned in the indigenous technology knowledge for watershed management in Indian upper north-west Himalayas, cultivation of several hilly crops is difficult due to the harm caused by natural bioenemies and pathogens [28]. As a result of that, development of resistance to pesticide and secondary pest outbreak was experienced [28]. This statement highlighted that importance of development of alternative plant protection technologies which favor the eco-friendly approaches in plant protection. The plant protection strategies should not be destructive to natural enemies but gradually remove sizeable proportions of pest populations and tend to keep their populations density below the economic threshold [28]. Physical or devices and procedures used to change physical environment of pest populations and mechanical or mitigating pest populations by cultural practices are the oldest of all such insect control methods.

Other than all these methods in plant protection, regional plant protection organizations have identified the importance of collaboration at the national, regional, and global levels. Specifically, there is a need to reinforce legislative and regulatory frameworks of contracting parties to promote collaboration between national stakeholders. Regional collaboration is considered necessary for better cooperation with neighboring countries [24]. Furthermore, importance of the policy aspects of national phytosanitary strategies and legislation, responses to national phytosanitary emergencies, resilience to the national political situations, and risk management of sea containers has been identified as key areas in plant protection by the international plant protection convention [24]. However, resource limitations are a common problem in plant protection activities in developing nations mainly in Asia and African regions. This condition adversely affects the sustainability of phytosanitary systems. These limitations include access to financial resources, training and development of skill full staff and retention, and infrastructure facilities to undertake phytosanitary activities. The National Plant Quarantine Station of

Sri Lanka is the national organization which has basic facilities to manage phytosanitary activities in Sri Lanka. Further capacity building through human resource development equipped with modern facilities is very important to strengthen the local plant protection activities.

Occurrence of pest in agricultural fields is quite obvious but crop monitoring or forecasting systems usually keep them in check. Total destruction or eradication of an insect pest will break the natural balance due to lack of food for their natural enemies. Therefore, plant protection technologies should be aimed to maintain the pest population below the economic threshold. That will help to maintain the natural balance in an ecosystem [29].

14.5 Pest and diseases

In general, plant pests are identified as an external and introduced factor to the crop production systems. We can't believe this statement at all the time because most of the time pest species emerge naturally within the agroecosystem. The associated species of pests such as predators, parasitoids, pathogens, pollinators, competitors, and decomposers are components of crop-associated agrobiodiversity. These organisms also have a wide range of responsibilities in ecosystem. There is a natural balance in an agroecosystem, due to various reasons, such as pest outbreaks or upsurges, this balance could be breakable. if so it would be yielded a serious losses [29].

For the betterment of the managerial aspect of pest, pests are classified as pest, quarantine pest, and regulated pest in the Plant Pest Surveillance report of FAO. According to that, pests are "any species, strain or biotype of plant, animal or pathogenic agent injurious to plants or plant products" [24]. Quarantine pest, "a pest of potential economic importance to the area endangered thereby and not yet present there, or present but not widely distributed and being officially controlled" [30]. And regulated pest means a quarantine pest or a regulated nonquarantine pest [30].

Plant pests and pathogens interfere with the growth, development, and reproduction of cultivated and naturally growing plants. As a result, crop plants would not be able to reach their genetic potential. Therefore, expected yield and postharvest quality of agricultural commodities will be reduced. Damage is continued from preharvest to postharvest [31]. Estimated crop loss in global agricultural production due to pest and disease infestation from field to fork is around 30–40%. That means, nearly half of the production will not be reached to the final consumer. Effort, energy, and money used for the production operations became a real waste. This preharvest and postharvest crop losses due to pest and diseases are greater in the area where there is more hunger. Inappropriate applications of these pesticides may have damaging environmental and human health impacts [31].

Plant pathology is the science that studies the cause of plant diseases and mechanisms by which diseases develop in individual plants or in a population. In addition to that, seek the possible ways and means how diseases can be controlled. Plants are living biological organisms, and they have interaction with microorganisms for their growth and development. Microbiota become pathobiota under the immune deficiency or adverse environmental conditions [31]. Plants have internal defense mechanism like in other animals that protect them against diseases. However, virulent pathogens compete with this defense mechanism and break it.

Weeds are troublesome in various ways. Weeds make disturbances for crop growth, development, and reproduction. Weeds are competitors for plant nutrients, space, and solar energy. In addition to that, weeds are alternative host for insects. Despite the myriad interactions of weeds and insects, many aspects of the relationship are predictable. Host specificity is the remarkable characteristic shown by certain group of insects. That means, only damage to single plant family. Some of the weeds can serve as a source of diseases including both diseases of insects and of crop plants. Rarely, weed contains entomopathogenic diseases but more frequently contains phytopathogenic diseases. Therefore, insect pest acts as a vector for some of the main plant diseases [32].

Stress is an external condition that adversely affects growth, development, or productivity of plants. Altered gene expression, changes in growth rates, cellular metabolism, and crop yields are few of the plant responses to the external stimuli. Those conditions are triggered by the stress that are generated by pest of crop plants. Plant stresses are of two types namely, abiotic stress and biotic stress. Abiotic stress is mainly due to environmental factors and their changes due to various interactions. And biotic stresses are generated by microorganisms and insects and non-insect pests [28]. Bacteria, fungi, nematodes, viruses, insects, arachnids, and weeds are the living organisms responsible for generating the biotic stress in crop plants. The agents causing biotic stress directly deprive their host of its nutrients and can lead to death of plants. Some of the fungal parasites secrete toxic substances to kill their host plant usually called as neurotropic. And feeding on living cells of host plant is another mechanism of fungus pathogenicity. Vascular wilts, leaf spots, and cankers in plants can be induced due to fungal invasion [33]. Nutrient deficiency, stunted growth, and wilting are some of the soil-borne diseases generated by nematodes [34]. Similarly, viruses are also capable of local and systemic damage resulting in chlorosis and stunting [35]. Mites and insects impair plants by either feeding by piercing and sucking on them or laying eggs. In contrast, insects might also act as carriers of other viruses and bacteria [36]. Biotic stress can occur during pre-harvest or postharvest life of a crop plant. Therefore, it is a major condition in plant life. Lack of adaptive immune system in plants can counteract biotic stresses by evolving themselves to certain sophisticated strategies. The defense mechanisms which act against these stresses are controlled genetically by plant's genetic code stored in them. The resistant genes against these biotic stresses present in plant genome are encoded in hundreds [37].

Physical barriers such as cuticles, wax, and trichomes act as a passive first line of defense in plant, which includes to avert pathogens and insects. Releasing chemical compounds to defend themselves from infecting pathogens is another strategy used by plants [38]. The first level of pathogen recognition encompasses pattern recognition receptors, which identify pathogen-associated molecular patterns (PAMPs). Such plant immunity is categorized as PAMP-triggered immunity [39]. Plants sense biological stress conditions, activate regulatory or transcriptional mechanisms, and ultimately produce appropriate responses. Although plant defenses against pathogen attacks are well understood, the interactions and effects of the various signals that generate a protective response to biological stress remain elusive [40]. Therefore, development of biotechnological approaches to mitigate the stress generated by biotic stress is a far better solution to improve the crop health.

14.6 Plant protection technologies

Plant protection technologies are aiming to minimize damage caused by main plant pest such as insect pest, plant pathogen, and weed. Other than these biotic agents, various other biotypes and environmental factors namely extreme soil pH, drought, salinity, and global climate change have negative impact on plant growth and development as abiotic stress factors. Physical, chemical, and biological methods have been implemented as a control measure of the stress generated by plant pests. As a control measure in plant pest, overreliance on pesticides is not suitable. It has negative impact on the natural ecosystem balance. Parasitoid and predator are natural enemies of plant pest but application of synthetic pesticide disrupts the life cycle of these potential bio-control agents. Therefore, development of secondary pests' outbreak could be resulted [41]. In addition to that, excessive exposure of farmers to various pesticides has crated serious health risks and has negative consequences to the environment. Often, less than 1% of pesticides applied actually reaches a target pest organism; the rest contaminates the air, soil, and water [42]. Application of these methods alone has less efficiency or may have side effect to the environment or final consumer. Therefore, novel plant protection technologies are aiming integrated approach to all technological processes, especially in agriculture [43]. The soil, weather conditions, and the population of insects living in the field under consideration are the main factors that determine the effectiveness of plant protection technologies. Therefore, plant protection is a process of application of chemical, biological, and agrotechnical measures to minimize the pest population in agro or natural ecosystem [44].

Effective use of crop protection techniques is one of the factors contributing to the gap between theoretical and actual yields. The pressure on crop protection products is now increasing due to the significant increase in resistance to pesticides, fungicides, pesticides, and herbicides. Therefore, new strategies or legislation mechanism need to be implemented to avoid the hazard rather than risk-based approach. However, prolonged use and successive pest control ability in chemical are difficult to be withdrawn from use. Therefore, new regulatory framework should be prepared in order to protect crop plants with the participation of private and public sector of the agricultural community [44].

The Food Research Partnership subgroup and the Biosciences Knowledge Transfer Network on a new and innovative approach to crop protection were held at the BIS Conference Center in London in January 2013 for 40 people from research, industry, organization, and the end. We held a workshop by experts: Users, census authorities, and government agencies [45]. The expert decision came out from this workshop was to use application for "Omics" technologies in crop protection.

The application of "omics"-based technologies to understand mechanisms of plant pathogen interactions and genetic variation among crop genotypes and populations of target organisms such as weeds, invertebrates, and pathogens has vital use in the development of innovative approaches in plant protection. Rather than conventional methods, "omics" technologies have linked to systems approaches. Therefore, application of this strategy now being increasingly used across the life science area to identify new potential "drugable" targets [45].

Compared to traditional methods of crop protection, "omics"-based techniques offer the following potential strategies: It uses a molecular approach to optimize the integrated use of pesticides and adapt to the mode of action of fungicides in host resistance of plant varieties with different genetic backgrounds. Identify new plant protection targets for interventions in pathogens, pests, and weeds that can form the basis for screening chemical and biological factors and how beneficial end fights and resistance inducers protect crops. Strategies, including less-studied interactions such as pest and disease resistance, allelopathy, to understand and thereby enhance their effectiveness and to understand and utilize "natural" crop protection. Understanding and methods of many mechanisms of resistance that reduce the ability to utilize existing toxophores, countering them, developing synergistic mixtures and formulations of pesticides, and direct bioactive natural products by bioprospecting identification [45]. According to these novel approaches in plant biotic and abiotic stress management, biotechnology has contributed immensely. Biotechnology is a branch of biology, which involves in development or makes product by using living systems or organisms.

14.7 Biotechnological approaches in plant stress management

SDGs work toward better future. That try to meet the needs of present generation without harming to fulfill the need of future generation. A-line with SDGs, plant biotechnology plays a major role as a sustainable tool to fulfill the future food demand without disturbing the natural ecosystem. Biotechnological tools have greatly contributed to many areas in crop production systems such as production and supply of improved quality seed and planting materials. Development of novel technological applications will benefit the farming community through utilization of disease-free planting materials, environmental friendly plant protection technologies, and good postharvest qualities in final harvest with higher economic potentials [46]. In addition to that, the following highlights the use of biotechnological approaches in plant stress management.

14.7.1 Biological control for plant stress management

Natural enemies: Organism/s such as parasites/parasitoids, pathogens, and predators can be used to manage the crop pests. These natural enemies provided the control of the pest called "Biocontrol or Biological control." Therefore, the use of natural enemies for biocontrol is all over the place, where the crops are cultivated, and it is extended further even in the forest and wildland weeds. Parasites, pathogens, and predators are the primary groups used in biocontrol of insects and mites. Not only these insect pests and mites but also all the plant pathogens, nematodes, and vertebrates have a lot of natural opponents, though it is difficult to be identified, poorly studied, or more challenging to succeed. Biological control could be possible in three major ways *viz.*, conservation, augmentation, and classical biological control. Generally, parasitoids and pathogens and predators are host-specific, and their host range almost a certain number of precise pest species.

14.7.1.1 Parasitoids

Parasitoids are animals or organisms that are completing their life cycle in or on another host animal or organism [47, 48]. Parasitoids are always insects and their larvae feed on other insects. There are a small number of parasitoids described against other arthropods and mollusks. Egg laying site of the parasitoids varies depending on the species, and it generally lays the eggs on the outer surface or within the insect body or rarely adjacent to the host body. Parasitoids larvae hatched from the laid eggs and started to feed the insect's body contents and subsequently kill the host completely [48, 49]. Sometimes parasitiods are called "parasite." However, true parasites (e.g., fleas and ticks) are different from the parasitiods since parasites do not normally kill their hosts but parasitiods ultimately kill their host. Therefore, parasitoids are useful for the biological control [50].

All the parasitoids belong to predominantly two orders, namely, Hymenoptera and Diptera, and few of them from the order Coleoptera. Insects such as sawflies, ants, bees, and wasps belong to te order Hymenoptera. The insects that belong to the suborder Symphyta are phytophagous, for example sawflies, even though family Orussidae (suborder Symphyta) insects larvae are parasitoids. The rest of Hymenoptera (the Aprocrita) are Parasitica or parasitoid wasps, which are entirely parasitoids. Among them, Ichneumonidae, Braconidae, and Chalcidoidea are the most important families. The subsequent most important group of parasitoids belongs to the order Diptera. Insects from the family Tachinidae are completely parasitoids. The most frequently encountered coleopteran beetle parasitoids are probably those in the rove beetle from the family Staphylinidae [48, 49].

Egg parasitoids are parasitoids that complete their development within the host eggs [47]; larval parasitoids complete their life cycle within larval hosts. Egg or larval parasitoids have been used for the biological control of pest. Globally, there are a number of parasitiods introduced for the pest control programmes and their information are included in the database of BIOCAT. For example, 91 egg parasitiods have been released to control more than 77 insect pests' species [51]. *Trichogramma* species have been used most frequently against almost all agricultural crops, such as cash crops, cereals, vegetables, and fruit crops [52–54].

14.7.1.2 Predators

Predators are the organisms that capture their prey and consume as their feedstuff; as a result of this phenomenon, they are killing many individuals during their lifespan. Predators of insect pests are several animal groups such as amphibians, birds, mammals, and reptiles. Coleopteran beetles, dipteran flies, lacewings, true bugs (order Hemiptera), and wasps are the insect predators that predate insects and mites. Almost all the spiders feed on insects. *Amblyseius* spp., *Neoseiulus* spp., and *Galendromus occidentalis* (western predatory mite) are examples of the predatory mites that are used against spider mites [50, 55].

14.7.1.3 Pathogens

Pathogens that are used for the biocontrol programs are commonly microbes including certain bacteria, fungi, nematodes, protozoa, and viruses. These microbes have the ability to infect and invade within the host body; ultimately, host will die due to their pathogenicity [56]. This type of pathogens is commonly used to control the insects and mites pest; therefore, these pathogens are normally known as "Entomopathogens" or "Acaropathogens" [57]. At the movement, there are plenty of beneficial pathogens that are commercially available as biological or microbial pesticides. These include entomopathogenic bacteria, entomopathogenic fungi, entomopathogenic nematodes, and entomopathogenic virus (granulosis viruses) [56–59].

Virus: There are many entomopathogenic virus (granulosis virus) identified and subsequently commercialized for the pest control programs all over the world to manage pests. For example, in the USA, AucaMNPV is against the insect host *Autographa californica* [60, 61]. AnfaMNPV (Syn. RaouMNPV) controlled the insect hosts such as *Anagrapha falcifera* and *Rachoplusia ou* [62, 63]. A biocontrol agent NeabSNPV is used in Canada against the insect pest *Neodiprion abietis* [64, 65], AdorGV explored in Japan and Europe for the management of the pest such as *Adoxophyes orana* [66].

Bacteria: Entomopathogenic bacteria are being used for the crop pest as well as household pest managements. Among them, some major and famous examples are indicated here such as *Bacillus thuringiensis* originally collected from the insect Mediterranean flour moth, *Ephestia kuhniella* in Germany [67], and it is subsequently commercialized to control many other pest. *B. t.* subsp. *kurstakia* originally isolated from the insect hosts *E. kuhniella*, *Pectinophora gossypiella* in Canada and the USA [57, 68]. *B. t.* subsp. *israelensisa* isolated and tested against a household pest *Culex quinquefasciatus* in Israel [69]. New Zealand researchers [70–72] reported *Serratia entomophila* bacterium against the pasture grub, *Entomophila zealandica*. An entomopathogenic bacterium *Chromobacterium subtsugae* isolated from the soils of the USA [73].

Fungi: Entomopathogenic fungi and acaropathogenic fungi (mycopathogens) are used to control many crops pest; mainly for the insect and mites [59]. At present, there are several commercialized bio pesticides available for the biological control of pest all over the world. Mycopathogens are plenty. Here are the some classical examples where the entomopathogenic fungi originally isolated and subsequently established for the commercialization [57]. An entomopathogenic fungus *Isaria fumosorosea* (Apopka strain 97) was isolated and tested against *Bemisia tabaci* biotype B in Florida (sweet potato whitefly), USA, by Lacey et al. [75]. And another most popular mycopathogen *Beauveria bassianaa* GHA was isolated from the infected cadaver of spotted cucumber beetle, *Diabrotica undecimpunctata* from the Oregon region of USA; subsequently, it is commercialized for the world population usage [76]. Another species of *Beauveria brongniartii* was isolated from the *Melolontha melolontha* by the Switzerland researchers [77, 78]. *Metarhizium anisopliaea* from the host *C. pomonella* [78], *Metarhizium acridum* (green muscle) isolated from insect host of *Ornithacris cavroisi* [79–81], *Lecanicillium muscariuma* from Whiteflies and thrips, and *Lecanicillium longisporuma* isolated from Aphids were reported in the UK and Sri Lanka [74, 82,

83]. *Neozygites tanajoae* is an acaropathogen isolated form Cassava green mite and *Mononychellus tanajoa* by the Brazil researchers [84, 85].

Entomopathogenic nematodes: Nematode genera, *Steinernema* (Steinernematidae) and *Heterorhabditidas* (Heterorhabditidae), are used with years of time for the biological control of insect pests, and they are typically called "entomopathogenic nematodes." Entomopathogenic nematodes enter into the insects' hemolymph through the natural opening and vomit their symbiotic bacteria into the insect body (haemolymph). Released symbiotic bacteria will multiply within the insect body and produce toxins. These toxins are lethal to the host insect and ultimately kill the insect host within a short period of time [86]. Recently, few researcher reported that there are a few other nematode families containing free-living nematodes [excluding these two typical entomopathogenic nematodes, *Steinernema* (Steinernematidae) and *Heterorhabditidas* (Heterorhabditidae)] that are shown lethal to the insect pests. For example *Acrobeloides* spp. is associated with insects [87], *Acrobeloides longiuterus* against the insects red flour beetle, *Tribollium castaneum* and sweet potato white grub larva *Phyllophaga ephilida* in Sri Lanka [88, 89], *Acrobeloides* (K29), against *Zeuzera pyrina* in Iran [90], *Oscheius* [91–93], and *Caenorhabditis* [94]. However, nematodes belonging to families Steinernematidae and Heterorhabditidae are considered potential insect biocontrol agents, and they are commercialized universally. The following are few examples of them: *Steinernema carpocapsae* (host insect *C. pomonella*) [95], *S. glaseri* (host Japanese beetle, *Popillia japonicum*) [96, 97], *S. riobravea* (host *Helicoverpa zea)* [98], *Heterorhabditis megadis* (host *P. japonicum*) [99], *H. bacteriophoraa* (host *Heliothis punctigera*) [100], and *H. marelatus* (host *Galleria mellonella*) [101].

14.7.1.4 Microbial antagonists

Application of natural antimicrobial substances or microorganism with such potential combined with novel technologies provides new inside into development of control measures for pathogenic microorganisms. A number of microbial antagonists have been identified by various research approaches such as bacteria, filamentous fungi, and yeasts [102]. These organisms have simple nutritional requirement, and they are not harmful for human or host organism [103]. These microbial antagonists use various mechanism to combat the pathogen. Among those strategies, the most common mechanism is competition for space and nutrients [104]. In addition to that, direct parasitism, production of siderophores, antibiotics, and volatile compounds are the other mechanisms used by microbial antagonist to suppress the growth, development, and reproduction of phytopathogens (Figure 14.1) [102].

As described by Díaz et al. [105], natural killer yeast is a potential source to manage postharvest diseases in fruits. These unicellular eukaryotes process various characteristic features to be qualified as microbial antagonist. Rapid utilization of

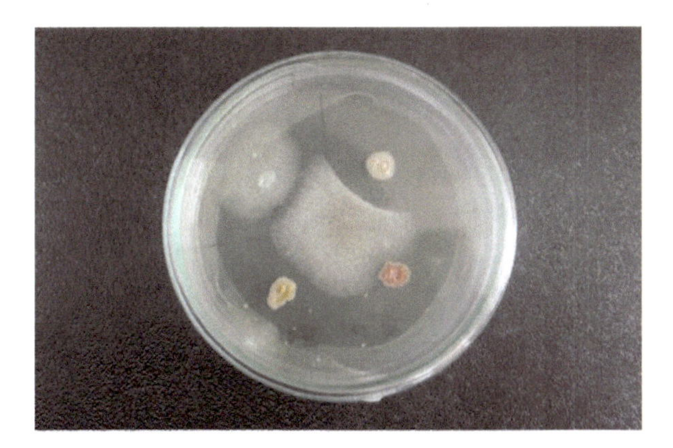

Figure 14.1: Some of the yeast isolates showing their antagonistic/killer activity against postharvest fungal pathogen in dual culture assay.

nutrients and lavish proliferation on fruit surfaces are some of the most common strategies. In addition to that, they can tolerate to the commonly applied pesticides, such as imazalil, tiabendazole, pyrimethanil, or fludioxonil, due to their extracellular polysaccharides [106]. Klassen et al. [107] have elaborated the mode of action of killer toxins in his paper. According to that, killer toxins can be divided into four categories based on the cell target. β-1,6-Glucan use as the primary cell wall receptor by the killer toxin type T1 and T2. T1-type toxins such as K1, K2, and PMKT bind to a membrane receptor and induce membrane pore formation. T3 and T4 toxin types have intracellular targets [108].

Perez et al. [109] described the use of native killer yeasts as biocontrol agents of postharvest fungal diseases in lemons. One of the main reasons for economic losses in citrus industry worldwide is the postharvest diseases. The "green mold" and "blue mold," caused by *Penicillium digitatum* and *P. italicum*, respectively, are the major diseases affecting citrus industry. Two strains of *Pichia* and one strain of *Wickerhamomyces* isolated from citrus peel depicted a significant protection ($p < 0.05$) from decay caused by *P. digitatum* in "in vivo" assessment. Therefore, microbial antagonistic ability showed by native killer yeasts through scientific researches will be a better alternative for the biocontrol of postharvest fungal diseases of lemons [109].

Papaya fruit anthracnose is a detrimental postharvest disease which is caused by *Colletotrichum gloeosporioides*. As estimated, in Malaysia, losses can be up to 62% and disease incidence ranging from 90% to 98% [110]. Hassan et al. [111] try to evaluate the microbial antagonistic ability of yeasts against *C. gloeosporioides* in papaya and elucidate the possible mechanism involved in plant abiotic stress tolerant through his resent research. The results revealed that isolated strain F001 had the strongest biocontrol activity and reduce disease incident and disease severity up to 66.7% and 25%, respectively, under in vivo condition. The strain F001 was confirmed as *Trichosporon asahii* through molecular biological identification [111].

Alternaria sp. is a fungus that causes rot in several crops with agricultural importance. Therefore, Bosqueiro et al. [112] have emphasized the use of yeast *Trichosporon asahii* (3S44) isolate as a microbial antagonist to control *Alternaria* rot of crops such as cereals, ornamental plants, broccoli, cauliflowers, potatoes, carrots, tomatoes, citruses, and apples as a biological control agents [113]. The social demand for modern agriculture has been increased with the application of biological control agents as an alternative to the chemical pesticides [112].

14.7.2 Protoplast fusion

The integration of genetic material from one organism into the genome of another organism can be successfully carried out by protoplasts fusion. Chemical removal of cell wall and either polyethylene glycol (PEG) treatment or electroporation allow transfer genetic material among genetically related organisms [114]. Particle bombardment is a potential alternative for transient delivery method for DNA-free gene-editing tools [115]. However, it may suffer from limitations in transformation efficiency and the regeneration of chimeric plants. However, protoplast fusion provide the nontransgenic approach and maintain higher precision and efficiency in gene delivery [116]. Chimerism or where only parts of the regenerated plant are descended from an edited cell would be a drawback in conventional, tissue-culture-based approaches with callus intermediate [117]. Chimerism can be avoided through protoplasts fusion as its potential in regenerating from a single cell. When nonselectable, nontransgenic approaches are used together with conventional tissue culture, chimerism can be a concern. Low editing efficiency is a limitation in such nonselectable strategies in the regenerated plants. In contrast, protoplast transformation efficiencies are much higher. The plants regenerated from protoplasts fusion can have successful edition of genome when transiently transformed with genome editing tools [114] (Figure 14.2).

The protoplast division and plant regeneration are basically depended upon the success of the protoplast culture media. The appropriate macronutrients, micronutrients, and additives, such as plant growth regulators, osmotic stabilizers, medium solidifiers, and supplements, are essential in protoplast culture. The division and microcallus formation potential of protoplasts highly depends on the protoplast culture conditions, such as the use of liquid or semisolid medium, temperature and light, cell density, or the presence of nurse cultures. The organogenesis or embryogenesis is the main route to regeneration of microcalli. PEG-mediated transformation is much popular than the electroporation as a method for protoplast transformation. The pulse voltage, pulse length, pulse number, cell number, DNA concentration, and electroporation buffer composition are factors to be considered during electroporation [118]. Lee et al. [118] found that when electroporation transformation was optimized for cabbage protoplasts, the transformation efficiency was

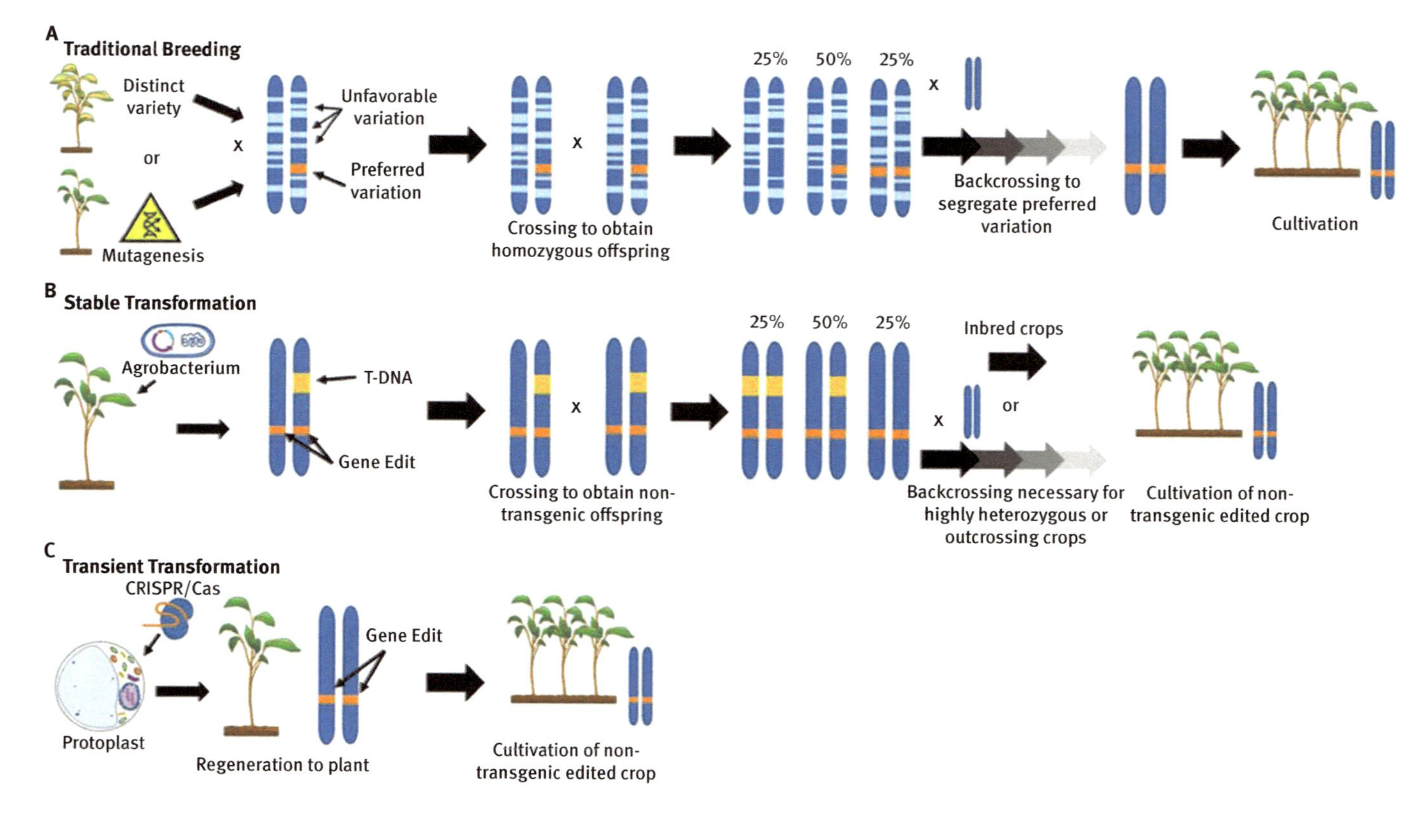

Figure 14.2: Avoiding the integration of foreign DNA into the host genome by gene editing through transient transformation and regeneration of protoplasts over conventional breeding or transgenic CRISPR/Cas9 approaches [114].

nearly double that of PEG-mediated delivery. Anyway, both transformation rates were low (3.4% and 1.8%, respectively) [119]. Transcriptomic analysis would be beneficial to gain fundamental knowledge of the transcriptional regulation of the regeneration process. Shulse et al. [120] demonstrated that single-cell transcriptome profiling, but it is quite difficult for the regeneration capability and the larger, doomed protoplast population.

14.7.3 Hybridization and marker-assisted selection

The sustainability of crop loss reduction and maintenance of food and nutritional security is mainly affected by biotic and abiotic stress generated on crop plants. Variable magnitudes can be seen in large populations in order to cope with new environmental changes. Plants with capability to adhere to these environmental stimuli mainly change their phenotype, anatomy, physiology, and genetic orientation. Therefore, changes against environmental conditions and yield potential, nutrient, and water-use efficiency against abiotic stress need to be identified, selected, and utilized in breeding programs so as to expedite the biotechnological enhancement of crop productivity. In contrast, molecular breeding techniques, particularly marker-assisted selection (MAS), deliver a huge opportunity to speed up the process of selecting plant populations having abiotic stress-tolerance traits with enhanced accuracy via the identification and utilization of DNA markers (distributed over the genome) linked with traits of interest [121].

The Snap bean is a valuable vegetable crop in Sri Lanka due to its nutritional properties. Snap beans which are consumed principally as mature seeds of dry beans and immature green pods [122]. Qualitative and quantitative yield losses in common bean resulted due to pests and pathogens attacks during pre-harvest stage. Bean rust, caused by the basidiomycete fungus *Uromyces appendiculatus*, is one of the most significant diseases. It was estimated that increase of 1% severity in bean rust can lead to a yield loss of approximately 19 kg/ha [123].

Induction of resistance through conventional breeding is identified as most cost-effective disease management strategy [124]. In common bean, rust resistance (RR) is controlled by major single dominant genes and 10 of these RR genes namely, "Ur-3, Ur-4, Ur-5, Ur-6, Ur-7, Ur-9, Ur-11, Ur-12, Ur-13, and Ur-14" [125]. Out of these RR genes, Ur-11gene has shown wide-spectrum resistance to reported races of the rust pathogen. Therefore, common bean cultivar having the Ur11 gene (PI 181996) usually use the RR donor parent in crop improvement program through MAS. Once after completion of the hybridization process, introgression of the resistant gene from the cultivar PI 181996 into the F1, F_2, and BC_1F_1 populations was evaluated using standard phenotypic screening method. The SCAR marker GT-2 and SAE 19 can be used to select the resistant lines in offspring. Introduction of Ur-11 gene in to the local rust susceptible cultivars would be beneficial to develop local races of RR.

The linkage mapping, molecular dissection of the complex agronomical traits, and marker-assisted breeding have been popularized during the last decade [126]. Accuracy and efficiency of plant breeding program can be accelerated by the molecular marker-based technology [127]. The varietal identification, diversity analysis, DNA fingerprinting, phylogenetic analysis, and marker-assisted breeding and map-based cloning of genes in rice have been supported by molecular markers both in basic and applied researches [128]. The microsatellite DNA markers such as simple sequence repeats (SSRs), amplified fragment length polymorphism, and ISSRs have been successfully used for genotype identification, diversity analysis, and gene/quantitative trait loci (QTL) analysis in rice including Basmati rice [129]. Stress-tolerant crop varieties have been bred by using transgenic and MAS approaches. A map-based marker-assisted in situ technology should be developed for precision breeding in important crops like rice, wheat, brassica, and chickpea. Molecular techniques can be used to evaluate hybrid vigor in breeding lines and identification of biosynthetic pathways in organisms. Landraces and progenitor species of crop plants are ideal sources to elucidate marker-associated gene block showing salt-tolerant genes. As mentioned by Flowers et al. [126], QTLs have been mapped for several important agronomical traits such as tolerance against abiotic stresses including drought and salinity in major crops like rice and wheat (Figure 14.2).

Shabanimofrad et al. [130] have developed a MAS strategy against most destructive plant pest brown planthopper (BPH) *Nilaparvata lugens*. They have identified MAS as an ideal approach to develop resistance against plant pest in order to manage them. Codominant maker such as SSR was used for the identification of suitable regions. As a result of that, 28 polymorphic SSR markers were analyzed in a progeny (108 F3) which was developed by crossing the Rathu Heenati and MR276 rice cultivars. This investigation was aimed to elucidate the association of BPH resistance against biotypes 2 and 3 [130]. The SSR markers, namely, RM210, RM217, RM401, RM5953, RM242, and RM1103 showed significant association with BPH resistance in biotypes 2 and 3. When considered about the total phenotypic variation against biotypes 2 and 3, these six SSR markers can be divided into two clear groups. Ultimately, researchers have highlighted that these six SSR markers as useful molecular marker to incorporate in crop improvement programs against BPH.

14.7.4 Recombinant DNA technology

Development of host plant resistance is the durable and long-lasting approaches to break the inappropriate use of chemical pesticides to manage the biotic stress generated by plant pest including insect, weed, and plant pathogenic microorganisms. In addition to that, generation of abiotic stress tolerant is the way to mitigate adverse effect of global climate change on agricultural development. The naturally acquired host plant resistance is the promising way to manage pre- and postharvest stress caused by biotic and abiotic agents. This eventually support for the reduction of harvest loss. The development of transgenic resistance is the next best alternative for lack of resistant germplasm for crop improvement technologies [131]. Protein-mediated resistance (PMR) and nucleic acid-mediated resistance (RMR) are the two branches of molecular mechanisms of pathogen-derived resistant [132]. Nucleic acid-mediated resistance (RMR) is highly specific and PMR can be considered as broad spectrum resistance. The genetic engineering or recombinant DNA technology is the alternative and rapid method for the transfer of resistant genes to the traditional crop cultivars [133].

Among microorganisms, viruses are becoming the most tedious types of plant pathogen all over the world [134]. The cotton leaf curl disease (CLCuD) caused by begomoviruses is a good example to prove the above statement. As highlighted by Rasool et al. [131], through their research activities, the cotton leaf curl disease is the most important limiting factor for the production of cotton across the subcontinent. This was started as epidemic but it was ignored as a major threat. However, after three years of time, the disease incidence and severity were very high in Punjab [131]. As a consequence of viral attack, post-transcriptional gene silencing (PTGS) or virus-induced gene silencing (VIGS) of the transgene can be resulted. As a result of that, the pathogen-derived transgenes which offer PMR may not work [134]. To eliminate the negative impact of gene silencing, some other technologies such as synthetic gene and nonpathogen-derived resistance (NPDR) have been proposed by various scientific communities [135].

Rasool et al. [131] described the development of transgenic expression of synthetic coat protein (CP) and synthetic replication-associated protein in order to manage biotic stress generated by Begomovirus in *Nicotiana benthamiana*. Their prime objective was to develop transgenic approaches for managing CLCuD. Resistant cultivars against CLCuD have not been accomplished yet. Therefore, they have used the synthetic genes approaches to induce host plant resistance, a line with the phenomenon call pathogen-derived resistance (PDR). The CP and replication-associated protein (Rep) genes were used to control CLCuKoV-Bu and associated betasatellite related to CLCuD [131]. The pJIT60 cloning vector having cauliflower mosaic virus (CaMV) 35S promoter was used for the study. The cloning sight for the synthesized codon-optimized CP (CPsyn) is in between the SalI and EcoRI sites of the cloning vector. This expression cassette of CPsyn (2.2 kb approx.) consisted of double CaMV 35S promoter and CaMV 35S terminator regions. After that, the expression cassette was cloned into a binary vector called pGreen0029. Once after the construction of binary vector system, it was transferred to the *Nicotiana benthamiana* plant tissues through *Agrobacterium*-mediated transformation approach. Abundance of trans genes CPsyn and Repsyn was identified by screening with virulent pathogen CLCuKoV-Bu. Furthermore, low virus accumulation in plant tissues was confirmed by the southern hybridization. Finally, the results reviled both CPsyn and Repsyn genes for the development of resistance against of CLCuKoV-Bu [131].

Rather introduction of exotic or foreign gene to a plant, genetic modification can be performed with the aid of gene originate only from a species itself or from a species that can be crossed conventionally with this species is referred as cisgenesis [136]. Introduction of a new gene is an extra copy to the existing genome and leading to a natural variant [137]. Use of genetic element from a same species or sexually compatible species is called as intragenesis or hybrid genes. They have different promoter or terminator for genes and loci [138]. This foreign gene or insert will create a new genetic arrangement leading to a modified or differentiated function from the intact genome [139]. To avoid the accidental insertion of vector sequence into the plant genome, *Agrobacterium*-mediated gene transfer method is performed in the process of intragenesis [140]. This strategy avoid the phenomenon called "linkage drag" or the transfer of other undesirable genes along with the gene of interest. This can happen in conventional plant breeding as classical introgression [141]. The cisgenes with potential to be used in crop improvement against abiotic and abiotic stress management can be elucidated from whole genome sequencing approaches. However, abundance of cisgenic promoters and efficient marker genes are the limiting factor in application of cisgenesis [37].

As described by Limera et al. [37], Cisgenesis/intragenesis has been applied in apple as woody perennial fruit crop. The fire blight disease caused by *Erwinia amylovora* is the most destructive disease in apple. Therefore, apple breeders have used this technology to develop host plant resistance against apple fire blight in blossoms. The cisgenic apple line C44.4.146 showing resistance to apple fire blight was developed from the cisgene FB_MR5 which was extracted from a fire blight susceptible cultivar "Gala Galaxy" from wild apple *Malus × robusta* 5 (Mr5) [142]. Cisgenesis and intragenesis have also been successfully applied to induce resistance to other diseases in both apple and other woody fruit tree and vines [37]. As an example for that, An et al. [143] described the development of cisgenic grapefruit (*Citrus paradisi*) line with "foreign DNA-free" intra-/cisgenic citrus cultivars. And powdery mildew (*Erysiphe necator*)-resistant grapevine (*Vitis vinifera* L.) has been developed by the research group, and Cisgenic plants showed a delay in powdery mildew disease development and decreased severity of black rot (*Guignardia bidwellii*) during field tests [144].

Chougule and Bonning [145] have mentioned that Hemiptera or the sap sucking insects such as aphids, whiteflies, plant bugs, and stink bugs have been emerging as a serious agricultural pest than ever. These sap-sucking insects have dual role like direct damage and indirect damages as a vector which disseminate the viral diseases throughout the crop population. Predominant management of insect pests is mainly focused on nonsustainable application of chemical insecticides. Continuous expression of entomopathogenic toxins like δ-endotoxin insecticidal proteins (Cry and Cyt toxins) secreted by *Bacillus thuringenesis* (Bt) in plant cells is becoming a more attractive biotechnological approach in plant stress management. Development of resistance against chemical insecticide in insect population is a common phenomenon but such resistance would not be developed in this biologically based, alternative management strategies [145].

14.7.5 RNA interference

New biotechnological approaches are used for modifying an existing DNA sequence in a plant, and this process includes insertion or deletion and gene replacement and silencing of a gene or promoter sequence [37]. The RNAi approach is really worthful to achieve the above-mentioned objectives. This phenomenon in plants was first discovered in 1990 [146]. RNAi is an endogenous cellular process that occurs naturally to "turn off" unwanted or harmful specific nucleic sequences or to regulate gene expression before translation [147]. RNAi is mediated by a complex of molecular mechanisms with the aid of double-stranded RNA molecules (dsRNAs). This mechanism mainly focuses on gene expression inhibition or suppression [148]. The discovery of this phenomenon led to the development of gene "knock-down" technology as a realistic biotechnological approach in plant stress management [37]. Simply speaking, RNAi utilizes the dsRNAs as active molecules which has homologous strand for the target transcriptome or mRNAs and eventually negatively regulate the transcriptional process [149]. Accordingly, RNAi-mediated gene silencing has bloomed as an ideal technology for targeting plant pests namely fungi [150], insects [151], bacteria [152], viruses [153], and plants [154]. PTGS [155], transcriptional gene silencing (TGS) [147], and microRNA silencing (miRNA) [156] are the different mechanism used for gene silencing in plants. As well as in plants, post-TGS is governed by miRNAs. As a result, transcript cleavage or translational repression accelerates the secondary siRNA production from Pol II-derived cleaved transcripts. The RNA-directed DNA methylation and TGS are mainly controlled by majority of siRNAs in plants [155].

All these pathways are manipulated by the presence of dsRNA molecules of different sizes. Those dsRNA molecules are converted into specific protein families such as Dicer or Dicer-like (DCL), Argonaute (AGO), and RNA-dependent RNA polymerases by the plant cell [157]. Understanding of the biogenesis of small RNAs in plants is prerequisite in biotechnological application of small RNA-mediated plant stress management attributes. The targeting dsRNA molecule needs to be single-strand or small RNAs in order to silence the gene expression. Therefore, this conversion is mediated by the Dicer enzymes [155]. Dicer molecule found in budding yeasts can be considered as the minimal functioning unit. And which is buildup with a ribonuclease III domain and a dsRNA-binding domain. However, that lacks the DExD box helicase and PAZ domains found in higher eukaryotes. The PAZ domain binds the 2-nucleotide 3′ overhang of dsRNAs and is connected to the catalytic domain through a α-helical structure that acts as a ruler to determine small RNA size. DCL protein elements were elucidated in *A. thaliana* through scientific experimentations, and it revealed that duplication of this DCL genes has occurred early in the plant evolution [155, 158].

As described by Weiberg et al. [159], in his experimentation, new discovery of the cross-talk or sharing RNA molecules among two different kingdom was resulted. Fungal microbes utilize RNAi to enhance their spread, whereas plants seem to use this mechanism to encounter infection by these pathogens. Final outcomes of the both situations were accomplished by the process of RNAi. Degradation of target messenger RNA was used as a strategy for disturb the gene [160]. In plants, endogenous gene expression is regulated by the RNA silencing. That is an evolutionary adaptation to overcome stress generated by plant pathogenic viruses [161]. According to Limera et al. [37], this RNAi-mediated gene silencing has been mainly applied in woody fruit species to induce resistance against phytopathogens and allow generation of pathogen genetic elements in host plant tissues leading to induce host plant immunity as a method of PDR [161].

RNAi has been ranked as the promising gene regulatory approach in functional genomics with relation to stress management in plants. This process leads to down-regulation of target gene expression while without affecting the expression of other genes [162]. Phytopathogens are harmful organisms that limit the growth, development, and reproduction of the crop plants globally while making economic impact. Plant pathologists and plant biotechnologists have used several methodologies to develop resistant genotypes of crop cultivars against pest and diseases. But, in the last decade, RNAi-induced gene silencing emerged as an effective tool to engineer pathogen resistant in plants [163]. Host gene silencing-hair pin RNAi (HGS-hpRNAi) is recognized as more stable gene silencing approch in plants [164].

A bacterial component called flagellin can stimulate the expression of specific miRNA in plants which induce the signal transduction pathways of disease resistance [165]. Research revealed that RNAi-mediated over-expression of a gene fatty acid amide hydrolase (AtFAAH) which is responsible for fatty acid *N*-acylethanolamines metabolism can alter phyto-hormone signals through overlapping with plant defense pathways to increase resistance against plant pathogenic bacterial in *Arabidopsis thaliana* [166]. Jiang et al. [167] highlighted that through his research findings, RNAi-mediated knockdown of rice fatty acid desaturate gene (OsSSI2) can increase the host resistance against rice leaf blight causing agent *Xanthomonas oryzae* pv. *oryzae* and rice blast fungus *Magnaporthe grisea*. In addition to that, siRNAs have shown greater effectiveness against the crown gall disease in *Arabidopsis*, *Nicotiana*, and *Lycopersicum* species caused by *Agrobacterium tumefaciens*. In the process of development of biotechnological approaches through siRNA technology to induce above-mentioned resistance, inverted repeats of gene ipt and iaaM which encode the precursors of biosynthesis of auxin and cytokinin have been transformed to the host cells [168]. Furthermore, Escobar et al. [168] explained the use of host-induced gene such as Avra10 as a gene silencing agent, but this phenomenon is limited to few fungal diseases in wheat (*Triticum aestivum*) and barley (*Hordeum vulgare*). The transient gene expression resistant to RNAi leading to silent point mutations is the mechanism involved in the above situation. To achieve the ultimate outcome, RNA molecule of the host plant transfers to the fungal pathogen such as *Blumeria graminis* in order to develop the RNAi-mediated plant protection.

The drought-responsive miR159, miR169g, and miRNA393 genes have been observed in *Arabidopsis thaliana* and *Oryza sativa* [169]. The gene expression pattern of the RACK1 gene under drought tolerance was studied by RNAi-mediated gene silencing. As a result, transgenic rice was shown a superior level of tolerance to drought than nontransgenic rice plants [170]. miR169, miR396, miR165, miR167, miR168, miR159, miR319, miR171, miR394, miR393, miR156, and miR158 were identified as drought-responsive micro-RNA in different plants such as *Arabidopsis*, *Populus trichocarpa*, and *Oryza sativa* [171]. In addition to the drought-responsive miRNAs, many miRNAs have been identified salinity stress in crop plants. The miR397, miR156, miR394, miR158, miR393, miR159, miR319, miR165, miR171, miR167, miR169, miR168, and miR398 were upregulated under the salinity stress in *Arabidopsis* [171]. The accumulation of miR159.2 and miRS1 was increased in *P. vulgaris* during the addition of NaCl [172]. As explained by Liu et al. [171], miR171l-n, miR530a, miR1446a-e, miR1445, and miR1447 were down-regulated and miR1450 and miR482.2 were upregulated in salt stress period in *P. trichocarpa*. The recent research findings elucidated the miR396, miR156, miR167, and miR164 groups and were downregulated, while miR474, miR162, miR395, and miR168 groups were upregulated in salinity-tolerant and a salt-sensitive line of maize [173].

Recent investigations reviled that variant expression patterns of miRNA in wheat under heat stress condition. The miRNAs originated under heat stress condition were cloned with the aid of Solexa high-throughput sequencing. Among 32 families of miRNA in wheat, 9 families were identified as heat-responsive [162]. The *P. trichcarpa*, Pt-miRNA levels of transcript were studied under stressed or tension-stressed xylem with the nonstressed xylem. Results highlighted that miR408 was upregulated, and miR156, miR48, miR162, miR475, miR164, and miR480 were downregulated under the stress condition. These findings revealed that miRNAs may be regulated in mechanical stress and could play a role in defense system for mechanical and structural fitness [174]. Aphids play a major role as an insect pest in most agricultural crops and make considerable amount of yield loss through direct damage and as a vector-borne disease disseminator. Plant-mediated RNAi or use of dsRNA molecule targeting specific genes of insect in transgenic plants have become a new tool in insect pest management [175]. Identification of the mechanism of RNAi in insects and effective RNAi target genes in aphids are studied by various plant molecular biologist worldwide. Whitefly (*Bemisia tabaci*) is an important agricultural pest and damage to most tropical and sub-tropical crops. Ibrahim et al. [176] reported that construction of RNAi-based plasmid containing an interfering cassette is designed to generate dsRNAs that target a novel v-ATPase transcript in whitefly. Transgenic lettuce lines were generated, and seven lines were tested with whiteflies over a period of 32 days. Results revealed 83.8–98.1% mortality rate during five days of feeding. Alien with that, when compared with control, 95% reduction in egg laying was observed in flies feeding on transgenic lettuce plants. Furthermore, QRT PCR confirmed that reduction of endogenous expression of v-ATPase gene in whiteflies feeding on transgenic plants. This biotechnological approach is a foundation for the development of whitefly-resistant commercial crops and improving agricultural sustainability and food security worldwide. In addition to that, this technology would be a beneficial to the reduced use of huge environmentally aggressive methods of pest control [176].

14.7.6 CRISPR/Cas9-mediated generation of stress tolerance

Since the advent of CRISPR/Cas9 and related gene-editing technology, direct modification of crop genomes has become the way of the future for advanced breeding techniques in agriculture [177] (Figure 14.3). These new plant breeding technologies have opened avenues of fundamental and translational research that were previously inaccessible. Abiotic stresses in plants namely drought, salinity, heavy metal toxicity, heat, and nutrients limitations significantly reduce the agricultural production worldwide [178]. The transcriptional activator-like effector nucleases and zinc finger nucleases can be considered as potential technology for genetic manipulation in plants. Over this technology, clustered regularly interspaced short palindromic repeats (CRISPR)/CRISPR-associated protein 9 (Cas9) technique has much more potential as a genome editing tool in plants to acquire stress tolerant against

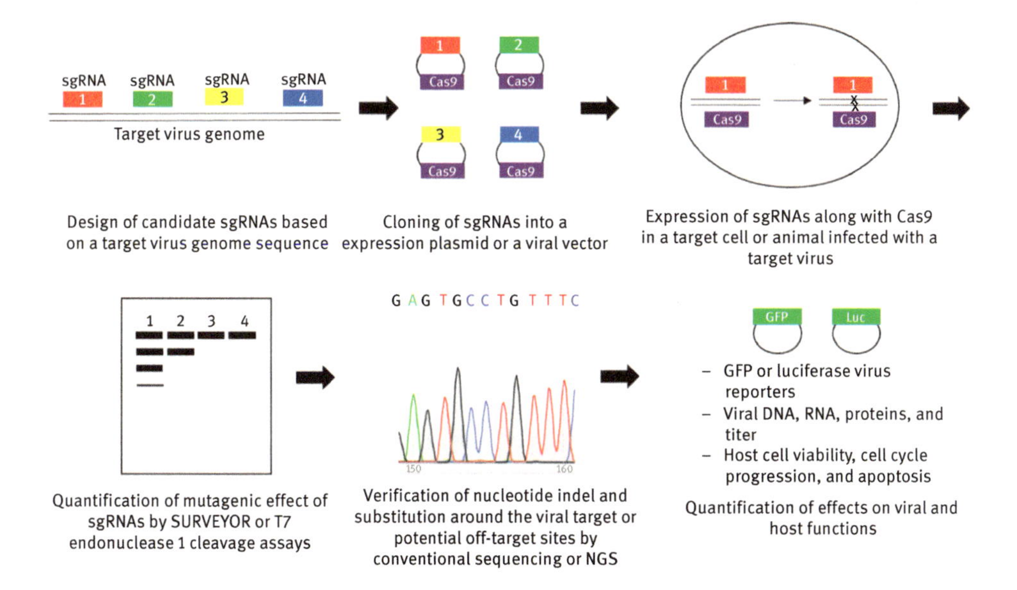

Figure 14.3: CRISPR/Cas9-based antiviral strategy [188].

biotic and abiotic factors [178]. Furthermore, CRISPR/Cas9-mediated genome editing technology provides more advantages for accomplish the objectives of sustainable agriculture over other conventional strategies.

The type II bacterial immune system engineered for genome editing purposes is referred as CRISPR/CRISPR-associated protein (CRISPR/Cas) [179]. Single-guided RNA (sgRNA) and Cas9 endonuclease are the two major components in the CRISPR/Cas-based genome editing system. The sgRNA–Cas9 complex search for the target genome site with the help of protospacer adjacent motif generates efficient DNA double-strand break. The CRISPR/Cas technology has been successfully implemented in generating crip cultivars with tolerant to biotic and abiotic stress [180].

Pramanik et al. [181] have demonstrated that successful utilization of CRISPR/Cas technology for the development of pathogen-resistant tomato against tomato Tomato yellow leaf curl virus (TYLCV) and powdery mildew caused by fungal *Oidium* sp. Tomato (*Solanum lycopersicum*) is one of the agriculturally important food crops consumed worldwide with good economic validity. According to the statistics of the Food and Agriculture Organization, it has generated USD 190.4 billion from the production tonnage in 188 million globally in 2018 [182]. However, significant yield losses occur due to a wide range of vulnerabilities to different pathogens. Tomato is infected by several phytopathogens, among that TYLCV plays a major role. This Begomovirus consists of a circular single-stranded DNA and belongs to Geminiviridae family [181]. TYLCV is a vector-borne disease and is transmitted with the help of phloem sap feeding whiteflies (*Bemisia tabaci*). In addition to that, accumulation of viral particles in seed also has potential to become a seed borne [183].

Several conventional strategies, such as stringent quarantine rules, integrated pest management, and conventional breeding, were used in TYLCV management [184]. With the help of omics technologies, the QTL associates with TYLCV resistance were recently identified in wild tomato varieties [185]. Out of these QTLs, Ty-2 gene has shown a potential hypersensitive response in *Nicotiana benthamiana* under TYLCV infection [186]. And the Ty-5 locus contains a gene called SlPelo that synthesizes a messenger RNA surveillance factor known as Pelota (PELO) [187]. The recycling of ribosome during the protein synthesis process is governed by Pelota (PELO). This activity was further confirmed by SlPelo knockout mutants and showed restricted TYLCV proliferation within the infected cells [181].

The PELO protein contains 387 amino acids and retains three conserved eukaryotic translation termination factor 1 (eRF1) domains named eRF1_1, eRF1_2, and eRF1_3. Out of these termination factors (eRF1_1, eRF1_2, and eRF1_3), the eRF1 domains playing a major role in ribosome recycling during the protein synthesis [13]. In the development process of TYLCV-resistant tomato line with the help of CRISPR/Cas technology, gRNAs were selected to disrupt the genome region of the eRF1_1 domain in the SlPelo locus that would eventually abolish the function of PELO protein. Pramanik et al. [180, 181] have revealed that sgRNA expression cassettes, plant selection marker (kanamycin), and functional SpCas9 expression cassette were combined into the T-DNA (binary transfer-DNA) vectors using Golden Gate cloning process [180]. Ultimately, *Agrobacterium*-mediated tomato transformation was carried out to deliver CRISPR/Cas9 components for the generation of genome-edited BN-86 lines [180]. As result of that, genome-edited tomato plants were developed. Callus regeneration and shoot induction were carried out by tissue culture technology. The genome-edited SlPelo-knockout mutants were tested for resistance against TYLCV. *Agrobacterium*-carrying TYLCV virulent factor was used to inoculate three biallelic knockout mutant plants. TYLCV symptoms were shown by WT-like plants three weeks of post-infection. The tested G1-41-40 and G1-41-42 plants exhibited no visible TYLCV symptom. The Disease Severity Index values were calculated according to the severity of visual TYLCV symptoms [181].

Tomato transmembrane protein MLO1 is involved in the interaction between tomato and powdery mildew fungal pathogen. Three independent SlMlo1-knockout lines, namely, homozygous (G1-22-31), biallelic (G1-22-32), and chimeric (G1-22-36) plants, along with WT-like control plants, were challenged for powdery mildew disease-resistance assay [181]. The fungal inoculum was sprayed in tested plants, and the fungal disease occurrence was monitored. All three SlMlo1 mutant plants showed complete resistance. The suppressed accumulation of TYLCV restricted the spread of disease to the noninoculated plant parts. Overall, the results demonstrate the efficiency of the CRISPR/Cas9 system to introduce targeted mutagenesis for the rapid development of pathogen-resistant varieties in tomato [181].

The precise gene mutations induction was carried out in several woody frit species with the help of CRISPR/Cas9 system. Peng et al. [189] demonstrated that the use of CRISPR/Cas9 system for development of disease resistance in citrus against *Xanthomonas citri* subsp. citri (Xcc) the causative agent of citrus canker. Outcome of this research was attained by targeting the modified effector binding element (EBEPthA4) of the susceptibility gene LATERAL ORGAN BOUNDARIES 1 (CsLOB1) promoter in Wanjincheng orange [189, 190]. The PthA4, transcription activator-like effector bind with the EBEPthA4 effector present on the promoter of CsLOB1 susceptibility gene during the infection. As result, Citrus canker disease would be developed [190]. CRISPR/Cas9 mediated editing of CsLOB1 gene promoter or deletion of the entire EBEPthA4 sequence from both CsLOB1 alleles was able to generate sufficient level of resistance in Wanjincheng orange against Citrus canker [189].

Molecular mechanisms involved in abiotic stress tolerance in plant are divers and complex in nature. Signal transduction pathway is the most prominent system which governs the cytoplasmic operations. As a result, many interactions and cross-talk are happening among molecules [191]. Individual function of molecules involved in signal transduction pathway can be studied extensively with the help of CRISPR-Cas9-mediated genome editing technology. CRISPR-Cas9-generated mutagenesis in mitogen-activated protein kinases3 (slmapk3) has shown to be drought-tolerant in tomato (*Solanum lycopersicum*) [192]. As well as CRISPR-Cas9-mediated genome editing, practice was carried out by research group with stress-ABA-activated protein kinase2 in rice to understand the mechanism involved in stress management [193]. The CRISPR-Cas9 approach was used to increase the expression level of the ARGOS8 gene to develop drought tolerance in maize (Figure 14.4). Results revealed that enhancement of the grain yields under drought conditions in mutant under the field conditions [194]. The above examples demonstrated that ability of CRISPR-Cas9 system is to mediate resistance against numerous abiotic stresses [195].

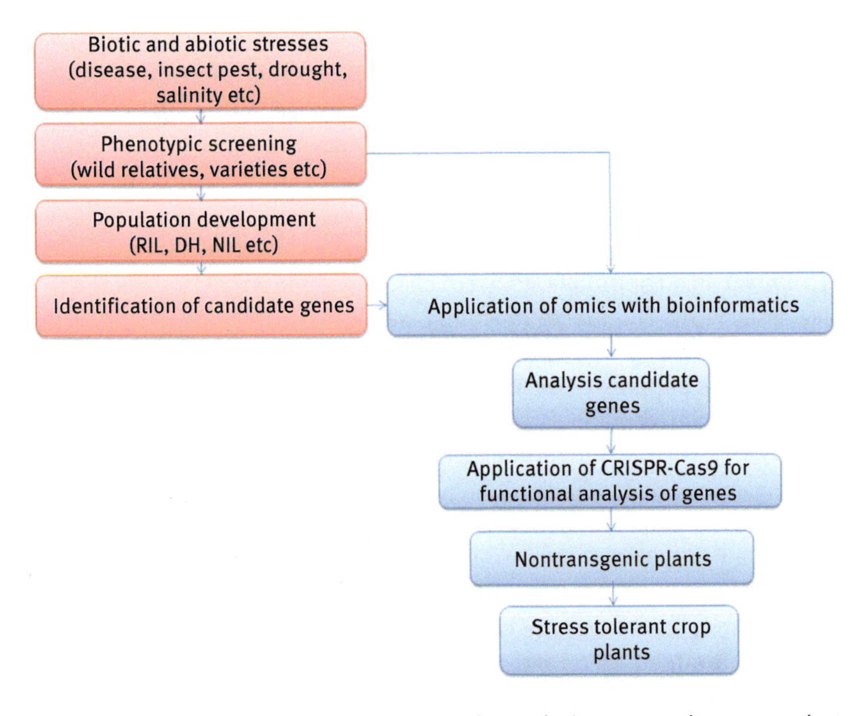

Figure 14.4: Omics and CRISPR-Cas9 strategies for producing stress-tolerant crop plants.

The competition with crops for growth space, sunlight, water, fertilizer, and spread pests and diseases directly or indirectly are the main consequences generated by weeds in farmland [196]. As a result of that, inhibition of crop growth, reduction of crop yields, and serious affect on the quality of crops can be seen [197]. Based on the economic and effective effects, chemical herbicides have been extensively used in agronomic practices to reduce the economic loss of weeds [198]. However, the extensive and recurrent use of the same herbicides for a long period of time, quick emergence of herbicide-resistant weeds can be resulted. Development of herbicide resistant in agricultural crops is ideal to control weeds and reduce biomagnification of toxicity in crops through food webs [199]. Dong et al. [200] reviewed on the development of herbicide resistance crops using CRISPR/Cas9-mediated gene editing and clearly mentioned that acetolactate synthase (ALS) gene has several target mutation sites to develop herbicide resistance crop plants (Figure 14.5).

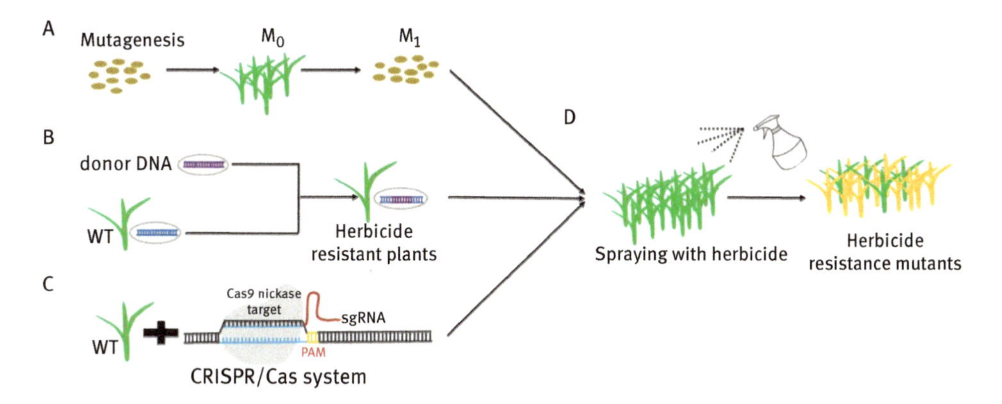

Figure 14.5: Different breeding approaches for development of herbicide resistant in crops: (A) mutation breeding; (B) transgenic approaches; (C) application of CRISPR/Cas system; and (D) screening and identification of herbicide resistant [200].

Mutation site Ala96 in rice, Pro165 in maize, Pro178 in soybean, Pro165 in maize, Pro174 in wheat, and Pro174 in watermelon of ALS gene have been used to generate herbicide-resistant crops in agriculture. Target site mutation, target site gene amplification, increased herbicide detoxification, and metabolism are the mechanisms which we can be used to achieve herbicide resistance in crops [201]. Among those, target site mutation has been used as most prominent approaches to generate herbicide resistant in many economically important agricultural crops [200].

References

[1] Jeger M, Beresford R, Bock C, et al. Global challenges facing plant pathology: Multidisciplinary approaches to meet the food security and environmental challenges in the mid-twenty-first century. CABI Agriculture and Bioscience 2021, 2, 20. doi:https://10.1186/s43170-021-00042-x

[2] Mahlein AK, Kuska MT, Thomas S, et al. Quantitative and qualitative phenotyping of disease resistance of crops by hyperspectral sensors: Seamless interlocking of phytopathology, sensors, and machine learning is needed! Current Opinion in Plant Biology 2019, 50, 156–162. doi:https://10.1016/j.pbi.2019.06.007

[3] Laine AL, Barrès B. Epidemiological and evolutionary consequences of life-history trade-offs in pathogens. Plant Pathology 2013, 62, 96–105. doi:https://10.1111/ppa.12129

[4] Pixley KV, Falck-Zepeda JB, Giller KE, et al. Genome editing, gene drives, and synthetic biology: Will they contribute to disease-resistant crops, and who will benefit?. Annual Review of Phytopathology 2019, 57, 165–188. doi:https://10.1146/annurev-phyto-080417-045954

[5] Newton AC, Torrance L, Holden N, et al. Climate change and defense against pathogens in plants. In: Gadd GM, Sariaslani S, ed. Advances in applied microbiology, vol. 81. Elsevier, 2012, 89–132. doi:https://10.1016/B978-0-12-394382-8.00003-4

[6] Showalter DN, Saville RJ, Orton ES, Buggs RJA, Bonello P, Brown JKM. Resistance of European ash (*Fraxinus excelsior*) saplings to larval feeding by the emerald ash borer (*Agrilus planipennis*). Plant People Planet 2020, 2, 41–46. doi:https://10.1002/ppp3.10077
[7] Carvajal-Yepes M, Cardwell K, Nelson A, et al. A global surveillance system for crop diseases. Science 2019, 364, 1237–1239. doi:https://10.1126/science.aaw1572
[8] McRoberts N, Thomas CS, Brown JK, Nutter FW, Stack JP, Martyn RD. The evolution of a process for selecting and prioritizing plant diseases for recovery plans. Plant Disease 2016, 100, 665–671. doi:https://10.1094/PDIS-04-15-0457-FE
[9] Sarkar D, Mandal R, Roy PC, Taradar J, Dasgupta B. Management of brown spot disease of rice by using safer fungicides and some bioagents. 2014. https://www.semanticscholar.org/paper/MANAGEMENT-OF-BROWN-SPOT-DISEASE-OF-RICE-BY-USING-Sarkar-Mandal/bf0b21e0414a935b42d048d57018a07737ccea1f#citing-papers
[10] Ashfaq B, Arshad HMI, Atiq M, Yousaf S, Saleem K, Arshad A. Biochemical profiling of resistant phenotypes against bipolaris oryzae causing brown spot disease in rice. Frontiers in Agronomy 2021, 3, 675895. doi:https://10.3389/fagro.2021.675895
[11] Biswas B, Scott PT, Gresshoff PM. Tree legumes as feedstock for sustainable biofuel production: Opportunities and challenges. Journal of Plant Physiology 2011, 168, 1877–1884. doi:https://10.1016/j.jplph.2011.05.015
[12] Katsantonis D, Koutroubas SD, Ntanos DA, Lupotto E. A comparison of three experimental designs for the field assessment of resistance to rice blast disease (Pyricularia oryzae). Journal of Phytopathology 2007, 155, 204–210. doi:https://doi.org/10.1111/j.1439-0434.2007.01218.x
[13] Song F, Goodman RM. Molecular biology of disease resistance in rice. Physiological and Molecular Plant Pathology 2001, 59, 1–11. doi:https://10.1006/pmpp.2001.0353
[14] Geetha NP, Amruthesh KN, Sharathchandra RG, Shetty HS. Resistance to downy mildew in pearl millet is associated with increased phenylalanine ammonia lyase activity. Functional Plant Biology: FPB 2005, 32, 267–275. doi:https://10.1071/FP04068
[15] Eliahu N, Igbaria A, Rose MS, Horwitz BA, Lev S. Melanin biosynthesis in the maize pathogen cochliobolus heterostrophus depends on two mitogen-activated protein kinases, Chk1 and Mps1, and the transcription factor Cmr1. Eukaryotic Cell 2007, 6, 421–429. doi:https://10.1128/EC.00264-06
[16] Gabbatiss J. Plants can see, hear and smell – And respond. The people who believe plants can talk. http://www.bbc.com/earth/story/20170109-plants-can-see-hear-and-smell-and-respond. Published 2017
[17] Trinkaus E. Early modern humans. Annual Review of Anthropology 2005, 34, 207–230. doi: https://10.1146/annurev.anthro.34.030905.154913
[18] Mannion AM. In: Montgomery DR. Dirt: The erosion of civilizations, vol. 12. Berkeley, Los Angeles, and London, University of California Press, Environ Hist Durh N C, 2007, 682–683. doi:https://10.1093/envhis/12.3.682
[19] Diamond J. Evolution, consequences and future of plant and animal domestication. Nature 2002, 418, 700–707. doi:https://10.1038/nature01019
[20] Stewart M. In: Vasey DE. An ecological history of agriculture, 10,000 B.C.-A.D. 10,000. Ames, Iowa State University Press, 1992 Environ Hist Durh N C, 1996, vol. 1, 91–92. doi:https://10.2307/3985162
[21] Dow GK, Olewiler N, Reed CG. The transition to agriculture: Climate reversals, population density, and technical change. Economic history 0509003. University Library of Munich, Germany, 2005, 42.
[22] Kremer M. Population growth and technological change: One million B.C. to 1990. The Quarterly Journal of Economics 1993, 108, 681–716. doi:https://10.2307/2118405

[23] Johns Hopkins Center for a Livable Future. First National Forum for Food Policy Councils. News. https://clf.jhsph.edu/about-us/news. Published 2021
[24] FAO, IPPC. Global Emerging Issues. 2017.
[25] FAO. International year of plant health. Rome: FAO, 2021. doi:https://10.4060/cb7056en.
[26] FAO, IPPC. 2020 IPPC annual report – Protecting the world's plant resources from pests. FAO on behalf of the Secretariat of the International Plant Protection Convention, 2021. doi: https://10.4060/cb3842en
[27] Government of Sri Lanka. Plant Protection Act, No 35 of 1999. Sri Lanka, 1999, 0–87. https://www.customs.gov.lk/wpcontent/uploads/2021/06/PLANT_PROTECTION_ACT_No._35_OF_1999.pdf
[28] Verma LR, Aparna N, Sharma PN. Indigenous technology knowledge for watershed management in upper north-west Himalayas of India. Participatory Watershed Management Training in Asia (PWMTA) Program, Kathmandu, Nepal. https://www.fao.org/3/x5672e/x5672e00.htm. Published 1998. Accessed December 26, 2021.
[29] FAO. Save and grow – A policymaker's guide to the sustainable intensification of smallholder crop production. Rome, FAO, 2011.
[30] PPC. Plant Pest Surveillance. 2016. https://www.ippc.int/static/media/files/publication/en/2018/06/Plant_Pest_Surveillance_Guide_Pr2Final_WEB_tzFeSDS.pdf.
[31] Thomas B, Murray BG, Murphy DJ. Encyclopedia of applied plant sciences, vols 1–3. Academic Press and Elsevier Ltd, Amsterdam Academic Press, 2017.
[32] Capinera JL. Relationships between insect pests and weeds: An evolutionary perspective. Weed Science 2005, 3, 892–901. doi:https://10.1614/WS-04-049R.1
[33] Sobiczewski P, Iakimova ET, Mikiciński A, Węgrzynowicz-Lesiak E, Dyki B. Necrotrophic behaviour of *Erwinia amylovora* in apple and tobacco leaf tissue. Plant Pathology 2017, 66, 842–855. doi:https://10.1111/ppa.12631
[34] Osman HA, Ameen HH, Mohamed M, Elkelany US. Efficacy of integrated microorganisms in controlling root-knot nematode *Meloidogyne javanica* infecting peanut plants under field conditions. Bulletin of the National Research Centre 2020, 44, 134. doi:https://10.1186/s42269-020-00366-0
[35] Pallas V, García JA. How do plant viruses induce disease? Interactions and interference with host components. The Journal of General Virology 2011, 92, 2691–2705. doi:https://10.1099/vir.0.034603-0
[36] Schumann GL, D'Arcy CJ. Essential plant pathology, 2nd edn. The American Phytopathological Society, 2010. doi:https://10.1094/9780890546710. ISBN: 9780890543818 / 089054381X.
[37] Limera C, Sabbadini S, Sweet JB, Mezzetti B. New biotechnological tools for the genetic improvement of major woody fruit species. Frontiers in Plant Science 2017, 8, 1–16. doi: https://10.3389/fpls.2017.01418
[38] Taiz L, Zeiger E. Secondary metabolites and plant defense. In: Plant physiology, 4th edn. Massachusetts U.S.A., Sinauer Associates Inc., Publishers Sunderland, 2006, 315–344.
[39] Monaghan J, Zipfel C. Plant pattern recognition receptor complexes at the plasma membrane. Current Opinion in Plant Biology 2012, 15, 349–357. doi:https://doi.org/10.1016/j.pbi.2012.05.006
[40] Iqbal Z, Iqbal MS, Hashem A, Abd Allah EF, Ansari MI. Plant defense responses to biotic stress and its interplay with fluctuating dark/light conditions. Frontiers in Plant Science 2021, 12, 631810. doi:https://10.3389/fpls.2021.631810
[41] Lewis WJ, van Lenteren JC, Phatak SC, Tumlinson JH. A total system approach to sustainable pest management. The Proceedings of the National Academy of Sciences 1997, 94, 12243–12248. doi:https://10.1073/pnas.94.23.12243

[42] Pimentel D, Levitan L. Pesticides: Amounts applied and amounts reaching pests. Bioscience 1986, 36, 86–91. doi:https://10.2307/1310108
[43] Belyakov NV, Nikolina NV. Plant protection technologies: From advanced to innovative. Journal of Physics. Conference Series 2021, 1942, 012072. doi:https://10.1088/1742-6596/1942/1/012072
[44] Belyakov N. Environmental educational project as a way of forming a healthy food culture. IOP Conference Series: Earth and Environmental Science 2019, 390, 012014. doi:https://10.1088/1755-1315/390/1/012014
[45] Goverment Office for Science. Government Office for Science Annual Review 2012–2013, UK, 2013. https://assets.publishing.service.gov.uk/government/uploads/system/uploads/attachment_data/file/275802/13-p95-government-office-for-science-annual-review-2012-2013.pdf.
[46] FAO. FAO Statement on Biotechnology. Agricultural Biotechnologies. http://www.fao.org/biotech/fao-statement-on-biotechnology/en/. Published 2000. Accessed December 28, 2021
[47] Mills N. Egg parasitoids in biological control and integrated pest management. In: Consoli FL, Parra JRP, Zucchi RA, eds. Egg parasitoids in agroecosystems with emphasis on *Trichogramma*. Dordrecht, Springer Netherlands, 2010, 389–411. https://10.1007/978-1-4020-9110-0
[48] Godfray HCJ. Parasitoids. In: Levin SA, ed. Encyclopedia of biodiversity. Oxford, Elsevier, 2007, 1–13. Online version doi:https://10.1016/B0-12-226865-2/00218-2
[49] Godfray HCJ. Parasitoids. In: Levin SA, ed. Encyclopedia of biodiversity, vol. 4. Oxford, Elsevier, 2001, 674–682. doi:https://10.1016/B978-0-12-384719-5.00152-0
[50] University of Maryland. Parasitoids, University of Maryland Extension. https://extension.umd.edu/resource/parasitoids. Published 2021. Accessed December 28, 2021.
[51] Greathead D, Greathead A. Filter by type search abstract biological control of insect pests by insect parasitoids and predators: The BIOCAT database. Biocontrol News and Information 1992, 13, N–68N. doi:https://10.1079/cabireviews/19921166435
[52] Hassan SA. The mass rearing and utilization of *Trichogramma* to control lepidopterous pests: Achievements and outlook. Journal of Pesticide Science 1993, 37, 387–391. doi:https://doi.org/10.1002/ps.2780370412
[53] Li L. Worldwide use of *Trichogramma* for biological control on different crops: A survey. In: Wajnberg E, Hassan S, eds. Biological control with egg parasitoids. Oxon, UK, CAB International, 1994, 37–51.
[54] Smith SM. Biological control with *Trichogramma*: Advances, successes, and potential of their use. Annual Review of Entomology 1996, 41, 375–406. doi:https://10.1146/annurev.en.41.010196.002111
[55] Australian Museum. Predators, parasites and parasitoids. Australian Museum. https://australianmuseum.net.au/learn/animals/insects/predators-parasites-and-parasitoids/%23:%7E: :text=A parasitoid is an organism,of both predators and parasites. &text=Parasitoids are very selective and,or several closely related species. Published 2019. Accessed December 28, 2021.
[56] Kaya HK, Vega FE. Scope and basic principles of insect pathology. In: Insect pathology, 2nd edn. Elsevier, 2012, 1–12. In: Vega F, Kaya HK, ed. Insect Pathology. 2nd edition. San Diego, CA, Academic Press, 2012, 1–12. doi:https://10.1016/B978-0-12-384984-7.00001-4
[57] Solter LF, Hajek AE, Lacey LA. Exploration for entomopathogens. In: Lacey LA, ed. Microbial control of insect and mite pests. Yakima, WA, United States, IP Consulting International, Elsevier, 2017, 13–23. doi:https://10.1016/B978-0-12-803527-6.00002-0. https://experts.illinois.edu/en/publications/exploration-for-entomopathogens

[58] Hajek AE, van Frankenhuyzen K. Use of entomopathogens against forest pests. In: Lacey LA, ed. Microbial control of insect and mite pests. Yakima, WA, United States, IP Consulting International, Elsevier, 2017, 313–330. doi:https://10.1016/B978-0-12-803527-6.00021-4
[59] Sinha KK, Choudhary AK, Kumari P. Entomopathogenic fungi. In: Omkar, ed. Ecofriendly pest management for food security. 125 London Wall, London EC2Y 5AS, UK, Elsevier, 2016, 475–505. doi:https://10.1016/B978-0-12-803265-7.00015-4
[60] Vail PV, Jay DL, Hunter DK, Staten RT. A nuclear polyhedrosis virus infective to the pink bollworm, *Pectinophora gossypiella*. Journal of Invertebrate Pathology 1972, 20, 124–128. doi:https://doi.org/10.1016/0022-2011(72)90092-4
[61] Vail PV, Jay DL. Pathology of a nuclear polyhedrosis virus of the alfalfa looper in alternate hosts. Journal of Invertebrate Pathology 1973, 21, 198–204. doi:https://doi.org/10.1016/0022-2011(73)90202-4
[62] Hostetter DL, Puttler B. A new broad host spectrum nuclear polyhedrosis virus isolated from a celery looper, *Anagrapha falcifera* (Kirby), (Lepidoptera: Noctuidae). Environmental Entomology 1991, 20, 1480–1488. doi:https://10.1093/ee/20.5.1480
[63] Harrison RL, Bonning BC. The nucleopolyhedroviruses of *Rachiplusia ou* and *Anagrapha falcifera* are isolates of the same virus. The Journal of General Virology 1999, 80, 2793–2798. doi:https://10.1099/0022-1317-80-10-2793
[64] Moreau G, Lucarotti CJ, Kettela EG, et al. Aerial application of nucleopolyhedrovirus induces decline in increasing and peaking populations of *Neodiprion abietis*. Biological Control 2005, 33, 65–73. doi:https://10.1016/j.biocontrol.2005.01.008
[65] Lucarotti CJ, Moreau G, Kettela EG. AbietiTM, a viral biopesticide for control of the balsam fir sawfly. In: Vincent C, Goettel MS, Lazarovits G, eds. Biological control: A global perspective. Oxfordshire, UK, CAB International, Oxfordshire, UK, 2007, 353–361.
[66] Sato T, Oho N, Kodomari S. A granulosis virus of the tea tortrix, H*omona magnanima* DIAKNOFF (Lepidoptera: Tortricidae): Its pathogenicity and mass-production method. Applied Entomology and Zoology 1980, 15, 409–415. doi:https://10.1303/aez.15.40
[67] Berliner E. Über die Schlaffsucht der Mehlmottenraupe (*Ephestia kühniella* Zell.) und ihren Erreger *Bacillus thuringiensis* n. sp. Zeitschrift Für Angewandte Entomologie 1915, 2, 29–56. doi:https://10.1111/j.1439-0418.1915.tb00334.x
[68] Dulmage HT. Insecticidal activity of HD-1, a new isolate of *Bacillus thuringiensis* var. *alesti*. Journal of Invertebrate Pathology 1970, 15, 232–239.
[69] Goldberg LJ, Margalit J. A bacterial spore demonstrating rapid larvicidal activity against *Anopheles sergentii, Uranotaenia unguiculata, Culex univitattus, Aedes aegypti* and *Culex pipiens*. Mosquito News 1977, 37, 355–358.
[70] Trought TET, Jackson TA, French RA. Incidence and transmission of a disease of grass grub (*Costelytra zealandica*) in Canterbury. New Zealand Journal of Experimental Agriculture 1982, 10, 79–82. doi:https://10.1080/03015521.1982.10427847
[71] Stucki G, Jackson TA, Noonan MJ. Isolation and characterisation of *Serratia* strains pathogenic for larvae of the New Zealand grass grub C*ostelytra zealandica*. New Zealand Journal of Science 1984, 27, 255–260.
[72] Jackson TA, Pearson JF, O'callaghan M, Mahanty HK, Willocks MJ. Pathogen to product-development of *Serratia entomophila* (Enterobacteriaceae) as a commercial biological control agent for the New Zealand grass grub (*Costelytra zealandica*). In: Jackson TA, Glare TR, eds. Use of pathogens in scarab pest management. Hampshire, England, Intercept, 1992, 191–198.
[73] Martin PAW, Gundersen-Rindal D, Blackburn M, Buyer J. *Chromobacterium subtsugae* sp. nov., a betaproteobacterium toxic to Colorado potato beetle and other insect pests. International Journal of Systematic and Evolutionary Microbiology 2007, 57, 993–999. doi: https://10.1099/ijs.0.64611-0

[74] Lacey LA, Wraight SP, Kirk AA. Entomopathogenic fungi for control of *Bemisia tabaci* biotype B: Foreign exploration, research and implementation. In: Gould J, Hoelmer K, Goolsby J, eds. Classical biological control of Bemisia Tabaci in the United States – A review of interagency research and implementation. Dordrecht, The Netherlands, Springer, 2008, 33–69.
[75] Lacey LA, Thomson D, Vincent C, Arthurs SP. Codling moth granulovirus: A comprehensive review. Biocontrol Science and Technology 2008, 18, 639–663. doi:https://10.1080/09583150802267046
[76] ARSEF. ARS collection of entomopathogenic fungal cultures database. USDA-Agricultural Research Service Entomopathogenic Collection. http://arsef.fpsnl.cornell.edu/4DACTION/W_List_Tables/param. Published 2014. Accessed December 28, 2021.
[77] Keller S. Infektionsversuche mit dem Pilz *Beauveria tenella* and adulten Maikäfern (*Melolontha melolontha* L.). Mitt Schweiz Entomol Gesell 1978, 51, 13–19.
[78] Zimmerman G. Use of the fungus, *Beauveria brongniartii*, for the control of European cockchafers, *Melolontha* spp. in Europe. In: Jackson TA, Glare TR, eds. Use of pathogens in scarab pest management. Andover, Intercept Press, 1992, 199–208.
[79] Prior C, Lomer CJ, Herren H, Paraiso A, Kooyman C, Smit JJ. The IIBC/IITA/DFPV collaborative research programme on the biological control of locusts and grasshoppers. In: Biological Control of locusts and grasshoppers: Proceedings of a Workshop held at the International Institute of Tropical Agriculture, Cotonou, Republic of Benin, 29 April–1 May 1991. Wallingford (United Kingdom), CAB International, 1992, 8–18. https://www.cabdirect.org/cabdirect/abstract/19921175466.
[80] Lomer CJ, Bateman RP, Johnson DL, Langewald J, Thomas M. Biological control of locusts and grasshoppers. Annual Review of Entomology 2001, 46, 667–702. doi:https://10.1146/annurev.ento.46.1.667
[81] C Lomer CJ, Prior C, Kooyman C. Deveolpment of *Metarhizium* spp. for the control of Grasshoppers and Locusts. Memoirs of the Entomological Society of Canada 1997, 129, 265–286. doi:https://10.4039/entm129171265-1
[82] Cuthbertson AGS, Walters KFA. Pathogenicity of the entomopathogenic fungus, Lecanicillium muscarium, against the sweetpotato Whitefly *Bemisia tabaci* under laboratory and glasshouse conditions. Mycopathologia 2005, 160, 315–319. doi:https://10.1007/s11046-005-0122-2
[83] Goettel MS, Koike M, Kim JJ, Aiuchi D, Shinya R, Brodeur J. Potential of *Lecanicillium* spp. for management of insects, nematodes and plant diseases. Journal of Invertebrate Pathology 2008, 98, 256–261. doi:https://10.1016/j.jip.2008.01.009
[84] Delalibera I, Gomez DRS, De Moraes GJ, De Alencar JA, Araujo WF. Infection of *Mononychellus tanajoa* (Acari: Tetranychidae) by the fungus *Neozygites* sp. (Entomophthorales) in northeastern Brazil. Florida Entomologist 1992, 75, 145–147. doi:https://10.2307/3495493
[85] Delalibera IJ, Hajek AE, Humber RA. *Neozygites tanajoae* sp. nov., a pathogen of the cassava green mite. Mycologia 2004, 96, 1002–1009.
[86] Askary TH. Nematodes as biocontrol agents. In: Lichtfouse E, eds. Sociology, Organic Farming, Climate Change and Soil Science. Sustainable Agriculture Reviews, vol 3. Dordrecht, Springer. https://doi.org/10.1007/978-90-481-3333-8_13
[87] Azizoglu U, Karabörklü S, Ayvaz A, Yilmaz S. Phylogenetic relationships of insect-associated free-living rhabditid nematodes from eastern mediterranean region of Turkey. Applied Ecology and Environmental Research 2016, 14, 93–103. doi:https://10.15666/aeer/1403_093103
[88] Thiruchchelvan N, Thirukkumaran G, Edgington S, Buddie A, Mikunthan G. Morphological characteristics and insect killing potential of a soil dwelling nematode, *Acrobeloides* cf. *longiuterus* from Sri Lanka. Plant Protection 2021, 5, 1–11. doi:https://10.33804/pp.005.01.3512

[89] Thiruchchelvan N, Thirukkumaran G, Mikunthan G. Mass production of the nematode *Acrobeloides longiuterus* using *Tribolium castaneum* and artificial solid media. Ruhuna Journal of Science 2021, 12, 14–25. doi:https://10.4038/rjs.v12i1.97
[90] Salari E, Karimi J, Fasihi Harandi M, Sadeghi Nameghi H. Comparative infectivity and biocontrol potential of *Acrobeloides* k29 and entomopathogenic nematodes on the leopard moth borer, *Zeuzera pyrina*. Biological Control 2021, 155, 104526. doi:https://doi.org/10.1016/j.biocontrol.2020.104526
[91] Tabassum AK, Shahina F. *Oscheius siddiqii* and *O. niazii*, two new entomopathogenic nematode species from Pakistan, with observations on *O. shamimi*. International Journal of Nematology 2010, 20, 75–84.
[92] Torrini G, Mazza G, Carletti B, et al. *Oscheius onirici* sp. N. (Nematoda: Rhabditidae): A new entomopathogenic nematode from an Italian cave. Zootaxa 2015, 3937, 533–548. doi: https://10.11646/zootaxa.3937.3.6
[93] Zhou G, Yang H, Wang F, et al. Oscheius microvilli n. sp. (Nematoda: Rhabditidae): A facultatively pathogenic nematode from Chongming Island, China. Journal of Nematology 2017, 49, 33–41. doi:https://10.21307/jofnem-2017-044
[94] Dillman AR, Chaston JM, Adams BJ, et al. An entomopathogenic nematode by any other name. PLoS Pathogens 2012, 8, e1002527–e1002527. doi:https://10.1371/journal.ppat.1002527
[95] Dutky SR, Hough WS. Note on a parasitic nematode from codling moth larvae, *Carpocapsa pamonetta* (Lepidoptera, Olethreutidae). In: Proceedig of Entomological Society 1955, 244.
[96] Glaser RW, Fox H. A nematode parasite of the japanese beetle (*Popillia japonica* Newm.). Science 1930, 71, 16–17. doi:https://10.1126/science.71.1827.16-b
[97] Gaugler R, Campbell JF, Selvan S, Lewis EE. Large-scale inoculative releases of the entomopathogenic nematode *Steinernema glaseri*: Assessment 50 years later. Biological Control 1992, 2, 181–187. doi:https://doi.org/10.1016/1049-9644(92)90057-K
[98] Cabanillas H, Poinar G, Railston J. *Steinernema riobravis* n. sp. (Rhabditia: Steinernematidae) from Texas. Fundamental and Applied Nematology 1994, 17, 123–131.
[99] Poinar GO, Jackson T, Klein M. Parasitic in the Japanese Beetle, *Popillia japonica* (Scarabaeidae: Coleoptera), in Ohio. Proceeding Helmintholgy Soc Washingt 1987, 54, 53–59
[100] Poinar GO. Description and biology of a new insect parasitic rhabditoid, *Heterorhabditis bacteriophora* n.gen., n.sp. (Rhabditida; Heterorhabditidae n.fam.). Nematologica 1975, 1, 463–470.
[101] Liu J, Berry RE. *Heterorhabditis marelatus* n.sp. (Rhabditida: Heterorhabditidae) from Oregon. Journal of Invertebrate Pathology 1996, 67, 48–54. https://doi.org/10.1006/jipa.1996.0008
[102] Vega FE. The use of fungal entomopathogens as endophytes in biological control: A review. Mycologia 2018, 110, 4–30. doi:https://10.1080/00275514.2017.1418578
[103] Liu J, Sui Y, Wisniewski M, Droby S, Liu Y. Review: Utilization of antagonistic yeasts to manage postharvest fungal diseases of fruit. International Journal of Food Microbiology 2013, 167, 153–160. doi:https://10.1016/j.ijfoodmicro.2013.09.004
[104] Spadaro D, Droby S. Development of biocontrol products for postharvest diseases of fruit: The importance of elucidating the mechanisms of action of yeast antagonists. Trends in Food Science and Technology 2016, 47, 39–49. doi:https://doi.org/10.1016/j.tifs.2015.11.003
[105] Díaz MA, Pereyra MM, Picón-Montenegro E, Meinhardt F, Dib JR. Killer yeasts for the biological control of postharvest fungal crop diseases. Microorganisms 2020, 8, 1680. doi: https://10.3390/microorganisms8111680
[106] Perez MF, Perez Ibarreche J, Isas AS, Sepulveda M, Ramallo J, Dib JR. Antagonistic yeasts for the biological control of *Penicillium digitatum* on lemons stored under export conditions. Biological Control 2017, 115, 135–140. doi:https://doi.org/10.1016/j.biocontrol.2017.10.006

[107] Klassen R, Schaffrath R, Buzzini P, Ganter PF. Antagonistic interactions and killer yeasts. In: Buzzini P, Lachance M, Yurkov A, eds. Yeasts in natural ecosystems: Ecology. Cham, Springer International Publishing, 2017, 229–275. doi:https://10.1007/978-3-319-61575-2_9
[108] Marquina D, Santos A, Peinado JM. Biology of killer yeasts. International Microbiology, the Spanish Society for Microbiology 2002, 5, 65–71. doi:https://10.1007/s10123-002-0066-z
[109] Perez MF, Contreras L, Garnica NM, et al. Native killer yeasts as biocontrol agents of postharvest fungal diseases in lemons. PLoS One 2016, 11, e0165590. doi:https://10.1371/journal.pone.0165590
[110] Rahman MA, Mahmud TMM, Kadir J, Rahman RA, Begum MM. Major postharvest fungal diseases of papaya cv. "Sekaki" in Selangor, Malaysia. Pertanika Journal of Tropical Agricultural Science 2008, 31, 27–34.
[111] Hassan H, Mohamed MTM, Yusoff SF, Hata EM, Tajidin NE. Selecting antagonistic yeast for postharvest biocontrol of *Colletotrichum gloeosporioides* in papaya fruit and possible mechanisms involved. Agronomy 2021, 11, 760. doi:https://10.3390/agronomy11040760
[112] Bosqueiro AS, Bizarria Júnior R, Rosa-Magri MM. Potential of *Trichosporon asahii* against *Alternaria* sp. and mechanisms of actions. Summa Phytopathologica 2020, 46, 20–25. doi: https://10.1590/0100-5405/220861
[113] Thomma BPHJ. *Alternaria* spp.: From general saprophyte to specific parasite. Molecular Plant Pathology 2003, 4, 225–236. doi:https://10.1046/j.1364-3703.2003.00173.x
[114] Reed KM, Bargmann BOR. Protoplast regeneration and its use in new plant breeding technologies. Frontiers in Genome Editing 2021, 3, 734951. doi:https://10.3389/fgeed.2021.734951
[115] Liang Z, Chen K, Zhang Y, et al. Genome editing of bread wheat using biolistic delivery of CRISPR/Cas9 *in vitro* transcripts or ribonucleoproteins. Nature Protocols 2018, 13, 413–430. doi:https://10.1038/nprot.2017.145
[116] Sauer NJ, Narváez-Vásquez J, Mozoruk J, et al. Oligonucleotide-mediated genome editing provides precision and function to engineered nucleases and antibiotics in plants. Plant Physiology 2016, 170, 1917–1928. doi:https://10.1104/pp.15.01696
[117] Charrier A, Vergne E, Dousset N, Richer A, Petiteau A, Chevreau E. Efficient targeted mutagenesis in apple and first time edition of pear using the CRISPR-Cas9 system. Frontiers in Plant Science 2019, 10, 40. https://www.frontiersin.org/article/10.3389/fpls.2019.00040
[118] Lee MH, Lee J, Choi SA, et al. Efficient genome editing using CRISPR–Cas9 RNP delivery into cabbage protoplasts via electro-transfection. Plant Biotechnology Reports 2020, 14, 695–702. doi:https://10.1007/s11816-020-00645-2
[119] Wójcik A, Rybczyński JJ. Electroporation and morphogenic potential of *Gentiana kurroo* (Royle) embryogenic cell suspension protoplasts. BioTechnologia 2015, 96, 19–29. doi: https://10.5114/bta.2015.54170
[120] Shulse CN, Cole BJ, Ciobanu D, et al. High-throughput single-cell transcriptome profiling of plant cell types. Cell Reports 2019, 27, 2241–2247. e4. doi:https://doi.org/10.1016/j.celrep.2019.04.054
[121] Gantait S, Sarkar S, Verma SK. Marker-assisted selection for abiotic stress tolerance in crop plants. In: Roychoudhury A, Tripathi D, eds. Molecular plant abiotic stress. The Atrium, Southern Gate, Chichester, West Sussex, UK, Wiley Online Books, Wiley, 2019, 335–368. doi: https://10.1002/9781119463665.ch18
[122] Broughton WJ, Hernández G, Blair M, Beebe S, Gepts P, Vanderleyden J. Beans (Phaseolus spp.) – Model food legumes. Plant and Soil 2003, 252, 55–128. doi:https://10.1023/A:1024146710611

[123] Stavely J, MA P-C. Rust. In: Schwartz HF, Pastor-Corrales MA, eds. Bean production problems in the tropics, 2nd edn. Cali, CO, Centro Internacional de Agricultura Tropical (CIAT), 1989, 159–194.
[124] Steadman JR, Pastor-Corrales MA, Beaver JS. An overview of the 3rd bean rust and 2nd bean common bacterial blight international workshops. 2002 1, 20–124. https://handle.nal.usda.gov/10113/IND23299719.
[125] Hurtado-Gonzales OP, Valentini G, Gilio TAS, Martins AM, Song Q, Pastor-Corrales MA. Fine mapping of Ur-3, a historically important rust resistance locus in common bean. G3: Genes, Genomes, Genetics 2017, 7, 557–569. doi:https://10.1534/g3.116.036061
[126] Flowers TJ, Koyama ML, Flowers SA, Sudhakar C, Singh KP, Yeo AR. QTL: Their place in engineering tolerance of rice to salinity. Journal of Experimental Botany 2000, 51, 99–106. doi:https://10.1093/jexbot/51.342.99
[127] Singh D, Kumar A, Ashok K, et al. Marker assisted selection and crop management for salt tolerance: A review. African Journal of Biotechnology 2011, 10, 14694–14698. doi:https://10.5897/AJB11.049
[128] Khush GS, Brar DS, Hardy B. Rice Genetics IV. In: Proceedings of the Fourth International Rice Genetics Symposium, 22–27 October 2000. Los Baños, Philippines. Enfield, NH (USA): Science Publishers, Inc., and Los Baños (Philippines): International Rice Research Institute, 2001, 488. doi:https://10.1142/6842
[129] Nagaraju J, Kathirvel M, Kumar RR, Siddiq EA, Hasnain SE. Genetic analysis of traditional and evolved Basmati and non-Basmati rice varieties by using fluorescence-based ISSR-PCR and SSR markers. Proceedings of the National Academy of Sciences of the United States of America 2002, 99, 5836–5841. doi:https://10.1073/pnas.042099099
[130] Shabanimofrad M, Yusop MR, Ashkani S, et al. Marker-assisted selection for rice brown planthopper (*Nilaparvata lugens*)resistance using linked SSR markers. Turkish Journal of Biology 2015, 39, 666–673. doi:https://10.3906/biy-1406-78
[131] Rasool G, Yousaf S, Ammara UE, et al. Transgenic expression of synthetic coat protein and synthetic replication associated protein produces mild symptoms and reduce begomovirus-betasatellite accumulation in *Nicotiana benthamiana*. Frontiers in Agronomy 2021, 3, 6768820. doi:https://10.3389/fagro.2021.676820
[132] Niu Q-W, Lin -S-S, Reyes JL, et al. Expression of artificial microRNAs in transgenic *Arabidopsis thaliana* confers virus resistance. Nature Biotechnology 2006, 24, 1420–1428. doi:https://10.1038/nbt1255
[133] Loriato VAP, Martins LGC, Euclydes NC, Reis PAB, Duarte CEM, Fontes EPB. Engineering resistance against geminiviruses: A review of suppressed natural defenses and the use of RNAi and the CRISPR/Cas system. Plant Science: An International Journal of Experimental Plant Biology 2020, 292, 110410. doi:https://doi.org/10.1016/j.plantsci.2020.110410
[134] Lucioli A, Sallustio DE, Barboni D, et al. A cautionary note on pathogen-derived sequences. Nature Biotechnology 2008, 26, 617–619. doi:https://10.1038/nbt0608-617
[135] Shepherd DN, Martin DP, Thomson JA. Transgenic strategies for developing crops resistant to geminiviruses. Plant Science: An International Journal of Experimental Plant Biology 2009, 176, 1–11. doi:https://doi.org/10.1016/j.plantsci.2008.08.011
[136] Schouten HJ, Krens FA, Jacobsen E. Cisgenic plants are similar to traditionally bred plants: International regulations for genetically modified organisms should be altered to exempt cisgenesis. EMBO Reports 2006, 7, 750–753. doi:https://10.1038/sj.embor.7400769
[137] Lusser M, Davies HV. Comparative regulatory approaches for groups of new plant breeding techniques. New Biotechnology 2013, 30, 437–446. doi:https://doi.org/10.1016/j.nbt.2013.02.004

[138] Rommens CM. Intragenic crop improvement: Combining the benefits of traditional breeding and genetic engineering. Journal of Agricultural and Food Chemistry 2007, 55, 4281–4288. doi:https://10.1021/jf0706631
[139] Conner AJ, Barrell PJ, Baldwin SJ, et al. Intragenic vectors for gene transfer without foreign DNA. Euphytica 2007, 154, 341–353. doi:https://10.1007/s10681-006-9316-z
[140] Rommens CM. All-native DNA transformation: A new approach to plant genetic engineering. Trends in Plant Science 2004, 9, 457–464. doi:https://10.1016/j.tplants.2004.07.001
[141] Jacobsen E, Schouten HJ. Cisgenesis strongly improves introgression breeding and induced translocation breeding of plants. Trends in Biotechnology 2007, 25, 219–223. doi:https://10.1016/j.tibtech.2007.03.008
[142] Kost TD, Gessler C, Jänsch M, Flachowsky H, Patocchi A, Broggini GAL. Development of the first cisgenic apple with increased resistance to fire blight. PLoS One 2015, 10, e0143980. doi:https://10.1371/journal.pone.0143980
[143] An C, Orbović V, Mou Z. An efficient intragenic vector for generating intragenic and cisgenic plants in citrus. American Journal of Plant Sciences 2013, 04, 2131–2137. doi:https://10.4236/ajps.2013.411265
[144] Dhekney SA, Li ZT, Gray DJ. Grapevines engineered to express cisgenic *Vitis vinifera* thaumatin-like protein exhibit fungal disease resistance. In Vitro Cellular & Developmental Biology 2011, 47, 458–466.
[145] Chougule NP, Bonning BC. Toxins for transgenic resistance to hemipteran pests. Toxins 2012, 4, 405–429. doi:https://10.3390/toxins4060405
[146] Metzlaff M, O'Dell M, Cluster PD, Flavell RB. RNA-mediated RNA degradation and chalcone synthase a silencing in *Petunia*. Cell 1997, 88, 845–854. doi:https://doi.org/10.1016/S0092-8674(00)81930-3
[147] Martínez de Alba AE, Elvira-Matelot E, Vaucheret H. Gene silencing in plants: A diversity of pathways. Biochimica Et Biophysica Acta – Gene Regulatory Mechanisms 2013, 1829, 1300–1308. doi:https://10.1016/j.bbagrm.2013.10.005
[148] Vaucheret H, Chupeau Y. Ingested plant miRNAs regulate gene expression in animals. Cell Research 2012, 22, 3–5. doi:https://10.1038/cr.2011.164
[149] Ipsaro JJ, Joshua-Tor L. From guide to target: Molecular insights into eukaryotic RNA-interference machinery. Nature Structural & Molecular Biology 2015, 22, 20–28. doi: https://10.1038/nsmb.2931
[150] Salame TM, Ziv C, Hadar Y, Yarden O. RNAi as a potential tool for biotechnological applications in fungi. Applied Microbiology and Biotechnology 2011, 89, 501–512. doi: https://10.1007/s00253-010-2928-1
[151] Scott JG, Michel K, Bartholomay LC, et al. Towards the elements of successful insect RNAi. Journal of Insect Physiology 2013, 59, 1212–1221. doi:https://10.1016/j.jinsphys.2013.08.014
[152] Navarro L, Dunoyer P, Jay F, et al. A plant miRNA contributes to antibacterial resistance by repressing auxin signaling. Science 2006, 312, 436–439. doi:https://10.1126/science.1126088
[153] Ding S-W. RNA-based antiviral immunity. Nature Reviews: Immunology 2010, 10, 632–644. doi:https://10.1038/nri2824
[154] Frizzi A, Huang S. Tapping RNA silencing pathways for plant biotechnology. Plant Biotechnology Journal 2010, 8, 655–677. doi:https://doi.org/10.1111/j.1467-7652.2010.00505.x
[155] Borges F, Martienssen RA. The expanding world of small RNAs in plants. Nature Reviews: Molecular Cell Biology 2015, 16, 727–741. doi:https://10.1038/nrm4085
[156] Jonas S, Izaurralde E. Towards a molecular understanding of microRNA-mediated gene silencing. Nature Reviews: Genetics 2015, 16, 421–433. doi:https://10.1038/nrg3965

[157] Molnar A, Melnyk C, Baulcombe DC. Silencing signals in plants: A long journey for small RNAs. Genome Biology 2011, 12, 215. doi:https://10.1186/gb-2010-11-12-219
[158] Henderson IR, Zhang X, Lu C, et al. Dissecting Arabidopsis thaliana DICER function in small RNA processing, gene silencing and DNA methylation patterning. Nature Genetics 2006, 38, 721–725. doi:https://10.1038/ng1804
[159] Weiberg A, Bellinger M, Jin H. Conversations between kingdoms: Small RNAs. Current Opinion in Biotechnology 2015, 32, 207–215. doi:https://10.1016/j.copbio.2014.12.025
[160] Cheng W, Song X-S, Li H-P, et al. Host-induced gene silencing of an essential chitin synthase gene confers durable resistance to *Fusarium* head blight and seedling blight in wheat. Plant Biotechnology Journal 2015, 13, 1335–1345. doi:https://10.1111/pbi.12352
[161] Carbonell A, Carrington JC. Antiviral roles of plant ARGONAUTES. Current Opinion in Plant Biology 2015, 27, 111–117. doi:https://10.1016/j.pbi.2015.06.013
[162] Younis A, Siddique MI, Kim C-K, Lim K-B. Interference RNA (RNAi) induced gene silencing: A promising approach of Hi-Tech plant breeding. International Journal of Biological Sciences 2014, 10, 1150–1158. doi:https://10.7150/ijbs.10452
[163] Duan CG, Wang CH, Guo HS. Application of RNA silencing to plant disease resistance. Silence 2012, 3, 5. doi:https://10.1186/1758-907X-3-5
[164] Senthil-Kumar M, Mysore KS. Virus-induced gene silencing can persist for more than 2years and also be transmitted to progeny seedlings in *Nicotiana benthamiana* and tomato. Plant Biotechnology Journal 2011, 9, 797–806. doi:https://doi.org/10.1111/j.1467-7652.2011.00589.x
[165] Fritz JH, Ferrero RL, Philpott DJ, Girardin SE. Nod-like proteins in immunity, inflammation and disease. Nature Immunology 2006, 7, 1250–1257. doi:https://10.1038/ni1412
[166] Kang JS, Frank J, Kang CH, et al. Salt tolerance of *Arabidopsis thaliana* requires maturation of N-glycosylated proteins in the Golgi apparatus. Proceedings of the National Academy of Sciences of the United States of America 2008, 105, 5933–5938. doi:https://10.1073/pnas.0800237105
[167] Jiang C-J, Shimono M, Maeda S, et al. Suppression of the Rice fatty-acid desaturase gene osssi2 enhances resistance to blast and leaf blight diseases in rice. Molecular Plant-Microbe Interactions 2009, 22, 820–829. doi:https://10.1094/MPMI-22-7-0820
[168] Escobar MA, Civerolo EL, Summerfelt KR, Dandekar AM. RNAi-mediated oncogene silencing confers resistance to crown gall tumorigenesis. The Proceedings of the National Academy of Sciences 2001, 98, 13437–13442. doi:https://10.1073/pnas.241276898
[169] Zhao Y, Srivastava D. A developmental view of microRNA function. Trends in Biochemical Sciences 2007, 32, 189–197. doi:https://10.1016/j.tibs.2007.02.006
[170] Jian X, Zhang L, Li G, et al. Identification of novel stress-regulated microRNAs from *Oryza sativa* L. Genomics 2010, 95, 47–55. doi:https://10.1016/j.ygeno.2009.08.017
[171] Liu -H-H, Tian X, Li Y-J, Wu C-A, Zheng -C-C. Microarray-based analysis of stress-regulated microRNAs in *Arabidopsis thaliana*. RNA 2008, 14, 836–843. doi:https://10.1261/rna.895308
[172] Arenas-Huertero C, Pérez B, Rabanal F, et al. Conserved and novel miRNAs in the legume *Phaseolus vulgaris* in response to stress. Plant Molecular Biology 2009, 70, 385–401. doi:https://10.1007/s11103-009-9480-3
[173] Ding D, Zhang L, Wang H, Liu Z, Zhang Z, Zheng Y. Differential expression of miRNAs in response to salt stress in maize roots. Annals of Botany 2009, 103, 29–38. doi:https://10.1093/aob/mcn205
[174] Lu PY, Xie F, Woodle MC. *In vivo* application of RNA interference: From functional genomics to therapeutics. Advances in Genetics 2005, 54, 117–142. doi:https://10.1016/S0065-2660(05)54006-9

[175] Yu X-D, Liu Z-C, Huang S-L, et al. RNAi-mediated plant protection against aphids. Pest Management Science 2016, 72, 1090–1098. doi:https://doi.org/10.1002/ps.4258
[176] Ibrahim AB, Monteiro TR, Cabral GB, Aragão FJL. RNAi-mediated resistance to whitefly (i*Bemisia tabaci*) in genetically engineered lettuce (*Lactuca sativa*). Transgenic Research 2017, 26, 613–624. doi:https://10.1007/s11248-017-0035-0
[177] Zhang Y, Malzahn AA, Sretenovic S, Qi Y. The emerging and uncultivated potential of CRISPR technology in plant science. Nature Plants 2019, 5, 778–794. doi:https://10.1038/s41477-019-0461-5
[178] Ahmed T, Noman M, Shahid M, et al. Potential application of CRISPR/Cas9 system to engineer abiotic stress tolerance in plants. Protein and Peptide Letters 2021, 28, 861–877. doi:https://10.2174/0929866528666210218220138
[179] Jinek M, Chylinski K, Fonfara I, Hauer M, Doudna JA, Charpentier EA. Programmable dual-RNA–guided DNA endonuclease in adaptive bacterial immunity. Science 2012, 337, 816–821. doi:https://10.1126/science.1225829
[180] Pramanik D, Shelake RM, Kim MJ, Kim J-Y. CRISPR-Mediated Engineering across the central dogma in plant biology for basic research and crop improvement. Molecular Plant 2021, 14, 127–150. doi:https://10.1016/j.molp.2020.11.002
[181] Pramanik D, Shelake RM, Park J, et al. CRISPR/Cas9-mediated generation of pathogen-resistant tomato against tomato yellow leaf curl virus and powdery mildew. International Journal of Molecular Sciences 2021, 22, 1878. doi:https://10.3390/ijms22041878
[182] FAO. Global Forest Products Facts and Figures (2018–2019). 2018. doi:http://www.fao.org/forestry/statistics/80938/en/
[183] Jones DR. Plant viruses transmitted by whiteflies. The European Journal of Plant Pathology 2003, 109, 195–219. doi:https://10.1023/A:1022846630513
[184] Prasad A, Sharma N, Hari-Gowthem G, Muthamilarasan M, Prasad M. Tomato yellow leaf curl virus: Impact, challenges, and management. Trends in Plant Science 2020, 25, 897–911. doi: https://doi.org/10.1016/j.tplants.2020.03.015
[185] Dhaliwal MS, Jindal SK, Sharma A, Prasanna HC. Tomato yellow leaf curl virus disease of tomato and its management through resistance breeding: A review. The Journal of Horticultural Science and Biotechnology 2020, 95, 425–444. doi:https://10.1080/14620316.2019.1691060
[186] Shen X, Yan Z, Wang X, et al. The NLR protein encoded by the resistance gene Ty-2 is triggered by the replication-associated protein Rep/C1 of tomato yellow leaf curl virus. Frontiers in Plant Science 2020, 11, 1384. https://www.frontiersin.org/article/10.3389/fpls.2020.545306
[187] Lapidot M, Karniel U, Gelbart D, et al. A novel route controlling begomovirus resistance by the messenger rna surveillance factor Pelota. PLoS Genetics 2015, 11, e1005538. doi: https://10.1371/journal.pgen.1005538
[188] Lee C. CRISPR/Cas9-based antiviral strategy: Current status and the potential challenge. Molecules 2019, 24, 1349. doi:https://10.3390/molecules24071349
[189] Peng A, Chen S, Lei T, et al.. Engineering canker-resistant plants through CRISPR/Cas9-targeted editing of the susceptibility gene CsLOB1 promoter in citrus. Plant Biotechnology Journal 2017, 15, 1509–1519. doi:https://10.1111/pbi.12733
[190] Hu Y, Zhang J, Jia H, et al. Lateral organ boundaries 1 is a disease susceptibility gene for citrus bacterial canker disease. The Proceedings of the National Academy of Sciences 2014, 111, E521–9. doi:https://10.1073/pnas.1313271111
[191] Mickelbart MV, Hasegawa PM, Bailey-Serres J. Genetic mechanisms of abiotic stress tolerance that translate to crop yield stability. Nature Reviews: Genetics 2015, 16, 237–251. doi:https://10.1038/nrg3901

[192] Wang L, Chen L, Li R, et al. Reduced drought tolerance by CRISPR/Cas9-Mediated SlMAPK3 mutagenesis in tomato plants. Journal of Agricultural and Food Chemistry 2017, 65, 8674–8682. doi:https://10.1021/acs.jafc.7b02745
[193] Lou D, Wang H, Liang G, Yu D. OsSAPK2 confers abscisic acid sensitivity and tolerance to drought stress in rice. Frontiers in Plant Science 2017, 8, 993. doi:https://10.3389/fpls.2017.00993
[194] Hirai MY, Sugiyama K, Sawada Y, et al. Omics-based identification of *Arabidopsis* Myb transcription factors regulating aliphatic glucosinolate biosynthesis. The Proceedings of the National Academy of Sciences 2007, 104, 6478–6483. doi:https://10.1073/pnas.0611629104
[195] Razzaq MK, Aleem M, Mansoor S, et al. Omics and CRISPR-Cas9 approaches for molecular insight, functional gene analysis, and stress tolerance development in crops. International Journal of Molecular Sciences 2021, 22, 1292. doi:https://10.3390/ijms22031292
[196] Quareshy M, Prusinska J, Li J, Napier R. A cheminformatics review of auxins as herbicides. Journal of Experimental Botany 2018, 69, 265–275. doi:https://10.1093/jxb/erx258
[197] Akbar N, Jabran K, Ali MA. Weed management improves yield and quality of direct seeded rice. Australian Journal of Crop Science 2011, 5, 688–694.
[198] Chauhan BS. Strategies to manage weedy rice in Asia. Crop Protection (Guildford, Surrey) 2013, 48, 51–56. doi:https://10.1016/j.cropro.2013.02.015
[199] Glick HL. Herbicide tolerant crops: A review of agronomic, economic and environmental impacts. In: The BCPC Conference: Weeds, 2001, Volume 1 and Volume 2. Proceedings of an International Conference Held at the Brighton Hilton Metropole Hotel, Brighton, UK, 12–15 November 2001. British Crop Protection Council, 2001, 359–366.
[200] Dong H, Huang Y, Wang K. The development of herbicide resistance crop plants using CRISPR/Cas9-mediated gene editing. Genes 2021, 12, 912. doi:https://10.3390/genes12060912
[201] Sammons RD, Gaines TA. Glyphosate resistance: State of knowledge. Pest Management Science 2014, 70, 1367–1377. doi:https://10.1002/ps.3743

Sikhamoni Bora, Rajib Kumar Borah, Krishna Giri

Chapter 15 Role of proteins and enzymes in plant disease control

Abstract: Plant disease is a disorder that occurs as a consequence of unusual changes in the morphology, physiology, integrity, or behavior of the plant. It is a harmful deviation from the regular functioning of biological processes of sufficient duration or intensity to cause disruption or cessation of vital activities, according to the American Phytopathological Society. Plant–pathogen interaction is a complex process where both the organisms play a significant role by synthesizing several proteins and enzymes. Plants have developed a complex defense system against diverse insects, pests, and pathogens. There are several proteins and enzymes such as membrane spanning protein, protein kinases, heat shock protein, peroxidase, chitinase, etc., which act against different plant pathogens through various signaling pathways. After pathogens overcome mechanical barriers to infection, plant receptors initiate multiple biochemical pathways that result in the activation of defense response genes. Plant immune systems completely depend on their ability to recognize foreign molecules and to respond defensively against the pathogen, through several biochemical pathways involving several genes and their products. The aim of this chapter is to highlight plant–pathogen interaction and role of protein and enzymes in disease suppression and control.

15.1 Introduction

The American Phytopathological Society and the British Mycological Society have both agreed that disease is a malfunctioning process induced by continual irritations that result in some discomfort and symptoms [1]. It can also be defined as a change in one or more of the tidy consecutive series of physiological processes, culminating in a loss of coordination of energy utilization in a plant as a result of the continuous irritation from the presence or absence of some external factor or agent. A pathogen is constantly associated with a disease and can be broadly defined as any agent or factor that can induce "pathos" or disease in an organism. So, the pathogen may be living, nonliving, or in between living and nonliving [2]. The interaction of the host, the pathogen, and the environment constitutes the disease triangle (Figure 15.1).

Plant diseases may be groped in various categories based on certain parameters. According to the pattern of their infection, the disease may be soilborne, seedborne, airborne, etc. Thus, the disease caused by animate or virus pathogens are often classified as [2]:

https://doi.org/10.1515/9783110771558-015

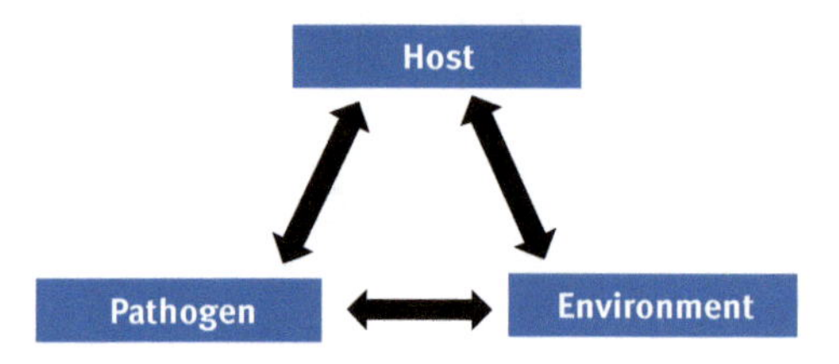

Figure 15.1: The disease triangle.

1. **Endemic disease:** If a disease is more or less continually present year to year in a moderate to acute form, in a particular place, it is classified as endemic.
2. **Epidemic disease:** It is described as a disease that arises widely but periodically, and may be present continually in the locality but assumes a critical form only on instances.
3. **Sporadic diseases:** Diseases that occur at random breaks and location and in relatively few instances.

Changes in plant growth or appearance in response to disease is called symptom of the disease. Symptoms of the disease are seen on the plant due to character and appearance of the visible pathogen or its structure or organs, or due to some effect upon or change in the host plant [2]. Some of them are:

a) Symptoms due to the character and appearance of the pathogen or its structure or organs, i.e., mildew, rusts, smuts, white blisters, scab, sclerotia, blotch, fruiting bodies, exudations, tar spots, etc.
b) Symptoms due to some effect on, or change in the host plant and as a result of disease there may be a visible change in the form, size, color, texture, attitude, or habit of the plant or some of its organs. Such symptoms are color changes, overgrowth or hypertrophy (i.e., galls, curl, pocket or bladder, witches broom, hairy root, intumescence), atrophy or hypoplasia, necrosis (spots, streaks or stripes, canker, blight, damping off, burn, scald or scorch, rot), wilts, or die-back [2].
c) Miscellaneous symptoms: It includes dropping of leaves, blossoms, fruits or twigs due to some physiological disorder or indirect effect of parasitic diseases; destruction of specific organs due to many parasitic diseases and transformation of organs or replacement of organs by new structure.

Categorization of plant ailment depends on numerous factors; sometimes they are classified on the basis of symptoms, i.e., root rots, cankers, wilts, leaf spots, rusts, scabs, blights, smuts, mosaics, and yellows, or plant part they affect (diseases of root, stem, foliage, fruit), or types of plants affected (diseases of field crop, vegetable, fruit tree, forest, turf, ornamental plants). The most useful criterion for classification of a disease, however, is the mode of pathogen that triggers the disease [3].

So, according to the type of pathogen that triggers plant diseases, they are classified as follows [3]:

1. **Infectious or biotic plant diseases**
   - 1.1 Disease caused by fungi
   - 1.2 Disease caused by prokaryotes i.e. bacteria and mycoplasmas
   - 1.3 Disease caused by parasitic higher plants
   - 1.4 Disease caused by viruses and viroids
   - 1.5 Disease caused by nematodes
   - 1.6 Disease caused by protozoa
2. **Noninfectious or abiotic diseases**
   - 2.1 Limited or excessive temperature
   - 2.2 Lack or excess of soil moisture and light
   - 2.3 Lack of oxygen
   - 2.4 Air pollution
   - 2.5 Nutrient deficiencies
   - 2.6 Mineral toxicities
   - 2.7 Soil acidity or basicity
   - 2.8 Toxicity of pesticides
   - 2.9 Improper cultural practices

There are several disease-causing organisms that attack the plants including microorganisms like fungi, bacteria, viruses, and nematodes, resulting in huge crop losses. Therefore, plant disease control is the prime focused area of plant breeding, plant pathology, and the agricultural chemical industry. There are various defense mechanisms through which plant defend themselves against pathogen, such as antimicrobial activity, secondary compounds, and by inducing defense responses [4, 5].

The prime causative agents in plant pathogenesis are fungi, which are known for their ability to infect, for more than 70% of major diseases in trees, agricultural crops, and landscapes [6,7]. Among several fungal diseases, common fungal diseases that occur in crop include wheat rust, powdery mildew of pulses, rice blast, rice sheath blight, etc. [7]. In general, fungal pathogens attack plants through natural openings or wounds, but true phytopathogenic fungi enter the plant cell by defeating the plant's structural defenses, such as the cuticle and epidermal cell wall, and secreting hydrolytic enzymes such as pectinases, chitinases, cellulases, and proteases. Pathogens must overcome the structural and biochemical defensive barriers of plants in order to successfully invade plant cells [6]. Cuticle, a structural component of plants, prevents fungal diseases from entering directly and spore germination by secreting wax [7]. Fungi can sometimes overcome plant defenses by employing a variety of approaches, such as secreting effector chemicals that interact with the plant's basic defense response [8]. Pathogens secrete chemical messengers to engage with the host plant's signaling mechanism, in order to compete with the plant's biochemical barriers [9].

15.2 Defense mechanism of plants against pathogen

In general, plants defend themselves against pathogens by employing structural or physical barrier and biochemical reactions [10]. In structural defense, the physical barriers impede the pathogen from entering and spreading the disease through the plant and, in biochemical reaction, they produce different secondary metabolites, toxin, or other substances, which directly or indirectly impede entry of the pathogen. The plant's defense system varies according to the age of the plants, kind of plant organ, and tissue attacked, climatic condition, nutritional state of the plants, as well as the defense system in various host pathogens and even in the same plant pathogen system [3]. Therefore, the outer surface of the plants play major role in the entrance of pathogen to a host tissue and cause infection. There are several structural defense systems present in the plants, i.e., thickness of wax, cuticle, structure of the epidermal cell walls, size, location, shapes of stomata and lenticels, and the presence of tissues made of thick-walled cells, which directly or indirectly hinder the rate of infection in plants [10, 3].

15.3 Structural defense mechanism or physical barrier

There are several types of structures that act as structural or physical barriers to prevent the plant from infection. Some of them are listed below:

15.3.1 Epidermal cell wall

The outermost wall of the cell known as epidermal cell wall, and it is a very significant part of plants. Plant resistance to pathogen completely relies on the thickness and toughness of the epidermal cell wall. The presence of tough or thick cell wall prevents the entry of the pathogen to host cell. Along with the thickness of the cell wall, the presence or absence of lignin and silicic acid affects the toughness of the cell wall, preventing disease spread [10, 3].

15.3.2 Cuticular wax

Cuticular wax is made up of cuticle and wax. Waxes form a water-repellent surface on leaf and fruit surfaces and, thereby, prevent the formation of a film of water on which pathogens might be accumulated and germinate. Cuticle thickness is not usually connected with resistance to infection, and many plant varieties with thick cuticles are easily invaded by directly invading pathogens [10, 3].

15.3.3 Natural openings

Stomata or lenticels are the natural openings present in plant cell wall through which many pathogenic fungi and bacteria enter the plant cell even in the closed stomata. However, in the case of stem rust of wheat, the pathogens can enter only when stomata are open [10, 3].

Defense structures formed due to infection by the pathogen

There are some defense structures formed due to infection by the pathogen, such as: [10, 3],

i) Histological defense structures
ii) Cellular defense structures
iii) Cytoplasmic defense reaction
iv) Necrotic defense reaction/ defense through hypersensitivity

i) Histological defense structures

Cork layer formation

The action of pathogen like fungi, bacteria, nematodes, and also the substances secreted by the pathogen stimulate the host cell to form numerous layers of cork cells. These cork layers impede the entry of the pathogen as well as block the spread of substances that may have toxic effect to the plant. It also prevents the movement of water and nutrients from healthy to infected area to stop the pathogen from gaining nourishment from the host tissue [10, 3].

Abscission layer formation

Abscission layers usually develop to separate the ripe fruits and old leaves in plants, but they are also formed in young, meristematic leaves of stone fruit trees after infection by any pathogen. It initiates gap between unhealthy and healthy tissue and because of this gap formation, the infected portion becomes unsupported, slowly dies, and falls down along with pathogen [10, 3].

Tyloses formation

Tyloses are considered as outgrowths of the protoplast of adjacent parenchymatous cells that protrude into xylem vessel through pits. The main function of tyloses is to stop the spread of pathogens by blocking the xylem vessel [10, 3].

Deposition of gums

After the infection by various pathogens or wound, different types of gums are produced by plants around the lesions. The defensive role provided by the gums is that they are quickly deposited in the intercellular spaces and within the cells adjacent to the point of infection, by forming a barrier that fully encloses the entry of pathogen into the host cell [10, 3].

ii) Cellular defense structures

The morphological changes in the host cell wall after invasion by the pathogen are referred to as cellular defense. In this defense mechanism, four structural changes have been observed, mainly: (a) Swelling of parenchyma cells because of the infection of incompatible bacteria. The infected or swelled cells then produce an amorphous fibrillar substance that surrounds and traps the pathogen and inhibits it from further spread; (b) Thickening of cell walls in response to some fungal and viral pathogens; (c) deposition of Callose papillae on inner side of the cell wall in defense of fungal pathogens; (d) Delaying penetration and imparting partial check of pathogen spread by hyphal sheathing [10, 3].

iii) Cytoplasmic defense reaction

In this defense reaction, cytoplasm plays the key role. In this type of defense, the cytoplasm and nucleus enlarge in the invaded cells. As a result, cytoplasm becomes granular and dense, and finally, the mycelium of the pathogen splits, and the invasion stops [10, 3].

iv) Necrotic defense reaction/defense through hypersensitivity

This type of defense reaction is common in diseases caused by obligate fungal parasites, viruses, and nematodes. In such cases, the necrotic tissue separates itself from the living substance on which it depends completely for its nutrition and, therefore, results in malnourishment and death [10, 3].

15.3.4 Metabolic (biochemical) defense

Apart from various structural defense systems, plants have definite metabolic processes in their cells, which generate resistance against pathogen attack. There are two types of biochemical defenses, as given below:

i. Pre-existing Biochemical Defense
ii. Post-inflectional biochemical defense mechanism

(i) Pre-existing biochemical defense

Inhibitors present in plants and inhibitor released by the plant in its environment
Plants secrete several substances through the surface of their aboveground parts and their root surfaces. There are certain compounds released by plants that appear to have an inhibitory action against pathogens. The plant cell possesses specific pre-existing inhibitory substances, which primarily play a defensive role against the particular pathogen. The pre-existing antimicrobial substances in plant cells include sulfur-containing compound, cynogenic glycosides, phenols and saponins (tomatin, solanin, etc.). The inhibitory substances secreted by the pathogen either directly affect micro-organisms or encourage certain groups to control the environment which may act as antagonist to the pathogen [10, 3].

(ii) Post-inflectional biochemical defense mechanism

After the pathogen attack, many biochemical substances like phytoalexins and hypersensitive defense are produced in plants. These substances that are produced around the site of injury are generally fungicidal. Subsequently, a series of biochemical reactions occur to heal the wound [10, 3].

– **Phytoalexins**

These are toxic antimicrobial compounds that are only produced in plants after being stimulated by certain phytopathogenic microorganisms or mechanical harm. Phytoalexins are produced by healthy cells, when they are in close proximity to damaged and necrotic cells [4].

– **Phenolic Compounds**

The molecules that are produced by plants in response to injury or infection are known as phenolic compounds. The “Shikimic acid pathway” or the “acetic acid pathway” is the mechanism used to synthesize phenolic chemicals. Some of the typical phenolic chemicals that are hazardous to pathogens are caffeic acid, chlorgenic acid, and ferulic acid. In resistant types, these phenolic chemicals are produced more quickly than in susceptible varieties. As a result, it is thought that the combined action of all phenolics that are present in the host or plants inhibit infection.

– **Enzymes in disease management**

Enzymes are the protein molecules that catalyze all the interrelated reactions in a cell. Most of the pathogens derive their energy from enzymatic breakdown of food matter from host tissue. There are numerous defense enzymes present in plants including peroxidase, ammonia lyase, polyphenol oxidase, phenylalanine, β-1,3-glucanase, and chitinase. These enzymes are directly or indirectly correlated to resistance induction in plants [11–13]. Among various defense related processes are suberization, hypersensitive response, lignification, cross-linking of phenolics and glycoproteins, and phytoalexin production, which are directly associated with the enzyme peroxidases [14, 15]. Increase in phenylalanine ammonia lyase (PAL) activity in infected plant tissues indicated the synthesis of phenolic compounds in plants to prevent insects and pathogens [16]. The hydrolysis of cell wall components in sequence such as chitin and β-1, 3-glucans are bought about by enzymes Chitinase and β-1, 3-glucanase [17]. Fungi employ enzymes like cellulases for degradation of plant cell walls. On detection of these fungal proteins and in response to that, plants react by producing enzyme inhibitors and depositing lignin and callose to brace the cell wall [18]. The level of protection observed in the plants is variable due to various reasons, such as specific action of the enzyme, its concentration and localization within the cell, the features of the fungal pathogen, and the nature of the host-pathogen interaction [19].

There are a number of enzymes related to plant defense mechanism as well as disease management:

Chitinases

It belongs to the glycosyl hydrolase group that has a molecular size of 20–90 k Da and accelerates the hydrolysis of glycosidic bonds present in protein chitin [20]. Several organisms such as bacteria, yeasts, plants, fungi, actinomycetes, arthropods, and humans produced chitinases naturally [21]. There are various reports that suggest that chitinases have several different functions, such as pathogenesis, parasitism, morphogenesis, nutrition, immunity, growth regulation, and defense [22].

Chitinolytic enzymes are considered to be the most important enzymes in disease management, as they are able to hydrolyze chitin that is usually present in fungal pathogens [23]. There are several reports, which suggest that plants produce pathogenesis-related enzymes against the action of pathogen and that they can strengthen the defense mechanism [24, 25]. Nowadays, chitinase genes are introduced in transgenic plants for cloning and then expressed into various plant species for improvement of disease resistance [26, 27]. For example, in tea, genome class I chitinase gene (AF153195) from potato was introduced, and it was found that its overexpression showed resistance against a disease called blister blight [28]. It was also reported that chitinase and β-1, 3-glucanase together can act synergistically to inhibit fungal growth [29]. Chitinase can degrade chitin and make it osmotically sensitive [30]. The chitinases have resilient antifungal activity against various pathogens such as *Rhizoctoni asolani, Alternaria spp., Bipolaris oryzae, Botrytis cinerea, Curvularia lunata, Fusarium oxysporum, F. udum, Mycosphaerella arachidicola,* and *Pestalotia theae* for tea leaf spot [31–33].

Peroxidases

It is a class of PR protein that belongs to PR protein 9 subfamily and is stimulated because of the action of pathogen infection in host tissues [34]. It is a major enzyme in the synthesis of lignin and suberin and is associated with a number of physiological functions. It contributes to resistance through oxidation of hydroxyl cinnamyl alcohol into free radical intermediates, phenol oxidation, hypersensitive responses, cross-linking of polysaccharide, and extension monomers. During resistance reaction, the phenolic material is deposited in the plant cell walls [35]. In higher plants, peroxidase activity is induced by fungi, bacteria, viruses, and viroids [36–39]. In oxidative coupling of lignin subunits, the cross-linking of phenolic monomers has been associated with peroxidase, using H_2O_2 as oxidant. Oxidative burst is the key reaction as an early response of host plant cells against the pathogen infection [40]. The peroxidase level increase due to induced systemic resistance initiates quick reactive oxygen derivative synthesis, leading to cell death of pathogens through oxidative burst reaction [41, 42]. Phenolic and quinines are oxidized by peroxidase enzyme and hydrogen peroxide produced in the cell. The antimicrobial nature of H_2O_2 and release of highly reactive free radicles increase the polymerization of phenolic compounds into lignin-like substances. These substances are deposited in the cell wall and papillae and hinder the proliferation of pathogens [3].

β-1,3-Glucanases

They belong to the PR-2 family of proteins and are the chief constituent of oomycetes group of fungi [43]. These enzymes decompose glucans like callose present in plant tissues for wall modification against disease resistance [44]. It has also been described that the antifungal β-glucanase-I seems to be tailored for defense against fungal diseases, while studies on β-glucanase I also implied that β-1,3-glucanases play a crucial role in viral pathogenesis through deficient mutants generated by antisense transformation [45]. The endotype β-1,3-glucanase enzyme, one of the most important for callosic wall degradation, and exotype β-1,3-glucanase are implicated in the further hydrolysis of released oligosaccharides. These glucanohydrolases function in two different ways: (i) directly by pathogen cell wall degradation, and (ii) by promoting the release of cell wall derived materials, which can act as elicitor in defense mechanism [46].

Phenylalanine ammonia lyase (PAL)

Phenylalanine ammonia lyase (PAL) is an enzyme that is responsible for linking primary metabolism of aromatic amino acids with secondary metabolic products [47]. It is the most extensively studied enzyme in plants, as it is accountable for the synthesis of many phenolic compounds as well as anthocyanin that are accountable for the resistance of pathogens [48]. It was reported that notable changes occur in PAL activity, during a pathological event [13]. It was found that the plant hormone ethylene and signaling molecules including jasmonic acid and salicylic acid can stimulate the PAL activity [49, 50]. Biotic and abiotic stresses such as pathogen invasion, wounding, chilling, and ozone can also induce PAL activity [51]. According to some reports, when strawberry plants were treated with abscisic acid and anthocyanin, PAL activity was seen to increase [52]. All phenylpropanoid compounds that are derivative of cinnamic acid formed from phenylalanine, by the activity of PAL. These compounds are accountable for crop development and mechanical support, disease resistance, as well as insect pest damages [53–56].

Polyphenol oxidase (PPO)

It is a group of enzymes containing copper in their structure, which catalyzes oxidation of hydroxy phenols to their quinone derivatives that possess antimicrobial activity [57]. Due to its reaction products and wound inducibility, it aids in defense against pathogen [58]. Pathogen-induced PPO activity has been documented in a variety of plant taxa, including monocots and dicots [59, 60]. The PPO activity of banana roots treated with *Fusarium oxysporum*-derived elicitor was found to increase [61]. High PPO levels in cultivars or lines with high pathogen resistance have also been linked in studies, indicating that PPO may play a role in pathogen defense [62]. It has also been observed that the correlation of the protective effects of rhizosphere bacteria with an induction of defense enzymes including PPO results in mixed success [59, 63]. Li and Steffens [64] suggested several possibilities, including general toxicity of PPO-generated quinones to pathogens and plant cells, accelerating cell death, alkylation, and reduced bioavailability of cellular proteins to the pathogen, cross-linking of quinones with protein or other phenolics, forming a physical barrier to pathogens in the cell wall, and quinone redox cycling, leading to H_2O_2 and other reactive oxygen species [64, 65].

15.3.5 Protein and its role in disease management

In 1839, the Dutch chemist G. J. Mulder coined the term protein (Greek: proteios, primary). Proteins are unbranched polymers constructed from 22 standard α-amino acids. Thus, proteins are the major armaments synthesized by both organisms in plant–pathogen interaction [66]. The structural blockades, such as epidermal cell wall, cutin, suberin, wax, lignin, cellulose, callose, and cell wall proteins are rapidly reinforced upon the infection. Plants also synthesized secondary metabolic compounds and antifungal proteins, which can function as antimicrobial compounds during their defense against microorganisms. These include saponins, phenolics, terpenoids, and steroids. There are reports of some preformed compounds that are directly toxic, while others occur as conjugates such as glycosides that are not toxic, but become poisonous, following disruption of the conjugate [67]. Over 500 naturally occurring proteins have been reported that are believed to interact with the fungal membrane, leading to pore formation, efflux of cellular components, and changes in the membrane potential [68].

There are various types of proteins that play a crucial role in disease management and perform various defensive roles. Some of them are discussed below:

– **R protein**

To detect proteins produced during infection by specific pathogens, plants have a well-evolved recognition system, and the protein produced is termed as effectors. The gene-for-gene interaction shows that plant disease resistance (R) proteins easily recognize the effectors in a specific manner [69].

– **PR protein**

In1980, Pathogenesis-related protein (PR) was coined to describe any protein that was coded for the host plant but induced in pathological conditions such as viral, fungal, or bacterial infections, parasitic attack by nematodes, phytophagous insects, and other higher forms of animals like herbivores [70]. Acidic and basic PRs are induced in response to abiotic stress components. Such induction in the absence of pathogenic stimuli suggests that they are involved in cellular structures protection, either physically by stabilizing membranes or by keeping hazardous bacteria in check [71].

Although some noninfectious physiological circumstances (e.g., toxin-induced chlorosis or necrosis) frequently prompt induction of specific PR proteins, abiotic stressors and disorders were not considered as inducers of PR proteins. [72]. Proteins that are constitutively present in low but detectable amounts in healthy tissues but are induced under pathological conditions are not considered PR proteins, according to research [73]. PR proteins were acidic having low molecular mass, extremely resistant to proteolytic degradation, and had low PH values located in the intercellular space of leaves, at first [74]. PR protein is able to directly alter pathogen integrity, and as a result, they can produce various signal molecules via their enzymatic activity, which works as an elicitor and induces various defense-related mechanisms [74–76]. The PR-like proteins are present in healthy plants, primarily induced in a developmentally controlled, tissue-specific manner. These proteins, not produced in response to pathogen infection or related stresses, are predominantly basic and localized intracellularly in the vacuole [77]. The expression of several PR proteins in pathological situations implies, but does not prove, that these proteins play a role in plant defense [78].

Therefore, although these proteins have generally been considered protective proteins that prevent the invasion and spread of pathogens, their involvement in disease resistance to early infections is usually small [79]. PR proteins are produced by plants during normal development or as part of guided defense against fungal pathogens [80]. PR proteins are induced in response to a variety of environmental stressors such as drought, salt, injury, heavy metals, treatment with endogenous and extrinsic triggers, and plant growth regulators [79–81]. The PR-2 protein family has β – 1, 3 glucanase activity and is abundant in plant tissues. It is also involved in the formation of callose, leaf, and stem hairs. Endo β – 1, 3 glucanases are found in a wide variety of plant species. Several β – 1, 3 glucanases have been found to be constitutively present and to be inducible by pathogens. [82]. Proteinase inhibitors belong to the PR-6 family of PRs. They are very stable defense proteins that are only induced in response to insect and pathogen attack and are developmentally controlled [83].

This protein family has endoproteinase activity. The amino acid sequence shows homology to subtlisin-like proteinase and plays a role in disease resistance in tomato plants in response to pathogen attack. [84]. PR-11 is a group of pathogenesis-related proteins that are activated in response to UV radiation and virus infection. They were first discovered in tobacco, and the only known molecular homologue is found in pepper [85, 86]. Defensins and thionins are two families of low-molecular-mass, cysteine-rich peptides found in mammals, insects, plants, and fungi [87]. Plant defensins (PR-12 proteins) and thionins (PR-13 proteins), present in both monocotyledonous and dicotyledonous plants, are toxic to fungi [88–90].

– **Antifungal proteins**

Antifungal proteins are known for their protective function against fungal invasion. They are reported to involve in constitutive and induced resistance to fungal invasion and are produced by a variety of organisms such as bacteria, fungi, gymnosperms, flowering plants, insects, molluscs, and mammals [91, 92]. Plant seeds are rich in antimicrobial proteins in comparison to those present in leaves or flowers [93].

– **Seed storage protein**

These are the proteins that are produced highly in plants during specific growing stages and accumulate in distinct vesicles called as protein storage vacuoles. Therefore, they are considered as a reservoir for surplus nitrogen, organic carbon, and sulfur [94]. There are a few storage proteins that have already been reported to contribute to plant defense mechanisms [95].

– **Vicilins**

Vicilins are considered as multifunctional proteins, because of their function, i.e., as a source of amino acids for the plant during growth and germination and also for being toxic to fungi and insects, at the same time [96–98].

– **Cyclophilins**

Cyclophilins are proteins that usually function as intracellular receptors for cyclosporin. For example, in mung bean, there is protein known as Mungin that inhibits α- and β-glycosidases in vitro [99] and Unguilin, a protein extracted from the seeds of the black-eyed pea (*Vigna unguiculata*) that possesses antiviral, antimitogenic, and antifungal activities towards specific fungi.

– **Plant lipid transfer proteins**

These types of proteins (LTPs; PR-14) are basic proteins that help in transfer of phospholipids between membrane layers and show antifungal activity, but the main reason responsible for antifungal activity still remains undiscovered; however, it was implied that the protein molecules fit snugly into the fungal cell membrane by forming a pore through their central hydrophobic cavity, thus allowing efflux of intracellular ions and leading to fungal cell death [92].These protein molecules incorporate themselves into the cell membrane of pathogen with their hydrophobic cavity by formation of a pore, permitting efflux of intracellular ions and leading to fungal cell death [92].

– **Killer proteins**

Several yeast are secreted killer proteins, which bind to specific surface receptors, and eventually, they can disrupt several mechanisms inside the cell, such as synthesis of DNA and K^+ channel activity, synthesis of cell wall, inhibition of β-1, 3-glucan synthesis and halting the cell cycle, leading to suppression of fungal growth and, ultimately, fungal cell death [100–104].

References

[1] Horsfall JG, Dimond AE. Plant Pathology– an advanced treatise, Vol. I, Academic Press, New York, San Francisco, London, 1959.
[2] Singh RS. Plant diseases, 6th edn, Oxford & IBH Publishing Co. Pvt. Ltd., New Delhi, Bombay, Calcutta, 1995, 2–30.
[3] Agrios GN. Plant pathology, 3rd edn, Academic Press, INC., San Diego, New York, Berkeley, Boston, London, Sydney, Tokyo, Toronto, 1988, 3–110.
[4] Hammond-Kosack K, Jones JDG. Response to plant pathogens. In: Buchanan B, Gruissem D, Jones, R, Rockville MD eds. Biochemistry and molecular biology of plants. Am. Soc. Plant Physiol 2000, 1102–1156.

[5] Heath MC. Nonhost resistance and nonspecific plant defenses. Current Opinion in Plant Biology 2000, 3, 315–319.
[6] Altieri MA. The ecological role of biodiversity in agroecosystems. In: Invertebrate biodiversity as bioindicators of sustainable landscapes. New York City, NY, USA, Elsevier, 1999, 19–31.
[7] Jones JD, Dangl JL. The plant immune system. Nature 2006, 444, 323–329.
[8] Chawade A, Alexandersson E, Bengtsson T, Andreasson E, Levander F. Targeted proteomics approach for precision plant breeding. Journal of Proteome Research 2016, 15, 638–646.
[9] Cui P, Zhang S, Ding F, Ali S, Xiong L. Dynamic regulation of genome-wide pre-mRNA splicing and stress tolerance by the Sm-like protein LSm5 In Arabidopsis. Genome Biology 2014, 15, R1.
[10] John V, Maurya AK, Murmu R. Plant defence mechanism against pathogen, Innovation in Agriculture, Environment and Health Research for Ecological Restoration, Society of Biological Sciences and Rural Development, 2019, 142–145.
[11] Prasannath K, De Costa DM. Induction of peroxidase activity in tomato leaf tissues treated with two crop management systems across a temperature gradient. Proceedings of the International Conference on Dry Zone Agriculture 2015. Sri Lanka, Faculty of Agriculture, University of Jaffna, 15th & 16th October 2015, 34–35.
[12] Gajanayaka GMDR, Prasannath K, De Costa DM. Variation of chitinase and –1,3-glucanase activities in tomato and chilli tissues grown under different crop management practices and agroecological regions. Proceedings of the Peradeniya University International Research Sessions. 4th – 5th July 2014, vol. 18, 519.
[13] Seneviratne DMAS, Prasannath K, De Costa DM. Quantification of phenylalanine ammonia lyase activity in tomato and chilli tissues grown under different crop management practices and agro-ecological regions in Sri Lanka. Proceedings of the 1st Faculty of Agriculture Undergraduate Research Symposium. 23rd December 2014.
[14] Nicholson RL, Hammerschmidt R. Phenolic compounds and their role in disease resistance. Annual Review of Phytopathology 1992, 30, 369–389.
[15] Wojtaszek P. The oxidative burst: A plant's early response against infection. Biochemical Journal 1997, 322, 681–692.
[16] Bi JL, Felton GW. Foliar oxidative and insect herbivory: Primary compounds, secondary metabolites, and reactive oxygen species as components of induced resistance. Journal of Chemical Ecology 1995, 21, 1511–1530.
[17] Ebrahim S, Usha K, Singh B. Pathogenesis Related (PR) Proteins in Plant Defense Mechanism. In: Science against microbial pathogens: communicating current research and technological advances, A. Mendez-Vilas (ed). 2011, 1043–1054.
[18] Bellincampi D, Cervone F, Lionetti V. Plant cell wall dynamics and wall-related susceptibility in plant–pathogen interactions. Frontiers of Plant Science 2014, 5, 228.
[19] Punja ZK, Zhang YY. Plant chitinases and their roles in resistance to fungal diseases. Journal of Nematology 1993, 25(4), 526–540.
[20] Hamid R, Khan MA, Ahmad M, Ahmad MM, Abdin MZ, Musarrat J, Javed S. Chitinases: An update. Journal of Pharmacy and Bioallied Sciences 2013, 5, 21–29.
[21] Henrissat B, Davies G. Structural and sequence-based classification of glycoside hydrolases. Current Opinion in Structural Biology 1997, 7, 637–644.
[22] Kuranda MJ, Robbins PW. Chitinase is required for cell separation during growth of *Saccharomycescerevisiae*. The Journal of Biological Chemistry 1991, 266, 19758–19767.
[23] Knowles J, Lehtovaara P, Teeri T. Cellulase families and their genes. Trends in Biotechnology 1987, 5, 255–261.
[24] Daizo K. Application of chitinase in agriculture. Journal of Metals Materials and Minerals 2005, 15, 33–36.

[25] Rathore AS, Gupta RD. Chitinases from bacteria to human: Properties, applications and future perspectives. Enzyme Research 2015, 2015, 791907.
[26] Fahmy A, Hassanein R, Hashem H, Ibrahim A, El Shihy O, Qaid E. Developing of transgenic wheat cultivars for improved disease resistance. Journal of Applied Biology and Biotechnology 2018, 6, 31–40.
[27] Jalil SU, Mishra M, Ansari MI. Current view on chitinase for plant defence. Trends Bioscience 2015, 8, 6733–6743.
[28] Singh HR, Deka M, Das S. Enhanced resistance to blister blight in transgenic tea (*Camellia sinensis* [L.] Kuntze) by overexpression of class I chitinase gene from potato (*Solanumtuberosum*). Functional and Integrative Genomics 2015, 15, 461–480.
[29] Mauch F, Mauch-Mani B, Boller T. Antifungal hydrolases in pea tissue. II. Inhibition of fungal growth by combinations of chitinase and β-1,3-glucanase. Plant Physiology 1988, 88, 936–942.
[30] Jach G, Gornhardt B, Mundy J, Logemann J, Pinsdorf E, Leah R, Schell J, Maas C. Enhanced quantitative resistance against fungal disease by combinatorial expression of different barley antifungal proteins in transgenic tobacco. The Plant Journal 1995, 8, 97–109.
[31] Chu KT, Ng TB. Purification and characterization of a chitinase-like antifungal protein from black turtle bean with stimulatory effect on nitric oxide production by macrophages. Biological Chemistry 2005, 386, 19–24.
[32] Saikia R, Singh BP, Kumar R, Arora DK. Detection of pathogenesis-related proteins– chitinase and β-1,3-glucanase in induced chickpea. Current Science 2005, 89(4), 659–663.
[33] Kirubakaran SI, Sakthivel N. Cloning and overexpression of antifungal barley chitinase gene in *Escherichia coli*. Protein Expression and Purification 2006, 52(1), 159–166.
[34] Passardi F, Cosio C, Penel C, Dunand C. Peroxidases have more functions than a Swiss army knife. Plant Cell Reports 2005, 24(5), 255–265.
[35] Thakker JN, Patel S, Dhandhukia PC. Induction of defense-related enzymes in banana plants: effect of live and dead pathogenic strain of *Fusariumoxysporum* f. sp. cubense. ISRN Biotechnology 2013. 10.5402/2013/601303.
[36] Sasaki K, Iwai T, Hiraga S. Ten rice peroxidases redundantly respond to multiple stresses including infection with rice blast fungus. Plant & Cell Physiology 2004, 45(10), 1442–1452.
[37] Lavania M, Chauhan PS, Chauhan SVS, Singh HB, Nautiyal CS. Induction of plant defense enzymes and phenolics by treatment with plant growth – Promoting rhizobacteria Serrtiamarcescens NBRI1213. Current Microbiology 2006, 52(5), 363–368.
[38] Diaz-Vivancos P, Rubio M, Mesonero V. The apoplastic antioxidant system in Prunus: Response to long-term plum pox virus infection. Journal of Experimental Botany 2006, 57(14), 3813–3824.
[39] Vera P, Tornero P, Conejero V. Cloning and expression analysis of a viroid-induced peroxidase from tomato plants. Molecular Plant-Microbe Interactions 1993, 6(6), 790–794.
[40] Almagro L, Gomez Ros LV, Belchi-Navarro S, Bru R, RosBarcelo A, Pedreno MA. Class III peroxidases in plant defence reactions. Journal of Experimental Botany 2009, 60(2), 377–390.
[41] Prasannath K, De Costa DM, Hemachandra KS. Quantification of Peroxidase activity in Chilli tissues Grown under Two Crop Protection Systems across a Temperature Gradient. Proceedings of the HETC Symposium 2014. 7th& 8th July 2014, 102.
[42] Halfeld-Vieira BA, Vieira JR Jr, Romeiro RS, Silva HAS, Baracat-Pereira MC. Introduction of systemic resistance in tomato by the autochthonous phylloplane resident Bacillus cereus. Pesquisa Agropecuária Brasileira 2006, 41(8), 1247–1252.
[43] Wessels JGH, Sietsma JH. Fungal cell walls: A survey. In: Tanner W, Loewus FA, eds. Encyclopedia of plant physiology. Berlin, Springer Verlag, 1981, 352–394.

[44] Smart MG. The plant cell wall as a barrier to fungal invasion. In: Cole GC, Hoch HC, eds. The fungal spore and disease initiation in plants and animals. New York, Plenum Press, 1991, 47–66.

[45] Beffa RS, Hofer RM, Thomas M, Meins F. Decreased susceptibility to virus disease of β-1,3-glucanasedeficient plants generated by antisense transformation. The Plant Cell 1996, 8, 1001–1011.

[46] Bowles DJ. Defense-related proteins in higher plants. Annual Review of Biochemistry 1990, 59, 873–907.

[47] Macdonold MJ, Dcunha GB. A modern view of phenylalanine ammonia lyase. Biochemistry and Cell Biology 2007, 85, 273–282.

[48] Dixon RA, Paiva NL, Bhattacharyya MK. Engineering disease resistance in plants: An overview. In: Singh RP, Singh US, eds. Molecular methods in plant pathology. Boca Raton, CRC Press, 1995, 249–270.

[49] Campos-Vargas R, Saltveit ME. Involvement of putative chemical wound signals in the induction of phenolic metabolism in wounded lettuce. Physiologia Plantarum 2002, 114, 73–84.

[50] Kim HJ, Fonseca JM, Choi JH, Kubota C. Effect on methyl jasmonate on phenolic compounds and carotenoids of Romaine Lettuce (Lactucasativa L.). Journal of Agricultural and Food Chemistry 2007, 55, 10366–10372.

[51] Lafuente MT, Zacarias L, Martinez-Telez MA, Sanchez-Ballesta MT, Granell A. Phenylalanine ammonia-lyase and ethylene in relation to chilling injury as affected by fruit age in citrus. Postharvest Biology and Technology 2003, 29, 308–317.

[52] Jiang YM, Joyce DC. ABA effects on ethylene production, PAL activity, anthocyanin and phenolic contents of strawberry fruit. Plant Growth Regulation 2003, 39, 171–174.

[53] Barber MS, Michell HJ. Regulation of Phenylalpropanoid metabolism in relation to lignin biosynthesis in plants. International Review of Cytology 1997, 172, 243–293.

[54] Chen AH, Chai YR, Li JN, Chen L. Molecular cloning of two genes encoding cinnamate 4-hydroxylase (C_4H) from oilseed rape (*Brassicanapus*). Journal of Biochemistry and Molecular Biology 2007, 40, 247–260.

[55] Harakava R. Genes encoding enzymes of the lignin biosynthesis pathway in eucalyptus. Genetics and Molecular Biology 2005, 28, 601–607.

[56] War AR, Paulraj MG, Ahmad T, Buhroo AA, Hussain B, Ignacimuthu S, Sharma HC. Mechanisms of plant defense against insect herbivores. Plant Signaling & Behavior 2012, 7 (10), 1306–1320.

[57] Chunhua S, Ya D, Bingle X, Xiao L, Yonshu X, Qinguang L. The purification and spectral properties of PPO I from *Nicotianantababcum*. Plant Molecular Biology 2001, 19, 301–314.

[58] Mayer AM, Harel E. Polyphenol oxidases in plants. Phytochemistry 1979, 18(2), 193–215.

[59] Chen C, Belanger R, Benhamou N, Paulitz TC. Defense enzymes induced in cucumber roots by treatment with plant growth-promoting rhizobacteria (PGPR) and *Pythiumaphanidermatum*. Physiological and Molecular Plant Pathology 2000, 56, 13–23.

[60] Deborah SD, Palaniswami A, Vidhyasekaran P, Velazhahan R. Time-course study of the induction of defense enzymes, phenolics and lignin in rice in response to infection by pathogen and non-pathogen. Journal of Plant Diseases and Protection 2001, 108, 204–216.

[61] Thakker JN, Patel N, Kothari IL. *Fusariumoxysporum* derived Elicitor-induced changes in Enzymes of Banana leaves against wilt disease. Journal of Mycology and Plant Pathology 2007, 37, 510–513.

[62] Raj SN, Sarosh BR, Shetty HS. Induction and accumulation of polyphenol oxidase activities as implicated in development of resistance against pearl millet downy mildew disease. Functional Plant Biology 2006, 33, 563–571.

[63] Ramamoorthy V, Raguchander T, Samiyappan R. Induction of defense-related proteins in tomato roots treated with Pseudomonas fluorescens Pf1 and *Fusariumoxysporum* f. sp. lycopersici. Plant and Soil 2002, 239, 55–68.
[64] Li L, Steffens JC. Overexpression of polyphenol oxidase in transgenic tomato plants results in enhanced bacterial disease resistance. Planta 2002, 215, 239–247.
[65] Jiang Y, Miles PW. Generation of H_2O_2 during enzymatic oxidation of catechin. Phytochemistry 1993, 33, 29–34.
[66] Ferreira R, Monteiro S, Freitas R, Santos C, Chen Z, Batista L, Duarte J, Borges A, Teixeira A. Fungal pathogens: The battle for plant infection. Critical Reviews in Plant Sciences 2006, 25, 505–524.
[67] Keen N. Mechanisms of pest resistance in plants. Bethesda, MD, Workshop on Ecological Effects of Pest Resistance Genes in Managed Ecosystems, 1999a.
[68] Tossi A, Sandri L, Giangaspero A. Amphipathic, alphahelical antimicrobial peptides. Biopolymers 2000, 55, 4–30.
[69] Flor HH. Current status of the gene-for-gene concept. Annual Review of Phytopathology 1971, 9, 275–296.
[70] Antoniw JF, Ritter CE, Pierpoint WS, van Loon LC. Comparison of three pathogenesis-related proteins from plants of two cultivars of tobacco infected with TMV. The Journal of General Virology 1980b, 47, 79–87.
[71] van Loon LC. Occurrence and properties of plant pathogenesis related proteins. In: Dutta SK, Muthukrishnan, S eds. Pathogenesis related proteins in plants. Boca Raton, CRC Press, 1999, 1–19.
[72] Jayaraj J, Anand A, Muthukrishnan S. Pathogenesis related proteins and their roles in resistance to fungal pathogens. In: Punja ZK, ed. Fungal disease resistance in plants. Biochemistry, molecular biology, and genetic engineering. New York, Haworth Press, 2004, 139–177.
[73] van Loon LC, van Kammen A. Polyacrylamide gel disc electrophoresis of the soluble leaf proteins from *Nicotianatabacum* var. 'Sansum' and 'Sansum NN'. Changes in protein constitution after infection with tobacco mosaic virus. Virology 1970, 4, 199–211.
[74] van Loon LC, Rep M, Pieterse CMJ. Significance of inducible defense-related proteins in infected plants. Annual Review of Phytopathology 2006, 44, 135–162.
[75] Linthorst HJM, van Loon LC. Pathogenesis-related proteins of plants. Critical Reviews in Plant Sciences 1991, 10, 123–150.
[76] Edreva A. Pathogenesis-related proteins: Research progress in the last 15 years. General and Applied Plant Physiology 2005, 31, 105–124.
[77] van Loon LC, Pierpoint WS, Boller T, Conejero V. Recommendations for naming plant pathogenesis-related proteins. Plant Molecular Biology Reporter 1994, 12, 245–264.
[78] van Loon LC. The nomenclature of pathogenesis-related proteins. Physiological and Molecular Plant Pathology 1990, 37, 229–230.
[79] Derckel JP, Legendre L, Audran JC, Haye B, Lambert B. Chitinases of the grapevine (*Vitisvinefera* L.): Five isoforms induced in leaves by salicylic acid are constitutively expressed in other tissues. Plant Science 1996, 119, 31–37.
[80] Xie Z, Staehelin C, Wiemken A, Broughton W, Muller J, Boller T. Symbiosis-stimulated chitinase isoenzymes of soybean (*Glycinemax* (L.) Merr.). Journal of Experimental Botany 1999, 50, 327–333.
[81] Yu XM, Griffith M, Wiseman SB. Ethylene induces antifreeze activity in winter rye leaves. Plant Physiology 2001, 126, 1232–1240.

[82] Metzger GL, Meins F. Functions and regulation of plant E 1–3 glucanases. In: Dutta SK, Muthukrishnan S, eds. Pathogenesis related proteins in plants. Boca Raton, CRC Press, 1999.
[83] Heitz T, Geoffrey YP, Frittig B, Legrand M. The PR-6 family of proteinase inhibitors in plant-microbe and plant-insect interaction. In: Dutta SK, Muthukrishnan S, eds. Pathogenesis related protein in plants. Boca Raton, CRC Press, 1999.
[84] Torenero P, Conejero V, Vera P. Identification of a new pathogen induced member of subtlisin like processing family from plants. Journal of Biological Chemistry 1997, 272, 14412–14419.
[85] Heitz T, Geoffrey YP, Frittig B, Legrand M. Molecular characterization of a novel tobacco PR protein: A new plant chitinase lysozyme. Molecular & General Genetics 1994, 245, 246–254.
[86] Bravo JM, Campo S, Murillo I, Coca M, Segundo B. Fungus and wound induced accumulation of mRNA containing class II chitinase of the pathogenesis related 4 family of maize. Plant Molecular Biology 2003, 52, 745–749.
[87] Theis T, Stahl U. Antifungal proteins: Targets, mechanisms and prospective applications. Cellular and Molecular Life Sciences 2004, 61, 437–455.
[88] Bohlmann H. The role of thionins in plant protection. Critical Reviews in Plant Sciences 1994, 13, 1–16.
[89] Broekaert WF, Terras FR, Cammue BP, Osborn RW. Plant defensins: Novel antimicrobial peptides as components of the host defense system. Plant Physiology 1995, 108, 1353–1358.
[90] Evans IJ, Greenland AJ. Transgenic approaches to disease protection: Applications of antifungal proteins. Pesticide Science 1998, 54, 353–359.
[91] Ng TB. Antifungal proteins and peptides of leguminous and non-leguminous origins. Peptides 2004, 25, 1215–1222.
[92] Selitrennikoff CP. Antifungal proteins. Applied and Environmental Microbiology 2001, 67, 2883–2894.
[93] Wang X, Bunkers GJ, Walters MR, Thoma RS. Purification and characterization of three antifungal proteins from cheeseweed (*Malvaparviflora*). Biochemical and Biophysical Research Communications 2001, 282, 1224–1228.
[94] Pernollet JC. Protein bodies of seeds: Ultrastructure, biochemistry and degradation. Phytochemistry 1978, 17, 1473–1480.
[95] Shewry PR, Napier JA, Tatham AS. Seed storage proteins: Structures and biosynthesis. The Plant Cell 1995, 7, 945–956.
[96] Macedo MLR, Da Andrade SLB, Moraes RA, Xavier-Filho J. Vicilin variants and the resistance of cowpea (Vignaunguiculata) seeds to the cowpea weevil (*Callosobruchusmaculatus*). Comparative Biochemistry and Physiology 1993, 105, 89–94.
[97] Sales MP, Gerhardt IR, Grossi-De-Sa MF, Xavier-Filho J. Do legume storage proteins play a role in defending seeds against bruchids? Plant Physiology 2000, 124, 515–522.
[98] Shutov AD, Kakhovskaya IA, Braun H, Baumlein H, Muntz K. Legumin-like and vicilin-like seed storage proteins: Evidence for a common single-domain ancestral gene. Journal of Molecular Evolution 1995, 41, 1057–1069.
[99] Ye XY, Ng TB. Mungin, a novel cyclophilin-like antifungal protein from the mung bean. Biochemical and Biophysical Research Communications 2000, 273, 1111–1115.
[100] Ahmed A, Sesti F, Ilan N, Shih TM, Sturley SL. Goldstein SA.A molecular target for viral killer toxin: TOK1 potassium channels. Cell 1999, 99, 283–291.
[101] Eisfeld K, Riffer F, Mentges J, Schmitt MJ. Endocytotic uptake and retrograde transport of a virally encoded killer toxin in yeast. Molecular Microbiology 2000, 37, 926–940.

[102] Kimura T, Kitamoto N, Kito Y, Iimura Y, Shirai T, Komiyama T, Furuichi Y, Sakka K, Ohmiya K. A novel yeast gene, RHK1, is involved in the synthesis of the cell wall receptor for the HM-1 killer toxin that inhibits beta-1,3-glucan synthesis. Molecular & General Genetics 1997, 254, 139–147.
[103] Kimura T, Komiyama T, Furuichi Y, Iimura Y, Karita S, Sakka K, Ohmiya K. N-glycosylation is involved in the sensitivity of *Saccharomyces cerevisiae* to HM-1 killer toxin secreted from *Hansenulamrakii* IFO 0895. Applied Microbiology and Biotechnology 1999, 51, 176–184.
[104] Suzuki C, Shimma YI. P-type ATPase spf1 mutants show a novel resistance mechanism for the killer toxin SMKT. Molecular Microbiolog 1999, 32, 813–823.

Bahman Fazeli-Nasab, Ramin Piri, Yamini Tak, Anahita Pahlavan, Farzaneh Zamani

Chapter 16
The role of PGPRs in phosphate solubilization and nitrogen fixation in order to promote plant growth parameters under salinity, drought, nutrient deficiency, and heavy metal stresses

Abstract: Sustainable agriculture is defined as appropriate management of agricultural resources in order to meet the changing needs of human, without harming natural resources. Management using sustainable energy contributes significantly to environmental sustainability, soil fertility, microorganism conservation, and reducing the use of chemical fertilizers. Plant growth-promoting microorganisms (PGPMs) are used in sustainable agriculture in order to deal with biological and nonbiological stresses. Among the performance-enhancing microorganisms in agriculture, are plant growth-promoting rhizobacteria (PGPRs) and plant growth-promoting bacteria (PGPB) including *Pseudomonas*, *Bacillus*, *Arthrobacter*, *Enterobacter*, *Alcaligenes*, *Burkholderia*, *Acinetobacter*, *Rhizobacter*, *Azotobacter*, *Azotobacter Azospirillum*, and *Beijerinckia*. PGPRs stimulate plant growth, both directly and indirectly, by modifying the rhizosphere microbial community and producing different substances. PGPRs will be effective in improving growth parameters, nutrient uptake, and photosynthesis under salinity stress by performance of osmotic adjustment, induction of ionic homeostasis and hormonal signaling, release of extracellular molecules, and the use of ACC deaminase producing bacteria. Moreover, PGPRs play a major role in improving growth parameters by coping with drought stress, heat, and heavy metal toxicity (by producing ACC deaminase and enhancing plant hormone activity), nutrient deficiency tolerance (by biological nitrogen fixation and phosphate solubility) and biological stresses (by production and release of nonvolatile antibiotics). In this review, the beneficial effects of PGPRs on plants and sustainable agriculture have been thoroughly investigated.

16.1 Introduction

Traditional agriculture is an agricultural style that includes the application of indigenous knowledge, traditional tools, and organic fertilizers [1] On the other hand, the indiscriminate use of chemical fertilizers, herbicides, and pesticides in modern and mechanized agriculture has led to pollution of the soil, environment, and

https://doi.org/10.1515/9783110771558-016

climate and human hazards. Moreover, the use of chemical fertilizers reduces nonrenewable resources and is expensive [2]. In contrast, sustainable agriculture is the proper management of agricultural resources to meet the changing needs of humans, without harming safe natural resources. It uses sustainable energy, maintains soil fertility and microorganisms, and reduces the usage of chemical fertilizers [3].

Sustainable agriculture makes use of PGPMs to deal with biological and nonbiological stresses. PGPMs live in the rhizosphere, which is the limited part of the soil that surrounds the roots or any specific volume of it that is related to the roots and plant materials [4]. Stress is the result of disruption of vital processes in plant cells, which is caused by one or more biological and environmental factors. It affects germination, growth, development as well as crop quality and yield and may even cause the death of part or all of the plant. Stresses have been classified into two: biotic and abiotic groups. Biotic stresses refer to plant pathogenic stresses such as viruses, fungi, bacteria, nematodes, insects, etc. Abiotic stresses involve soil heavy metal content, drought, nutrient deficiency, salinity, temperature, etc. [5].

Growth-promoting bacteria have enhanced drought tolerance in Arabidopsis and soybeans [6]. Increased germination and seedling indices of cumin have been reported by *Pseudomonas* and *Bacillus* growth-promoting bacteria under drought and salinity stresses [7, 8]. Growth-promoting bacteria can affect plant growth by producing and releasing secondary metabolites, so that increased amount of available substances and nutrients in the soil due to their activity results in an increase in the production rate and improvement in the quality of agricultural products [9].

PGPMs are either endophyte or epiphyte. Not only do they not cause disease in the host plant [10], they also help improve plant health and productivity [11]. Root microorganisms can form colonies in the form of free-living organisms, parasites, or saprophytes and are very diverse in this regard. They convert substances such as amino acids, monosaccharides, and organic acids into nutrients, which are sent to the roots as the primary source. They support the function of various microorganisms [12].

Among the growth-promoting microorganisms, PGPRs [13] can be used to deal with various biological and nonbiological stresses (Figure 16.1). PGPRs are one of the growth-promoting microorganisms [13], which can be used to deal with various biotic and abiotic stresses (Figure 16.1). Bacteria make up the largest group of soil microorganisms. On average, there are 10^8 to 10^9 bacterial cells per gram of soil [14].

Among the growth-promoting bacteria that improve agricultural performance are *Pseudomonas*, *Bacillus*, *Arthrobacter*, *Enterobacter*, *Alcaligenes*, *Burkholderia*, *Acinetobacter*, *Azotobacter*, *Rhizobium*, *Erwinia*, *Flavobacterium*, *Azospirillum*, and *Beijerin* [15].

PGPRs can boost plant strength and growth as follows [16–19]:

1) providing nutritional elements such as phosphorus, nitrogen, and other nutrients by increasing the solubility of the nutrients
2) helping or enhancing nitrogen fixation by either nonsymbiotic or symbiotic nitrogen fixation, nodule occupancy or nodulation

3) affecting plant growth by production of growth promoters and phytohormones such as auxins, cytokinins, and gibberellins
4) synthesis of antibiotics, fungicides, and various factors that protect plants against different diseases
5) enhancing plant tolerance to improper environmental conditions and stresses
6) increasing plant yield and growth by ACC-deaminase–containing rhizobacteria

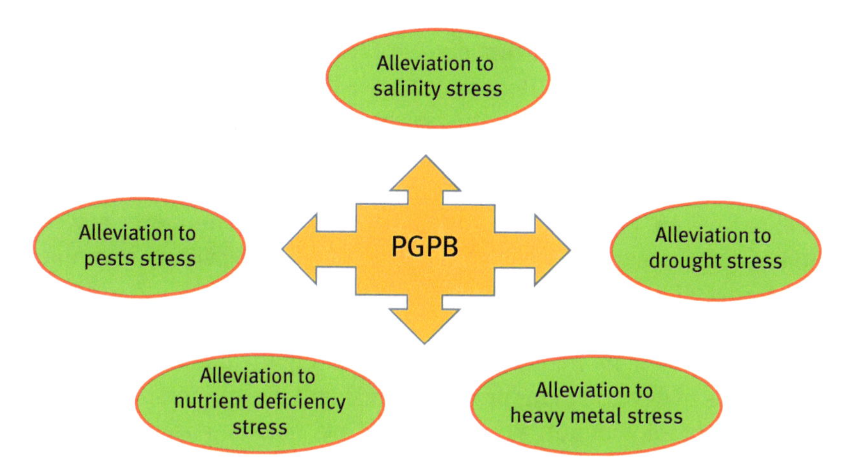

Figure 16.1: The role of PGPB in dealing with biotic and abiotic stresses.

PGPRs stimulate plant growth by changing the rhizosphere microbial community through the production of various substances. These bacteria affect different models of biological control either directly (biological fertilizer) by facilitating the absorption of resources (nitrogen, phosphorus, and other essential minerals) and modulating the level of plant hormones, or indirectly (biological control) by reducing the effects of various pathogens on plant growth and development [20]. The direct mechanism is based on promoting plant growth through nitrogen fixation, phosphate solubility, production of phytohormones (auxins, cytokinins and gibberellins), enzymatic activity of 1-aminocyclopropane 1-carboxylic acid (ACC) deaminase and iron complex by bacterial siderophores. The indirect mechanism is based on the synthesis of antibiotics, cell wall-destroying enzymes, or induced systemic resistance (ISR), which is responsible for inhibition of pathogenic organisms in plants [21]. The mechanisms performed by PGPBs and PGPRs depend on plant type and species and, in turn, are affected by biotic (such as plant defense processes and genotype) and abiotic (such as climatic conditions and soil composition) factors [22].

To produce high-yielding stress-resistant plants, it is necessary to comprehend the physiological, biochemical, and molecular mechanisms of PGPRs, plant type and species, and their interactions. In the following, we have discussed some physiological and ecophysiological efficiency of PGPB and their functional roles in plants.

16.2 The role of PGPRs in dealing with plant salinity stress

Salinity reduces crop productivity and destroys the sustainability of the world's crop systems. Salinity decreases the potential of soil minerals and creates osmotic stress leading to reduced hydraulic conductivity and greater water and minerals uptake by plants [23]. Increased water and minerals uptake will cause drought; and, drought and salinity simultaneously disrupt diverse agricultural systems [24]. Moreover, salinity stress leads to plant nutrient deficiencies by directly interfering with ion transporters in the root plasma membrane (selective K^+ channels) and by inhibiting root growth. Salinity affects plants in both direct (regulated water uptake, reduced stomatal conductance, osmotic tension of root surface, food restriction, ion toxicity due to excessive Na^+) and indirect (reduction of available nutrients as well as accumulation of organic materials) ways [25]. Furthermore, soil salinity may take place due to improper distribution of cationic (Ca^{+2}, Mg^{+2} and Na^+) and anionic (Cl^-, $SO4^{-2}$ and HCO^{-3}) ions [26].

Farmers use environmentally friendly methods including the use of PGPRs and bio-priming techniques to develop plant protection against salinity stress. There are other methods including biotechnology, plant breeding, genetic engineering, or conventional breeding methods in which salinity tolerance traits are transferred to high-yield cultivars. Among salinity tolerant PGPRs that are commonly used in salinity tolerance, *Rhizobium sp., Acinetobacter, Pseudomonas sp., Azotobacter, Serratia sp.* and *Bacillus sp.* can be mentioned [9, 27].

Inoculation of plant seeds with PGPRs improves plant growth and salinity tolerance. Some of the mechanisms that are used to reduce salinity stress by boosting water absorption are as follows: increased essential nutrients uptake, accumulation of osmolytes (e.g., proline (Pro), glutamate (Glu), glycine-betaine, soluble sugars, choline, O-sulfate, and polyols), increased AE activity (e.g., superoxide dismutase, peroxidase, catalase, ascorbate peroxidase, monohydroascorbate reductase, dehydroascorbate reductase, glutathione reductase. and nonenzymatic antioxidants (NEAs) (e.g., ascorbate (ASC), glutathione (GSH), tocopherols (TCPs), carotenoids (Car). and polyphenols (PPs) in plant tissues [28, 29]. Among all types of soil salinity, sodium chloride (NaCl) has the greatest effect, due to its high solubility. On the other hand, Na^+ is the main cause of damage in many plant species, especially cereals, although high concentration of Cl^- anion is also reported to be toxic to plants [30].

PGPRs enhance nitrogen fixation, phosphate, and mineral potassium solubility and the synthesis of plant hormones and siderophores, which are specifically attached to specific metal ions, in order to facilitate their absorption [31]. During salinity stress, oxidative stress is also caused by reactive oxygen species (ROS), which cause damage to the chloroplast, cell membrane, and nucleic acids [32]. Another process that takes place during salinity stress is ion imbalance in which the exchange of

Na^+ and K^+ ions is disrupted, and Na^+ accumulation inside the cell leads to less water uptake and reduced plant growth [33]. Antioxidant enzymes such as peroxidase, superoxide dismutase, catalase, lipoxygenase, and ascorbate peroxidase are produced by plants to destroy ROS and protect plants against osmotic and oxidative stresses. It has also indicated that antioxidant response of plants is correlated with stress tolerance [34].

PGPRs can be effective in improving growth parameters, nutrient uptake, and photosynthesis in seeds, seedlings, and plants under salinity stress by regulating osmotic pressure, ionic homeostasis, hormonal signaling, and the release of extracellular molecules.

16.2.1 The role of PGPRs in creating osmotic balance under salinity stress conditions

PGPRs play a major role in salinity tolerance by regulating stomatal conductance, water potential, transpiration rate, and hydraulic conductivity. Inoculation of plants with *Bacillus megaterium* regulates two ZMP IP genes, which are responsible for synthesis of aquaporins in the plasma membrane [35]. *Bacillus subtilis* increases the expression of proline biosynthesis stimulating genes in the transgenic plant, Arabidopsis thaliana (harboring BA pro BA genes) and enhancing plant tolerance under stress condition [36].

PGPRs stimulate carbohydrate metabolism, which can play an important role in resource-reservoir relation, CO2 stabilization, and increased biomass allocation at high growth rates [9]. In salinity stress, inoculation of annual bell pepper with *Pantoea dispersa* and *Azospirillum brasilense* increased plant dry matter biomass after 36 days, due to elevated photosynthesis rate and stomatal conductance, but had no effect on photochemical efficiency of photosystem II and chlorophyll content [37]. Under salinity stress, more osmolytes accumulate. These osmolytes protect the plant against oxidative damage and preserve osmotic balance. Glycine, betaine, proline, and trehalose osmolites are synthesized by PGPRs more rapidly in bacterial inoculated plants than uninoculated plants [38]. On the other hand, increased salinity tolerance gene expression has been indicated in plant nodes, resulting in plant resistance due to the secretion of trehalose (an extracellular carbohydrate) [39].

16.2.2 The role of PGPRs in ionic homeostasis under salinity conditions

PGPRs make plants resistant by producing complex cations and exopolysaccharides on top of plant roots and controlling the expression of ion transporter genes. Bacteria balance the transport of large ions and micronutrients. Infiltration of Na^+ and Cl^-

ions leads to toxicity and a balance. This balance is maintained by changes in the rhizosphere pH due to the release of organic acids by PGPRs, the chelation of metals using siderophores, and the release of mineral enzymes [40].

Polyphenols, proline, and chlorophyll content in corn have increased under salinity stress, due to inoculation with different PGPR strains, including *Azotobacter sp. C9 and C5* [41]. In *Puccinellia tenuiflora*, PtHKT1 and PtSOS1 gene expression were also increased, but PtHKT2 gene expression was decreased in roots under salinity stress, when inoculated with GB03 strain of *B. subtilis* (200 mM NaCl) [42].

16.2.3 The role of PGPRs in creating hormonal signaling under ion stress condition

Plants respond to salinity stress by changing the amount of exogenous metabolites, phytohormones, and enzymes. On the other hand, microbes provide these metabolites and phytohormones as microbial-plant interactions [43]. In addition, different phytohormones have diverse roles and are part of major hormones in plant tolerance.

Auxin is synthesized by plant roots using tryptophan as the main precursor and is converted to indole-3-acetic acid by PGPRs in plant roots. Exogenous and endogenous IAAs act simultaneously to initiate a signaling pathway. This pathway may have an inhibitory, stimulating, or neutralizing effect on plant growth [44].

Ethylene is synthesized in response to stress in plants, increasing tolerance and precocity (aging in plants) [45]. Ethylene limits the response to auxin and increases plant adaptation during growth and development, while PGPRs synthesize 1-aminocyclopropane-1-carboxylase (ACC) diaminase, which is in contrast with ethylene synthesis. Stress tolerance increases with ACC deaminase activity in plants with PGPRs [46].

Not much information is available on the use of exogenous ABA, bacterial ABA, and plant ABA in increasing plant tolerance. However, findings have shown that PGPRs produce some amount of ABA, which plays a role in signaling, where the level of ABA synthesis is constantly changing in response to salinity [9].

16.2.4 The role of PGPRs in the release of extracellular molecules under salinity stress condition

Extracellular molecules, including polyamines, proteins, volatiles, hormones, and other free compounds are released by PGPRs, altering the plant's response to stress. On the other hand, bacteria regulate activities such as plant growth, defense response, and pathogen tolerance and increase plant tolerance to various biotic and abiotic stresses by developing different signaling pathways [47]. They produce exopolysaccharides that interact with other microbes and plant roots to promote cation exchange and water molecule binding [48]. These exopolysaccharides, in addition to supporting microbial colony formation and biofilm synthesis, help the plant to fight environmental damage, retain water, nutrients, and exogenous colonies [49]. Salinity tolerance depends on the composition and synthesis of exopolysaccharides [49, 50].

When plants make flavonoids in response to abiotic stresses, PGPRs are involved in signaling by synthesizing lipocytoligosaccharides (LCOs), causing plant nodule formation and nitrogen fixation in legumes [51]. Inoculation of soybean (Glycine max) with *Bradyrhizobium japonicum* has increased nitrogen fixation rate under salinity stress [52]. Further, growth of chickpeas with root formation and higher soil density has been reported under salinity stress (up to 200 mM NaCl) due to inoculation with *Rhizobacteria Planococcus rifietoensis RT4* and *HT1 Halomonas variabilis* [53].

16.2.5 The role of ACC deaminase–containing PGPRs under salinity stress condition

Although ethylene synthesis during germination has been reported in plants such as wheat, rice, soybeans, and corn under salinity stress, its biosynthesis varies under stress conditions. For example, an increase in ACC concentration, which eventually led to an increase in ethylene synthesis, inhibited alfalfa seedling growth at high concentrations of ethephon and ethylene, under salinity stress conditions. On the other hand, PGPRs have increased the tolerance of peanut seedlings to salinity stress using a low Na^+ / K^+ ratio maintenance mechanism [54, 55]. Na^+ concentration in saline environment is generally greater than K^+, so that these two elements compete with each other [56]. At high salinity concentrations, sodium (Na^+) reduces K^+ uptake, inducing toxic sodium levels in plants [57].

Under salinity stress condition, some PGPR strains have the ability to produce exopolysaccharides that bind cations to Na^+, leading to low Na^+ uptake [58]. Under this condition, high concentrations of Cl^- and Na^+ reduces K^+ and Ca^{2+}availability and their transfer to different plant parts, so that it negatively affects the vegetative and reproductive organs and ultimately decreases their quality [59]. The ability of plants to tolerate restricted salinity stress conditions depends on Na^+ uptake as well as the continuous K^+ uptake by plant roots [60]. PGPRs have positive effects on regulating nutrient uptake and maintaining nutritional balance in plants growing in saline conditions [61]. As salinity tolerance in plants mainly depends on high Na^+ / K^+ ratio, inoculation of plants with specific PGPRs strains may be able to adjust Na^+ / K^+ uptake and keep nutritional balance in plants. This is because salinity tolerance in plants generally depends on Na^+ / K^+ ratio [62].

16.2.6 The role of PGPRs in improving growth parameters, nutrient uptake, and photosynthesis in seeds, seedlings, and plants under salinity stress condition

Seed is the main contributor to the production potential and yield in agriculture. Seed germination is the first and most important stage of the plant life cycle [63]. Uniform germination is one of the basic criteria used to evaluate seed vigor [64]. Climate change exposes seeds to environmental stress and decreases the percentage and rate of germination, resulting in seedling failure and reduced yield. Seeds and seedlings are always more sensitive to salinity as compared to mature plants [65].

Salinity and drought restrain protein synthesis and disrupt the germination process [66]. Alpha-amylase is the key factor in starch hydrolysis during seed germination, because it provides the energy needed for seed germination in the early growth stages before photosynthesis begins[63]. Salinity diminishes water absorption and alpha-amylase activity and, thus, delays the germination time [67, 68]. Salinity decreases germination and seedling indices including average germination time, seed vigor index, root and shoot length, as well as seedling fresh and dry weight [68].

Seed bio-priming with PGPRs stimulates germination rate and uniformity, promotes rapid establishment, and, thus, improves fruit /seed yield and quality under both stressed and nonstressed conditions [69]. Inoculation of rice seeds with *Enterobacter sp.* P23 strain increased germination percentage (GP) and seedling vigor index (SVI), so that the level of soluble protein, superoxide dismutase, catalase, polyphenol oxidase, and malondialdehyde reached its maximum under salinity stress condition [70]. 130 mM NaCl treatment affected the germination rate of alfalfa (Medicago sativa L.) seeds significantly, as compared to control seeds. On the other hand, inoculation of seedlings with PGPBs has increased vegetative parameters such as PH, RL, NL, TLA, and TPDW [71].

Pleiotropic biological activities such as growth regulation [72] and antioxidant properties [73] have been extensively used as a promising tool in diminishing salinity stress in plants. The combination of *Rhizobium leguminosarum* (N-stabilizing bacteria) and *Azotobacter crococum* (EPS-producing bacteria) inoculums has increased salinity stress tolerance in beans [24]. Bean seeds (*Vicia faba*) treatment with 100 μM inoculums reached the Chla, Chlb, Car and Pro content to the highest level, showing the synergistic effect of seeds and PGPRs in improving growth and other physiological aspects of the beans. Bean growth, an increase in PhoPs, Pro and N-P-K uptake, and a decrease in Na^+ / K^+ ratio were obtained as a result of seed inoculation using PGPR, in saline conditions. Increasing GRA and SVI indices in primary seeds compared to unprimed seeds has also been one of the achievements of bacterial inoculation [74] throughout the growth period. Inoculation of corn (*Zea mays* L.) and beans (*Vicia faba* cv. Giza3) seeds with *Azotobacter chroococcum* as an

EPS-producing bacterium decreased Na^+ and Cl^- concentrations and increased N, P, and K concentrations in plant tissues [24].

PGPRs can improve plant growth and nutrient uptake and stimulate PhoPs synthesis in stress-free environments. Strong antioxidants such as PPs remove free radical species including O_2, O_{2-}, OH^-, and H_2O_2 [67]. Total PPs content including phenolic acids, flavonoids, anthocyanins, and proanthocyanidins have increased in young and mature corn leaves with increasing salinity [75]. High PCs concentrations were inversely related to H_2O_2 content and LP level in leaves, showing the inhibitory activity of endogenous PCs against free radicals [76]. PPs in *Azotobacter chroococcum*-inoculated maize seedlings were higher than uninoculated maize seedlings proteins at any salt concentration. Furthermore, PPs reached the highest level when NaCl was at the highest level, which was 5.85 g NaCl / kg soil [41]. Investigating the effects of endophytic *B. subtilis* (BERA71) in chickpea plants (*Cicer arietinum* cv. Giza 1) showed a positive correlation between Pro accumulation and adaptation to salinity stress. On the other hand, PGPRs have reduced Na^+ accumulation and increased N, Ca^{2+}, Mg^{2+}, K^+, and nutrient uptake in chickpea plants [77].

The increase in Mg^{2+} uptake as the main component of Chl was due to inoculation of rice and potatoes with *Bacillus subtilis* and *Bacillus pomilus*, along with increasing PhoP content [78]. P fixation in tomatoes [79] or N fixation in soybeans [80] was due to the presence of *B. subtilis* in the soil. However, sodium showed limited uptake in rice inoculated with *B. pumilus* [78]. In contrast, a large accumulation of Na^+ was observed in the stems of halophyte Arthrocnemum macrostachyum inoculated with *bacillus* under high NaCl concentration (1030 mm) [81].

16.3 The role of PGPRs in dealing with drought stress in plants

Among nonbiological factors, drought is the major cause of declining productivity and growth, worldwide. More than 50% of the cultivated area will probably be subjected to drought by 2050 [82]. Plants resist drought stress by synthesizing osmolites and increasing osmotic potential and, hence, minimize drought stress indices in plants [83]. Plants produce glycine-betaine osmolite under stress conditions. On the other hand, glycine-betaine is released by rhizosphere bacteria, which is involved in drought tolerance, together with glycine-betaine produced by the plant [84].

Phytohormones such as gibberellin, auxin, and ethylene are synthesized by PGPRs, which augment the number of primary and lateral roots and meet the plant's water needs by going deep into the soil, under water stress conditions. Another process to counteract water stress by PGPRs is the production of trehalose saccharide, which protects biomolecules and other compounds against degradation induced by osmotic and oxidative stresses [85]. Increased synthesis of antioxidant compounds and chlorophyll pigment has been reported in Ocimum basilicum L. plants inoculated with *Azospirillum brasilense*, *Pseudomonas sp.*, and *Bacillus lentus* PGPRs under drought stress conditions [86].

Bacillus sp. family members promote plant growth and protect the plant against stress conditions by production of PO_4^{-2} solubilizing enzymes, indole acetic acid, and release of NH3 and hydrogen cyanide(HCN). In *Zea mays* L., they also help plants tolerate harsh conditions by synthesizing biofilm-forming exopolysaccharides and improve plant traits such as stem growth, leaf area, and stem dry weight [87]. PGPRs synthesize polysaccharides in plants and cause drought tolerance, a process that contributes to the normal plant growth [88]. *Azospirillum* enhances soil fertility by nitrogen fixation. Another mechanism through which *azospirillum* acts as PGPRs is augmenting plant growth by producing biocidal biologic compounds such as HCN, siderophores, proteolytic enzymes, and bacteriocins, which destroy pathogenic microbes and protect the plant from the risks of biotic and abiotic stresses [89].

More phosphatidylcholine and less phosphatidyl ethanolamine were observed in wheat seedlings in dry conditions, an observation that was inhibited by wheat inoculation with *Rhizobacterium azospirillum* [90]. The WRKY and MRB families are responsible for regulating 93 drought-related genes in the sugarcane crop. Inoculation of plants with *Gluconacetobacter diazotrophicus* and *Herbaspirillum spp* in response to biotic stress results in salicylic acid formation [91]. PGPRs inoculation increases drought tolerance in many crops including mung bean, sunflower, sugarcane, wheat, as well as corn. Therefore, the first necessary step to prevent abiotic stress in plants using PGPR is bacterial contact with the plant [92].

Achromobacter piechaudii-inoculated pepper and tomato plants had higher fresh and dry weight than uninoculated plants. This is due to elevated ACC deaminase enzyme activity, which decreases the activity of ethylene [93].

16.3.1 The role of PGPRs in production of ACC deaminase for drought tolerance

Drought stress causes specific reactions as well as plant damage by affecting the relationship of water with the plant cell, along with the entire plant surface. Bacteria act as protectors against drought stress and enhance plant viability under different stress conditions, by producing exopolysaccharide (EPS), which is associated with higher water retention as well as regulation of organic carbon emissions [94]. On the other hand, boosting ethylene synthesis not only inhibits root and stem growth and promotes aging of immature plant parts, but also changes the plasma membrane integrity, pigment content, and leaf dry weight, promotes disease symptoms, and reduces endogenous resistance [95].

EPS helps microorganisms attach irreversibly to plant roots and form colonies. On the other hand, EPS-producing microbes protect the plant from drying [53]. ACC deaminase enzymes also have drought tolerance capacity and their inoculation into plants increases drought tolerance capacity in arid / semi-arid regions [96]. ACC is the ethylene precursor and is converted to ethylene as a result of the ACC oxidase activity. ACC deaminase-containing strains can use ACC as the only source of nitrogen and control its negative effects on root growth by decreasing ethylene concentration. Some growth-promoting rhizosphere bacteria have the ability to produce ACC deaminase enzyme. Under drought stress conditions, this enzyme prevents ethylene formation and, therefore, inoculation of plants with bacteria containing this enzyme reduces ethylene production under stress conditions and, accordingly, balances plant growth and development [97]. The mechanism of ACC deaminase-containing bacteria in drought stress conditions is that it is converted to alpha-ketobutyrate (KBα) and ammonium by ACC deaminase, before oxidation by ACC oxidase occurs [98].

16.4 The role of PGPRs in dealing with heat stress (High temperature tolerance)

Extreme temperatures can include very high or very low temperatures, affecting the natural metabolism of plants, in both cases. High temperatures alter the amount of phytohormones, particularly ethylene. High temperatures cause denaturation or accumulation of proteins in the cell, and if this comes about on a large scale, it leads to cell necrosis. Furthermore, if the levels of ABA and cytokinin are not balanced, grain immaturity or infertility takes place [99]. Under high temperature stress conditions, changes in transcription and translation level in plant cells result in more heat shock proteins synthesis in order for the thermal response by plants to start [100]. Raising the temperature to a certain extent linearly increases the root elongation in plants [101].

At suboptimal temperatures, bacterial strains increase the growth of soybean plants through further nitrogen fixation and other non-nitrogen-dependent pathways [102]. The rate of biomass reduction and electrolyte leakage in *Vitis vinifera* was obtained under cold conditions (4 °C) when inoculated with *B. phytofirmans PsJN* bacterial strain [103].

16.5 The role of PGPRs in tolerating nutrient deficiencies

Many nutrients, such as phosphate, potassium, iron and zinc (Zn), have limited solubility. PGPRs are used to resolve the problem of nutrient deficiency by releasing enzymes and hydrolytic acids and dissolving precipitated ions, thus avoiding nutrient deficiencies. Another mechanism that works with the help of bacteria is the excretion of siderophore. Siderophore chelates metal ions and causes chelate ions to be absorbed by plants. Siderophore chelates metal ions to be absorbed by plants. Siderophore-producing bacteria use plant nutrients by releasing organic acids in a small environment and modifying the oxidation state of metal ions such as copper(Cu), iron, Zn, etc. [9].

Siderophores are low molecular weight organic molecules with a strong affinity for iron ferric ions [104]. This action not only restricts the access of iron to pathogenic microbes, but also indirectly improves plant health by making soluble iron available to plants [105]. Iron absorption also affects nitrogenase enzyme as iron is an integral part of the nitrogenase, and it directly regulates NIF gene expression in plants [106]. The most common examples of nutrient deficiencies studies have been carried out on diazotrophic bacteria that can live free or coexist and help fix nitrogen and solve nitrogen deficiency in plants. In addition, PGPRs can diminish stress levels by neutralizing the effects of stress hormones produced by plants [107]. On the other hand, this is attributed to plants that, in the absence of nutrients, produce more ACC deaminase, inhibit ethylene synthesis, and root growth. Overall, plants use diverse mechanisms to alter the physiology of the roots and subsequently absorb more nutrients, thus protecting themselves against abiotic stresses [9].

The release of more carbohydrates by plants indirectly raises the number of rhizosphere bacteria and helps absorb nutrients, creating more bacterial colonies for nutrient bioavailability [108]. Reduction of cadmium (Cd) in barley due to inoculation with *Klebsiella mobilis* strain CIAM 880 has doubled grain yield. The reason for this could be the movement of free Cd ions from the rhizoplan to the rhizosphere and, in the second stage, the formation of a complex with PGPRs, which provides less free Cd for absorption by the plant [109].

PGPRs alter plant nutrient uptake by quantitatively changing hydrolyzing enzymes or the levels of modified phytohormones. However, it is not obvious that the elevated nutrient uptake was due to more root growth or more mineral uptake, but plants generally compensate the lack of nutrients in this way [110]. PGPRs can make phosphorus available to plants through mineral phosphate solution or organic phosphate mineralization [111]. *Bacillus* and *Pseudomonas* bacteria have the ability to work in a wide pH range and to use various carbohydrate sources to dissolve potassium through producing numerous exopolysaccharides and organic acids, especially oxalic acid and citric acid [112].

16.6 The role of PGPRs in nitrogen fixation in plants

Microorganisms can convert nitrogen to ammonia through the nitrogen fixation process. About two-thirds of the total nitrogen amount is produced by the well-known enzyme nitrogenase through a complex mechanism [113]. There are two groups of microbes that make atmospheric nitrogen usable: (a) symbiotic nitrogen-fixing bacteria [114] and (b) nonsymbiotic nitrogen-fixing form such as *cyanobacteria* [115].

The host plant, which is associated with nonsymbiotic nitrogen-fixing bacteria, can fix small amounts of nitrogen. Molybdenum nitrogenase containing diazotrophs are nitrogen-fixing microbes that are responsible for the biological nitrogen fixation and other related activities [116]. Free-ranging bacteria, such as *Azospirillum spp.* and *Rhizobia spp.*, are able to fix nitrogen and provide it to the plant [117]. It should be noted that free living bacteria produce only a small amount of fixed nitrogen and provide it to the plant. Nitrogenases (Nifs) require structural genes, protein-iron activating genes, molybdenum cofactor genes, and electron donation to fix nitrogen. Moreover, the presence of regulatory genes is crucial for enzymes function and synthesis [118]. Just like Nif genes, they usually belong to a group of 7 operons of 10 to 20 Kbps, encoding 20 proteins [119].

16.7 The role of PGPR in phosphate solubilization

Phosphorus (P) is the second most important nutrient necessary for plant growth in soil after nitrogen, which is present in both organic and inorganic forms [120]. Soil phosphorus is insoluble and becomes soluble only when monounsaturated ions (H_2PO_4) and bipolar ions are absorbed by plants. Phosphorus is available in plants in little amounts [115]. Phosphorus deficiency is observed in field soils, as plants absorb less phosphate fertilizers, and other residual complexes rapidly become insoluble when they react with other phosphate fertilizers and soil components. Phosphate fertilizer treatment is both expensive and unnecessary [121]. Thus, a suitable alternative to phosphate fertilizers are phosphate solubilizing microorganisms (PSMs) that can provide the phosphorus sources required by the plant [122]. The most potent biofertilizers are various PSMs residing in the rhizosphere and phosphate-solubilizing bacteria (PSB), though plants can simply acquire the right amount of phosphorus through other biological pathways [123]. Many crops such as radishes, potatoes, tomatoes, and wheat are associated with microbial species that dissolve phosphorus [124].

PGPRs can dissolve mineral phosphate and provide it to the plant to enhance the growth [125]. PSB, which are plentiful in the rhizosphere, are able to make insoluble phosphate compounds available to plants by secreting organic acids and phosphatases (Figure 16.2). Inoculation of plants with phosphate-solubilizing microorganisms results in phosphorus uptake and consequently improves plant growth. Furthermore,

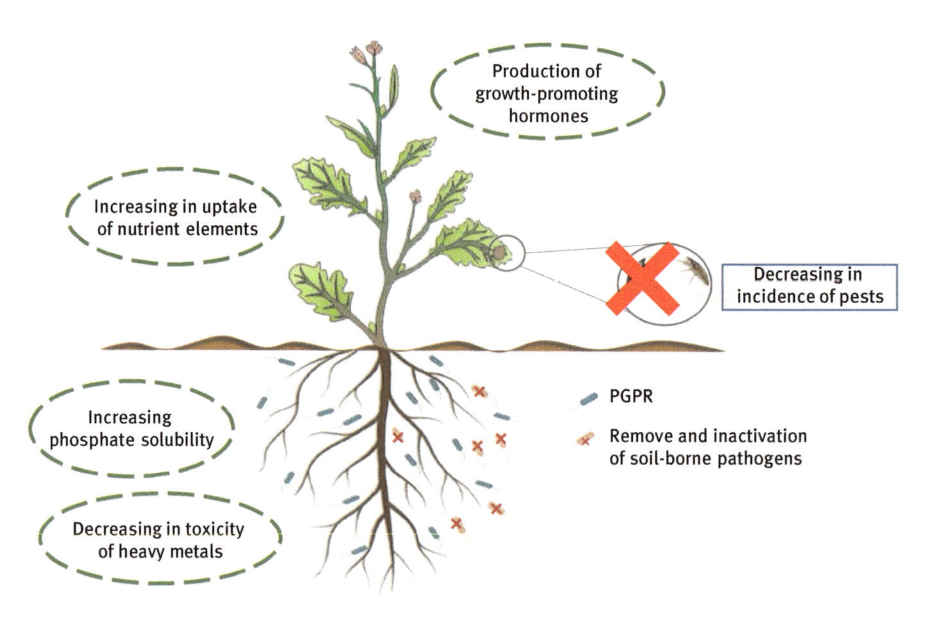

Figure 16.2: Direct and indirect PGPRs roles in plants.

the synthesis of organic acid by rhizosphere microbes can be the reason for the dissolution of mineral phosphorus [126]. Commercial use of PGPB phosphate solubilization got unfortunately limited due to inconsistent results [127].

Phosphate-solubilizing microorganisms are able to dissolve phosphorus from calcium phosphate complexes by lowering the pH of their environment through the release of organic acids or the excretion of protons. The released organic acids are able to dissolve mineral phosphorus directly and can chelate aluminum (Al) and iron ions, as well. Phosphate-solubilizing microorganisms are able to produce different organic acids, such as acetate, lactate, oxalate, tartrate, glycolate, succinate, citrate, gluconate, and ketogluconate [128].

Some PGPRs strains such as *Azotobacter chroococcum*, *Bacillus sp.* and *Pseudomonas sp.* can dissolve phosphorus [124]. PGPRs inoculation enhances the solubility of phosphorus and, thus, enhances plant growth [9]. There are several reports that different strains of PGPRs have the ability to dissolve insoluble phosphate mineral compounds. They reduce the pH of the rhizosphere soil by secretion of organic acids such as gluconic acid, oxalic acid, and citric acid and, thus, increase the solubility of insoluble phosphorus[129]. The ability of a number of soil microorganisms to convert insoluble phosphorus forms into an accessible form to improve plant yield has been studied. Phosphate-solubilizing bacteria that have been isolated, screened and identified through the production of organic acid have been found to have the highest phosphorus release in tricalcium phosphate in culture medium, at the pH of 4.9 [128].

16.8 The role of PGPRs in tolerating heavy metal toxicity in plants

Heavy metals enter the environment through different sources including mines, burning fossil fuels, sewage, sludge, and chemical effluents. These metals involve Cd, Pb, Zn, and Ni. Their characteristics include lack of biodegradation and long shelf life, which damage plants, seas, and animals by entering the food chain [130]. Heavy metals remain in the soil and are transported from one trophic level to another through the food chain. Zn (even in small amounts) is the most dangerous element among heavy metals and is classified as a micronutrient, according to the needs of plants [131].

Studying nickel metal toxicity in *R. leguminosarum bv. trifoli* showed the negative effect of this metal on nodule formation [132]. On the other hand, the presence of Zn in large quantities has been indicated to be toxic to living organisms [133], as Zn disrupts the natural metabolism of plants and inhibits plant growth [134]. PGPRs play a major role in neutralizing heavy metal toxicity and their negative effects on plants. Plants use a variety of mechanisms to counteract the effects of heavy metals (Figure 16.2). Among these factors is the increased activity of ACC deaminase, which leads to the production of ethylene and high auxin levels. This leads to increased lateral root growth, with a larger surface area and chelation of metal ions using plant-produced siderophores and PGPRs. PGPRs enhance plant growth by detoxifying metal ions using antioxidant enzymes and nonenzymatic metabolites [135]. Higher levels of H_2O_2 and malondialdehyde (MDA) were observed in PGPRs-inoculated plants compared to uninoculated plants, under Zn stress conditions [136]. *P. aeruginosa*-inoculated plants have significantly increased nonenzymatic antioxidant metabolites content such as total phenolics and ascorbic acid, compared to uninoculated plants [137].

Enhancing the capacity of phytotherapy in plants inoculated with *Bacillus sp. MN3-4* (lead-resistant strain) indicated that lead is released from the rhizosphere by intracellular accumulation and extracellular deposition [138]. Similarly, higher mineralization and metal uptake have been reported in inoculation with rhizobacteria, leading to increased phytometal extraction, along with increased heavy metals and nutrients uptake [139].

Pseudomonas sp. A3R3, of increased *Brassica juncea* biomass through growth-promoting mechanisms such as IAA biosynthesis, phosphorus accumulation, ACC deaminase synthesis, hydrolyzing enzymes, plant biomolecules, and siderophore production in nickel contaminated soil. Inoculation of *B. napus* with PGPB has increased Cu resistance, Cu transfer, and plant biomass using plant growth-promoting traits such as IAA synthesis, siderophore production, ACC diaminase synthesis, arginine decarboxylase function [140]. *Neorhizobium huautlense T1-17* and *B. megaterium H3* have reduced Cd availability in soil and Cd accumulation in rice [64].

PGPRs degrade, remove, and precipitate toxic metals from the soil and help purify plants [141]. Purification is carried out by stimulating plant growth, enhancing plant tolerance to toxic heavy metals, boosting the transfer of metal ions inside the plant, changing the availability of different metal ions, and metal accumulation in the plant. These changes include metal resistance, accumulation, detoxification, and precipitation, which lead to less access to metal ions in contaminated soils [142].

16.8.1 The role of PGPRs in ACC deaminase mechanism in tolerating heavy metal toxicity in plants

Heavy metal-contaminated terrestrial and aquatic environments are a major problem, worldwide. All living organisms such as plants, animals, and microorganisms are affected by the toxicity of the heavy metals around them. Common toxic metals involve mercury (Hg), lead (Pb), Cd, Cu, chromium (Cr), manganese (Mn), Zn, and Al. In addition, a few metals such as antimony (Sb) and arsenic (As) are considered toxic. The US Environmental Protection Agency (US EPA) has identified 20 hazardous substances, including several heavy metals like Pb, As, Cd, and Hg [143].

Some cell wall bacteria can participate in the bacterial accumulation of heavy metals such as Cd, due to their ability to bind to metals [144]. Microbial phytoremediation is an economical and environmentally friendly technology for detoxifying heavy metal-contaminated soils. Soil conditions affect metal capacities; for example, microorganisms affect metals bioavailability by acidifying the environment and affecting redox potential changes [145]. The release of chelators can cause metal movement though metal adsorption by cell components, and the consequent intracellular separation or deposition as insoluble organic or inorganic compounds reduces heavy metal mobility [145]. Rhizobacteria can bind free Cd ions to undesirable biological complex forms that cannot be absorbed by the plant roots [146]. ACC deaminases-containing rhizobacteria are released in the rhizosphere, where metal chelating siderophores affect iron, Zn, and Cu uptake, ultimately increasing biomass [147].

16.9 The role of PGPRs in increasing the activity of plant hormones in plants

Auxin (indole-3-acetic acid (IAA)), gibberellic acid (GA), cytokinin (CK), and ethylene are organic chemical compounds that have been shown to regulate plant growth and development. These chemical molecules are known as the major biochemical, physiological, and morphological hormones required for plant growth. PGPRs species including *Azospirillum, Pseudomonas, Xanthomonas, Rhizobium,* and *Bradyrhizobium* can cause the formation of phytohormones [148]. Most plant processes are controlled by auxin either directly or indirectly [132]. Growth-promoting bacteria can directly affect plant growth through the production of hormones. *Azospirillum brasilense* has been shown to promote auxin production in plants [149].

The success of plant hormones depends on their exogenous usage. Auxin results from the differentiation of vascular tissue, the appearance of lateral roots, the stimulation of cell division, and the elongation of stems and roots. The type of cultivated species and strains, developmental conditions and availability of nutrients in the rhizosphere are important for proper and applicable IAA production [110, 132].

Elevated L-tryptophan levels promote the biochemical and metabolic activities of bacterial BIPs or APBs and modify root length and architecture, with subsequent reactions. The main metabolism pathways include tryptophol, tryptamine, indole-3-pyruvic acid (IPA), and indole-3-acetamide. Evidence suggests that species that make less auxin due to L-tryptophan deficiency produce more auxin by increasing L-tryptophan level, especially in the presence of a live Rhizobium strain. IAA alone may not be sufficient to achieve maximum plant yield, but it does help plant growth [150, 151]. Increased production of the cytokinin hormone by *Bacillus subtilis* has been reported [152].

GA can enhance the tolerance of biotic and abiotic stresses in many plant products. Exogenous application of these growth-regulating hormones can be beneficial in soil and in improvement of crop yields. GA can be produced both endogenously and exogenously by plants or PGPRs [153]. One study reports that inoculation of seeds with *Citricoccus zhacaiensis* increased the growth rates of onion seedlings under stress conditions by secreting growth-regulating hormones such as gibberellin and auxin [154]. Gibberellin production has also been reported in *Rhizobium phaseoli, Bacillus cereus,* and *Acinetobacter calcoaceticus* [155].

16.10 The role of PGPRs in dealing with biological stresses in plants

One of the biggest threats to food security around the world is the removal of crops affected by plant diseases. Plant diseases cause a broad spectrum of damages ranging from a small reduction in plant growth to irreparable damage to the plant, finally leading to reduced yield and plant death. Many factors are involved in the practicality of a biological control system to deal with pathogen, one of which is the ability to establish itself in the host plant. Another factor is the crop environment because it has an important role in determining the competition of the existing microflora, with the effective population level of an antagonist. The environment affects the choice of antagonist, as well. For example, yeasts can survive on leaves faster under unfavorable moisture conditions than non-spore-forming bacteria. It is, therefore, indispensable to make out the basic mechanism [153, 156–158].

Understanding the biological control mechanisms of plant diseases and the interactions between antagonists and pathogens provides the best and most effective control agent by manipulating the soil environment to create favorable conditions for biological control against disease [159].

PGPB are known to promote plant growth and vigor in diverse culture systems and to develop plant resistance to pathogens and insects [160]. Some PGPB may confront soil pathogens by producing siderophores and antimicrobial metabolites, or by competing for food and ecological nests.

PGPRs are responsible for a large part of crop protection, growth, and soil health [9]. Most of these bacteria, including *Pseudomonas*, promote plant growth and development by reducing or eliminating the harmful effects of pathogens through various mechanisms such as induction of plant resistance system against pathogens [161]. *Pseudomonas, Bacillus, Azospirillum,* and *Rhizobium* strains are involved in inhibiting or killing pathogens by making antibiotic mixtures. Microbial antagonist is another tool used to suppress pathogens (Figure 16.2). PGPRs regulate a broad range of pathogens including bacteria, fungi, viruses, and nematic diseases [162, 163]. *Serratia spp.* and *Paenibacillus spp.* have controlled verticillium wilt disease in cauliflower and canola plants, both in pots and in vitro condition [164]. The incidence of Fusarium wilt disease in tomatoes has been reduced by *Pseudomonas fluorescens* [165]. *Pseudomonas flouorescens* and *Trichoderma asperellum* have promoted chickpea plants resistance to Erysiphe pisi [166].

16.11 Major antibiotics produced by PGPRs

Plant pests and diseases are the most important factors that decrease the quantity and quality of crops and yields. In the last decade, the use of PGPB with the ability to induce systemic resistance to different plant diseases has become one of the most useful methods for the biological control of plant pests and diseases, in the direction of sustainable management of the agroecosystems [167, 168].

PGPRs play a key role in the production of a wide range of antibiotics and the inhibition or suppression of pathogenic microbes. *Bacillus* and *Pseudomonas fluorescens* species are effective in eliminating pathogens, producing inhibitory metabolites and antagonists, and defending against pathogens. In addition, plant-ISR antibiotics play an essential role in antagonistic action. Antibiotics as a heterogeneous community of low molecular weight organic complexes can impair the production or metabolism of various microorganisms [167, 169, 170].

The most important aspect of plant growth that enhances the growth of rhizo pathological bacteria and resistance to other pathogens is production of one or more antibiotics. Antibiotics show potentially beneficial properties such as antimicrobial, antiviral, and antioxidant properties; on the other hand, they enhance plant growth [167, 171, 172].

They are classified into volatile and nonvolatile groups. Aldehydes, alcohols, sulfides, ketones, and HCN are in the volatile group, while heterocyclic nitrogen is in the nonvolatile group [22, 172].

16.11.1 The role of PGPRs in the production of nonvolatile antibiotics in plants

The polyketide metbolite (2,4-Diacetylphloroglucinol (DAPG or Phl)) is a phenolic polyoxide compound produced by *Pseudomonas fluorescens* with antibacterial, antifungal, and antioxidant activities. DAPG is the main determinant of the biocontrol activity of PGPRs [173]. The antibiotic 2,4-DAPG can impair *Gaeumannomyces graminis* var. tritici, the causative agent of wheat disease. *P. fluorescens* strains produce this antibiotic with their nematicidal activities [174].

In plants, ISR microorganisms extract Phl, which is involved in the management of plant diseases and plant cell development. Pyoluteorin is a natural antibiotic that is biosynthesized via the nonribosomal peptide synthetase (NRPS) and polyketide synthase (PKS) pathways [175, 176]. Plt was isolated from *Pseudomonas aeruginosa* T359 and IFO 3455 strains in the 1950s. It was toxic to bacterial, fungal, and plant pathogens [177]. Most fungal pathogens such as *Pythium ultimate* have been inhibited by Plt-producing *Pseudomonas* strains [178]. *Glomerella tucumanensis* causes red root rot disease in sugarcane. It has shown that *P. putida* resist this disease by producing piolutorin [179].

Bacterial species such as *Pseudomonas, Burkholderia, Brevibacterium,* or *Streptomyces* produce low molecular weight pigments called phenazines, which are, basically, heterocyclic nitrogen compounds. Some bacterial strains produce different phenazine derivatives as a mixture simultaneously [180]. Some *Pseudomonas* strains have antibiotic and antitumor characteristics and are active in suppressing plants pathogens including fungi and nematodes [181]. *P. fluorescens* 2–79 and *P. aureofaciens* 30–84 produce a compound called phenazine-1-carboxylic acid (PCA), which provides biological control of G. graminis var. tritici in wheat [182]. *P. aeruginosa* produces pyocyanin and phenazine-1-carboxylic acid, which show antagonistic activity against F. oxysporum, Aspergillus niger, and other pathogens [183].

HCN is one of the volatile antibiotics produced by *Chromobacterium violaceum, P. aeruginosa,* and *P. fluorescens* as secondary metabolites [184] and is a key determinant of the biological control [185]. *P. chlororaphis* O6 bacterial strains showed nematicidal activity against Meloidogyne hapla by producing HCN [186].

16.12 Conclusion

Due to the growing trend in world's population, food shortage, excessive use of natural resources to provide food, and ample use of fertilizers and chemical toxins, application and promotion of alternative methods (PGPMs) seems necessary for environmental sustainability and organic production. PGPRs promote plant growth and vigor in a variety of ways such as providing bio-elements, enhancing nitrogen fixation, antibiotic synthesis, etc. Moreover, PGPRs stimulate plant growth by changing the microbial community of the rhizosphere through the production of different substances. Therefore, more comprehensive and meticulous study to promote our knowledge on the use of PGPMs, such as PGPRs, and their molecular and physiological roles in dealing with biotic and abiotic stresses is necessary to achieve sustainable agriculture worldwide and to protect the environment.

Abbreviation

Al	aluminum
Cd	cadmium
Cu	copper
GA	gibberellic acid
HCN	hydrogen cyanide
IAA	auxin (indole-3-acetic acid
ISR	induced systemic resistance
PGPB	plant growth-promoting bacteria
PGPMs	plant growth-promoting microorganisms
PGPRs	plant growth-promoting rhizobacteria
PSB	phosphate-solubilizing bacteria
Zn	Zinc

References

[1] Fazelienasab B, Omidi M, Amiritokaldani M. Effects of abscisic acid on callus induction and regeneration of different wheat cultivars to mature embryo culture. News directions for a diverse planet: Proceedings of the 4th International Brisbane, Australia, 2004, 26.

[2] Fazeli-Nasab B, A-R S, Azad H. Effect of titanium dioxide nanoparticles on essential oil quantity and quality in *Thymus vulgaris* under water deficit. Journal of Medicinal Plants and By-product 2018, 7, 125–133.

[3] Umesha S, Singh PK, Singh RP. Microbial biotechnology and sustainable agriculture. In: Singh RL, Mondal S (eds) Biotechnology for sustainable agriculture. Elsevier, Amsterdam, Netherlands, 2018, 185–205.

[4] Yadav AN, Rastegari AA, Yadav N, Kour D. Advances in plant microbiome and sustainable agriculture. Singapore, Springer, 2020.

[5] Vejan P, Abdullah R, Khadiran T, Ismail S, Nasrulhaq Boyce A. Role of plant growth promoting rhizobacteria in agricultural sustainability – a review. Molecules 2016, 21, 573.

[6] Liu W, Sikora E, Park S-W. Plant growth-promoting rhizobacterium, *Paenibacillus polymyxa* CR1, upregulates dehydration-responsive genes, RD29A and RD29B, during priming drought tolerance in arabidopsis. Plant Physiology and Biochemistry 2020, 156, 146–154.

[7] Piri R, Moradi A, Balouchi H. Improvement of salinity stress in cumin (*Cuminum cyminum*) seedling by inoculation with *Rhizobacteria*. Indian Journal of Agricultural Sciences 2020, 90, 371–375.

[8] Piri R, Moradi A, Balouchi H, Salehi A. Improvement of cumin (*Cuminum cyminum*) seed performance under drought stress by seed coating and biopriming. Scientia Horticulturae 2019, 257, 108667.

[9] Fazeli-Nasab B, Sayyed RZ. Plant growth-promoting rhizobacteria and salinity stress: A journey into the soil. In: Sayyed RZ, Arora, NK, Reddy, MS, eds. Plant growth promoting rhizobacteria for sustainable stress management: Volume 1: Rhizobacteria in abiotic stress management. Singapore, Springer Singapore, 2019, 21–34.

[10] Kuklinsky-Sobral J, Araújo WL, Mendes R, Geraldi IO, Pizzirani-Kleiner AA, Azevedo JL. Isolation and characterization of soybean-associated bacteria and their potential for plant growth promotion. Environmental Microbiology 2004, 6, 1244–1251.

[11] Berg G, Grube M, Schloter M, Smalla K. Unraveling the plant microbiome: Looking back and future perspectives. Frontiers in Microbiology 2014, 5, 148.

[12] Hamid S, Lone R, Mohamed HI. Production of antibiotics from PGPR and their role in biocontrol of plant diseases. In: Mohamed HI, Saad El-Beltagi HE, Abd-Elsalam KA (eds) Plant growth-promoting microbes for sustainable biotic and abiotic stress management. Springer Cham, 2021, 441–461.

[13] Nadeem SM, Ahmad M, Zahir ZA, Javaid A, Ashraf M. The role of mycorrhizae and plant growth promoting rhizobacteria (PGPR) in improving crop productivity under stressful environments. Biotechnology Advances 2014, 32, 429–448.

[14] Dunbar J, Barns SM, Ticknor LO, Kuske CR. Empirical and theoretical bacterial diversity in four Arizona soils. Applied and Environmental Microbiology 2002, 68, 3035–3045.

[15] Kumar A, Maurya B, Raghuwanshi R, Meena VS, Islam MT. Co-inoculation with Enterobacter and Rhizobacteria on yield and nutrient uptake by wheat (*Triticum aestivum* L.) in the alluvial soil under Indo-Gangetic plain of India. Journal of Plant Growth Regulation 2017, 36, 608–617.

[16] Basu A, Prasad P, Das SN, Kalam S, Sayyed R, Reddy M, El Enshasy H. Plant growth promoting rhizobacteria (PGPR) as green bioinoculants: Recent developments, constraints, and prospects. Sustainability 2021, 13, 1140.

[17] Sagar A, Shukla P, Sayyed R, Ramteke P. Stimulation of seed germination and growth parameters of rice var. Sahbhagi by enterobacter cloacae in the presence of ammonium sulfate as substitute of ACC. In: Sayyed RZ, Reddy MS, Antonius S (eds) Plant growth promoting rhizobacteria (PGPR): Prospects for sustainable agriculture. Springer, Singapore, 2019, 117–124.

[18] Shaikh S, Sayyed R. Role of plant growth-promoting rhizobacteria and their formulation in biocontrol of plant diseases. In: Plant microbes symbiosis: Applied facets. Springer, 2015, 337–351.

[19] Shaikh S, Wani S, Sayyed R. Impact of interactions between rhizosphere and rhizobacteria: A review. Journal of Bacteriology and Mycology 2018, 5, 1058.

[20] Glick BR. Bacteria with ACC deaminase can promote plant growth and help to feed the world. Microbiological Research 2014, 169, 30–39.
[21] Olanrewaju OS, Glick BR, Babalola OO. Mechanisms of action of plant growth promoting bacteria. World Journal of Microbiology & Biotechnology 2017, 33, 1–16.
[22] Gouda S, Kerry RG, Das G, Paramithiotis S, Shin H-S, Patra JK. Revitalization of plant growth promoting rhizobacteria for sustainable development in agriculture. Microbiological Research 2018, 206, 131–140.
[23] Farouk S, Elhindi KM, Alotaibi MA. Silicon supplementation mitigates salinity stress on *Ocimum basilicum* L. via improving water balance, ion homeostasis, and antioxidant defense system. Ecotoxicology and Environmental Safety 2020, 206, 111396.
[24] El-Ghany A, Mona F, Attia M. Effect of exopolysaccharide-producing bacteria and melatonin on faba bean production in saline and nonsaline soil. Agronomy 2020, 10, 316.
[25] Waśkiewicz A, Muzolf-Panek M, Goliński P. Phenolic content changes in plants under salt stress. In: Ecophysiology and responses of plants under salt stress. Springer, 2013, 283–314.
[26] Bothe H. Arbuscular mycorrhiza and salt tolerance of plants. Symbiosis 2012, 58, 7–16.
[27] Amozadeh S, Fazeli-Nasab B. Improvements methods and mechanisms to salinity tolerance in agricultural crops. The first national agricultural conference in difficult environments. 2012, Ramhormoz Branch, Islamic Azad University.
[28] Jogawat A. Osmolytes and their role in abiotic stress tolerance in plants. In: Molecular plant abiotic stress: Biology and biotechnology. 2019, 91–104.
[29] Abd El-Azeem SA, Elwan MW, Sung J-K, Ok YS. Alleviation of salt stress in eggplant (*Solanum melongena* L.) by plant-growth-promoting rhizobacteria. Communications in Soil Science and Plant Analysis 2012, 43, 1303–1315.
[30] Tian S, Guo R, Zou X, Zhang X, Yu X, Zhan Y, Ci D, Wang M, Wang Y, Si T. Priming with the green leaf volatile (Z)-3-hexeny-1-yl acetate enhances salinity stress tolerance in peanut (*Arachis hypogaea* L.) seedlings. Frontiers in Plant Science 2019, 10, 785.
[31] Nadeem SM, Ahmad M, Naveed M, Imran M, Zahir ZA, Crowley DE. Relationship between in vitro characterization and comparative efficacy of plant growth-promoting rhizobacteria for improving cucumber salt tolerance. Archives of Microbiology 2016, 198, 379–387.
[32] Ambede JG, Netondo GW, Mwai GN, Musyimi DM. NaCl salinity affects germination, growth, physiology, and biochemistry of bambara groundnut. Brazilian Journal of Plant Physiology 2012, 24, 151–160.
[33] Hanin M, Ebel C, Ngom M, Laplaze L, Masmoudi K. New insights on plant salt tolerance mechanisms and their potential use for breeding. Frontiers in Plant Science 2016, 7, 1787.
[34] Abogadallah GM. Insights into the significance of antioxidative defense under salt stress. Plant Signaling & Behavior 2010, 5, 369–374.
[35] Marulanda A, Azcón R, Chaumont F, Ruiz-Lozano JM, Aroca R. Regulation of plasma membrane aquaporins by inoculation with a *Bacillus megaterium* strain in maize (*Zea mays* L.) plants under unstressed and salt-stressed conditions. Planta 2010, 232, 533–543.
[36] Chen M, Wei H, Cao J, Liu R, Wang Y, Zheng C. Expression of *Bacillus subtilis* proBA genes and reduction of feedback inhibition of proline synthesis increases proline production and confers osmotolerance in transgenic Arabidopsis. BMB Reports 2007, 40, 396–403.
[37] Del Amor FM, Cuadra-Crespo P. Plant growth-promoting bacteria as a tool to improve salinity tolerance in sweet pepper. Functional Plant Biology 2011, 39, 82–90.
[38] König P, Averhoff B, Müller V. K+ and its role in virulence of Acinetobacter baumannii. International Journal of Medical Microbiology 2021, 151516.

[39] Manaf HH, Zayed MS. Productivity of cowpea as affected by salt stress in presence of endomycorrhizae and *Pseudomonas fluorescens*. Annals of Agricultural Sciences 2015, 60, 219–226.
[40] Lugtenberg BJ, Malfanova N, Kamilova F, Berg G. Plant growth promotion by microbes. Molecular Microbial Ecology of the Rhizosphere 2013, 2, 561–573.
[41] Rojas-Tapias D, Moreno-Galván A, Pardo-Díaz S, Obando M, Rivera D, Bonilla R. Effect of inoculation with plant growth-promoting bacteria (PGPB) on amelioration of saline stress in maize (*Zea mays*). Applied Soil Ecology 2012, 61, 264–272.
[42] Niu S-Q, Li H-R, Paré PW, Aziz M, Wang S-M, Shi H, Li J, Han -Q-Q, Guo S-Q LJ. Induced growth promotion and higher salt tolerance in the halophyte grass *Puccinellia tenuiflora* by beneficial rhizobacteria. Plant and Soil 2016, 407, 217–230.
[43] Egamberdieva D, Wirth S, Bellingrath-Kimura SD, Mishra J, Arora NK. Salt-tolerant plant growth promoting rhizobacteria for enhancing crop productivity of saline soils. Frontiers in Microbiology 2019, 10, 2791.
[44] Kunkel BN, Johnson JM. Auxin plays multiple roles during plant–pathogen interactions. Cold Spring Harbor Perspectives in Biology 2021, 13, a040022.
[45] Li X, Kong X, Zhou J, Luo Z, Lu H, Li W, Tang W, Zhang D, Ma C, Zhang H. Seeding depth and seeding rate regulate apical hook formation by inducing *GhHLS1* expression via ethylene during cotton emergence. Plant Physiology and Biochemistry 2021, 164, 92–100.
[46] Bharti N, Barnawal D. Amelioration of salinity stress by PGPR: ACC deaminase and ROS scavenging enzymes activity. In: Singh AK, Kumar, A, Singh, PK, eds. PGPR amelioration in sustainable agriculture. Woodhead Publishing, Sawston, United Kingdom, 2019, 85–106.
[47] Zhou C, Ma Z, Zhu L, Xiao X, Xie Y, Zhu J, Wang J. Rhizobacterial strain *Bacillus megaterium* BOFC15 induces cellular polyamine changes that improve plant growth and drought resistance. International Journal of Molecular Sciences 2016, 17, 976.
[48] Khan N, Bano A. Exopolysaccharide producing rhizobacteria and their impact on growth and drought tolerance of wheat grown under rainfed conditions. PloS One 2019, 14, e0222302.
[49] Banerjee A, Sarkar S, Cuadros-Orellana S, Bandopadhyay R. Exopolysaccharides and biofilms in mitigating salinity stress: The biotechnological potential of halophilic and soilinhabiting PGPR microorganisms. In: Giri B, Varma A (eds) Microorganisms in saline environments: strategies and functions. Springer, Cham, 2019, 133–153.
[50] Mohammed AF. Effectiveness of exopolysaccharides and biofilm forming plant growth promoting rhizobacteria on salinity tolerance of faba bean (*Vicia faba* L.). African Journal of Microbiology Research 2018, 12, 399–404.
[51] Kirova E, Kocheva K. Physiological effects of salinity on nitrogen fixation in legumes–a review. Journal of Plant Nutrition 2021, 44(17), 1–10.
[52] Duzan H, Zhou X, Souleimanov A, Smith D. Perception of *Bradyrhizobium japonicum* Nod factor by soybean [*Glycine max* (L.) Merr.] root hairs under abiotic stress conditions. Journal of Experimental Botany 2004, 55, 2641–2646.
[53] Qurashi AW, Sabri AN. Bacterial exopolysaccharide and biofilm formation stimulate chickpea growth and soil aggregation under salt stress. Brazilian Journal of Microbiology 2012, 43, 1183–1191.
[54] Ji J, Yuan D, Jin C, Wang G, Li X, Guan C. Enhancement of growth and salt tolerance of rice seedlings (*Oryza sativa* L.) by regulating ethylene production with a novel halotolerant PGPR strain Glutamicibacter sp. YD01 containing ACC deaminase activity. Acta Physiologiae Plantarum 2020, 42, 1–17.
[55] Tester M, Davenport R. Na+ tolerance and Na+ transport in higher plants. Annals of Botany 2003, 91, 503–527.

[56] Maathuis FJ, Verlin D, Smith FA, Sanders D, Fernandez JA, Walker NA. The physiological relevance of Na+-coupled K+-transport. Plant Physiology 1996, 112, 1609–1616.
[57] Saqib M, Akhtar J, Qureshi R, Aslam M, Nawaz S. Effect of salinity and sodicity on growth and ionic relations of different wheat genotypes. Pakistan Journal of Soil Science (Pakistan) 2000, 18, 99–104.
[58] Ashraf M, Hasnain S, Berge O, Mahmood T. Inoculating wheat seedlings with exopolysaccharide-producing bacteria restricts sodium uptake and stimulates plant growth under salt stress. Biology and Fertility of Soils 2004, 40, 157–162.
[59] Kohler J, Caravaca F, Carrasco L, Roldan A. Contribution of *Pseudomonas mendocina* and *Glomus intraradices* to aggregate stabilization and promotion of biological fertility in rhizosphere soil of lettuce plants under field conditions. Soil Use and Management 2006, 22, 298–304.
[60] Solaimanifar S, Asemaneh T. Investigation of silicon effects on some growth and biochemical parameters of lens culinaris medik under salt stress. Journal of Plant Nutrition 2021, 44(15), 1–15.
[61] Nadeem SM, Zahir ZA, Naveed M, Arshad M. Rhizobacteria containing ACC-deaminase confer salt tolerance in maize grown on salt-affected fields. Canadian Journal of Microbiology 2009, 55, 1302–1309.
[62] Yun P, Xu L, Wang -S-S, Shabala L, Shabala S, Zhang W-Y. Piriformospora indica improves salinity stress tolerance in *Zea mays* L. plants by regulating Na+ and K+ loading in root and allocating K+ in shoot. Plant Growth Regulation 2018, 86, 323–331.
[63] Thu HPT, Thu TN, Thao NDN, Le Minh K, Do Tan K. Evaluate the effects of salt stress on physico-chemical characteristics in the germination of rice (*Oryza sativa* L.) in response to methyl salicylate (MeSA). Biocatalysis and Agricultural Biotechnology 2020, 23, 101470.
[64] Liu T, Li R, Jin X, Ding J, Zhu X, Sun C, Guo W. Evaluation of seed emergence uniformity of mechanically sown wheat with UAV RGB imagery. Remote Sensing 2017, 9, 1241.
[65] Zheng H, Zhou X, He J, Yao X, Cheng T, Zhu Y, Cao W, Tian Y. Early season detection of rice plants using RGB, NIR-GB and multispectral images from unmanned aerial vehicle (UAV). Computers and Electronics in Agriculture 2020, 169, 105223.
[66] Fercha A, Capriotti AL, Caruso G, Cavaliere C, Stampachiacchiere S, Zenezini Chiozzi R, Laganà A. Shotgun proteomic analysis of soybean embryonic axes during germination under salt stress. Proteomics 2016, 16, 1537–1546.
[67] Ha-Tran DM, Nguyen TTM, Hung S-H, Huang E, Huang -C-C. Roles of plant growth-promoting rhizobacteria (PGPR) in stimulating salinity stress defense in plants: A review. International Journal of Molecular Sciences 2021, 22, 3154.
[68] Rajabi Dehnavi A, Zahedi M, Ludwiczak A, Cardenas Perez S, Piernik A. Effect of salinity on seed germination and seedling development of sorghum (*Sorghum bicolor* (L.) Moench) genotypes. Agronomy 2020, 10, 859.
[69] Mahmood A, Turgay OC, Farooq M, Hayat R. Seed biopriming with plant growth promoting rhizobacteria: A review. FEMS Microbiology Ecology 2016, 92, 1–14.
[70] Sarkar A, Pramanik K, Mitra S, Soren T, Maiti TK. Enhancement of growth and salt tolerance of rice seedlings by ACC deaminase-producing Burkholderia sp. MTCC 12259. Journal of Plant Physiology 2018, 231, 434–442.
[71] Zhu Z, Zhang H, Leng J, Niu H, Chen X, Liu D, Chen Y, Gao N, Ying H. Isolation and characterization of plant growth-promoting rhizobacteria and their effects on the growth of *Medicago sativa* L. under salinity conditions. Antonie van Leeuwenhoek 2020, 113, 1263–1278.
[72] Sarropoulou V, Dimassi-Theriou K, Therios I, Koukourikou-Petridou M. Melatonin enhances root regeneration, photosynthetic pigments, biomass, total carbohydrates and proline

content in the cherry rootstock PHL-C (*Prunus avium× Prunus cerasus*). Plant Physiology and Biochemistry 2012, 61, 162–168.
[73] Tal O, Haim A, Harel O, Gerchman Y. Melatonin as an antioxidant and its semi-lunar rhythm in green macroalga Ulva sp. Journal of Experimental Botany 2011, 62, 1903–1910.
[74] Uçarlı C. Effects of salinity on seed germination and early seedling stage. In: Abiotic stress in plants. IntechOpen, 2020.
[75] Hichem H, Mounir D. Differential responses of two maize (*Zea mays* L.) varieties to salt stress: Changes on polyphenols composition of foliage and oxidative damages. Industrial Crops and Products 2009, 30, 144–151.
[76] Ksouri R, Megdiche W, Debez A, Falleh H, Grignon C, Abdelly C. Salinity effects on polyphenol content and antioxidant activities in leaves of the halophyte *Cakile maritima*. Plant Physiology and Biochemistry 2007, 45, 244–249.
[77] Abd_Allah EF, Alqarawi AA, Hashem A, Radhakrishnan R, Al-Huqail AA, Fon A-O, Malik JA, Alharbi RI, Egamberdieva D. Endophytic bacterium *Bacillus subtilis* (BERA 71) improves salt tolerance in chickpea plants by regulating the plant defense mechanisms. Journal of Plant Interactions 2018, 13, 37–44.
[78] Khan A, Zhao XQ, Javed MT, Khan KS, Bano A, Shen RF, Masood S. *Bacillus pumilus* enhances tolerance in rice (*Oryza sativa* L.) to combined stresses of NaCl and high boron due to limited uptake of Na+. Environmental and Experimental Botany 2016, 124, 120–129.
[79] Nassal D, Spohn M, Eltlbany N, Jacquiod S, Smalla K, Marhan S, Kandeler E. Effects of phosphorus-mobilizing bacteria on tomato growth and soil microbial activity. Plant and Soil 2018, 427, 17–37.
[80] Gopalakrishnan S, Srinivas V, Samineni S. Nitrogen fixation, plant growth and yield enhancements by diazotrophic growth-promoting bacteria in two cultivars of chickpea (*Cicer arietinum* L.). Biocatalysis and Agricultural Biotechnology 2017, 11, 116–123.
[81] Navarro-Torre S, Barcia-Piedras J, Mateos-Naranjo E, Redondo-Gómez S, Camacho M, Caviedes M, Pajuelo E, Rodríguez-Llorente I. Assessing the role of endophytic bacteria in the halophyte *Arthrocnemum macrostachyum* salt tolerance. Plant Biology 2017, 19, 249–256.
[82] Vinocur B, Altman A. Recent advances in engineering plant tolerance to abiotic stress: Achievements and limitations. Current Opinion in Biotechnology 2005, 16, 123–132.
[83] Farooq M, Wahid A, Kobayashi N, Fujita D, Basra S. Plant drought stress: Effects, mechanisms and management. In: Sustainable agriculture. 2009, 153–188.
[84] Yuwono T, Handayani D, Soedarsono J. The role of osmotolerant rhizobacteria in rice growth under different drought conditions. Australian Journal of Agricultural Research 2005, 56, 715–721.
[85] Khan Z, Rho H, Firrincieli A, Hung SH, Luna V, Masciarelli O, Kim S-H, Doty SL. Growth enhancement and drought tolerance of hybrid poplar upon inoculation with endophyte consortia. Current Plant Biology 2016, 6, 38–47.
[86] Heidari M, Golpayegani A. Effects of water stress and inoculation with plant growth promoting rhizobacteria (PGPR) on antioxidant status and photosynthetic pigments in basil (*Ocimum basilicum* L.). Journal of the Saudi Society of Agricultural Sciences 2012, 11, 57–61.
[87] Kavamura VN, Santos SN, da Silva JL, Parma MM, Ávila LA, Visconti A, Zucchi TD, Taketani RG, Andreote FD, de Melo IS. Screening of Brazilian cacti rhizobacteria for plant growth promotion under drought. Microbiological Research 2013, 168, 183–191.
[88] Nocker A, Fernández PS, Montijn R, Schuren F. Effect of air drying on bacterial viability: A multiparameter viability assessment. Journal of Microbiological Methods 2012, 90, 86–95.
[89] Vacheron J, Renoud S, Muller D, Babalola OO, Prigent-Combaret C. Alleviation of abiotic and biotic stresses in plants by *Azospirillum*. Handbook for Azospirillum 2015, 1, 333–365.

[90] Dehghani Bidgoli R. Investigating the possibility of increasing the physiological function of (*Lippia citriodora* L.) using biological stimuli under salinity stress conditions. Eco-phytochemical Journal of Medicinal Plants 2019, 7, 77–88.
[91] Rocha FR, Papini-Terzi FS, Nishiyama MY, Vêncio RZ, Vicentini R, Duarte RD, de Rosa VE, Vinagre F, Barsalobres C, Medeiros AH. Signal transduction-related responses to phytohormones and environmental challenges in sugarcane. BMC Genomics 2007, 8, 1–22.
[92] Daim SB, Meijer J, Kasim WA, Osman ME, Omar MN, Islam A, Abd E. Control of drought stress in wheat using plant-growth-promoting bacteria. Journal of Plant Growth Regulation 2013, 32, 122–130.
[93] Mayak S, Tirosh T, Glick BR. Plant growth-promoting bacteria confer resistance in tomato plants to salt stress. Plant Physiology and Biochemistry 2004, 42, 565–572.
[94] Nadeem SM, Ahmad M, Tufail MA, Asghar HN, Nazli F, Zahir ZA. Appraising the potential of EPS-producing rhizobacteria with ACC-deaminase activity to improve growth and physiology of maize under drought stress. Physiologia Plantarum 2021, 172, 463–476.
[95] Ayub MA, Ahmad HR, Ali M, Rizwan M, Ali S, Ur Rehman MZ, Waris AA. Salinity and its tolerance strategies in plants. In: Tripathi DK et al (eds) Plant life under changing environment. Elsevier, Amsterdam, Netherlands, 2020, 47–76.
[96] Balota M, Cristescu S, Payne W, te Lintel Hekkert S, Laarhoven L, Harren F. Ethylene production of two wheat cultivars exposed to desiccation, heat, and paraquat-induced oxidation. Crop Science 2004, 44, 812–818.
[97] Saleem M, Arshad M, Hussain S, Bhatti AS. Perspective of plant growth promoting rhizobacteria (PGPR) containing ACC deaminase in stress agriculture. Journal of Industrial Microbiology & Biotechnology 2007, 34, 635–648.
[98] Glick BR, Todorovic B, Czarny J, Cheng Z, Duan J, McConkey B. Promotion of plant growth by bacterial ACC deaminase. Critical Reviews in Plant Sciences 2007, 26, 227–242.
[99] da Rosa TC, Carvalho IR, Hutra DJ, Bradebon LC, da Rosa Sarturi MV, da Rosa JAG, Szareski VJ. Maize breeding for abiotic stress tolerance: An alternative to face climate changes. Agronomy Science and Biotechnology 2020, 6, 1–13.
[100] Krishna P. Brassinosteroid-mediated stress responses. Journal of Plant Growth Regulation 2003, 22, 289–297.
[101] Zhang J, Poudel B, Kenworthy K, Unruh JB, Rowland D, Erickson JE, Kruse J. Drought responses of above-ground and below-ground characteristics in warm-season turfgrass. Journal of Agronomy and Crop Science 2019, 205, 1–12.
[102] Chen D, Liu X, Bian R, Cheng K, Zhang X, Zheng J, Joseph S, Crowley D, Pan G, Li L. Effects of biochar on availability and plant uptake of heavy metals–A meta-analysis. Journal of Environmental Management 2018, 222, 76–85.
[103] Ait Barka E, Nowak J, Clément C. Enhancement of chilling resistance of inoculated grapevine plantlets with a plant growth-promoting rhizobacterium, *Burkholderia phytofirmans* strain PsJN. Applied and Environmental Microbiology 2006, 72, 7246–7252.
[104] Rani A, Tokas J, Punia H. Chapter-9 amelioration of abiotic stresses using PGPRs. Chief Editor Dr RK Naresh 2020, 91, 163.
[105] Pahari A, Mishra B. Characterization of siderophore producing Rhizobacteria and Its effect on growth performance of different vegetables. International Journal of Current Microbiology and Applied Sciences 2017, 6, 1398–1405.
[106] Rosconi F, Davyt D, Martínez V, Martínez M, Abin-Carriquiry JA, Zane H, Butler A, de Souza EM, Fabiano E. Identification and structural characterization of serobactins, a suite of lipopeptide siderophores produced by the grass endophyte *Herbaspirillum seropedicae*. Environmental Microbiology 2013, 15, 916–927.

[107] Shaharoona B, Naveed M, Arshad M, Zahir ZA. Fertilizer-dependent efficiency of *Pseudomonads* for improving growth, yield, and nutrient use efficiency of wheat (*Triticum aestivum* L.). Applied Microbiology and Biotechnology 2008, 79, 147–155.
[108] Wittenmayer L, Merbach W. Plant responses to drought and phosphorus deficiency: Contribution of phytohormones in root-related processes. Journal of Plant Nutrition and Soil Science 2005, 168, 531–540.
[109] Pishchik V, Vorobyev N, Chernyaeva I, Timofeeva S, Kozhemyakov A, Alexeev Y, Lukin S. Experimental and mathematical simulation of plant growth promoting rhizobacteria and plant interaction under cadmium stress. Plant and Soil 2002, 243, 173–186.
[110] Khan N, Bano A, Ali S, Babar MA. Crosstalk amongst phytohormones from planta and PGPR under biotic and abiotic stresses. Plant Growth Regulation 2020, 90, 189–203.
[111] Alori ET, Glick BR, Babalola OO. Microbial phosphorus solubilization and its potential for use in sustainable agriculture. Frontiers in Microbiology 2017, 8, 971.
[112] Keshavarz Zarjani J, Aliasgharzad N, Oustan S, Emadi M, Ahmadi A. Isolation and characterization of potassium solubilizing bacteria in some Iranian soils. Archives of Agronomy and Soil Science 2013, 59, 1713–1723.
[113] Gulati R. Legume symbiosis under abiotic stresses-a review. Molecular Physiol Abiotic Stress Plant Production 2018, 186, 1–12.
[114] Ahemad M, Khan M. Toxicological assessment of selective pesticides towards plant growth promoting activities of phosphate solubilizing *Pseudomonas aeruginosa*. Acta Microbiologica Et Immunologica Hungarica 2011, 58, 169–187.
[115] Bhattacharyya PN, Jha DK. Plant growth-promoting rhizobacteria (PGPR): Emergence in agriculture. World Journal of Microbiology & Biotechnology 2012, 28, 1327–1350.
[116] Navarro-Rodríguez M, Buesa JM, Rubio LM. Genetic and biochemical analysis of the azotobacter vinelandii molybdenum storage protein. Frontiers in Microbiology 2019, 10.
[117] Wisniewski-Dyé F, Vial L, Burdman S, Okon Y, Hartmann A. Phenotypic variation in Azospirillum spp and other root-associated bacteria. Hoboken, Biological nitrogen fixation Wiley, 2015, 1047–1054.
[118] Bruto M, Prigent-Combaret C, Muller D, Moënne-Loccoz Y. Analysis of genes contributing to plant-beneficial functions in plant growth-promoting rhizobacteria and related proteobacteria. Scientific Reports 2014, 4, 1–10.
[119] Glick BR. Plant growth-promoting bacteria: Mechanisms and applications. Scientifica 2012, 2012. Article ID: 963401.
[120] Khan MS, Zaidi A, Wani PA, Oves M. Role of plant growth promoting rhizobacteria in the remediation of metal contaminated soils. Environmental Chemistry Letters 2009, 7, 1–19.
[121] Kaur G, Reddy MS. Influence of P-solubilizing bacteria on crop yield and soil fertility at multilocational sites. European Journal of Soil Biology 2014, 61, 35–40.
[122] Tennakoon P, Rajapaksha R, Hettiarachchi L. Tea yield maintained in PGPR inoculated field plants despite significant reduction in fertilizer application. Rhizosphere 2019, 10, 100146.
[123] Yadav J, Verma JP, Jaiswal DK, Kumar A. Evaluation of PGPR and different concentration of phosphorus level on plant growth, yield and nutrient content of rice (*Oryza sativa*). Ecological Engineering 2014, 62, 123–128.
[124] Kumar V, Behl RK, Narula N. Establishment of phosphate-solubilizing strains of *Azotobacter chroococcum* in the rhizosphere and their effect on wheat cultivars under green house conditions. Microbiological Research 2001, 156, 87–93.
[125] Guo JK, Ding YZ, Feng RW, Wang RG, Xu YM, Chen C, Wei XL, Chen WM. Burkholderia metalliresistens sp. nov., a multiple metal-resistant and phosphate-solubilising species isolated from heavy metal-polluted soil in Southeast China. Antonie van Leeuwenhoek 2015, 107, 1591–1598.

[126] Barea J-M, Richardson AE. Phosphate mobilisation by soil microorganisms. In: Lugtenberg B (ed) Principles of plant-microbe interactions. Springer, Cham, 2015, 225–234.
[127] Ghosh UD, Saha C, Maiti M, Lahiri S, Ghosh S, Seal A, Mitra Ghosh M. Root associated iron oxidizing bacteria increase phosphate nutrition and influence root to shoot partitioning of iron in tolerant plant *Typha angustifolia*. Plant and Soil 2014, 381, 279–295.
[128] Chen Y, Rekha P, Arun A, Shen F, Lai W-A, Young CC. Phosphate solubilizing bacteria from subtropical soil and their tricalcium phosphate solubilizing abilities. Applied Soil Ecology 2006, 34, 33–41.
[129] Khan AA, Jilani G, Akhtar MS, Naqvi SMS, Rasheed M. Phosphorus solubilizing bacteria: Occurrence, mechanisms and their role in crop production. Journal of Agriculture and Biological Sciences 2009, 1, 48–58.
[130] Rajkumar M, Vara Prasad MN, Freitas H, Ae N. Biotechnological applications of serpentine soil bacteria for phytoremediation of trace metals. Critical Reviews in Biotechnology 2009, 29, 120–130.
[131] Taniguchi J, Hemmi H, Tanahashi K, Amano N, Nakayama T, Nishino T. Zinc biosorption by a zinc-resistant bacterium, Brevibacterium sp. strain HZM-1. Applied Microbiology and Biotechnology 2000, 54, 581–588.
[132] Agarwal P, Singh PC, Chaudhry V, Shirke PA, Chakrabarty D, Farooqui A, Nautiyal CS, Sane AP, Sane VA. PGPR-induced OsASR6 improves plant growth and yield by altering root auxin sensitivity and the xylem structure in transgenic *Arabidopsis thaliana*. Journal of Plant Physiology 2019, 240, 153010.
[133] He Y, Chen H-Y, Hou J, Li Y. Indene– C60 bisadduct: A new acceptor for high-performance polymer solar cells. Journal of the American Chemical Society 2010, 132, 1377–1382.
[134] John R, Ahmad P, Gadgil K, Sharma S. Effect of cadmium and lead on growth, biochemical parameters and uptake in *Lemna polyrrhiza* L. Plant, Soil and Environment 2008, 54, 262–270.
[135] Nadgórska-Socha A, Kafel A, Kandziora-Ciupa M, Gospodarek J, Zawisza-Raszka A. Accumulation of heavy metals and antioxidant responses in Vicia faba plants grown on monometallic contaminated soil. Environmental Science and Pollution Research 2013, 20, 1124–1134.
[136] Turan M, Yildirim E, Ekinci M, Argin S. Effect of biostimulants on yield and quality of cherry tomatoes grown in fertile and stressed soils. HortScience 2021, 56, 414–423.
[137] Gururani MA, Upadhyaya CP, Baskar V, Venkatesh J, Nookaraju A, Park SW. Plant growth-promoting rhizobacteria enhance abiotic stress tolerance in *Solanum tuberosum* through inducing changes in the expression of ROS-scavenging enzymes and improved photosynthetic performance. Journal of Plant Growth Regulation 2013, 32, 245–258.
[138] Shin M-N, Shim J, You Y, Myung H, Bang K-S, Cho M, Kamala-Kannan S, Oh B-T. Characterization of lead resistant endophytic Bacillus sp. MN3-4 and its potential for promoting lead accumulation in metal hyperaccumulator *Alnus firma*. Journal of Hazardous Materials 2012, 199, 314–320.
[139] Chen L, Luo S, Li X, Wan Y, Chen J, Liu C. Interaction of Cd-hyperaccumulator *Solanum nigrum* L. and functional endophyte *Pseudomonas* sp. Lk9 on soil heavy metals uptake. Soil Biology & Biochemistry 2014, 68, 300–308.
[140] Sun L-N, Zhang Y-F, He L-Y, Chen Z-J, Wang Q-Y, Qian M, Sheng X-F. Genetic diversity and characterization of heavy metal-resistant-endophytic bacteria from two copper-tolerant plant species on copper mine wasteland. Bioresource Technology 2010, 101, 501–509.
[141] Sobariu DL, Fertu DIT, Diaconu M, Pavel LV, Hlihor R-M, Drăgoi EN, Curteanu S, Lenz M, Corvini PF-X, Gavrilescu M. Rhizobacteria and plant symbiosis in heavy metal uptake and its implications for soil bioremediation. New Biotechnology 2017, 39, 125–134.

[142] Ma Y, Rajkumar M, Zhang C, Freitas H. Beneficial role of bacterial endophytes in heavy metal phytoremediation. Journal of Environmental Management 2016, 174, 14–25.
[143] Pandey S, Ghosh PK, Ghosh S, De TK, Maiti TK. Role of heavy metal resistant Ochrobactrum sp. and *Bacillus* spp. strains in bioremediation of a rice cultivar and their PGPR like activities. Journal of Microbiology 2013, 51, 11–17.
[144] Sinha S, Mukherjee SK. Cadmium–induced siderophore production by a high Cd-resistant bacterial strain relieved Cd toxicity in plants through root colonization. Current Microbiology 2008, 56, 55–60.
[145] Banala UK, Das NPI, Toleti SR. Microbial interactions with uranium: Towards an effective bioremediation approach. Environmental Technology & Innovation 2020, 21, 101254.
[146] Peng W, Li X, Song J, Jiang W, Liu Y, Fan W. Bioremediation of cadmium-and zinc-contaminated soil using rhodobacter sphaeroides. Chemosphere 2018, 197, 33–41.
[147] Chen J, Li N, Han S, Sun Y, Wang L, Qu Z, Dai M, Zhao G. Characterization and bioremediation potential of nickel-resistant endophytic bacteria isolated from the wetland plant *Tamarix chinensis*. FEMS Microbiology Letters 2020, 367, fnaa098.
[148] Mohamed H, Gomaa E. Effect of plant growth promoting bacillus subtilis and pseudomonas fluorescens on growth and pigment composition of radish plants (*Raphanus sativus*) under NaCl stress. Photosynthetica 2012, 50, 263–272.
[149] Martínez-Morales LJ, Soto-Urzúa L, Baca BE, Sánchez-Ahédo JA. Indole-3-butyric acid (IBA) production in culture medium by wild strain *Azospirillum brasilense*. FEMS Microbiology Letters 2003, 228, 167–173.
[150] Guzmán-Albores JM, Bojórquez-Velázquez E, De León-Rodríguez A, de Jesús Calva-cruz O, de la Rosa APB, Ruíz-Valdiviezo VM. Comparison of *Moringa oleifera* oils extracted with supercritical fluids and hexane and characterization of seed storage proteins in defatted flour. Food Bioscience 2021, 40, 100830.
[151] Agathokleous E, Kitao M, Calabrese EJ. Biphasic effect of abscisic acid on plants: An hormetic viewpoint. Botany 2018, 96, 637–642.
[152] Arkhipova T, Veselov S, Melentiev A, Martynenko E, Kudoyarova G. Ability of bacterium *Bacillus subtilis* to produce cytokinins and to influence the growth and endogenous hormone content of lettuce plants. Plant and Soil 2005, 272, 201–209.
[153] Singh RP, Jha PN. The PGPR stenotrophomonas maltophilia SBP-9 augments resistance against biotic and abiotic stress in wheat plants. Frontiers in Microbiology 2017, 8, 1–15.
[154] Selvakumar G, Bhatt RM, Upreti KK, Bindu GH, Shweta K. *Citricoccus zhacaiensis* B-4 (MTCC 12119) a novel osmotolerant plant growth promoting actinobacterium enhances onion (*Allium cepa* L.) seed germination under osmotic stress conditions. World Journal of Microbiology & Biotechnology 2015, 31, 833–839.
[155] Kang S-M, Joo G-J, Hamayun M, Na C-I, Shin D-H, Kim HY, Hong J-K, Lee I-J. Gibberellin production and phosphate solubilization by newly isolated strain of *Acinetobacter calcoaceticus* and its effect on plant growth. Biotechnology Letters 2009, 31, 277–281.
[156] Kumar A, Singh VK, Tripathi V, Singh PP, Singh AK. Chapter 16 – plant growth-promoting rhizobacteria (PGPR): Perspective in agriculture under biotic and abiotic stress. Editor(s): Ram Prasad, Sarvajeet S. Gill, Narendra Tuteja. Crop Improvement Through Microbial Biotechnology 2018, 2018, 333–342.
[157] Khan A, Sayyed R, Seifi S. Rhizobacteria: Legendary soil guards in abiotic stress management. In: Sayyed,RZ, Arora NK, Reddy MS (eds) Plant growth promoting rhizobacteria for sustainable stress management. Springer, Singapore, 2019, 327–343.
[158] Xia AY, Farooq MA, Javed MT, Kamran MA, Mukhtar T, Ali J, Tabassum T, Rehman S, Munis MFH, Sultan T, Chaudhary HJ. Multi-stress tolerant PGPR *Bacillus xiamenensis* PM14

activating sugarcane (*Saccharum officinarum* L.) red rot disease resistance. Plant Physiology and Biochemistry 2020, 151, 640–649.
[159] Vega FE. The use of fungal entomopathogens as endophytes in biological control: A review. Mycologia 2018, 110, 4–30.
[160] Jia R, Li M, Zhang J, Addrah ME, Zhao J. Effect of low temperature culture on the biological characteristics and aggressiveness of *Sclerotinia sclerotiorum* and *Sclerotinia minor*. Ocl 2021, 28, 20.
[161] Singh SK, Singh PP, Gupta A, Singh AK, Keshri J. Chapter twelve – tolerance of heavy metal toxicity using PGPR strains of pseudomonas species. In: Singh AK, Kumar, A, Singh, PK, eds. PGPR amelioration in sustainable agriculture. Food Security and Environmental Management. 2019, 239–252.
[162] Jiao X, Takishita Y, Zhou G, Smith DL. Plant associated rhizobacteria for biocontrol and plant growth enhancement. Frontiers in Plant Science 2021, 12.
[163] Liu X-M, Zhang H. The effects of bacterial volatile emissions on plant abiotic stress tolerance. Frontiers in Plant Science 2015, 6.
[164] Rybakova D, Schmuck M, Wetzlinger U, Varo-Suarez A, Murgu O, Müller H, Berg G. Kill or cure? The interaction between endophytic *Paenibacillus* and *Serratia* strains and the host plant is shaped by plant growth conditions. Plant and Soil 2016, 405, 65–79.
[165] Srivastava R, Khalid A, Singh U, Sharma A. Evaluation of arbuscular *mycorrhizal* fungus, *fluorescent Pseudomonas* and *Trichoderma harzianum* formulation against *Fusarium oxysporum* f. sp. lycopersici for the management of tomato wilt. Biological Control 2010, 53, 24–31.
[166] Patel JS, Sarma BK, Singh HB, Upadhyay RS, Kharwar RN, Ahmed M. *Pseudomonas fluorescens* and *Trichoderma asperellum* enhance expression of Gα subunits of the pea heterotrimeric G-protein during *Erysiphe pisi* infection. Frontiers in Plant Science 2016, 6, 1206.
[167] Kenawy A, Dailin DJ, Abo-Zaid GA, Abd Malek R, Ambehabati KK, Zakaria KHN, Sayyed R, El Enshasy HA. Biosynthesis of antibiotics by PGPR and their roles in biocontrol of plant diseases. In: Sayyed RZ (ed) Plant growth promoting rhizobacteria for sustainable stress management. Springer, Singapore, 2019, 1–35.
[168] Suriani NL, Suprapta DN, Nazir N, Parwanayoni NMS, Darmadi AAK, Dewi DA, Sudatri NW, Fudholi A, Sayyed R, Syed A. A mixture of piper leaves extracts and rhizobacteria for sustainable plant growth promotion and bio-control of blast pathogen of organic bali rice. Sustainability 2020, 12, 1–18.
[169] Sagar A, Sayyed R, Ramteke P, Sharma S, Marraiki N, Elgorban AM, Syed A. ACC deaminase and antioxidant enzymes producing halophilic Enterobacter sp. PR14 promotes the growth of rice and millets under salinity stress. Physiology and Molecular Biology of Plants 2020, 26, 1847–1854.
[170] Kannojia P, Choudhary KK, Srivastava AK, Singh AK. PGPR bioelicitors: Induced systemic resistance (ISR) and proteomic perspective on biocontrol. In: Singh AK, Kumar, A, Singh, PK, eds. PGPR amelioration in sustainable agriculture. Woodhead Publishing, Sawston, United Kingdom, 2019, 67–84.
[171] Sagar A, Sayyed RZ, Ramteke PW, Sharma S, Marraiki N, Elgorban AM, Syed A. ACC deaminase and antioxidant enzymes producing halophilicEnterobacter sp. ameliorates salt stress and promotes the growth of rice and millets under salt stress. Physiology and Molecu lar Biology of Plants (In press) 2020
[172] Fernando WGD, Nakkeeran S, Zhang Y, Savchuk S. Biological control of sclerotinia sclerotiorum (Lib.) de bary by *Pseudomonas and Bacillus* species on canola petals. Crop Protection 2007, 26(9), 1847–1854, 100–107.

[173] Gaur R. Diversity of 2, 4-diacetylphloroglucinol and 1-aminocyclopropane 1-carboxylate deaminase producing rhizobacteria from wheat rhizosphere. PhD thesis, GB Pant University of Agriculture and Technology, Pantnagar, 2002.
[174] Weller DM, Landa BB, Mavrodi OV, Schroeder KL, De La Fuente L, Blouin Bankhead S, Allende Molar R, Bonsall RF, Mavrodi DV, Thomashow LS. Role of 2,4-diacetylphloroglucinol-producing fluorescent *Pseudomonas* spp. in the defense of plant roots. Plant Biol (Stuttg) 2007, 9, 4–20.
[175] Shahryari F, Esmaeili M. Recent developments in pseudomonas biocontrol mechanisms. Journal of Biosafety 2019, 12, 27–50.
[176] Magarvey NA, Beck ZQ, Golakoti T, Ding Y, Huber U, Hemscheidt TK, Abelson D, Moore RE, Sherman DH. Biosynthetic characterization and chemoenzymatic assembly of the cryptophycins. Potent anticancer agents from nostoc cyanobionts. ACS Chemical Biology 2006, 1, 766–779.
[177] Lukkani NJ, Reddy EC. Screening of ACC-deaminase and antifungal metabolites producing fluorescent *pseudomonads* isolated from rhizosphere soil of groundnut. Current Trends in Biotechnology and Pharmacy 2019, 13, 309–316.
[178] Gurney J, Aldakak L, Betts A, Gougat-Barbera C, Poisot T, Kaltz O, Hochberg ME. Network structure and local adaptation in co-evolving bacteria–phage interactions. Molecular Ecology 2017, 26, 1764–1777.
[179] Hassan MN, Afghan S, Hafeez FY. Biological control of red rot in sugarcane by native pyoluteorin-producing *Pseudomonas putida* strain NH-50 under field conditions and its potential modes of action. Pest Management Science 2011, 67, 1147–1154.
[180] Guttenberger N, Blankenfeldt W, Breinbauer R. Recent developments in the isolation, biological function, biosynthesis, and synthesis of phenazine natural products. Bioorganic & Medicinal Chemistry 2017, 25, 6149–6166.
[181] Zhou L, Jiang HX, Sun S, Yang DD, Jin KM, Zhang W, He YW. Biotechnological potential of a rhizosphere *Pseudomonas aeruginosa* strain producing phenazine-1-carboxylic acid and phenazine-1-carboxamide. World Journal of Microbiology & Biotechnology 2016, 32, 50.
[182] Yu JM, Wang D, Pierson LS, Pierson EA. Disruption of MiaA provides insights into the regulation of phenazine biosynthesis under suboptimal growth conditions in *Pseudomonas chlororaphis* 30-84. Microbiology 2017, 163, 94–108.
[183] Abo-Zaid GA, Soliman NA-M, Abdullah AS, El-Sharouny EE, Matar SM, Sabry SA-F. Maximization of siderophores production from biocontrol agents, *Pseudomonas aeruginosa* F2 and *Pseudomonas fluorescens* JY3 using batch and exponential fed-batch fermentation. Processes 2020, 8, 455.
[184] Blom D, Fabbri C, Eberl L, Weisskopf L. Volatile-mediated killing of arabidopsis thaliana by bacteria is mainly due to hydrogen cyanide. Applied and Environmental Microbiology 2011, 77, 1000–1008.
[185] Anderson AJ, Kim YC. Biopesticides produced by plant-probiotic *Pseudomonas chlororaphis* isolates. Crop Protection 2018, 105, 62–69.
[186] Kang BR, Anderson AJ, Kim YC. Hydrogen cyanide produced by *Pseudomonas chlororaphis* O6 exhibits nematicidal activity against *Meloidogyne hapla*. Plant Pathology Journal 2018, 34, 35–43.

Jaagriti Tyagi, Parul Chaudhary, Jyotsana, Upasana Bhagwati, Geeta Bhandari, Anuj Chaudhary

Chapter 17 Impact of endophytic fungi in biotic stress management

Abstract: Biotic stress is known to cause numerous harmful effects on agricultural harvests worldwide. Many biotic/living microbes, such as bacteria, fungi, virus, weeds, insects, nematodes, and pests, are responsible for causing biotic stress in several crops. When biotic stress crosses a specific value in crops, it causes lethal harm. Practices such as spraying, integrated pest management, and tiling are used for basic control of biotic stress when recognized at an early stage. At the very initial phase of infection, due to the small size of weeds, this makes it hard for crops to recognize. It is essential to develop new control strategies toward plant pathogens. Soil microorganisms play a vital role in plant growth productivity in a stressful environment. Fungal endophytes are the solution to overcome the tasks faced with conventional farming. These are environment friendly microbial commodities that colonize in plant tissues without causing any damage. Endophytes help in increasing the growth of plants. They also help in abiotic and biotic stress tolerance in host plants in addition to being used as biocontrol agents. They are able to diminish the injury triggered by pathogens via antibiosis, production of lytic enzymes, and hormone activation. Thus, this chapter highlights the fungal-mediated improvement in crops that are under pathogen attack.

17.1 Introduction

Biotic stress can be defined as the condition in which the normal metabolic function of plants is disturbed by living organisms [1]. It causes a hazardous impact on crops due to the different types of living microbes, such as bacteria, virus, and fungi. These are most common pathogens that result in plant disease and show a variety of symptoms, such as formation of local lesion, stunting, rotting, and chlorosis [2]. These microorganisms unite, resulting in severe symptoms of disease instead of appearing as individual pathogens, causing the infection. Biotic stress depends on several factors such as environment, types of causal agent, host specificity, and causal agent [3].

Environmental factors play a vital role in biotic stress and help us know whether it results in yield or loss of crop quality, as it differs from region to region or even from one country to another. For example, Barley foliar diseases were considered to be the

https://doi.org/10.1515/9783110771558-017

major cause of biotic stress in Australia, which resulted in increasing yield and quality losses. In some regions of Australia that are known for the barley, nematodes, such as root lesion nematode (*Pratylenchus* spp.), play an important role as barley pest [4]. Another factor is climate change that plays a significant part in the spread of disease. In the near future, with the increase in temperatures, as expected, diseases caused by thermophilic bacteria are anticipated to occur. Rapid change in races, caused by insects and pathogens, is also considered a factor for biotic stress, which results in a non-durable process of resistance. However, it is a difficult job for biotic stress to provide resistance to plants in case of non-availability of sources of durable resistance. Generally, plants are provided resistance against biotic stress via one or more gene. Thus, in crop plants, genetic basis can be exploited by plant breeders for the development of resistance against various pests and diseases.

Endophytes, in the form of potential biocontrol agents, represent as a good alternative to harmful chemicals, such as fertilizers, herbicides, pesticides, and insecticides. Unfortunately, the use of pesticides still dominates over biocontrol products, which is estimated to be only 3.5% of the pesticide market in the world despite being versatile in nature [5]. Thus, growth-promoting endophytes are drawing special attention for the management of phytopathogens belonging to the group of bacteria, fungi, virus, aphids, and nematodes. Fungal endophytes offer ecological support to the host plant and get nutrients and protection in return [6]. Fungal endophytes can be found in various parts of the plant – stem, branches, roots, and leaves; they reside without causing any adverse effect to the host [7]. When found in close association with the host, fungal endophytes inculcate many benefits in the host. They help in developing pest and disease resistance by disrupting the life cycle and reproductive phases of the pest. Examples include maize, coffee, tomato, and banana plants. The process of hindering pest survival inside a host is possible due to the production of secondary metabolites by the fungal endophyte, which are toxic to the pest [8]. Fungal endophyte also helps the plant to tolerate abiotic and biotic stresses. They produce plant growth, promoting factors such as gibberellic acid [9]. Examples include *Penicillium aurantiogriseum*, *Piriformospora indica*, *Aspergillus oryzae*, and *A. awamori*. Several other fungal strains, including *Beauveria bassiana*, *Beauveria metarhizium*, *M. robertsii*, *Chaetomium globosum*, and *Acremonium* spp., are successful in plant protection [10]. With a wide host range, endophytic fungus is advantageous over other biocontrol agents. Particularly, *Trichoderma viride*, isolated from *Spilanthes paniculata*, showed a broad range of activities toward *Colletotrichum capsici*, *Fusarium solani*, and *Pythium aphanidermatum*. All the same, a deeper understanding on their mechanism and mode of action is vital for exploiting endophytes as better biocontrol agents. In this chapter, we discuss the types of biotic stress and few mechanisms employed by fungal endophytes in controlling diseases in plants.

17.2 Types of biotic stress

17.2.1 Bacterial

Most bacteria that are pathogenic to plants are mostly saprophytic in nature and are known for not damaging the plant itself. But there are about 100 bacterial species that are known for causing damage in plants. As per reports, most bacterial diseases are observed in tropical regions of the world [11]. Plants are mostly infected by rod-shaped bacteria. These diseases spread in plants through specific pathogenic factors that include phytohormones, cell wall degrading enzymes, effector protein, exopolysaccharides, and toxins [12]. Some bacterial diseases in plants include wildfire of tobacco, granville wilt, blight of beans, crown gall, and soft rot that are caused due to *Pseudomonas syringae*, *Pseudomonas solacearum*, *Xanthomonas campestris*, *Agrobacterium tumifaciens*, and *Erwinia carotovora*.

17.2.2 Virus

Virus is considered as a big threat to many plants, decreasing the crop yield. Geminiviridae and Reoviridae are two main groups of viruses that cause plant diseases. The infected seeds and pest-carrying virus act as means for invading the infected plants [13]. The injured or the damaged parts of the plants provide the viral genome a direct contact for entering the plants cell. So, these parts are known for spreading virus infection. The viral vector should be controlled for reducing the viral infection by (a) enhancing agricultural practices, (b) implementing practices such as crop trapping, and (c) use of transgenic crops [14].

17.2.3 Nematodes

Various nematodes are well known for causing disease in plants, including citrus plants. *Tylenchulus semipenetrans* is the most common citrus nematode found across the world. It causes a slow decline disease in citrus growth [15]. Nematodes damage the plants by causing various diseases, such as stubby root, root knots and lesion, and strings [16]. Legume plants are mostly affected by parasitic nematodes [17]. Pea cyst nematode is responsible for causing damage in the plant tissues of pea and broad beans. Root-knots and lesion are some diseases that are caused by ring and stunt nematodes, which destroy the peanut plants [18]. Potato rot, lesion, and root knots are some of the nematodes that are known for causing damage in potatoes [19]. For prevention of these infections, plants form various metabolic products, proteins, and some specific body structures, which create repulsive and harming effect on the attacking nematodes [20].

17.2.4 Weeds

Weeds are provided with some distinct habits for growth and adaption, which helps them grow in various ecological niches where it is not possible for the growth of other plants. Weeds compete with crop plants [21]. The most common adaptation practices that are related to competitive benefits of weeds include growth and rapid establishment of seedlings, rapid growth of plants that result in tolerating the shading effects by some valuable plants, adaption to harsh climatic conditions and to edaphic regimes, treating the post seeding troubles by providing them with the relative's immunity, lowering the use of herbicides, and using agricultural practices [22]. A strong competition was seen between weeds and crops during the initial stage when weeds invade the ecological niche for the available resources [23]. Plant competition is considered a natural phenomenon where there is a strong correlation between the growth and yield of crops, and weeds.

17.2.5 Fungi

Fungi are parasitic to plants and are divided into biotrophs and necrotrophs on the basis of their existence [24]. Biotrophs are those that are directly associated with the host of living tissues for deriving food, while necrotrophs are those that severely damage the plant tissue and are dependent on dead tissues, which act as source of food for these necrotrophs [25]. But, hemibiotroph is a group of fungi that are considered as plant pathogens and depending on the environmental conditions, they act as biotrophs or necrotrophs [26].

17.2.6 Insects

Insects are considered as important biotic stress factors, which cause crop losses throughout the world. Insect pests, such as wheat aphid, fruit fly, American bollworm, Brown plant hopper, and Gall midge, are responsible for causing disease in plants. Man-made manipulated habitats are mostly adaptable by insect pests, where crops containing high nutritional value, large size, and high yield are selected. Insect pests are also able to adapt to new challenges, such as insecticides, resistance toward host, or changes in climate; they are capable of evolving to biotypes. Response by plants is the most important mechanism for controlling the attack of insects. Plants are able to secrete various toxins against insect attacks; in plants, there are variations in the ingredients used in defense, and toxins impact the health of the insect pest, thereby offering resistance to plants [20].

17.3 Role of microbes in stress management

Plants benefit expansively via protecting the allied symbiotic microorganisms, as they encourage growth of plants and offer improved resistance to pathogens through antibiotics production [27–29]. Habitat-imposed biotic stress is a thoughtful state for land deprivation in arid and semi-arid areas, and interferes in crop productivity. Some plant populations have successfully adapted their strategy of stress tolerance while some become sensitive under such stressed conditions. Roots of plants are bound by a thin layer of soil known as rhizosphere. This is the main site where nutrient uptake, physiological and biological activities occur [30]. Bacteria, such as *Bacillus, Pseudomonas, Rhizobium, Sinorhizobium, and Azotobacter,* are the most abundant microorganisms present in the rhizosphere. Some fungi, such as *Serendipita indica, Aspergillus niger, Aspergillus flavus, Fusarium,* and *Mucor,* are also present in the rhizosphere areas [31–32].

Nowadays, safety measures are needed for commercial agriculture production to protect plants from microbial pathogens as they decrease the quality and yield of crops. Alternatives to synthetic substances have been a vigorous part of studies, with the arrival of organic agriculture, but they also affect crop productivity and soil fertility. Apart from this, biologically approachable, harmless approaches for plant protection have been adopted, specifically biocontrol methods that employ beneficial microbes [33–34].

Plant roots and soil are habitats for establishment of variability of soilborne pathogens as well as beneficial microbes [35–36]. Plant root triggers microbial multiplicities that also invite plant pathogens and pests, causing massive destruction of crop yield. The commonly seen influences of these issues are hormonal/nutrient imbalance and physiological illness. So, the usage of beneficial microbes as biological control agents and as plant growth promoting microorganisms has been observed to be a substitute and as a supportable method to change pesticides as well as chemical composts [37–39]. The plant-microbe association in natural habitats is very important for proper growth and development. They play a crucial role in nutrient utilization and release chemicals that activate the biochemical and physiological transitions in plants [40–41].

17.4 How fungi provide protection to plants and their mechanism

The role of microbes and their interactions are important not only for the host but also for fungal existence under stress conditions. Plants fight against different nematodes, bacteria, and weeds. Many microorganisms have the capability to encourage microbial metabolites production, nutrient acquisition, and phytohormones production that help in the improvement of plant growth (direct mechanism) [42–43]. Defensive responses are initiated by microbes, which occur via diverse mechanisms; they induce systemic and systemically acquired resistance (Indirect mechanism) [44]. Induced response might be toughened by nonpathogenic root linked growth promoting microbes, whereas SAR may involve alteration in gene expression and pathogen-related proteins [45]. Plants exhibit different defense tackles, which can be easily observed in reactions to biotic stress. A system of inter-related signaling pathways regulates the defense mechanism toward plant pathogens. The important components of the system are plant salicylic, jasmonic, and ethylene acids, which are signaling hormones [46]. Reduction in fungal infection is caused by *Alternaria alternate* and the endophytic fungi associated with it are *Penicillium citrinum* and *Aspergillus terreus*. Microorganisms evoke an immune response in plants by generating a systematic transmission signal through the help of systemic acquired resistance (SAR), in which salicylic and jasmonic acids (SA and JA) plays a very significant role in the activation of these responses [47].

17.4.1 Nutrient attainment

Endophytic fungi expand the uptake of nutrients such as phosphorus, iron, zinc, copper, etc. from soil to the host plant. *Serendipita indica* is an endophyte that is involved in nutrient transference, including the distribution of phosphate compounds to plants [48]. According to many studies, it has been proved that endophytic interaction helps in the improvement of phosphorus acquisition [49]. Symbiotic association of *Glomus intraradices* and *Trichoderma atrroviride* results in an increase in productivity of plants by 14%-70%, thereby helping in enhancing the yield and uptake of nutrient for crops, such as lettuce, by acting as a bio stimulant [50]. Inoculation of AFM enhances the P and N concentration in the tissues of *Chrysanthemum morifolium* [51]. The *B. bassiana* and *M. robertsii* deliver N to plants that is integrated during insect parasitization. Upregulation of the phosphate transporter activated genes in cotton plants when treated with *Rhizophagus irregularis* and showed biocontrol activity [52].

17.4.2 Phytohormone production

Endophytic fungi secrete auxins, gibberellins, and cytokines, which are vital in chemical signaling for growth of plants under unfavorable situations [53]. Auxins are the main plant growth regulators and show many positive and negative effects, such as cell division and differentiation, beginning of root progression, etc. [54]. Gibberellins are involved in seed germination, fruit formation and senescence etc. *Penicillium citrinum* IR-3-3 produces gibberellins, which enables plant growth. IAA production by *T. harzianum* showed biocontrol activity toward anthracnose disease in sorghum plants [55].

17.4.3 Antibiotics and secondary metabolites production

In plants, the production of antibiotics and secondary metabolites is considered as the main approach for endophytic fungi in providing defense in various stressful environmental conditions. Endophytes are able to produce various secondary metabolites, such as alkaloids, peptides, polyketones, lipids, etc., which provide protection against pathogens [56]. *Epichloë* species is an endophytic fungus that produces alkaloids, a secondary metabolite, which helps in repelling insects and pests as it is toxic to them. Fungal endophytes, during abiotic stress, activate the ISR and ASR pathways, which release metabolites that are lethal to pathogens. Furthermore, competition can arise to circumvent herbivory and disease [57]. There are various studies that prove that endophytic colonization provides protection to plants against biotic stress. Colonization of cotton plants by endophytic fungus, *Phialemonium inflatum*, conquers the penetration of *Meloidogyne incognita* nematodes in roots, and this also affected their reproduction. Alkaloids have the potential to inhibit the spread of microbes. Fungal endophytes, such as *Clavicipitaceae* sp., isolated from the grass family, showed the production of alkaloids that are harmful to aphids [58]. Altersetin, a new alkaloid that was isolated from *Alternaria* spp., showed a strong antibacterial consequence toward pathogenic bacteria, as reported by Hellwig et al. [59]. Production of thermostable metabolites, such as d-Norandrostane and Longifolenaldehyde, by endophytic fungus *Alternaria alternata* AE1, isolated from neem leaves, showed antibacterial activities. Endophytic fungi cause plant colonization that benefits the host plant (Figure 17.1).

17.4.4 Lytic enzymes

Extracellular enzymes that exhibit biocontrol activity are being increasingly explored as potential antimicrobials to target pathogenic microbes. Numerous endophytes have been reported to produce different lytic enzymes, such as lipases, chitinase, cellulose, proteases, hemicelluloses, and amylase, which aid the hydrolysis of polymers. Chitinases enzymes, β 1–3 glucanases, and proteases, secreted from *Trichoderma herzianum* and *Trichoderma viride*, significantly reduced the incidence of collar rot disease caused by *Aspergillus niger* [60]. The production of phospholipase C by *M. brunneum* hydrolyzes phosphodiester bonds and destroys the cell membrane of insects, permitting the fungus to go in hemocel and infect the tissues

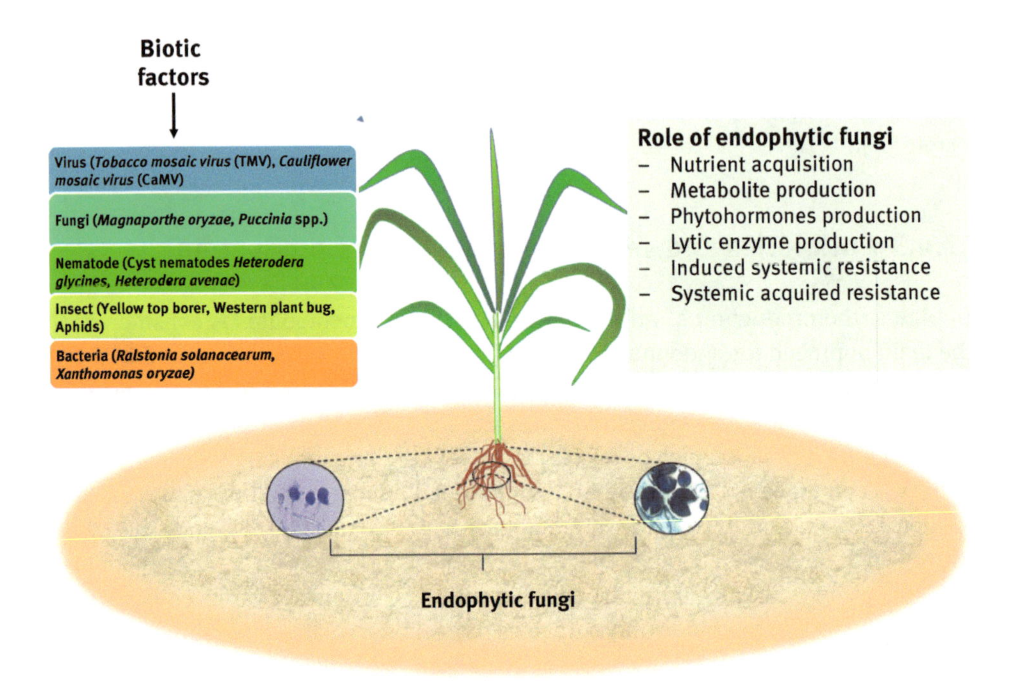

Figure 17.1: Role of fungal endophyte to mitigate the biotic stress.

of insects [61]. Tomato plants exhibited higher chitinase and glucanase activities against *Alternaria solani*, when treated with *Funneliformis mosseae* [62].

17.4.5 Stimulation of systemic resistance toward biotic stress

Endophytic fungi activate the ISR and ASR pathways as a part of the plant defense system against pathogens and pests. Activation of the systemic resistance pathway is initiated by the presence of pathogen-associated molecular patterns (PAMP's). For microbes, the term used is MAMPs (microbe-associated molecular patterns), which can be identified via the receptor of plants [63–64]. MAMPs activate the ISR and ASR pathways, producing signaling molecules (such as salicylic acid and ethylene). These pathways promote many defense responses, such as cell wall thickening of plants, creation of protein-related (PR) molecules and cause cell death. Endophytic colonization causes a primer consequence, making the plant far more infection-prone by infective microorganisms and nematodes [65]. Inoculating a fungal endophyte, A. *lancea*, with *Gilmaniella sp* resulted in various metabolic activities, such as modulating antioxidant system, improved photosynthesis to cope with biotic stress by inducing salicylic acid (SA), Nitric acid (HNO_3), and hydrogen peroxide (H_2O_2), which act as signaling molecules in stressful environmental conditions [31].

ISR pathway has been elucidated moderately in numerous models of plant systems, such as *Arabidopsis*, and it is considered complex. In induced resistance system, there are three pathways that are generally recognized in *Arabidopsis*, in which two of them are related to the creation of pathogenesis-related (PR) protein. In one of these pathways, due to attack by pathogenic microorganism, it results in the production of PR protein. On the other hand, in the other pathway, production of PR proteins generally takes place due to wounding or due to the presence of plant pathogen that induces necrosis. AMF increase PR proteins to reduce cotton disease (*Verticillium* wilt) caused by *V. dahlia* [66]. However, in both the pathways, induction takes place by an alternative mechanism. Salicylic acid (SA), a signaling molecule produced by plants, plays an important role. The pathogen encouraged by the pathway relies on SA, while the hurtful pathway depends on jasmonic acid (JA), which acts as a signaling molecule [67].

It is assumed that plant health is benefitted due to the association between the plant roots and the microbial communities present in the suppressive soil. In the host plant, there are numerous biocontrol microbes that provoke ISR, which help plants to withstand the attack of pathogen on roots or leaves, although total protection is not provided [68]. There are various biocontrol PGPR, irrespective of the production of antibiotics that elicits ISR [69]. In *Arabidopsis thaliana*, due to the investigation through transcriptome, the effects of three different strains of *Pseudomonas* spp. were seen to mediate ISR. *Fusarium oxysporum* strain, Fo47 via endophyte-mediated-resistance (EMR), was found to suppress various wilt diseases in tomato, flax, watermelon, and pepper [70].

SA produces antibacterial composites that are of low molecular weight and these are able to express pathogenesis-related proteins at the site of the pathogen infection as well as nearby tissues [71]. SA plays a significant role in providing resistance in plants against diseases even though it does not act as a transmitter of signal in SAR. In the case of inoculation of *Arabidopsis* with mycorrhizal fungi *P. indica*, in the early phase of recognition of the host, activation of a slight defense response by SA and JA takes place when there is no establishment of physical content [72]. Similarly, when the roots of *Arabidopsis* are treated with *Sebacina vermifera* (fungi), there is an increase in the concentration of glucosinolate and jasmonate, on addition of SA [73]. In the case of various poaceae species, synthesis of alkaloids takes place in a fungal endophyte – *Epichloe*, which results in an increase in the defense of antiherbivore of symbiotic plants, when treated along with JA [74].

Jasmonic acid (JA) is a set plant hormones that contains signaling molecules, such as oxylipins (oxidized lipid molecules) [75]. The JAs in plants comes from compounds (linolenic/raughanic acid) that are finely dispersed in algae, angiosperms, and certain fungi. JAs play a crucial part in farming by initiating the defense response toward phytopathogens. *Lasiodiplodia theobrome* is a commonly known plant pathogenic fungus that is proficient in manufacturing JAs in significant amounts [76–77]. The probability of the existence of JA-serine/threonine conjugates was confirmed from fungal sp. in a fermentation broth using HPLC-ESI [78]. These conjugating enzymes were first separated from the flowering plant, *Arabidopsis thaliana*, and compared with the peptidase activity of *L. theobromae*. This enzyme has the ability to determine the conjugates of JAs along with amino acids. It was cleansed by hydrophobic chromatography and identified as glycoprotein (107 kDa). It was observed that its amidohydrolase action was exactly same as JA [79]. Therefore, fungi may require Jasmonic acid at the time of the host plant infection to alter plant progressions, such as nutrient release and senescence, which is possibly helpful for the growth of fungal sp. [80–81].

The association of endophytic microbes with plants contributes in various ways, such as in the promotion of the growth of plants and tolerating biotic stress against pathogens, which play an important role in sustainable agriculture. The various other functions involving soil fungi in nutrient cycling are decomposition of organic matter, protection against pathogen, and plant growth promotion [82] In plants, stress tolerance is conferred once the endophytic fungi associated with the root activates the pathways of jasmonate/ethylene signaling [83]. Agricultural management practices show an imperative role in influencing the fungal communities present in soil. In agriculture, arbuscular mycorrhizal fungi act as biofertilizer and is involved in the growth of plant and in the improvement of the quality of plants. AMF offers several benefits to organic farming, such as providing resistance against soilborne pathogen, improving quality of plants by contributing in its nutritional uptake, etc.

17.5 Conclusion

The presence of different kinds of pest and pathogen in crops leads to a decrease in the yield of fruits and vegetables, resulting in heavy crop losses every year. To reduce the loss of crop yield and to control diseases, different effective methods should be used. The use of fungal endophytes as biocontrol agents and biofertilizers is an auspicious approach in the agriculture field to overcome biotic stress in plants. The association of endophytic fungi with plants provides protection to plants from the attack of various pests, insects, pathogens, and nematodes. It also, helps them to survive by adaptation in harsh biotic/abiotic stress conditions.

References

[1] Yu X, Zhang W, Lang D, Zhang X, Cui G, Zhang X. Interactions between endophytes and plants: Beneficial effect of endophytes to ameliorate biotic and abiotic stresses in plants. Journal of Plant Biology 2019, 62, 1–13.

[2] Sapre S, Gontia-Mishra I, Thakur VV, Sikdar S, Tiwari S. Molecular techniques used in plant disease diagnosis. In: Food security and plant disease management. Woodhead Publishing, 2021, 405–421.

[3] Pandey P, Ramegowda V, Senthil-Kumar M. Shared and unique responses of plants to multiple individual stresses and stress combinations: Physiological and molecular mechanisms. Frontiers of Plant Science 2015, 6, 723.

[4] Angessa TT, Li C. Exploration and utilization of genetic diversity exotic germplasm for barley improvement. In: Exploration, identification and utilization of barley germplasm. Academic Press, Cambridge, United States, 2016, 223–240.

[5] Kumar A, Sharma A, Chaudhary P, Gangola S. Chlorpyrifos degradation using binary fungal strains isolated from industrial waste soil. Biologia 2021, 76, 3071–3080.

[6] Tyagi J, Sultan E, Mishra A, Kumari M, Pudake RN. The impact of AMF symbiosis in alleviating drought tolerance in field crops. In: Ajit Varma, Ram Prasad, Narendra Tuteja eds. Mycorrhiza-nutrient uptake, biocontrol, ecorestoration. Cham, Springer, 2017, 211–234.

[7] Tyagi J, Varma A, Pudake RN. Evaluation of comparative effects of arbuscular mycorrhiza (*Rhizophagus intraradices*) and endophyte (*Piriformospora indica*) association with finger millet (*Eleusine coracana*) under drought stress. European Journal of Soil Biology 2017, 81, 1–10.

[8] Jaber LR, Araj SE. Interactions among endophytic fungal entomopathogens (Ascomycota: Hypocreales), the green peach aphid *Myzus persicae Sulzer* (Homoptera: Aphididae), and the aphid endoparasitoid Aphidius colemani Viereck (Hymenoptera: Braconidae). Biological Control 2018, 116, 53–61.

[9] Yao YQ, Lan F, Qiao YM, Wei JG, Huang RS, Li LB. Endophytic fungi harbored in the root of Sophora *tonkinensis Gapnep*: Diversity and biocontrol potential against phytopathogens. Microbiol Open 2017, 6, e00437.

[10] Grabka R, d'Entremont TW, Adams SJ, Walker AK, Tanney JB, Abbasi PA, Ali S. Fungal endophytes and their role in agricultural plant protection against pests and pathogens. Plants 2022, 11, 384.

[11] Dass A, Shekhawat K, Choudhary AK, Sepat S, Rathore SS, Mahajan G, Chauhan BS. Weed management in rice using crop competition-a review. Crop Protection 2017, 95, 45–52.

[12] Gagne-Bourgue F, Aliferis KA, Seguin P, Rani M, Samson R, Jabaji S. Isolation and characterization of indigenous endophytic bacteria associated with leaves of switchgrass (*Panicum virgatum* L.) cultivars. Journal of Applied Microbiology 2013, 114, 836–853.

[13] Fereres A, Moreno A. Behavioural aspects influencing plant virus transmission by homopteran insects. Virus Research 2013, 141, 158–168.

[14] Prins M, Laimer M, Noris E, Schubert J, Wassenegger M, Tepfer M. Strategies for antiviral resistance in transgenic plants. Molecular Plant Pathology 2008, 9, 73–83.

[15] Safdar A, Javed N, Khan SA, Safdar H, Haq IU, Abbas H, Ullah Z. Synergistic effect of a fungus, Fusarium semitectum and a nematode, *Tylenchulus semipenetrans* on citrus decline. Pakistan Journal of Zoology 2013, 45, 643–651.

[16] Etebu E, Nwauzoma AB. A review on sweet orange (Citrus sinensis L Osbeck): Health, diseases and management. American Journal of Research Communication 2014, 2, 33–70.

[17] Engstrom MT, Karonen M, Ahern JR, Baert N, Payré B, Hoste H, Salminen JP. Chemical structures of plant hydrolyzable tannins reveal their in vitro activity against egg hatching and motility of *Haemonchus contortus* nematodes. Journal of Agricultural and Food Chemistry 2016, 64, 840–851.

[18] Eschen R, Roques A, Santini A. Taxonomic dissimilarity in patterns of interception and establishment of alien arthropods, nematodes and pathogens affecting woody plants in Europe. Diversity and Distributions 2015, 21, 36–45.

[19] Balasubramanian N, Hao YJ, Toubarro D, Nascimento G, Simões N. Purification, biochemical and molecular analysis of a chymotrypsin protease with prophenoloxidase suppression activity from the entomopathogenic nematode *Steinernema carpocapsae*. International Journal for Parasitology 2009, 39, 975–984.

[20] War AR, Sharma HC, Paulraj MG, War MY, Ignacimuthu S. Herbivore induced plant volatiles: Their role in plant defense for pest management. Plant Signaling and Behavior 2011, 6, 1973–1978.

[21] Ramesh K, Rao AN, Chauhan BS. Role of crop competition in managing weeds in rice, wheat, and maize in India: A review. Crop Protect 2017, 95, 14–21.

[22] Sardana V, Mahajan G, Jabran K, Chauhan BS. Role of competition in managing weeds: An introduction to the special issue. Crop Protect 2017, 95, 1–7.

[23] Bürger J, Darmency H, Granger S, Guyot SH, Messéan A, Colbach N. Simulation study of the impact of changed cropping practices in conventional and GM maize on weeds and associated biodiversity. Agricultural Systems 2015, 137, 51–63.

[24] Kabbage M, Yarden O, Dickman MB. Pathogenic attributes of *Sclerotinia sclerotiorum*: Switching from a biotrophic to necrotrophic lifestyle. Plant Science 2015, 233, 53–60.

[25] Campe R, Loehrer M, Conrath U, Goellner K. Phakopsora pachyrhizi induces defense marker genes to necrotrophs in Arabidopsis thaliana. Physiol and Molecular Plant Patho 2014, 87, 1–8.

[26] Mang HG, Laluk KA, Parsons EP, Kosma DK, Cooper BR, Park HC, AbuQamar S, Boccongelli C, Miyazaki S. Consiglio Fand Chilosi G. The *Arabidopsis* RESURRECTION1 gene regulates a novel antagonistic interaction in plant defense to biotrophs and necrotrophs. Plant Physiology 2009, 151, 290–305.

[27] Agri U, Chaudhary P, Sharma A. In vitro compatibility evaluation of agriusable nanochitosan on beneficial plant growth-promoting rhizobacteria and maize plant. National Academy Science Letters 2021, 44, 555–559.

[28] Agri U, Chaudhary P, Sharma A, Kukreti B. Physiological response of maize plants and its rhizospheric microbiome under the influence of potential bioinoculants and nanochitosan. Plant and Soil 2022, 474, 451–468. https://doi.org/10.1007/s11104-022-05351-2.

[29] Chaudhary P, Chaudhary A, Agri U, Khatoon H, Singh A. Recent trends and advancements for agro-environmental sustainability at higher altitudes. In: Goel R, Soni R, Suyal DC, Khan M, eds. Survival strategies in cold-adapted microorganisms. Singapore, Springer, 2022, 425–435.

[30] Chaudhary P, Sharma A, Chaudhary A, Khati P, Gangola S, Maithani D. Illumina based high throughput analysis of microbial diversity of rhizospheric soil of maize infested with nanocompounds and Bacillus sp. Applied Soil Ecology 2021, 159, 103836.

[31] Mishra D, Kumar A, Tripathi S, Chitara MK, Chaturvedi P. Endophytic fungi as biostimulants: An efficient tool for plant growth promotion under biotic and abiotic stress conditions. In: Biostimulants for crops from seed germination to plant development. Academic Press, 2021, 365–391.

[32] Chaudhary P, Khati P, Chaudhary A, Maithani D, Kumar G, Sharma A. Cultivable and metagenomic approach to study the combined impact of nanogypsum and *Pseudomonas taiwanensis* on maize plant health and its rhizospheric microbiome. PLoS One 2021, 16, e0250574.

[33] Chaudhary P, Parveen H, Gangola S, Kumar G, Bhatt P, Chaudhary A. Plant growth-promoting rhizobacteria and their application in sustainable crop production. In: Bhatt P, Gangola S, Udayanga D, Kumar G, eds. Microbial technology for sustainable environment. Singapore, Springer, 2021, 217–234.
[34] Chaudhary A, Parveen H, Chaudhary P, Khatoon H, Bhatt P. Rhizospheric microbes and their mechanism. In: Bhatt P, Gangola S, Udayanga D, Kumar G, eds. Microbial technology for sustainable environment. Singapore, Springer, 2021, 79–93.
[35] Khati P, Bhatt P, Nisha KR, Sharma A. Effect of nanozeolite and plant growth promoting rhizobacteria on maize. 3 Biotech 2018, 8, 141.
[36] Chaudhary P, Chaudhary A, Bhatt P, Kumar G, Khatoon H, Rani A, Kumar S, Sharma A. Assessment of soil health indicators under the influence of nanocompounds and *Bacillus* spp. in field condition. Frontiers of Environmental Science 2022, 9, 769871.
[37] Chaudhary P, Sharma A. Response of nanogypsum on the performance of plant growth promotory bacteria recovered from nanocompound infested agriculture field. Environment and Ecology 2019, 37, 363–372.
[38] Chaudhary A, Chaudhary P, Upadhyay A, Kumar A, Singh A. Effect of gypsum on plant growth promoting rhizobacteria. Environment and Ecology 2021, 39, 1248–1256.
[39] Kumari S, Sharma A, Chaudhary P, Khati P. Management of plant vigor and soil health using two agriusable nanocompounds and plant growth promotory rhizobacteria in Fenugreek. 3 Biotechnology 2020, 10, 461.
[40] Chaudhary P, Khati P, Chaudhary A, Gangola S, Kumar R, Sharma A. Bioinoculation using indigenous *Bacillus* spp. improves growth and yield of *Zea mays* under the influence of nanozeolite. 3 Biotechnology 2021, 11, 11.
[41] Chaudhary P, Khati P, Gangola S, Kumar A, Kumar R, Sharma A. Impact of nanochitosan and *Bacillus* spp. on health, productivity and defence response in *Zea mays* under field condition. 3 Biotechnology 2021, 11, 237.
[42] Kukreti B, Sharma A, Chaudhary P, Agri U, Maithani D. Influence of nanosilicon dioxide along with bioinoculants on *Zea mays* and its rhizospheric soil. 3 Biotechnology 2020, 10, 345.
[43] Chaudhary P, Chaudhary A, Parveen H, Rani A, Kumar G, Kumar A, Sharma A. Impact of nanophos in agriculture to improve functional bacterial community and crop productivity. BMC Plant Biology 2021, 21, 519.
[44] Pieterse CM, Zamioudis C, Berendsen RL, Weller DM, Van Wees SC, Bakker PA. Induced systemic resistance by beneficial microbes. Annual Review of Phytopathol 2014, 52, 347–375.
[45] Lee BD, Dutta S, Ryu H, Yoo SJ, Suh DS, Park K. Induction of systemic resistance in Panax ginseng against Phytophthora cactorum by native *Bacillus amyloliquefaciens* HK34. Journal of Ginseng Research 2015, 39, 213–220.
[46] Hardoim PR, Van Overbeek LS, Berg G, Pirttilä AM, Compant S, Campisano A, Döring M, Sessitsch A. The hidden world within plants: Ecological and evolutionary considerations for defining functioning of microbial endophytes. Microbiology and Molecular Biology Reviews 2015, 79, 293–320.
[47] Waqas M, Khan AL, Hamayun M, Shahzad R, Kang SM, Kim JG, Lee IJ. Endophytic fungi promote plant growth and mitigate the adverse effects of stem rot: An example of *Penicillium citrinum* and *Aspergillus terreus*. Journal of Plant Interactions 2015, 10, 280–287.
[48] Hiruma K, Kobae Y, Toju H. Beneficial associations between Brassicaceae plants and fungal endophytes under nutrient-limiting conditions: Evolutionary origins and host–symbiont molecular mechanisms. Current Opinion in Plant Biology 2018, 44, 145–154.
[49] Baron NC, Costa NTA, Mochi DA, Rigobelo EC. First report of *Aspergillus sydowii* and *Aspergillus brasiliensis* as phosphorus solubilizers in maize. Annals of Microb 2018, 68, 863–870.

[50] Rana KL, Kour D, Kaur T, Devi R, Yadav AN, Yadav N, Dhaliwal HS, Saxena AK. Endophytic microbes: Biodiversity, plant growth-promoting mechanisms and potential applications for agricultural sustainability. Antonie Van Leeuwenhoek 2018, 113, 1075–1107.
[51] Wang Y, Wang M, Li Y, Wu A, Huang J. Effects of arbuscular mycorrhizal fungi on growth and nitrogen uptake of *Chrysanthemum morifolium* under salt stress. PLoS One 2018, 13, e0196408.
[52] Gao X, Guo H, Zhang Q, Guo H, Zhang L, Zhang C, Gou Z, Liu Y, Wei J, Chen A, Chu Z, Zeng F. Arbuscular mycorrhizal fungi (AMF) enhanced the growth, yield, fiber quality and phosphorus regulation in upland cotton (*Gossypium hirsutum* L.). Scientific Reports 2020, 10, 2084–2084.
[53] Khan N, Bano A, Ali S, Babar M. Crosstalk amongst phytohormones from planta and PGPR under biotic and abiotic stresses. Plant Growth Regulation 2020, 90, 189–203.
[54] Numponsak T, Kumla J, Suwannarach N, Matsui K, Lumyong S. Biosynthetic pathway and optimal conditions for the production of indole-3-acetic acid by an endophytic fungus *Colletotrichum fructicola* CMU-A109. PLoS One 2018, 13, e0205070.
[55] Saber WIA, Ghoneem KM, Rashad YM, Al-Askar AA. *Trichoderma harzianum* WKY1: An indole acetic acid producer for growth improvement and anthracnose disease control in sorghum. Biocontrol Science and Technology 2017, 27, 654–676.
[56] Kaddes A, Fauconnier ML, Sassi K, Nasraoui B, Jijakli MH. Endophytic fungal volatile compounds as solution for sustainable agriculture. Molecules 2019, 24, 1065.
[57] Chitnis VR, Suryanarayanan TS, Nataraja KN, Prasad SR, Oelmüller R, Shaanker RU. Fungal endophyte-mediated crop improvement: The way ahead. Frontiers of Plant Science 2020, 11, 1588.
[58] Panaccione DG, Beaulieu WT, Cook D. Bioactive alkaloids in vertically transmitted fungal endophytes. Functional Ecology 2014, 28, 299–314.
[59] Hellwig V, Grothe T, Mayer-Bartschmid A, Endermann R, Geschke FU, Henkel T, Stadler M. Altersetin, a new antibiotic from cultures of endophytic Alternaria spp. Taxonomy, fermentation, isolation, structure elucidation and biological activities. The Journal of Antibiotics 2002, 55, 881–892.
[60] Gajera HP, Vakharia DN. Production of lytic enzymes by trichoderma isolates during in vitro antagonism with aspergillus niger, the causal agent of collar ROT of peanut. Brazilian Journal of Microbiology 2012, 43, 43–52.
[61] Santi L, Beys da Silva WO, Berger M. Conidial surface proteins of *Metarhizium anisopliae*: Source of activities related with toxic effects, host penetration and pathogenesis. Toxico 2010, 55, 874–880.
[62] Song Y, Chen D, Lu K, Sun Z, Zeng R. Enhanced tomato disease resistance primed by arbuscular mycorrhizal fungus. Frontiers in Plant Science 2015, 6, 786.
[63] Latz MA, Jensen B, Collinge DB, Jørgensen HJ. Endophytic fungi as biocontrol agents: Elucidating mechanisms in disease suppression. Plant Ecology & Diversity 2018, 11, 555–567.
[64] Yan L, Zhu J, Zhao X, Shi J, Jiang C, Shao D. Beneficial effects of endophytic fungi colonization on plants. Applied Microbiol and Biotechnology 2019, 103, 3327–3340.
[65] Poveda J, Abril-Urias P, Escobar C. Biological control of plant-parasitic nematodes by filamentous fungi inducers of resistance: *Trichoderma*, mycorrhizal and endophytic fungi. Frontiers in Microbiology 2020, 11, 992.
[66] Liu RJ. Effect of vesicular-arbuscular mycorrhizal fungi on Verticillium wilt of cotton. Mycorrhiza 1995, 5, 293–297.
[67] Pieterse CMJ, Ton J, van Loon LC. Cross-talk between plant defence signalling pathways: Boost or burden?. Agricultural Biotechnology Net 2001, 3, 1–18.

[68] Harman GE, Howell CR, Viterbo A, Chet I, Lorito M. Trichoderma species – opportunistic, avirulent plant symbionts. Nature Reviews Microbiology 2004, 2, 43–56.

[69] Ongena M, Duby F, Rossignol F, Fauconnier ML, Dommes J, Thonart P. Stimulation of the lipoxygenase pathway is associated with systemic resistance induced in bean by a nonpathogenic Pseudomonas strain. Molecular Plant-Microbe Interactions 2004, 17, 1009–1018.

[70] Trouvelot S, Olivain C, Recorbet G, Migheli Q, Alabouvette C. Recovery of *Fusarium oxysporum* Fo47 mutants affected in their biocontrol activity after transposition of the Fot1 element. Phytopathology 2002, 92, 936–945.

[71] Oliveira MDM, Varanda CMR, Félix MRF. Induced resistance during the interaction pathogen x plant and the use of resistance inducers. Phytochem Letters 2016, 15, 152–158.

[72] Thürich J, Meichsner D, Furch AC, Pfalz J, Krüger T, Kniemeyer O, Brakhage A, Oelmüller R. *Arabidopsis* thaliana responds to colonisation of *Piriformospora indica* by secretion of symbiosis-specific proteins. PLoS One 2018, 13, e0209658.

[73] Lahrmann U, Strehmel N, Langen G, Frerigmann H, Leson L, Ding Y, Scheel D, Herklotz S, Hilbert M, Zuccaro A. Mutualistic root endophytism is not associated with the reduction of saprotrophic traits and requires a noncompromised plant innate immunity. New Phytologist 2015, 207, 841–857.

[74] Bastías DA, Martínez-Ghersa MA, Newman JA, Card SD, Mace WJ, Gundel PE. Jasmonic acid regulation of the anti-herbivory mechanism conferred by fungal endophytes in grasses. Journal of Ecology and Environment 2018, 106, 2365–2379.

[75] Wasternack C, Feussner I. The oxylipin pathways: Biochemistry and function. Annual Review of Plant Biology 2018, 69, 363–386.

[76] Eng F, Haroth S, Feussner K, Meldau D, Rekhter D, Ischebeck T, Brodhun F, Feussner I. Optimized jasmonic acid production by *Lasiodiplodia theobromae* reveals formation of valuable plant secondary metabolites. PloS One 2016, 11, e0167627.

[77] Salvatore MM, Alve A, Andolfi A. Secondary metabolites of *Lasiodiplodia theobromae*: Distribution, chemical diversity, bioactivity, and implications of their occurrence. Toxins 2020, 12, 457.

[78] Eng F, Marin JE, Zienkiewicz K, Gutiérrez-Rojas M, Favela-Torres E, Feussner I. Jasmonic acid biosynthesis by fungi: Derivatives, first evidence on biochemical pathways and culture conditions for production. Peer Journal 2021, 9, e10873.

[79] Andolfi A, Maddau L, Cimmino A, Linaldeddu BT, Basso S, Deidda A, Serra S, Evidente A, Lasiojasmonates A–C. three jasmonic acid esters produced by *Lasiodiplodia* sp., a grapevine pathogen. Phytochemical 2014, 103, 145–153.

[80] Cimmino A, Cinelli T, Masi M, Reveglia P, da Silva MA, Mugnai L, Michereff SJ, Surico G, Evidente A. Phytotoxic lipophilic metabolites produced by grapevine strains of *Lasiodiplodia* species in Brazil. Journal of Agricultural and Food Chemistry 2017, 65, 1102–1107.

[81] Matsui R, Amano N, Takahashi K, Taguchi Y, Saburi W, Mori H, Kondo N, Matsuda K, Matsuura H. Elucidation of the biosynthetic pathway of cis-jasmone in *Lasiodiplodia theobromae.* Scientific Reports 2017, 7, 1–9.

[82] Kour D, Rana KL, Yadav N, Yadav AN, Singh J, Rastegari AA, Saxena AK. Agriculturally and industrially important fungi: Current developments and potential biotechnological applications. In: Recent advancement in white biotechnology through fungi. Cham, Springer, 2019, 1–64.

[83] Lahlali R, McGregor L, Song T, Gossen BD, Narisawa K, Peng G. *Heteroconium chaetospira* induces resistance to clubroot via upregulation of host genes involved in jasmonic acid, ethylene, and auxin biosynthesis. PLoS One 2014, 9, e94144.

G. K. Dinesh, B. Priyanka, Archana Anokhe, P. T. Ramesh, R. Venkitachalam, K. S. Keerthana Sri, S. Abinaya, V. Anithaa, R. Soni

Chapter 18
Ecosystem services and ecological role of birds in insect and pest control

Abstract: Agricultural ornithology is an emerging science that renders great ecosystem services. Most of the bird species play a vital role in agriculture by managing pest and rodent populations. In agricultural countries like India, this field has a special importance in protecting agricultural crops from invading insect pests and, thereby, preventing economic losses to farmers in an ecofriendly manner. In addition, birds aid farmers by reducing invertebrate crop pests, yet they are seldom used in Integrated Pest Management (IPM) practices. Nonetheless, several bird species have a substantial influence on agricultural productivity. To develop practical IPM approaches, it is necessary to understand bird flocking and their behavior at different stages of crops, so that the agricultural yield will be enhanced. Developing strategies on overall aspects help complete essential knowledge gaps around the complex roles of birds in agricultural systems. Although many better works have been started, a lot is yet to be done in order to stabilize the ecological approaches. This book chapter describes the ecosystem services provided by birds and agricultural ornithology advancements, highlights difficulties and information shortages, and suggests future research possibilities.

18.1 Introduction

Agricultural ornithology deals with the role of birds in pollination, seed dispersal, crop pest control, crop grain management, and associated ecosystem services. Birds play a significant part in crop insect management in all agricultural ecosystems, whether dry or irrigated; it primarily focuses on bird ecology and management in agroecosystems. It also collects scientific data on birds and controls insects [1]. Agriculture crops offer birds a concentrated and reliable food supply such as insects, other arthropods, rodents, grains, etc. Birds are the most effective insect predators among vertebrates, because they can swiftly gather

Acknowledgment: The authors would like to acknowledge their affiliated institutes. They would like to acknowledge Dr. R. Venkitachalam, C. Vijaya Kumar, R. Suryaprahasan, V. Ilaiya Bharathi, and G. K. Dinesh for bird photo plates.

https://doi.org/10.1515/9783110771558-018

in great numbers upon a pest outbreak and protect various agriculture crops, and hence, they serve as excellent natural insect pest managers [2]. In agroecosystems, most bird species are insectivorous and play an important role in maintaining a population of insect pests, and thereby, they are beneficial to farmers [3]. Another study estimated that the prey biomass consumed by insectivorous birds ranges from 400–500 million metric tons per year [4] Birds are one of the key pollinator groups in the animal kingdom, rendering service to about 80% of plants that depend on other animals for pollination. Seed dispersal is another important ecosystem service carried out by birds that helps sustain green plants and trees. However, granivorous bird species cause severe damage to crops, leading to crop loss. Thus, crops, birds, and insects have mutualistic and antagonistic relationships and a significant ecological connection that supports biological cycles and food chains. Few studies have examined this complex and delicate relationship, and more thorough studies should be done. This book chapter deals with the role of birds in the biological control of insects and pests in various agroecosystems.

18.2 Ecosystem services of birds in the agroecosystem

Nature delivers both massive ecosystem services (ES) and ecosystem disservices (EDS). However, the affinity between the two is less well-recognized [5]. As in the case of birds, they are flying vertebrates and recognized as sources of various ES [6] and function as an effective alternative to chemical pesticides [7]. Based on the Millennium Ecosystem Assessment, ecosystem services have been categorized into four categories as provisioning services, regulating services, supporting services, and cultural services [8]. Birds provide multifaceted environmental services, including nutrition cycling through droppings (supportive services), pest and bug management (regulatory services), scavenging, and seed dispersion (cultural services), besides their behavior-driven services [9] (Figure 18.1). The potential ecosystem services and disservices by birds and their ecological relationship between various organisms are complex (Figure 18.2).

18.2.1 Regulating services of birds

Birds are important biocontrol agents in regulating pests and insects in agricultural landscapes. Insectivorous birds such as Black-backed Puffback (*Dryoscopus cubla)*, the Cape White-eye (*Zosterops virens*), the Tawny-flanked Prinia (*Prinia subflava*), the Gorgeous Bush-shrike (*Telophrous viridis*), and the Brown-hooded Kingfisher (*Halcyon*

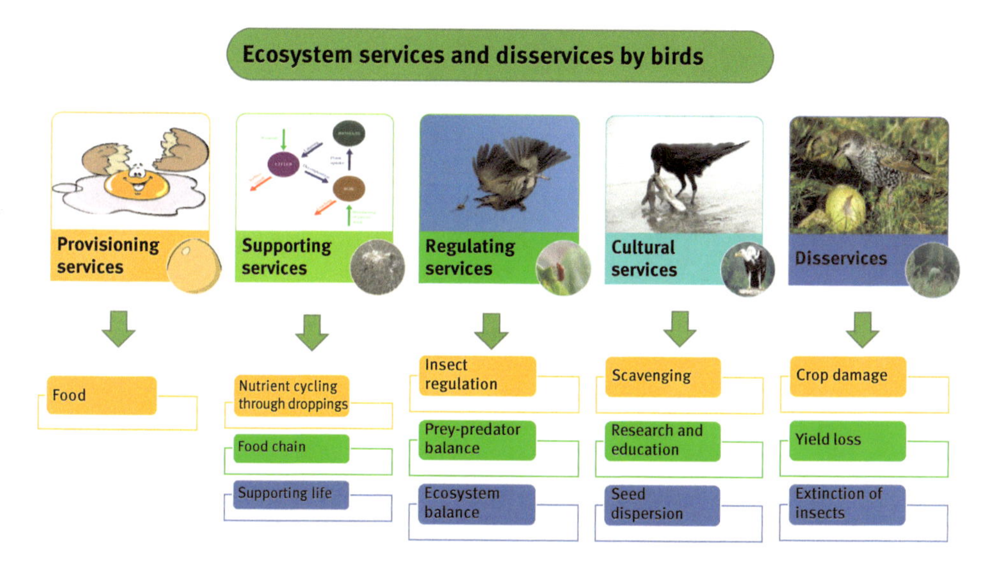

Figure 18.1: Ecosystem services and disservices provided by the bird.

albiventris), Cattle egret (*Bubulcus Ibis)*, Common wood shrike (*Tephrodornis pondicerianus)*, Hoopoe (*Upupa epops*), Black drongo (*Dicrurus macrocercus)*, Indian roller (*Coracias benghalensis*), and Asian green bee-eater (*Merops orientalis*) are well-known insectivorous bird species across agroecosystems [10–12]. Some of the birds are shown in Figure 18.3. These well-identified insect predators have not been documented well on their economic benefits in different agroecosystems. Only a few scientific reports are available on the economic effects of birds on biological pest management. Both ES and EDS are subject to change in response as the landscape structure changes [5, 13], and farmers often see surviving natural areas as lost land and the locations where pests thrive [14]. Loss of natural habitat for birds affects these ecosystem service providers, resulting in the increase of the pest population. Thus, a trade-off between ES and EDS may exist, mediated by landscape composition. Other agricultural systems have benefited significantly from closeness to the forest areas in bird variety and predator hunting success [15]. Crop raiding by birds occurred exclusively in near-natural vegetation and caused a yield loss of around 26% [5]. However, bird biocontrol was more effective in the natural vegetation habitat. According to the research done in Macadamia farms of South Africa, biological management by birds and bats had an economic effect of about USD 5,000 ha/year. Another study points out that the revenue losses were reduced by bird pest management to the value of USD 1,595/ha as income gain. Bats and birds enhance the ecosystem services when released near artificial roosts and nest sites, but species-specific preference studies of birds are yet to take off well. It is critical to educate farmers since many are unaware of the advantages of birds [5]. To boost the bird populations as biological control, artificial nesting and roosting places may be built to attract birds to agricultural regions with little

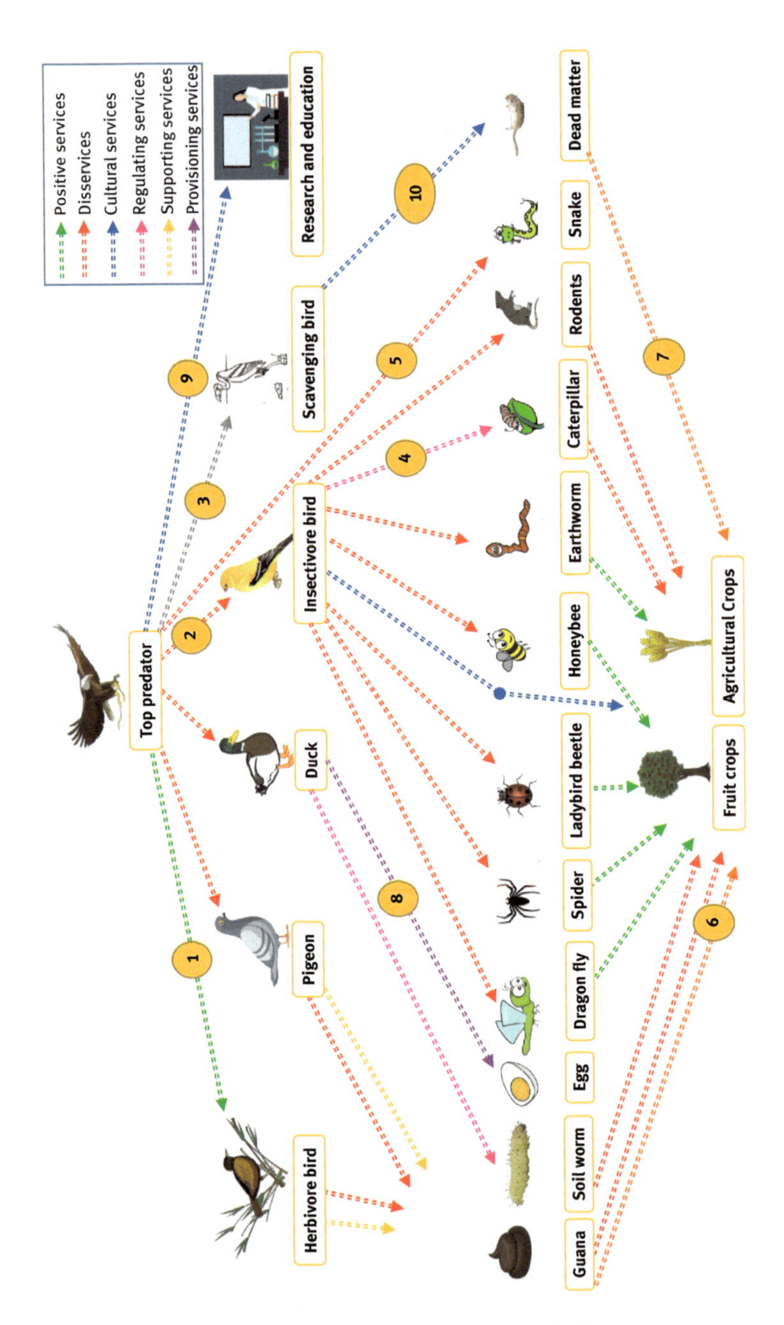

Figure 18.2: The potential services and disservices by birds in an agroecosystem.
Notes: 1. Top predator predating herbivore birds that are harmful for crops; 2. Top predator predating insectivore birds that are beneficial for crops; 3. Top predators not having any guild on scavenging birds, hence they are neutral; 4. Insectivore bird eating caterpillar, by which it is reducing the population and they are classified as regulating services; 5. Top predators predating snakes, which it is indirectly leading to yield losses by rodents; 6. Guana or faecal

Figure 18.3: Various Ecosystem services and disservices provided by the common agricultural field birds.

Figure 18.2 (continued)
matter by birds are both harmful and beneficial, the faecal matter on economic parts may be considered as poor quality, and if the faecal matter added to the soil it is beneficial to soil for nutrient cycling and they are classified as supporting services; 8. some farm birds such as duck may eat the harmful soil worms and larvae, also it provides nutritious meat and egg, they are classified as provisioning services; 9. Birds are useful for research and education, it also have aesthetic value, hence it is classified as cultural services; 10. Some scavenging birds may clean up the ecosystem by eating dead matter, it is classified into cultural services.

natural habitat nearby [7]. This would be more cost-efficient than using chemical pesticides to control pest species below an economic threshold level [16]. Biodiversity research in specific agricultural regions is incumbent to developing suitable designs for birds as biological pest control.

18.3 Birds' niche in paddy ecosystem

Wetlands play a vital role in our natural environment. Wetlands are described as the "kidney of the landscape" because they function as the downstream receivers of water and waste from both natural and human resources [17]. Wetlands provide shelter for roosting, nesting, and foraging environments to waterfowl, fish, amphibians, reptiles, and many plant species at critical life phases and as protection from harsh weather [10]. They also serve as stepping stones or corridor habitats for migratory bird species such as waterfowl and waders. Wetlands offer habitat and shelter for many species and aid resilience of species as a means to overcome population decline [9]. Studies on birds in rice fields and the number of reported species are listed in Table 18.1.

Table 18.1: Studies on birds in rice fields and the number of reported bird species [9, 10, 21].

Location	Cultivation system	No. of species	Source
Indian subcontinent			
Ahmedabad, Gujarat, India	Seasonal	28	[21]
Bangalore district, Karnataka, India	All-year	131	[21]
Etawah and Manipuri districts, Uttar Pradesh, India	Seasonal	114	[21]
Kerala, India	All-year	133	[21]
Kheda district, Gujarat, India	Seasonal	65	[37]
Ludhiana, Punjab, India	All-year	68	[26]
Ludhiana, Punjab, India	All-year	23	[27]
Mysore and Coorg districts, Karnataka, India	All-year	68	[21]
Nellore district, Andhra Pradesh, India	All-year	62	[21]
Pondicherry, India	All-year	34	[31]
Ranga Reddy district, Andhra Pradesh, India	All-year	56	[35]
Tamil Nadu Agricultural University, Coimbatore, Tamil Nadu, India (Conventional rice fields)	All-year	26	[9]

Table 18.1 (continued)

Location	Cultivation system	No. of species	Source
Tamil Nadu Agricultural University, Coimbatore, Tamil Nadu, India (Organic rice fields)	All-year	25	[9]
The Indian subcontinent	All-year	89	[32]
The Indian subcontinent	All-year	59	[39]
Uttaranchal, India	Seasonal	58	[21]
West Bengal, India	Seasonal	67	[21]
Other countries			
Bangladesh	Flooded	88	[21]
Kurunegala district, Sri Lanka	All-year	61	[38]
Nepal	Seasonal	175	[21]
Pakistan	Seasonal	60	[33, 34]
Southwestern Louisiana, USA	All-year	37	[25]
Sri Lanka	All-year	73	[36]

The Indian subcontinent is home to approximately 1300 bird species. Rice fields serve as artificial wetlands and are home to various bird species. Rice is an important food crop in India, with 81 lakh hectares planted nationwide [18]. The Indian subcontinent has the highest farmland cover per unit area globally, with rice (*Oryza sativa*) as the second most significant crop; Irrigated paddy fields in tropical Asia are the source of one-third of the world's rice [19]. Paddy fields in south India may have three rice seasons (e.g., in Tamil Nadu, India); other locations may have one (e.g., Uttar Pradesh (India), Pakistan, and Bangladesh) or two harvests in a year (e.g., parts of Karnataka in India). At least close to 351 species utilize rice fields in the subcontinent, even though just 2.7% of the subcontinent's birds nest in these fields [21]. Diverse aquatic creatures inhabit the paddy field environment, and they are considered as ecosystem engineers, since they play an essential part in the food web's dynamics [20, 22]. Algae, macroinvertebrates, vertebrates, microorganisms, and mosquito vector species thrive in this environment [23].

Birds have an ecological and functional value in the rice wetlands that act as a habitat to live and survive. Paddy fields are occupied mainly by waterfowls and ducks throughout the season, with certain species from nearby habitats only visiting the fields on rare occasions. For example, Lapwings, Baya weaver, Munia, and Larks require nesting season in moist habitats of rice-growing areas. Farmers' intervention will be less in the fields during the initial rice-growing season after transplanting, since it will be maintained in a submerged condition for two to three months. At this time, is the fields are best suited as bird nesting grounds. According to a study in Louisiana (USA), 37% of the birds recorded (37 species) inbreed in paddy fields. In contrast, just 2.7% of birds have been observed to breed in paddy fields [24, 25].

Paddy fields are being used by wintering birds in the northern subcontinent, although migratory birds use paddy fields throughout their stay. The rice is harvested after the fields have been drained. Many bird species exploit the dried fields, ripened grains, and grain leftovers left after harvesting. After the dry summer season, fields are flooded and plowed, attracting diverse bird species in large numbers. The observations made from the studies [9] show that the plowing followed by irrigation exposes many grubs, cocoons, and pests from the topsoil layer in rice fields; hence, the birds are attracted more during tillage practices when they effectively forage on the soil insects [21]. The utilization of fields by the birds appears to be less in machine-tilled fields, but this difference has not been documented scientifically.

18.3.1 Relationship between birds and insects in rice wetlands

A comparative study on the role of birds in organic and inorganic paddy ecosystems by Tamil Nadu Agricultural University, India, revealed that 26 species of birds were found in the inorganic rice ecosystem, compared to 25 species in the organic rice fields. However, the bird population in the organic environment was 29.69% greater than in the inorganic environment [9, 10]. Granivorous birds were seen during harvesting stages in both organic and inorganic rice. The common birds seen in the organic rice fields include Indian Pond Heron, Egret, Common Sandpiper, Red Wattled Lapwing, and White-browed Wagtail, whereas, in inorganic rice fields, the Indian Pond Heron, Red-wattled Lapwing, Egret, Common myna, and Black drongo were the most common birds [9, 10]. A similar study in 1990 studied the bird population of an intensively maintained area in Ludhiana and found 68 species [26]. Among them, 10 were grain eaters, 12 were omnivores, 38 were insectivores, and 8 had mixed foraging habits. The top four species (two omnivores and two grain eaters) accounted for 47% of the occurrence of 68 bird species. Also, 38 insectivorous species accounted for barely 30% of the birds. Another study was carried out to find agricultural bird diversity in the Ludhiana district's rice fields [27]. Nearly 23, with 12 non-passerine and 11 passerine species, belonging to 6 orders, 17 families, and 21 genera were found. Another study in the United States compared the count and diversity of bird species on organic and conventional farms in Florida [28]. They discovered that mixed field plantings, margins, adjacent matrix of woodland, and hedgerow had the highest bird abundances. This is because the insectivorous birds rely on the insect food available in fields [29, 30]. Major studies on birds in rice fields are enlisted in Table 18.1.

The feeding pattern of birds in the diversified agroecosystems varies according to the habitats [40]. Weaverbirds and Munia choose rice fields with more vegetation complexity than resource availability. The findings imply that predator avoidance rather than resources influence the reported foraging pattern by avians found in the agroecosystems. Some scientists reported that a few common species dominate disturbed habitats, and agricultural areas are considered highly disturbed habitats [41]. Hence, diversity is less in artificial wetlands than natural wetlands. The overall abundance of birds and richness in agroecosystems does not differ between open and partially wooded regions [42]. However, trees in the surrounding landscape altered the abundance of several common species. Studies by Indian scientists summarized the data of birds in paddy fields found in the Indian subcontinent [21]. The rice ecosystem birds and their target insect species are enlisted in Table 18.2.

Table 18.2: Rice ecosystem birds and their target insect species [9, 10].

Common name	Scientific name	Family	Order	Target insect species
Asian Open Bill Stork	*Anastomus oscitans* (Boddaert) 1783	Ciconiidae	Ciconiiformes	Wide range of insects
Black Drongo	*Dicrurus macrocercus* (Vieillot) 1817	Dicruridae	Passeriformes	Grasshopper *Atractomorphalata* (Mochulsky, 1866), Green leafhopper *Nephotetixvirescens* (Distant, 1908)
Black Kite	*Milvus migrans* (Boddaert) 1783	Accipitridae	Accipitriformes	Grasshopper *Atracto morphalata* (Mochulsky, 1866) Green leafhopper *Nephotetix virescens* (Distant, 1908)
Blue Rock Pigeon	*Columba livia* (Gmelin) 1789	Columbidae	Columbiformes	Ant *Solenopsis geminate* (Fabricius, 1804)
Blue-tailed Bee-eater	*Merops philippinus* (Linnaeus) 1766	Meropidae	Coraciiformes	Grasshopper *Atractomorphalata* (Mochulsky, 1866)
Brahminy Kite	*Haliaster Indus* (Boddaert) 1783	Accipitridae	Accipitriformes	Rice bugs *Leptocorisaoratoria* (Fabricius, 1764)
Bronze-winged Jacana	*Metopidius indicus* (Latham) 1790	Jacanidae	Charadriiformes	Larva, worms, and ants
Cattle Egret	*Bubulcus ibis* (Linnaeus) 1758	Ardeidae	Pelecaniformes	Rice leaf folder *Cnaphalocrocismedinalis* (Guenee, 1859)

Table 18.2 (continued)

Common name	Scientific name	Family	Order	Target insect species
Common Myna	*Acridotheres tristis* (Linnaeus) 1758	Sturnidae	Passeriformes	Swarming caterpillar *Spodoptera mauritia* (Boisduval) 1833
Common Sandpiper	*Acititishy poleucos* (Linnaeus) 1758	Scolopacidae	Charadriiformes	Worms, larvae
Darter	*Plotus anhinga* (Linnaeus) 1766	Anhingidae	Suliformes	Larvae
House Crow	*Corvus splendens* (Vieillot) 1817	Corvidae	Passeriformes	Ant *Solenopsis geminate* (Fabricius, 1804), Brown planthopper *Nilaparvatalugens* (Stal, 1854), Stink bugs *Nezaraviridula* (Linnaeus, 1758)
Eurasian Collared-Dove	*Streptopelia decaocto* (Frivaldszky) 1838	Columbidae	Columbiformes	Ant *Solenopsis geminate* (Fabricius, 1804)
Common Greenshank	*Tringa nebularia* (Gunnerus) 1767	Scolopacidae	Charadriiformes	Larvae
Grey Wagtail	*Motacilla cinerea* (Tunstall) 1771	Motacillidae	Passeriformes	Ant *Solenopsis geminate* (Fabricius, 1804), larvae, worms
Indian Pond Heron	*Ardeola grayii* (Sykes) 1832	Ardeidae	Pelecaniformes	Damselfly *Agriocnemispygmaea* (Rambur, 1842), Dragonfly *Sympetrum flaveolum* (Selys, 1854)
Indian Roller	*Coracias benghalensis* (Linnaeus) 1758	Coraciidae	Coraciiformes	Grasshopper *Atractomorphalata* (Mochulsky, 1866) Green leafhopper *Nephotetixvirescens* (Distant, 1908)
Little Egret	*Egretta garzetta* (Linnaeus) 1766	Ardeidae	Pelecaniformes	Damselfly *Agriocnemispygmaea* (Rambur, 1842), Dragonfly *Sympetrum flaveolum* (Selys, 1854)
Indian Peafowl	*Pavo cristatus* (Linnaeus) 1758	Phasianidae	Galliformes	Cutworm *Spodoptera litura* (Fabricius, 1775), Rice caseworm *Nymphuladepunctalis* (Guenee) 1854, Rice hispa *Dicladispaarmigera* (Oliver) 1808

Table 18.2 (continued)

Common name	**Scientific name**	**Family**	**Order**	**Target insect species**
Grey-headed Swamphen Moorhen	*Porphyrio poliocephalus* (Latham) 1801	Rallidae	Gruiformes	Worms, larvae
Red-wattled Lapwing	*Vanellus indicus* (Boddaert) 1783	Charadriidae	Charadriiformes	Cutworm *Spodoptera litura* (Fabricius, 1775), Rice caseworm *Nymphuladepunctalis* (Guenee) 1854, Rice hispa*Dicladispaarmigera* (Oliver) 1808
Spotted Munia	*Lonchura punctulata* (Linnaeus) 1758	Estrildidae	Passeriformes	Ants and small insects
Spotted Owlet	*Athene brama* (Temminck) 1821	Strigidae	Strigiformes	Cutworm *Spodoptera litura* (Fabricius, 1775), Grasshopper *Atracto morphalata* (Mochulsky, 1866)
Tricolored Munia	*Lonchura malacca* (Linnaeus) 1766	Estrildidae	Passeriformes	Ants and small insects
Eurasian Whimbrel	*Numenius phaeopus* (Linnaeus) 1758	Scolopacidae	Charadriiformes	Larvae
White-breasted Waterhen	*Amaurornis phoenicurus* (Pennant) 1769	Rallidae	Gruiformes	Larvae
White-browed Wagtail	*Motacilla maderaspatensis* (Gmelin) 1789	Motacillidae	Passeriformes	orthopterans, caterpillars, and spiders
White-throated Kingfisher	*Halcyon smyrnensis* (Linnaeus) 1758	Alcedinidae	Coraciiformes	Ants, larvae, and small insects

18.3.2 Birds–insects relationship in other crop fields

Insectivorous birds have a broad range of eating preferences, from chasing insects to digging in the dirt and woodlands. Approximately 60% of known species are insectivorous [43]. About 545 species of birds use agricultural areas for food and other activities, and 25 species have been observed to harm crops and fruits. The majority of the identified species are insectivorous [44]. Sturnidae and Turdidae starlings and thrushes eat a lot of insects and other arthropods [4]. These insects provide vital protein and other nutrients to newborn birds. The bird population in the Ludhiana district was studied while sowing maize, in which the results showed that species diversity (H'), species richness, and uniformity (E) were 2.53, 17, and 0.89, respectively [45]. Similar research was done in the boll formation stages of the cotton field, in which 26 bird species were documented [27]. Fourteen insectivorous/omnivorous species used T-perches to forage insect pests in the adjacent cotton plants. In the wheat fields of Ludhiana, Punjab, 19 different species of birds from five orders, 11 families, and 13 genera were identified. Bird species recorded in various states of India through agroecosystem studies were elucidated as a map (Figure 18.4).

During wheat harvest, the bird population showed successive patterns that connected to plant types [46]. In 1762, the Common Myna *Acridotheres tristis* was imported in Mauritius to control the red locust, *Nomadacris septemtasciata*. Several investigations on red locust gut content analyses have demonstrated that many birds rely on insect pests for food, which includes Brahminy Starling Rosy Starling, Gray-headed Starling, Yellow Wagtail, Spotted Owlet, etc. Some omnivores, including the Black-throated weaver bird, Baya Weaver, Large Grey Babbler, House sparrow, and Babbler species, eat insects and perform a dual function. For example, the barn owl eats only rodents, but the Spotted Owlet mostly feeds on Coleopteran and Lepidopteran adults. Several studies have attempted to measure the function of birds in pest management [47].

There are 14 bird species that were identified as *Helicoverpa armigera* predators in chickpea cropping systems, reducing the larval population by 73% and increasing crop output. Aside from direct predation, these birds cause natural epizootics by spreading germs and viruses in the host insect habitat. *Bacillus popillae* var. *Holotrichiea* causes milky sickness in *Holotrichia consanguinea*. It was also found that Myna and house sparrow feed on both healthy and infected H. armigera larvae, aiding the spread of NPV [48]. Drongo feces contain high amounts of NPV poly occlusion bodies (POBs) @ 7×10^8 POBs per gram of feces. Raptors eat 60–70% orthopterans. Adults with young ones may hunt on 20 mice per day; hence, raptors are the best rodent pest-controlling bird. The Spotted Owlet eats insects (54%), followed by rodents (36%), micro mammals (10%), and unidentified vertebrates (10%). The relative number of insects was greatest, with Orthoptera (24%) followed by Coleoptera (19%) and Dermaptera (11%) [49].

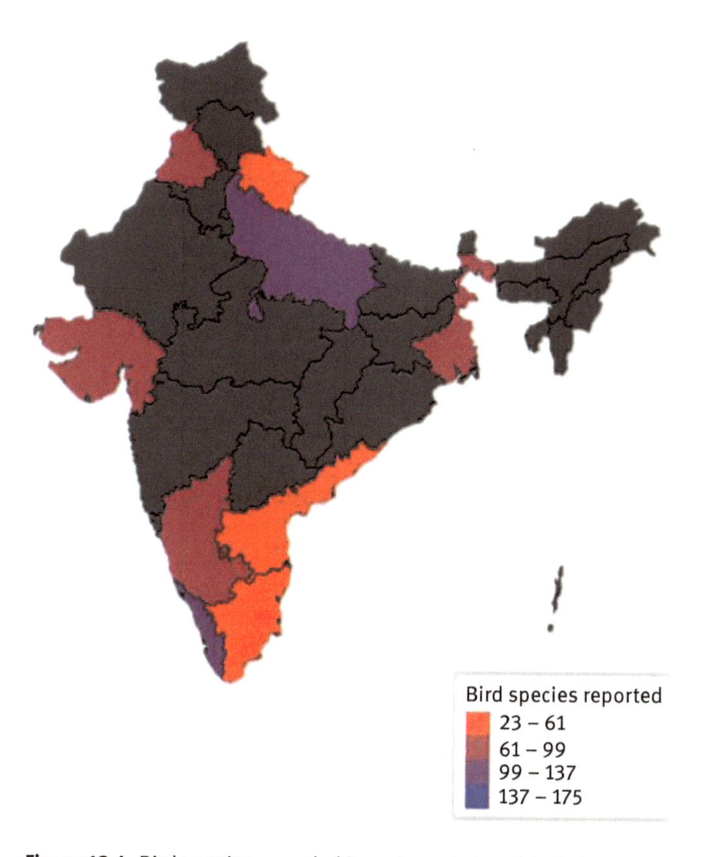

Figure 18.4: Bird species recorded in various states of India by agroecosystem studies.

An exclusive study on strawberries found that 3.8% of insect damage was reduced by an insectivorous bird [50]. In a comparative study of food preference among plowed and unplowed crop fields of pigeon pea, sorghum, maize, and cow pea, the maximum number of birds was recorded in unplowed fields and the least insectivorous birds were recorded in sorghum. In sorghum, H. armigera was the most preferred species [51]. Jungle babbler (*Turdoides striatus*), a widely spread subtropical insectivorous passerine, is considered beneficial to an agroecosystem, as they devour voraciously on insect matter, especially *Helicoverpa armigera*. The Manila tamarind, Tamarind *Pithecolobium dulce* attracts significant numbers of Rosy Starling, which suppressed *H. armigera* and *Spodoptera litura* in summer groundnut, but the adjacent region without manila trees showed an increased prevalence of such pests, requiring pesticide treatment. Mulberry fruits also have an ecological role in keeping insectivorous (beneficial) birds in the neighboring agricultural system. The *Salvadora persica* fruit attracts 18 bird species and aids in the natural management of *H. armigera* [52]. The Cattle egret is an essential bird predator in agricultural cropping systems, capturing insect pests, forming an integral part of IPM. In wheat crop, the role of birds in the natural regulation of a polyphagous lepidopteran pest, *H. armigera*, was studied in Gujarat, which concluded that the birds preferred insects over maturing wheat grains [53]. The potential of insectivorous birds to regulate the pests in woody and fruit crops has been evaluated in Spain by exclusion study and artificial nesting. The result shows that Great tit (*Parus major*) and sparrows (*Passer domesticus and P. montanus*) increased over time and controlled the Greater wax moth and sentinel caterpillars. It was found that Great tits (*Parus major*) reduced the insect damage by 19% and increased fruit yield by 66% in commercial apple orchards [53]. By providing nest boxes to insectivorous birds in Californian vineyards, the population of Beet armyworms (*Spodoptera exigua*) was reduced by 3.5 times [54]. In Costa Rica, birds reduced half an infestation of coffee berry borer beetles (*Hypothenemus hampei*) [55]. In Indonesia's bird and bat exclusion study, 31% increased pest abundance, decreasing final crop yield [56]. Several workers have also reported that natural enemies like predatory birds and insectivorous birds help in control of fruit pests such as aphids, mites, psyllids, and leaf miners by natural enemies [57].

Insectivorous birds, which decrease the number of leaf-eating insects, can enhance forest tree growth by reducing leaf damage [58]. Insectivorous birds and bats are considered effective arthropod predators in the tropical agricultural ecosystem [59]. Insect pests from the order Coleoptera and Lepidoptera (especially immature form) are effectively reduced by the bird population buildup across different habitat types [60]. Thirty-eight species of insectivorous birds belonging to 17 families under 26 genera were recorded at the maize agroecosystem in the Bengaluru region [61]. Insectivorous bird activity was also recorded in Bt and Non-Bt maize crops, and it was observed that higher insectivorous birds are under organic systems, large woodlots, dense trees, and thick hedgerows. The well-known insect pests of cacao, the cacao pod borer, and *Helopeltis bugs* were not affected by bat and bird exposure. The main reason appeared to be that the pod borer spends most of its immature stages hidden inside the pod and, therefore, is not a visible prey [62]. Nocturnal arthropods strongly declined in abundance due to exposure to nocturnal predators, including birds [63]. Likewise, organic farming is associated with increased insectivorous bird numbers, largely due to high spatial and habitat heterogeneity in the cultivated units benefiting insectivorous over granivorous birds [64]. Generalist species are favored because conventional farming reduces habitat quality by homogenizing landscapes [65]. Insectivorous bird species and their target insects in other crops are listed in Table 18.3.

18.4 Economic worthiness of insectivorous birds

Alternative use of synthetic chemicals and insectivorous birds may provide an environmentally sound and economically viable alternative in many agroecological cropping systems [70]. Birds can economically benefit crop plants across the tropical and

Table 18.3: Insectivorous bird species and their target insects in other crops [9, 10, 61, 66–69].

Common insectivorous birds	Scientific name	Target insects	Crops
Ashy Prinia	*Prinia socialis*	*Aphis gossypii*; Hymenoptera and Hemiptera insects	Guava; Cotton
Asian Brown Flycatcher	*Muscica padauurica*	Different species of insects	Different species of crops
Asian Palm-Swift	*Cypsiurus balasiensis*	Aerial insectivore; Preys insects on flight	Different species of crops
Asian Paradise (Flycatcher)	*Terpsiphone paradisi*	Different species of insects	Different species of crops
Barn Swallow	*Hirundo rustica*	Different species of flies, including horn flies, face flies, horse flies; leafhoppers, grasshoppers, wasps, crickets, moths, and beetles	Horn flies, face flies, and horse flies are domestic animals' pests, whereas crickets are household pests. Moths, beetle, leafhoppers, and polyphagous are pests of different crops.
Black-Backed Kingfisher or the oriental dwarf kingfisher	*Ceyx erithaca*	katydid	Household pest
Black-Billed Cuckoo	*Coccyzus erythropthalmus*	Tent caterpillar nest, Malacosoma	Pest of an urban and foreign tree.
Blackbirds	*Turdus merula*	Diamond-Back Moth	Cruciferous vegetable
Black-Cheeked Woodpecker	*Melanerpes pucherani*	Wasp	Generalized predator
Black Drongo	*Dicrurus macrocercus*	Pod borer larvae – *Helicoverpa armigera*	Pigeon pea, chickpea, maize, and sorghum
Black-Rumped Flameback	*Dinopium benghalense*	Stem borers (moths and beetles)	Cardamom, coffee, and cocoa
Blue-Cheeked Bee-Eater	*Merops superciliosus*	*Vanessa* adult	Generalized predator

Table 18.3 (continued)

Common insectivorous birds	Scientific name	Target insects	Crops
Blue-Throated Flycatcher	*Cyornisru beculoides*	Wide range of insect species	Different species of crops
Boat-Tailed Blackbird (Green-Tailed Grackle)	*Quiscalus major*	Cabbage Looper	Cabbage
Botteri's Sparrow	*Peucaea botterii*	*Helicoverpa Spp*,	Polyphagous pest
Brain fever bird or Indian Hawk Cuckoo	*Hierococcyx varius*	Cricket – *Gryllodes melanocephalus*, Weaver ant – *Oecophylla smaragdina*, Cotton stainer – *Dysdergus superstitious*, Sand burrowing insect – *Schyzodactylus monstruosu*, Adult and larva of *Agrotis ipsilon*, Cotton stainer nymphs beetle – *Anomala*, Mole cricket – *Gryllotalpa*, Tobacco grasshopper – *Attractographa*	Guava, jackfruit, sapota (chiku), paddy, ragi, sorghum, tobacco, and cotton
Brewer's Blackbird, Baltimore Oriole, English Sparrow (house sparrow), Warbling Vireo	*Euphagus cyanocephalus, Icterus galbula, Passer domesticus, Vireo gilvus*	Rose Weevil, Canker Worm	Rose, Deciduous forest, shade forest, and ornamental trees
Bronze Grackle (Common Grackle), Yellow-Headed Blackbird, English Sparrow (house sparrow), Vesper Sparrow, Migrant Shrike (Loggerhead Shrike)	*Quiscalus quiscula, Xanthocephalus xanthocephalus, Passer domesticus, Pooecetes gramineus, Lanius ludovicianus*	Armyworm	Polyphagous
Cattle Egret	*Bubulcus ibis*	Spodoptera, Helicoverpa	Sunflower, groundnut, and potato
California Gull	*Larus californicus*	Mormon Cricket	Household pest

Table 18.3 (continued)

Common insectivorous birds	Scientific name	Target insects	Crops
Cedar Bird (Cedar Waxwing)	*Bombycilla cedrorum*	Elm Leaf Beetle	Leaf-chewing insect of an elm tree
Chin-Spot Puffback	*Batis molitor*	Tree looper	Pest of forest tree and fruit crop
Chippy (Chipping Sparrow)	*Spizella passerina*	Cabbage Worm, Pea Louse	Cabbage, pea
Common Hoopoe/ Eurasian Hoopoe	*Upupa epops*	Mole cricket – *Gryllotalp*, Sand burrowing insect – *Schyzodactylus monstruosu*, Adult and larva of *Agrotis ipsilon* and *spilosoma obliqua*	Pulse crops like chickpea, pigeon pea; oilseed crops like castor, sunflower, and groundnut
Common Iora	*Aegithina tiphia*	Variety of insect species and spiders	Different species of crops
Common tailorbird	*Orthotomus sutorius*	Aphids	Mustard, guava, cotton, cowpea, and beans
Crow (Am. Crow), Robin (American Robin)	*Turdus migratorius*	White Grub (June beetle)	Sugarcane, groundnut, and many other crops
Crow Blackbird (Common Grackle)	*Quiscalus mexicanus*	Periodical Cicada	Loudest insect
Eurasian Blackbird	*Turdus merula*	Wide species range	Different species of crops
Forest Wagtail	*Dendronanthus indicus*	Wide insect species range	Different species of crops
Grasshopper Sparrow, Greater Roadrunner, American Kestrel, Common Black Hawk, Corn Bunting, Lesser Kestrel, Lilac Breasted Roller	*Ammodramus savannarum, Geococcyx californianus, Falco sparverius, Buteogallus anthracinus, Emberiza calandra, Falco naumanni, Coracias caudatus*	Grasshopper	Pest of many crops

Table 18.3 (continued)

Common insectivorous birds	Scientific name	Target insects	Crops
Great Grey Shrike	*Lanius excubitor*	Insect species belongs to Orthoptera, Hymenoptera, Lepidoptera, Coleoptera, and Heteroptera	Different species of crops
Great Tit	*Parus major*	*Helicoverpa armigera* and other Lepidoptera insects; Also insects belonging to Orthoptera, Neuroptera, Coleoptera, Heteroptera, Myriapoda, Formicidae, Gasteropoda; larvae belonging to Symphita	Cotton, sorghum, pulses, groundnut, tomato
Greenish Warbler	*Phylloscopus trochiloides*	Diverse insect targets	Different species of crops
Grey Wagtail	*Motacilla cinerea*	*Plutellama culipennis*, *Brevicoryne brassicae* and *Aphis craccivora*,	Cotton, Cabbage, peas, field bean, and guava
Hairy Woodpecker	*Leuconotopicus villosus*	Cecropia Moth, Tussock Moth	Polyphagous, Blueberry
Indian Crow	*Corvus plendens*	Borer – *Adisuraatkinsoni*, sand borrowing insect – *Schyzodactylusmonstruosu*, *Cricket – Brachytripsachattinu*, *Darkling beetle – Opatrum sp.*, *Mole cricket – Gryllotalpa*	Tomato, field beans, groundnut, sorghum, millets, cotton, pulses
Indian roller	*Coracias benghalensis*	Rhinoceros beetle – Oryctus rhinoceros, Mole cricket – Gryllotalpa	Groundnut, ragi, pigeon pea, cotton, cucurbits, Guava
Indian wren warbler, fledglings	*Prinia inornata*	Hairy caterpillars, aphids, defoliators	Ragi, millets, groundnut, pulses

Table 18.3 (continued)

Common insectivorous birds	Scientific name	Target insects	Crops
Jack snipe (Wilson's Snipe)*, Curlews, Upland plover (Upland Sandpiper), Plovers, Quail (No. Bobwhite), Prai-Rie Chicken, Blackbirds, Yellow-Headed Blackbird, Bobolink, Western meadow lark, Orioles, Sparrows, Robin (American Robin)	*Lymnocryptes minimus, Numenius spp, Bartramia longicauda, Colinus virginianus, Tympanuchus cupid, Turdus merula, Xanthocephalus xanthocephalus, Dolichonyx oryzivorus, Sturnella neglecta, Icterus, Turdus migratorius*	Rocky Mountain Locust	Polyphagous
Large-Billed Leaf Warbler	*Phylloscopus magnirostris*	Many species of insects	
Little Green Bee-Eater	*Merops orientalis*	*Aphis craccivora, Pieris brassicaae; Apis florea, A. mellifera, A. dorsata and A. cerana.*	Pea, cabbage, cauliflower
Little Swift	*Apus affinis*	Aerial insectivore; Insects belong to Diptera, Hymenoptera, Hemiptera, and Homoptera	Different species of crops
Magnolia Warbler	*Setophaga magnolia*	Spruce budworm,	Pest of Spruce
Myna adults	*Acridotheres tristis*	Ash weevil–*Myllocerous discolor,* Weaver ant – *Oecophylla smaragdina,* Driver ant – *Dorylu,* Rice grasshopper, beetle – *Anomola,* Darkling beetle	Pigeon pea, paddy, jackfruit, guava, sapota (chiku), custard apple
Myna Fledglings	*Acridotheres tristis*	Insect species of Orthoptera, Lepidoptera, Coleoptera, Hymenoptera, *Trichoplusia,* and other defoliators	Sunflower, cotton, pigeon pea, maize, sorghum

Table 18.3 (continued)

Common insectivorous birds	Scientific name	Target insects	Crops
Myrtle Warbler, Blackpoll Warbler, Oregon Chickadee (Black-Capped Chickadee)	*Setophaga coronata, Setophaga striata, Poecile atricapilla*	Plant lice (Aphids)	Polyphagous
Oriental Honey Buzzard	*Pernis ptilorhyncus*	Larvae of Paper wasps – *Polistes* spp.	Generalized predator
Paddy Field Pipit	*Anthus rufulus*	Diverse insect species	Different species of crops
Paddy field Warbler	*Acrocephalusa gricola*	Diverse insect species	Different species of crops
Pied Bushchat	*Saxicola caprata*	*Helicoverpa armigera* and other insects	Sorghum, Pulses, cotton, groundnut, and tomato
Red-Rumped Swallow	*Hirundo daurica*	Various insect species belong to Hymenoptera, Homoptera, Coleoptera, and Diptera	Different species of crops
Red-Shafted Flicker (N. Flicker)	*Colaptes auratus cafer*	Codling Moth	Different species of crops
Red-Throated Pipit	*Anthus cervinus*	Different insects	Different species of crops
Road Runner	*Geococcyx californianus* (the greater roadrunner) *Geococcyx velox* (The lesser roadrunner)	Passion-Vine Caterpillar	Passion flower
Rose-Breasted Grosbeak Cliff Swallow	*Pheucticus ludovicianus*	Potato Beetle	Potato and sweet potato
Rufous Woodpecker	*Celeus brachyurus*	Diversified target species	Different species of crops
Oriental Magpie Robin	*Copsychus saularis*	*Helicoverpa armigera* and other insects	Sorghum, pulses, cotton, groundnut, and tomato

Table 18.3 (continued)

Common insectivorous birds	**Scientific name**	**Target insects**	**Crops**
Starling, Western Meadowlark	*Sturnus vulgaris, Sturnella neglecta*	Cutworms	Cereals and vegetables
Tawny Flanked Prinia	*Prinia subflava*	Aerial insectivore; Insects belonging to Hemiptera and Hymenoptera	Different species of crops
Thick-Billed Warbler	*Acrocephalus aedon*	Diversified insect species	Different species of crops
Tickell's Blue Flycatcher	*Cyornis tickelliae*	Diverse insect species	Different species of crops
Valley Quail (CA Quail)	*Callipepla californica*	Black Olive Scale	Olive
Verditer Flycatcher	*Eumyias thalassinus*	Wide range of insects	Different species of crops
Western Crow (Am. Crow)	*Corvus brachyrhynchos*	Climbing Cutworm	Maize
Western Meadowlark	Sturnella neglecta	Coulee Cricket	Household pest
White-Bellied Fantail	*Rhipidura euryura*	Wide range of insects	
White-Breasted Nuthatch	*Sitta carolinensis*	Pear Psylla	Pear
White-Browed Fantail	*Rhipidura aureola*	Wide range of insects	
White-Crowned Sparrow	*Zonotrichia leucophrys*	Aphis (Rose Aphid)	Rose
Wire-Tailed Swallow	*Hirundo smithii*	Flies including horn flies, face flies, horse flies, leafhoppers, grasshoppers, crickets, moths, beetles, wasps.	Different species of crops
Wood shrike	*Tephrodornis pondicerianus*	Diverse insect targets	Different species of crops
Yellow-Billed Cuckoo	*Coccycus americanus*		

Table 18.3 (continued)

Common insectivorous birds	Scientific name	Target insects	Crops
Yellow-Billed CuckooBaltimore Oriole, English Sparrow (House Sparrow), Cedar Bird (Cedar Waxwing), Yellow Warbler, Robin (American Robin), Blue Jay, Orchard Oriole	*Coccyzus americanus, Icterus galbula, Passer domesticus, Bombycilla cedrorum, Setophaga petechial, Turdus migratorius*	Walnut Caterpillar, Forest Tent Caterpillar, Catalpa Sphinx, Orchard Tent Caterpillar	Walnut, forest tree, Catalpa, trees
Yellow-Billed Cuckoo, Kingbird (E. Kingbird), Great-Crested Flycatcher, Phoebe (E. Phoebe), Wood-Pewee (E. Wood-Pewee), Orchard Oriole, Baltimore Oriole, English Sparrow (House Sparrow), Chippy (Chipping Sparrow), Field Sparrow, Song Sparrow, Chewink (E. Towhee), Cardinal, Scarlet Tanager, Cedar Bird (Cedar Waxwing), Red-Eyed Vireo, Warbling Vireo, Yellow Warbler, Catbird (Grey Catbird), Carolina Wren	*Colaptes auratus cafer, Tyrannus tyrannus, Myiarchus crinitus, Sayornis phoebe, Contopus sordidulus, Icterus spurius, Passer domesticus, Spizella passerine, Spizella pusilla, Melospiza melodia, Pipilo erythrophthalmus, Cardinalis cardinalis, Piranga olivacea, Bombycilla cedrorum, Vireo olivaceus, Vireo gilvus, Setophaga petechial, Dumetella carolinensis, Thryothorus ludovicianus*	Leaf-miner	Polyphagous
White-Browed Wagtail	*Motacilla madaraspatensis*	*Aphis craccivora, Plutellama culipennis* and *Brevicoryne brassicae*	Peas, field beans, cabbage, guava, and cotton
White-Naped Tit	*Machlolophus nuchalis*	Pod Borer/American Bollworm – *Helicoverpa armigera*	Sorghum, cotton, groundnut, and pulses
White Wagtail	*Motacilla alba*	*Plutellama culipennis, Aphis craccivora* and *Brevicoryne brassicae,*	Peas, field beans, cabbage, guava, and cotton

temperate zones [71]. It has also been estimated that the expenditure on bird ranges adds between 100–400 billion USD per annum to the world's economy and showed birds were responsible for increased production of coffee worth US$310/ ha by controlling coffee berry borer, *Hypothenemus hampei* through bird predating [55]. In addition, birds saved farmers from post-harvest costs of removing mummified almonds and gave a positive net return of AUD 25–275/ ha averaged across almond orchards [72]. Based on gut contents analysis, it was observed that each bird consumes around 2,100 Casebearer larvae, estimated to add the worth of US$2900 to the pecan industry [73].

Tropical studies revealed that the success of birds as predators of insect pests in agricultural landscapes is correlated with native forest proximity. Lemon-bellied white-eye (*Zosterops chloris*) occurs in both forest habitats and coffee plantations, and it plays an important role in suppressing insect pests population in a coffee plantation, and the pest control is enhanced when it is planted at forest proximity [15]. The cabbage looper population has been reduced by 21%, and in broccoli too, insectivorous birds have reduced the two-caterpillar species [74]. Birds killed an average of 41% of overwintering codling moths, *Cydia pomenella,* in walnut orchards [75]. Various studies reported that between 13–99% of the overwintering stage of Codling Moth were consumed by birds, especially near the habitat. The setting of nest boxes increases the density of Great Tits (a relative of chickadees), thereby increasing apple yields by 66% [7, 76]. Birds help reduce olive fruitflies in pupae stages. They consume 65–71% of the pupae in the soil. Dark-Eyed Juncos in large flocks of 50 to 150 birds consume large numbers of insects, as it was found that 23,000 to 70,000 Pear Psylla females, with a potential to produce around 7 to 23 million eggs, were preyed on by the birds. Another study of gut contents analysis of Black-Capped Chickadees, Red-Breasted, and Golden-Crowned Kinglets Nuthatches contained large numbers of winter psyllas [77]. Installation of bird nest boxes has greatly increased the abundance of Western Bluebirds and their ability to insects predation in the vineyards. The highest removal of insects, i.e., 59%, was seen closest to the bird boxes. DNA analysis of fecal matter showed that birds were not consuming natural enemy insects as only 3% of the natural enemies were reported in their diet [54]. Songbirds reduced the population of alfalfa weevils by over 33% [78]. Birds reduced corn insect pests by 34–98% in different studies. Birds were estimated to reduce 20–26% of leaf damaging grasshoppers in millet [79]. Examination of fecal matter collected from the active mud cavity previously nested by birds reported 18–84% of pest insects and around 34–70% of cutworms in the fecal analysis [80]. In feeding trials using nonparasitized and parasitized Armyworms, preference of birds was tested, and it was seen that birds strongly preferred the larger nonparasitized insects than the parasitized ones [28]. Birds reduced grasshoppers selectively in different sites by 25–55% [81, 82]. When pest abundance ranged from medium to high, European Starlings ate 40–60% of pasture grubs [83]. Releasing wild falcons in New Zealand vineyards reduces pests by 78–83% and grape damage by 55–95% [84]. Pest reduction by various birds and crops is listed in Table 18.4.

Table 18.4: Pest insects and other invertebrates managed with beneficial birds.

Crop	Insectivorous birds	Target insect and other arthropod pests	Benefits by insectivores bird	Reference
Vegetable crops				
Broccoli	Red-Crested Cardinals	*Trichoplusia ni* and *Artogeia larvae*	More insectivores bird activity	[74]
	Swallows and Sparrows	Aphids, Caterpillars, and Flea Beetles		[85]
	Wild Birds in South Korea	Aphids		[86]
	Savanna Sparrows	Experimental Cabbage Loopers	Pest reduction by 49%	[87]
Beetroot	W. Bluebirds, Chipping Sparrows, Am. Goldfinches, and other insect-eating birds	Beet Armyworms	Reduced Pest by 58%; and consumed around 3% natural Enemies	[54]
Dried Beans	Sparrows, Baya's, Mynahs, and Black Drongos	Bollworm/Pod Borer	Insects damage by 84%, thereby Increasing in yield by 71%;	[88]
Fruit crops				
Apples	CA Scrub and Stellar's Jays, Am. Robins, EU Starlings, Ruby-Crowned Kinglets, N. Flickers, Downy Woodpeckers, Oak Titmice, Brewer's Blackbirds, and Chestnut-Backed Chickadees	Live Experimental Codling Moth Larvae	Reduction of overwintering egg 77–91%;	
	EU Robins, Common Blackbirds, EU Blackcaps, EU Wrens, Great and EU Blue Tits	Aphids, Apple Blossom Weevils, and other arthropods	Reduction in Aphid and natural enemy population	[89]
	Great Tits	Codling Moths	Reduction in pest damage by 11–14% and overwintering pest by 90% thereby increasing yield by 66%	[76]

Table 18.4 (continued)

Crop	Insectivorous birds	Target insect and other arthropod pests	Benefits by insectivores bird	Reference
Cacao	Indonesia Birds Including Flowerpeckers, Sunbirds, and White Eeyes	Arthropods	Yield increased by 31%	[76]
Olives	Birds of Greece	Olive Fruit Flies	Reduced overwintering Pupae by 65–71	[90]
Strawberries	Insect-eating birds (e.g., Black Phoebes, Pacific-Slope Flycatchers), and fruit-eating birds (e.g., House Finches, Am. Robin, EU Starling)	Mainly Lygus Bugs, but also Leaf-rollers and Slugs	Reduced insect damage by 3.8% but created 3.2% birds Damage	[50]
Wine Grapes	Great Tits, House and Tree Sparrows	Greater Wax Moth Caterpillars.	Reduced one-third pest population	[53]
Foodgrain crops				
Corn	Red-Winged Blackbirds	EU Corn Borers	Reduced damage 20% and overwintering Pest by 64–82%	[91]
	Red-Winged Blackbirds	Corn Rootworm Beetles	Reduced insect damage by 50% in early stage of planting but caused 80% crop damage later on	[92]
	Downy Woodpeckers	Corn Borers	Reduction in overwintering insects stages by 34%	[93]
Millet	Senegalese Birds, Including Cattle Egrets, Sparrows, Rollers, Buzzards	Senegalese Grasshoppers	20–26% Pest population has been reduced	[79]
	Grackles, Yellow-Headed Blackbirds, Chipping Sparrows, Bluebirds, Prairie Hens, and EU Starlings	Armyworms	40 avian Species with a significant amount of Armyworms in their gut had been removed	[94]

Table 18.4 (continued)

Crop	Insectivorous birds	Target insect and other arthropod pests	Benefits by insectivores bird	Reference
Wheat	Horned Larks and McCown's Longspur	Pale Western Cut Worms, Grasshoppers, Ants, and Beetles	3 Gut contents revealed 4–70% of Caterpillars	[95]
Commercial crops				
Coffee	Guatemalan Birds Foraging on Insects	Coffee-Berry Borers and other arthropods	Reduced insect pest by 64–80% and foliage damage by 28%;	[96]
Cotton	Ugandan Birds	Experimental Caterpillars	Catches 2–4% insects per day	[97]
	Most important birds: Orioles, Black Bird, Meadlowlarks, Painted Buntings, Quail, Morning Doves, and Mockingbirds	Cotton boll Weevils	28 bird species found with significant amount of boll Weevils in their gut	[97]
Rapeseed and Oilseeds	Ethiopian Birds	Cabbage Flea Beetles, Aphids, Lepidopteran Skeletonizers and Chewing Larvae	Birds decrease leaf damage	[98]
	Common Swifts, Barn Swallows, and House Martins	Cabbage Seedpod Weevils and Pollen Beetle Pests	18–84% of prey consumed crop-damaging pest	[99]
Oil Palm	Oriental Magpie Robin, Ashy Tailor Birds, and Greater Coucal	Several Caterpillars	Reduced damage at an early stage in plants	[98]
Tea	Indian Birds (Asian Pied Starling, Chestnut-Tailed Starling, Jungle Myna, and Red-Vented Bulbul	Geometrid Looper Caterpillars	Reduction in pest populations	[98]
Tobacco	Common Grackles	Tobacco Hornworm	Eaten away 50–100% of insects pest	[98]
	Am. Crows, Mockingbirds, E. Bluebirds, and House Sparrows	Tobacco Hornworm	Reduction in 40–50% of insect populations	[100]

18.5 Improving the population of insectivorous birds

The availability of various resources such as perching trees, nesting trees, and substrates near human inhabitation may encourage insectivorous bird species to inhabit and reproduce. Nest boxes may greatly enhance the insectivorous bird population in the absence of cavities. Using palm and coconut leaf stumps as bird perches in rice fields has boosted insectivorous bird visits. Birds that nest in cavities have specific nest box needs. Similarly, the widespread use of nest boxes may minimize the requirement for chemical spraying and its associated costs in an integrated pest control system. Species-specific box placement should be varied. Various studies created nest boxes for Indian Myna, Brahminy Starling, Bank Myna, Blue Jay, Spotted Owlet, Magpie Robin, and Indian Robin. Bird nest boxes for bug-eating birds may help reduce agricultural pest numbers and avoid pest out breaks. Fixing perches @50/ha in chickpea boosted the effectiveness of predatory birds, including Drongo, Mynas, and sparrow. Some nest specifications for various species are listed in Table 18.5.

Table 18.5: Nest specifications of various bird species (inches).

Species	Diameter of entrance	Entrance height above the bottom	Depth of cavity	Bottom of cavity
Kestrel	3.0	12–14	14–18	8 × 8
Woodpecker	1.25	6–8	8–10	4 × 4
Myna	2.5	14–16	16–18	7 × 7
Blue Jay	2.0	9–12	12–15	6 × 6
House Sparrow	1.0	5–6	6–8	4 × 4
Indian Robin	1.5	7–8	8–10	5 × 5
Swallows	1.5	4–5	6	5 × 5

To improve the insectivorous bird population and replace the chemical usage, peak insect infestation periods may use a combo of bird releases and bio-pesticides. A long-term, comprehensive strategy is needed in this respect, requiring more scientific studies. Chemical recommendations should be based on detailed data on residual levels. Collaborative work of entomologists, ornithologists, and extension agents is needed in technology transmission.

18.6 Birds as a menace to crops and their control

While insectivorous birds have more beneficial and less detrimental consequences, granivorous birds have a more detrimental effect on economic crop and fruit yield. Some farmers employ illegal methods of killing birds using agrochemicals. Other strategies like the wrapping method, reflective bird scaring ribbon, high-density cropping, mechanical bird scare, and herbal repellents (Figure 18.5) are also used not directly to harm bird life. A description of various bird control methods is listed in Table 18.6.

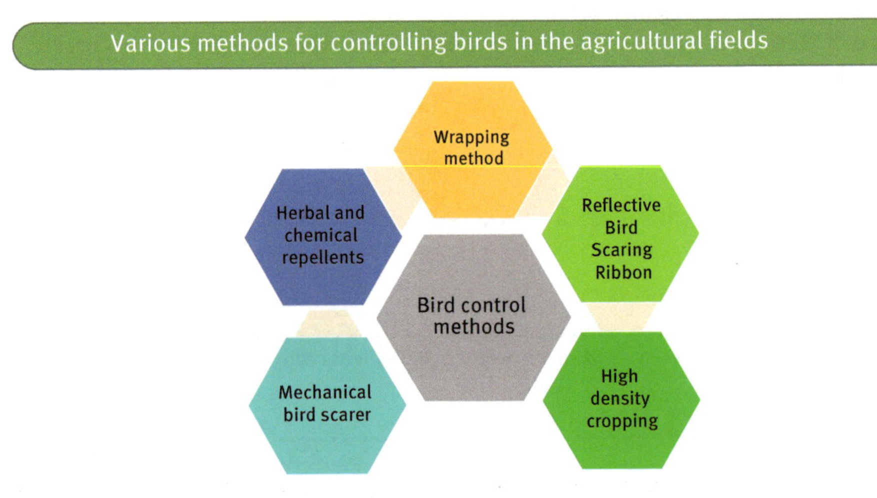

Figure 18.5: Methods for controlling birds in agricultural fields.

18.7 Constraints and uncertainties in the use of birds as insect control

The use of various birds as an agent to control the insects is unreliable, in some cases, as it can cause crop damage, which includes in the ecosystem disservices by birds [101] (Figure 18.6). Ecosystem disservices, or those features of an ecosystem that have a detrimental effect on people, may disproportionately influence conservation choices of farmers. Birds, in particular, might have broader impacts on crops, ranging from beneficial to detrimental. Therefore, it is critical to measure them scientifically. For example, birds may be profuse on farms next to natural areas and help the ecology by eating pests. Birds are opportunistic feeders, and they show facultative responses to a sudden increase in the prey [102]. The major constraint is that the bird species would indirectly consume the beneficial and the potential agents of natural controllers of pest arthropods like a dragonfly, damselfly, and yellow banded wasps

Table 18.6: Various methods for controlling birds in agricultural fields.

Methods	Methodology	Advantages	Crops
Wrapping method	A leaf was wrapped around the cob in two or three loops and secured by a loose knot in the last loop near the top. The objective was to disguise a cob in a leaf.	This procedure is less tedious than scaring and does not need any materials.	Maize
Reflective bird scaring ribbon	The reflective ribbon is a polyester film with a red and silver metallic coating. It is made of a polyester sheet, 1.5 cm wide and 15–20 m long strips. They are attached parallel to the crop at 0.5 m height above the crop using bamboo poles and strings at 5 m intervals. It should be orientated north or south to gain more sunshine.	The use of ribbons to scare birds is extremely successful and acceptable to farmers.	Cumbu and other millets
High-density cropping	Planting the crops with close row and plant spacing	High-density sorghum (fodder crop) and maize planting minimized the grain damage caused by parakeets and other birds. This approach increased the yield, while simultaneously providing fodder.	Sorghum and maize
Mechanical bird scare	It is small machinery, and a basic sound generator works on 1 kilogram of calcium carbide and water for 24 h. This approach efficiently covers 1-hectare areas.	To prevent habituation, vary the firing frequency, location, and direction.	Maize and other grains
Herbal and chemical repellents	Bird damage was reduced in maize and sorghum cropping systems using 200 ml/L Neem cake mixture and 10% Tobacco leaf decoction. A 10% sorghum spray during the lactation stage reduced avian harm. Copper-oxychloride (3 gm/kg seed) reduces bird seedling losses in maize, chickpea, sunflower, and peanut. During the milky stage of sunflower, to repel the insects, the heads are covered with aluminum foil-covered paper plates and sprayed with egg solution @20 ml/lit. Using organic manures like FYM enhances the microclimate and attracts insectivorous birds.	Easy to use	Sorghum, maize, pulses

[103]. When birds prey more on beneficial predatory arthropod insects than on pest species (intraguild predation), they may harm farmers by reducing crop productivity. A soybean field study conducted in Illinois (USA) discovered that plants with birds have far less insect damage to their leaves than in control plots, but their influence on yield was not significant. Thus, the data indicated no net benefit or loss on the productivity of the crop [104].

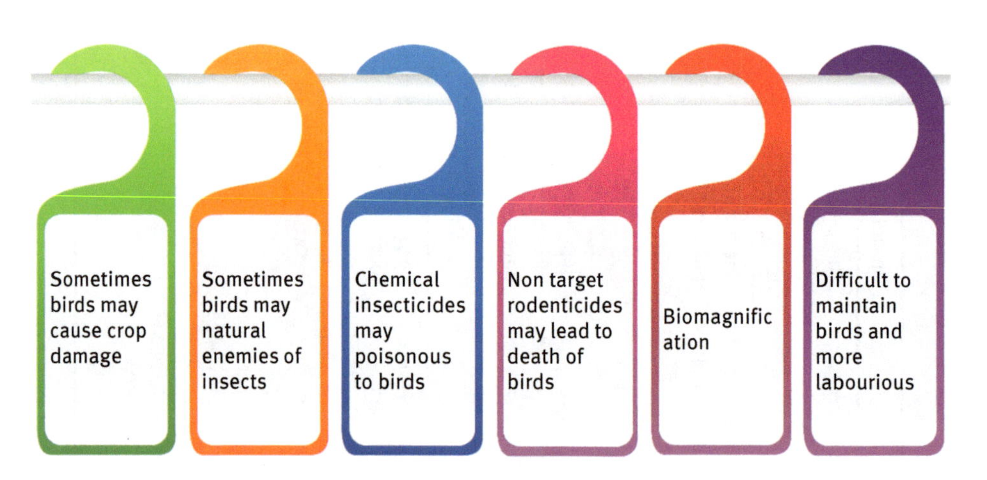

Figure 18.6: Constraints and uncertainties in using birds as biocontrol agents.

Furthermore, studies conducted in Tamil Nadu Agricultural University farms, India, suggested employing the birds, an effective pest biocontrol agent, but not much information is available on their yield losses quantification [9]. Hence, more scientific studies on the ecosystem services and disservices of birds must be done in agricultural ecosystems, with various cropping systems in different agroclimatic regions and seasons.

Syrphid flies and lady beetles are natural enemies of the pest aphid of cereal crops, especially at the high-density times of aphids. These are natural enemies eaten away by insectivorous Eurasian Tree Sparrows (*Passer montamus*) [105]. Birds are considered as vectors of disease-causing enteric pathogens such as *Salmonella enterica*. It was evident that migrating Sandhill cranes were causative agents of the outbreak of Campylobacter in peas [106]. Bird species may adversely affect the cultivars by damaging the crop intraguild predators, leading to economic losses [107].

18.8 Impact of agriculture expansion on bird population

The expansion and intensification of agriculture, such as increased mechanization and high use of chemical pesticides, ultimately affect the birds of croplands. Studies conducted by Tamil Nadu Agricultural University, India, found that many bird visits occurred in organic rice fields rather than conventional chemical-sprayed rice fields [9, 10]. Several studies showed that agriculture-dependent birds are harmed due to the use of pesticides. It is evidenced that insecticides used on rice seeds in nurseries killed Ruddy Shelduck, Lesser-whistling Duck, and Spot-billed ducks. The Indian Peafowl have faced a high degree of threats due to poison baits. The extinction of scavenging birds, such as the vulture and decline of crow and granivorous birds may be attributable to the lingering effects of agrochemicals. The disappearance of several insectivorous birds (Drongos, Bee-eaters) and nesting failure in raptors have been documented at pesticide-contaminated locations in Ambala, Bharatpur, Morena, Thrissur, Parambikulam, Coimbatore, and Andhra Pradesh [108].

Similarly, the pelican population in Karnataka decreased by 2000 [109], and the breeding population of Sarus crane in Bharatpur, Rajasthan, was also found to decline. The presence of different residual levels of organochlorine were detected in several tissues and eggs of bird species, including large Egret, large cormorant, Indian shag, Darter, Grey heron, Cattle egret, and Painted stork [110, 111]. In Corbett National Park, the Himalayan Grey-headed Fishing Eagle has failed to reproduce due to varying levels of pesticide residues [111, 112]. Between 1987 and 1990, Aldrin application on the wheat fields killed 50 collared doves and a few blue rock pigeons in Bharatpur, Madhya Pradesh, and Aldrin residues were found in the gastrointestinal tract of a Sarus crane and a Collared dove. Embryonic maldevelopment, behavioral aberrations in breeding, and physiological disturbances are all impacts. Organophosphate and carbamate compounds could affect singing, attentiveness, and disrupted nesting behavior. So maintaining the balance between deterring predators and encouraging insectivorous birds may be implemented. Several key practices are briefly mentioned below. Some recent studies revealed that these birds were contaminated by chemical residues such as polychlorinated hydrocarbons and dioxins [113–116]. This leads to thinning of eggshells, immature growth of young ones, habitat destruction, food availability, and breeding resources reduction, and finally, mortality [117]. The extensive use of Second Generation Anticoagulant Rodenticide (SGARS) affects nontarget species like Red Kites and barn owls by transfer through the food chain, causing potentially lethal levels of nearly about 30% [118]. According to the studies, extensive large-scale ranching and shifting cultivation of slash and burn agriculture practices adversely affect the birds and allow migration to various habitats other than particular agricultural croplands [119]. This decline in bird population may also occur due to the intentional monoculture of commercial cultivation like biofuel crops [120]. The positive effects of birds as insect control on farmlands can be improved by promising methods such as bird netting, natural habitats of hedgerows, etc. However, the major implied constraint is that these methods are expensive and make it difficult for widespread application by the farmers.

18.9 Research gaps

18.9.1 Need for advanced studies

Molecular methods should be used in ecological studies of bird and insect relationships. For example, quantifying net impacts and characterizing bird diets using molecular methods may help us understand agroecosystem ecology and its relationship with birds and insects [70, 103]. However, lack of funding and fewer researchers on this science are also major problems in conducting advanced research.

18.9.2 Need for crop-specific and regional studies

Very little research on net impacts has been done on various crop species and locations. It is uncertain that previous generalizable studies on birds across various agroecosystems are usually exaggerate the ecosystem services by birds, but it can be quantified when additional research is undertaken across croplands and geographies [103]. Researchers will not make universal advice but must, instead, adjust their results to each agroecosystem [103, 121].

18.9.3 Need for realistic studies

Studies from the period of economic ornithology typically tell us the role of birds as biocontrol agents, as interpreted by the presence of a pest in a bird's stomach as proof of pest population control [70, 103]. The role of a few bird species in pest control is uncertain to farmers. Conversely, several species implicated as pest control agents may harm crops or devour beneficial arthropods. Many diet studies (molecular or otherwise) will be required to understand better, how different species adjust their diets in various agricultural environments throughout the year.

18.9.4 Need for economic studies

However, a farmer may opt to repel hazardous birds, if the cost of the damage surpasses the expense of the deterrent strategy. In this situation, netting barriers, falconry, or raptor perches may help reduce bird damage. However, before using bird deterrents, a farmer must determine their cost-effectiveness [103, 122]. Birds may soon be included in regional IPM programs, as more research is being done on their benefits and drawbacks. The emerging science of molecular food analysis will aid researchers in determining which birds contribute to net impacts and in what agricultural scenarios.

18.9.5 Need for collaborative work

Researchers must work closely with stakeholders to acquire relevant data for critical decision-making phases (e.g., adoption, planning, intervention, assessment) [103, 121]. The decline of economic ornithology is due to a lack of practical advice and stakeholder participation [69, 70]. To maintain stakeholder support, admit when to use and when birds cannot be used as biological control agents. For example, blueberries and cherries are high-energy foods for various fruit-eating birds [103, 123]. In some cases, rather than improving services, it may be required to investigate techniques to reduce disservices (e.g., bird netting or falconry). Generalizing some bird species as beneficial or pests may not be fair, without any scientific assessment. Collaborating with farmers, landowners, and other stakeholders to convey this context-specific knowledge is critical. Scientists must work closely with farmers, landowners, and other stakeholders to guarantee the delivery of ecosystem services by birds (e.g., insect control) [103, 124].

18.9.6 Planning and policy to incorporate in IPM

Though many studies on birds have already been done, it is unclear how avians as biological control agents fit into IPM plans. Using natural enemy measures to integrate arthropod natural enemies into IPM decision-making in wheat, soybean, and walnut cropping systems may give promising results [103, 125, 126]. From economic ornithology to bird net effects and ecosystem services, the study of birds in agroecosystems has to be evolved. In addition, this research might provide valuable information on how certain bird populations affect specific crops in specific places [103].

18.10 Future scope and perspectives

Studying the agroecosystem of birds through ecosystem services and disservices allows us to revisit the insights given in economic ornithology developed in earlier days. However, the current studies on birds in insect control need several changes.

First, there is a strong need to conduct advanced studies in molecular biology, regionally specific, crop-specific, and agroecosystem-specific studies. The studies should also report the realistic scenario of using birds as biocontrol of insects without bias and exaggeration. Finally, economic analysis should be done to know the feasibility of the birds as biocontrol agents, especially in changing climatic conditions. After the economic analysis, the planning and policy to incorporate in IPM should be done, if it is economically feasible (Figure 18.7).

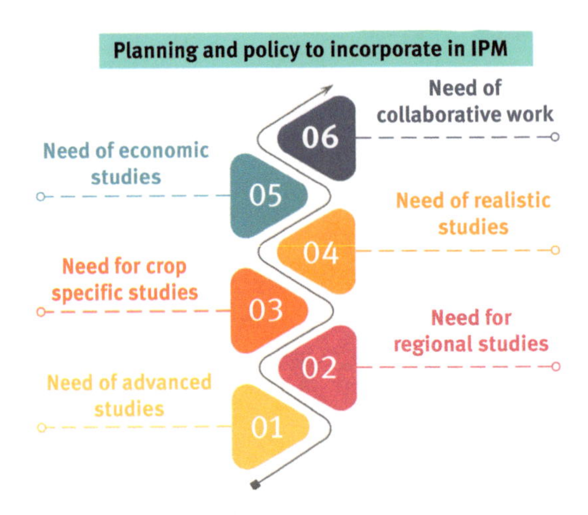

Figure 18.7: Future scope, perspectives, and research gaps in biocontrol of insects by birds.

18.11 Conclusion

Birds provide various ecosystem services; among them, one of the most important services is insect and pest regulation. As a consequence of their usefulness to farmers, it is advocated that most birds in the agro-environment should be protected, since they are beneficial in one way or another. It is vital to organize periodic awareness programs on the importance of birds to benefit farmers in maximizing their profits. It should also be made clear to people that they should not cause harm to perch and nesting areas. An artificial perch will be placed amid crops to attract more birds to the region, and this will be done in future. Maintaining biofencing in the agricultural farm lands may shelter many bird species that serve as insect predators. In order to minimize reliance on chemical fertilizers and pesticides, it is necessary to encourage the use of biological fertilizers and biological control agents, as this approach will not cause harm to birds. More research on nesting performance of commercially important bird species, breeding time, and the coincidence of these with the crop stages is the need of the hour, as this vital ecosystem service is highly underexplored scientifically. It is the right time for fruitful research to be carried out on this aspect in order to include insectivorous birds as a vital component in the Integrated Pest Management system.

Work contribution

G. K. Dinesh, Writing – first draft, data visualization, Software; **B. Priyanka,** Writing – first draft; **Archana Anokhe,** Writing – first draft; **P. T. Ramesh,** Writing – review and editing; **R. Venkitachalam,** Writing – review and editing; **K. S. Keerthana Sri,** Writing – review and editing; **S. Abinaya,** Writing – review and editing; **V. Anithaa,** Writing – review and editing; **R. Soni,** Writing – review and editing.

References

[1] Ali S. Ornithology in India: its past, present and future (Sunder Lal Hora memorial lecture). Proceedings of Indian National Science Academy 1971, B37, 99–113.

[2] Dhindsa MS, Saini HK. Agricultural ornithology: An Indian perspective. Journal of Biosciences 1994, 19, 391–402.

[3] Asokan S, Ali AMS, Manikannan R. Diet of three insectivorous birds in Nagapattinam District, Tamil Nadu, India-a preliminary study. Journal of Threatened Taxa 2009, 1(6), 327– 330.

[4] Nyffeler M, Şekercioğlu ÇH, Whelan CJ. Insectivorous birds consume an estimated 400–500 million tons of prey annually. Science Nature 105, 472018.

[5] Linden VMG, Grass I, Joubert E, Tscharntke T, Weier SM, Taylor PJ. Ecosystem services and disservices by birds, bats and monkeys change with macadamia landscape heterogeneity. Journal of Applied Ecology 2019, 56, 2069–2078.

[6] Sekercioglu CH. Increasing awareness of avian ecological function. Trends in Ecology & Evolution 2006, 21, 464–471.

[7] Mols CMM, Visser ME, Great tits (Parus major) reduce caterpillar damage in commercial apple orchards PLoS One 2007, 2, e202.

[8] Millennium Ecosystem Assessment, MEA. Ecosystems and human well-being. *Synthesis (Stuttg)*, 2005.

[9] Dinesh GK, Ecology of birds and insects in organic and inorganic rice ecosystem. Coimbatore, Tamil Nadu Agricultural University, 2017.

[10] Dinesh GK, Ramesh PT, Chitra N, Sugumaran MPP. Ecology of birds and insects in organic and conventional (In-Organic) rice ecosystem. International Journal of Current Microbiology and Applied Sciences 2018, 7, 1769–1779.

[11] Symes CT, Venter SM, Perrin MR. Afromontane forest avifauna of the eastern Soutpansberg mountain range, Northern Province, South Africa. South African For Journal 2000, 189, 71–80.

[12] Symes CT, Perrin MR. The habitat and associated bird assemblages of the grey-headed parrot poicephalus fuscicollis suahelicus in Limpopo Province, South Africa. Ostrich-Journal African Ornithol 2008, 79, 9–22.

[13] Rusch A, Chaplin-Kramer R, Gardiner MM, Hawro V, Holland J, Landis D, Thies C, Tscharntke T, Weisser WW, Winqvist C. Agricultural landscape simplification reduces natural pest

control: A quantitative synthesis. Agriculture Ecosystems and Environment 2016, 221, 198–204.
[14] Tscharntke T, Karp DS, Chaplin-Kramer R, Batáry P, DeClerck F, Gratton C, Hunt L, Ives A, Jonsson M, Larsen A. When natural habitat fails to enhance biological pest control–Five hypotheses. Biological Conservation 2016, 204, 449–458.
[15] Maas B, Tscharntke T, Saleh S, Dwi Putra D, Clough Y. Avian species identity drives predation success in tropical cacao agroforestry. Journal of Applied Ecology 2015, 52, 735–743.
[16] Puig-Montserrat X, Torre I, López-Baucells A, Guerrieri E, Monti MM, Ràfols-García R, Ferrer X, Gisbert D, Flaquer C. Pest control service provided by bats in Mediterranean rice paddies: Linking agroecosystems structure to ecological functions. Mammalian Biology 2015, 80, 237–245.
[17] Mitsch WJ, Gosselink JG. The value of wetlands: Importance of scale and landscape setting. Ecological Economics 2000, 35, 25–33.
[18] Government of India. State of Indian agriculture 2015-16. New Delhi, 2016.
[19] Frolking S, Yeluripati JB, Douglas E. New district-level maps of rice cropping in India: A foundation for scientific input into policy assessment. Field Crops Research 2006, 98, 164–177.
[20] Edirisinghe JP, Bambaradeniya CNB. Rice fields: An ecosystem rich in biodiversity. Journal of the National Science Foundation Sri Lanka 2006, 34(2), 57–59
[21] Sundar KSG, Subramanya S. Bird use of rice fields in the Indian subcontinent. Waterbirds 2010, 33, 44–70.
[22] Bambaradeniya CNB, Fonseka KT, Ambagahawatte CL. A preliminary study of fauna and flora of a rice field in Kandy, Sri Lanka. Ceylon Journal of Science (Biological Sciences) 1998, 25, 1–22.
[23] Ponraman G, Anbalagan S, Dinakaran S. Diversity of aquatic insects in irrigated rice fields of South India with reference to mosquitoes (Diptera: Culicidae). Journal of Entomology and Zoology Studies 2016, 4, 252–256.
[24] Pierluissi S. Breeding waterbirds in rice fields: A global review. Waterbirds 2010, 33, 123–132.
[25] Pierluissi S, King SL, Kaller MD. Waterbird nest density and nest survival in rice fields of southwestern Louisiana. Waterbirds 2010, 33, 323–330.
[26] Dhindsa MS, Toor HS. Feeding ecology of three sympatric species of Indian weaverbirds in an intensively cultivated area. In: Granivorous birds in agricultural landscape. Pinowski J, Summers-Smith JD, eds. Warsaw, PWN-Polish Scientific Publishers, 1990, 217–236.
[27] Kler TK, Singh T. Potential of reflective ribbons and cob wrapping methods as birds deterrents in maturing maize. Pestology 2007, 31, 62–64.
[28] Jones GA, Sieving KE, Jacobson SK. Avian diversity and functional insectivory on north-central Florida farmlands. Conservation Biology 2005, 19, 1234–1245.
[29] Dickson JG, Segelquist CA. Breeding bird populations in pine and pine-hardwood forests in Texas. Journal of Wildlife Management 1979, 43, 549–555.
[30] Ali S. The book of Indian birds (Salim Ali centenary edition). Oxford University press, 1996.
[31] Nathan SPC, Rajendran B. Bird fauna of the rice crop ecosystem in Pondicherry region. Journal of the Bombay Natural History Society 1982, 79, 204–205.
[32] Ali S, Ripley SD. Handbook of the birds of India and Pakistan Compact edition Oxford UnivPress BNHS, MumbaiAli, SSD Ripley (1995)PictGuidto Birds Indian Subcontinent Oxford UnivPress BNHS, Mumbai Baskaran, ST (1992)Sighting Dusky Horned Owl Newsl Birdwatchers. 1983, 32, 10.
[33] Roberts TJ. The Birds of Pakistan (Volume 1): Regional Studies and Non-passeriformes. USA, Oxford University Press, 1991.

[34] Brooke RK, Crowe TM. The birds of Pakistan Volume 2 Passeriformes, 1994.
[35] Srinivasulu B, Srinivasulu C, Rao VV, Koteshwarulu C, Nagulu V. Avian use of paddy agro-ecosystem. Pavo 1997, 35, 75–84.
[36] Henry GM. A guide to the birds of Sri Lanka. Oxford University Press, 1998.
[37] Borad CK, Aeshita M, Parasharya BM. Conservation of the avian biodiversity in paddy (Oryza sativa) crop agroecosystem. Indian Journal of Agricultural Sciences 2000, 70, 378–381.
[38] Bambaradeniya CNB, Edirisinghe JP, De Silva DN, Gunatilleke CVS, Ranawana KB, Wijekoon S. Biodiversity associated with an irrigated rice agro-ecosystem in Sri Lanka. Biodiversity Conservation 2004, 13, 1715–1753.
[39] Grimmett R, Inskipp C, Inskipp T. Birds of the Indian subcontinent: India, Pakistan, Sri Lanka, Nepal, Bhutan, Bangladesh and the Maldives. Bloomsbury Publishing, 2016.
[40] Subramanya S. Non-random foraging in certain bird pests of field crops. Journal of Biosciences 1994, 19, 369–380.
[41] MacArthur RH. Geographical ecology: Patterns in the distribution of species. Princeton University Press, 1984.
[42] Pithon JA, Moles R, O'Halloran J. The influence of coniferous afforestation on lowland farmland bird communities in Ireland: Different seasons and landscape contexts. Landscape Urban Plan 2005, 71, 91–103.
[43] Morse DH. The insectivorous bird as an adaptive strategy. Annual Review of Ecology Evolution and Systematics 1971, 2, 177–200.
[44] Losey JE, Vaughan M. The economic value of ecological services provided by insects. Bioscience 2006, 56, 311–323.
[45] Kler TK, Kumar M, Dhatt JS. Study on the population of house sparrow passer domesticus in urban and rural areas of Punjab. International Journal of Advanced Research 2015, 3, 1339–1344.
[46] Kler TK. Adaptability of 'T'-perches by insectivorous birds in Bt cotton fields. Pestology 2007, 31, 52–55.
[47] Singh M. Bird composition, diversity and foraging guilds in agricultural landscapes: A case study from eastern Uttar Pradesh, India. Journal of Threatened Taxa 2021, 13, 19011–19028.
[48] Shera PS, Arora R. Biointensive integrated pest management for sustainable agriculture. In: Singh B, Arora R, Gosal SS, eds. Biological and Molecular Approaches in Pest Management, Jodhpur, India, Scientific Publishers, 373–429.
[49] Rutz C. Assessing the breeding season diet of goshawks Accipiter gentilis: Biases of plucking analysis quantified by means of continuous radio-monitoring. Journal of Zoology 2003, 259, 209–217.
[50] Gonthier DJ, Sciligo AR, Karp DS, Lu A, Garcia K, Juarez G, Chiba T, Gennet S, Kremen C. Bird services and disservices to strawberry farming in Californian agricultural landscapes. Journal of Applied Ecology 2019, 56, 1948–1959.
[51] Bharucha B, Padate GS. Assessment of beneficial role of an insectivorous bird, jungle babbler (Turdoides striatus) predation, on Helicoverpa armigera infesting pigeon pea (Cajanus cajan). Crop Acta Agronomica 2010, 59, 228–235.
[52] Sri, NR, in: Chandel, BS ed. Latest trends in zoology and entomology sciences. New Delhi, India, AkiNik publications, 2018, 145.
[53] Rey Benayas JM, Meltzer J, de Las Heras-bravo D, Cayuela L. Potential of pest regulation by insectivorous birds in Mediterranean woody crops. PLoS One 2017, 12, e0180702.
[54] Jedlicka JA, Greenberg R, Letourneau DK. Avian conservation practices strengthen ecosystem services in California vineyards. PLoS One 2011, 6, e27347.
[55] Karp DS, Mendenhall CD, Sandí RF, Chaumont N, Ehrlich PR, Hadly EA, Daily GC. Forest bolsters bird abundance, pest control and coffee yield. Ecology Letters 2013, 16, 1339–1347.

[56] Maas B, Clough Y, Tscharntke T. Bats and birds increase crop yield in tropical agroforestry landscapes. Ecology Letters 2013, 16, 1480–1487.
[57] Saunders ME, Peisley RK, Rader R, Luck GW. Pollinators, pests, and predators: Recognizing ecological trade-offs in agroecosystems. Ambio 2016, 45, 4–14.
[58] Marquis RJ, Whelan CJ. Insectivorous birds increase growth of white oak through consumption of leaf-chewing insects. Ecology 1994, 75, 2007–2014.
[59] Morrison EB, Lindell CA. Birds and bats reduce insect biomass and leaf damage in tropical forest restoration sites. Ecological Applications 2012, 22, 1526–1534.
[60] Bael S, Philpott SM, Greenberg R, Bichier P, Barber NA, Mooney KA, Gruner DS. Birds as predators in tropical agroforestry systems. Ecology 2008, 89, 928–934.
[61] Rajashekara S, Venkatesha MG. Insectivorous bird communities of diverse agro-ecosystems in the Bengaluru region. India Journal of Entomology and Zoology Studies 2014, 2, 142–155.
[62] Wielgoss A, Clough Y, Fiala B, Rumede A, Tscharntke T. A minor pest reduces yield losses by a major pest: Plant-mediated herbivore interactions in Indonesian cacao. Journal of Applied Ecology 2012, 49, 465–473.
[63] Meyer CFJ, Schwarz CJ, Fahr J. Activity patterns and habitat preferences of insectivorous bats in a West African forest–savanna mosaic. Journal of Tropical Ecology 2004, 20, 397–407.
[64] Chamberlain DE, Wilson JD, Fuller RJ. A comparison of bird populations on organic and conventional farm systems in southern Britain. Biological Conservation 1999, 88, 307–320.
[65] Wilson JD, Evans AD, Grice PV. Bird conservation and agriculture. Cambridge University Press Cambridge, 2009.
[66] Sridhara S. Vertebrate pests in agriculture scientific publishers. New Delhi, India, 2016.
[67] Gandhi T, Birds W. Animals and agriculture: Conflict and coexistence in India the Orient Blackswan. New Delhi, India, 2015.
[68] CSIR. The Wealth of India : Birds, 1948.
[69] Evenden MD. The laborers of nature: Economic ornithology and the role of birds as agents of biological pest control in North American agriculture, ca1880–1930. For Conservation History 1995, 39, 172–183.
[70] Whelan CJ, Şekercioğlu ÇH, Wenny DG. Why birds matter: From economic ornithology to ecosystem services. Journal of Ornithology 2015, 156, 227–238.
[71] Classen A, Peters MK, Ferger SW, Helbig-Bonitz M, Schmack JM, Maassen G, Schleuning M, Kalko EKV, Böhning-Gaese K, Steffan-Dewenter I. Complementary ecosystem services provided by pest predators and pollinators increase quantity and quality of coffee yields. Proceedings of the Royal Society B: Biological Sciences 2014, 281, 20133148.
[72] Luck GW, Spooner PG, Watson DM, Watson SJ, Saunders ME. Interactions between almond plantations and native ecosystems: Lessons learned from north-western Victoria. Ecological Management and Restoration 2014, 15, 4–15.
[73] Whitcomb WH. Tall Timbers Conf Ecol Anim Control Habitat Manage Proc, 1971.
[74] Hooks CRR, Pandey RR, Johnson MW. Impact of avian and arthropod predation on lepidopteran caterpillar densities and plant productivity in an ephemeral agroecosystem. Ecological Entomology 2003, 28, 522–532.
[75] Heath SK. Avian diversity, pest-reduction services, and habitat quality in an intensive temperate agricultural landscape: how effective is local biodiversity enhancement? University of California, Davis 2018.
[76] Mols CMM, Visser ME. Great tits can reduce caterpillar damage in apple orchards. Journal of Applied Ecology 2002, 39, 888–899.
[77] Odell TT. Bulletin: Number 549: The Food of Orchard Birds with Special Reference to the Pear Psylla, 1927.

[78] Kross SM, Kelsey TR, McColl CJ, Townsend JM. Field-scale habitat complexity enhances avian conservation and avian-mediated pest-control services in an intensive agricultural crop. Agriculture Ecosystems and Environment 2016, 225, 140–149.
[79] Axelsen JA, Petersen BS, Maiga IH, Niassy A, Badji K, Ouambama Z, Sonderskov M, Kooyman C. Simulation studies of Senegalese grasshopper ecosystem interactions II: The role of egg pod predators and birds. International Journal of Pest Management 2009, 55, 99–112.
[80] Bumelis KH. Niche segregation among three sympatric species of swallows in southern Ontario, 2020.
[81] Joern A. Experimental study of avian predation on coexisting grasshopper populations (Orthoptera: Acrididae) in a sandhills grassland. Oikos 1986, 46(2), 243–249.
[82] Joern A. Variable impact of avian predation on grasshopper assemblies in sandhills grassland. Oikos 1992, 64(3), 458–463.
[83] East R, Pottinger RP. Starling (Sturnus vulgaris L.) predation on grass grub (Costelytra zealandica (White), Melolonthinae) populations in Canterbury. New Zealand Journal of Agricultural Research 1975, 18, 417–452.
[84] Kross SM, Tylianakis JM, Nelson XJ. Effects of introducing threatened falcons into vineyards on abundance of passeriformes and bird damage to grapes. Conservation Biology 2012, 26, 142–149.
[85] Smith RK, Pullin AS, Stewart GB, Sutherland WJ. Effectiveness of predator removal for enhancing bird populations. Conservation Biology 2010, 24, 820–829.
[86] Martin-Albarracin VL, Amico GC, Simberloff D, Nuñez MA. Impact of non-native birds on native ecosystems: A global analysis. PLoS One 2015, 10, e0143070.
[87] Strandberg JO. Predation of cabbage looper, Trichoplusia ni, pupae by the striped earwig, Labidura riparia, and two bird species. Environmental Entomology 1981, 10, 712–715.
[88] Gopali JB, Raju T, Mannur DM, Suhas Y. Bird perches for sustainable management of pod borer, Helicoverpa armigera (Hubner) in chickpea ecosystem. Karnataka Journal of Agricultural Sciences 2009, 22, 541–543.
[89] García D, Miñarro M, Martínez-Sastre R. Birds as suppliers of pest control in cider apple orchards: Avian biodiversity drivers and insectivory effect. Agriculture Ecosystems and Environment 2018, 254, 233–243.
[90] Ramadan-Jaradi G, Ramadan-Jaradi M. Olive Groves' avifauna in Lebanon: The composition of bird species and the importance of the inter-relation olive ecosystem and bird diversity. Chief Ed ProfrKhaled HAbu-Elteen 2021, 8, 23–28.
[91] Bendell BE, Weatherhead PJ, Stewart RK. The impact of predation by red-winged blackbirds on European corn borer populations. Canadian Journal of Zoology 1981, 59, 1535–1538.
[92] Bollinger EK, Caslick JW. Red-winged Blackbird predation on northern corn rootworm beetles in field corn. Journal of Applied Ecology 1985, 22(1), 39–48.
[93] Fye RE. Temperatures near the surface of cotton leaves. Journal of Economic Entomology 1972, 65, 1209–1210.
[94] Walton WR. The True Army Worm and Its Control US Government Printing Office, 1916.
[95] Mcewen LC, Deweese LR, Schladweiler P. Bird predation on cutworms (Lepidoptera: Noctuidae) in wheat fields and chlorpyrifos effects on brain cholinesterase activity. Environmental Entomology 1986, 15, 147–151.
[96] Railsback SF, Johnson MD. Effects of land use on bird populations and pest control services on coffee farms. Proceedings of the National Academy of Sciences 2014, 111, 6109–6114.
[97] Howe N, Della Porta S, Recchia H, Funamoto A, Ross H. "This bird can't do it 'cause this bird doesn't swim in water": Sibling teaching during naturalistic home observations in early childhood. Journal of Cognition and Development 2015, 16, 314–332.

[98] Mcewen LC, Deweese LR, Schladweiler P, Stewart RE, Thurston R, Prachuabmoh O, Sinu PA, Koh LP, Orłowski G, Karg J, Karg G, Lemessa D, Hambäck PA, Hylander K. Avian pest control in tea plantations of sub-Himalayan plains of Northeast India: Mixed-species foraging flock matters. Biology Control 2011, 58, 362–366.
[99] Orłowski G, Karg J, Karg G. Functional invertebrate prey groups reflect dietary responses to phenology and farming activity and pest control services in three sympatric species of aerially foraging insectivorous birds. PLoS One 2014, 9, e114906.
[100] Stewart RE. Breeding Birds of North Dakota Tri-college center for environmental studies Fargo, ND, 1975.
[101] Smith OM, Kennedy CM, Echeverri A, Karp DS, Latimer CE, Taylor JM, Wilson-Rankin EE, Owen JP, Snyder WE. Complex landscapes stabilize farm bird communities and their expected ecosystem services. Journal of Applied Ecology 2022, 59(4), 927–941.
[102] Avery ML. Birds in Pest management. In: Pimental D, Ed. Encyclopedia of Pest management. New York, Marcel Dekker, 104-106.
[103] Garcia K, Olimpi EM, Karp DS, Gonthier DJ. The Good, the bad, and the risky: Can birds be incorporated as biological control agents into integrated pest management programs? Journal of Integrated Pest Management 2020, 11, 11.
[104] Garfinkel M, Fuka ME, Minor E, Whelan CJ. When a pest is not a pest: Birds indirectly increase defoliation but have no effect on yield of soybean crops. Ecological Applications 2022, 32(4), e2527.
[105] Pejchar L, Clough Y, Ekroos J, Nicholas KA, Olsson OLA, Ram D, Tschumi M, Smith HG. Net effects of birds in agroecosystems. Bioscience 2018, 68, 896–904.
[106] Ormerod SJ, Watkinson AR. Birds and agriculture-editor's introduction. Journal of Applied Ecology 2000, 37(5), 699–705.
[107] Martin EA, Reineking B, Seo B, Steffan-Dewenter I. Natural enemy interactions constrain pest control in complex agricultural landscapes. Proceedings of the National Academy of Sciences 2013, 110, 5534–5539.
[108] Hamisi RA, Warui CM, Njoroge P. Nesting success of sharpe's longclaw (Macronyx sharpei Jackson, 1904) around the grasslands of lake Ol'bolossat Nyandarua, Kenya. Journal of Threatened Taxa 2022, 14, 20461–20468.
[109] Kannan V, Manakadan R. The status and distribution of Spot-billed Pelican Pelecanus philippensis in southern India. Forktail 2005, 21, 9.
[110] Dhananjayan V, Jayakumar S, Muralidharan S. Assessment of exposure and effect of organochlorine pesticides on birds in India. Research Review AJ Toxicology 2011, 1, 17–23.
[111] Dhananjayan V, Ravichandran B. Organochlorine pesticide residues in foodstuffs, fish, wildlife, and human tissues from India: historical trend and contamination status. In Environmental deterioration and human health. Dordrecht, Springer, 2014, 229–262.
[112] Naoroji R. Contamination in egg shells of Himalayan grey-headed fishing eagle ichthyaetus nana plumbea in Corbett National Park, India. Journal of the Bombay Natural History Society 1997, 94, 398–400.
[113] Tela M, Cresswell W, Chapman H. Pest-removal services provided by birds on subsistence farms in south-eastern Nigeria. PLoS One 2021, 16, e0255638.
[114] Tue NM, Goto A, Fumoto M, Nakatsu S, Tanabe S, Kunisue T. Nontarget screening of organohalogen compounds in the liver of wild birds from Osaka, Japan: specific accumulation of highly chlorinated POP homologues in raptors. Environmental Science & Technology 2021, 55(13), 8691–8699.
[115] Arya AK, Singh A, Bhatt D. Pesticide applications in agriculture and their effects on birds: an overview. Contaminants. In Agriculture and Environment: Health Risks and Remediation 2019, 5(10), 129–137.

[116] Reindl AR, Falkowska L. Food source as a factor determining birds' exposure to hazardous organic pollutants and egg contamination. Marine and Freshwater Research 2019, 71, 557–568.
[117] Morelli F. Quantifying effects of spatial heterogeneity of farmlands on bird species richness by means of similarity index pairwise. International Journal of Biodiversity 2013, 2013.
[118] Langford KH, Reid M, Thomas KV. The occurrence of second generation anticoagulant rodenticides in non-target raptor species in Norway. Science of the Total Environment 2013, 450, 205–208.
[119] Grass I, Lehmann K, Thies C, Tscharntke T. Insectivorous birds disrupt biological control of cereal aphids. Ecology 2017, 98, 1583–1590.
[120] Uden DR, Allen CR, Mitchell RB, McCoy TD, Guan Q. Predicted avian responses to bioenergy development scenarios in an intensive agricultural landscape. Gcb Bioenergy 2015, 7, 717–726.
[121] Chaplin-Kramer R, O'Rourke M, Schellhorn N, Zhang W, Robinson BE, Gratton C, Rosenheim JA, Tscharntke T, Karp DS. Measuring what matters: Actionable information for conservation biocontrol in multifunctional landscapes. Frontiers in Sustainable Food Systems 2019, 60.
[122] Spurr EB, Coleman JD. Cost-effectiveness of bird repellents for crop protection. In Proceedings of the 13th Australasian Vertebrate Pest Conference, 2–6 May 2005, Wellington, New Zealand, 227–233.
[123] Hannay MB, Boulanger JR, Curtis PD, Eaton RA, Hawes BC, Leigh DK, Rossetti CA, Steensma KMM, Lindell CA. Bird species and abundances in fruit crops and implications for bird management. Crop Protection (Guildford, Surrey) 2019, 120, 43–49.
[124] Geertsema W, Rossing WAH, Landis DA, Bianchi FJJA, Van Rijn PC, Schaminée JHJ, Tscharntke T, Van Der Werf W. Actionable knowledge for ecological intensification of agriculture. Frontiers in Ecology and the Environment 2016, 14, 209–216.
[125] Giles KL, McCornack BP, Royer TA, Elliott NC. Incorporating biological control into IPM decision making. Current Opinion in Insect Science 2017, 20, 84–89.
[126] Mace KC, Mills NJ. Connecting natural enemy metrics to biological control activity for aphids in California walnuts. Biology Control 2017, 106, 16–26.

G. Venkatesh, P. Sakthi Priya, V. Anithaa, G. K. Dinesh,
S. Velmurugan, S. Abinaya, P. Karthika, P. Sivasakthivelan,
R. Soni, A. Thennarasi

Chapter 19
Role of entomopathogenic fungi in biocontrol of insect pests

Abstract: Chemical pesticides have an adverse impact on non-target organisms, and it leads to biodiversity loss, loss of food safety, development of insect resistance and resurgence in newer areas. All these have led scientists to create more ecofriendly alternatives, such as the use of entomopathogenic fungi against insect pests. Entomopathogenic fungus is a promising alternative to chemical insecticides that provides biological plant protection against insect pests in a sustainable pest control approach. Insect-infecting fungi are now classified into 90 genera and roughly 800 entomopathogenic fungal species have been documented. However, most commercial mycoinsecticides target just three genera: *Beauveria bassiana*, *Metarhizium anisopliae*, and *Isaria fumosoroseus*. They cause about 60 percent of insect diseases. These fungi are key contributors to soil insect population dynamics. Hence, entomopathogenic fungi are important biocontrol agents against insect populations. Insect-infecting fungi are found in several distinct groupings. Insect fungal pathogens include those from the phyla Chytridiomycota, Zygomycota, Oomycota, Ascomycota, and Deuteromycota, which are known to be the best entomopathogens against various insect pests. Entomopathogenic fungi kill or inactivate insects by attacking and infecting their insect hosts. Entomopathogenic fungi are soil-dwelling fungi that infect and kill insects by breaching their cuticle. Most insect-infecting fungi work through penetration. Entomopathogens produce these extracellular enzymes (protease and lipase) and toxins in their adaptive response. Together with a mechanical process via appressoria growth, these enzymes break the insect cuticle and enter the body of the insect to infect and kill it by getting their nourishment from the insect tissues. On the other hand, insects have developed many defense against these fungal pathogens. Insect pests are effectively killed by the soil fungus, *Beauveria bassiana*, and are easy to use in the field. Now mass manufacturing of new fungal formulations are possible. Further, modern genetic engineering and biotechnology approaches may assist in increasing the bioactivity of entomopathogenic fungi. This chapter discusses entomopathogenic fungi and their detailed usage description in the current scenario. It also explains the mode of infection, approaches, plans, and policies for entomopathogenic fungi.

Acknowledgment: The authors would like to acknowledge their affiliated institutes.

https://doi.org/10.1515/9783110771558-019

19.1 Introduction

Insects are omnipresent, vast in their diversity, and make up almost three-quarters of all living species. Insects make for 4–8 million of the estimated 5.57–9.8 million animal species worldwide. They are exploited in the manufacture of silk, honey, lac, medicinal entomological medications, forensics, and biocontrol agents [1]. At the same time, they are potentially responsible for damaging crops, humans, and livestock, mostly as a vector for pathogens. Pest control is advancing with modern technologies while remaining eco-friendly. This led to the employment of biological species, particularly microorganisms such as bacteria, fungi, protozoa, and viruses, generally referred to as entomopathogens, to manage such insects. Entomopathogens, rather than broad-spectrum insecticides, can be employed to control insects. They contribute to the natural control of arthropod population and are preferable to pesticides in many instances. In addition to their efficiency, using entomopathogens has several advantages. These include preservation of natural enemies, reduced pesticide residues in food, and greater biodiversity in a controlled environment [2], and can be used in various ways, including as beneficial PGPRs and biofertilizers, in addition to their traditional role as bio-insecticide [3]. Fungi, as entomopathogenic, have a wider range of hosts and can infect both above-ground and underground pests, including soil-dwelling nematodes, and are most advisable for managing the pests of the soil, whereas the bacteria and viruses are more specific to their host [4]. This is one of the reasons why entomopathogenic fungi are preferred. Various fungi are entomopathogens. They range from severe specialists with restricted host ranges to generalists with relatively broad host ranges [5]. As both living and resting spores, fungi survive in the environment and infect insects when they contact them. Entomopathogenic fungi do not kill all insects but rather maintains their population below the economic threshold level. Entomopathogenic fungi are commercially cultured on a large scale as interest in these fungi is gradually increasing among the farmers. Over 170 different strains of mycopesticides are now used in commercial applications [3]. Several research studies are also being carried out everyday in this field to ensure its efficiency and for more beneficial exploitation. This chapter will give a brief summary of entomopathogenic fungi, their classification, mode of action, formulations, and finally, the new policies and plans for importing and exporting them.

19.2 Classification of entomopathogenic fungi

Among the living organisms, fungi range between 1.5–5.1 million globally. Fungi invading the dead and living insects are classified as saprophagous and entomophagous, respectively [6]. Entomophagous fungi consist of 100 genera. Out of the 100 genera, 750–1000 species are entomopathogens [6], [7]. Recently, insect-infecting fungi have been discovered in more than 700 species and 90 genera [8–10]. Entomopathogenic are categorized into 12 groups and six phyla within the kingdom fungi [11]. In these six phyla, true fungi are categorized into four phyla, namely, Basidiomycota, Ascomycota, Zygomycota (subphylum: Entomophthoromycotina), and Chytridiomycota. Artificial phylum Deuteromycota, which includes a filamentous fungus that exists in asexual forms (anamorph) [12] and the newly discovered Glomeromycota are the only phylum in which entomopathogens are not present [13]. Most entomopathogenic fungi that are discovered belong to the Zygomycota (class Entomophtohorales) and Deuteromycota (class Hyphomycetes). Entomopathogenic fungi never form one monophyletic group. Thus, 12 Oomycetes species, 65 Chytridiomycota species, 339 Microsporidia species, 474 Entomophthoramycota species, 238 Basidiomycota species, and 476 Ascomycota species have been reported [14]. Alternaria, Cladosporium, Aspergillus, and Penicillium are the common fungi found in insect cadavers. Insect orders affected by entomopathogenic fungi include Diptera, Lepidoptera, Orthoptera, Hymenoptera, Coleoptera, and Hemiptera [15]. The classification of phylum is shown in Figure 19.1. Classification of various entomopathogenic fungi and their family are listed in Table 19.1.

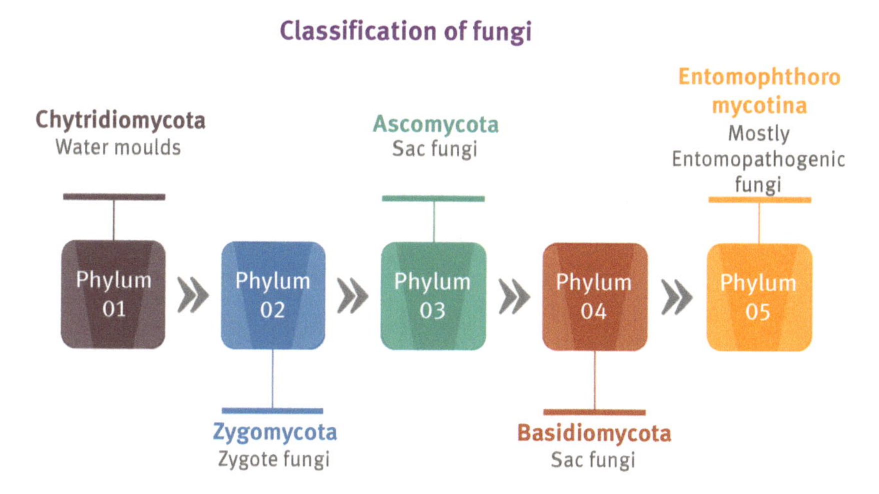

Figure 19.1: Classification of the fungal phylum.

19.2.1 Oomycota

Oomycota behave like a fungus because they have cellulose-based cell walls and plant-like biochemical properties. They are categorized as part of the kingdom Stramenophila rather than the kingdom Fungi [12]. Sexual reproduction can occur on the same or distinct hyphae between antheridia and archegonia. Lagenidiales and

Table 19.1: Classification of entomopathogenic fungi [6, 14, 19–37].

No.	Order	Family	Genus	Examples
Phylum: Chytridiomycota				
1.	Blastodadiales	Coelomomycetaceae	Coelomomyces	*Coelomomyces indicus*
2.	Chytridiales	Achlyogetonaceae	Myiophagus	*Myiophagus ucrainicus*
Phylum: Zygomycota				
1.	Mucorales	Mucoraceae	Mucor Rhizopus	*Mucor racemosus* *Rhizopus stolonifer*
Phylum: Ascomycota				
1.	Eurotiales	Trichocomaceae	Isaria	*Isaria farinosa* *Isaria fumosorosea*
2.	Onygenales	Ascosphaeraceae	Ascosphaera	*Ascosphaera apis* *Ascosphaera aggregate*
3.	Capnodiales	Capnodiaceae	Cladosporium	*Cladosporium cladosporioides*
4.	Hypocreales	Plectophaerelaceae	Verticillium	*Verticillium lecanii*
		Ophiocordycipitaceae	Hirsutella, Hymenostilbe, Tolypocladium	*Hirsutella citriformis* *Hymenostilbe dipterigena* *Tolypocladium cylindrosporum*
		Clavicipitaceae	Aschersonia, Metarhizium, Nomuraea, Nigelia	*Aschersonia aleyroids* *Metarhizium anisopliae* *Nomuraea rileyi* *Nigelia aurantiaca*
		Cordycipitaceae	Cordyceps, Beauveria, Gibellula	*Cordyceps lloydii* *Cordyceps militaris* *Beauveria brongniarti* *Beauveria bassiana* *Gibellula pulchra*
5.	Moniliales	Tuberculariaceae	Fusarium	*Fusarium coccophilum*

Table 19.1 (continued)

No.	Order	Family	Genus	Examples
Phylum: Basidiomycota				
1.	Septobasidiales	Septobasidiaceae	Septobasidium	*Septobasidium bogoriense*
Phylum: Entomophthoromycotina				
1.	Entomophthorales	Entomophthoraceae	Erynia, Entomophaga, Entomophthora, Batkoa, Furia, Massospora, Zoophthora	*Erynia aquatic* *Entomophaga grylli* *Entomophthora muscae* *Batkoa apiculate* *Furia americana* *Massospora cicadina* *Zoophthora radicans*
		Ancylistaceae	Conidiobolus, Pandora	*Conidiobolus coronatus* *Pandora blunckii*
		Neozygitaceae	Neozygites	*Neozygites floridana*

Saprolegniales contain entomophagous fungi. The genus Lagenidium consists of entomopathogenic fungi under the order Lagenidiales. The pathogen, *Lagenidium giganteum*, which affects mosquito larvae, has been investigated extensively [16, 17]. Crabs and aquatic crustaceans are also poisoned by several Lagenidium species [18]. These mosquito larvae are also infected by the Saproleginales species [19].

Domain: Eukarya
Kingdom: Fungi
Phylum: 5

19.2.2 Zygomycota

Zygomycota include multicellular, non-septate hyphae. The merging of gametangia forms zygospores. Molecular investigations have not revealed Zygomycota to be monophyletic [38]. EPF is categorized into two classes, Trichomycetes and Zygomycetes. Some trichomycetes EPF species can infect water insects such as mosquito larvae [39, 40]. Mucorales and Entomophthorales consist of EPF. Entomophthorales are major pathogens of both epigeal and soil-inhabiting insects. Around 200 species of entomogenous fungi have been reported in Entomophthorales [16], and Mucorales are a specific type of pathogens that infect only the weak insects.

19.2.3 Ascomycota

Ascomycota are septate haploid hyphae, along with yeasts. Sexual reproduction occurs by fusing modified hyphae or yeast-like cells, which leads to the formation of asci, which produce ascospores in groups of eight [41]. Cordyceps, the best-known ascomycete, contain over 300 insect-infecting species. The genus Ascosphaera has sexual dimorphism that causes chalkbrood disease in bees. The most common insect-infecting genera are *Beauveria, Metarhizium, Hirsutella, Paecilomyces, Aschersonia, Culicinomyces, Sorosporella, Lecanicillium*, and *Tolypocladium*. Biological or molecular investigations demonstrating the genetic link between teleomorphs and anamorphs can be utilized to confirm that these genera are related to one or more other genera [6, 42].

19.2.4 Basidiomycota

Basidiomycota includes fungi with dolipore septate hyphae as well as yeasts [41]. Hyphae generate sexual reproductive cells or basidia. Basidiospores are generated during nuclear fusion and meiosis on each basidium, commonly in groups of four [43]. Asexual reproduction is found apparently in a few species that produce conidia [41]. Probably, few Basidiomycetes have been discovered to be insect pathogens. The genera, Uredinella and Septobasidiales, have been described by certain scientists. Septobasidium is an infection-causing agent in insects, although it has a symbiotic connection with insects, namely, scales [40].

19.2.5 Chytridiomycota

Chytridiomycetes are categorized by their sexual reproduction system, including the unification of motile gametes and the synthesis of asexual, uniflagellate propagative spores [41]. EPF species are reported in Blastocladiales and Chytridiales. These are significant entomopathogens of aquatic insects. In Blastocladiales, the genera Coelomomyces is made up of more than 70 insect pathogenic species [44]. Coelomomyces cause infection to insects, including mosquitoes, black flies, midges, and backswimmers. In Chytridiales, the genus Myiophagus is pathogenically found on dipteran pupae with a specific affinity for insect-protective scale [40]. Various entomopathogenic fungi and their insect hosts are listed in Table 19.2.

Table 19.2: An overview of entomopathogenic fungal genera along with instances of insect hosts.

Fungus	Features	Target pest	References
Beauveria bassiana	Conidia are globous or broadly ellipsoid, ≤ 3.5 μm in diameter.	Coleoptera (Lepidoptera, Scarabaeidae, Castniidae, Curculionidae) Hymenoptera (Formicidae), Diptera (Tipulidae), Hemiptera (Lygaeidae, Cercopidae, Cicadellidae, Aleyrodidae, Aphididae, Pseudococcidae, Psyllidae) Lepidoptera (Noctuidae) Thysanoptera (Thripidae)	[41, 45]
Beauveria amorpha	Conidia is a curved or short cylinder with a flattened shape, and its dimensions are 3.5–5 × 1.5–2.0 μm.	Homoptera, Coleoptera, Hymenoptera, Lepidoptera	[46, 47]
Beauveria caledonica	Conidia is cylindrical or ellipsoidal and its size 3.7–5.2 × 1.9–2.3 μm	Coleoptera	
Beauveria asiatica	Conidia is oblong or ellipsoidal and its size 2.5–4 × 2 – 3 μm	Coleoptera: Cerambycidae Scarabaeidae	
Beauveria australis	Conidia is Sub-globose, moderately ellipsoid or ellipsoid, and globose are less common and its size 2–2.5 × 1.5–2.5 μm	Orthoptera: Acrididae	
Beauveria brongniartii	Conidia is ellipsoidal to sub-cylindrical and its size 2.5–4.5 μm.	Coleoptera: Scarabaeidae, Cerambycidae, European cockchafer, and other scarab beetle species	[48–49]
Beauveria kipukae	Conidia is globose or rarely ellipsoid, and its dimensions are 2–3.5 × 1.5–3 μm	Homoptera	[46, 47]
Beauveria malawiensis	Cylindrical and its dimensions 3.7–4.5 × 1.3–1.9 μm	Coleoptera: Scarabaeidae	
Beauveria sungii	Oblong or ellipsoidal and measures 2.5–3.5 × 1.5–2.5 μm	Coleoptera	
Beauveria varroae	Globose or broadly ellipsoid and measures 2–3.5 × 2–3 μm	Coleoptera: Curculionidae Acari: Varroidae	

Table 19.2 (continued)

Fungus	Features	Target pest	References
Beauveria lii	Cylindrical to ellipsoidal, occasionally obovoid, and its measures 3.1–10.1 × 1.4–3.6 µm	Coleoptera: Coccinellidae	[50]
Beauveria sinensis	Ellipsoidal to cylindrical and its measures 3–5 × 1.5–2 µm	Lepidoptera: Geometridae	[51]
Beauveria hoplocheli	Cylindrical to sub cylindrical and 3.5–4.5 × 1.5–2.5 µm	Coleoptera: Melolonthidae	
Beauveria rudraprayagi	Globose to sub-globose and its measures 2.5–4.0 × 2.5–4.0 µm	Skill worm	[55]
Metarhizium anisopliae	Conidia cells are cylinder and 9 µm length	Isoptera: Kalotermitidae, Rhinotermitidae, Termopsidae) Hemiptera: Aleyrodidae, Aphididae) Coleoptera: Curculionidae, Scarabaeidae), Hemiptera Blattodea: (Blattellidae, Blattidae), Thrips, fruit flies, mealybugs Orthoptera (Acrididae), Aphids, Sugarcane spittle bug	[45, 49, 53]
Metarhizium flavoviride	Conidia are ellipsoid greyish-green in masses, 7–11 µm in length.	Scarab larvae, Termites, Red-headed cockchafer	[45, 54–56]
Metarhizium chaiyaphumense	Perithecia are pyriform to oviform, and 320–380 µm wide, 550–670 µm long, Ascospores are hyaloid, filiform	Adult cicadas of the genus Platypleura.	[60]
Verticillium fusiporum	Fusoid conidia	Scale insects, aphids, and other insects	
Verticillium lecanii	Conidia ovoid to cylindrical with rounded apices 2–10 µm long and usually 1–1.7 µm wide.	Cabbage aphid, Mustard Aphid	[58]
Lecanicillium longisporum		Aphids	[45, 54–56]
Lecanicillium muscarium		Thrips and Whiteflies	
Lecanicillium lecanii		Whiteflies and thrips	[49]

Table 19.2 (continued)

Fungus	Features	Target pest	References
Isaria farinosa	Conidia are short fusoid to lemon-shaped ≤ 3 µm long, Synnemata present	Apple moth, Siberian pine caterpillar, and larch caterpillar	[59]
Isaria fumosorosea	Synnemata are usually smooth, uncolored long ovoid, with conidiophores and phialides, ≤ 4 µm rosy to tan to smoky pink in mas	Whiteflies, aphids, thrips, psyllids, mealybugs, and fungus gnats	
Isaria lilacinus	Conidia are fusoid to ellipsoid and 2–3 µm long	Wax moth	[57]
Hirsutella thompsonii	Conidia globose with a wrinkled or smooth but no visible slime layer	Mites, Acari	[45, 54–56]
Hirsutella citriformis	Synnemata are usually long, with numerous browns or grey with many short lateral branches.	Leaf and planthoppers, Hemiptera	[57]
Hirsutella rhosiliensis	Conidiophores cells are conidia with short, narrow neck and orange segments straight on one side and curved on the other or ellipsoid. Not forming synnemata	Mites	
Cordyceps militaris	Stromata with a swollen fertile section at the tip, < 10 cm high, densely clavate, and unbranched orange.	Lepidopterans	
Cordyceps lloydii	Stromata, white to cream-colored, ≤ 1 cm tall, with the discoid apical fertile region and slightly immersed perithecia.	Ants	
Cordyceps mrciensis	Stromata black to dark brown, < 8 cm tall with an elongated and slightly enlarged fertile part.	Spiders	
Cordyceps tuberculate	Perithecia are slightly saturated sulfur to intense yellow and dispersed towards apices, Stromata off-white	Lepidoptera	

Table 19.2 (continued)

Fungus	Features	Target pest	References
Cordyceps takaomontana	Fruit bodies are tubular or clavate, pale yellow, and shaped from pseudo sclerotium and 1–5 cm long, diameter up to 3.5 mm	Lepidoptera	
Entomophaga aulicae	Conidiogenous cells are non-elongated with narrow-necked subtending conidia.	Lepidoptera	
Entomophaga maimaiga	Conidia are obclavate or pyriform, hyaline, 16–28 μm width, and 20–36 μm length.	Gypsy moths	
Entomophaga calopteni	Formation of resting spores but no primary conidia	Melanopline (grasshopper)	
Entomophaga grylli	Conidiophores and conidia are formed by hyphae having cell walls.	Diverse acridids	
Entomophthora culicis	Binucleate Conidia	Black flies and mosquitoes	
Entomophthora muscae	Primary conidia and minute secondary conidia are present.	Muscoid flies	
Entomophthora planchoniana	Primary conidia are bell-shaped plurinucleate with a sharp-pointed tip and flattened papilla. Conidia dimensions are 15–20 to 12–16 μm.	Aphids	
Erynia aquatic	Conidia is clavate and measuring 30–40 × 15–18 μm	Diptera	
E. rhizospora	Conidia are straight to lunate and measure 30–40 × 8–10 μm in length	Trichoptera	
E. conica	Conidia are curved, tapering to a sub-acute apex and measuring 30–80 × 12–15 μm in length.	Diptera	
E. ovispora	Conidia are ovoid to ellipsoid and measures 23–30 × 12–14 μm.	Nematoceran dipterans	

Table 19.2 (continued)

Fungus	Features	Target pest	References
Cladosporium cladosporioides	Conidiophores brownish color, irregularly branched, 40–350 µm long 2–6 µm wide.	Aphids	
Septobasidium bogoriense	Basidiomata are branches and measure between 1–5 cm wide, 2–15 cm long, whitish-grey to greyish-brown	Scale insects	
Batkoa apiculata	Papilla often with pointed exterior and Conidia 30–40 µm diameter.	Homopterans and flies	
Tolypocladium cylindrosporum	Conidia are cylindrical or straight, or slightly curved.	Small Dipterans	
Ascosphaera aggregate	Ascospores 4–7 µm long ovoid to cylindrical.	Leaf-cutting bees	
Fusarium coccophilum	Macroconidia with transverse septa	Scale and other insects	
Aschersonia aleyroids	Stroma 2 × 2 mm tall. Thin halo hyphae spread throughout the leaf surface, ranging from orange to pink to cream in color.	Coccicds & aleyroidids	
Gibellula pulchra	Synnemata orange-yellow, Conidiophores are bulging from the surface of white lilac	Spiders	
Hymenostilbe furcate	Creamy white synnemata, conidiogenous cells two to seven forked denticles	Hemipteran insects,	
Nomuraea rileyi	Conidia are ovoid, Conidial mass grey-green covering a host	Lepidoptera	
Paecilomyces fumosoroseus		Mustard aphids, Diamondback moth	[45, 54–56]
Peacilomyces lilcinus		Mustard Aphids	

Table 19.2 (continued)

Fungus	Features	Target pest	References
Nigelia aurantiaca	Stromata up righted was cylindrical to club-shaped, yellowish or brownish. Perithecia submerged, wobbly, or closely composed, 320–440 µm wide, 520–680 µm long.	Lepidoptera	[57]
Conidiobolus thromboides	Conidia is pyriform, papilla emerges into spore, No capilliconidida or microconidia.	Aphids/Thrips/ Whiteflies	
Furia Americana	Conidia is obovoid and measures 28–35 × 14–16 µm.	Cyclorrhaphan muscoid flies	
Massospora cicadina	Conidia 1–6 nucleate	17-year Cicada	
Neozygites floridana	Zygospores sub globose dark brown, and Conidia are 10–14 µm diameter	Tetranychid mites	
Pandora blunckii	Conidia pyriform and measuring 15–20 × 7–11 µm	Lepidoptera	
Pandora dephacis	Conidia is clavate and measures 30–35 × 12–18 µm.	Hemiptera	
Zoophthora phytonomi	Conidia cylindrical, resting spores colorless	Coleoptera on alfalfa	
Coelomomyces indicus	Sporangia are anastomosing and measuring 25–65 × 30–40 µm	Mosquitoes	
Myiophagus ucrainicus	Zoospores are rarely present. Sporangia is 20–30 µm diameter and golden brown globose reticulated surface.	Beetle larvae, Scale insects, Mealybugs, Weevils, and Lepidoptera.	

19.3 Mode of action

Insect-pathogenic fungi must overcome various host challenges to produce sufficient new infectious spores in every generation to maintain a healthy population. Unlike other fungi, entomopathogens may infect their hosts directly through the exoskeleton or cuticle, where the non-feeding stages, such as eggs and pupae, can also be affected. High humidity encourages germination; so invasion occurs more readily between the mouthparts and the intersegmental folds and through the cuticle, when unsclerotized [60], [61]. The cuticle is the first point of contact between the fungus and insect. The insect dies through various causes due to this invasion, including tissue injury, nutritional loss, mycoses, and bodily toxins. Entomopathogenic fungi have evolved host surface attachment and recognition mechanisms. The adaptive reactions in EPF include the production of infectious structures such as appressoria or penetration tubes, secondary metabolites, hydrolytic, assimilatory, and detoxifying enzymes. Various steps involved in EPF infection are depicted in Figure 19.2.

Figure 19.2: Steps involved in the infection process of entomopathogenic fungi.

On the other hand, insects have evolved several defense mechanisms against illnesses, including creating epicuticular antimicrobial lipids, proteins, and metabolites, cuticle shedding during growth, induced fever, burrowing, and grooming, which are examples of biochemical-environmental adaptations. Various defense mechanisms of insects against EPF are shown in Figure 19.3. These features help insects prevent infections from penetrating the cuticle [65]. There is a coevolutionary arms race between pathogens and the target insects. Current research implies that the cuticular surface plays a role in the pathogen-host coevolutionary arms race. Surface interactions cause the infection to produce Mycosis or the host to protect itself [63]. The life cycle of entomopathogenic fungi in insects is shown in Figure 19.4.

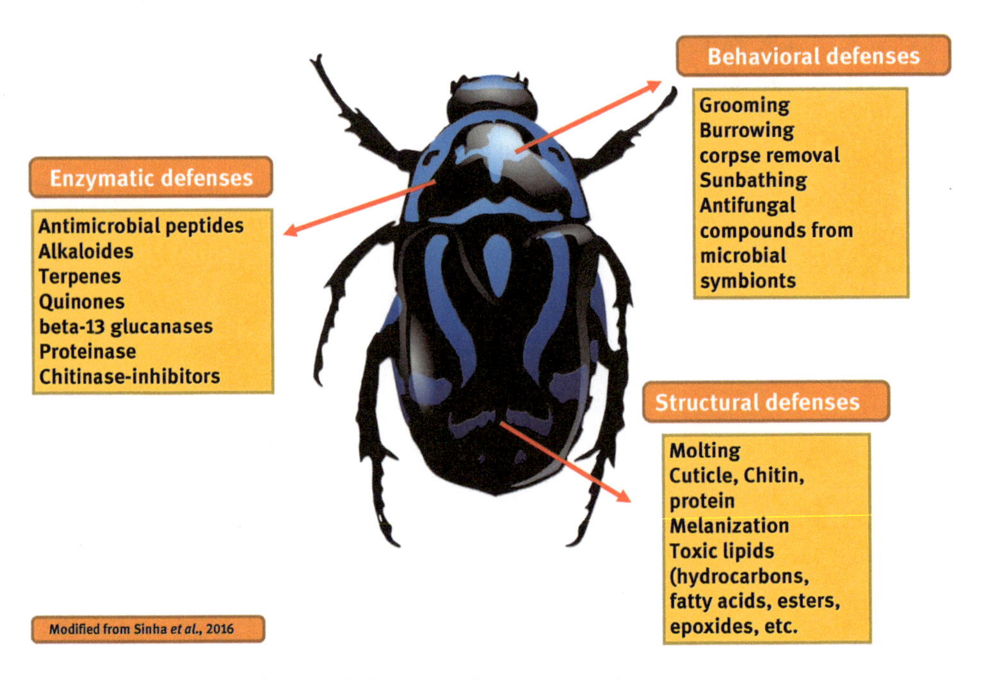

Figure 19.3: Defense mechanisms in insects against entomopathogens.

19.3.1 Adhesion and germination of conidia

Wind or water might aid attachment, which is a passive technique. Mycosis begins with spores (conidia or blastospores) attaching to the cuticle surface of a susceptible host. However, certain insects have preferred spots to enter the host's cuticle [64]. Insect cuticle are complex structures that change their composition with time. Epicuticle is the cuticle's outermost layer, followed by the procuticle, which is further divided into Exo, meso, and endo-cuticular layers. Finally, the epidermis surrounds the inner structures and is the innermost layer.

A layer of interwoven fascicles of hydrophobic rodlets made up of protein hydrophobins was identified in the dry spores of *B. bassiana* [65]. This rodlet layer is only seen in conidial cells and not vegetative cells. Rodlets exert non-specific hydrophobic forces on the cuticle, causing dry spore adhesion [66]. In *B. bassiana*, two hydrophobins (Hyd1 and Hyd2) influence the rodlet layer formation, contributing to adhesion to hydrophobic surfaces, pathogenicity, and cell-surface hydrophobicity [65, 67]. *M. anisopliae* has Mad1 and Mad2 adhesion genes [68]. Mad1 deficiency reduced cuticle adhesion, germination, blastospore production, and pathogenicity.

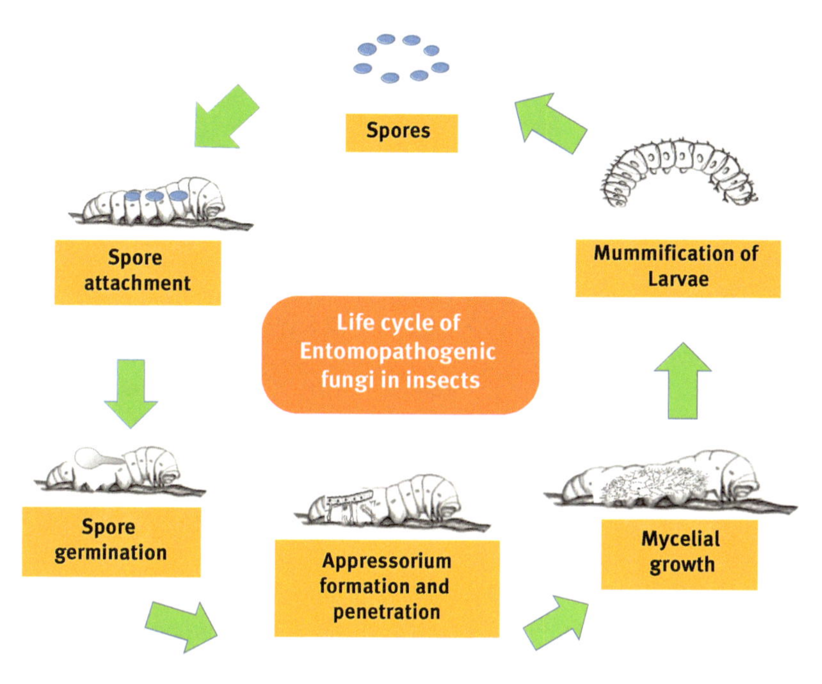

Figure 19.4: Life cycle of entomopathogenic fungi in insects.

The non-specific (passive) adsorption stage is attributed to hydrophobins, whereas the target-specific (active) stage is attributed to Mad2 adhesions [63]. Once a pathogen penetrates and attaches to the host cuticle, it germinates, and the availability of nutrients, oxygen, water, pH, temperature, and the impact of toxic host surface chemicals all play a role in its continuing growth. Fungi, with extensive host ranges, germinate in cultures in response to a wide range of non-specific nitrogen and carbon sources. [69] On the other hand, entomopathogenic fungi with a limited host range appear to have particular germination demands. For instance, *B. bassiana* has been proven to thrive on insect hydrocarbons, including methyl-branched and aliphatic alkanes [70].

19.3.2 Formation of infectious structure

The fungus infects the host by penetrating the cuticle. The epicuticle (outer cuticle) lacks chitin but it is structurally complex. It includes phenol-stabilizing proteins and coated by a waxy coating of fatty acids, lipids, and sterols [71]. The procuticle (inner cuticle) contains most of the cuticle's components. A protein matrix holds chitin fibrils, lipids, and quinones. For the fungus to penetrate the cuticle, it has to form a germ tube or an appressorium [71]. This appressorium formation focuses its physical and chemical energy across a narrow region and ensures efficient entry.

The host's surface topography and intracellular second messengers, such as Ca2 + and cAMP, impact appressorium formation [65].

19.3.3 Penetration of host cuticle

Entomopathogenic fungi penetrates the insect body through the cuticle to feed on nutrition. Fungal invasion requires both mechanical and enzymatic pressure. The site in the epicuticle appears as a black, melanotic lesion after penetration [72]. Proteins are an insect's recyclable resource and an important cuticle component. Proteases, lipases, and chitinase degrade the cuticle during the entomopathogenic fungi penetration. Proteases include trypsin, chymotrypsin, esterase, collagenase, and chymoelastase [73]; [74]. Endoproteases (PR1 and PR2) and aminopeptidase are the first cuticle enzymes that are generated and generally associated with appressoria formation [75]. Active endoprotease in the fungi helps to penetrate the cuticle, as protein constitutes up to 70% of the insect cuticle.

Since the insect cuticle is complete and is among the enzymes involved in the cuticle penetration, N-acetyl glucosaminidase is produced more slowly than proteolytic enzymes. Hence, it requires the coordinated activity of several enzymes to penetrate the cuticle. In *B. bassiana*, cuticular lipid breakdown involves eight CYP genes, four catalases, three lipases/esterase, long-chain alcohol and aldehyde dehydrogenases, and a putative hydrocarbon transporter protein [76,77]. Temperature, humidity, and light also affect insect cuticle adherence, germination, development, and penetration. Insect hemocoel is immediately penetrated by the fungus via the cuticle, employing extracellular enzymes (chitinases, lipases, esterase, and numerous proteases), and many enzymes work together to penetrate the cuticle. The fungus then disperses into the hemolymph through blastospores or a yeast-like structure, reaching the insect's respiratory system for optimal feeding. Finally, the insect dies via a combination of causes, including mechanical harm from tissue invasion, nutrition loss, and toxin formation (toxicosis) [62].

19.3.4 Production of toxins

Entomopathogenic fungi produce major toxins such as destruxins, efrapetins, beauvericin, leucinostatines, and bassianolide. Various toxins produced by entomopathogenic fungi are described in Figure 19.5.

Deuteromycetes pathogens create a variety of fungal toxins associated with insect health. The effects of these toxins on diverse insect tissues have been shown (Table 19.1). Cytotoxins may be involved in the breakdown of cells prior to hyphae entry. Neuromuscular toxins cause paralysis, sluggishness, and reduced excitability in insects infected with fungi, according to [78]. *B.bassiana*

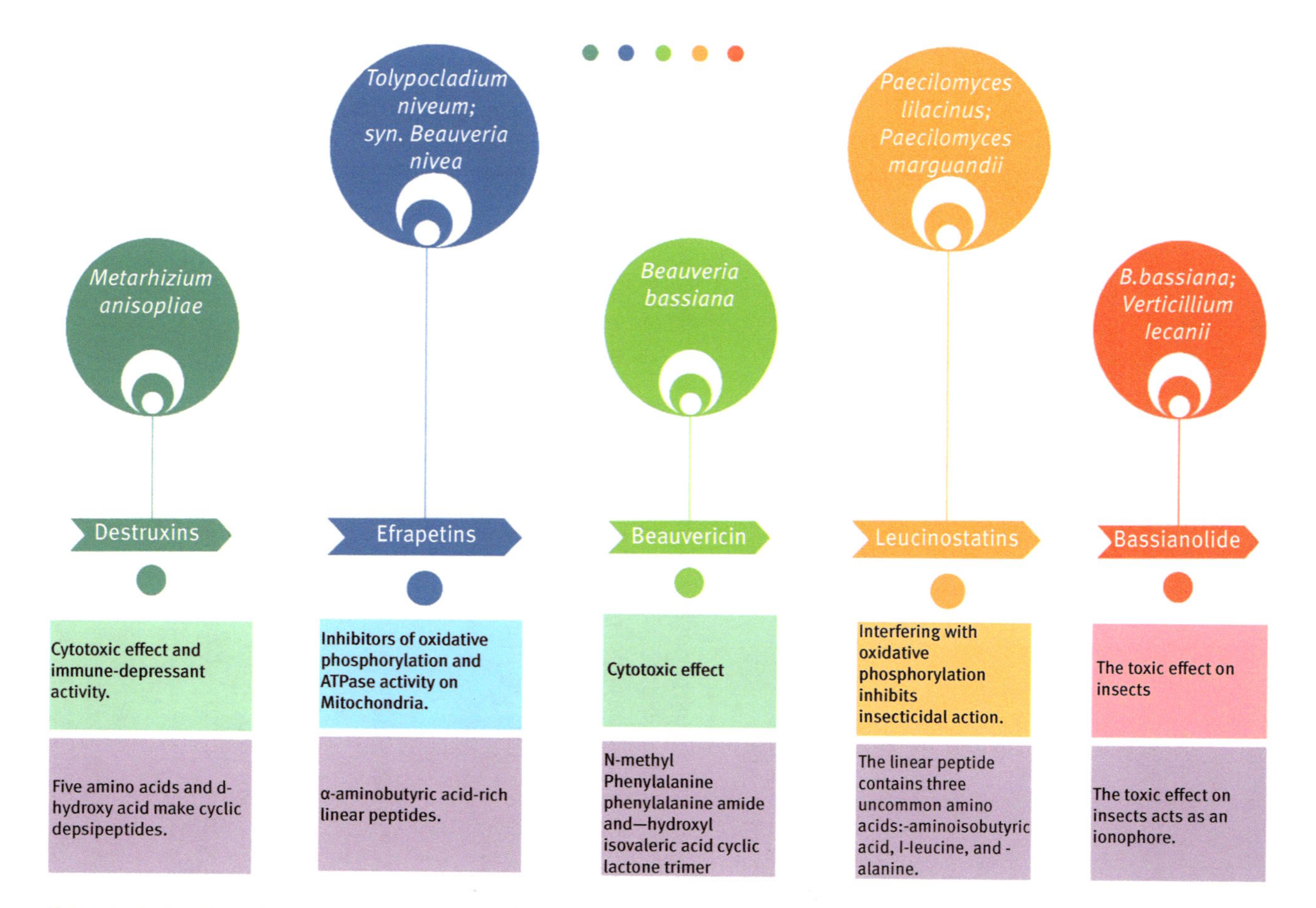

Figure 19.5: Toxins produced by entomopathogenic fungi [62].

toxins include beauvericin, beauverolides, bassianolide, and isarolides [79]. *Metarhizium anisopliae* infestation forces insects to produce destruxins and cytochalasins. DTX depolarizes lepidopteran muscle membranes and changes hemocyte function by activating calcium channels [80].

19.3.4.1 Destruxins

Destruxins are cyclic depsipeptides that comprise five amino acids and a D-hydroxyl acid. It was first found in *M.anisopliae* [81–83]. They are produced after the mycelial growth and suppress the insect's immune system. In addition, Destruxins may help the pathogen to establish itself in the host. These poisons kill insects after a fungus infestation [84].

19.3.4.2 Efrapeptins

Soil hyphomycetes, *Tolypocladium niveum* and *Beauveria nivea*, produce efrapeptins, a complex peptide antibiotic. Colorado potato beetle *Leptinotarsa decemlineata* (Coleoptera) is poisoned by efrapeptins. All peptides are strong mitochondrial oxidative phosphorylators and ATPase inhibitors when tested against entomopathogenic fungi (*M.anisopliae* and *T.niveum*) preparations. These peptides are presumably catalytic-site competitive inhibitors that bind to the soluble F′ component of the mitochondrial ATPase [85].

19.3.4.3 Bassianolide

Bassionolide is an ionophore of divalent cation. It is cytotoxic and insecticidal against mosquito larvae, blowflies, and the Colorado potato beetle. It has been found in *B. bassiana*, *P. fumosoroseus*, *Fusarium semitectum*, *Fusarium moniliforme* var. *subglutinans*, and *Polyporus sulphureus*, a plant pathogenic basidiomycetous fungi [62].

19.4 Alternate modes of action by entomopathogenic fungi

Insect infection by microbial ingestion is typical when the entomopathogen is a virus, bacterium, or protozoa; however, it has been reported that entomopathogenic fungi can utilize oral and respiratory routes as an alternative to cuticle penetration for their entry [88–91]. These strategies may provide an opportunity to boost efficiency against fungal-resistant arthropods by embedding antifungal compounds in their cuticle [92], [76], [93]. The first reports on entomopathogenic fungus infection pathways were published [94–96]. After several years, with no new understanding, researchers began to look into the molecular processes behind these infection pathways. Next-generation technology enabled the collection of vast volumes of genomic and transcriptome data from fungus and arthropods [90]. Alternative routes of entry of entomopathogenic fungus into the host are depicted in Figure 19.6.

19.4.1 Oral infection route by entomopathogenic fungi in terrestrial insects

Researchers have long been intrigued by the idea of entomopathogenic fungi invading the oral/gastrointestinal tract; however, much remains unknown. A century ago, scientists advocated that *B.bassiana* infect pine weevils' through the oral cavity [96]. From the mid-1940s to the mid-1980s, a discontinued study series found *M. anisopliae* hyphae surrounding the implantation of the mandibles and oesophagus of *Ephestia kuhniella* Zeller (Lepidoptera: Pyralidae) and non-germinated *M. anisopliae* spores in the stomach of *Oryctes* larvae. Scientists concluded that oral infection was frequent on maxillary palps and the head of *Schistocerca gregaria* by feeding on leaves infected with *M. anisopliae* spores [95]. Some researchers reported substantial mortality and hyphal development in all regions of the digestive tube and no apparent evidence of germination in the stomach, suggesting fungal conidia invade through the beetle mouthparts [88].

More evidence that fungal conidia adhere to, germinate on, and enter through the buccal apparatus to kill insects, rather than penetrating the gut that is protected by its microbiota, comes from *S. gregaria* fed *M. anisopliae* conidia [78]. In recent research on *Sitophilus granarius* Linnaeus (Coleoptera: Curculionidae), it was found that conidia of *B. bassiana* and *M. anisopliae* could infect the insects' digestive tube cuticles. They were disinfected to prevent cuticular breaching and killing of the beetle. However, this study found inadequate histological data to support the stated idea [91].

According to research on terrestrial insects, the mechanisms by which they swallowed spores and kill the host remain unknown. Because conidia do not germinate in the stomach, it appears that fungal spores preferentially adhere to the buccal cavity rather than the digestive system. Furthermore, studies are needed to better understand the physiological and molecular changes that occur when entomopathogenic fungus spores are eaten [87].

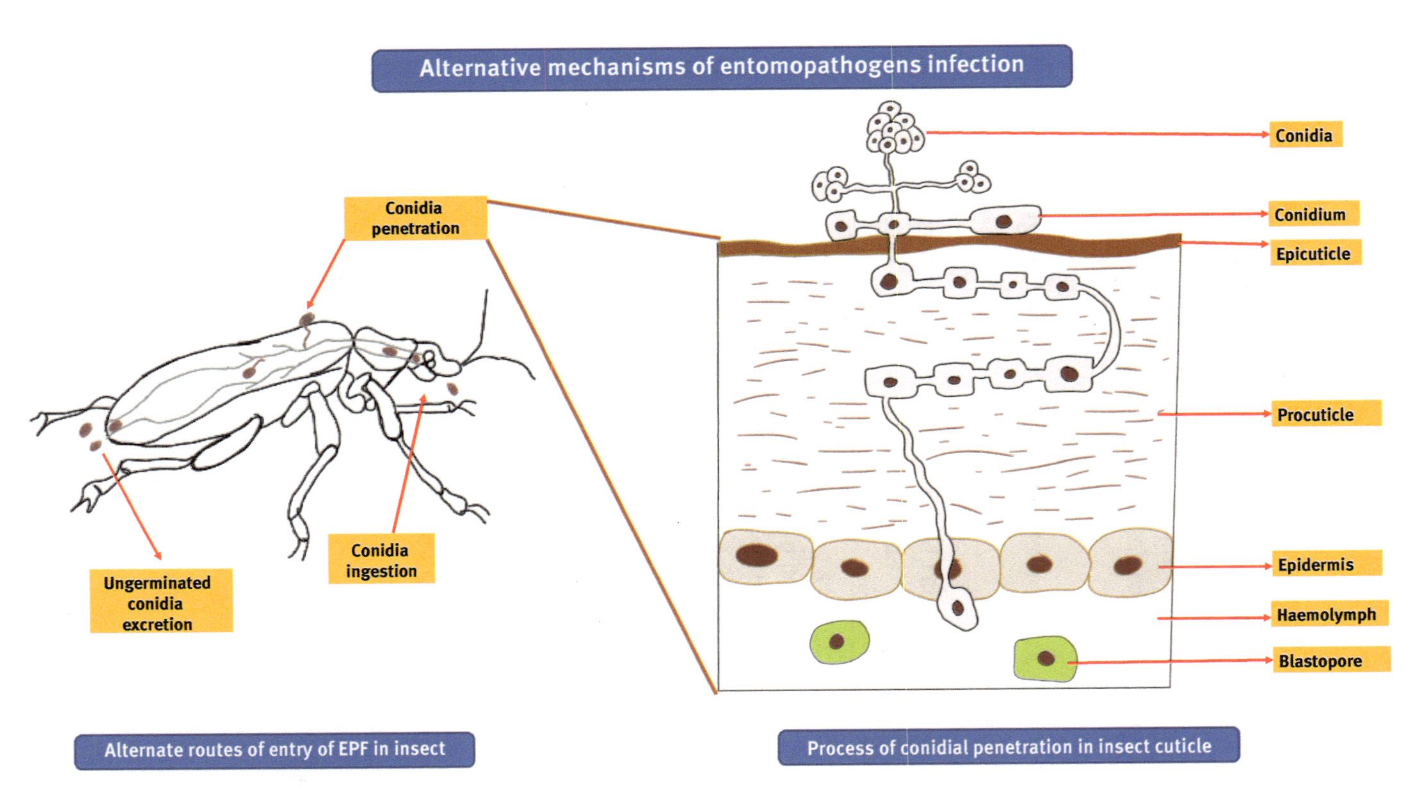

Figure 19.6: Alternative routes of entry of entomopathogenic fungus into the host.

19.4.2 Oral infection route by entomopathogenic fungi in aquatic insects

Infection by entomopathogenic fungi that have adapted to terrestrial hosts appears to be possible in mosquito larvae. While entomopathogenic fungi have pathogenic traits that kill hosts, they do not infect as actual aquatic fungal infections [97]. Conidia of *M. anisopliae* have a hydrophobic surface and float. Conidia connect to the siphon tip when the larvae open their perispiracular valves for air intake; hyphae develop into the trachea and suffocate the insect. Lesser conidia can be given to larvae after treating a non-ionic deterrent. Conidia live in insect's guts and kill them with poisons, but they do not infect the rest of the animal [94].

Because conidia are indigestible for larvae and occupy the digestive tract, nutritional support is essential in this last entry route [98]. Ingestion of conidia happens; however, how the infection develops beyond the entrance point differs, depending on the host and pathogen. However, *M. anisopliae* conidia in the gut of *S.gregaria* showed fungitoxicity, which is dependent on the gut flora [78, 99]. The head and anal areas were the most common infection sites for *Aedes aegypti* Linnaeus (Diptera: Culicidae) when administered with *B. bassiana* conidia [100]. Insect mortality appears to be connected with autolysis that is induced by caspases, which are protease enzymes implicated in apoptotic processes. Some scientists reported that the infected larvae's Hsp70 modulates the protease inhibitor mechanism [101]. Aquatic larvae of culicid dipterans have anal papillae that adhere to the anal papillae, rather than the cuticle. This fungus may infect the host through the intestinal cavity, but does not adhere to the cuticle [102], [103]. Few researchers tested the insecticidal activity of *Culicinomyces clavisporus* (Hypocreales: Cordycipitaceae) isolates on *Aedes aegypti* larvae, eggs, and adults[104]. They observed that they killed faster at a lower dose after several serial repassages of *Aedes aegypti* larvae [105].

19.4.3 Molecular evidence for oral infection

Researchers revealed that some entomopathogenic fungi contain a repertoire of genes that allow them to have oral toxicity. For example, *B. bassiana* has at least 13 heat-labile bacteria-like enterotoxins, whereas *M. robertsii* has six. Additionally, *B. bassiana* has eight Cry-like delta enterotoxins and three bacteria-like zeta toxins proteins; the rest of the entomopathogenic fungi have one [67, 106, 107]. Various genes that are involved in the oral infection of Entomopathogenic fungi are mentioned in Table 19.3.

Table 19.3: Genes of entomopathogenic fungi involved in the oral infection [106].

Gene family	No. of genes in EPF		Description of genes
	Metarhizium anisopliae	***Beauveria bassiana***	
Zeta toxins, bacterial-like	0	3	Bacterial heat-labile enterotoxin IIB, A chain (enzymatic), and IIA A
Cry-like delta enterotoxins	0	8	Bacterial delta endotoxin, N-terminal
Heat labile bacterial-like Toxins	6	13	Bacterial toxin

19.5 Mass production

A large load of inoculum is required to manage insect pests at the field level. It is not feasible in natural settings because the inoculum is less in fields. So, the EPF should be manufactured in large quantities in vitro and applied in the field to control insect pests. The success of deploying entomopathogenic fungi in pest control is dependent on reliable and cheap methods for mass production.

19.5.1 Methods of mass production

Different fungi have diverged nutrient requirements. The host range of fungi that belong to the class oomycetes, zygomycetes, and chytridiomycetes is very narrow and are specific in their nutrient specificity. On the other hand, fungi that belong to the class ascomycetes enjoy an extensive host range so that the fungi can be grown in various substrates; a complex mixture of nutrients is best suited. The major concern is to mass-produce entomopathogenic fungi cheaply without compromising their efficacy [108].

19.5.1.1 Solid substrate fermentation

This is the primary production method with no free moisture, which simulates the natural conditions and results in aerial conidia production by the fungi [109, 110]. The other name, the solid-state fermentation, is used interchangeably. Most commonly used media include rice, wheat bran, cracked barley, millets, corn, rye, sorghum, and peat soil [62, 109]. Sugarcane bagasse has scope to be used as a supporting matrix to boost growth and sporulation [111]. The media should be moistened to the optimum level to support fungal growth. Take the moistened media in an autoclavable polyethylene cover and autoclave at 121 °C at 15 psi for 15–20 min. Inoculate the media with seedling inoculum prepared in an agar plate. Transfer it to an Erlenmeyer flask containing 100 mL liquid and it serves as the starting material for solid substrate fermentation. The fungi-inoculated polyethylene bag should be incubated at 25 °C and 12 h light: 12 h dark photoperiod for 10–14 days. Then, dry the culture media until the moisture level comes down to 5% [62, 112]. This method is labor-intensive, requires a long incubation period, and may involve inevitable contaminations. [110].

19.5.1.2 Liquid fermentation

Liquid fermentation is of two types, namely, submerged fermentation and stationary liquid fermentation. In submerged fermentation, continuous agitation and aeration are provided to the media, favoring blastospore formation, micro-cycle conidia, and microsclerotia. Submerged fermentation is a commonly deployed method to mass-produce fungi such as *Beauveria, Metarhizium, Isaria,* and *Lecanicillium.* However, Conidia is not possible in this method for obtaining 100% blastospores or unblended micro-cycle. The second method is stationary liquid fermentation, in which fungi grow and conidiate on the standstill liquid media, in which mycelia and conidia are formed. In large-scale production, such as in industries, the fermentation is accomplished in a fermenter [109]. This method has the added advantages such as automation of the process and scaling up [110].

19.5.1.3 Biphasic system

The seedling inoculum is prepared by liquid fermentation in the biphasic system, which is the starting material for solid fermentation. Therefore, this method is also referred to as liquid-solid fermentation. Using blastospores produced from liquid fermentation as inoculum for the solid substrate, fermentation yields higher conidia than conidia as inoculum [111].

19.5.2 Formulations of EPF

EPF Formulation is a blend of viable conidia meant for successful pest control. Formulating EPF prolongs the shelf life, handling, application, storage, safety, and effectiveness. EPFs come in many different formulations. Formulations are of two types, i.e., solid and liquid formulation. Formulations are usually described using abbreviations [113].

19.5.2.1 Dust

The active ingredient is mixed with finely ground solid inert material such as clay, talc, etc. The particle size of dust formulation ranges from 50–100 micrometers. The dust formulation is then applied directly to the target insect. The choice of inert material is determined by its characteristics such as anticaking, UV protectant, and adhesiveness. Usually, the titter of microorganisms will be 10% [114].

19.5.2.2 Granular

A ready-to-use free-flowing solid formulation with a predetermined granule size range is available[115]. Formulations have larger and heavier particles with sizes ranging from 100–1000 micrometers for granules; micro granules range from 100–600 micrometers in size. The inert materials used in these formulations include kaolin, silica, polymers, starch, attapulgite, ground plant residues, and dry fertilizers. The concentration of an *a.i.* is 5–20%. The granular formulation is mostly applied to the soil for controlling nematodes, weeds, and soil-dwelling insects. The *a.i.* is released slowly from the granules. A specific soil moisture level is required while using granular formulation to act effectively [114].

19.5.2.3 Wettable powder (WP)

A powder formulation is applied as a suspension after dispersion in water [115]. It is free from moisture. The active ingredients are mixed with adjuvants such as a surfactant, dispersing agent, wetting agents, or inert material, crushed to a fine powder with a particle size of 5 microns, and applied as a suspension by mixing with water. Extended storage stability, efficient water dispersal, and ease of application with standard sprayers are just a few of the benefits [114].

19.5.2.4 Water dispersible granule (WDG)

Same as wettable powder, this formulation is developed to overcome the dustiness problem in wettable powder [114].

19.5.2.5 Emulsion

The emulsion contains an immiscible liquid in which droplets of size 0.1–1 micrometer are dispersed. It is of two types, namely, normal emulsion (oil in water) and invert emulsion (water in oil). However, it should be combined with water [114].

19.5.2.5 Suspension concentrate (SC)

A steady suspension of active ingredients with water is to be diluted before usage [12]. These are solid active substances that have been extensively crushed and disseminated in water. Agitation is required before application to keep the particles evenly distributed. Many ingredients are used in SC formulations, including wetting agents, dispersion agents, foam inhibitors, gelling agents, and so on. They are created using a wet grinding process, with particle sizes ranging from 1 to 10 µm.

19.5.2.6 Oil Dispersion

Solid active ingredient is dispersed in a liquid other than water, most often in oil; however, plant oil is predominantly used. This formulation has excellent penetration and spreading activity and is also very helpful in delivering water-sensitive ingredients [114].

19.5.2.7 Aerosol

One or other active ingredient is mixed with a solvent. Mostly, aerosol formulation contains a lower percentage of active ingredients. Aerosols are delivered as fine droplets. They are employed in greenhouses, indoors, and in localized outdoor areas [116].

19.6 Methods of application

Several factors are to be considered before applying EPF. They include spore concentration and environmental conditions. Currently, four common methods are plant or root dipping, foliar application, soil application, and vector transmission. Foliar spray is a promising and convenient application method among all those methods. Root dipping can be followed in crops that are transplanted. Soil application of EPF is used for targeting soil-dwelling insects. Major entomopathogenic fungi formulations and their host along with trade names are provided in Table 19.4.

Table 19.4: Popular entomopathogenic fungi, host and their commercial names.

Fungi	Host	Commercial name
Lagenidium giganteum	Mosquito	Lagenex
Coelomyces psorophorae	Mosquito	Lagenex
Conidiobolus Coronatus	Aphids, flies, caterpillars	–
Pandora	BPH and GLH of Rice	–
Beauveria bassiana	Cotton bollworms, coffee berry borer, DBM	Beveroz, Sun bio Bevigaurd, Dr. Bacto's Brave
Metarhizium anisopliae	Sugarcane pyrilla, Rhinoceros beetle	Metarhoz, Almid, Dr. Bacto's Meta, Sun Bio Meta
Lecanicillium lecanii	Whiteflies, scales, aphids, mealybugs, planthopper, thrips	Biogreen, Vertilac, Biocatch, Verticare
Isaria fumosorocea	Whitefly	Priority
Hirsutella thompsonii	Phytophagous mites	No mite

19.7 Approaches toward the application of entomopathogenic fungi

EPF is a natural enemy of insect pests and regulates their ecosystem population. In recent years, EPF-based biological control programs followed the same inundative spraying pattern and inoculative biological control strategy [117]. In addition, entomopathogenic fungi have several properties that make them an excellent alternative or adjunct to synthetic pesticides [118]. The main approaches in the application of EPF for pest management are described in Figure 19.7. They are classical control/introduction, augmentation, and conservation.

19.7.1 Classical control

In this approach, natural enemies are used to control exotic hosts that become invasive in new environments/localities where those specific natural enemies are absent [119]. So the correct identification of pests and their locality is necessary for a

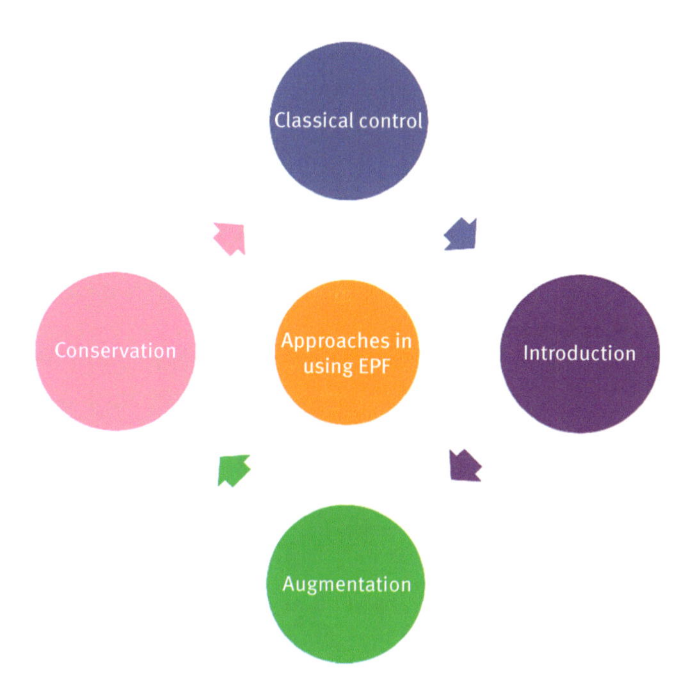

Figure 19.7: Approaches toward the application of entomopathogenic fungi.

successful classical biological control program. The early history of the application of EPF using this approach starts with the introduction of *Entamophaga maimaiga* (Entamophthorales: Entamophthoraceae) that was isolated in Japan and used in the US for the control of larvae of the Gypsy moth, *Lymantria dispar (shah)*. Subsequently, a successful permanent introduction of EPF was reported owing to the conidial movement of fungi and human manipulation [120]. Following this, another entomopathogenic *Zoophthora radicans* (Zygomycetes: Entamophthothorales) from Israel had been successfully introduced in Australia for the management of spotted alfalfa aphid *Therioaphis trifolii* (Homoptera: Drepanosiphidae) [3]. At the same time, standards for collection and importation/exportation of entomopathogen should be framed by IPPC.

19.7.2 Augmentation

In augmentation, control is brought about by manipulating within organisms. This technique is carried out by a mass production technology and genetic improvement to promote efficiency and population [5]. In the case of fungi, augmentation usually involves applying in vitro-produced conidia or mycelia in aqueous suspensions in a greenhouse or field, often along with formulations to improve its persistence and infectivity [6]. Hypocreales are more generalized, while entomopathoromycota are more target-specific; both have little influence on natural enemies [118].

19.7.3 Conservation

In the case of conservation, manipulation is carried out in the environment. In this technique, the habitat is modified to favor an organism's potentiality, virulence, and the successful establishment of epizootics. Conservation should be the first consideration in biocontrol programs [5]. Usually, entomopathogenic fungi have two phases in their life cycle: a typical mycelia development phase that occurs primarily well outside the host body and a budding phase that occurs mainly in the hemocoel of the insect host [122]. Enhancement or conservation can be done by

- Maintaining the hosts and diversity in that ecosystem to overcome the unsuitable situation of the biocontrol agent
- Manipulating alternate host
- Manipulating suitable environmental conditions such as temperature and humidity while building a greenhouse/polyhouse/glasshouse [121]
- Infectivity of *B. bassiana* in the soil can be improved by radiation of the modified hyphal strands [123]

19.7.4 Persistence of entomopathogenic fungi

Fungi require specific circumstances that favor the growth, development, and persistence of pathogens to control insect pests effectively. According to many studies, the change in the organism's reaction in laboratory and field conditions is the reason for the non-persistence of many fungal entomopathogens, and hence stabilizers should be used. When there is a lack of a host and a non-favorable environmental condition sustains, most entomopathogenic fungi produce either meiotic resting spore (zygospore) or mitotic resting spore (azygospore) and persists in the soil until a favorable time arises [49, 124].

19.7.5 Safety

There is fear that using entomopathogenic Hyphomycetes as biocontrol agents might be dangerous to the applicator or the environment, just like other pathogens. Despite their facultative character and rather extensive host ranges, entomopathogenic Hyphomycetes appear to provide negligible harm to humans, domesticated pets, fauna, and the ecosystem [125]. A no-risk situation cannot be assured for chemical pesticides and biopesticides [126, 127]. As *B. bassiana* and *B. brangroftii* are soil-dwelling organisms, they show effects in soil pests and soil-inhabiting non-target species [127]. Field observations did not show any possible adverse effects on bees, natural enemies, and earthworms. In contrast, laboratory studies negatively affected the carabids, cicindellids, and collembolans when pushed to stress conditions [125, 127]. With respect to the effects of *Beauveria* species on mammals and humans, there are no pathogenic, allergic, or toxic hazards. [125, 126, 128–131]. A report documented that a person under immunosuppressive therapy was infected with *B. bassiana* in deep tissue [132]. *B. brangroftii* showed non-pathogenicity over warm-blooded animals and not even a single negative report over decades of use [133, 134]. Investigations into the behavior of a transgenic *M. anisopliae* ARSEF 1080 strain in the soil under field conditions revealed that the transgenic strain did not suppress the culturable native fungal microbiota [135]. *M. anisopliae* was first experimented with inhalation on mice, guinea pigs, and rats, resulting in no evidence of allergy [136–138]. An isolate of *M. anisopliae* var *acridium* showed severe dermal allergic response in humans [131]. Studies beyond allergic reactions on humans, such as carcinogenic, serological, and more genetic studies, have also been commenced [139]. Mitosporic fungi are typically safe and nontoxic, with no or mild mammalian toxicity and persistent toxicity [140]. Various advantages and disadvantages in using entomopathogenic fungi as a biocontrol agent are described in Figure 19.8.

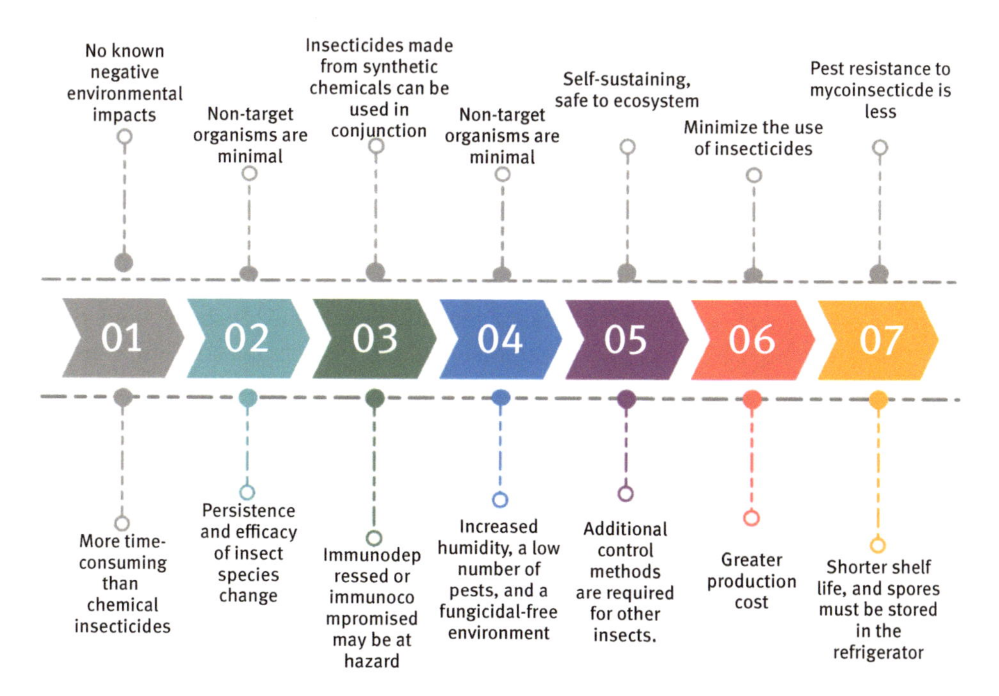

Figure 19.8: Advantages and disadvantages of using entomopathogenic fungi as a biocontrol agent [146, 147].

19.8 Policy and plans in the import and export of EPF

Though there are almost 170 strains of mycopesticides formulated and available for commercial use, some exotic pests that are not indigenous to that country will be effectively controlled by such a specific biocontrol agent present in that exotic area itself. In such cases where the existing biocontrol agents fail to act effectively, considering the pest status and its impacts on yield, importing/exporting of specific biocontrol agents can be encouraged. Certain rules and guidelines for risk management related to the export, shipment, import, and release of biocontrol agents and other beneficial organisms have been laid out. These rules and guidelines are called International Standards for Phytosanitary Measures (ISPMs), which are available on the International Phytosanitary Portal (IPP). It lists the related responsibilities of contracting parties to the International Plant Protection Convention (IPPC), Convention on Biological Diversity (CBD), National Plant Protection Organization (NPPO), or other responsible authorities, importers, and exporters. The scope of these standards do not include living modified organisms, issues related to registration of biopesticides, or microbial agents intended for vertebrate pest control. These contents and data are taken from the ISPM 3 guidelines of the Food and Agricultural Organization of the United Nations for the export, shipment, import, and release of biological control agents and other beneficial organisms produced by the International Plant Protection Convention (IPPC) [141].

19.9 Background of IPPC

The setting up of IPPC was based on the need for a common and effective action to prevent the introduction and spread of pests and for promotion of control measures. In this context, the provisions of IPPC extend to any organism that is capable of harboring or spreading plant pests, mainly where international transportation is involved (Article 1). Section 4.1 of ISPM 20 (*Guidelines for a phytosanitary import regulatory system*) contains a reference to the regulation of biological control agents. It states: Imported commodities that may be regulated include articles that may be infested or contaminated with regulated pests. Examples of regulated articles are pests and biological control agents. Phytosanitary concerns may include the possibility that newly introduced biological control agents may primarily affect other non-target organisms. It may be a potential pest itself; it may act as a carrier/pathway for pests, hyperparasitoid, and hyperpathogens. In this sense, they may be regulated articles according to Article VII.1 of IPPC and ISPM20. In addition, the IPPC considers globally accepted environmental standards (Preamble), and contracting parties should also examine the potential for more significant environmental implications from releasing biological control agents and other beneficial creatures (for example, impacts on non-target invertebrates).

19.10 Purpose of the standard and some important standards

The objective of the standard is
- To ease the safe shipment, import, export, and release of biocontrol agents by providing guidelines, mainly through the development of national legislation where it does not exist.
- To describe the need for cooperation between importing and exporting countries so that benefits can be obtained with minimal adverse effects.
- To promote practices that will ensure safe and efficient use while minimizing environmental risks due to improper handling.

Some important standards are:

ISPM 2 – Framework for pest risk analysis.
ISPM 3 – Code of conduct for import and release of exotic biocontrol agents.
ISPM 11 – Pest risk analysis for quarantine pests (this includes pest risk assessment in relation to environmental risks, and this aspect covers environmental concerns related to the use of biological control agents).
ISPM 19 – Guidelines on lists of regulated pests.
ISPM 20 – Guidelines for a phytosanitary import regulatory system.

19.11 Requirements

The basic requirements for import and export Entomopathogenic fungi were already framed and proposed by the International Plant Protection Convention in 2005 and the document was published in 2017. The basic requirements involved in importing and exporting entomopathogenic fungi are mentioned in Figure 19.9.

19.12 Designation of responsible authority and description of general responsibilities

19.12.1 Contracting parties

Contracting parties should designate an authority with appropriate competencies (usually their NPPO) to be responsible for export certification and regulate the import or release of biological control agents and other beneficial organisms, subject to relevant phytosanitary measures and procedures.

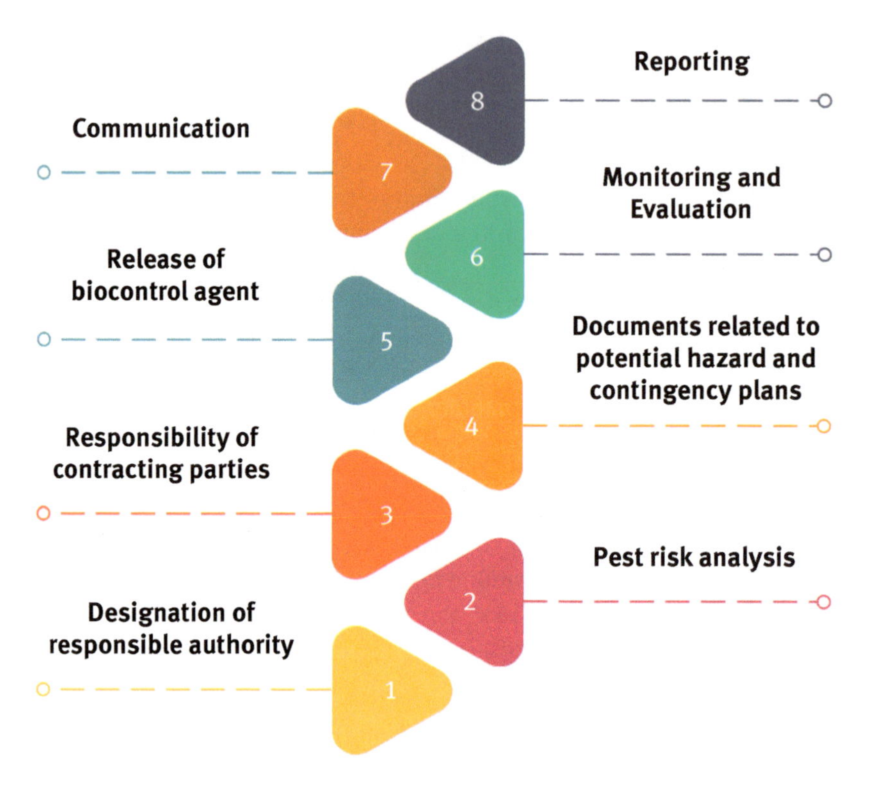

Figure 19.9: Basic requirements involved in importing and exporting of entomopathogenic fungi.

19.12.2 General responsibilities

NPPO or other responsible authorities should access and audit the import/export documentation; should carry out a pest risk analysis before import or release of biological control agents, labeling; should ensure that biocontrol agents are taken directly to either quarantine station or mass rearing unit; and should also consider possible impacts on the environment.

19.12.3 Pest risk analysis

This analysis is required to estimate the likelihood or successful invasion of a plant pest, hyperparasitoid, and specificity toward a target pest that may occur due to the import/export of biocontrol agents. In addition, the analysis is required to assess the potential impact and also the options to mitigate. This step helps the decision on the control of import/export.

19.12.4 Responsibility of contracting parties

19.12.4.1 Importing country's responsibility

If the biological control agent or another beneficial organism already exists in the country, regulation may be required solely to guarantee that the organism is not contaminated or infested; and to ensure that interbreeding with indigenous genotypes of the same species does not introduce new phytosanitary hazards. For these reasons, inundative release may be limited. Documentaries about the organism's biology, origin, distribution, ecology, economic value, environmental impacts, and other enemies should all be kept on hand.

19.12.4.2 Exporting country's responsibility

If the biological control agent or other beneficial organism is already present in the country, regulation may only be needed to ensure there is no contamination or infestation of this organism or that interbreeding with local genotypes of the same species does not result in new phytosanitary risks. Inundative release may be restricted for these reasons.

19.12.4.3 Documents related to the potential hazard and contingency plans

Before the first importation, the importer should provide the documents related to the potential health hazard and risks posed to staff operatives while handling. NPPO or other responsible authority develops or adopts contingency plans or procedures. When the problem is identified, they should consider the possible emergency actions and implement them in an appropriate situation.

19.12.4.4 Release

NPPO/responsible authority should analyze the release requirement only in a specific area. In addition, they should have sufficient documents to allow traceback of the released biocontrol agents.

19.12.4.5 Monitoring and evaluation

The authorities should evaluate and respond to the impact on the target and non-target organisms.

19.12.4.6 Communication

It is recommended that the NPPO or other responsible authority ensure that local users and suppliers of biological control agents or other beneficial organisms, farmers, farmer organizations, and other stakeholders are kept sufficiently informed and educated on the appropriate measures for their use.

19.12.4.7 Reporting

The contracting party should abide by any reporting obligations under the IPPC; for example, where an organism used as a biological control agent or beneficial organism has shown pest characteristics.

19.13 Authorities, acts, and conventions in India

9.1. National Biodiversity Authority (NBA): It is a statutory autonomous body under the Ministry of Environment, Forests and Climate Change, Government of India, established in 2003 to implement the provisions under the Biological Diversity Act, 2002, after India signed the Convention on Biological Diversity in 1992.

9.2. Salient provisions related to biocontrol agents in Biological Diversity Act, 2002.

Section 3: All foreign nationals require NBA approval to obtain Biological Resources.

Section 4: Indian individuals/entities to seek approval before transferring knowledge/research and material to foreigners.

Section 6: Prior approval of NBA before applying for any IPR based on research conducted on biological material and/or associated knowledge obtained from India.

9.3. The Cartagena protocol: The Biosafety Protocol under CBD (Convention on Biological Diversity) requires parties to make decisions on the import of LMOs for intentional introduction into the environment for risk assessments (Article 15). Risk assessment in Annexure III. Parties are also required to take measures to prevent unintentional transboundary movements.

9.4. The Nagoya–Kuala Lumpur Supplementary Protocol on Liability and Redress to the Cartagena Protocol on Biosafety: The Supplementary Protocol also includes provisions about civil liability.

19.14 Future prospects

In general, our knowledge of fungal entomopathogens has moved from simple observations of fungus killing insects to a determined attempt to learn how to employ these organisms as biological control agents. Most previous studies on endophytic fungi and other associated investigations have concentrated on co-culturing methods in an in vitro dual plate experiment to assess endophytic fungi's antagonistic effects against key targets. Anti-pest proteins produced by recombinant endophytic microbes can successfully invade the host employed in insect pest management [142]. However, it is to be noted that recombinant endophytic fungi with increased virulence can be a risk to pollinators and other beneficial insects. Therefore, the integrated use of EPF, such as *B. bassiana*, in combination with other chemical pesticides, has been investigated. According to certain scientists, the integration of various entomopathogens might aid in improving resistance management tactics and reducing ecosystem damage caused by the overuse of inorganic pesticides [143]. According to a study, one of the significant areas to be concentrated on is the ecology of fungal entomopathogens. Though there have been some significant developments in this part [123], in-depth knowledge of the most common genera of fungal entomopathogens should reveal the significant insights that make our understanding better so that it can be employed in an ecological approach for biological control. Some scientists advocated that several research areas should be used to understand fungal entomopathogens better [123]. Still, there are many other areas in need of research; for example, the impact of climate change on fungal entomopathogens and spore surface properties of fungi. All studies should include most entomopathogenic species and not only about *M. anisopliae* and *B. bassiana*. These investigations should give us new knowledge about manufacturing, storage, maintainance and utilization of fungal entomopathogens in the field. Further studies can also focus on the following:

i. Plant-microbe associations for stress tolerance and adaptation
ii. Symbiosis effect on plant's secondary metabolism
iii. Symbiosis impact on secondary metabolism of microbes
iv. The use of metagenomics and bioinformatics tools to determine entomopathogens diversity and phylogeny [144].

Important information on pathogenicity, genes, proteins, metabolites, and genetic linkages will be revealed by this endeavour. Furthermore, including EPF into IPM systems entails a thorough understanding of the abiotic and biotic factors that determine EPF's insecticidal effect and endophytic activity. Apart from that, to establish effective management techniques, it is critical to evaluate inoculation methods for extended colonization [145]. Even though biopesticides account for just 3% of the crop protection market worldwide currently, they are consistently expanding at the rate of 10% each year. Mycoinsecticide comes in second (27 percent) in the worldwide biopesticide market, next to *Bacillus thuringiensis* products [10].

19.15 Conclusion

Biopesticides, based on entomopathogenic fungi, are safe and effective in managing insect pests. However, environmental and food safety issues drive the use of entomopathogenic fungus in the myco-biocontrol of insects. Also, contamination with mycotoxins (aflatoxins, trichothecenes, zearalenone, fumonisins, citrinin, etc.) that are generated by saprophytic fungi cannot be ruled out. Currently, much research is being done on entomopathogenic fungus identification and characterization. However, it is critical to find new EPF strains and commercialize those that have previously shown efficiency against certain insects. The current pathogenesis mechanism of entomopathogens is sluggish and has to be improved. Modern molecular biology approaches may alter the beneficial features of this fungus to increase field activity. The ecology of entomopathogens must also be better understood to create sensible techniques for enhancing their efficacy in field applications. It is also exciting to see biotechnology being used, but it should be more rigorous. A multi-faceted approach using all available integrated pest management (IPM) techniques offers a higher chance of controlling pests affordably. Hence, entomopathogenic fungi will likely become a major IPM component. The future of EPF-based biopesticides in IPM relies on scientists and other stakeholders working together.

Work contribution

G. Venkatesh, Writing – first draft; **P. Sakthi Priya,** Writing – first draft; **V. Anithaa,** Writing – first draft; **G. K. Dinesh**, Data Visualization, Software, Writing – first draft, review and editing; **S. Vel Murugan,** Writing – first draft; **Abinaya S.,** Writing – review and editing; **P. Karthika,** Writing – first draft, review, and editing; **P. Sivasakthivelan,** Writing – review and editing; **R. Soni,** Writing – review and editing; **A. Thennarasi,** Writing – review and editing

References

[1] Lokeshwari RK, Shantibala T. A review on the fascinating world of insect resources: Reason for thoughts. Psyche (Stuttg) 2010, 2010, 207570.

[2] Mwamburi LA. Beauvaria. In: Amaresan N, Senthil Kumar M, Annapurna K, Kumar K, Sankaranarayanan A-BT-BM in A.-E. eds. Beneficial microbes in agro-ecology. Amsterdam, Academic Press, Elsevier, 2020, 727–748.

[3] Bamisile BS, Akutse KS, Siddiqui JA, Xu Y. Model application of entomopathogenic fungi as alternatives to chemical pesticides: Prospects, challenges, and insights for next-generation sustainable agriculture. Frontiers of Plant Science 2021, 12. doi:10.3389/fpls.2021.741804.

[4] Bahadur A. Entomopathogens: Role of insect pest management in crops. Trends Hortic 2018, 1. doi:10.24294/th.v1i4.833.

[5] Vega FE, Meyling NV, Luangsa-ard JJ, Blackwell M. Insect pathology. In: Vega FE, Kaya HKBT-IP. 2nd edn, San Diego, Academic Press, 2012, 171–220.

[6] Hibbett DS, Binder M, Bischoff JF et al. A higher-level phylogenetic classification of the Fungi. Mycological Research 2007, 111, 509–547.

[7] St. Leger RJ, Wang C. Genetic engineering of fungal biocontrol agents to achieve greater efficacy against insect pests. Applied Microbiology and Biotechnology 2010, 85, 901–907.

[8] Batta YA. Control of main stored grain insects with new formulations of entamopathogenic fungi in diatomaceous earth dusts. International Journal of Food Engineering 2008, 4, 1–16.

[9] Ramanujam B, Poornesha B, Yatish KR, Renuka S. Evaluation of pathogenicity of different isolates of *Metarhizium anisopliae* sorokin against maize stem borer, *Chilo partellus* using laboratory bioassays. Biopesticides International 2015, 11, 89–95.

[10] Um M, Zakaria D, Galadima IB, Gambo FM, Maina UM, Zakaria D. A review on the use of entomopathogenic fungi in the management of insect pests of field crops. Journal of Entomology and Zoology Studies 2018, 6, 27–32.

[11] Abdelgany TM. Entomopathogenic fungi and their role in biological control. Omi. Gr. eBooks 2015, 46.

[12] Alexopoulos CJ, Mims CW, Blackwell M. Introductory mycology. New York, John Wiley & Sons, 1996.

[13] Mora MAE, Castilho AMC, Fraga ME. Classification and infection mechanism of entomopathogenic fungi. Arquivos Do Instituto Biologico (Sao. Paulo) 2018, 84. doi:10.1590/1808-1657000552015.

[14] Araújo JPM, Hughes DP. Diversity of entomopathogenic fungi. Which groups conquered the insect body? Advances in Genetics 2016, 94, 1–39.

[15] Ramanujam B, Rangeshwaran R, Sivakumar G, Mohan M, Yandigeri MS. Management of insect pests by microorganisms. Proc. Indian Natl. Sci. Acad. 2014, 80, 455–471.

[16] Glare T. Entamopathogenic fungi and their role in regulation of insect populations. Insect Control Biological Synthetic Agent 2010, 6, 387–419.

[17] Shah PA, Pell JK. Entomopathogenic fungi as biological control agents. Applied Microbiology and Biotechnology 2003, 61, 413–423.

[18] Hatai K, Roza D, Nakayama T. Identification of lower fungi isolated from larvae of mangrove crab *Scylla Serrate* in Indonesia. Mycoscience 2000, 41(6), 565–572.

[19] Tanada Y, Kaya HK. Insect pathology. Amsterdam, Acadenic Press, 2012, 665.

[20] Abdelaziz O, Senoussi MM, Oufroukh A et al. Pathogenicity of three entomopathogenic fungi to the aphid species, *Metopolophium dirhodum* Walker (Hemiptera: Aphididae) and their alkaline protease activities. Egyptian Journal of Biological Pest Control 2018, 28, 1–5.

[21] Torres Acosta RI, Humber RA, Sánchez-Peña SR. *Zoophthora radicans* (Entomophthorales), a fungal pathogen of *Bagrada hilaris* and *Bactericera cockerelli* (Hemiptera: Pentatomidae and Triozidae): Prevalence, pathogenicity, and interplay of environmental influence, morphology, and sequence data on fungal identification. Journal of Invertebrate Pathology 2016, 139, 82–91.

[22] Aung MO, Kang JC, Liang ZQ, Soytong K, Hyde KD. A new entamopathogenic species *Hymenostilbe furcata* parasitic on a hemipteran nymph in Northern Thailand. Mycotaxon 2006, 97, 241–245.

[23] Bensch K, Groenewald JZ, Dijksterhuis M et al. Species and ecological diversity within the *Cladosporium cladosporoides* (Davidiellaceae, Capnodiales). Studies in Mycology 2010, 67, 1–94.
[24] Ben FI, Boukhris-Bouhachem S, Eilenberg J, Allagui MB, Jensen AB. The occurrence of two species of entomophthorales (entomophthoromycota), pathogens of *Sitobion avenae* and *Myzus persicae* (Hemiptera: Aphididae), in Tunisia. BioMed Research International 2013, 2013. doi:10.1155/2013/838145.
[25] Kepler RM, Luangsa-Ard JJ, Hywel-Jones NL et al. A phylogenetically-based nomenclature for Cordycipitaceae (Hypocreales). IMA Fungus 2017, 8, 335–353.
[26] Lacey LA, Frutos R, Kaya HK, Vail P. Insect pathogens as biological control agents: Do they have a future? Biological Control 2001, 21, 230–248.
[27] Lacey LA. Manual of techniques in insect pathology. Amsterdam, Academic press, 1997.
[28] Luangsa-ard JJ, Mongkolsamrit S, Thanakitpipattana D, Khonsanit A, Tasanathai K, Noisripoom W, Humber RA. Clavicipitaceous entamopathogens: New species in Metarhizium and a new genus Nigelia. Mycological Progress 2017, 16, 369–391.
[29] Luangsa-Ard J, Houbraken J, van Doorn T, Hong SB, Borman AM, Hywel-Jones NL, Samson RA. Purpureocillium, a new genus for the medically important *Paecilomyces lilacinus*. FEMS Microbiology Letters 2011, 321, 141–149.
[30] Mains EB. The genus Gibellula on spiders in North America. Mycologia 1950, 42, 306–321.
[31] Sanchez-Pena SR, Peña S. In vitro production of hyphae of the grasshopper pathogen Entomophaga grylli (Zygomycota: Entomophthorales): Potential for production of conidia. Florida Entomol 2005, 88, 332.
[32] Rumbos CI, Mendoza A, Kiewnick S. Effect of *Paecilomyces lilacinus* strain 251 on the survival and virulence of entamopathogenic nematodes under laboratory conditions. Nematologia Mediterranea 2007, 35, 103–107.
[33] Soper RS, Shimazu M, Humber RA, Ramos ME, Hajek AE. Isolation and characterization of entomophaga *Maimaiga* Sp. Nov., a fungal pathogen of gypsy moth *Lymantria Dispar* from Japan. Journal of Invertebrate Pathology 1988, 51(3), 229–241.
[34] Taheri-Talesh N, Horio T, Araujo-Bazán L et al. The tip growth apparatus of *Aspergillus nidulans*. Molecular Biology of the Cell 2008, 19, 1439–1449.
[35] Wynns AA, Jensen AB, Eilenberg J, James R. *Ascosphaera subglobosa*, a new spore cyst fungus from North America associated with the solitary bee *Megachile rotundata*. Mycologia 2012, 104, 108–114.
[36] Zekeya N, Mtambo M, Ramasamy S, Chacha M, Ndakidemi PA, Mbega ER. First record of an entomopathogenic fungus of tomato leafminer, *Tuta absoluta* (Meyrick) in Tanzania. Biocontrol Science and Technology 2019, 29, 626–637.
[37] Zimmermann G. The entomopathogenic fungi *Isaria farinosa* (formerly *Paecilomyces farinosus*): Biology, ecology and use in biological control. Biocontrol Science and Technology 2008, 18, 865–901.
[38] Tanabe Y, O'Donnell K, Saikawa M, Sugiyama J. Molecular phylogeny of parasitic Zygomycota (Dimargaritales, Zoopagales) based on nuclear small subunit ribosomal DNA sequences. Molecular Phylogenetics and Evolution 2000, 16, 253–262.
[39] Glare TR, Milner RJ. Ecology of entomopathogenic fungi. In: Arora DK, Ajello L, Mukerji KG. 2nd edn, New York, Marcel Dekker, 1991, 547–612.
[40] Samson RA, Evans HC, Latge JP. Atlas of entamopathogenic fungi. Heidelberg, Springer Berlin, 1998.
[41] Deacon JW. Modern mycology. Cambridge, Blackwell Science ltd, 1997.
[42] Bischoff JF, Rehner SA, Humber RA. A multilocus phylogeny of the *Metarhizium anisopliae* lineage. Mycologia 2009, 101, 512–530.

[43] Lefeuvre P, Calmes C, Reynaud B, Nibouche S, Costet L. Description and phylogenetic placement of *Beauveria hoplocheli* sp. nov., used in the biological control of the sugarcane white grub, *Hoplochelus marginalis* in Reunion Island. Mycologia 2015, 107, 1221–1232.
[44] Barr DJS. Chytridiomycota. In: Systematics and evolution. Berlin, Heidelberg, Springer, 2001, 93–112.
[45] De Faria MR, Wraight SP. Mycoinsecticides and Mycoacaricides: A comprehensive list with worldwide coverage and international classification of formulation types. Biological Contro 2007, 43, 237–256.
[46] Rehner SA, Buckley EP. A Beauveria phylogeny inferred from nuclear ITS and EF1-alpha sequences evidence for cryptic diversification and links to Cordyceps teleomorphs. Mycologia 2005, 97, 84–98.
[47] Rehner SA, Posada E, Buckley P, Infante F, Castillo A, Vega E. Phylogenetic origins of African and neotropical *Beauveria bassiana* pathogens of the coffee berry borer, *Hypothenemus hampei*. Journal of Invertebrate Pathology 2006, 93, 11–21.
[48] Wraight S, Jackson M, De Kock S. Fungi as biocontrol agents: progress, problems and potential. Central Agricultural Biosecurity International 2001, 253–287.
[49] Shah PA, Pell JK. Entomopathogenic fungi as biological control agents. Applied Microbiology and Biotechnology 2003, 61, 413–423.
[50] Zhang SL, He LM, Chen X, Hueng B. B*eauveria lii* sp. nov., isolated from *Henosepilachna vigintiopunctata*. Mycotaxon 2012, 121, 199–206.
[51] Chen MJ, Huang B, Li ZZ, Spatafora JW. Morphological and genetic characterisation of *Beauveria sinensis* sp. nov, from China. Mycotaxon 2013, 124, 301–308.
[52] Agrawal Y, Mual P, Shenoy BD. Multi gene genealogies reveal cryptic species *Beauveria rudraprayagi* sp. nov. from India. Mycosphere 2014, 5, 719–736.
[53] Zimmermann G. Review on safety of the entomopathogenic fungus *Metarhizium anisopliae*. Biocontrol Science and Technology 2007, 17, 879–920.
[54] Milner JR. Current status of Metarhizium as a mycoinsecticide in Australia. Biocontrol News Information 2000, 21, 47–50.
[55] Mishra J, Tewari S, Singh S, Arora NK. Biopesticides: Where We Stand? In: Arora N, eds. Plant Microbes Symbiosis: Applied Facets. New Delhi, Springer. 2015. https://doi.org/10.1007/978-81-322-2068-8_2.
[56] Kachhawa D. Microorganism as a biopesticides. Journal of Entomology and Zoology Studies 2017, 5, 468–473.
[57] Rajula J, Rahman A, Krutmuang P. Entomopahogenic fungi in southeast Asia and Africa and their possible adoption in biological control. Biological Contro 2020, 151, 104–399.
[58] Ujjan AA, Shahzad S. Use of entomopathogenic fungi for the control of mustard aphids (*Lipaphis erysimi*) on canola(*Brassica napus*). Pakistan Journal of Botany 2012, 44, 2081–2086.
[59] Weng Q, Zhang X, Chen W, Hu Q. Secondary metabolites and the risks of *Isaria fumosorosea* and *Isaria farinose*. Molecules 2019, 24, 664.
[60] Clarkson JM, Charnley AK. New insights into the mechanisms of fungal pathogenesis in insects. Trends in Microbiology 1996, 4, 197–203.
[61] Hajek AE, St. Leger RJ. Interactions between fungal pathogens and insect hosts. Annual Review of Entomology 1994, 39, 293–322.
[62] Sinha KK, Choudhary AK, Kumari P. Entomopathogenic fungi. 2016.
[63] Ortiz-Urquiza A, Keyhani NO. Action on the surface: Entomopathogenic fungi versus the insect cuticle. Insects 2013, 4, 357–374.

[64] Gabarty A, Salem HM, Fouda MA, Abas AA, Ibrahim AA. Pathogencity induced by the entomopathogenic fungi *Beauveria bassiana* and *Metarhizium anisopliae* in *Agrotis ipsilon* (Hufn.). Journal of Radiation Research and Applied Sciences 2014, 7, 95–100.
[65] Cho EM, Kirkland BH, Holder DJ, Keyhani NO. Phage display cDNA cloning and expression analysis of hydrophobins from the entomopathogenic fungus *Beauveria (Cordyceps) bassiana*. Microbiology 2007, 153, 3438–3447.
[66] Boucias DG, Pendland JC, Latge JP. Nonspecific factors involved in attachment of entomopathogenic Deuteromycetes to host insect cuticle. Applied and Environmental Microbiology 1988, 54, 1795–1805.
[67] Gao Q, Jin K, Ying H et al. Genome sequencing and comparative transcriptomics of the model entomopathogenic fungi *Metarhizium anisopliae* and *M. acridum*. PLoS Genetics 2011, 7. doi:10.1371/journal.pgen.1001264.
[68] Wang C, St. Leger RJ. The MAD1 adhesion of Metarhizium anisopliae links adhesion with blastospore production and virulence to insects, and the MAD2 adhesin enables attachment to plants. Eukaryotic Cells 2007, 6, 808–816.
[69] St.Leger RJ, Butt TM, Goettel MS, Staples RC, Roberts DW. Production in vitro of appressoria by the entomopathogenic fungus *Metarhizium anisopliae*. Experimental Mycology 1989, 13, 274–288.
[70] Napolitano R, Juárez MP. Entomopathogenous fungi degrade epicuticular hydrocarbons of *Triatoma infestans*. Archives of Biochemistry and Biophysics 1997, 344, 208–214.
[71] Hackman RH. Biology of the Integument. Berlin, Heidelberg, Springer, 1984, 583–610.
[72] Zacharuk RY. Penetration of the cuticular layers of elaterid larvae (Coleoptera) by the fungus *Metarrhizium anisopliae*, and notes on a bacterial invasion. Journal of Invertebrate Pathology 1973, 21, 101–106.
[73] Smith RJ, Grula EA. Toxic components on the larval surface of the corn earworm *(Heliothis zea)* and their effects on germination and growth of *Beauveria bassiana*. Journal of Invertebrate Pathology 1982, 39, 15–22.
[74] Sánchez-Pérez LDC, Barranco-Florido JE, Rodríguez-Navarro S, Cervantes-Mayagoitia JF, Ramos-López MÁ. Enzymes of entomopathogenic fungi, advances and insights. Advanced Enzyme Research 2014, 02, 65–76.
[75] Beli WJ. Comprehensive insect physiology, biochemistry and pharmacology. The International Journal of Biochemistry 1985, 17, 1281–1282.
[76] Pedrini N, Zhang S, Juárez MP, Keyhani NO. Molecular characterization and expression analysis of a suite of cytochrome P450 enzymes implicated in insect hydrocarbon degradation in the entomopathogenic fungus *Beauveria bassiana*. Microbiology 2010, 156, 2549–2557.
[77] Pedrini N, Ortiz-Urquiza A, Huarte-Bonnet C, Zhang S, Keyhani NO. Targeting of insect epicuticular lipids by the entomopathogenic fungus *Beauveria bassiana*: Hydrocarbon oxidation within the context of a host-pathogen interaction. Frontiers in Microbiology 2013, 4, 1–18.
[78] Dillon RJ, Charnley AK. Inhibition of *Metarhizium anisopliae* by the gut bacterial flora of the desert locust, *Schistocerca gregaria*: Evidence for an antifungal toxin. Journal of Invertebrate Pathology 1986, 47, 350–360.
[79] Eyal J, Fischbein AMKL, Grace R. Assessment of *Beauveria bassiana* which produces a red pigment for microbial control. World Journal of Microbiology and Biotechnology 1994, 44, 2263–2268.
[80] Bradfisch GA, Harmer SL. ω-Conotoxin GVIA and nifedipine inhibit the depolarizing action of the fungal metabolite, destruxin B on muscle from the tobacco budworm (*Heliothis virescens*). Toxicon 1990, 28, 1249–1254.

[81] Suzuki A, Taguchi H, Tamura S. Isolation and structure elucidation of three new insecticidal cyclodepsipeptides, destruxins C and D and desmethyldestruxin B, produced by *metarrhizium anisopliae*. Agricultural and Biological Chemistry 1970, 34, 813–816.
[82] Kodaira Y. Toxic substances to insects, produced by *aspergillus ochraceus* and *oopsra destructor*. Agricultural and Biological Chemistry 1961, 25, 261–262.
[83] Kodaira Y. Studies on the new toxic substances to insects, destruxin a and b, produced by *oospora destructor*. Agricultural and Biological Chemistry 1962, 26, 36–42.
[84] Samuels RI, Reynolds SE, Charnley AK. Calcium channel activation of insect muscle by destruxins, insecticidal compounds produced by the entomopathogenic fungus *Metarhizium anisopliae*. Comparative Biochemistry and Physiology Part C: Comparative 1988, 90, 403–412.
[85] Ebel RE, Lardy HA. Stimulation of rat liver mitochondrial adenosine triphosphatase by anions. The Journal of Biological Chemistry 1975, 250, 191–196.
[86] Gelardi M, Angelini C, Costanzo F, Dovana F, Ortiz-Santana B, Vizzini A. *Neoboletus antillanus* sp. nov. (Boletaceae), first report of a red-pored bolete from the dominican republic and insights on the genus neoboletus. MycoKeys 2019, 49, 73–97.
[87] Mannino MC, Huarte-Bonnet C, Davyt-Colo B, Pedrini N. Is the insect cuticle the only entry gate for fungal infection? Insights into alternative modes of action of entomopathogenic fungi. Journal of Fungi 2019, 5, 33.
[88] Schabel HG. Oral infection of Hylobius pales by *Metarrhizium anisopliae*. Journal of Invertebrate Pathology 1976, 27, 377–383.
[89] Rafaluk-Mohr C, Wagner S, Joop G. Cryptic changes in immune response and fitness in *Tribolium castaneum* as a consequence of coevolution with *Beauveria bassiana*. Journal of Invertebrate Pathology 2018, 152, 1–7.
[90] Biswas T, Joop G, Rafaluk-Mohr C. Cross-resistance: A consequence of bi-partite host-parasite coevolution. Insects 2018, 9. doi:10.3390/insects9010028.
[91] Batta YA. Efficacy of two species of entomopathogenic fungi against the stored-grain pest, sitophilus granarius l. (curculionidae: Coleoptera), via oral ingestion. Egyptian Journal of Biological Pest Control 2018, 28, 1–8.
[92] Jayasiri SC, Hyde KD, Ariyawansa HA, Bhat J et al. The faces of fungi database: Fungal names linked with morphology, phylogeny and human impacts. Fungal Diversity 2015, 74, 3–18.
[92] Lord JC, Hartzer K, Toutges M, Oppert B. Evaluation of quantitative PCR reference genes for gene expression studies in *Tribolium castaneum* after fungal challenge. Journal of Microbiological Methods 2010, 80, 219–221.
[94] Keilin D. On a new *Saccharomycete monosporella unicuspidata* gen. n. nom., n.sp., parasitic in the body cavity of a dipterous larva (d*asyhelea obscura winnertz*). Parasitology 1920, 12, 83–91.
[95] Van Der Veen KJ, Willebrands AF. Isoenzymes of creatine phosphokinase in tissue extracts and in normal and pathological sera. Clinica Chimica Acta 1966, 13, 312–316.
[96] Peirson HB. The life history and control of the pales weevil (*Hylobius Pales*). Petersham, Harvard Forest, 1921.
[97] Pedrini N. Molecular interactions between entomopathogenic fungi (Hypocreales) and their insect host: Perspectives from stressful cuticle and hemolymph battlefields and the potential of dual RNA sequencing for future studies. Fungal Biology 2018, 122, 538–545.
[98] Lacey CM, Lacey LA, Roberts DR. Route of invasion and histopathology of *Metarhizium anisopliae* in *Culex quinquefasciatus*. Journal of Invertebrate Pathology 1988, 52, 108–118.
[99] Goettel MS. Viability of *Tolypocladium cylindrosporum* (Hyphomycetes) conidia following ingestion and excretion by larval *Aedes aegypti*. Journal of Invertebrate Pathology 1988, 51, 275–277.

[100] Miranpuri GS, Khachatourians GG. Infection sites of the entomopathogenic fungus *Beauveria bassiana* in the larvae of the mosquito *Aedes aegypti*. Entomologia Experimentalis Et Applicata 1991, 59, 19–27.
[101 Butt TM, Coates CJ, Dubovskiy IM, Ratcliffe NA. Entomopathogenic Fungi: New insights into host-pathogen interactions. Advances in Genetics 2016, 94, 307–364.
[102] Sweeney AW. Infection of mosquito larvae by *Culicinomyces* sp. through anal papillae. Journal of Invertebrate Pathology 1979, 33, 249–251.
[103] Rodrigues J, Campos VC, Humber RA, Luz C. Efficacy of Culicinomyces spp. against *Aedes aegypti* eggs, larvae and adults. Journal of Invertebrate Pathology 2018, 157, 104–111.
[104] Rodrigues ML, Casadevall A. A two-way road: Novel roles for fungal extracellular vesicles. Molecular Microbiology 2018, 110, 11–15.
[105] Rodrigues J, Luz C, Humber RA. New insights into the in vitro development and virulence of *Culicinomyces spp*. as fungal pathogens of *Aedes aegypti*. Journal of Invertebrate Pathology 2017, 146, 7–13.
[106] Xiao G, Ying SH, Zheng P et al. Genomic perspectives on the evolution of fungal entomopathogenicity in *Beauveria bassiana*. Scientific Reports 2012, 2. doi:10.1038/srep00483.
[107] Hu X, Xiao G, Zheng P. Trajectory and genomic determinants of fungal-pathogen speciation and host adaptation. Proceedings of the National Academy of Sciences of the United States of America 2014, 111, 16796–16801.
[108] Bartlett MC, Jaronski ST. Mass Production of Entomogenous Fungi for Biological Control of insects. In: Burge MN, ed. Fungi in Biological Control Systems. Manchester UK, Manchester University Press, 1988, 61–85.
[109] Jaronski ST. Mass production of entomopathogenic fungi: State of the art. Sydney, USA. 2013.
[110] Lopes RB, Faria M, Glare TR. A nonconventional two-stage fermentation system for the production of aerial conidia of entomopathogenic fungi utilizing surface tension. Journal of Applied Microbiology 2019, 126, 155–164.
[111] Santos P, Abati K, Mendoza NVR, Mascarin GM, Delalibera Júnior I. Nutritional impact of low-cost substrates on biphasic fermentation for conidia production of the fungal biopesticide *Metarhizium anisopliae*. Bioresource Technology Reports 2021, 13, 100619.
[112] Pham TA, Kim JJ, Kim K. Optimization of solid-state fermentation for improved conidia production of B*eauveria bassiana* as a Mycoinsecticide. Mycobiology 2010, 38, 137.
[113] McWhorter CG. Pesticide Formulations. Journal of Environmental Quality 1974, 3, 94–95.
[114] Gasic S, Tanovic B. Biopesticide formulations, possibility of application and future trends. Pesticidi I Fitomedicina 2013, 28, 97–102.
[115] IRAC. Technical monograph no 2. 6th ed, Catalogue of pesticide formulation types and international coding system 2008.
[116] McWhorter CG. Pesticide Formulations. Journal of Environmental Quality 1974, 3, 94–95.
[117] Eilenberg J, Hajek A, Lomer C. Suggestions for unifying the terminology in biological control. Bio Control 2001, 46, 387–400.
[118] Dara SK, Montalva C, Barta M. Microbial control of invasive forest pests with entomopathogenic fungi: A review of the current situation. Insects 2019, 10. doi:10.3390/insects10100341.
[119] Samways MJ. Classical biological control and insect conservation: Arethey compatible? Environmental Conservation 1988, 15, 349–354.
[120] Hajek AE, Elkinton JS, Witcosky JJ. Introduction and spread of the fungal pathogen Entomophaga maimaiga (Zygomycetes: Entomophthorales) along the leading edge of gypsy moth (Lepidoptera: Lymantriidae) spread. Environmental Entomology 1996, 25, 1235–1247.

[121] Gautam RD. Biological pest suppression. New Delhi, Westvile publication house, 1994.
[122] Khan S, Guo L, Maimaiti Y, Mijit M, Qiu D. Entomopathogenic fungi as microbial biocontrol agent. Molecular Plant Breeding 2012. doi:10.5376/mpb.2012.03.0007.
[123 Vega FE, Goettel MS, Blackwell M et al. Fungal entomopathogens: New insights on their ecology. Fungal Ecology 2009, 2, 149–159.
[124] Glare TR, Milner RJ. Handbooks of applied mycology. Volume 2: Humans, animals and insects. In: Arora DK, Ajello L, Mukerj KG, ed. New York, Marcel Dekker, 1991, 547–612.
[125] Vestergaard S, Cherry A, Keller S, Goettel M. Environmental impacts of microbial insecticides. Dordrecht, Springer Netherlands, 2003, 35–62.
[126] Otieno WA. Microbial control of pest and plant diseases 1970–1980. In: Burges, ed. London, Academic Press, 1981.
[127] Zimmermann G. Review on safety of the entomopathogenic fungi *Beauveria bassiana* and *Beauveria brongniartii*. Biocontrol Science and Technology 2007, 17, 553–596.
[128] Steinhaus EA. Microbial diseases of insects. Annual Review of Microbiology 1957, 11, 165–182.
[129] Ignoffo CM, Hostetter DL, Sikorowski PP, Sutter G, Brooks WM. Inactivation of representative species of entomopathogenic viruses, a bacterium, fungus, and protozoan by an ultraviolet light source. Environmental Entomology 1977, 6, 411–415.
[130] Goettel MS, Douglas Inglis G. Manual of techniques in insect pathology. New York, Elsevier, 1997, 213–249.
[131] Goettel MS, Hajek AE, Siegel JP, Evans HC. In: Butt TM, Jackson C, Magan N, ed. Fungi as biocontrol agents: Progress, problems and potential. Wallingford, CAB International, 2001, 347–376.
[132] Henke MO, de Hoog GS, Gross U, Zimmermann G, Kraemer D, Weig M. Human deep tissue infection with an entomopathogenic *Beauveria* species. Journal of Clinical Microbiology 2002, 40, 2698–2702.
[133] Semalulu SS, MacPherson JM, Schiefer HB, Khachatourians GG. Pathogenicity of *Beauveria bassiana* in Mice. Journal of Veterinary Medicine Series B 1992, 39, 81–90.
[134] Strasser H, Vey A, Butt TM. Are there any risks in using entomopathogenic fungi for pest control, with particular reference to the bioactive metabolites of metarhizium, tolypocladium and beauveria species? Biocontrol Science and Technology 2000, 10, 717–735.
[135] Hu G, St. Leger RJ. Field studies using a recombinant mycoinsecticide (*Metarhizium anisopliae*) Reveal that it is rhizosphere competent. Applied and Environmental Microbiology 2002, 68, 6383–6387.
[136] Schaerffenberg B. Untersuchungen über die Wirkung der Insektentötenden PilzeBeauveria bassiana (Bals. Vuill.) undMetarrhizium anisopliae (Metsch.) Sorok. auf Warmblütler. Entomophaga 1968, 13, 175–182.
[137] El-Kadi MK, Xará LS, De Matos PF, Da Rocha JVN, De Oliveira DP. Effects of the Entomopathogen *Metarhizium anisopliae* on Guinea Pigs and Mice 1. Environmental Entomology 1983, 12, 37–42.
[138] Little LM, Shadduck JA. Pathogenesis of rotavirus infection in mice. Infection and Immunity 1982, 38, 755–763.
[139] Ferron P. Biological control of insect pests by entomogenous fungi. 1978.
[140] Copping LG. The manual of biocontrol agents. Alton, U.K, British crop protection council, 2004.
[141] International Plant Protection Convention. Guidelines for the export, shipment, import and release of biological control agents and other beneficial organisms. Rome, Italy, 2017.
[142] Fadiji AE, Babalola OO. Exploring the potentialities of beneficial endophytes for improved plant growth. Saudi Journal of Biological Sciences 2020, 27, 3622–3633.

[143] Aguilar-Marcelino L et al. Formation, resistance, and pathogenicity of fungal biofilms: Current trends and future challenges. In: Liliana Aguilar-Marcelino, Laith Khalil Tawfeeq Al-Ani, Filippe Elias de Freitas Soares, André Luís Elias Moreira, Maura Téllez-Téllez, Gloria Sarahi Castañeda-Ramírez, Ma. de Lourdes Acosta-Urdapilleta, Gerardo Díaz-Godínez, Jesús Antonio Pineda-Alegria, eds. Recent trends in mycological research. Cham, Springer, 2021, 411–438.
[144] Fadiji AE, Ayangbenro AS, Babalola OO. Unveiling the putative functional genes present in root-associated endophytic microbiome from maize plant using the shotgun approach. Journal of Applied Genetics 2021, 62, 339–351.
[145] Mantzoukas S, Eliopoulos PA. Endophytic entomopathogenic fungi: A valuable biological control tool against plant pests. Applied Science 2020, 10. doi:10.3390/app10010360.
[146] Singh D, Kour Raina T, Singh J. Entomopathogenic fungi: An effective biocontrol agent for management of insect populations naturally. Journal of Pharmaceutical Sciences Research 2017, 9(6), 883.
[147] Bamisile BS, Siddiqui JA, Akutse KS, Aguila LCR, Xu Y. General limitations to endophytic entomopathogenic fungi use as plant growth promoters, pests and pathogens biocontrol agents. Plants 2021, 10. doi:10.3390/plants10102119.

Debiprasad Dash, Pallabi Mishra

Chapter 20 Indigenous practices for pest control and marketability of the produce for development of sustainable agriculture

Abstract: The increasing trend of using chemical inputs in our Indian agriculture is reducing soil fertility, enhancing resistance to chemical inputs in pest and diseases, and polluting environment day by day. Hence, it is important to find alternative measures to control plant diseases that do not harm the environment and at the same time increase yield and improve the quality of product. Despite the fact that the importance of nutrients in disease control has been recognized for some of the most severe diseases, the correct management of nutrients in order to control disease for a sustainable agriculture has received little attention. Keeping in view the above issues, the following objectives have been kept for the study: indigenous technical knowledge (ITK) adopted by farmers for controlling plant diseases and marketability of the indigenous knowledge. The approach for this study is case-oriented across different crops.

20.1 Introduction

Practice of sustainable agriculture is the main aim of farmers and the governments of all countries. Sustainable development relates to meeting the needs of people with proper utilization and management of resources in a harmless manner. This involves social, economical, and environmental development without causing any harm to our resources [18]. There has always been a conflict between mankind and plant pests for superior yield and continued existence of crop since the advent of agriculture. The conventional farming system is an age-old sustainability-based farming method that has been developed by primordial farmers through generations of their interface with nature and its resources in the form of food, fodder, and fibers [12]. This self reliant method of farming using resources available locally, without external inputs, is an indigenous way of cultivation. The production of grains in India has steadily increased due to technological advances, but there is a 10% constant loss postharvest. Indigenous practices are more eco-friendly, less hazardous to biomass, and very close to nature. Indigenous plant protection practices (IPPP) ensemble awareness and knowledge of a range of factors linked to management of pests that are developed by farmers over an extended period, by practice and experiments. They continue to increase and are disseminated from every corner of different societies and groups. The employment of chemical pesticides has its own adverse effects

https://doi.org/10.1515/9783110771558-020

on plants and the ecosystem. Artificial and chemical pesticides used by farmers have an unfavorable effect on health and growth of plants that imbalance the ecology. They are gradually diminishing the natural sustainability of the environment. There is a concern about global warming and agriculturists, farmers and consumers too wish to save our earth and the mankind from destruction. The use of chemical pesticides is prevalent across the globe because of its immediate effect and ready availability. Though chemical pesticides provide immediate control of pests but the traditional knowhow positions well in consideration to ecology and sustainability [16]. There is need for a change from the traditional agriculture system that uses chemical pesticides to an integrated approach. Keeping in view the above issues, the following objectives have been kept for the study: indigenous technical knowledge (ITKs) adopted by farmers for controlling plant diseases and the issues involved in marketability of the indigenous technical knowledge produce.

20.2 Pesticides overview

More than 50% of the pesticides used globally are used in Asia. "India positions 12th in pesticide use globally and ranks 3rd in Asia after China and Turkey". The most important sector of India economy is agriculture, which employs around 70% of the entire population. Pesticides and fertilizers are vital elements of contemporary agriculture. Fungicides, insecticides, and herbicides are the most widely used pesticides for suitable management of uncontrolled weeds and pests. However, in the consumption of total pesticides, insecticides holds the topmost market share in India, which is only 1% of the global usage of pesticide. "As per the data of FAO, India has utilized around 58,160 tonnes of pesticide in 2018. The per hectare application rate of pesticide was only 0.31 kg in 2017 while consumption in China, Japan, and America was around 13.07, 11.76 and 3.57 kg per hectare of pesticides, respectively". Hence, it is apparent that India uses a lesser quantity of pesticides per hectare of crop. Inspite of the low share, in relation to global usage, the uncontrolled and haphazard pesticide treatment results in high pesticide residues in both soil as well as the natural environment. To reduce the hazardous effect of chemical pesticides, bio-pesticides have been proven as the best alternative for a sustainable development of agriculture. They also reduce pollution when compared to the use of chemical pesticides. A lot of bio-pesticides are being formulated in India that will possibly be outstanding substitutes to chemical pesticides. Several ready-made environment-friendly plant-based microbial bio-pesticides are being sold in the market. Bio-pesticides consumption amounts to 8% in India. For "sustainable agricultural development" and environment fortification from unfavorable effects of chemical pesticides, formulation and usage of bio-pesticides needs to be promoted [21].

Pesticides are generally used to kill the pests, thereby increasing agricultural productivity, but their indiscriminate use pollute the biota. Some insecticides are even hazardous to human health and the environment. It has been observed that roughly about 0.1% of chemical pesticides come in contact with the targeted organisms, and others pollute and damage the environment [4, 10]. Crops are treated frequently with pesticides than before. "The amount of pesticide used across the globe, in tonnes of active ingredient, has increased by 46% between 1996 and 2016, according to the FAOSTAT database of WHO" (2019). Pesticides damage the focused pests as well as the non-targeted plants and animals. "Organophosphate, carbamate, and pyrethroid insecticides are the most often used pesticides [11]. At present, around four million tonnes of pesticides are used per year on a global basis; most of which are herbicides (56%), followed by insecticides (19%), fungicides (25%), and other types such as rodenticides and nematicides" [9].

20.2.1 Importance of ITKs

Indigenous knowledge of farmers includes their relationship with the religious and natural environment, utilization of natural resources, community association, values, organization, and laws that create a base for a radical scientific system. The basic strategies for solving problems of the poor underutilized local resource communities are provided by this indigenous knowledge. Sustainability of agriculture arises from the traditional methods of agriculture, which is for the long term than focusing on short term yield maximization. Sustainability of agriculture is safe for the health of mankind and the environment. Traditional farmers adopted safe indigenous methods acquired from nature that is soothing to corporeal and socio-economic environments of an agroecosystem. Indigenous practices adopted by crop growers are considered as the lynchpin of the society. The experience and knowledge derived by farmers from these indigenous practices pass on from generation to generation. In current times, the change in climate has an apparent effect on the facets of agriculture. Due to the change in climate, knowledge of the ecology and biology of varied pests for the protection of plants has become imperative. These unnatural changes have made pest control mechanisms all the more thorny and composite. Moreover, the arbitrary use of chemical pesticides on crops has resulted in development of pest resistance, impacted insect pollinators, destroyed natural foes of crop pests, impacted farming enterprises, and has led to the degradation of the environment by soil and water pollution. The deteriorating soil holds less water that causes farmers to strain the already depleted water reservoirs. Adopting indigenous ways of managing pest is economical, effective, and they also do not have a deteriorating effect on the environment. Indigenous knowledge application in agriculture reduces ecosystem disturbance and also sustains the growth of plants, crop production, and protection from crop pests. Indigenous knowledge intensification through awareness of the provisions provided, farmer training, and information sharing strengthen the application among farmers, mostly in tropical African countries. Based upon application and attainment of benefits, farmers use the indigenous knowledge techniques. But there is a difference between the techniques adopted, as some of them are overriding, easily reachable, and safe and sound for man, animals while also promoting social solidity due to the dissemination mechanism, whereas some of them are inefficient in their dissemination of indigenous knowledge methods. Integration of some major indigenous practices with contemporary research creates a readiness amongst the farmers to participate and act in response to global prospects and challenges. Biological resources should be harnessed to protect the crop, as chemical pesticides are expensive and hazardous to the environment and crops [30]. Generations of experience, watchful observations, trials and errors have led to the development of indigenous knowledge (ITK) systems. The majority of the indigenous technologies and practices are similar to contemporary technologies that have provided the expected results. The dynamism in ITK is due to the exogenous knowledge and endogenous creativity. The nature of ITK thinking is intuitive. ITK is mostly qualitative and requires a holistic approach. It provides priceless insights into the reserve, procedures, potentialities, and problems in a particular area, if correctly tapped. ITK is recorded, transferred traditionally, and learned through observation and hands-on experience [19]. Indigenous knowledge is effective, locally available, relatively cheap, and less destructive to local environments [8]. In sustainable agricultural development, this knowledge associates with farmers to natural, physical, and socioeconomic environment from any agroecosystem [25].

20.2.2 Approaches of ITKs

One of the foremost biological constraints that hamper crop production globally is insect pests. To eradicate pests, various control methods and measures are adopted.

A study on Malayali tribal farmers of Kolli hills of Namakkal district of Tamil Nadu, India, was conducted to focus on the Indigenous Crop Protection Practices (ICPPs). The adoption level of the farmers for ICPPs was studied. To prevent the attack of thrips, 73.33% of the tribes apply *Calotropis gigantea* (L.) in the nursery with a rationality score of 2.80. Likewise, 6 kg of *Melia azadirachta* L. (6 kg) kernel powder mixed with 200 L of water, left undisturbed whole night, filtered, and sprayed the next morning for the control of brown plant hopper and green leaf hopper by 66.67% of the farmers was found to be rationale (3.21). Moreover, the rationality score was 2.60 towards pressing and incorporating the leaves of *Calotropis gigantea* (L.) to the soil in the interspace available in the field for controlling brown plant hopper by 70% of the farmers. All the same, to control the invasion of hopper, 76.67% respondents expressed planting of *Calotropis gigantea* (L.) at an interval of 12 feet along all sides of the paddy fields, which was found irrational (2.34). About 66.67% of the farmers practiced dusting of ash on the standing crop of paddy [5] and 50% of the farmers sprayed a solution containing a mix of 3 kg of fish, 5 kg of neem leaf, or table salt to control sucking pest was found irrational (2.41). Oil of lemon grass along with the leaf extract of *Ocimum tenuiflorum* L., cow urine, and butter milk was mixed with water and sprayed to control the sap feeder by 26.67% of the respondents, which has a rationality score of 2.46. About 66.67% of the farmers applied a powder of neem seed kernel 2–3 times after transplanting to control ear head bug in paddy. It was found to be rational (3.21). To control damage due to sucking pest, 83.33% of the farmers followed dusting of 15–20 kg ash /ha over Italian millet (*Setaria italica*), Little millet (*Panicum miliare*), or Finger millet (*Eleucine coracana*) crops. This was found to have a scientific rationale score of 2.60 [20].

To control shedding of flowers and premature fall of pods (Morales, 2002), 80% of the farmers applied a mixture leaf extracts of nochi (Vitex negundo L.) and neem cake, which was sprayed on a field of bean (Lablab purpureus L.). This was found to be rational (2.98). To swear of white fly and aphids in tapioca (*Manihot esculenta*), 73.33% of farmers (followed the practice of mixing latex (Calotropis + Jatropha + Mango) with hot water. However, this practice was found to be irrational (2.29). All together, there were 11 ICPPs on sucking insect management. Of these four ICPPs were irrational, but had been adopted by more than 50% of the farmer respondents. To reduce the damage caused by sucking pest, 83.33% of the farmers following dusting of 15–20 kg ash /ha over crops. This had a scientific rationale score of 2.60 [20]. Such ICCPs may further be taken for on-farm trial for further assessment.

In paddy crop, the main problem is with the attack of stem borer and leaf folder. To control stem borer in paddy fields, 70% of the farmers spread the stems and leaves of *Datura stramonium* L. in stagnant and puddled waters in the field. This was found rational (2.80) as the bitterness and smell of Datura leaves keep the stem-borer away from the area. About 30% of the farmers also mixed neem oil and water @ 30 mL/L to make a spray to control stem borer. This practice has a rationality score of 3.73. To control leaf folder in paddy, 36.37% of the farmers sprayed leaf extract of Notchi (Vitex negundo L.) with buttermilk. This was found to be rational (2.78). About 43.33% of the farmers sprayed cow dung with leaf extract of *Adhatoda vasica* Nees. This was also found rational (2.60), and 66.67% also controlled leaf folder by spraying crops (15–20 kg/acre), which was followed by 83.33% of the farmers to kill sucking pests. This practice has a scientific rationale score of 2.60 [20].

A paste of pounded neem leaves was made by boiling 10 kg leaves for half-an-hour in 1 L of water and left overnight. This solution is sprayed the next morning by mixing with 200 liters of water for controlling leaf folder. This was practiced by 30% of the farmers and was found to be rational (2.73). Moreover, stem borer and leaf folder in paddy was controlled by 30.00% of them by mixing 5 L of kerosene and soap solution and spraying the mixed solution in 1 ha of paddy field. It is scientifically irrational (1.88), perhaps thinking that the spray kerosenated soap water would suffocate and kill the larvae of stem borer and leaf folder. Clipping the top portion of seedlings before transplanting to prevent the seedling from stem borer and hispa eggs was followed by 43.33% of the farmers and was found to be rational (3.65). About 90% of the farmers grew castor growing in the field bunds of red gram (*Cajanus cajan*). Since the pod borer is attracted by the castor crops, which prevent the damage to the main crop, this practice has a rationality score of 3.24. Moreover, 83.33% of the farmers followed the use of 'Naithulasi' (*Ocimum sanctum*) having linalool (2.95), which is a strong insecticidal with fumigant action against the damage of pod borer. In the category of pest management of borer, there were about ten ICPPs. Eight ICPPs were adopted for paddy crop and two ICPPs for red gram crop. It was found that less than 30% adopted three ICPPs. Moreover, among these, one ICPP, that is, spraying kerosene along with soap solution in paddy fields, was found irrational, with a score of 1.88. Such ICPPs may be inhibited from further dissemination. Leaves of pungam (*Pongam pinnata* L.), Notchi (*Vitex negundo* L.), and neem (*Azadirachta indica* L.) are mixed with seeds of paddy so that pests do not attack those that are under storage. This was being practiced by 83.33% of the farmers and has a rationality score of 3.60. About 33.33% of the farmers used the leaf extract of *Vitex negundo* L, alone or with Vitex + neem, as a spray to control rice weevil. This has a scientific rationality score of 3.51. About 66.67% of the farmers mixed turmeric powder with paddy grains to controls weevil in stored grains. This has a rationality score of 3.43. About 50% of the farmers stored paddy seeds on a floor coated with the slurry of cow dung to avoid insect attack. This practice was found irrational (2.37). About 63.33% of the farmers filled water in vessels that were kept inside the store room, for attracting the insects and thereby reducing damage. This practice was found irrational (2.07). Leaves of *Cipadessa baccifera* (Roth) were spread over paddy seeds in the storage structure to repel the storage pests. This practice is scientifically rational (3.00). About 46.67% of the farmers prevented stored grain from being attacked by pests by placing 20 to 30 red chillies in one quintal of rice. The pungent odor of chillies was expected to prevent the attack by pests (2.95 R). About 43.33% used pepper powder to ward away the storage pests in paddy. This practice was found to be irrational (2.34). Since 14% or less moisture content is preferred for storing little millet (*Panicum* miliare) to reduce the risk of damage due to storage pest, it is sun dried. This practice was followed by 73.33% of the farmers and was found rational (2.61). About 80% of the farmers used the leaves of Neem (*Azadirachta indica* L.) and Thumbai (*Leucas aspera* L.) while storing finger millets. It is scientifically rationale (3.07) as the pests are kept away from the storage due to the strong odor coming from these leaves [6], reducing attacks from toothed beetle (*Oryzaephilus surinamensis*), flat grain beetle (*Cryptolestes minutus*), and grain borers (*Rhyzopertha dominica*). The seeds of red gram were dried and stored in gunny bags. Subsequently, dried leaves of 'Naithulasi' were placed inside the bag along with red gram seeds. About 90% of the tribal farmers spread the leaves of Neem over the red gram seeds and was found rational (3.39) due to its repelling action against the pests [24]. About 90% of the farmers smeared oil from neem seed over *Cajanus cajan* seeds as it checks the entry of larvae, which was found rational (3.39). Thus, damage is avoided, and 200 g of common salt per one kilo gram of red gram seeds is mixed manually and stored in jute gunny bags and stitched to the storage to keep away insects away from the stored grains. This practice was adopted by 83.33% of the farmers and is rational (2.85). The abrasive action of salt on the insect's skin prevents its movement [5]. This practice makes the seeds safe for storage for a period of 6–8 months (short-term duration). Dusting of turmeric powder or powder from the leaves of *Vitex negundo* L. controls pulse beetle, a storage pest. This practice was rational (3.05). About 83.33% of the respondents adopted the practice of using dried chilly pods in the containers of red gram, which has a rationality score of 2.95, since they understood that this practice may avoid the attack of bruchids (beetle). Before storing, mixing dried leaves of neem or nochi (*Vitex negundo* L.) with red gram seeds is seen to show a repellent action against pests under storage, which was rational (3.56). About 50 kg of red gram seeds is mixed with 2.5 kg of red earth slurry and dried for storage. This practice has a scientific rationality score of 3.34, as the coated seeds of red earth repel pests. Management of pest under storage with ICPPs was predominant in Kolli hills. All the 17 ICPPs on storage pest management were found to be rational. Only one ICPP, the use of pepper powder to ward away storage pest in paddy grains, was found to be irrational. Likewise, though the rationality scoring of the use of Vitex negundo L. leaf extract spray was 3.51, it was adopted only by 33.33% of the farmers. Such ICPPs may be further refined and replicated in other areas, for better performance. To prevent pests and diseases from attacking paddy fields, leaves of neem are used by about 73.33% of the farmers and applied in the field. This practice has a rationality score of 3.02 [3]. About 63.33% sprayed Notchi (Vitex negundo L.) leaf extract to control rice tungro virus. This was found to be rational (3.22) as Vitex has pesticidal property. To control brown spot disease in paddy crop, 40% of the farmers mixed 5 kg of common salt with 15 kg of sand. The mixture was then incorporated in one acre of paddy field. This was rather irrational (1.63) [18, 28].

Dead frogs or crabs or snails are crushed, fermented and placed randomly in the paddy field during the milking stage, to act as repellents as their foul smell keeps away the Gundhi bug from paddy crops. To control the stem borer in rice, Citrus or pumello fruits are cut into small pieces and are kept in paddy fields with the help of bamboo sticks to act as insect repellents. Setting up bird perches (made up of bamboo and having many branches) in rice, brinjal, cruciferous crops, millets, and maize crops attract carnivorous birds, which act as predators and control the stem borer of rice and brinjal fruit, and shoot borer and stem borer in millets and maize fields, and with different lepidepterous larvae from infesting cruciferous crops. Pest management strategy is also boosted due to the presence of hydrocyanic acid content and larvicidal properties of bamboo. Obliteration of alternating hosts eliminate the microhabitats of insects during the off season and controls brown plant hopper in rice through keeping the field bunds clean from weeds. Placing of

1 m high branches of crofton weeds (*Eupatorium adenophorusm*) (100/ha) in paddy fields also control brown plant hopper due to the presence of alkanoid compounds, terpinoids, etc. [26].

A foremost threat to the productivity and trade of mango in sub-Saharan Africa is posed by the invasive fruit fly *Bactrocera dorsalis*. Different innovations are devised by farmers for pest management with the purpose of reducing yield loss and cost of production to maximize revenues. Mexican marigold and pepper as bitter herbs are used to create smoke. This indigenous practice was reported by respondents during a survey on managing the tephritid fruit flies of mango (*Bactrocera dorsalis*) [29]. The use of silica in plants helps in strengthening the cell wall of the plants, enhancing resistance to diseases caused by pests both in the field and during storage of the crops. Insect pest, such as brown planthopper, green leafhopper, stem borers, and whitebacked planthopper, and spider mites-like non-insect pests were suppressed by the use of silica [24, 17]. Application of silicon to sugarcane creates resistance to pest damage. Sugarcane varieties containing highest solidity of silicon per unit area in the leaf sheath are tolerant to the shootborer, Chilo infuscatelus. A study conducted in Florida showed that a considerable decrease in leaf freckling of sugarcane was seen after the application of 20 t/ha of silica slag to muck soil. An increase in silica to plants resulted in better resistance to the stem borer Diatraea saccharalis F in sugarcane. Another study in Taiwan showed the use of diverse forms of silicon use, including blast furnace ash and silica slag, decreased the appearance of borer pest than on untreated plant [15]. Some studies in South Africa have focused on the correlation between the absorption of silicon and plant resistance to Eldana saccharina. A green house sugarcane variety was artificially inoculated with *E. Saccharina* and treated with calcium silicate. A 30% reduction of damage by borer and a 20% reduction in borer mass was seen when the plant was treated with 10 t/ha calcium silicate [13].

Application of silicon to wheat also resulted in resistance to pest damage. It was found that when wheat (Triticum aestivum L.) plants were grown in a solution of sodium silicate, there was no injury due to the larvae of Hessian fly (Phytophaga destructer Say.). Reports reveal that silicon application to vulnerable wheat improved crop resistance and decreased the infestation of pest in the field as well as in storage [15]. Rice crop also showed an increase in resistance to pest by the application of silicon. A negative correlation was found between the rice (Oryza sativa) applied with silicon and the number of stems infected by the African striped borer (Chilo zacconius). It was further observed that yellow rice borer (Scirpophaga incertulas Walker) larvae did not have an effect on the silicon-treated rice plants. Paddy grown in silicon-rich soil showed a high resistance to the stem borer Chillo suppressalis. The quantity of larvae that penetrated the stem and the feces amount showed a negative correlation (Ma and Takahashi, 2002). At high levels of silica, fewer nymphs of brown plant hopper became adults along with a decrease in the longevity of adults and reduction in female productivity. In the case of maize (Zea mays), it was found that higher the silicon content, lower is the manifestation of stalk borer (Chilo zonellus Swinhoe). The application of sodium metasilicate at a rate of 0, 0.56, and 0.84 gm Si/plant reduced the larvae survival of the borer Sesamia calamistis from 26.0% (control) to 4.0% at 0.56 g Si/plant.

The red spiter mite (Tetranychus urticae, Koch) infests dicotyledonous vegetables such as brinjal, cucumber, tomato, and green bean. A test was performed on the brinjal plant. A combination of silicon and biocontrol agent Beauveria bassiana enhanced the control of red smiter mite. The underlying reason is that silicon, which is soluble and absorbed by plants, catalyzes the mechanism of resisting pests. For this reason, pests that feed on silicon-treated plants will freely feed on the plants and will succumb to the effect [15].

Applying wood ashes to and around crops of Amaranthus, vegetable crops, kitchen garden, and Legume crop storage kills chewing pests such as blister beetle, Mylabris phalerata, termites (Odonotermes spp.) and Bruchids. Further, smoke produced by burning coconut husk or paddy husk or maize cobs in citrus orchards act as a repellent to fruit sucking moth (Eudocima fullonia). Foams of the soaps and detergents applied to vegetable plants make their surface slippery, affecting sucking pests such as whitefy (Bemisia tabaci Gennadius), leaf hopper (Amrasca biguttula biguttula Ishida), etc. After consuming hookah, the left-over hookah water is sprayed on kitchen garden, which repels pod borers (Bhendi pod borer, Earias vitella Fab., sucking bugs [Riptorus spp., Clavigralla spp.]) of vegetables, etc. Placing the middle portion of the jackfruit (shaft) in gourd-type crops, and also in mango and guava orchards repels fruit fly (Bactrocera spp.). Application of ash to Amaranthus, vegetable crops, and kitchen garden plants repels chewing pests such as beetles (e.g., blister beetle Mylabris phalerata) and termites (Odonotermes spp.). Tin boxes are cut open and placed near the stem on the ground region to capture coconut rat, Bandicota indica (Burrowing rat), and squirrels (Funambulus spp.). The slippery surface of the substrate reduces the frictional force, causing a barrier for climbing [26].

In the Garhwal hills of Himachal, farmers reported that they prepare a solution of cow urine with the meshed leaves of Kandali (Nettle leaves), Daikan leaves (wild Azadirachta), Tulsi leaves (Holy Basil), and spray it on infected plants. Farmers were very affirmative of this practice owing to its effectiveness and at nearly no cost [21]. A sample size of 75 respondents of the indigenous society of Magar, Gurung, and Newar in Tanahun, Lamjung, and Kaski districts of Nepal were interviewed with a semi-structured questionnaire. It was revealed from the study that commercial farmers were not common users of ITK whereas semicommercial and subsistence farmers widely used ITK. About 85% of the subsistence farmers, followed by 60% of semicommercial farmers, and lastly 10% of the commercial farmers used ITK for pest management [20].

20.3 Marketability of indigenous crops and vegetables

Indigenous crops and vegetables are generally cultivated by farmers for self-consumption. With a growing demand in the market, indigenous vegetables are being widely sold in local markets. Market surveys of Kenyan urban and sub-urban markets, such as the Nakuru market, show that the there is hypo-supply of indigenous vegetables in comparison to the market demand. There is a demand for indigenous vegetables in the United Kingdom and the United States, which shows the potential of regional and international markets [1]. In bigger cities, such as Nairobi, authentic traditional dishes are made from indigenous vegetables by hotels and restaurants. A small but significant restaurant chain in Nairobi, called the Amaica, uses indigenous cowpea leaves and spider plant, which is bought from the local women farmers. Amaica dispenses the vegetables to restaurants in Nairobi, including Jomo at the Kenyatta International Airport. Generally, indigenous vegetables are bought by middlemen who then resell it to customers by setting their own price. This occurs in cases where the farmer fails to directly sell to his customers. It has been observed that cultivation, procurement, and marketing of indigenous vegetables are generally carried out by women. The profits are considerably higher than chemically treated vegetables (well over 75%). In the Naivasha area, farmers having excess indigenous vegetables sell them via intermediaries to supermarkets in Kasarani, Nairobi, and Naivasha, depending on the demand on a particular day [14]. The indigenous technology used in farming is a substantial innovation for the rural economy. This technology would meet the requirements of farmers at a lower cost and higher profits [27]. It has been seen that there is a growing demand for organic vegetables that are grown indigenously, in both domestic as well as in international markets due to the increase in purchasing power and health consciousness of the consumers. In international markets, organic products fetch 50% to 200% higher price than conventional inorganic produce. Farmers who have adopted integrated organic farming system (IOFS) have reaped better crops that are sold on a local basis and on highways, with a margin of 10 to 15% as compared to conventional produce. Marketing and value addition are treated as the most important factors for organic produce [7, 12]. Some other issues faced by the farmers are inadequate transportation to cities and towns. Further studies have revealed that growing indigenous fruit trees has helped in improving the socio-economic conditions of farmers, leading to improvement in health, better education, and sustainable agriculture [14].

Marketing of indigenous leafy vegetables has become a major source of income for women farmers of Nigeria. Some marketing issues such as good communication network between buyers and sellers, an efficient marketing system, and effective distribution channels are to be addressed. Other challenges include setting of the right price, lack of reliable market information and provision of market advisory service, high perishability of leafy vegetables, and inconvenience during rainy seasons in reaching the market place in time. Low level of productivity due to erratic investments and poor financial conditions also hamper crop productivity, and thereby the marketability of indigenous crops. Inadequate and poor ways of sorting, grading, and packaging of vegetables pose a threat to marketability [2].

20.4 Case studies results

In the context the adoption of ITKs, the following case studies have been undertaken in Odisha under the North Eastern Coastal Plain Zone of the Agroclimatic Zone of Odisha and Agroecological Situation (Table 20.1). The crops are canal irrigated in alluvial soil. The annual rainfall is 1427 mm and the mean maximum temperature and mean minimum temperature are 32.4 °C and 21.4 °C, respectively.

Table 20.1: Details of agroclimatic zone [33].

Sl. no.	Item	Information
1	Major farming system/ enterprise	Rice, black gram/green gram/mustard/sunflower/ vegetable/sugarcane, Pisciculture, Dairy, Poultry, Mushroom
2	Agroclimatic Zone	North Eastern Coastal Plain Zone
3	Agroecological situation	AES(3) – Alluvial canal irrigated – Low lying and flood prone – Saline soil group
4	Soil type	Alluvial soil: 83,209 ha, Saline soil: 20,200 ha, Sandy soil: 19,146 ha
5	Mean yearly rainfall & temperature of the district	1427 mm, mean max temp: 32.4 °C and min temp: 21.5 °C

Farmers have adopted various organic methods along with indigenous technology. Due to the adoption of these various natural ways of controlling insect pests and maintaining soil health, the cost of cultivation has been reduced and better quality produced along with improving soil health. Initially, they started using neem cake and vermicompost in paddy crops along with two sprays of neem oil to reduce fertilizer dose and control sucking pests, respectively. The use of neem cake and vermicompost has also shown control over sheath blight and bacterial blight in paddy. When they realized the impact of using organic inputs on paddy crop and soil, fermented organic liquid was further used in the paddy crop to control other insect pests in paddy crop.

20.4.1 Case studies on paddy crop

Initially, one of the farmers started using vermicompost and neem cake in paddy crop as basal dose along with one spray of neem oil at 20 DAT to reduce the fertilizer dose and control sucking pests, respectively. The use of neem cake and vermicompost also resulted in control over sheath blight and bacterial blight in paddy, and 39% of the fertilizers were also applied as basal dose. When he realized the impact of using organic inputs on paddy crop and soil, fermented organic liquid-1 & 2 were further used in the paddy crop to control other insect pests in the paddy crop at 40 & 60 DAT, respectively. Murette of potash (MOP) at 40 & 60 DAT and vermiwash at 60 DAT were also applied with the fermented organic liquid. Usage of various ITKs reduced the cost of cultivation and saved 15% over conventional cost of cultivation of paddy. Moreover, 71% and 87% were saved from fertilizers and pesticides over conventional paddy. The cost of cultivation was Rs. 15,500, with a gross profit of Rs. 35,200, net return of Rs. 19,700, and B:C ratio of 2.27 due to the adoption of these practices, as given in Table 20.2. In comparison, the B:C ratio was 2.10 in conventional paddy, which might be due to the high cost of cultivation from using fertilizer and pest management. The cost of cultivation was Rs. 18,300, with a gross profit of Rs. 38,400 and a net return of Rs. 20,100.

Another farmer applied vermicompost along with 73% of fertilizers, and neem oil at 20 DAT with fermented-1 was applied at 20, 40, and 60 DAT. The cost of cultivation was Rs. 14,380 and 2.2% of cost was saved over conventional method of paddy cultivation. Gross profit was Rs. 31,500, with a net profit of Rs. 17,120, and B:C ratio of 2.19 in the case of paddy cultivation using ITKs. The gross profit was Rs. 30,000, with a net profit of Rs. 15,300 and B:C ratio of 2.04 in case of conventional method of paddy cultivation. The overall saving of cost from fertilizer and pesticides were 27 and 86%, respectively, over conventional paddy cultivation.

One of the farmers practiced the application of vermicompost along with 69% of fertilizers, as compared to doses of fertilizer in conventional paddy cultivation. Pest management was done through neem oil along with fermented liquid-1 applied at 20 DAT, and fermented liquid-3 at 40 and 60 DAT. Due to the practice of using ITKs, he saved 8.4% of cost over the cost of cultivation through conventional methods. Similarly, 31 and 93% of cost through fertilizer and pesticides were saved when compared to the cost incurred due to the adoption of conventional methods. The cost of cultivation and gross profit were Rs. 15,750 and 33,000, respectively and the net return was Rs. 17,250, with a B:C ratio of 2.10. The cost of cultivation was Rs. 17,200, with a gross profit of Rs. 27,000, net return of Rs. 9,800, and B:C ratio of 1.57 in conventional paddy cultivation. Though the net return was more from the conventional practices of paddy cultivation, the B:C ratio is more in ITKs-based paddy cultivation in all the above cases. Moreover, the cost towards fertilizers and pesticides were saved. This might be due to lowering of the pest problems through the use of various ITKs, which kept most of the pests away from the field and increased the beneficial insects in the paddy field.

a. Fermented Liquid 1:
Neem (*Azadirachta indica*) leaves: 3 kg
Karanja (*Millettia pinnata*) leaves: 2 kg
Arakha/Aak/Madar (*Calotropis gigantean*): 2 kg
Pokasungha (*Ageratum conyzoides*): 2 kg
Cow urine: 10 L

Method of preparation:
All leaves and the cow urine were placed in a 50 L plastic drum and kept for 30 days for fermentation and were stirred thrice a week. After 30 days, the fermented liquid was filtered and stored in bottles.

Uses: 500 mL in 15 L of water for spray

Benefits of use: leaf-eating and sucking insects
Practice of the farmer in paddy crop: Before using the above fermented liquid, he prepares a solution of murette of potash (MOP) by mixing 500 g of MOP in 1 L water, which is later filtered. A 100 mL of this filtered solution of MOP and 500 mL of fermented liquid is then mixed with 15 L of water for spraying the paddy crop after 40 DAT (days after transplanting). This practice has controlled the menace of sucking pests such as BPH, GLH, WBPH, and leaf eating and other lepidopteron insects.

b. Fermented Liquid 2:
Ingredients
Cow dung: 1 kg
Cow urine: 2 L
Molasses/curd: 50 to 100 g
Leaves of bitter taste (Neem/Karanj): 1 kg
Leaves of strong smell (Pokasungha – *Ageratum conyzoides*/Pasaruni – *Paederia foetida*/Ban tulsi – *Croton bonplandianus*): 1 kg
Sticky leaves (Madar – *Calotropis gigantean*/Banyan – *Ficus benghalensis*): 1 kg
Garlic paste: 50 g
Turmeric powder: 50 g

Method of preparation:
Leaves are cut to small pieces and pestled. Then, cow urine and dung, along with pestled leaves, are taken in a 20 L bucket with a cover. One L of water is added to it, and garlic paste, turmeric powder, and molasses or Gur are mixed and the bucket is covered. This bucket is kept under shade for 30 days and stirred once in 3 to 4 days. After a month, the mixture is diluted with water; the quantity of water being two times the volume of the mixture, and filtered by wearing a polythene cover in hand as gloves. The filtered liquid is stored in a bottle after adding one tea spoon of fresh milk and half tea spoon of turmeric powder. This liquid can be used as pesticides for up to six months.

Uses: After a week of filtration, the liquid is used for foliar spray after dilution (4 to 7 times water is used for dilution, depending on the vegetable crop).

Benefits of use: Sucking pests, Borer, fungal infection, leaf blight are under control

Vermiwash: 300 mL in 15 L of water

c. Fermented Liquid 3:
Neem (*Azadirachta indica*) leaves:1 kg
Karanja (*Millettia pinnata*) leaves: 1 kg
Arakha/Aak/Madar (*Calotropis gigantean)* leaves: 1 kg
Tobacco (*Ageratum conyzoides*) leaves: 1 kg
Chiraitah (Swertia chirayita) leaves: 1 kg
Cow urine: 5 L

Method of preparation
The leaves and cow urine were mixed in a 20 L plastic drum and kept for 30 days to allow for fermentation, while stirring thrice a week (Figures 20.1 and 20.2 here). After 30 days, the fermented liquid is filtered and stored in bottles.

Uses: 100 mL in 15 L of water for spray

Benefits of use: Sucking insects, leaf eating caterpillar, gundhi bug, blast, sheath rot, bacterial blight

The results are represented in Table 20.2.

The benefit to cost ratio of the same paddy treated with three different types of ITKs and chemical pesticides are represented in Figure 20.3. It can be seen that ITK1 has the highest benefit/cost ratio, followed by ITK2 and then ITK3. When compared with chemically treated pesticides, it was observed that the benefit/cost ratio of ITKs-treated paddy was better.

Figure 20.1: Cow urine used as ITK.

Figure 20.2: Preparation of vermiwash as ITK.

Table 20.2: Economics of case studies on paddy crop by practicing farmers.

S. No.	Crop	ITKs	Benefits experience over pests and diseases	Cost of cultivation (Rs.) Chemical	Cost of cultivation (Rs.) ITKs	Gross profit (Rs.) chemical	Gross Profit (Rs.) ITKs	Net return (Rs.) chemical	Net return (Rs.) ITKs	B:C Ratio chemical	B:C ratio ITKs	Savings over chemical fertilizers and pesticides
1.	Paddy	1. Neem cake + Vermicompost + 39% fertilizer 2. Neem oil at 20 DAT 3. Fermented liquid-1 + MOP solution at 40DAT 4. Fermented liquid-2 + MOP solution + Vermiwash at 60 DAT	1. Establishment of healthy seedlings controlling Sheath rot, bacterial blight & Sheath blight. 2. Sucking pests 3. Sucking pests, leaf eating caterpillar 4. Sucking pest, Blast, Lepidopteron insects, gundhi bug	18,300	15,500	38,400	35,200	20,100	19,700	2.10	2.27	Fertilizer:71% Pesticides:87% Cost of cultivation:15%
2.	Paddy	1. Vermicompost + 73% fertilizers 2. Neem oil + Fermented liquid-1 at 20 DAT 3. Fermented liquid-1 at 40 & 60 DAT	1. Establishment of healthy seedlings 2. Sucking pest, Blast, Lepidopteron insects, leaf eating caterpillar, Bacterial blight & Sheath blight, Sheath rot 3. Sucking pest, Blast, gundhi bug	14,700	14,380	30,000	31,500	15,300	17,120	2.04	2.19	Fertilizer:27% Pesticides:86% Cost of cultivation:2.2%

3.	Paddy	1. Vermicompost + 69% fertilizer 2. Neem oil + Fermented liquid-1 at 20, Fermented liquid-3 at 40 & 65 DAT	1. Establishment of healthy seedlings 2. Sucking insects, leaf eating caterpillar, gundhi bug, blast, sheath rot, bacterial blight	17,200	15,750	27,000	33,000	9800	17,250	1.57	2.10	Fertilizer:31% Pesticides:93% Cost of cultivation:8.4%

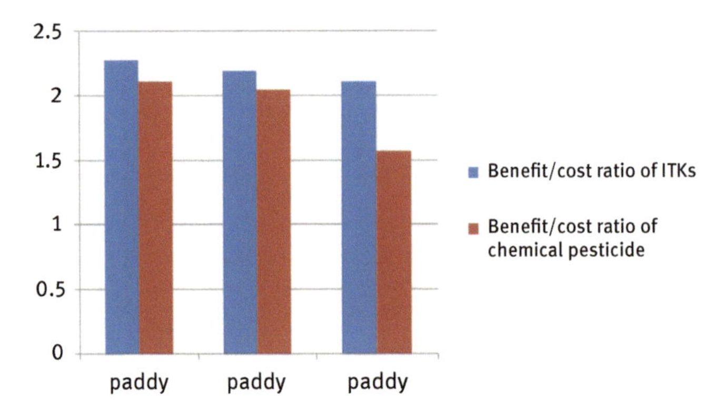

Figure 20.3: Benefit to cost ratio of paddy after using ITKs.

20.4.2 Practices of the farmers in vegetables

Various vegetables were cultivated only with soil and foliar application as mentioned above without the application of fertilizers or pesticides (Figure 20.4).

Soil application: Vermicompost along with neem cake is applied while transplanting seedlings of vegetables. Further, vermicompost is used for top dressings. Use of neem cake controls most of the soilborne diseases and some of the sucking pests due to the bitterness in neem cake, which has a repellent action.

Foliar application: Use of neem oil at the initial stage is helpful in controlling sucking pest and leaf-eating insects because of the bitter taste and repellent action. Fermented organic liquid-1 and 2 are applied in each at intervals of 12 to 15 days, alternately, due to which sucking pest, leaf eating caterpillar, fruit rot, and fruit borer are controlled. The population of beneficial insects increased by the application of ITKs because no fertilizer and pesticides were used, which helped in controlling sucking pest and caterpillars.

Use of cow urine in tomato

1 L of cow urine in 5 liters of water controls wilting of tomato crop.

The negative impact of fertilizers and pesticides on soil, insect pests, and environment in the cultivation of vegetables are gradually being felt by many farmers and during this pandemic situation of covid-19, most farmers are opting to produce and eat chemical-free vegetables. Our government also promotes production of healthy foods through integrated farming and organic farming at-scale. In vegetable cultivation, the major problems are nutrients and pest management. Imbalanced use of fertilizers and the improper manner of controlling pest and diseases through the increasing

Figure 20.4: Results of ITKs use on vegetables.

using of pesticides are the major causes of increased cost of cultivation with low net return harming soil health and environment. As depicted in Table 20.3, as per the farmers' experience on various economics, the cost of cultivations in all the vegetables were more in ITKs, which might be due to use of vermicompost and the requirement of more number of labor hours. Erratic use of pesticides enhances pest loading whereas ITKs reduce pest loading, enhancing the population of beneficial insects that help in controlling pests while maintaining an ecological balance. Though production of vegetables were less while using ITKs, the higher price due to non application of fertilizers and pesticides were afforded by the nearby consumers for availing good quality vegetables, which enhanced the gross profit and net return in all vegetables, except brinjal. It might be due to lack of pest-specific management through ITKs. Increased integrated pest management along with the use of ITKs is to be focused and studied for nutrient and pest management. Nutrient management is also very essential for a healthy crop through which pest problem can be reduced. Application of vermicompost supplies macro and micro nutrients in a balanced manner to vegetables, which helps in growing healthy crops, since a balanced nutrition keeps diseases and pests away from any living beings.

The benefit to cost ratio of different vegetables treated with ITKs and chemicals are represented in Figure 20.5. As seen from the graph, the benefit /cost ratio of chemical pesticide is better in the case of brinjal, pointed gourd, and tomato; same in the case of okra, and less than in ITKs in the case of ridge gourd, cowpea, bitter gourd, and cucumber.

Table 20.3: Economics of case studies on vegetable crops by practicing farmers.

S. No.	Crop	Cost of cultivation (Rs.) chemical	Cost of cultivation (Rs.) ITKs	Gross cost (Rs.) chemical	Gross cost (Rs.) ITKs	Net return (Rs.) chemical	Net return (Rs.) ITKs	B:C ratio chemical	B:C ratio ITKs	ITKs	Benefits experience over pests & diseases
1	Brinjal	46,750	52,300	112,500	105,000	65,750	52,750	2.41	2.01	1. Neem cake + Vermicompost 2. Neem oil + Fermented liquid-1 3. Neem oil + Fermented liquid-2 + vermiwash (Applied above two fermented liquid alternately in every 12 to 15 days interval) 4. Increase in population of beneficial insects	1. Establishment of healthy seedlings controlling Wilting, damping off 2. Sucking pest, leaf eating caterpillar 3. Sucking pest, leaf eating caterpillar, fruit rot, fruit borer 4. Controlled sucking pests and caterpillars
2	Ridge gourd	36,200	44,500	120,000	135,000	83,800	90,500	3.31	3.03		
3	Okra	41,500	46,500	95,000	105,000	53,500	58,500	2.29	2.26		
4	Cowpea	27,200	31,500	60,000	70,000	32,800	38,500	2.20	2.22		
5	Bitter gourd	52,600	59,500	147,000	180,000	94,400	120,500	2.79	3.02		
6	Cucumber	43,400	51,500	103,500	122,000	60,100	70,500	2.38	2.37		
7	Pointed gourd	74,600	82,500	18,3000	232,000	108,400	149,500	2.45	2.81		
8	Tomato	47,400	56,500	96,000	114,400	48,600	57,900	2.02	2.02		

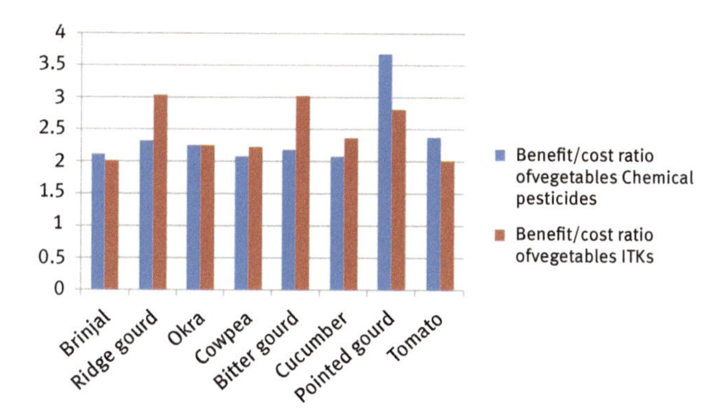

Figure 20.5: Benefit to cost ratio of vegetables.

20.5 Discussion and conclusion

The above study tries to find out the importance of using indigenous techniques of pest management. The main objective is to minimize the use of chemical and synthetic pesticides, which not only kill the targeted pest but also the beneficial insects. Moreover, the hazardous effect on the plant, soil, and environment has made agriculture unsustainable. In order to maintain the sustainability of agriculture and the ecosystem, the use of ITKs is necessary. The case studies on the use of home-made ITKs on paddy and vegetable pests have been studied. The observations show that farmers were satisfied with the use of ITKs but in some cases, the cost incurred was more than by using chemical pesticides. The reason being the small scale use of ITKs as many farmers are skeptical about the use of ITKs. Moreover, commercially grown crops are not much treated with ITKs as the price would be high and the farmers are not getting their desired rates. It was found that the ITKs produce were sold locally by farmers in the neighborhood to customers who valued chemical-free products. They were ready to pay the premium too. Farmers faced issues such as fetching the right price for their produce, lobbying by middlemen, lack of proper transportation to the nearby city or town, and lack of other resources. Inspite of these challenges, some of them are adopting ITKs demonstrating a hope for sustainable agriculture, which shows their love and compassion for nature and ecology.

The developing strategy that can be adopted is an awareness of the benefits of consuming indigenous produce by the consumers. This would create a demand in the market for such a produce. Certification and branding of indigenous produce can increase the adoption rate. On the other hand, farmers should be encouraged by the government and other bodies to use ITKs for pest management. Proper training regarding the same should be imparted to them. A value chain can be developed, starting from providing high quality seeds, improving cultivation practices by new techniques and innovations, followed by development of low cost technology for harvesting and post-harvesting practices, and last but perhaps the most crucial part being the creation of demand and facilitating value addition to market the produce. Finally, to enhance sustainable agriculture, one should avoid using chemical pesticides and adopt indigenous pest management techniques.

References

[1] Abukutsa-Onyango MO, Adipala E, Tusiime G, Majaliwa JGM. Strategic repositioning of African indigenous vegetables in the horticulture sector. In: Second RUFORUM biennial regional conference on building capacity for food security in Africa. Uganda, Entebbe, 2010, 1413–1419.
[2] Agbugba IK, Okechukwu FO, Solomon RJ. Challenges and strategies for improving the marketing of Indigenous leaf vegetables in Nigeria. Journal of Home Economics Association of Nigeria (HERAN) 2011, 15, 11–20.
[3] Arora S, Sharma JP, Chakravorty S, Sharma N, Joshi P. Indigenous technologies in plant protection. New Delhi, India, ICAR – National Research Centre for Integrated Pest Management, 2016, 15–20.
[4] Carriger JF, Rand GM, Gardinali PR, Perry WB, Tompkins MS, Fernandez AM 2006. Pesticides of potential ecological concern in sediment from South Florida Canals: An ecological risk prioritization for aquatic arthropods. Soil and Sediment Contamination: An International Journal 2006, 15(1), 21–45.
[5] Chandola M, Rathore S, Kumar B. Indigenous pest management practices prevalent among hill farmers of Uttarakhand. Indian Journal of Traditional Knowledge 2011, 10(2), 311–315.
[6] Chhetry G, Belbahri L. Indigenous pest and disease management practices in traditional farming systems in north east India. A Review. Journal of Plant Breeding and Crop Science 2009, 1(3), 28–38.
[7] Das A, Layek J, Ramkrushna GI, Babu S. Integrated organic farming system in North East India. Conservation Agriculture for Advancing Food Security in Changing Climate 2018, 1, 301–318.
[8] Elango K, Sobhana E, Sujithra P, Bharath D, Ahuja A. Traditional agricultural practices as a tool for management of insects and nematode pests of crops: An overview. 2020, 8(3), 237–245.
[9] FAO. Pesticide Use Data-FAOSTAT. 2018. (Accessed January 2, 2022 at http://www.fao.org/faostat/en/#data/RP).
[10] Gill HK, Garg H. Pesticides: Environmental impacts and management strategies. Pesticides: Environmental impacts and management strategies. Pesticides Toxic Aspects Marcelo L. Larramendy and Sonia Soloneski, IntechOpen, 2014, 8, 187.
[11] Gilbert S. A small dose of toxicology: The health effects of common chemicals. Boca Raton, FL, USA, Healthy World Press, 2012.
[12] Karthikeyan C, Veeraragavathatham DK, Ayisha SF. Traditional storage practices. Indian Journal of Traditional Knowledge 2009, 8(4), 564–568.
[13] Keeping M, Meyer J. Silicon-mediated resistance of sugarcane to Eldana saccharina Walker (Lepidoptera: Pyralidae): Effects of silicon source and cultivar. Journal of Applied Entomology 2006, 130, 410–420.
[14] Knaepen H. Towards a sustainable food system in the Lake Naivasha Basin. Kenya, ecdpm, 2018, 230, 10–26.

[15] Laing MD, Gatarayiha MC, Adandonon A. Silicon use for pest control in agriculture: A review. In Proceedings of the South African Sugar Technologists' Association. 2006, 80, 278–286.
[16] Lal C, Verma LR. Use of certain bio-products for insect-pest control. Indian Journal of Traditional Knowledge 2006, 5(1), 79–82.
[17] Ma JF, Takahashi E. Soil, fertilizer, and plant silicon research in Japan. Amsterdam, The Netherlands, Elsevier Science, 2002.
[18] Mishra P, Dash D. Rejuvenation of biofertilizer for sustainable agriculture and economic development. Consilience 2014, 11, 41–61.
[19] Mushtaq A, Pathania SS, Khan ZH, Ahmad MO. Indigenous technical knowledge in pest management. 2020, 8(5), 296–302.
[20] Naharki K, Jaishi M. Documentation of indigenous technical knowledge and their application in pest management in western mid hill of Nepal. SAARC Journal of Agriculture 2020, 18(1), 251–261.
[21] Nayak P, Solanki H. Pesticides and Indian agriculture – a review. International Journal of Research GRANTHAALAYAH 2021, 9(5), 250–263.
[22] Papnai G, Nautiyal P, Joshi N, Supyal V. Traditional knowledge and indigenous practices still in vogue among rural populace of Garhwali hills, Uttarakhand, India. J Pharmacogn. Phytochem 2020, 9, 145–147.
[23] Sah U, Dubey SK, Saxena H, Kumar R 2019. Scientific rationality of indigenous technical knowledge related to pulse production: Researchers' perception. Journal of Community Mobilization and Sustainable Development 2019, 14(1), 20–24.
[24] Savant NK, Korndorfer GH, Datnoff LE, Snyder GH. Silicon nutrition and sugarcane production: A review. Journal of Plant Nutrition 1999, 22, 1853–1903.
[25] Sharma IP, Kanta C, Dwivedi T, Rani R. Indigenous agricultural practices: A supreme key to maintaining biodiversity. Microbiological Advancements for Higher Altitude Agro-Ecosystems and Sustainability 2020, 91–112.
[26] Singh S, Das B, Das A, Majumder S, Devi HL, Godara RS, Sahoo MR. Indigenous plant protection practices of Tripura, India. Journal of Ethnobiology and Ethnomedicine 2021, 17(1), 1–19.
[27] Umoh EEA. Indigenous technological innovations: Crossing the valley of death to the marketplace. Conference paper, Federal Polytechnic, Kaura Namoda, Zamfara State, Nigeria, 21st – 23rd November 2017.
[28] Venkatesan P, Sundaramari M, Ahire LM. Adoption of rationale indigenous crop protection measures by "Malayali" Tribes of Kolli Hills, Tamil Nadu. Journal of Community Mobilization and Sustainable Development 2021, 16(1), 316–322.
[29] Wangithi CM, Muriithi BW, Belmin R. Adoption and dis-adoption of sustainable agriculture: A case of farmers' innovations and integrated fruit fly management in Kenya. Agriculture 2021, 11(4), 338.
[30] Yigezu G, Wakgari M 2020. Local and indigenous knowledge of farmers management practice against fall armyworm (Spodoptera frugiperda)(JE Smith)(Lepidoptera: Noctuidae): A review. Journal of Entomology and Zoology Studies 2020, 8, 765–770.
[31] Annual report of KVK Bhadrakh. 2020–21, (Accessed on January 20, 2022 http://www.kvkbhadrak.org/uploads/fck/APR%202020-21.pdf,).

Thajudeen Thahira, Kinza Qadeer, Julie Sosso,
Mohamed Cassim Mohamed Zakeel

Chapter 21
Role of plant microbiome in crop protection

Abstract: With the rapid increase in global population, the demand for more food is also escalating. This demand not only increases the need for extensive crop production but also the capacity to protect crops from the hazards of biotic and abiotic stresses that threaten agricultural products. These stresses include the widespread distribution of pests and diseases, heavy metal induction due to rapid industrialization, and chemical fertilizer use. Nature has developed a security system to protect crops from such stressors in the form of microbes. These microbes are predominantly present in the rhizosphere where they enhance the growth and development of plants by mitigating the negative effects of environmental stress. They reduce the exposure of plants to stress in many ways. These mechanisms include the production of secondary metabolites and the induction of plant hormones. They also resist plant phytopathogens, such as nematodes and fungus, by using induced systemic resistance against them. In addition, they act as biofertilizers for plants as they improve the uptake of nutrients, nitrogen fixation, and solubilization of unavailable phosphorus from the soil. Another way that soil microbes are beneficial to plants is the mitigation of heavy metal stress by transforming these compounds into less toxic forms and so making them safely available to the plant through biosorption and bioaccumulation. The processes of acidification, bioleaching, and detoxification are also microbial mechanisms used for the transformation of not just heavy metals but for other toxic compounds such as DDTs, PCB, etc. These microbes are natural plant protectors and using them commercially as an alternative to chemical fertilizers and pesticides will be more beneficial for the crops, the environment, and human health.

21.1 Introduction

According to the Food and Agricultural Organization (FAO) of the United Nations report (2009), the global population will reach more than 9.3 billion by 2050; so the demand for food, feed, and fiber will also increase [1]. Along with this population increase, there is an ongoing need to improve food quality and quantity [2]. Crop production must increase significantly on the limited arable land available, by using novel crop production techniques to overcome future food scarcity [3].

Plants face different stresses under field conditions. Plant stress can be referred to as an external factor that affects the growth and development of the plant. These stresses are mainly caused by biotic and abiotic factors. Plants respond to stress in

https://doi.org/10.1515/9783110771558-021

different ways, such as altering gene expression, decreasing the crop yield and growth rate, modifying cellular metabolism, and increasing senescence [4]. Biotic agents cause injury and reduce the vigor of the host plant, resulting in increased plant mortality. This process is known as biotic stress. These biological agents include bacteria, fungi, viruses, nematodes, and weeds [5].

In addition, abiotic stress refers to the physicochemical factors that negatively affect plant growth, development, yield, and reproduction. Such factors include drought, excessive water, extreme temperature, salinity, alkalinity, and heavy metals [4]. Abiotic and biotic stresses lower the productivity of crops, which leads to reductions in global food and fiber production [6, 7]. Pandey et al. [8] noted that in previous studies, the spread of biotic agents has been influenced by abiotic factors such as drought, high and low temperature, and salinity. Therefore, combining the studies of abiotic and biotic stress on plants is necessary.

Although microorganisms such as extremophiles can survive in extreme environmental conditions [9], the plant microbiome is a community of microorganisms occupying a reasonably well-defined habitat. These microbes have distinct physicochemical properties that have the potential for both beneficial and harmful effects on plants. These communities include bacteria, viruses, and fungi. Plant microbiomes can be divided into three types: rhizospheric microbes, epiphytic microbes, and endophytic microbes. These microbe groups live in the soil near the roots, establish colonies on the phyllosphere, and live inside the plant tissues, respectively [9]. Some beneficial microorganisms, called plant growth-promoting microbes, can protect plants from stress. They produce phytohormones, such as Indole-3-acetic acid (IAA), gibberellin, abscisic acid (ABA), and cytokinin, to regulate the plants' growth and prevent diseases by producing antibiotics, antifungal agents, and metabolites [5]. Hashem et al. [5] stated that *Bacillus* species can increase plant growth and prevent diseases by producing long-lived stress-tolerant spores and secreting the metabolites. Yadav [10] stated that the species of *Azotobacter*, *Flavobacterium*, *Bacillus*, *Burkholderia*, *Pseudomonas*, *Methylobacterium*, and *Serratia* have reduced stress (drought, alkalinity, acidity, and low/high temperature) and enhanced the yield of crops. Furthermore, *Microbacterium* sp., *Pseudomonas* sp., *Flavobacterium*, *Kluyver* sp., *Curtobacterium* sp., *Clavibacter* sp., *Bacillus* sp., and *Alcaligenes* sp. have the potential to act as biocontrol agents against different plant pathogens.

Understanding the association between microbes and plants can lead to valuable applications in crop protection as they can act as biofertilizers, biopesticides, and weedicides [7, 10].

21.2 Abiotic stress

Plants are subjected to a variety of abiotic stresses, such as drought, salinity, heat, heavy metals, cold, water logging, and acidity. These stressors are interrelated and cause osmotic stress, dysfunction of ion regulation, and plants homeostasis [4]. Soil salinity is the amount of all soluble ions in the soil water and is one of the threats to global agriculture. The level of salinity in irrigated agricultural land has increased by 37% and has caused 10–25% loss yield in the last two decades [11]. Major effects due to salinity are osmotic stress and ion toxicity, and it has reduced the growth rates of plants. Salinity also reduces cell expansion, membrane function, and cytolytic metabolism [12, 13]. Drought is one of the major issues and it depends on low precipitation and global climatic changes. Droughts can limit plant growth and productivity by increasing abscisic acid (ABA) levels, which leads to closure of stomata, thereby greatly reducing the availability of carbon dioxide (CO_2) for photosynthesis, and limiting the release of accumulated reactive oxygen species (ROS). Further, drought stress reduces plant biomass and inhibits flower formation [10]. Increase in global temperatures has the most severe effect on plants. It can affect the hormone activities and cellular metabolism. This leads to growth retardation and disturbs homeostasis. The combined effect of drought and heat stress harms the reproductive process of plants by decreasing pollen fertility [4, 13]. Extreme low temperatures lead to cold stress, which limits plant growth. The effect is determined by the severity and exposure duration. The cold temperature leads to physiological and biochemical changes in plants by accumulating reactive oxygen species (ROS) and increasing cellular Ca^{2++} ions. In addition, sub-optimal temperatures cause internal desiccation by reducing the water uptake by plants and, ultimately, freezing can cause physical damage and protein denaturation. Heavy metals contamination occurs due to industrialization and anthropogenic activities that can affect soil surface and soil water. They affect the metabolic reactions of plants by denaturing enzymes and lowering the photosynthetic activity [14].

21.3 Biotic stress

Plants face a variety of biotic stresses generated by fungi, viruses, bacteria, and nematodes. These biotic agents cause diseases and disorders that damage the plants and decrease crop production and yield [4].

Fungi attack plants in two ways: necrotrophic fungi kill the host by producing toxins and biotrophic fungi consume the host while living on it, which leads to vascular wilts, leaf spots, and cankers in plants [15]. Pandey et al. [8], and Noman et al. [16] reported that fungal pathogens such as *Fusarium* sp., *Alternaria* sp., *Cladosporium* sp., *Colletotrichum* sp., and *Phomopsis* sp., together with a bacterium, *Xanthomonas arboricola*, cause brown necrosis in walnut (*Juglans regia*). Significant pathogenic bacteria, such as *Pseudomonas syringae*, *Dickeya solani*, *Rostonia solonacrarum*, *Agrobacterium tumefaciens*, and *Erwinia amylovora*, also cause major crop losses [17].

21.4 Microbial capacity to protect crops against stress

Earlier in this chapter, there was a brief discussion on how various environmental stresses limit the production of crops. Plants have a diverse group of microorganisms associated with them, known as the plant microbiomes, which counter the harmful effects produced by these stresses. These microbiomes are considered holobiont, that is, species living in or around the plant, and mainly include bacteria and fungi [18]. These microorganisms mostly occupy the rhizospheric region of plants but also exist in the phyllosphere and endosphere regions [19–21]. They have a significant effect on the growth and development of plants as they provide resistance to both biotic and abiotic stress [22] (Table 21.1). In addition, this plant-microbe interaction increases the crop yield and enhances the nutritional value of crops [23, 24]

Table 21.1: Microbial role in protecting crops against stress.

Microbial strain	Crop	Stress	Process used to mitigate stress effects	Reference
Pseudomonas libanensis strain EU-LWNA-33	Wheat	Drought	Increases crop biomass, improves accumulation of proline, and enhances the amount of Glycine betine in the cell	[25]
Enterobacter sp. UPMR18 & UPMR2	Okra	Salt	Improves the rate of seed germination, increases chlorophyll content, and enhances the activity of ROS scavenging enzymes	[26]
Pseudomonas fluorescens, *Pseudomonas migulae* strain 8R6	Tomato	Salt	Increases tolerance against salt stress, increases fresh and dry weight, and enhances chlorophyll content	[27]
Bacillus sp. strain QX8 & QX13	*Solanum nigrum*	Cadmium	Mitigates the negative effects of cd and improves enzymatic activity	[28]

Table 21.1 (continued)

Microbial strain	Crop	Stress	Process used to mitigate stress effects	Reference
Kocuria erythromyxa strain EY43, *Staphylococcus kloosii* strain EY37	Strawberry	Salt	Increases the relative water content, increases yield and higher chlorophyll content as compared to a non-inoculated plant	[29]
Pseudomonas aeruginosa	*Lycopersicon esculentum*	Cadmium	Increases nutrient uptake and tolerance against cadmium stress	[30]
Agrobacterium fabrum strain $CdtS_5$, *Stenotrophomonas maltophilia* strain $CdtS_7$	Wheat	Cadmium	Alleviates cd stress, improves chlorophyll content, and increases rate of seed germination	[31]
Bacillus megaterium strain MCR-8	*Vinca rosea*	Nickel	Increases total soluble sugar, photosynthetic pigments and proline content, and improves activity of antioxidant enzymes	[32]
Pseudomonas fluorescens strain P22, *Pseudomonas* sp. strain Z6	Maize, Sunflower	Copper	Increases stem height and plant biomass, increases uptake of nutrients, and enhances protection against copper stress	[33]
Pseudomonas sp. strain PS3	White radish	Fungicide residues	Improves growth and development, increases sugar and proline content, and enhances activity of antioxidant enzymes	[34]
Pseudomonas putida	Eruca sativa	Nickel	Increases plant biomass, increases activity of photosynthetic pigments, and increases uptake of minerals	[35]
Paenibacillus polymyxa, Bacillus circulans	Maize	Copper	Decreases activity of ROS scavenging enzymes, reduces the amount of MDA content, and enhances crop protection against stress	[36]
Alcaligenes faecalis, Bacillus amylolique faciens	Spinach	Lead	Increases root growth, alleviates lead stress, and increases potassium uptake	[37]

Table 21.1 (continued)

Microbial strain	Crop	Stress	Process used to mitigate stress effects	Reference
Pseudomonas sp. strain Str-1 & Str-2	Okra	Chromium	Improves rate of photosynthesis and transpiration, increases growth rate, and decreases oxidative stress	[38]
P. fluorescens strain K23, *Luteibacter* sp. *Variovorax* sp.	*Lathyrus sativus*	Lead	Increases plant biomass and nodule formation, enhances phenolic content, decreases amount of MDA, and increases membrane stability	[39]
Bacillus subtilis strain HAS31	Potato	Drought	Increases the number of tubers, increases enzymatic activity and chlorophyll content, and increases crop yield	[40]
Rhizobium sp. strain Thal-8, *Pseudomonas* sp. strain 54RB	Zea mays	Salt	Decreases electrolyte leakage, increases accumulation of proline, and improves potassium uptake	[41]
Bacillus subtilis strain DHK & B1N1	Maize	Drought	Increases the amount of osmolytes, increases nutrient uptake, and enhances siderophore production	[42]
Stenotrophomonas maltophilia, Agrobacterium fabrum	Bitter gourd	Cadmium	Improves tolerance against stress, increases yield, and improves uptake of nutrients	[43]
Agrobacterium fabrum, Leclercia adecarboxylata	Zea mays	Chromium	Increases uptake of nitrogen, potassium, and phosphorus, and increases plant height	[44]
P. fluorescens strain A506, *P. gessardii* strain BLP141, *P. fluorescens* strain LMG 2189	Sunflower	Lead	Improves antioxidant activity, increases accumulation of proline, and improves yield than a non-inoculated plant	[45]
Bacillus cereus	Tomato	Heat	Increases plant fresh and dry weight, improves phosphate solubilization, and decreases activity of hydrogen cyanide	[46]

Table 21.1 (continued)

Microbial strain	Crop	Stress	Process used to mitigate stress effects	Reference
Azotobacter chroococcum strain CAZ3,	Maize	Copper, Lead	Increases crop production, improves kernel attributes, including number and protein content, and improves accumulation of osmoprotectants	[47]
Burkholderia phytofirmans strain PsJN	Grapevine	Cold stress	Increases the amount of soluble sugar and starch, and enhances stomatal conductance	[48]
Bacillus cereus strain SA1	Soybean	Heat	Increases resistance to heat stress, increases the level of salicylic acid, decreases the activity of abscisic acid, and improves the activity of antioxidant enzymes	[49]
Pseudomonas frederiksbergensis strain OS211, *Flavobacterium glaciei* strain OB146, P. vancouverensis strain OB155	Tomato	Cold	Decreases electrolyte leakeage and reduces the activity of ROS scavenging enzymes	[50]
Enterobacter sp. strain EG16, *Enterobacter ludwigii* strain DJ3	Tomato	Cadmium	Improves stress tolerance, decreases translocation of cadmium to plant aerial parts, and reduces the availability of cadmium	[51]
Serratia proteamaculans, Pseudomonas putida, P. fluorescens	Tomato	Root knot nematode	Decreases ration of galls and juveniles, improves growth rate as compared to non-inoculated plants, and suppresses disease	[52]
Pantoea agglomerans stain SWg2	Mulberry	Pseudomonas syringae	Protects against Mulberry blight, restricts spread of disease, and increases plant growth and development	[53]

Table 21.1 (continued)

Microbial strain	Crop	Stress	Process used to mitigate stress effects	Reference
Stenotrophomonas maltophilia strain SBP-9	Wheat	*Fusarium graminearum*, Salt	Decreases the amount of proline, increases the antioxidant activity, and enhances resistance to fungal infection	[54]
Pseudomonas fluorescens strain Pf.SS101	*Arabidopsis thaliana*	*Pseudomonas syringae*	Induces systemic resistance against disease and increases the number of lateral roots	[55]
Glomus intraradices, *Acinetobacter* sp.	Oat	Petroleum	Reduces MDA accumulation, increases tolerance against harmful effects of petroleum residues, and improves growth and development	[56]
Bacillus xiamenensis strain PM14	Sugarcane	Red rot disease	Induces systemic resistance through hyperparasitism, increases the activity of antioxidants, and decreases electrolyte leakage	[57]
Bacillus firmus	Tomato	*Meloidogyne incognita*	Increases plant biomass, decreases nematode population, and reduces the formation of galls and eggs	[58]
Trichoderma viridae strain ES-1, *Pseudomonas fluorescens* strain Bak-150	Potato	*Phytophthora infestans*	Increases resistance to disease, and lowers extension of fungal micellium	[59]

21.5 Mechanisms of microbiome-mediated stress tolerance in plants

There are certain biotic and abiotic stresses in nature that plants face. They disintegrate the cellular membrane, release reactive oxygen species (ROS), and reduce the rate of photosynthesis, leading to decreased crop yield [60]. Plants contain an inbuilt system that mitigates the effects of stress, but they also rely heavily on the supporting microbial community to alleviate lasting damage [61]. The plant microbiome improves crop yield and protects the plant from both biotic and abiotic stress using direct and indirect mechanisms, as depicted in Figure 21.1 [62]. Directly, they fix the nitrogen from the soil, produce siderophores, develop the roots, and degrade pollutants to make them unavailable to the plant [63]. Indirectly, they induce systematic resistance, produce antibiotics, and compete with pathogens [64].

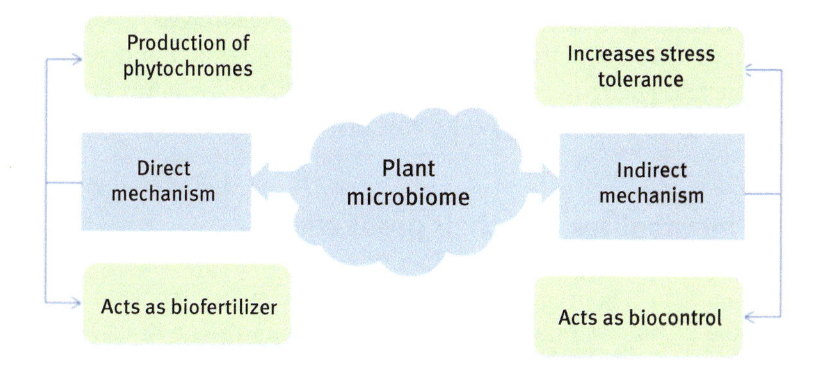

Figure 21.1: Depiction of microbial mechanisms used to mitigate stress for plants.

21.5.1 Nitrogen fixation

Nitrogen is considered one of the most important micronutrients for plants [64] and plays a major role in plant growth and development, and directly affects crop yield [65]. It is one of the most abundant elements in nature (78%) but is unavailable to plants in its atmospheric form [66]. However, it is made accessible to plants by nitrogen-fixing bacteria that convert the atmospheric nitrogen to the usable ammonia form [67]. These bacteria include the ones that form symbiotic associations in the plant rhizosphere and free-living endophytes [68].

During stress conditions, the plant's ability to uptake nitrogen gets reduced. Exposure of plants to heavy metals can reduce the assimilation process while salt stress greatly affects the process of nodulation by reducing the activity of the nitrogenase enzyme [69, 70]. It has been found that inoculation of plants with nitrogen-fixing microbial species helps manage diseases by balancing the nitrogen level available to the plant [71].

21.5.2 Phosphate solubilization

Apart from nitrogen, phosphorus is also a key element in the growth and development of plants. Its major role in a plant is signal transduction at the cellular level [72]. Soil contains phosphorus, both in organic and inorganic forms, but despite its abundance, plants cannot utilize it. The reason for this nonavailability to plants is the presence of phosphorus in insoluble and immobilized forms [8]. The plant can only utilize phosphorus when it is available in monobasic and diabasic forms [73]. The plant microbiota, especially rhizobacteria, strategically converts the nonavailable forms to available ones by mineralizing the phosphate [74]. The most-reported phosphate-solubilizing bacteria include *Rhizobium*, *Bacillus*, *Rhodococcus*, *Pseudomonas*, and *Enterobacter* [64]. They also play a role in the suppression of disease and can facilitate an increase in plant growth. The bacterial genera, including *Bacillus* and *Rhizobium*, are reported to be effective against plant pathogens [75].

21.5.3 Microbial metabolites and their production

A large quantity of metabolites is produced by rhizospheric microbes. They do not play a direct role in the growth and development of plants but act as a line of defense against various stresses faced by plants [76]. There are several groups of metabolite-producing microbes, but the one that is more beneficial to plants is plant growth-promoting rhizobacteria (PGPR) [77]. For instance, in a study presented by Jain et al. [78] maximum root exudates (phenolic compounds) were produced by *Pisum sativum* when inoculated with *Bacillus subtilis*. These exudates help the plant to sustain against biotic and abiotic stress [79]. Based on function, microbial metabolites are categorized into two groups as shown in Figure 21.2.

Various amino acids, vitamins, and organic acids come under the category of primary metabolites and they are released by the microbes when they are in their growth phase of development [80]. These primary metabolites do not directly help in the defense against stress but play an important role in plant growth and development, while secondary metabolites that are released by the primary metabolites act directly to protect the plant against biotic and abiotic stress [81].

21.5.3.1 Siderophores

Iron is regarded as a crucial element that helps to sustain life forms and is necessary for all living organisms [82]. This element is plentiful in the environment but is not easily accessible mainly due to its presence in the Fe^{3+} form [83]. Plant-associated microbes are helpful in this regard as they release siderophores. Siderophores are defined as low molecular weight peptides that can bind iron to them [84]. Microbial siderophores chelate the insoluble iron and provide it to the plant in the soluble form. They not only enhance the growth and development of plants by providing iron but also limit the development of phytopathogens by making the iron that is not available to them [85]. Siderophores not only sequester the iron but also make complexes with other metals, such as zinc, cadmium, lead, and copper, mitigating the heavy metal stress in the plant [86]. Some of the common siderophore-producing

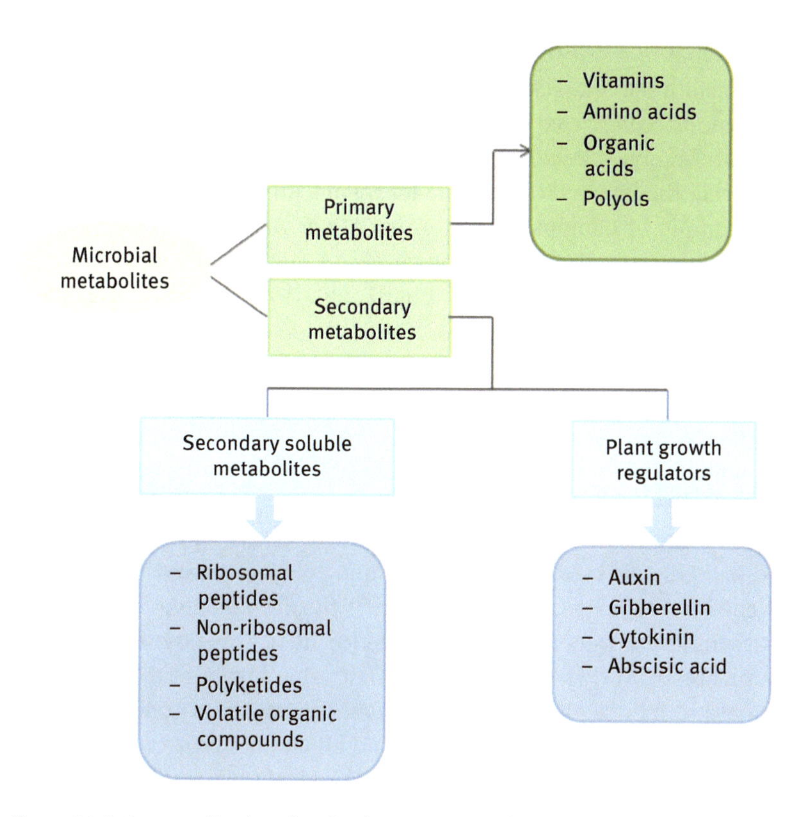

Figure 21.2: A generalized outline for the categories of microbial metabolites.

bacterial genera are *Azotobacter*, *Bacillus*, *Nocardia*, *Pseudomonas*, and *Rhizobium* [87–89]. Furthermore, Jacob et al. [90] elaborated in their study that stem and root disease in groundnut was prevented when the plant was inoculated with *Streptomyces* sp. RP1A-12. Another study indicated that inoculation of chickpea plants with *Pseudomonas* strain was found effective in limiting root disease [91]. Many siderophore-producing bacterial species are also reported to have the ability to help plants fight against nematode infection [92].

21.5.3.2 Osmolytes

When a plant faces stress, especially drought and salinity stress, the amount of reactive oxygen species is increased [93]. This increase causes an osmotic imbalance as there is a leakage of electrolytes and H_2O from the cell. To maintain the cell osmotic balance, the plant accumulates lower molecular weight compounds called osmolytes that are released from microbial cells [94]. These osmolytes include trehalose, amino acids, polyols, glycine betaine, and polyamines [95]. Glycine betaine is an important quaternary osmolyte that is produced by bacteria. During stress conditions, it protects the thylakoid membrane and accumulates a large number of K^+ ions, while trehalose is another osmoprotectant that stabilizes the protein structure [36, 96, 97]. The chemical structure of osmolytes released by microbes is compatible with the protein channels in a plant cell; that is why it is feasible for them to uptake these compounds directly [98]. An example of osmoprotectant stimulation: *Arthrobacter*-inoculated wheat plants were reported to activate signaling pathways that synthesized osmolytes against salt stress [99]. Similarly, according to Bharti et al. [100], in PGPR-inoculated *Bacopa monnieri*, there was an enhanced release of Bacoside A compound to defend against salt stress.

21.5.3.3 ACC deaminase

Ethylene is a hormone necessary for the growth and development of plants, and the amount produced is highly affected by various physiological factors and stresses. An elevation in ethylene level above a certain limit restricts the development of the plant. It also damages tissues and lowers the plant biomass [65, 101]. This phytohormone is produced by the precursor ACC (1-Aminocyclopropane-a-carboxylate) that uses the enzyme ACC synthase to induce ethylene biosynthesis. During stress conditions, microbes release ACC deaminase that degrades ACC into ammonia and 2-oxobutanoate [102]. In a study by Zhu et al. [103], an ACC deaminase-producing microbial stain *Varivorax paradoxus* 5C-2 was inoculated in a pea plant that helped the plant survive extreme salt stress. In another study presented by Vimal et al. [104], inoculation of *Oryza sativa* with *Curtobacterium albidum*, in the presence of salt stress, showed enhanced growth because of the release of ACC deaminase enzymes.

21.5.3.4 Induced systemic resistance

Induced systemic resistance is the induction of a defense mechanism by microbes against phytopathogens. This type of resistance is not specific for any pathogen but can be induced against all kinds of biotic stress [105]. In this regard, beneficial microbes present in the rhizosphere produce certain enzymes such as catalase, superoxide dismutase, ascorbate peroxidase, and peroxidase [106]. These enzymes that are produced increase the number of phenolic compounds in plant cells and thus resist the effect of the disease.

Induced systemic resistance is not only shown by bacterial species; some fungal genera present in the rhizosphere also contain this defense system. The fungal species deposits lignin in the plant's cell wall that acts as a barrier against any plant pathogen [107]. This fungal-induced systemic resistance also showed the potential for defense against various bacteria. For example, angular leaf spot caused by *Pseudomonas syringae* was successfully controlled by plant growth-promoting rhizobacteria [108].

Similarly, many microbes release salicylic acid, and its accumulation acts as a signal that triggers the NPR1 gene. This protein is responsible for the activation of both induced systemic resistance and systemic acquired resistance [101]. NPR-1 is the main protein behind the phenomena of induced systemic resistance that is encoded by the NPR1 gene. According to Hallmann et al. [109], *Rhizobium elite* acted as a defense against the infection caused by cyst nematodes in potatoes through induction of systemic resistance. Furthermore, treating tomatoes with a microbial consortium increased the resistance against nematodes by increasing the number of defense enzymes such as phenylalanine [106].

21.5.3.5 Volatile organic compounds

Volatile organic compounds and their relationship with plants against stress were first elucidated in the findings of Ryu et al. [110]. These organic compounds contain sulfur and are generally released by *Bacillus* species. They help plants in growth and development [107], and defense against nematodes and fungal diseases [111]. These compounds mainly include indole, terpenes, sulfide compounds, pyrrols, aniline, and pyrazines [76]. In the 2004 study by Ryu et al. [110], results indicated that *Bacillus* sp. produced 2, 3 butanediol, which increased the biomass of *Arabidopsis thaliana*. Some strains of *Pseudomonas* were reported to produce hydrogen cyanide that acts as a biocontrol agent [107]. Another volatile organic compound that restricts the spread of *Erwinia carotovora* in the tobacco plant is 3R-butanediol that is released by *Pseudomonas* [72].

21.5.3.6 Exopolysaccharides

Extracellular polysaccharides are produced by microbial species such as *Bacillus* and *Rhizobium*. They are important as they form a biofilm, colonize the root, and protect from phytopathogens. Crown rot disease in peanuts was successfully eradicated by *Paenibacillus polymyxa* [107].

21.5.3.7 Production of lytic enzymes

Microbial cells release different enzymes that inhibit the growth of pests, especially nematodes. These enzymes mainly include chitinase, protease, peroxidase, and lipases. Mhatre et al. [106] reported that the release of chitinase enzyme from *Corynebacterium* sp. restricted the eggs of nematode from hatching. Similarly, Insunza et al. [112] isolated three bacterial stains from plants that have nematicidal ability and it decreased the population of nematode on potatoes by half.

21.5.3.8 Phytochromes

When a plant faces any stress, the microbial community associated with it releases certain proteins that are known as phytochromes [113]. These phytochromes are one type of secondary metabolites that act as chemical messengers to regulate various cellular and physiological processes in plants [114]. The most common phytochromes include auxin, gibberellin, cytokinin, and abscisic acid.

Auxin
Auxin is also known as IAA (Indole-3-acetic acid) and is one of the widely reported phytochromes from PGPR. This phytochrome not only acts as a line of defense against phytopathogens but also plays its role in growth and development of the plant [115]. There are a total of six pathways that are reported to synthesize auxin. These pathways include the indole-3-pyruvate pathway, the tryptamine pathway, the acetonitrile pathway, the indole-3-acetonitrile pathway, the tryptophan-independent pathway, and the tryptophan side-chain oxidase pathway [116, 117]. In a study presented by Egamberdieva et al. [113] showing the beneficial effects of PGPR, it was reported that inoculation of wheat with *Pseudomonas aurantiaca* decreased the saline and sodic stress within the crop.

Cytokinin
Cytokinin is the phytochrome produced by both plants and microbes for better seed germination and fruit development [91]. It is commonly produced by two bacterial genera, *Bacillus* and *Pseudomonas* [118]. Zeatin is the common cytokinin that is required by seeds to increase their germination rate. Cytokinin is generated by two microbial pathways, the tRNA pathway and the adenosine monophosphate pathway [119]. Großkinsky et al. [120] reported that inoculation of *Arabidopsis* with cytokinin-producing *Pseudomonas fluorescens* prevented the infection of *P. syringae* in the controlled samples. The common cytokinin-producing strains include *Pseudomonas*, *Azospirillum*, *Bacillus*, and *Arthrobacter* [121].

Gibberellin
Gibberellins are tetracyclic carboxylic acids that are necessary for plant development. They play an important role in the germination of seeds and elongation of the stem [70]. Gibberellins are also produced by bacterial species, of which the most common genera include *Azotobacter*, *Azospirillum*, *Bacillus*, *Rhizobacteria*,

and *Acinetobacter* [122]. According to Tuna et al. [123], maize plants, when exposed to Gibberellin (GA_3), sustained the negative effects produced by lead and cadmium.

Abscisic acid

One of the negative effects of stress on plants is that it damages the protein structure and thus various metabolic processes in plants get disturbed. Abscisic acid (ABA), released by the rhizospheric microbes against stress, especially abiotic stress, acts as the first line of defense. For instance, Bücker-Neto et al. [124] reported that brassica plants, inoculated with *Bacillus* sp., released abscisic acid under a high amount of lead toxicity.

21.5.3.9 Antibiotics

Antibiotics are chemical substances that are released by some bacterial species. They are low in molecular weight, and compete with and parasitize pests [106]. Microbial species produce these antibiotics that act as toxins to reduce/inhibit the reproduction of nematodes [125].

Most of the bacterial species that produce antibiotics are ammonifying bacteria. They decompose nitrogenous compounds, producing ammonia, as a result of which the survival rate of nematodes is lowered [126]. Some of the bacterial species also produce hydrogen cyanimide around the roots that also helps in reducing the plant pest [127]. One study also reported that *Pseudomonas fluorescens* reduces the number of nematodes around plants and acts on the phenomena of antibiosis [128].

21.5.3.10 Hyperparasitism

Hyperparasitism is one of the microbial modes of action against phytopathogens in which both organisms are parasites. Most of this interaction is observed between fungal species although there are rare cases of bacteria. In bacterial species, the strains of *Bdellovibrio bacteriovorus* are the ones that hunt down other bacterial pathogens, mainly *Agrobacterium tumefaciens* and *Pseudomonas syringae* [129]. In fungal genera, hyperparasitism is mostly investigated. For example, more than 30 species have so far been reported as mycoparasites for a fungal species *Rhizoctonia solani*. A vast number of studies reported two common genera involved in hyperparasitism, *Trichoderma* and *Clonostachys* [130, 131]. Similarly, it was reported that, including *Cladosporium* sp., there are more than 30 species of fungi that act against rust pathogens [132]. All of the species involved in hyperparasitism release lytic enzymes that kill their hosts.

21.5.3.11 Biotransformation of toxic compounds

The toxic compounds that plants usually face are from heavy metals. These heavy metal compounds decrease the yield of the crop in many ways as discussed earlier in this chapter. The microbial community associated with plants has developed several mechanisms to mitigate the negative effects of heavy metals. These mechanisms are discussed in detail below:

Detoxification of heavy metals

Nature has presented plants with microbiota that can remediate heavy metals [133]. This microbiota includes a wide range of bacterial genera that includes *Bacillus*, *Pseudomonas*, *Enterobacter*, and *Rhizophagus* [134]. These genera are not only capable of remediating heavy metals but also have the ability to detoxify multiple pollutants. In these genera, *Bacillus* and *Pseudomonas* are the ones that can remediate almost every heavy metal [135]. Not only bacterial species but some of the fungal species are also reported to alleviate the heavy metal stress in plants, such as *Penicillium*, *Trichoderma* and *Aspergillus*. According to [136], when chickpea plants were inoculated with *Trichoderma*, it mitigated the harmful effects of arsenic. Another study [137] indicated that *Humicola* sp. strain 2WS detoxified arsenic in *Bacopa monnieri*. Amongst all the soil microbes, the ones present in the plant rhizosphere are highly effective in heavy metal detoxification, and the process is called rhizoremediation [138]. These microbial species mainly use two strategies to lower the effect of heavy metals in plants, biosorption and bioaccumulation [139]. Apart from these, there are other mechanisms as well, such as acidification and biotransformation that microbes use to lower the toxicity of heavy metals.

Biosorption

Biosorption is the remediation strategy that sequesters heavy metals from soil. It has the advantage over other strategies in that it is not dependent upon metabolism and is faster than other methods. Within microbes, bacteria commonly biosorbs heavy metal because it has the advantage of being extremely small. The most common genera of bacteria that have the capacity of biosorption are *Micrococcus*, *Pseudomonas*, *Escherichia*, and *Bacillus* [140, 141]. Many fungal genera can also sequester heavy metals. These fungal genera include *Penicillium*, *Rhizopus*, *Aspergillus*, and *Phanerochaete* [142].

These species have chemosorption sites on which cations of heavy metals bind. These cations further pass through the cell membrane using binding proteins and react with anions present in the microbial cell. In the end, these are volatilized by certain enzymes present inside the cell [143].

Bioaccumulation

Bioaccumulation, known as active biosorption, is also the accumulation of heavy metals in the cell, but is different from biosorption in that this process does not maintain equilibrium and depends upon the metabolism of cells, unlike the biosorption process [144, 145]. It also takes more time to complete and is only performed by living cells or biomass [146].

In bioaccumulation, microbes use protein complexes through which the heavy metal easily gets translocated into the cell. Inside the cell, there are specific ligands that cloister these heavy metals and make them unavailable for the plant to uptake [147, 148]. Bioaccumulation processes are carried out by different bacterial and fungal species such as *Saccharomyces cerevisiae*, *Candida tropicalis*, *Candida* sp., *Pichia guilliermondii*, *Geobacillus thermantarcticus*, and *Pseudomonas aeruginosa* [145, 149–151].

Acidification

Microorganisms, especially the ones that are present in the rhizosphere, mobilize the heavy metal compounds and make them unavailable to plants. They secrete organic ligands that make the heavy metal insoluble. They also produce organic acids, such as oxalate, malate, and succinate that lower the pH around roots, which in turn influences the speciation of heavy metals [152, 153]. The heavy metal ions present in the soil bind with the organic acids released by microbes and form complexes. There is a wide range of microbial species that releases organic acids but the most important is *Pseudomonas fluorescens*. According to the finding of Sheng et al. [154], the acids produced by *Microbacterium* sp. G16 mobilized lead effectively. Similarly, Wani et al. [155] presented that the *Bacillus* strains mobilized both zinc and lead by releasing organic acids. Therefore, it can be concluded that lower pH increases the mobility of heavy metals [156].

Transformation of heavy metals

In microbial cells, heavy metals are transformed by redox reactions that make them bioavailable in less toxic forms [157]. The lowering of oxidation states also makes the heavy metal more soluble. According to Khan and Bano [158], *Geobacillus* sp. was successful in transforming arsenite into arsenate, which is the least toxic form of arsenic. In another study by Jobby et al. [159], *Sinorhizobium* sp. SAR1 converted the toxic form of chromium (i.e. hexavalent) into the least toxic trivalent form.

21.6 Conclusion and future perspective

It can be concluded from the above review that because of the increases in environmental change, crop production is facing serious threats. Such threats can be minimized by the microbiome associated with plants. Plant microbiome plays a major role in the protection of crops as they release certain enzymes, metabolites, and hormones against both biotic and abiotic stress. They also develop Induced Systemic Resistance against phytopathogens. Now is the time to utilize microbes commercially for the betterment of crops. They can be used as biopesticides to protect crops and increase their yield.

Recently, more work has been done on the indigenous microbial species and their interaction with plants at the molecular level. There are more advances in this category because of the use of model plants that helps in elaborating the plant-microbiome interaction. Apart from the old approaches, there is also an inclusion of new techniques, such as liquid-chromatography, mass spectrophotometry, and mass spectrophotometry imaging, that helps to carry out the study of microbial metabolites at an advanced level. This type of new instrumentation is further required to assess the methodology through which these microbes are used at commercial levels against plant stress.

From a future perspective, there is a need to design studies that involve the evolutionary relationship between plants and microbes, as there is a huge amount of data already available on the physiological interaction of plants and their microbiome. Epigenetics can be used to assess the compatibility between plants and microbes at a genetic level over time. This will help to ascertain whether the close and long-term biological relationship between plants and microbes is either mutualistic or commensalistic.

References

[1] Le Mouël C, Forslund A. How can we feed the world in 2050? A review of the responses from global scenario studies. European Review of Agricultural Economics 2017, 44(4), 541–591. doi:https://10.1093/erae/jbx006.

[2] Tian L, Lin X, Tian J, et al. Research advances of beneficial microbiota associated with crop plants. International Journal of Molecular Sciences 2020, 21(5), 1–18. doi:https://10.3390/ijms21051792.

[3] Timmusk S, Zucca C. The plant microbiome as a resource to increase crop productivity and soil resilience: A systems approach. Journal of the Cameroon Academy of Sciences 2019, 14(3), 181. doi:https://10.4314/jcas.v14i3.2.

[4] Gull A, Ahmad Lone A, Ul Islam Wani N. Biotic and Abiotic Stresses in Plants. Abiotic and Biotic Stress in Plants 2019, 1–6. doi:https://10.5772/intechopen.85832.

[5] Hashem A, Tabassum B, Fathi Abd_Allah E. *Bacillus subtilis*: A plant-growth promoting rhizobacterium that also impacts biotic stress. Saudi Journal of Biological Sciences 2019, 26(6), 1291–1297. doi:https://10.1016/j.sjbs.2019.05.004.

[6] Ke J, Wang B, Yoshikuni Y. Microbiome engineering: Synthetic biology of plant-associated microbiomes in sustainable agriculture. Trends in Biotechnology 2021, 39(3), 244–261. doi:https://10.1016/j.tibtech.2020.07.008.

[7] Srivastava AK, Kashyap PL, Santoyo G, Newcombe G. Editorial: Plant microbiome: Interactions, mechanisms of action, and applications. Frontiers in Microbiology 2021, 12(July), 10–13. doi:https://10.3389/fmicb.2021.706049.

[8] Pandey P, Irulappan V, Bagavathiannan MV, Senthil-Kumar M. Impact of combined abiotic and biotic stresses on plant growth and avenues for crop improvement by exploiting physio-morphological traits. Frontiers in Plant Science 2017, 8(April), 1–15. doi:https://10.3389/fpls.2017.00537.

[9] Nath Yadav A, Yadav N, Author C. Stress-adaptive microbes for plant growth promotion and alleviation of drought stress in plants bioremediation and waste management for environmental sustainability view project special issue on "Microbes for agricultural and environmental sustainability. 2018(May), 2–6. https://www.researchgate.net/publication/325537515

[10] Yadav AN. Beneficial plant-microbe interactions for agricultural sustainability. Journal of Applied Biology and Biotechnology 2021, 9(1), i–iv. doi:https://10.7324/JABB.2021.91ed.

[11] Waqas MA, Kaya C, Riaz A, et al. Potential mechanisms of abiotic stress tolerance in crop plants induced by thiourea. Frontiers in Plant Science 2019, 10(October), 1–14. doi:https://10.3389/fpls.2019.01336.

[12] Rajput VD, Minkina T, Kumari A, et al. Coping with the challenges of abiotic stress in plants: New dimensions in the field application of nanoparticles. Plants 2021, 10(6), 1–25. doi:https://10.3390/plants10061221.

[13] Suzuki N, Rivero RM, Shulaev V, Blumwald E, Mittler R. Abiotic and biotic stress combinations. The New Phytologist 2014, 203(1), 32–43. doi:https://10.1111/nph.12797.

[14] Imran QM, Falak N, Hussain A, Mun BG, Yun BW. Abiotic stress in plants, stress perception to molecular response and role of biotechnological tools in stress resistance. Agronomy 2021, 11, 8. doi:https://10.3390/agronomy11081579.

[15] Iqbal Z, Iqbal MS, Hashem A, Abd_Allah EF, Ansari MI. Plant defense responses to biotic stress and its interplay with fluctuating dark/light conditions. Frontiers in Plant Science 2021, 12(March). doi:https://10.3389/fpls.2021.631810.

[16] Noman M, Ahmed T, Ijaz U, et al. Plant–microbiome crosstalk: Dawning from composition and assembly of microbial community to improvement of disease resilience in plants. International Journal of Molecular Sciences 2021, 22(13). doi:https://10.3390/ijms22136852.

[17] Compant S, Samad A, Faist H, Sessitsch A. A review on the plant microbiome: Ecology, functions, and emerging trends in microbial application. Journal of Advanced Research 2019, 19, 29–37. doi:https://10.1016/j.jare.2019.03.004.

[18] Trivedi P, Leach JE, Tringe SG, Sa T, Singh BK. Plant–microbiome interactions: From community assembly to plant health. Nature Reviews: Microbiology 2020, 18(11), 607–621. doi:https://10.1038/s41579-020-0412-1.

[19] Finkel OM, Burch AY, Lindow SE, Post AF, Belkin S. Geographical location determines the population structure in phyllosphere microbial communities of a salt-excreting desert tree. Applied and Environmental Microbiology 2011, 77(21), 7647–7655. doi:https://10.1128/AEM.05565-11.

[20] Inceoğlu Ö, Al-Soud WA, Salles JF, Semenov AV, van Elsas JD. Comparative analysis of bacterial communities in a potato field as determined by pyrosequencing. PLoS One 2011, 6(8), e23321. doi:https://10.1371/journal.pone.0023321.

[21] Rodriguez RJ, White JF, Arnold AE, Redman RS. Fungal endophytes: Diversity and functional roles: Tansley review. The New Phytologist 2009, 182(2), 314–330. doi:https://10.1111/j.1469-8137.2009.02773.x.

[22] Schuhegger R, Ihring A, Gantner S, et al. Induction of systemic resistance in tomato by N-acyl -L-homoserine lactone-producing rhizosphere bacteria. Plant, Cell & Environment 2006, 29(5), 909–918. doi:https://10.1111/j.1365-3040.2005.01471.x.

[23] Goicoechea N, Antolín MC. Increased nutritional value in food crops. Microbial Biotechnology 2017, 10(5), 1004–1007. doi:https://10.1111/1751-7915.12764.

[24] Ye L, Zhao X, Bao E, Li J, Zou Z, Cao K. Bio-organic fertilizer with reduced rates of chemical fertilization improves soil fertility and enhances tomato yield and quality. Scientific Reports 2020, 10(1), 1–11. doi:https://10.1038/s41598-019-56954-2.

[25] Kour D, Rana KL, Sheikh I, et al. Alleviation of drought stress and plant growth promotion by Pseudomonas libanensis EU-LWNA-33, a drought-adaptive phosphorus-solubilizing bacterium. Proceedings of the National Academy of Sciences, India, Section B: Biological Sciences 2020, 90(4), 785–795. doi:https://10.1007/s40011-019-01151-4.

[26] Habib SH, Kausar H, Saud HM. Plant growth-promoting rhizobacteria enhance salinity stress tolerance in okra through ROS-scavenging enzymes. BioMed Research International 2016, 2016, doi:https://10.1155/2016/6284547.

[27] Ali S, Charles TC, Glick BR. Amelioration of high salinity stress damage by plant growth-promoting bacterial endophytes that contain ACC deaminase. Plant Physiology and Biochemistry: PPB / Societe Francaise de Physiologie Vegetale 2014, 80, 160–167. doi:https://10.1016/j.plaphy.2014.04.003.

[28] He X, Xu M, Wei Q, et al. Promotion of growth and phytoextraction of cadmium and lead in Solanum nigrum L. mediated by plant-growth-promoting rhizobacteria. Ecotoxicology and Environmental Safety 2020, 205(June), 111333. doi:https://10.1016/j.ecoenv.2020.111333.

[29] Karlidag H, Yildirim E, Turan M, Pehluvan M, Donmez F. Plant growth-promoting rhizobacteria mitigate deleterious effects of salt stress on strawberry plants (*Fragaria× ananassa*). HortScience 2013, 48(5), 563–567.

[30] Khanna K, Jamwal VL, Gandhi SG, Ohri P, Bhardwaj R. Metal resistant PGPR lowered Cd uptake and expression of metal transporter genes with improved growth and photosynthetic pigments in Lycopersicon esculentum under metal toxicity. Scientific Reports 2019, 9(1), 1–14. doi:https://10.1038/s41598-019-41899-3.

[31] Zafar-ul-hye M, Shahjahan A, Danish S, Abid M, Qayyum MF. Mitigation of cadmium toxicity induced stress in wheat by ACC-deaminase containing PGPR isolated from cadmium polluted wheat rhizosphere. Pakistan Journal of Botany 2018, 50(5), 1727–1734.

[32] Khan WU, Ahmad SR, Yasin NA, Ali A, Ahmad A, Akram W. Application of Bacillus megaterium MCR-8 improved phytoextraction and stress alleviation of nickel in Vinca rosea. International Journal of Phytoremediation 2017, 19(9), 813–824. doi:https://10.1080/15226514.2017.1290580.

[33] Abbaszadeh P, Farhad D, Atajan A, Omidvari M, Tahan V, Kariman K. Mitigation of copper stress in maize (Zea mays) and sunflower (Helianthus annuus) plants by copper-resistant Pseudomonas strains. Current Microbiology 2021, 78(4), 1335–1343. doi:https://10.1007/s00284-021-02408-w.

[34] Khan S, Shahid M, Khan MS, Syed A, Bahkali AH. Fungicide-tolerant plant growth-promoting rhizobacteria mitigate physiological disruption of white radish caused by fungicides used in the field cultivation. International Journal of Environmental Research and Public Health 2020, 17(19), 7251. doi:https://doi.org/10.3390/ijerph17197251.

[35] Aqeel M, Ali S, Akber M, et al. Ecotoxicology and environmental safety bioaccumulation of nickel by E. sativa and role of plant growth promoting rhizobacteria (PGPRs) under nickel stress. Ecotoxicology and Environmental Safety 2016, 126(256–263). doi:https://10.1016/j.ecoenv.2016.01.002.
[36] Abdelrahman M, Jogaiah S, Burritt DJ, Tran LSP. Legume genetic resources and transcriptome dynamics under abiotic stress conditions. Plant, Cell & Environment 2018, 41(9), 1972–1983. doi:https://10.1111/pce.13123.
[37] Zafar-Ul-hye M, Naeem M, Danish S, et al. Alleviation of cadmium adverse effects by improving nutrients uptake in bitter gourd through cadmium tolerant rhizobacteria. Environments – MDPI 2020, 7(8), 1–16. doi:https://10.3390/environments7080054.
[38] Mushtaq Z, Asghar HN, Zahir ZA. Comparative growth analysis of okra (Abelmoschus esculentus) in the presence of PGPR and press mud in chromium contaminated soil. Chemosphere 2021, 262, 1–8. doi:https://10.1016/j.chemosphere.2020.127865.
[39] Abdelkrim S, Jebara SH, Saadani O, Jebara M. Potential of efficient and resistant plant growth-promoting rhizobacteria in lead uptake and plant defence stimulation in Lathyrus sativus under lead stress. Plant Biology 2018, 20(5), 857–869. doi:https://10.1111/plb.12863.
[40] Batool T, Ali S, Seleiman MF, et al. Plant growth promoting rhizobacteria alleviates drought stress in potato in response to suppressive oxidative stress and antioxidant enzymes activities. Scientific Reports 2020, 10(1), 1–19. doi:https://10.1038/s41598-020-73489-z.
[41] Bano A, Fatima M. Salt tolerance in Zea mays (L). following inoculation with Rhizobium and Pseudomonas. Biology and Fertility of Soils 2009, 45(4), 405–413. doi:https://10.1007/s00374-008-0344-9.
[42] Sood G, Kaushal R, Sharma M. Alleviation of drought stress in maize (Zea mays L.) by using endogenous endophyte Bacillus subtilis in North West Himalayas. Acta Agriculturae Scandinavica, Section B – Soil & Plant Science 2020, 70(5), 361–370. doi:https://10.1080/09064710.2020.1743749.
[43] Zafar-ul-hye M, Naeem M, Danish S, et al. Plants effect of cadmium-tolerant rhizobacteria on growth attributes and chlorophyll contents of bitter gourd under cadmium toxicity. Plants 2020, 9(10), 1386. doi:https://doi.org/10.3390/plants9101386.
[44] Danish S, Kiran S, Fahad S, et al. Alleviation of chromium toxicity in maize by Fe fortification and chromium tolerant ACC deaminase producing plant growth promoting rhizobacteria. Ecotoxicology and Environmental Safety 2019, 185(July), 109706. doi:https://10.1016/j.ecoenv.2019.109706.
[45] Saleem M, Asghar HN, Zahir ZA, Shahid M. Impact of lead tolerant plant growth promoting rhizobacteria on growth, physiology, antioxidant activities, yield and lead content in sunflower in lead contaminated soil. Chemosphere 2018, 195, 606–614. doi:https://10.1016/j.chemosphere.2017.12.117.
[46] Mukhtar T, Rehman S, Smith D, Sultan T. Mitigation of heat stress in Solanum lycopersicum L. by ACC – Deaminase and exopolysaccharide producing Bacillus cereus: Effects on biochemical profiling. Sustainability 2020(May), 12(6), 2159. doi:https://10.3390/su12062159.
[47] Rizvi A, Khan MS. Ecotoxicology and environmental safety heavy metal induced oxidative damage and root morphology alterations of maize (*Zea mays* L.) plants and stress mitigation by metal tolerant nitrogen fixing *Azotobacter chroococcum*. Ecotoxicology and Environmental Safety 2018, 157(March), 9–20. doi:https://10.1016/j.ecoenv.2018.03.063.
[48] Fernandez O, Theocharis A, Bordiec S, et al. *Burkholderia phytofirmans* PsJN acclimates grapevine to cold by modulating carbohydrate metabolism. Molecular Plant-Microbe Interactions 2012, 25(4), 496–504.

[49] Khan MA, Asaf S, Khan AL, et al. Thermotolerance effect of plant growth-promoting *Bacillus cereus* SA1 on soybean during heat stress. BMC microbiology 2020, 20(1), 1–14.
[50] Subramanian P, Kim K, Krishnamoorthy R. Cold stress tolerance in psychrotolerant soil bacteria and their conferred chilling resistance in tomato (Solanum lycopersicum Mill.) under low temperatures. PLoS one 2016, 11(8), 1–17. doi:https://10.1371/journal.pone.0161592.
[51] Li Y, Zeng J, Wang S, Lin Q, Ruan D. Effects of cadmium-resistant plant growth-promoting rhizobacteria and Funneliformis mosseae on the cadmium tolerance of tomato (Lycopersicon esculentum L.). International Journal of Phytoremediation 2020, 22(5), 451–458. doi: https://10.1080/15226514.2019.1671796.
[52] Zhao D, Zhao H, Zhao D, et al. Isolation and identification of bacteria from rhizosphere soil and their effect on plant growth promotion and root-knot nematode disease. Biological Control 2018, 119(January), 12–19. doi:https://10.1016/j.biocontrol.2018.01.004.
[53] Xie J, Shu P, Strobel G, et al. Pantoea agglomerans SWg2 colonizes mulberry tissues, promotes disease protection and seedling growth. Biological Control 2017, 113(9–17). doi:https://10.1016/j.biocontrol.2017.06.010.
[54] Singh RP, Jha PN. The PGPR stenotrophomonas maltophilia SBP-9 augments resistance against biotic and abiotic stress in wheat plants. Frontiers in Microbiology 2017, 8, 1945. doi:https://10.3389/fmicb.2017.01945.
[55] Cheng X, Etalo DW, van de Mortel JE, et al. Genome-wide analysis of bacterial determinants of plant growth promotion and induced systemic resistance by Pseudomonas fluorescens. Environmental Microbiology 2017, 19(11), 4638–4656. doi:https://10.1111/1462-2920.13927.
[56] Xun F, Xie B, Liu S, Guo C. Effect of plant growth-promoting bacteria (PGPR) and arbuscular mycorrhizal fungi (AMF) inoculation on oats in saline-alkali soil contaminated by petroleum to enhance phytoremediation. Environmental Science and Pollution Research 2015, 22(1), 598–608. doi:https://10.1007/s11356-014-3396-4.
[57] Amna, Xia Y, Farooq MA, et al. Multi-stress tolerant PGPR Bacillus xiamenensis PM14 activating sugarcane (Saccharum officinarum L.) red rot disease resistance. Plant Physiology and Biochemistry: PPB / Societe Francaise de Physiologie Vegetale 2020, 151(April), 640–649. doi:https://10.1016/j.plaphy.2020.04.016.
[58] Terefe M, Tefera T, Sakhuja PK. Effect of a formulation of Bacillus firmus on root-knot nematode Meloidogyne incognita infestation and the growth of tomato plants in the greenhouse and nursery. Journal of Invertebrate Pathology 2009, 100(2), 94–99. doi:https://10.1016/j.jip.2008.11.004.
[59] Zegeye ED, Santhanam A, Gorfu D, Kassa B. Biocontrol activity of *Trichoderma viride* and *Pseudomonas fluorescens* against *Phytophthora infestans* under greenhouse conditions. Journal of Agricultural Technology 2011, 7(6), 1589–1602.
[60] Wang Z, Li G, Sun H, et al. Effects of drought stress on photosynthesis and photosynthetic electron transport chain in young apple tree leaves. Biology Open 2018, 7(11). doi:https://10.1242/bio.035279.
[61] Carmen B, Roberto D. Soil bacteria support and protect plants against abiotic stresses. Abiotic Stress in Plants – Mechanisms and Adaptations 2011, 22, 143–170. doi:https://10.5772/23310.
[62] Choudhary DK, Varma A. Microbial-mediated induced systemic resistance in plants 2016, USA, Springer, 1–226. doi:https://10.1007/978-981-10-0388-2.
[63] Glick BR. Plant growth-promoting bacteria: Mechanisms and applications. Scientifica *Cairo* 2012, 2012, 1–15. http://www.hindawi.com/journals/scientifica/2012/963401/
[64] Ahemad M, Kibret M. Mechanisms and applications of plant growth promoting rhizobacteria: Current perspective. Journal of King Saud University – Science 2014, 26(1), 1–20. doi:https://10.1016/j.jksus.2013.05.001.

[65] Zhang X, Davidson EA, Mauzerall DL, Searchinger TD, Dumas P, Shen Y. Managing nitrogen for sustainable development. Nature 2015, 528(7580), 51–59. doi:https://10.1038/nature15743.
[66] Mustafa S, Kabir S, Shabbir U, Batool R. Plant growth promoting rhizobacteria in sustainable agriculture: From theoretical to pragmatic approach. Symbiosis 2019, 78(2), 115–123. doi:https://10.1007/s13199-019-00602-w.
[67] Amoo AE, Babalola OO. Ammonia-oxidizing microorganisms: Key players in the promotion of plant growth. The Journal of Soil Science and Plant Nutrition 2017, 17(4), 935–947. doi:https://10.4067/S0718-95162017000400008.
[68] Defez R, Andreozzi A, Bianco C. The overproduction of Indole-3-Acetic Acid (IAA) in endophytes upregulates nitrogen fixation in both bacterial cultures and inoculated rice plants. Microbial Ecology 2017, 74(2), 441–452. doi:https://10.1007/s00248-017-0948-4.
[69] Nascimento F, Brígido C, Alho L, Glick BR, Oliveira S. Enhanced chickpea growth-promotion ability of a Mesorhizobium strain expressing an exogenous ACC deaminase gene. Plant and Soil 2012, 353(1-2), 221–230. doi:https://10.1007/s11104-011-1025-2.
[70] Vishal B, Kumar PP. Regulation of seed germination and abiotic stresses by gibberellins and abscisic acid. Frontiers in Plant Science 2018, 9(June), 838. doi:https://10.3389/fpls.2018.00838.
[71] Govind Gupta SS. Plant Growth Promoting Rhizobacteria (PGPR): Current and future prospects for development of sustainable agriculture. Journal of Microbial & Biochemical Technology 2015, 07(02), 96–102. doi:https://10.4172/1948-5948.1000188.
[72] Khan MS, Zaidi A, Ahemad M, Oves M, Wani PA. Plant growth promotion by phosphate solubilizing fungi – Current perspective. Archives of Agronomy and Soil Science 2010, 56(1), 73–98. doi:https://10.1080/03650340902806469.
[73] Bhattacharyya PN, Jha DK. Plant growth-promoting rhizobacteria (PGPR): Emergence in agriculture. World Journal of Microbiology & Biotechnology 2012, 28(4), 1327–1350. doi:https://10.1007/s11274-011-0979-9.
[74] Sharma SB, Sayyed RZ, Trivedi MH, Gobi TA. Phosphate solubilizing microbes: Sustainable approach for managing phosphorus deficiency in agricultural soils. Springerplus 2013, 2(1), 1–14. doi:https://10.1186/2193-1801-2-587.
[75] Panhwar QA. Isolation and characterization of phosphate-solubilizing bacteria from aerobic rice. African Journal of Biotechnology 2012, 11(11), 2711–2719. doi:https://10.5897/ajb10.2218.
[76] Ullah A, Bano A, Janjua HT. Microbial secondary metabolites and defense of plant stress. Elsevier Inc., 2020. doi:https://10.1016/B978-0-12-819978-7.00003-8.
[77] Etalo DW, Jeon JS, Raaijmakers JM. Modulation of plant chemistry by beneficial root microbiota. Natural Product Reports 2018, 35(5), 398–409. doi:https://10.1039/c7np00057j.
[78] Jain A, Singh A, Singh S, Singh HB. Phenols enhancement effect of microbial consortium in pea plants restrains Sclerotinia sclerotiorum. Biological Control 2015, 89, 23–32. doi:https://10.1016/j.biocontrol.2015.04.013.
[79] Bai Y, Müller DB, Srinivas G, et al. Functional overlap of the Arabidopsis leaf and root microbiota. Nature 2015, 528(7582), 364–369. doi:https://10.1038/nature16192.
[80] Zaynab M, Fatima M, Sharif Y, Zafar MH, Ali H, Khan KA. Role of primary metabolites in plant defense against pathogens. Microbial Pathogenesis 2019, 137(July), 103728. doi:https://10.1016/j.micpath.2019.103728.
[81] Zaynab M, Fatima M, Abbas S, et al. Role of secondary metabolites in plant defense against pathogens. Microbial Pathogenesis 2018, 124(198–202). doi:https://10.1016/j.micpath.2018.08.034.

[82] Rout GR, Sahoo S. Role of iron in plant growth and metabolism. Reviews in Agricultural Science 2015, 3, 1–24. doi:https://10.7831/ras.3.1.

[83] López-Millán AF, Grusak MA, Abadía A, Abadía J. Iron deficiency in plants: An insight from proteomic approaches. Frontiers in Plant Science 2013, 4(Jul), 1–8. doi:https://10.3389/fpls.2013.00254.

[84] Goswami D, Thakker JN, Dhandhukia PC. Portraying mechanics of plant growth promoting rhizobacteria (PGPR): A review. Cogent Food & Agriculture 2016, 2(1), 1–19. doi:https://10.1080/23311932.2015.1127500.

[85] dos Santos RM, Diaz PAE, Lobo LLB, Rigobelo EC. Use of plant growth-promoting rhizobacteria in maize and sugarcane: Characteristics and applications. Frontiers in Sustainable Food Systems 2020, 4(September), 1–15. doi:https://10.3389/fsufs.2020.00136.

[86] Gururani MA, Upadhyaya CP, Baskar V, Venkatesh J, Nookaraju A, Park SW. Plant growth-promoting rhizobacteria enhance abiotic stress tolerance in Solanum tuberosum through inducing changes in the expression of ROS-scavenging enzymes and improved photosynthetic performance. Journal of Plant Growth Regulation 2013, 32(2), 245–258. doi:https://10.1007/s00344-012-9292-6.

[87] Deori M, Jayamohan NS, Kumudini BS. Production, characterization and iron binding affinity of hydroxamate siderophores from rhizosphere associated fluorescent Pseudomonas. Journal of Plant Protection Research 2018, 58(1), 36–43. doi:https://10.24425/119116.

[88] Pourbabaee AA, Shoaibi F, Emami S, Alikhani HA. The potential contribution of siderophore producing bacteria on growth and Fe ion concentration of sunflower (Helianthus annuus L.) under water stress. Journal of Plant Nutrition 2018, 41(5), 619–626. doi:https://10.1080/01904167.2017.1406112.

[89] Romero-Perdomo F, Abril J, Camelo M, et al. Azotobacter chroococcum as a potentially useful bacterial biofertilizer for cotton (Gossypium hirsutum): Effect in reducing N fertilization. Revista Argentina de Microbiologia 2017, 49(4), 377–383. doi:https://10.1016/j.ram.2017.04.006.

[90] Jacob S, Sajjalaguddam RR, Kumar KVK, Varshney R, Sudini HK. Assessing the prospects of Streptomyces sp. RP1A-12 in managing groundnut stem rot disease caused by Sclerotium rolfsii Sacc. Journal of General Plant Pathology 2016, 82(2), 96–104. doi:https://10.1007/s10327-016-0644-0.

[91] Akhtar MS, Siddiqui ZA. Use of plant growth-promoting rhizobacteria for the biocontrol of root-rot disease complex of chickpea. Australasian Plant Pathology 2009, 38(1), 44–50. doi:https://10.1071/AP08075.

[92] El-Sayed WS, Akhkha A, El-Naggar MY, Elbadry M. In vitro antagonistic activity, plant growth promoting traits and phylogenetic affiliation of rhizobacteria associated with wild plants grown in arid soil. Frontiers in Microbiology 2014, 5(Dec), 1–12. doi:https://10.3389/fmicb.2014.00651.

[93] Sharma A, Zheng B. Melatonin mediated regulation of drought stress: Physiological and molecular aspects. Plants 2019, 8(7), 190. doi:https://10.3390/plants8070190.

[94] Sadak MS. Physiological role of trehalose on enhancing salinity tolerance of wheat plant. Bulletin of the National Research Centre 2019, 43(1), 1–10. doi:https://10.1186/s42269-019-0098-6.

[95] Anjum SA, Ashraf U, Tanveer M, et al. Drought induced changes in growth, osmolyte accumulation and antioxidant metabolism of three maize hybrids. Frontiers in Plant Science 2017, 8(February), 1–12. doi:https://10.3389/fpls.2017.00069.

[96] Annunziata MG, Ciarmiello LF, Woodrow P, Dell'aversana E, Carillo P. Spatial and temporal profile of glycine betaine accumulation in plants under abiotic stresses. Frontiers in Plant Science 2019, 10(March), 1–13. doi:https://10.3389/fpls.2019.00230.

[97] Kosar F, Akram NA, Sadiq M, Al-Qurainy F, Ashraf M. Trehalose: A key organic osmolyte effectively involved in plant abiotic stress tolerance. Journal of Plant Growth Regulation 2019, 38(2), 606–618. doi:https://10.1007/s00344-018-9876-x.
[98] Ghorai AK, Patsa R, Jash S, Dutta S. Microbial secondary metabolites and their role in stress management of plants. 2021. UK, Woodhead Publishing doi:https://10.1016/b978-0-12-822919-4.00012-0.
[99] Safdarian M, Askari H, Shariati JV, Nematzadeh G. Transcriptional responses of wheat roots inoculated with Arthrobacter nitroguajacolicus to salt stress. Scientific Reports 2019, 9(1), 1–12. doi:https://10.1038/s41598-018-38398-2.
[100] Bharti N, Yadav D, Barnawal D, Maji D, Kalra A. Exiguobacterium oxidotolerans, a halotolerant plant growth promoting rhizobacteria, improves yield and content of secondary metabolites in Bacopa monnieri (L.) Pennell under primary and secondary salt stress. World Journal of Microbiology & Biotechnology 2013, 29(2), 379–387. doi:https://10.1007/s11274-012-1192-1.
[101] Olanrewaju OS, Glick BR, Babalola OO. Mechanisms of action of plant growth promoting bacteria. World Journal of Microbiology & Biotechnology 2017, 33(11), 1–16. doi:https://10.1007/s11274-017-2364-9.
[102] del Carmen Orozco-Mosqueda M, Glick BR, Santoyo G. ACC deaminase in plant growth-promoting bacteria (PGPB): An efficient mechanism to counter salt stress in crops. Microbiological Research 2020, 235, 126439. doi:https://10.1016/j.micres.2020.126439.
[103] Hussain I, Puschenreiter M, Gerhard S, Schöftner P, Yousaf S, Wang A, Syed JH, Reichenauer TG. Rhizoremediation of petroleum hydrocarbon-contaminated soils: improvement opportunities and field applications. Environmental and Experimental Botany 2018, 147, 202–219.
[104] Vimal SR, Patel VK, Singh JS. Plant growth promoting Curtobacterium albidum strain SRV4: An agriculturally important microbe to alleviate salinity stress in paddy plants. Ecological Indicators 2019, 105(July 2017), 553–562. doi:https://10.1016/j.ecolind.2018.05.014.
[105] Beneduzi A, Ambrosini A, Passaglia LMP. Plant growth-promoting rhizobacteria (PGPR): Their potential as antagonists and biocontrol agents. Genetics and Molecular Biology 2012, 35, 1044–1051. doi:https://10.1590/S1415-47572012000600020.
[106] Mhatre PH, Karthik C, Kadirvelu K, et al. Plant growth promoting rhizobacteria (PGPR): A potential alternative tool for nematodes bio-control. Biocatalysis and Agricultural Biotechnology 2019, 17(November 2018), 119–128. doi:https://10.1016/j.bcab.2018.11.009.
[107] Arora NK. Plant microbe symbiosis: Fundamentals and advances 2013. India, Springer, doi:https://10.1007/978-81-322-1287-4.
[108] Gamalero E, Glick BR. Bacteria in agrobiology. In: Maheshwari DKK, ed. Bacteria in Agrobiology: Plant Nutrient Management. Berlin/Heidelberg, Germany, Springer, 2011, 17–47. doi:https://10.1007/978-3-642-21061-7.
[109] Hallmann J, Quadt-Hallmann A, Miller WG, Sikora RA, Lindow SE. Endophytic colonization of plants by the biocontrol agent Rhizobium etli G12 in relation to Meloidogyne incognita infection. Phytopathology 2001, 91(4), 415–422. doi:https://10.1094/PHYTO.2001.91.4.415.
[110] Ryu CM, Farag MA, Hu CH, Reddy MS, Kloepper JW, Paré PW. Bacterial volatiles induce systemic resistance in Arabidopsis. Plant Physiology 2004, 134(3), 1017–1026. doi:https://10.1104/pp.103.026583.
[111] Garbeva P, Hordijk C, Gerards S, de Boer W. Volatiles produced by the mycophagous soil bacterium Collimonas. FEMS Microbiology Ecology 2014, 87(3), 639–649. doi:https://10.1111/1574-6941.12252.
[112] Insunza V, Alström S, Eriksson KB. Root bacteria from nematicidal plants and their biocontrol potential against trichodorid nematodes in potato. Plant and Soil 2002, 241(2), 271–278. doi:https://10.1023/A:1016159902759.

[113] Egamberdieva D, Wirth SJ, Alqarawi AA, Abd-Allah EF, Hashem A. Phytohormones and beneficial microbes: Essential components for plants to balance stress and fitness. Frontiers in Microbiology 2017, 8(Oct), 1–14. doi:https://10.3389/fmicb.2017.02104.

[114] Lymperopoulos P, Msanne J, Rabara R. Phytochrome and phytohormones: Working in tandem for plant growth and development. Frontiers in Plant Science 2018, 9(July), 1–14. doi:https://10.3389/fpls.2018.01037.

[115] Denancé N, Sánchez-Vallet A, Goffner D, Molina A. Disease resistance or growth: The role of plant hormones in balancing immune responses and fitness costs. Frontiers in Plant Science 2013, 4(May), 1–12. doi:https://10.3389/fpls.2013.00155.

[116] Duca DR, Glick BR. Indole-3-acetic acid biosynthesis and its regulation in plant-associated bacteria. Applied Microbiology and Biotechnology 2020, 104(20), 8607–8619. doi:https://10.1007/s00253-020-10869-5.

[117] Imada EL, de Paiva Rolla dos Santos AA, de Oliveira ALM, Hungria M, Rodrigues EP. Indole-3-acetic acid production via the indole-3-pyruvate pathway by plant growth promoter Rhizobium tropici CIAT 899 is strongly inhibited by ammonium. Research in Microbiology 2017, 168(3), 283–292. doi:https://10.1016/j.resmic.2016.10.010.

[118] Cortleven A, Leuendorf JE, Frank M, Pezzetta D, Bolt S, Schmülling T. Cytokinin action in response to abiotic and biotic stresses in plants. Plant, Cell & Environment 2019, 42(3), 998–1018. doi:https://10.1111/pce.13494.

[119] Amara U, Khalid R, Hayat R. Soil bacteria and phytohormones for sustainable crop production BT – Bacterial metabolites in sustainable agroecosystem. In: Maheshwari D, eds. Bacterial Metabolites in Sustainable Agroecosystem. Sustainable Development and Biodiversity, vol 12. Cham, Springer. https://doi.org/10.1007/978-3-319-24654-3_5.

[120] Großkinsky DK, Tafner R, Moreno MV, et al. Cytokinin production by Pseudomonas fluorescens G20-18 determines biocontrol activity against Pseudomonas syringae in Arabidopsis. Scientific Reports 2016, 6(September 2015), 1–11. doi:https://10.1038/srep23310.

[121] Tsukanova KA, Chebotar V, Meyer JJM, Bibikova TN. Effect of plant growth-promoting Rhizobacteria on plant hormone homeostasis. South African Journal of Botany 2017, 113, 91–102. doi:https://10.1016/j.sajb.2017.07.007.

[122] Dodd IC, Zinovkina NY, Safronova VI, Belimov AA. Rhizobacterial mediation of plant hormone status. The Annals of Applied Biology 2010, 157(3), 361–379. doi:https://10.1111/j.1744-7348.2010.00439.x.

[123] Tuna AL, Kaya C, Dikilitas M, Higgs D. The combined effects of gibberellic acid and salinity on some antioxidant enzyme activities, plant growth parameters and nutritional status in maize plants. Environmental and Experimental Botany 2008, 62(1), 1–9. doi:https://10.1016/j.en vexpbot.2007.06.007.

[124] Bücker-Neto L, Paiva ALS, Machado RD, Arenhart RA, Margis-Pinheiro M. Interactions between plant hormones and heavy metals responses. Genetics and Molecular Biology 2017, 40(1), 373–386. doi:https://10.1590/1678-4685-gmb-2016-0087.

[125] Siddiqui ZA, Mahmood I. Role of bacteria in the management of plant parasitic nematodes: A review. Bioresource Technology 1999, 69(2), 167–179. doi:https://10.1016/S0960-8524(98)00122-9.

[126] Siddiqui IA, Shaukat SS. Suppression of root-knot disease by Pseudomonas fluorescens CHA0 in tomato: Importance of bacterial secondary metabolite, 2,4-diacetylpholoroglucinol. Soil Biology & Biochemistry 2003, 35(12), 1615–1623. doi:https://10.1016/j.soilbio.2003.08.006.

[127] Tian B, Yang J, Lian L. Role of an extracellular neutral protease in infection against nematodes by Brevibacillus laterosporus strain G4. Applied microbiology and biotechnology 2007, 74(2), 372–380. doi:https://10.1007/s00253-006-0690-1.

[128] Rizvi R, Mahmood I, Tiyagi SA, Khan Z. Conjoint effect of oil-seed cakes and Pseudomonas fluorescens on the growth of chickpea in relation to the management of plant-parasitic nematodes. 2012, 55 December, 801–808.
[129] McNeely D, Chanyi RM, Dooley JS, Moore JE, Koval SF. Biocontrol of Burkholderia cepacia complex bacteria and bacterial phytopathogens by Bdellovibrio bacteriovorus. Canadian Journal of Microbiology 2017, 63(4), 350–358. doi:https://doi.org/10.1139/cjm-2016-0612.
[130] Gautam AK, Avasthi S. Fungal endophytes: Potential biocontrol agents in agriculture. In: Kumar A, Singh AK, Choudhary KK, eds. Role of plant growth promoting microorganisms in sustainable agriculture and nanotechnology 2019, USA, Woodhead Publishing, 241–283. doi:https://10.1016/b978-0-12-817004-5.00014-2.
[131] Grabka R, D'entremont TW, Adams SJ, et al. Fungal endophytes and their role in agricultural plant protection against pests and pathogens. Plants 2022, 11(3), 1–29. doi:https://10.3390/plants11030384.
[132] Zheng L, Zhao J, Liang X, Zhan G, Jiang S, Nosanchuk JD. Identification of a novel alternaria alternata strain able to Hyperparasitize Puccinia striiformis f. sp. tritici, the causal agent of wheat stripe rust. Frontiers in Microbiology 2017, 8(January), 1–10. doi:https://10.3389/fmicb.2017.00071.
[133] Nacer A, Boudjema S, Boudouaia N. Bioremediation of hexavalent chromium by an indigenous bacterium Bacillus cereus S10C1: optimization study using two level full factorial experimental design. Comptes Rendus. Chimie. 2021, 24(S1), 57–70. doi:https://10.5802/crchim.81.
[134] Devi R, Behera B, Raza B, Mangal V, Ahsan M, Ravinder A. An insight into microbes mediated heavy metal detoxification in plants: A review. The Journal of Soil Science and Plant Nutrition 2022, 22, 914–936. doi:https://10.1007/s42729-021-00702-x.
[135] Hassan TU, Bano A, Naz I. Alleviation of heavy metal toxicity by the application of plant growth promoting rhizobacteria(PGPR) and effects on wheat grown in saline sodic field. International Journal of Phytoremediation 2016, 19(6), 522–529.
[136] Oladipo OG, Olufemi O, Olayinka A, Carlos C, Steve M. Heavy metal tolerance traits of filamentous fungi isolated from gold and gemstone mining sites. Brazilian Journal of Microbiology 2017, 49(1), 29–37. doi:https://10.1016/j.bjm.2017.06.003.
[137] Tripathi P, Khare P, Barnawal D, et al. Science of the Total Environment Bioremediation of arsenic by soil methylating fungi: Role of Humicola sp. strain 2WS1 in amelioration of arsenic phytotoxicity in Bacopa monnieri L. The Science of the Total Environment 2020, 716, 136758. doi:https://10.1016/j.scitotenv.2020.136758.
[138] Hoang SA, Lamb D, Seshadri B, Sarkar B, Kirkham MB, Bolan NS. Rhizoremediation as a green technology for the remediation of petroleum hydrocarbon-contaminated soils. Journal of Hazardous Materials 2021, 401, 123282. doi:https://10.1016/j.jhazmat.2020.123282.
[139] Ma Y, Prasad MNV, Rajkumar M, Freitas H. Plant growth promoting rhizobacteria and endophytes accelerate phytoremediation of metalliferous soils. Biotechnology Advances. 2011, 29(2), 248–258. doi:https://10.1016/j.biotechadv.2010.12.001.
[140] Priyadarshanee M, Das S. Biosorption and removal of toxic heavy metals by metal tolerating bacteria for bioremediation of metal contamination: A comprehensive review. Journal of Environmental Chemical Engineering. 2021, 9(1), 104686.
[141] Verma S, Kuila A. Environmental technology & innovation bioremediation of heavy metals by microbial process. Environmental Technology & Innovation 2019, 14, 100369. doi:https://10.1016/j.eti.2019.100369.
[142] Dusengemungu L, Kasali G, Gwanama C. Recent advances in biosorption of copper and cobalt by filamentous fungi. 2020, 11(December), 1–16. doi:https://10.3389/fmicb.2020.582016.

[143] Aryal M. A comprehensive study on the bacterial biosorption of heavy metals: materials, performances, mechanisms, and mathematical modellings. Reviews in Chemical Engineering 2021, 37(6), 715–754. doi:https://doi.org/10.1515/revce-2019-0016.
[144] Chojnacka K. Biosorption and bioaccumulation – The prospects for practical applications. Environment International 2010, 36(3), 299–307. doi:https://10.1016/j.envint.2009.12.001.
[145] Aksu Z, Donmez G. The use of molasses in copper (II) containing wastewaters: Effects on growth and copper (II) bioaccumulation properties of Kluy 6 eromyces marxianus. Process Biochemistry 2000, 36(5), 451–458.
[146] Timkov I, Sedl J, Pristaš P. Biosorption and bioaccumulation abilities of actinomycetes / streptomycetes isolated from metal contaminated sites. Separations 2018, 5(4), 54. doi:https://10.3390/separations5040054.
[147] Diep P, Mahadevan R, Yakunin AF. Heavy metal removal by bioaccumulation using genetically engineered microorganisms. Frontiers in Bioengineering and Biotechnology 2018, 29(6), 157. doi:https://10.3389/fbioe.2018.00157.
[148] Mishra A, Malik A. Recent advances in microbial metal bioaccumulation recent advances in microbial metal. Critical Reviews in Environmental Science 2013, 43(11), 37–41. doi: https://10.1080/10934529.2011.627044.
[149] Aslam F, Yasmin A, Sohail S. Bioaccumulation of lead, chromium, and nickel by bacteria from three different genera isolated from industrial effluent. International Microbiology 2020, 23(2), 253–261.
[150] Donmez G, Aksu Z. The effect of copper (II) ions on the growth and bioaccumulation properties of some yeasts. Process Biochemistry 1999, 35, 135–142.
[151] De Silóniz M, Balsalobre L, Alba C, Valderrama M, Peinado JM. Feasibility of copper uptake by the yeast Pichia guilliermondii isolated from sewage sludge. Research in Microbiology 2002, 153, 173–180.
[152] Renella G, Landi L, Nannipieri P. Degradation of low molecular weight organic acids complexed with heavy metals in soil. Geoderma 2004, 122, 311–315. doi:https://10.1016/j.geoderma.2004.01.018.
[153] Seneviratne M, Seneviratne G, Madawala HM, Vithanage M. Role of rhizospheric microbes in heavy metal uptake by plants. In: Agro-environmental sustainability, Cham, Springer, 2017, 147–163. doi:https://10.1007/978-3-319-49727-3.
[154] Sheng X, Xia J, Jiang C, He L, Qian M. Characterization of heavy metal-resistant endophytic bacteria from rape (Brassica napus) roots and their potential in promoting the growth and lead accumulation of rape. Environmental Pollution (Barking, Essex: 1987) 2008, 156(3), 1164–1170. doi:https://10.1016/j.envpol.2008.04.007.
[155] Wani PA, Khan MS, Zaidi A. Chromium reduction, plant growth – Promoting potentials, and metal solubilization by Bacillus sp. Isolated from Alluvial Soil 2007, 54, 237–243. doi:https://10.1007/s00284-006-0451-5.
[156] Khan N, Bano A, Rahman MA, Guo J, Kang Z, Babar A. Comparative physiological and metabolic analysis reveals a complex mechanism involved in drought tolerance in Chickpea (Cicer arietinum L.). Induced by PGPR and PGRs 2019(July 2018), 1–19. doi:https://10.1038/s41598-019-38702-8.
[157] Amstaetter K, Borch T, Larese-casanova P, Kappler A. Redox transformation of arsenic by Fe (II)-activated goethite (r-FeOOH). Environmental Science & Technology 2010, 44(1), 102–108.
[158] Khan N, Bano A. Modulation of phytoremediation and plant growth by the treatment of PGPR, Ag nanoparticle and untreated municipal wastewater. International Journal of Phytoremediation 2016, 18(12), 1258–1269. doi:https://10.1080/15226514.2016.1203287.
[159] Jobby R, Jha P, Gupta A, Gupte A, Id ND. Biotransformation of chromium by root nodule bacteria Sinorhizobium sp. SAR1. PLoS one 2019, 14(7), 1–16.

List of contributing authors

Chapter 1
Sanjay Kumar Joshi
Department of Agri-Business and Rural Management
Indira Gandhi Krishi Vishwavidyalaya, Raipur
India
sanjay29872@gmail.com

Rashmi Upadhyay
Department of Plant Molecular Biology and Biotechnology
Indira Gandhi Krishi Vishwavidyalaya, Raipur
India

Chapter 2
Himani
Department of Chemistry
College of Basic Sciences and Humanities
G.B. Pant University of Agriculture and Technology
Pantnagar
U.S. Nagar 263145
Uttarakhand
India

Pringal Upadhyay
Department of Chemistry
College of Basic Sciences and Humanities
G.B. Pant University of Agriculture and Technology
Pantnagar
U.S. Nagar 263145
Uttarakhand
India

Sonu Kumar Mahawer
Department of Chemistry
College of Basic Sciences and Humanities
G.B. Pant University of Agriculture and Technology
Pantnagar
U.S. Nagar 263145
Uttarakhand
India

Ravendra Kumar
Department of Chemistry
College of Basic Sciences and Humanities
G.B. Pant University of Agriculture and Technology
Pantnagar
U.S. Nagar 263145
Uttarakhand, India
ravichemistry.kumar@gmail.com

Om Prakash
Department of Chemistry
College of Basic Sciences and Humanities
G.B. Pant University of Agriculture and Technology
Pantnagar
U.S. Nagar 263145
Uttarakhand, India

Chapter 3
Parul Chaudhary
Govind Ballabh Pant University of Agriculture and Technology
Pantnagar
Uttarakhand
India
parulchaudhary1423@gmail.com

Shivani Singh
Govind Ballabh Pant University of Agriculture and Technology
Pantnagar
Uttarakhand
India

Anuj Chaudhary
School of Agriculture and Environmental Science
Shobhit University
Gangoh
Uttar Pradesh
India

https://doi.org/10.1515/9783110771558-022

Upasana Agri
Govind Ballabh Pant University of Agriculture and Technology
Pantnagar
Uttarakhand
India

Geeta Bhandari
Swami Rama Himalayan University
Dehradun
Uttarakhand
India

Chapter 4

Nitika Negi
Forest Pathology Discipline
Forest Protection Division
Forest Research Institute
Dehradun

Siya Sharma
Forest Pathology Discipline
Forest Protection Division
Forest Research Institute
Dehradun

Nitika Bansal
Forest Pathology Discipline
Forest Protection Division
Forest Research Institute
Dehradun

Aditi Saini
Forest Pathology Discipline
Forest Protection Division
Forest Research Institute
Dehradun

Ratnaboli Bose
Forest Pathology Discipline
Forest Protection Division
Forest Research Institute
Dehradun

M. S. Bhandari
Genetics and Tree Improvement Division
Forest Research Institute
Dehradun

Amit Pandey
Forest Pathology Discipline
Forest Protection Division
Forest Research Institute
Dehradun

Shailesh Pandey
Forest Pathology Discipline
Forest Protection Division
Forest Research Institute
Dehradun
pandeysh@icfre.org; shailesh31712@yahoo.co.in

Chapter 5

Hari Narayan
Department of Agricultural Microbiology
College of Agriculture
Indira Gandhi KrishiVishwaVidyalaya
Raipur
Chhatisgarh
India

Pragati Srivasatava
Department of Microbiology
G. B. Pant University of Agriculture and Technology
Pantnagar 263 145
Uttarakhand
India

Divya Joshi
Department of Microbiology
BFIT
Dehradun
Uttarakhand
India

Santosh Chandra Bhatt
Department of Soil Science
D.K.S.G. Akal College of Agriculture
Eternal University
Baru Sahib
Sirmaur-173101
Himachal Pradesh
India

Ravindra Soni
Department of Agricultural Microbiology
College of Agriculture
Indira Gandhi KrishiVishwaVidyalaya
Raipur
Chhatisgarh
India
rs31693@gmail.com

Chapter 6

Sonu Kumar Mahawer
Govind Ballabh Pant University of Agriculture and Technology
Pantnagar 263145
Uttarakhand
India

Sushila Arya
Govind Ballabh Pant University of Agriculture and Technology
Pantnagar 263145
Uttarakhand
India

Tanuja Kandpal
Govind Ballabh Pant University of Agriculture and Technology
Pantnagar 263145
Uttarakhand
India

Ravendra Kumar
Govind Ballabh Pant University of Agriculture and Technology
Pantnagar 263145
Uttarakhand
India
ravichemistry.kumar@gmail.com

Om Prakash
Govind Ballabh Pant University of Agriculture and Technology
Pantnagar 263145
Uttarakhand
India

Manoj Kumar Chitara
Govind Ballabh Pant University of Agriculture and Technology
Pantnagar 263145
Uttarakhand
India

Puspendra Koli
Post Harvest Biosecurity
Murdoch University
Perth, WA 6150
Australia

Chapter 7

Amir Khan
Plant Pathology and Nematology Section
Department of Botany
Aligarh Muslim University
Aligarh 202002
Uttar Pradesh
India

Mohd Shahid Anwar Ansari
Department of Plant Protection
Faculty of Agriculture Science
Aligarh Muslim University
Aligarh 202002
Uttar Pradesh
India

Irsad
Department of Plant Protection
Faculty of Agriculture Science
Aligarh Muslim University
Aligarh 202002
Uttar Pradesh
India

Touseef Hussain
Plant Pathology and Nematology Section
Department of Botany
Aligarh Muslim University
Aligarh 202002
Uttar Pradesh
India

And
Division of Plant Pathology
Indian Agricultural Research Institute
New Delhi 110012
India
hussaintouseef@yahoo.co.in

Abrar Ahmad Khan
Plant Pathology and Nematology Section
Department of Botany
Aligarh Muslim University
Aligarh 202002
Uttar Pradesh
India

Chapter 8

Hemant Sharma
Department of Botany
Sikkim University
Gangtok
Sikkim 737102
India

Binu Gogoi
Department of Botany
Sikkim University
Gangtok
Sikkim 737102
India

Arun Kumar Rai
Department of Botany
Sikkim University
Gangtok
Sikkim 737102
India
akrai@cus.ac.in

Chapter 9

Sougata Ghosh
Department of Physics
Faculty of Science
Kasetsart University
Bangkok
Thailand
and
Department of Microbiology
School of Science
RK University
Rajkot
Gujarat
India
ghoshsibb@gmail.com

Bishwarup Sarkar
College of Science
Northeastern University, Boston
MA, USA

Sirikanjana Thongmee
Department of Physics
Faculty of Science
Kasetsart University
Bangkok
Thailand

Chapter 10

Bahman Fazeli-Nasab
Research Department of Agronomy and Plant Breeding
Agricultural Research Institute
University of Zabol
Zabol
Iran
bfazeli@uoz.ac.ir, bfazelinasab@gmail.com

Ramin Piri
Department of Agronomy and Plant Breeding
Faculty of Agriculture
University of Tehran
Tehran
Iran

Ahmad Farid Rahmani
Department of Horticulture
Faculty of Agriculture
Herat University
Herat
Afghanistan

Chapter 11
Viphrezolie Sorhie
Applied Environmental Microbial
Biotechnology Laboratory
Department of Environmental Science
Nagaland University
Zunheboto
Nagaland
India

Pranjal Bharali
Applied Environmental Microbial
Biotechnology Laboratory
Department of Environmental Science
Nagaland University
Zunheboto
Nagaland
India
prangenetu@gmail.com

Alemtoshi
Applied Environmental Microbial
Biotechnology Laboratory
Department of Environmental Science
Nagaland University
Zunheboto
Nagaland
India

Chapter 12
Neela Gayathri Ganesan
Department of Chemical Engineering
Birla Institute of Technology and Science-
Pilani
K.K. Birla Goa Campus
Zuarinagar
Goa 403726
India

Subhranshu samal
Department of Chemical Engineering
Birla Institute of Technology and Science-
Pilani
K.K. Birla Goa Campus
Zuarinagar
Goa 403726
India

Vinoth Kannan
Department of Chemical Engineering
Birla Institute of Technology and Science-
Pilani
K.K. Birla Goa Campus
Zuarinagar
Goa 403726
India

Senthil Kumar Rathnasamy
School of Chemical and Biotechnology
Green Separation Engineering Laboratory
School of Chemical and Biotechnology
SASTRA University
Thanjavur 613 401
Tamil Nadu
India

Vivek Rangarajan
Department of Chemical Engineering
Birla Institute of Technology and Science-
Pilani
K.K. Birla Goa Campus
Zuarinagar
Goa 403726
India
vivekr@goa.bits-pilani.ac.in

Chapter 13
Pragati Srivasatava
Department of Microbiology
G. B. Pant University of Agriculture and
Technology
Pantnagar 263 145
Uttarakhand
India

Chapter 14
U. M. Aruna Kumara
Department of Agricultural Technology
Faculty of Technology
University of Colombo
Homagama
Sri Lanka
umarunakumara@at.cmb.ac.lk

N. Thiruchchelvan
Department of Agricultural Biology
Faculty of Agriculture
University of Jaffna
Kilinochchi
Sri Lanka

Chapter 15
Sikhamoni Bora
Rain Forest Research Institute
Jorhat
Assam
India
borahsikhamoni95@gmail.com

Rajib Kumar Borah
Rain Forest Research Institute
Jorhat
Assam
India

Krishna Giri
Rain Forest Research Institute
Jorhat
Assam
India
and
Biodiversity and Climate Change Division
Indian Council of Forestry Research and Education
Dehradun
Uttarakhand
India

Chapter 16
Bahman Fazeli-Nasab
Research Department of Agronomy and Plant Breeding
Agricultural Research Institute
University of Zabol
Zabol
Iran
bfazeli@uoz.ac.ir, bfazelinasab@gmail.com

Ramin Piri
Department of Agronomy and Plant Breeding
Faculty of Agriculture
University of Tehran
Tehran
Iran

Yamini Tak
Department of Biochemistry
Agriculture University
Kota
Rajasthan 324001
India

Anahita Pahlavan
Department of Horticultural
Faculty of Agriculture
University of Zabol
Zabol
Iran

Farzaneh Zamani
Department of Agronomy and Plant Breeding
Faculty of Agriculture
University of Tehran
Tehran
Iran

Chapter 17
Jaagriti Tyagi
Department of Microbial Biotechnology
Amity University
Noida
Uttar Pradesh
India

Parul Chaudhary
Govind Ballabh Pant University of Agriculture and Technology
Pantnagar
Uttarakhand
India

Jyotsana
Department of Microbial Biotechnology
Amity University
Noida
Uttar Pradesh
India

Upasana Bhagwati
Department of Microbial Biotechnology
Amity University
Noida
Uttar Pradesh
India

Geeta Bhandari
Swami Rama Himalayan University
Dehradun
Uttarakhand
India

Anuj Chaudhary
School of Agriculture and Environmental Sciences
SUG
Uttar Pradesh
India
anujchaudharysvp@gmail.com

Chapter 18

G. K. Dinesh
Division of Environment Science
ICAR – Indian Agricultural Research Institute
New Delhi 110012
India
gkdineshiari@gmail.com

B. Priyanka
Department of Environmental Science
Dr.Y.S. Parmar University of Horticulture and Forestry
Nauni, Solan
Himachal Pradesh 173230
India

Archana Anokhe
Division of Agricultural Entomology
ICAR – Indian Agricultural Research Institute
New Delhi 110012
India

P. T. Ramesh
Agricultural College and Research Institute
Agricultural University
Killikulam
Tamil Nadu
India

R. Venkitachalam
Department of Zoology
Kongunadu Arts and Science College (Autonomous)
Coimbatore 641029
India

K. S. Keerthana Sri
Faculty of Agricultural Sciences
University of Göttingen
Germany

S. Abinaya
Department of Entomology
University of Agricultural Sciences
Raichur
Karnataka 584104
India

V. Anithaa
Department of Agricultural Entomology
Palli Siksha Bhavana
Visva Bharati University
Bolpur 731236
West Bengal
India

R. Soni
Indira Gandhi Krishi Vishwavidyalaya
Raipur 492012
India

Chapter 19

G. Venkatesh
Department of Entomology
Annamalai University
Chidambaram 608002
India

P. Sakthi Priya
Department of Plant Pathology
School of Crop Protection
Umiam 793103
Meghalaya
India

V. Anithaa
Department of Agricultural Entomology
Palli Siksha Bhavana
Visva Bharati University
Bolpur 731236
West Bengal
India

G. K. Dinesh
Division of Environment Science
ICAR – Indian Agricultural Research Institute
New Delhi 110012
India
+91-8428719080

S. Velmurugan
Division of Plant Pathology
ICAR – Indian Agricultural Research Institute
New Delhi 110012
India

S. Abinaya
Department of Entomology
University of Agricultural Sciences
Raichur
Karnataka 584104
India

P. Karthika
Department of Microbiology
Chaudhary Charan Singh Haryana
Agricultural University
Hisar 125004
India

P. Sivasakthivelan
Department of agricultural microbiology
Annamalai University
Chidambaram 608001
India

R. Soni
Indira Gandhi Krishi Vishwavidyalaya
Raipur 492012
India
rs31693@gmail.com

A. Thennarasi
Department of Agri Biology
Sree Ramu College of Arts and Science
Pollachi 642 002
Coimbatore
India

Chapter 20

Debiprasad Dash
KVK
Bhadrak
OUAT
Bhubaneswar
India

Pallabi Mishra
Department of Business Administration
Utkal University
Bhubaneswar
India

Chapter 21

Thajudeen Thahira
Postgraduate Institute of Science
University of Peradeniya
Peradeniya
Sri Lanka

Kinza Qadeer
Postgraduate Institute of Science
University of Peradeniya
Peradeniya
Sri Lanka

Julie Sosso
School of Agriculture and Food Sciences
The University of Queensland, Gatton
Campus
QLD 4345
Australia

Mohamed Cassim Mohamed Zakeel
Centre for Horticultural Science
Queensland Alliance for Agriculture and Food Innovation
The University of Queensland
GPO Box 267
Brisbane
QLD 4001
Australia
And
Department of Plant Sciences
Faculty of Agriculture
Rajarata University of Sri Lanka
Puliyankulama
Anuradhapura
Sri Lanka
m.mohamedzakeel@uqconnect.edu.au;
zakeel@agri.rjt.ac.lk

Index

https://doi.org/10.1515/9783110771558-023